P9-DTQ-699

...otter © Frederick Warne PLC 1984 Licensed by ©opyrights

HUMAN PHYSIOLOGY

ARTHUR J. VANDER

JAMES H. SHERMAN

DOROTHY S. LUCIANO

University of Michigan

HUMAN PHYSIOLOGY
THE MECHANISMS OF BODY FUNCTION

SECOND EDITION

McGRAW-HILL BOOK COMPANY

New York St. Louis San Francisco Auckland Düsseldorf Johannesburg Kuala Lumpur London Mexico Montreal
New Delhi Panama Paris São Paulo Singapore Sydney Tokyo Toronto

This book was set in Vega by York Graphic Services, Inc. The editors were Thomas A. P. Adams and Carol First; the designer was Barbara Ellwood; the production supervisor was Sam Ratkewitch. New drawings were done by Eric G. Hieber Associates Inc. R. R. Donnelley & Sons Company was printer and binder.

HUMAN PHYSIOLOGY
THE MECHANISMS OF BODY FUNCTION

3 4 5 6 7 8 9 0   DODO   7 9 8 7 6 5

Library of Congress Cataloging in Publication Data

Vander, Arthur J   date
  Human physiology: the mechanisms of body function.
  Bibliography: p.
  1. Human physiology. I. Sherman, James H.,
date,   joint author. II. Luciano, Dorothy S.,
joint author. III. Title. [DNLM: 1. Physiology.
QT104 V228h]
QP34.5.V36 1975   612   74-13431
ISBN 0-07-066954-6

Contents

Preface

The primary purpose of this book remains what it was in the first edition: to present the fundamental mechanisms of human physiology. Our aim has been to tell a story, not to write an encyclopedia. The book is intended for undergraduate students, regardless of their scientific background. The physics and chemistry requisite for an understanding of the physiology are presented where relevant in the text. Students with little or no scientific training will find this material essential, while others, more sophisticated in the physical sciences, should profit from the review of basic science oriented toward specifically biological applications. Thus this book is suitable for most introductory courses in human physiology, including those taken by students in the health professions.

The overall organization and approach of the book is based upon a group of themes which are developed in the Introduction and form the framework for our descriptions: (1) All phenomena of life, no matter how complex, are ultimately describable in terms of physical and chemical laws; (2) certain fundamental features of cell function are shared by virtually all cells and, in addition, constitute the foundation upon which specialization is built; (3) the body's various coordinated functions—circulation, respiration, etc.—result from the precise control and integration of specialized cellular activities, serve to maintain relatively constant the internal composition of the body, and can be described in terms of control systems similar to those designed by engineers.

In keeping with these themes, the book progresses from the cell to the total body, utilizing at each level of increasing complexity the information and principles developed previously. Part 1 is devoted to an analysis of basic cellular physiology and the essential physics and chemistry required for its understanding. Part 2 analyzes the concept of the body's internal environment, the nature of biological control systems, and the properties of the major specialized cell types—nerve, muscle, and gland—which comprise these systems. Part 3 then analyzes the coordinated body functions in terms of the basic concepts and information developed in Parts 1 and 2. In this way we have tried to emphasize the underlying unity of biological processes.

This approach has resulted in several characteristics of the book: (*1*) Cell physiology has received extensive coverage. (*2*) We have been willing to spend a considerable number of pages logically developing a single cellular concept (such as the origin of membrane potentials) required for the understanding of total-body physiological processes, and which we have found to offer considerable difficulties for the student. (*3*) We have made every effort not to mention facts simply because they happen to be known, but rather to use facts as building blocks for general principles and concepts. In this last regard, we confess that even an introductory course in physiology must teach a rather frightening number of facts, and our book is no exception. We have tried to keep in mind, however, the words of John Hunter, the eighteenth-century British anatomist and surgeon: "Too much attention can not be paid to facts; yet too many facts crowd the memory without advantage, any further than they lead us to establish principles." (*4*) We have employed a very large number of figures as an aid to developing concepts and explanations. These figures also provide an excellent summary of the most important

material in each chapter. (*5*) We have not shied away from pointing out the considerable gaps in current understanding.

In summary, our book may be long in areas where other texts are brief, and vice versa; moreover, our approach requires the student to think rather than simply memorize. However, our experience over a number of years of presenting this material to students with varying backgrounds has fully confirmed our belief that students do respond to this approach with considerable enthusiasm and excitement.

In this second edition, the level, scope, and emphasis remain unchanged. The text has been completely updated, a process involving changes too numerous to list here. We have also done considerable rewriting in order to improve clarity of organization and presentation. In this regard, we wish to thank the many students and faculty for their comments and suggestions.

In response to requests from many users of the first edition, and incorporating many of their suggestions, there will be an instructor's manual to accompany this second edition. We hope it will be a useful aid to those instructors adopting the new edition. Once again, we thank Mrs. Helen L. Mysyk, who performed superhuman feats in typing the manuscript, and the staff of McGraw-Hill for their continuous help and advice. Above all, we gratefully acknowledge the support and encouragement of Dr. Horace W. Davenport, whose scientific work, teaching, and writings have set an example for us to follow.

ARTHUR J. VANDER
JAMES H. SHERMAN
DOROTHY S. LUCIANO

One cannot meaningfully analyze the enormously complex activities of the human body without a framework upon which to build, a set of viewpoints to guide one's thinking. It is the purpose of this introduction to establish those viewpoints and to orient the reader to our general approach to the subject.

Introduction

Mechanism and vitalism

The *mechanist* view of life holds that all phenomena, no matter how complex, are ultimately describable in terms of physical and chemical laws and that no "vital force" distinct from matter and energy is required to explain life.

This view has predominated in the twentieth century because virtually all information gathered from observation and experiment has agreed with it. But *vitalism,* its opposite, is not completely dead, nor is it surprising that it lingers in fields (like brain physiology) where we are almost entirely lacking in hypotheses to explain such phenomena as thought and consciousness in physicochemical terms. We believe that even these areas will ultimately yield to physicochemical analysis, but we also feel that it would be unscientific, on the basis of present knowledge, to dismiss the problem out of hand, and we shall analyze it further in Chap. 18. Man, then, is a machine—an enormously complex machine, but a machine, nevertheless.

Cells: the basic units

Individual cells are the basic units of both the structure and the function of living things. One of the crucial unifying generalizations of biology is that certain fundamental activities are common to almost all cells and represent the minimal requirements for maintaining the integrity and life of the cell. Thus, a human liver cell and an amoeba are remarkably similar in their means of exchanging materials with their immediate environments, of obtaining energy from organic nutrients, of synthesizing complex proteins, and of duplicating themselves.

This is not to say that there are no significant differences between an amoeba and a liver cell or between the liver cell and a nerve cell. However, a second crucial generalization is that these differences in cell function generally represent specializations of one or more of the fundamental common properties. For example, the excitability of nerve cells represents a specialization of electrical phenomena common to the membranes of virtually all cells; the secretion of protein hormones by certain gland cells of the body is a specialized form of the genetically controlled protein synthesis found in all cells; the transport of food molecules across the cells forming the intestinal wall results from a specialized orientation of transport mechanisms remarkably similar in most cells. These specializations have all occurred as a part of evolution and have resulted in the adaptation of certain cells for specific roles.

A society of cells

The human organism begins as a single cell, the fertilized ovum, which gives rise to the entire body by cell division and differentiation. The result is the formation of various types of specialized cells, each type differing in structure and function from the others. Cells which have a similar origin and structure and subserve the same general function are frequently found grouped together in sheetlike masses, or *tissues,* of which four main types are generally recognized: (*1*) muscle tissue, (*2*) nerve tissue, (*3*) connective tissue, and (*4*) epithelial (or lining) tissue. However, each of these four major categories actually consists of specific cell types which may differ considerably from other cell types of the same category; e.g., there are three distinct types of muscle cells: skeletal muscle, heart muscle, and smooth muscle.

Sometimes a single cell or tissue may function fairly independently of all others, but more commonly a number of different cell types are intimately associated with each other to form larger functional units called *organs:* heart, liver, kidney, pancreas, etc. The kidney, for example, consists largely of (*1*) a series of small tubes each composed of a single layer of epithelial cells; (*2*) blood vessels, whose walls consist of an epithelial lining and varying quantities of smooth muscle and connective tissue; (*3*) nerve processes; and (*4*) an enclosing connective-tissue capsule.

Finally, the last order in classification is that of the *organ system,* a collection of organs which together subserve an overall function. Thus the kidneys, the bladder, and the tubes leading from kidneys to bladder and from the bladder to the exterior constitute the urinary system.

In essence, then, the human body can be viewed as a complex society of cells of many different types which are structurally and functionally combined and interrelated in a variety of ways to carry on the functions essential to the survival of the organism as a whole. Yet the fact remains that individual cells still constitute the basic units of this society and that almost all these cells individually exhibit the fundamental activities common to all forms of life. Indeed, many of the body's different cell types can be removed from the body and maintained in test tubes as free-living cells.

There is a definite paradox in this analysis. If each individual cell performs the fundamental activities required for its own survival, what contributions do the different organ systems make? How can we refer to a system's functions as being "essential to the survival of the organism as a whole" when each individual cell of the organism seems to be capable of performing its own fundamental activities? The resolution of this paradox is found in the isolation of most of the cells of a multicellular organism from the environment surrounding the body (*external environment*). An amoeba and a human liver cell both obtain most of their required energy by the breakdown of certain organic nutrients; the chemical reactions involved in this intracellular process are remarkably similar in the two types of cells and involve the utilization of oxygen and the production of carbon dioxide. The amoeba picks up required oxygen directly from its environment and eliminates the carbon dioxide into it. But how can the liver cell obtain its oxygen and eliminate the carbon dioxide when, unlike the amoeba, it is not in direct contact with the external environment? Supplying oxygen to the liver is the function both of the respiratory system (comprising the lungs and the airways leading to them), which takes up oxygen from the environment, and of the circulatory system, which distributes oxygen to all parts of the body. Conversely, the circulatory system carries the carbon dioxide generated by the liver cells and all the other cells of the body to the lungs, which eliminate it to the exterior.

Similarly, the digestive and circulatory systems, working together, make nutrients from the external environment available to all the body's cells. Wastes, other than carbon dioxide, are carried by the circulatory system from the cells which produced them to the kidneys, which excrete them from the body.

Thus, the overall effect of the activities of organ systems is to create *within* the body the environment required for all cells to function. This concept of an *internal environment,* to be developed in detail in Chap. 5, supplies the basis for understanding the common denominator of the functions carried out by the organ systems, namely, *maintaining stable conditions within the internal environment.*

The total activities of every individual cell in the body fall into two categories: (*1*) Each cell performs for itself all those fundamental basic cellular processes (movement of materials across its membrane, extraction of energy, protein synthesis, etc.) which represent the minimal requirements for maintaining its individual integrity and life; (*2*) each cell simultaneously performs one or more specialized activities which in concert with the other cells of its tissue or organ system contribute to the survival of the total organism by helping maintain the stable internal environment required by all cells. These latter specialized activities together constitute the coordinated body processes (circulation, respiration, digestion, etc.) typical of multicellular organisms like man.

Clearly, the society of cells which constitutes the human body bears many striking similarities to a society of persons (although the analogy must not be pushed too far). Each person in a complex society must perform for himself a set of fundamental activities (eating, excreting, sleeping, etc.), which is virtually the same for all persons. In addition, because the complex organization of a society makes it virtually impossible for any individual within the society to raise his own food, arrange for the disposal of his wastes, and so on, each individual participates in the performance of one of these supply-and-disposal operations required for the survival of all. A specialized activity, therefore, becomes an *additional* part of his daily routine, but it never allows him to cease or to reduce his performance of the fundamental activities required for his survival.

The importance of control

Implicit in life is *control*. Regardless of its level of organizational complexity, no living system can exist without precise mechanisms for controlling its various activities.

Every one of the fundamental processes performed by any single cell (amoeba or liver cell) must be carefully regulated. What determines how much sugar is to be transported across the cell's membrane? What proportion of this sugar, once inside the cell, is to be utilized for energy or transformed into fat or protein? How much protein of each type is to be synthesized and when? How large is the cell to grow, and when is it to divide? The list is almost endless. Thus, an understanding of cellular physiology requires not only a knowledge of the basic processes but also of the mechanisms which control them. Indeed, the two are inseparable.

In any multicellular organism like man, these basic intracellular regulators remain, but the existence of a multitude of different cells organized into specialized tissues and organs obviously imposes the need for overall regulatory mechanisms. Information about all important aspects of the external and internal environments must be monitored continuously; this information must be integrated, and on the basis of its content "instructions" must be sent to the various tissue and organ cells (particularly muscle and gland cells) directing them to increase or decrease their activities.

This transmission and integration of information is performed primarily (although not exclusively) by the nervous and hormonal systems. Thus these two systems contribute to the survival of the organism by controlling the activities of the various bodily components so that any change (or impending change) in the body's internal environment automatically initiates a chain of events wiping out the change or leading to a state of readiness should the impending change actually occur. For example, when (for any reason) the concentration of oxygen in the body significantly decreases below normal, the nervous system detects the change and increases its output to the skeletal muscles responsible for breathing movements; the result is a compensatory increase in oxygen uptake by the body and a restoration of normal internal oxygen concentration.

The above example is analogous to certain systems in engineering, say for maintaining constant oxygen concentration in a submarine. Similarly, the body temperature is regulated by a system whose underlying principles are nearly identical to the thermostatically controlled system which keeps a house at some specified temperature. These are known as *control systems*.

A summation and rationale

With this framework in mind, the overall organization and approach of this book should easily be understood. Because the fundamental features of cell function are shared by virtually all cells and, in addition, constitute the foundation upon which specialization develops, we devote the first section of this book on the human animal to an analysis of basic cellular physiology. We also emphasize cell physiology at the start because the analysis of cellular phenomena in terms of physicochemical principles has proved very successful of late. Indeed, much of cell biology is now referred to as *molecular biology* in recognition of the ultimate goal of explaining all cellular processes in terms of interactions between molecules of known structure.

At the other end of the organizational spectrum, the third part of the book describes how the body's various coordinated functions (circulation, respiration, etc.) result from precisely controlled and integrated activities of specialized cells grouped together in tissues and organs. The theme of these descriptions is that each of these coordinated functions (with the obvious exception of reproduction) serves to keep some important aspect of the body's internal environment relatively constant and can be therefore described in terms of a control system similar to those familiar in engineering.

The second section of the book provides the principles and information required to bridge the gap between these two organizational levels, the cell and the body. First, the evolutionary origin and physicochemical composition of the internal environment are presented. Second, control systems are analyzed in general terms to emphasize that the basic principles governing virtually all control systems are the same. Finally, the bulk of this section is concerned with the major components of the body's control systems (nerve cells, muscle cells, and gland cells); the physicochemical basis for their specialized cellular activities and the interactions between them are emphasized. Once acquainted with the cast of major characters—nerve, muscle, and gland cells—and the theme—the maintenance of a stable internal environment through the interactions of these characters—the reader will be free to follow the specific plot lines—circulation, respiration, etc.—of Part 3.

The human aspect of human physiology

The principles discussed thus far apply to almost any complex multicellular organism, and most of the material in Parts 2 and 3 applies to all mammals (indeed, most of it comes from observations and experiments on non-human mammals). There are, of course, significant differences between the mammals with respect to the detailed mechanisms by which their various organs function. No human organ is precisely like its counterpart in the dog or rat, and we have made every attempt to give the specifically human mechanism whenever it is known (such information is frequently lacking).

With one exception, these differences are of relatively minor significance, but the exception, the phenomena which we shall group together as conscious experience, is of enormous importance and constitutes the unique aspect of human beings. It is not consciousness alone which distinguishes man, for all mammals clearly manifest this characteristic. It is the *content* of our consciousness which seems to differentiate us, although biologists certainly have little knowledge, at present, of what the content of any other mammal's consciousness is. We attempt to analyze this problem in Chap. 18. and have introduced it here by way of explaining the use of

the word "almost" in our previous assertions that the common denominator of bodily activities, other than reproduction, is the maintenance of a stable internal environment. Does the appreciation of a Mozart concerto really fit this description? Perhaps one could squeeze it into the mold by pushing, but why try? Better to admit that we understand nothing of such things.

Teleology and causality

We have emphasized that a common denominator of physiological processes is their contribution to survival. Unfortunately, it is easy to misunderstand the nature of this relationship. Consider, for example, the statement that "during exercise a person sweats *because* his body *needs* to get rid of the excess heat generated." This type of statement is an example of *teleology,* the explanation of events in terms of purpose. But it is not an *explanation* at all, in the scientific sense of the word. It is somewhat like saying, "The furnace is on because the house needs to be heated." Clearly, the furnace is on not because it senses, in some mystical manner, the house's *needs* but because the temperature has fallen below the thermostat's set point and the electric current in the connecting wires has turned on the heater and blower.

Is it not true to say the sweating actually serves a useful purpose because the excess heat, if not eliminated, might have caused sickness or even death? Correct, but this is totally different from stating that a "need" to avoid injury *caused* the sweating to occur. The *cause* of the sweating, in reality, was an automatically occurring sequence of events initiated by the increased heat generation; increased heat generation—increased blood temperature—increased activity of specific nerve cells in the brain—increased activity, in turn, of a series of nerve cells, the last of which stimulate the sweat glands—increased production of sweat by the sweat-gland cells. Each of these steps occurs by means of physicochemical changes in the cells involved. In science to *explain* a phenomenon is to reduce it to a sequence of physicochemical events. This is the scientific meaning of causality, of the word "because."

Of course, that the phenomenon is beneficial to the person is of considerable interest and importance. It is attributable to evolutionary processes, which result in the selecting out of those responses having survival value. Evolution is the key to understanding why most bodily activities do indeed appear to be purposeful. Throughout the book we emphasize how a particular process contributes to survival, but the reader must never confuse this survival value of a process with the explanation of the mechanisms by which the process occurs.

BASIC
CELL
FUNCTIONS PART

1

Chemical composition of the body

Atoms and molecules

Like all matter, the human body is composed of atoms and molecules, and its moment-to-moment functioning depends upon the millions of interactions occurring between the various atoms and molecules it contains. The wide diversity of structure and function in the body is a reflection of the properties of the molecules found in various tissues and their chemical and physical interactions. Before we can consider the properties of cells, tissues, organs, and organ systems, we must have some understanding of the types of atoms and molecules which compose the substance of the body and form the foundation upon which all other levels of body structure and function are based. Biologists are beginning to relate the infinitesimal realm of the atom to some of the most complex properties of living systems, such as disease, heredity, and memory. We shall begin with a brief description of those basic properties of atoms and molecules that are most relevant to the types of interactions that occur in living systems.

Atoms

Atoms are, of course, very small, but how small is small? To describe the dimensions of such small objects we must use appropriately small units of measurement. The dimensions of atoms and molecules are measured in angstrom units of length, where one angstrom (Å) is equal to about 0.000000004 in. Hydrogen, the smallest atom, is a sphere about 1 Å in diameter. A cell, the smallest unit of living matter, is typically about 100,000 Å in diameter, considerably larger than individual atoms. However, many important structures within a cell have dimensions

1

TABLE 1-1
English and metric units

	English	Metric
Length	1 foot = 0.305 meter	1 meter = 39.37 inches
	1 inch = 2.54 centimeters	1 centimeter (cm) = 1/100 meter
		1 millimeter (mm) = 1/1000 meter
		1 micrometer (μm) = 1/1000 meter
		1 nanometer† (nm) = 1/1000 micrometer
		1 angstrom (Å) = 1/10 nanometer
Weight	1 pound = 433.59 grams	1 kilogram (kg) = 1,000 grams = 2.2 pounds
	1 ounce = 27.1 grams	1 gram (g) = 0.037 ounce
Volume	1 gallon = 3.785 liters	1 liter = 1,000 cubic centimeters = 0.264 gallon
	1 quart = 0.946 liter	1 liter = 1.057 quarts
		1 milliliter (ml) = 1/1000 liter

† In the past the term millimicron (mμ), one-thousandth of a micron (now termed micrometer), was used for this unit. A more consistent terminology results from the use of separate prefixes for the subdivision of basic units; thus, milli- = one-thousandth, micro- = one-millionth, and nano- = one-billionth.

which indicate that they must be composed of relatively few atoms. Thus cell membranes which cover the surfaces of cells and surround many intracellular organelles are only about 75 Å thick.

The angstrom unit belongs to the metric system of measurement which is used by scientists throughout the world in reporting the magnitudes of their observations of natural phenomena. This system was introduced in France in 1790 and was spread throughout Europe by Napoleon's conquests. The metric system of meters, grams, and liters is now used as the basic units of measurement for both science and commerce in most countries of the world with the exception of the United States, which has retained the English units of feet, pounds, and gallons. The metric system of units will be used throughout this book. Table 1-1 provides a comparison of the basic English and metric units.

Atoms, the building blocks of matter, are composed of three fundamental particles: electrons, protons, and neutrons. These subatomic particles can be distinguished on the basis of their mass (weight) and electric charge (Table 1-2). The electron has a negative charge and almost no mass; the proton has a positive charge and about 2,000 times the mass of an electron. The neutron has about the same mass as the proton but is not electrically charged. The major effect of an electric charge is to exert a force on other electric charges in its vicinity. An attractive force occurs between a positive and a negative charge, and a repulsive force occurs between two similar charges, two positive or two negative charges.

Attractive and repulsive forces between these charged particles are responsible for forming chemical bonds between atoms and molecules and are ultimately responsible for the physical and chemical interactions of matter. (A further discussion of electric charge will be found in Chap. 6 when we consider the electrical properties of cells.)

Protons and neutrons are clustered together in a very small volume at the center of an atom, forming a positively charged nucleus; negatively charged electrons revolve about the nucleus much as the planets revolve about the sun (Fig. 1-1). The positive protons, which would normally repel each other, are held together in the nucleus of the atom by very powerful short-range nuclear forces, which are responsible for the tremendous amounts of energy released when the nucleus of an atom is split in an atomic reaction such as occurs when a nuclear weapon is exploded. These powerful nuclear energies which stabilize atomic structure play no role in

TABLE 1-2
Subatomic particles

Particle	Discoverer	Date	Electric charge	Relative mass
Electron	Thomson	1897	Negative	1
Proton	Rutherford	1919	Positive	1,836
Neutron	Chadwick	1932	Neutral	1,838

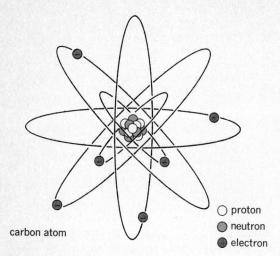

proton
neutron
electron

FIGURE 1-1
Atomic structure. Six electrons (negative) revolve around the nucleus of a carbon atom containing six protons (positive) and six neutrons (neutral).

the chemical and physical interactions of atoms in the body.

The atom as a whole is electrically neutral since it has as many negative electrons orbiting the nucleus as it has positive protons in its nucleus. The chemical properties of the elements are determined by the number of electrons associated with a particular atom. If we rank the 103 known elements on the basis of the number of electrons they contain, we find an unbroken sequence from hydrogen, the smallest atom (1 electron), to lawrencium (103 electrons). The number of electrons orbiting the nucleus of an atom is known as the *atomic number*. Since atoms are electrically neutral, the atomic number is also equal to the number of protons in the nucleus of the atom (Table 1-3).

Since neutrons and protons are about 2,000 times

heavier than electrons, most of the mass of the atom is located in the nucleus. Hydrogen is the lightest element, with only a single proton in its nucleus, and lawrencium is the heaviest with 103 protons and 154 neutrons. Because of the negligible weight of the electrons, we would expect the lawrencium atom to be $103 + 154 = 257$ times heavier than the hydrogen atom. By comparing the weights of the different atoms, one can establish a relative weight scale. A carbon atom is 12 times heavier than a hydrogen atom; thus we assign a relative weight of 12 to carbon, a relative weight of 1 to hydrogen, and so on for the other atoms. The relative weight of an atom is known as the *atomic weight*. No actual unit, such as grams or pounds, is associated with atomic weight; the atomic weight of carbon, for example, means that the carbon atom is 12 times heavier than the hydrogen atom with an atomic weight of 1. How useful the relative atomic weight scale is can be seen by considering the types of numbers needed to refer to the actual weight of atoms. The actual weights in grams are:

Weight of 1 carbon atom
$$= 0.0000000000000000000000199$$
Weight of 1 hydrogen atom
$$= 0.0000000000000000000000017$$

The relative weight of a molecule, composed of two or more atoms, is known as the *molecular weight* and is equal to the sum of the atomic weights of all the atoms in the molecule. Thus, a molecule of water, H_2O, which contains two hydrogen atoms and one oxygen atom, has a molecular weight of $1 + 1 + 16 = 18$. The quantity of a chemical compound present in a test tube, a cell, or the total body can be represented by stating the number of grams of the substance or the total number of molecules present. An additional unit for measuring the amount of a substance present in a system, which combines both the weight of the molecule and the number of molecules present, is known as a *mole*. The total number of moles of a compound present in a system is equal

TABLE 1-3
Composition of atoms

Atom	Atomic number	Number of electrons	Number of protons	Number of neutrons	Atomic weight
Hydrogen	1	1	1	0	1
Carbon	6	6	6	6	12
Nitrogen	7	7	7	7	14
Oxygen	8	8	8	8	16
Lawrencium	103	103	103	154	257

to the weight of the compound in grams divided by its molecular weight:

$$\text{Moles} = \frac{\text{weight in grams}}{\text{molecular weight}}$$

Thus one mole of a compound is the weight of the compound in grams equal to its molecular weight; e.g., 18 g of water which has a molecular weight of 18 is 1 mole of water, and 9 g of water is $\frac{1}{2}$ mole. The number of grams of most compounds present in biological systems is generally much less than 1 mole, and the millimole (mmole, 0.001 mole) is the most common unit used for measuring the amount of substances present in biological systems. Since the mole is based upon the molecular weight of the substance, which is proportional to the actual weight of the molecule, 1 mole of any compound must contain the same number of molecules as 1 mole of any other compound. The actual number of molecules in 1 mole is a very large number, about 6×10^{23} molecules per mole.

It is often more useful to describe the amount of substance present in a specified volume of the system than to give the total amount present in the entire system. The term *concentration* refers to the amount of a substance present in a specified unit volume of the system. The primary unit of volume in the metric system is the liter; thus concentrations are given as grams per liter, molecules per liter, or moles per liter. By using concentration rather than total amounts, it is possible to compare the chemical compositions of systems which have quite different volumes, such as the composition of a single cell having a volume of 0.000000000000001 liter with the composition of the blood having a volume of about 5 liters. Although the total amount of a substance in the body may vary with the size of an individual, the concentration of many biological substances is approximately the same in all individuals, independent of the body size.

Although 103 different chemical elements are known, only a few of them play an important role in living organisms. In fact, just four elements, hydrogen, carbon, nitrogen, and oxygen, account for 96 percent of the body weight and over 99 percent of the atoms in the body (Table 1-4). In pure form, oxygen, hydrogen, and nitrogen are gases at room temperature, and carbon is ordinary charcoal; yet most of the body's substance is built out of these four basic building blocks. From Table 1-4 we see that 65 percent of the body weight is composed of oxygen and 10 percent of hydrogen. However, since one oxygen atom weighs 16 times as much as one hydrogen atom, there is actually a greater number of hydrogen atoms in the body than oxygen atoms. Thus, the percent-

TABLE 1-4
Atomic composition of the body: four most numerous elements

Element	Atomic number	Total body weight, %	Total atoms in body, %
Hydrogen, H	1	10	63
Carbon, C	6	18	9
Nitrogen, N	7	3	1
Oxygen, O	8	65	26
Total		96	99

age composition of the body based upon weight does not represent the relative *numbers* of the different atoms in the body. Hydrogen, the simplest of all atoms, not only is the most numerous atom in the body but accounts for over 99 percent of all atoms in the universe.

Although the four elements—hydrogen, oxygen, carbon, and nitrogen—account for over 99 percent of the atoms in the body, most of the remaining 1 percent of the atoms play extremely important roles and are essential for the life of the organism. Table 1-5 lists the 24 elements that are presently known to be essential for the normal functioning of the body. The four major elements provide the basic building blocks for the compounds of the body and for water. Most of the seven elements listed as major minerals exist as electrically charged particles (ions, see below) and, along with water, provide the intracellular and extracellular media in which the materials of the body are suspended. We shall see that these elements play very important roles in the electrical activity of the body and in the transfer and utilization of chemical energy by cells. The 13 trace elements listed in Table 1-5 are those elements that are present in very small quantities but are essential for the maintenance of normal growth and functioning in mammals. There is a growing amount of indirect evidence that these same elements are essential for normal human growth and function. In future years it is possible that additional trace elements will be added to this list.

In addition to the 24 elements listed in Table 1-5, many other elements can be detected in the body. These elements enter the body through the foods we eat and the air we breathe but at the present time do not appear to have any essential chemical function. Some elements, for example lead and mercury, not only are not essential for normal function but may be extremely toxic and damaging to the normal chemical processes in the body.

In 1936 one could purchase from a chemical sup-

TABLE 1-5
Essential elements in the body

Symbol	Element
1 Major elements: 99.3% total atoms	
H	Hydrogen
O	Oxygen
C	Carbon
N	Nitrogen
2 Major minerals: 0.7% total atoms	
Ca	Calcium
P	Phosphorus
K (Latin, *kalium*)	Potassium
S	Sulfur
Na (Latin, *natrium*)	Sodium
Cl	Chloride
Mg	Magnesium
3 Trace elements: less than 0.01% total atoms	
Fe (Latin, *ferrum*)	Iron
I	Iodine
Cu (Latin, *cuprum*)	Copper
Zn	Zinc
Mn	Manganese
Co	Cobalt
Cr	Chromium
Se	Selinium
Mo	Molybdenum
F	Fluorine
Sn (Latin, *stannum*)	Tin
Si	Silicon
V	Vanadium

plier all the elements in the body for 98 cents. (Today, with inflation, the cost of these elements would be about $3.50.) The important question is not so much what elements are present in the body but how these elements are assembled to form molecules which interact with other molecules to produce the variety of functions carried out by the human body.

Molecules

Molecules are formed when atoms are linked together by chemical bonds. It is during the process of forming and breaking chemical bonds between atoms that the chemical potential energy of a molecule is released. A chemical bond may be compared to a spring joining two atoms, the stretch of the spring representing the stored potential energy, which is released when the bond is broken.

Forming a chemical bond involves an exchange or sharing of electrons between two atoms. Since atoms are electrically neutral, any transfer of an electron from one atom to another results in one atom's becoming more positive and the other more negative. The attraction between the positive and negative charges forms the basis of the chemical bond.

Hydrogen atoms can react with atoms of carbon, nitrogen, and oxygen to form molecules of methane, CH_4, ammonia, NH_3, and water, H_2O (Fig. 1-2). Wherever a chemical bond has been formed, the electron associated with the hydrogen atom has become partially incorporated into the electron structure of the opposite atom. With the formation of the chemical bond, the electrons are shared by both atoms, and electric interaction holds the atoms together. The number of bonds that can be formed with a given atom depends upon its electron structure. Thus, hydrogen is capable of forming only a single bond, whereas oxygen can form two bonds, nitrogen three, and carbon four. Sometimes a double bond may be formed between two atoms, as in carbon dioxide, CO_2, $O=C=O$, where carbon still forms four chemical bonds and oxygen only two. In methane the four chemical bonds of carbon are formed with four separate hydrogen atoms. A chemical bond is generally represented as a line joining two atoms, two lines indicating a double bond.

Although individual atoms are spherical, when atoms combine, the shape of the resulting molecule is usually not a sphere. When a single atom, such as carbon, can form four separate bonds, a molecule made up of many carbon atoms can have many branches, as illustrated in Fig. 1-3. The shape of a molecule often plays a very important role in its function in living organisms. Even when only a few atoms form a molecule, as in methane, ammonia, and water, the result has a definite geometrical configuration (Fig. 1-4). When more than one chemical bond is formed with a given atom, the bonds are distributed about the atom in a definite pattern which may or may not be symmetrical. Thus, the four bonds of a carbon atom occur at the four corners of a tetrahedron with the carbon atom at the center. The three bonds of the nitrogen atom form the three legs of a tripod, and the two bonds of the oxygen atom are not on a straight line but form an angle of 105°.

Ions and polar molecules

Just as the atoms of oxygen and hydrogen are themselves electrically neutral, the molecule of water, H_2O, formed from them is electrically neutral. The total number of electrons in a water molecule is equal to the total number of protons. However, the negative electron of

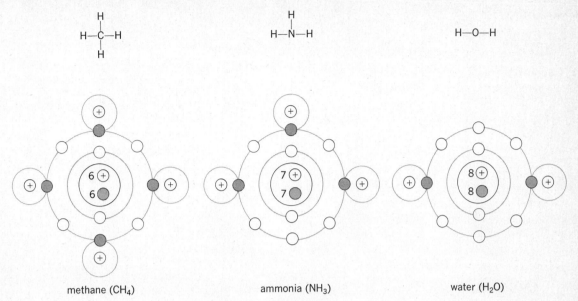

methane (CH₄)   ammonia (NH₃)   water (H₂O)

FIGURE 1-2
Chemical bonds formed between hydrogen atoms and atoms of carbon, nitrogen, and oxygen.

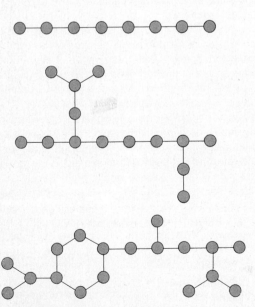

FIGURE 1-3
Combination of spherical atoms to form molecules of various shapes.

each hydrogen atom is more strongly attracted to the positive oxygen nucleus, containing eight protons, than to the single proton of the hydrogen nucleus, and therefore the oxygen atom is slightly negative and the two hydrogen atoms slightly positive. The hydrogen-oxygen bonds of the water molecule are polarized, and water is said to be a polar molecule (Fig. 1-5). In general, when different atoms are joined by a chemical bond, the electrons are not equally distributed and a polar bond results. In contrast, when two similar atoms such as carbon are joined, the electrons are shared equally and the bond is not polarized. The presence of polar bonds in a molecule provides sites for strong electric interactions with other molecules.

Some interactions between atoms result in the complete transfer of an electron from one atom to another, producing atoms having a net charge and known as *ions*. Thus, an atom of sodium transfers an electron to an atom of chlorine, forming a positive sodium ion containing 11 protons but only 10 electrons. Upon gaining an electron the chlorine atom becomes a negative chloride ion containing 18 electrons but only 17 protons (Fig. 1-6). Ion formation is an extreme case of a polarized chemical bond in that there has been a complete transfer

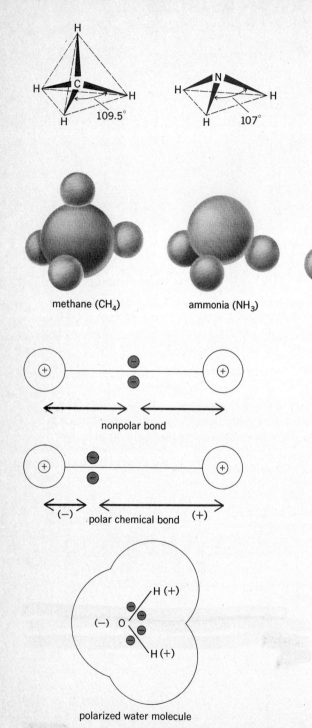

FIGURE 1-4
Geometrical configuration of chemical bonds around the carbon, nitrogen, and oxygen atoms.

methane (CH_4)   ammonia (NH_3)   water (H_2O)

nonpolar bond

(−)   polar chemical bond   (+)

polarized water molecule

FIGURE 1-5
Formation of polar chemical bonds resulting from the unequal distribution of electrons between two atoms.

of charge between two atoms. Some atoms are able to lose or gain more than one electron. For example, calcium and magnesium atoms can lose two electrons to become the positively charged calcium, Ca^{2+}, and magnesium, Mg^{2+}, ions.

One of the most important ions in the body is the hydrogen ion, H^+. When the single electron of the hydrogen atom is completely transferred to another atom, only the positive proton of the hydrogen nucleus remains; this constitutes the hydrogen ion. Molecules which give rise to hydrogen ions are known as *acids*. The acidity of a solution is determined by the concentration of hydrogen ions present. Thus, hydrochloric acid, HCl, dissolved in water produces a hydrogen ion, H^+, and a chloride ion, Cl^-, the electron of the hydrogen atom being transferred to the chlorine atom. Likewise acetic acid, CH_3COOH, can form a hydrogen ion, H^+, and an acetate ion, CH_3COO^- (Fig. 1-7). Here the hydrogen-oxygen bond has broken, and the oxygen atom has completely captured the hydrogen's electron. When the hydrogen-oxygen bond is broken, the molecule is said to undergo *ionization*, the resulting products being the hydrogen and acetate ions.

Although many chemical bonds are able to undergo ionization and thus produce charged molecules, most chemical bonds involving carbon and hydrogen atoms do not ionize and remain electrically neutral. Thus methane, CH_4, cannot produce hydrogen ions and become a meth-

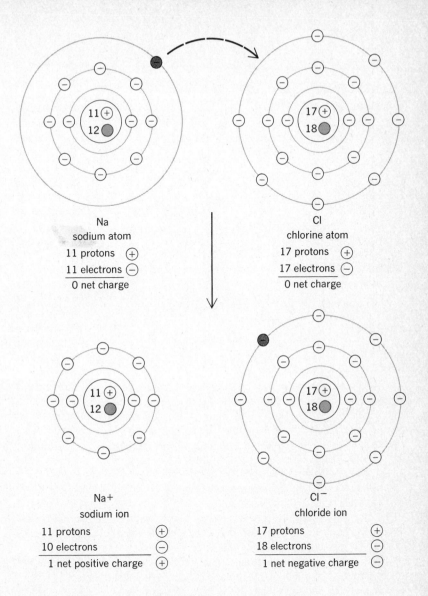

Na
sodium atom
11 protons   ⊕
11 electrons   ⊖
0 net charge

Cl
chlorine atom
17 protons   ⊕
17 electrons   ⊖
0 net charge

Na+
sodium ion
11 protons   ⊕
10 electrons   ⊖
1 net positive charge   ⊕

Cl⁻
chloride ion
17 protons   ⊕
18 electrons   ⊖
1 net negative charge   ⊖

FIGURE 1-6
Formation of charged ions resulting from the transfer of electrons from one atom to another.

ane ion. In this molecule the forces attracting the hydrogen electron to the carbon atom are not sufficiently strong to allow carbon to completely remove the electron from hydrogen and form a hydrogen ion. The most commonly encountered polar and ionic groupings in organic molecules will be considered later (Fig. 1-14).

In summary, molecules consist of atoms held together by chemical bonds, which are formed by the sharing of electrons between two atoms. The electric forces holding atoms together are a source of potential energy, which can be released when the bonds are broken.

Depending upon the types of chemical bonds in a molecule, the molecule may be electrically neutral, polarized, or ionized.

States of matter

When a number of molecules are brought together, they form larger aggregates of matter. Three general states of matter are recognized: gaseous, liquid, and solid. The ability of matter to exist in any of these three states depends upon the forces of interaction between molecules. In all three states the molecules are in continuous motion. The

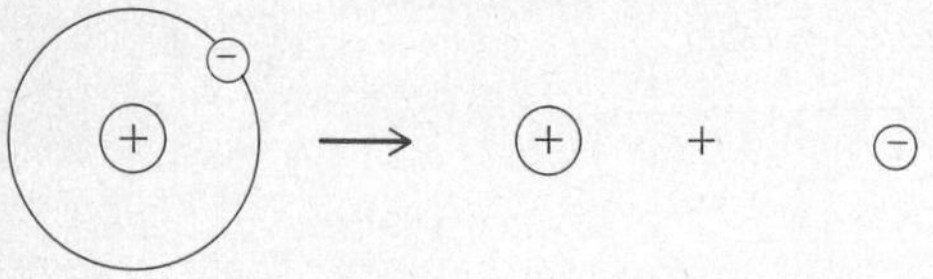

H $\longrightarrow$ H$^+$ + e$^-$
hydrogen atom   hydrogen ion   electron
                    (proton)

HCl $\longrightarrow$ H$^+$ + Cl$^-$
hydrochloric acid   chloride ion

$$CH_3-\overset{\overset{\textstyle O}{\|}}{C}-OH \longrightarrow H^+ + CH_3-\overset{\overset{\textstyle O}{\|}}{C}-O^-$$
acetic acid   acetate ion

FIGURE 1-7
Ionization of molecules of an acid leads to the formation of hydrogen ions.

higher the temperature, the faster is their motion. The three states of matter differ in the degree to which the individual molecules are free to move about. Like the forces which hold atoms together, the forces between molecules are electric forces arising from ionized groups on a molecule or polarized chemical bonds. A negative ion is attracted to a positive ion, and the negative portion of a polarized chemical bond on one molecule is attracted to the positive portion of a polarized bond of another molecule. In general, the electric forces interacting between molecules are considerably weaker than those holding atoms together within a molecule.

The electric force of attraction, which draws molecules together, is offset by the motion of the molecules, which tends to separate them. If the attractive force between molecules is greater than the average kinetic energy of the moving molecules, their movement is severely restricted and a solid results. In the solid state the molecules vibrate in relatively fixed positions with little freedom of independent movement relative to each other. In the gaseous state the forces of attraction between molecules are very weak relative to the kinetic energy of the molecules, and there is little restriction of the movement of individual molecules. The liquid state is intermediate in that there is greater freedom of movement than in the solid state but not the unrestricted motion characterizing

the gaseous state. A transition from the solid to the liquid or from the liquid to the gaseous state can often be achieved by raising the temperature, which increases the kinetic energy of the individual molecules so that they are able to break away from the attracting forces holding them together. Thus, warming solid ice converts it to liquid water, and further warming of water converts it to gaseous water vapor.

All three states of matter are found in the body. The air we breathe is a mixture of gases; bones, teeth, fingernails, and hair are solids. The major portion of the body, however, exists in the liquid state, and the molecule responsible for this liquid state is water.

Water Water forms the medium in which all living processes occur, and life as we know it is inconceivable in the absence of this deceptively simple little molecule; 60 percent of the body weight is water and 80 percent of the weight of a typical living cell is water. Just as hydrogen is the most numerous atom in the body, water is the most numerous molecule in the body; 99 out of every 100 molecules in the body are water molecules.

We noted earlier that the hydrogen-oxygen bonds of the water molecule are polarized so that the oxygen atom is slightly negative and the two hydrogen atoms are slightly positive. This polar structure is responsible for water's being a liquid at room temperatures. The slightly negative oxygen atom of one water molecule is electrically attracted to the slightly positive hydrogen atom of a second water molecule, forming an electrostatic attraction between two water molecules that is much weaker than the chemical bond between hydrogen and oxygen atoms within the water molecule. It does not involve any exchange of electrons between the molecules. Each water molecule is able to form four such electrostatic bonds, one with each of its two hydrogen atoms and two with its oxygen atom. This pattern is repeated over and over between adjacent water molecules until a three-dimensional lattice structure is formed. At low temperatures water exists as solid ice in which almost all the water molecules are bound to each other and there is relatively little freedom of movement for the individual molecules. As the temperature is raised, the kinetic energy of the molecules is increased, and a number of these weak bonds are broken, transforming the solid ice into liquid water. In the liquid state water probably exists as a number of clusters of associated water molecules (Fig. 1-8). At higher temperatures more and more of the weak electric bonds are broken, and the freedom of movement of individual water molecules is increased, allowing them

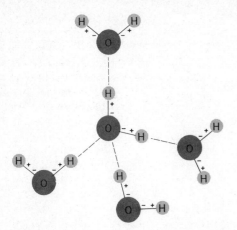

FIGURE 1-8
Electric attraction between polar water molecules is responsible for the liquid state of water at body temperature.

to escape from their neighbors into the gaseous state. A molecule dissolved in a liquid is known as a *solute*, and the liquid in which it dissolves is known as the *solvent*. Water is able to act as a solvent for a wide variety of solutes because water molecules can form weak electric bonds with the solute molecule. For example, common table salt (sodium chloride, NaCl) is a solid crystalline substance because of the strong electric attraction between the positive sodium ions, Na^+, and the negative chloride ions, Cl^-. Salt dissolves in water because the polar water molecules are attracted to the highly charged ions of sodium and chloride and surround them with a cluster of water molecules (Fig. 1-9). The cluster of water molecules around the ions decreases the electric force of attraction between the sodium and chloride ions, allowing them to separate and go into solution in the water medium.

For a molecule to dissolve in water it must be able to form electrostatic bonds with individual water molecules. Thus it must possess a certain number of polar or ionic groups in its molecular structure which can become associated with the polar groups on the water molecule. Molecules which do not dissolve in water have few if any polar or ionic groups. Although most of the molecules we shall be considering can dissolve in water because they contain such groups, one very important class of biological molecules, which includes the fats, is not soluble in water. The fact that these molecules cannot associate with the polar water molecule is an important aspect of their functional role in living systems.

Because so many different molecules can dissolve in water, it is an ideal medium for chemical reactions. In order for two molecules to undergo a chemical reaction they must come close enough to each other to interact by forming new chemical bonds. In the solid state there is not enough movement of the individual molecules to bring many of them together in a chemical reaction. In a gas the molecules have freedom of movement, but the concentration (number of molecules in a given volume) is so small that the probability of their coming in contact with each other is small. When molecules are dissolved in water, a relatively high concentration can be achieved in a small volume with enough freedom of movement to allow frequent collisions between the molecules.

Water is not only the medium in which chemical reactions occur, but it is frequently a participant. Its atoms may be incorporated into another molecule in the course of a chemical reaction

$$R—O—\overset{O}{\overset{\|}{C}}—R' + HOH \longrightarrow R—OH + HO—\overset{O}{\overset{\|}{C}}—R'$$

or a molecule of water may be formed during the course of a reaction

$$R—OH + HO—\overset{O}{\overset{\|}{C}}—R' \longrightarrow R—O—\overset{O}{\overset{\|}{C}}—R' + HOH$$

where R and R' represent the remaining structure of the molecule.

We have discussed only a few of the basic properties of water, but succeeding chapters will show that water participates in practically every process in a living organism. The circulation of blood and body fluids would be impossible without a liquid medium such as water, and the regulation of body temperature depends, in part, on the evaporation of water from the skin. The living state must maintain a proper water balance, and the body has developed numerous mechanisms for regulating the supply, distribution, and elimination of this critical substance.

The structure and properties of organic molecules

The chemistry of carbon

Despite the importance of water in living systems, the chemistry of life is centered around the chemistry of the carbon atom. If we remove all the water from the body (60 percent of the body weight), the remaining 40 percent is the dry weight of the body. Although hydrogen is the most numerous atom in the body on the basis of either wet or dry weight, carbon, rather than oxygen, is the

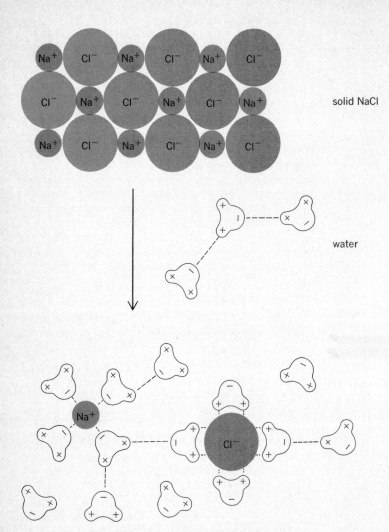

solid NaCl

water

solution of sodium and chloride ions

FIGURE 1-9
The ability of water to dissolve sodium chloride crystals depends upon the electric attraction between the polar water molecules and the charged sodium and chloride ions.

second most numerous atom among the elements that compose the dry weight.

Until the nineteenth century chemists had made little progress in understanding the chemical reactions involving carbon. It was known that molecules containing carbon were found primarily in living organisms, whereas nonliving matter was composed primarily of mineral elements such as iron, copper, zinc, calcium, magnesium, phosphorus, sulfur, chlorine, sodium, and potassium as well as oxygen, nitrogen, and hydrogen. Mineral elements readily undergo chemical reactions with each other to form simple molecular structures consisting of relatively few atoms, but the carbon compounds present in living organisms were found to have very complex structures, containing many atoms (often thousands). They did not appear to undergo chemical reactions as readily as the mineral elements. It was thought that they were basically different from the compounds found in the nonliving world and that they could be synthesized within living cells only under the influence of some mystical "vital force." In the early nineteenth century chemists finally succeeded in synthesizing carbon-containing molecules in a test tube, independent of any living matter, and were able to show that these molecules obey the same basic chemical laws

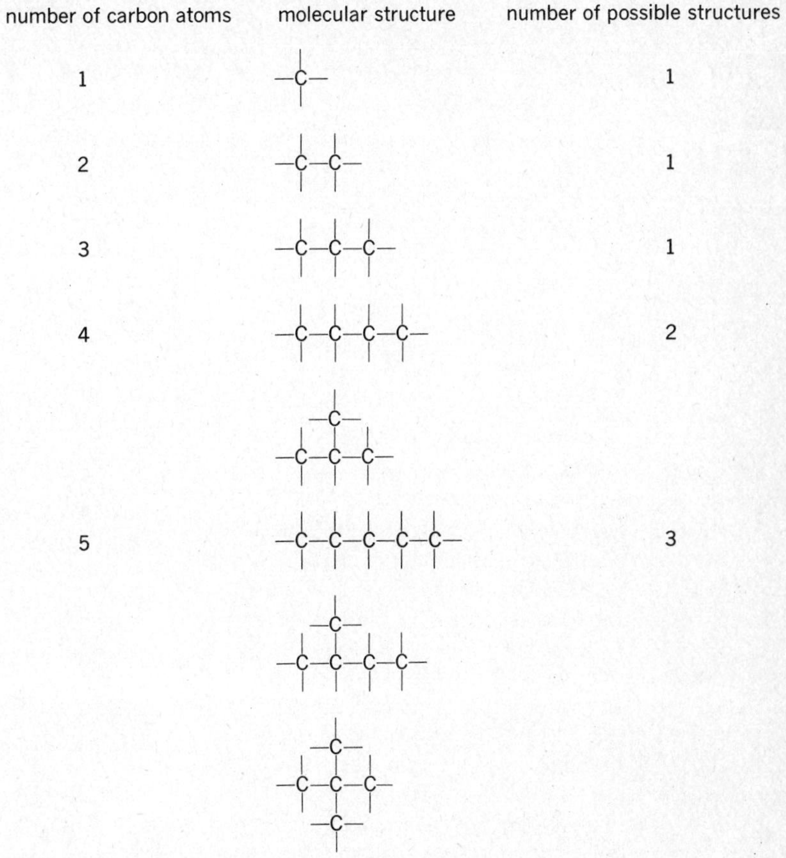

number of carbon atoms	molecular structure	number of possible structures
1	$-\overset{\mid}{\underset{\mid}{C}}-$	1
2	$-\overset{\mid}{\underset{\mid}{C}}-\overset{\mid}{\underset{\mid}{C}}-$	1
3	$-\overset{\mid}{\underset{\mid}{C}}-\overset{\mid}{\underset{\mid}{C}}-\overset{\mid}{\underset{\mid}{C}}-$	1
4	$-\overset{\mid}{\underset{\mid}{C}}-\overset{\mid}{\underset{\mid}{C}}-\overset{\mid}{\underset{\mid}{C}}-\overset{\mid}{\underset{\mid}{C}}-$	2

FIGURE 1-10
Molecular structures that can be formed from one, two, three, four, and five carbon atoms.

as the mineral elements. No vital forces unique to living organisms were necessary to explain the structure and properties of organic molecules.

Because most of the naturally occurring carbon molecules are found in living organisms, the study of the chemistry of carbon compounds became known as *organic chemistry,* the chemistry of noncarbon molecules being known as *inorganic chemistry.* Once chemists were able to manipulate the reactions of carbon molecules in a test tube, the science of organic chemistry expanded rapidly. Many products of our technology, such as plastics, synthetic rubber, nylon fibers, paints, and detergents, have developed out of the science of organic chemistry although they have little in common with the molecules present in living organisms. Organic chemistry grew out of the study of molecules found in living organisms, but the chemistry of living substances, *biochemistry,* now forms only a portion of the broad field of organic chemistry.

One property of the carbon atom makes it possible for an entire field of chemistry to be devoted to its study and makes life, with its variety of structures and functions, possible. This is the ability of carbon atoms to form four separate chemical bonds with other atoms, in particular with other carbon atoms, so that larger and larger molecules can be formed having a variety of structures. Figure 1-10 shows the number of ways in which one, two, three, four, and five carbon atoms can be combined. With four carbon atoms two different carbon structures can be formed; with five carbon atoms three molecular structures are possible. With each addition of one carbon atom the number of possible arrangements increases. There is no limit to the number of carbon atoms that can be combined in this way. If we also consider the possibility of carbon atoms combining with other atoms, such as hydrogen, oxygen, nitrogen, and sulfur, the variety of molecular structures that can be formed with relatively few atoms

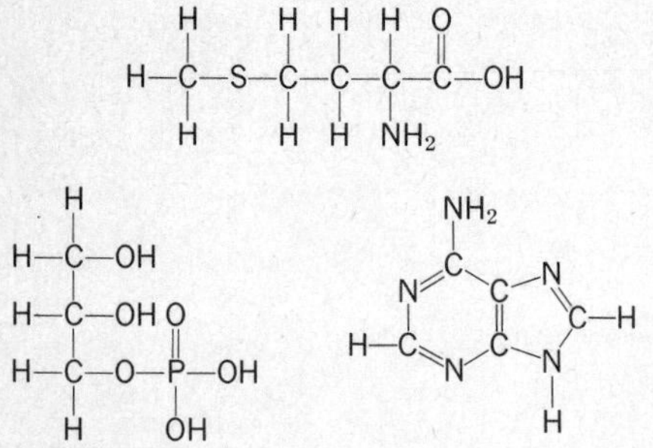

FIGURE 1-11
Three organic molecules: Methionine, glycerolphosphate, and adenine.

is considerable. Some examples of simpler organic molecules are shown in Fig. 1-11.

Although we draw diagrammatic structures on flat sheets of paper, we must remember that molecules are three-dimensional. Recall that the four bonds to a carbon atom form a tetrahedron surrounding the carbon atom. Thus, when a number of carbon atoms are linked together in succession, they do not form a perfectly straight line since no two bonds on the same carbon atom lie in a straight line with each other. Other atoms, such as hydrogen, occupying the remaining carbon bonds are oriented toward the sides of the carbon chain.

The three-dimensional shape of an organic molecule is not a rigid structure, however. Within limits, molecules are flexible and can change their shape without changing any of their chemical bonds because the atoms can rotate about their bonds. A chemical bond is similar to an axle joining two atoms about which the atoms can rotate independently of each other. Figure 1-12 shows

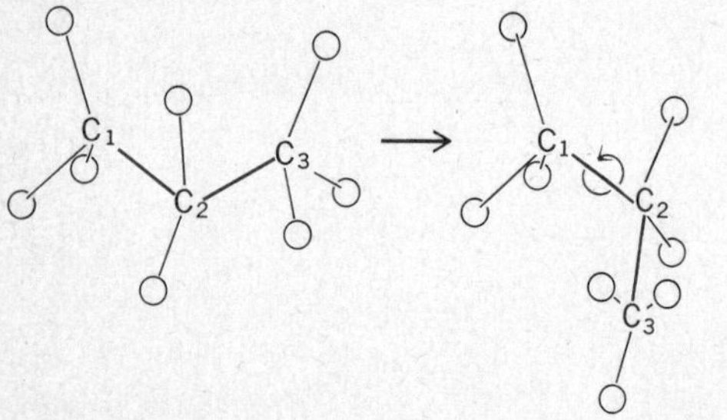

FIGURE 1-12
Rotation of parts of a molecule around a chemical bond can alter the shape of a molecule.

the change in shape that rotation makes possible. Carbon atom 1 remains fixed in position while atoms 2 and 3 alter their position as they rotate about the carbon bond joining atoms 1 and 2. Each carbon-to-carbon bond can rotate in a similar fashion. Thus, a sequence of six carbon atoms can assume a number of different shapes, some of which are shown in Fig. 1-13. Each shape represents the same molecule since no destruction, alteration, or formation of bonds has occurred.

In addition to carbon and hydrogen, atoms of oxygen, nitrogen, and occasionally sulfur and phosphorus are found in organic molecules, where they form polar or ionic bonds. Figure 1-14 lists some of the most common chemical groups and indicates their polar or ionic structure. In the polar hydroxyl group, —OH, the oxygen is slightly negative and the hydrogen slightly positive, as in water. Molecules containing carboxyl groups, —COOH, are acids because of the ability of one of the oxygen atoms to ionize, forming a hydrogen and a carboxyl ion. The amino group, —NH_2, can become ionized by combining with a free hydrogen ion. The phosphate group, —H_2PO_4, is generally found in its ionic form due to the ionization of two of its oxygen atoms. The presence of these groups in an organic molecule increases its solubility in water because their ionic and polar structures associate with the polar water molecules.

Classification of molecules found in living systems

The various classes of molecules found in the body are given in Table 1-6. We have already discussed the general properties of water. The minerals constitute the class of inorganic substances, most of which exist as ions such as Na^+, K^+, Mg^{2+}, Ca^{2+}, and Cl^-. The functions of these minerals in the body range from the formation of the crystalline substance of bones to generating electric currents in nerve and muscle cells, and most of these functions depend upon the fact that these ions are relatively small, highly charged particles.

Water and the mineral elements constitute the inorganic components of the body; the remaining classes, the proteins, lipids, carbohydrates, intermediates, and nucleic acids, constitute the organic molecules. We shall characterize each of these classes briefly, expanding upon the properties of particular molecules in each class in later chapters.

Proteins   The term protein comes from the Greek *proteios,* of the first rank, which aptly describes their importance. From Table 1-6 we see that proteins account for 17

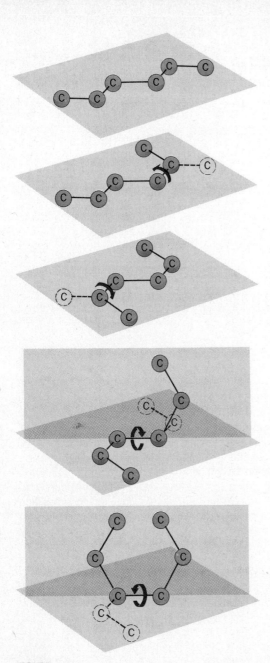

FIGURE 1-13
Changes in molecular shape occur as the molecule rotates around different carbon-to-carbon bonds, transforming it from a relatively straight chain into a ring.

TABLE 1-6
Molecular composition of the body

Constituent	Body wet weight, %
Water	60
Protein	17
Lipid	15
Minerals (Na, K, Cl, Ca, Mg, etc.)	5
Intermediates and nucleic acids	2
Carbohydrate	1

percent of the body weight and about 50 percent of the organic material in the body. No other class of organic molecules plays so many functional roles in living organisms. Proteins participate in both the static and dynamic functions of an organism. They are the basic structural units of cellular architecture, giving shape and form to cells and organelles. Protein molecules are the strings tying the organism together and giving it structural unity. The connective tissue of the body, which forms a structural matrix throughout all tissues and includes such specialized structures as the skin, hair, the ligaments connecting the bones together, and the tendons linking the muscles to the bones, is composed primarily of protein molecules. This static structural role constitutes only one of the functions of protein in the body. Proteins catalyze most of the chemical reactions in the body, involving both the synthesis and breakdown of organic molecules. The organism's ability to regulate its chemistry and derive energy from chemical reactions depends on the properties of protein molecules which facilitate these reactions, as we shall see in Chap. 3. The ability of muscle to contract depends upon the presence of specific contractile proteins within these cells. Many of the chemical messengers in the body, the hormones, such as insulin, are proteins. Many diseases are the result of foreign proteins that enter the body as components of bacteria or viruses, which in turn are often combated by special proteins known as antibodies. The red color of blood is due to the protein hemoglobin, which is involved in carrying oxygen from the lungs to the tissues. This partial list can only begin to suggest the enormous variety of functions with which proteins are associated.

Although a protein molecule may consist of thousands of atoms, the basic structure of the molecule is deceptively simple. Just as a large molecule can be formed by linking many carbon atoms together, a protein molecule is made by linking together a large number of

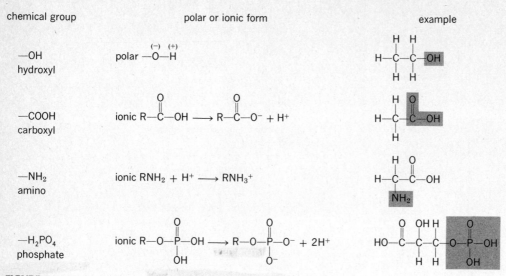

chemical group	polar or ionic form	example

FIGURE 1-14

Commonly encountered chemical groups in organic molecules. R corresponds to the remainder of the molecule.

similar molecular subunits. Large molecules made up of repeating subunits are known as *polymers* (Greek: many parts). The molecular subunit of protein architecture is the *amino acid,* proteins being polymers of amino acids. All amino acids have one structural property in common, namely, both an amino group, —NH$_2$, and a carboxyl (acid) group, —COOH, attached to the terminal carbon of the molecule, as shown in Fig. 1-15. The remainder of the molecule, represented by R, may have a number of different molecular forms and is often referred to as the side chain of the amino acid. About 20 different amino acids, that is, 20 different R groups, have been found in proteins, but not all 20 amino acids need be present in any one protein. Figure 1-16 shows a few examples of amino acid side chains. Some of the side chains are neutral, others are polar, and others contain ionized

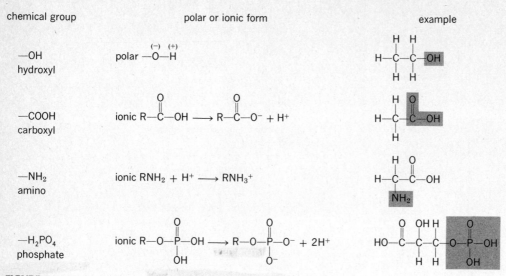

amino group

amino acid

FIGURE 1-15

General structure of amino acids. The remainder of the molecule, symbolized by R, is what distinguishes one amino acid from another.

groups and thus have either a positive or negative electric charge. These charged groups play an important role in the functioning of the protein molecule.

Protein molecules are formed by linking the amino group of one amino acid to the carboxyl group of another amino acid through a bond known as a *peptide bond.* A molecule of water is formed during the process (Fig. 1-17). Note that the molecule that results from the linking of two amino acids has a carboxyl group at one end and an amino group at the other. Each of these groups can form further peptide bonds with other amino acids and thus lengthen the chain of amino acids indefinitely. The peptide bonds between the amino and carboxyl groups form the backbone of the protein, and the R groups of the amino acids stick out to the sides of the chain.

The primary structure of a protein molecule is thus simply a chain of amino acids held together by peptide bonds. How do proteins manage to perform so many functions in the body when their molecular structure appears to be rather repetitive? To begin with, the sequence of amino acids in a protein is not a sequence of identical molecular units. There are 20 different amino acids and thus 20 possible different side chains branching off from the protein backbone. If we consider a very small protein molecule only three amino acids long and assume that only two different amino acids A and B can occur in the protein, it is possible to build eight different

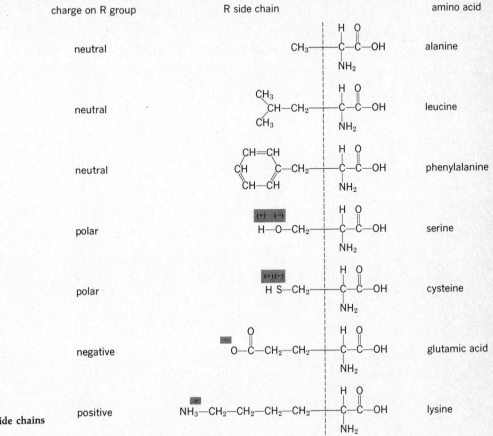

FIGURE 1-16
Some of the types of side chains present in amino acids.

protein molecules having the amino acid sequences shown in Fig. 1-18. One of the two amino acids A or B can occupy the first, second, and third position; thus there is a possibility of $2 \times 2 \times 2 = 8$ different molecules. If we go one step further and allow any of the 20 different amino acids to occupy any of the three positions in our simple protein, the number of possible molecules only 3 amino acids long becomes $20 \times 20 \times 20 = 8,000$. If we consider proteins that are 6 amino acids long, we find that $20^6 = 64,000,000$ possible protein molecules that can be formed from 20 amino acids. But even this number is infinitesimal when we consider that, with a few exceptions, even the smallest proteins found in the body contain a sequence of about 50 amino acids. The number of different proteins 50 amino acids long and built up of

20 amino acid structures is 20^{50}, or 1 followed by 65 zeros. But even 50 amino acids makes a very small protein. Hemoglobin, the protein found in red blood cells, contains 574 amino acids, and myosin, one of the contractile proteins of muscle, contains over 4,500 amino acids. Thus, starting with the 20 amino acids, an almost unlimited variety of protein molecules can be constructed by simply rearranging the sequence and altering the total number of amino acids, thus changing the physical and chemical properties of the molecule. This amounts to saying that it is possible to build a protein structure to fit almost any particular function in the body. We shall see examples later of how changing one single amino acid in a sequence of several hundred can completely alter the function of a protein molecule.

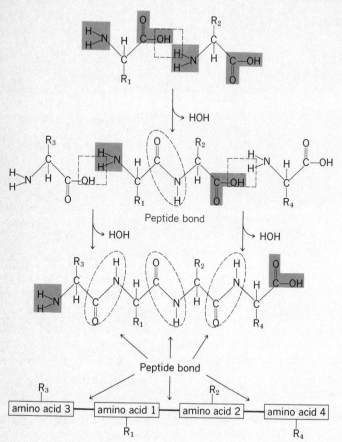

FIGURE 1-17
Proteins consist of amino acids linked together by peptide bonds; these bonds are formed between the carboxyl group of one amino acid and the amino group of another.

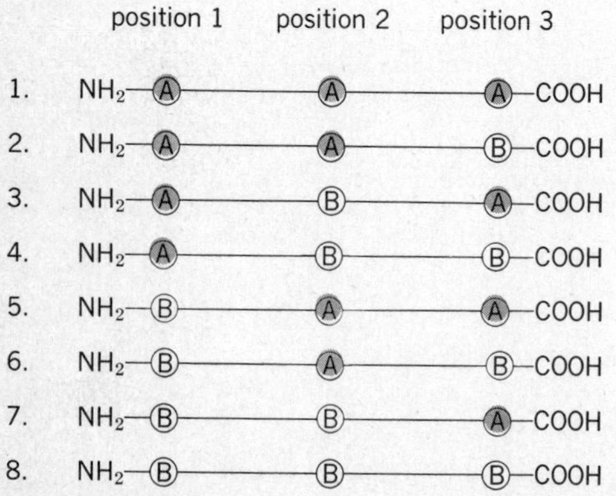

position 1   position 2   position 3

1.   NH₂—Ⓐ————————Ⓐ————————Ⓐ—COOH

2.   NH₂—Ⓐ————————Ⓐ————————Ⓑ—COOH

3.   NH₂—Ⓐ————————Ⓑ————————Ⓐ—COOH

4.   NH₂—Ⓐ————————Ⓑ————————Ⓑ—COOH

5.   NH₂—Ⓑ————————Ⓐ————————Ⓐ—COOH

6.   NH₂—Ⓑ————————Ⓐ————————Ⓑ—COOH

7.   NH₂—Ⓑ————————Ⓑ————————Ⓐ—COOH

8.   NH₂—Ⓑ————————Ⓑ————————Ⓑ—COOH

FIGURE 1-18
Eight different protein molecules composed of three linked amino acids can be formed from two different amino acids A and B.

When we say that we can build a protein molecule to "fit" almost any particular function, it is almost literally true, since many protein reactions involve fitting the geometrical shapes of two molecules together, much like the pieces of a jigsaw puzzle. Thus far, we have considered only the *primary* structure of proteins, i.e., the sequence of amino acids in the protein chain. Since a protein molecule can rotate about each of its chemical bonds, the molecule is not a rigid rod but more like a flexible piece of rope that can be twisted into a number of configurations. The bonds to the oxygen and nitrogen atoms adjacent to the peptide bonds linking the amino acids together are polarized, the oxygen being slightly negative and the hydrogen attached to the nitrogen of the amino group slightly positive. Electric attraction between these polarized groups in different parts of the molecule tends to draw the molecule together. Because the peptide bonds are regularly spaced along the protein backbone, a regu-

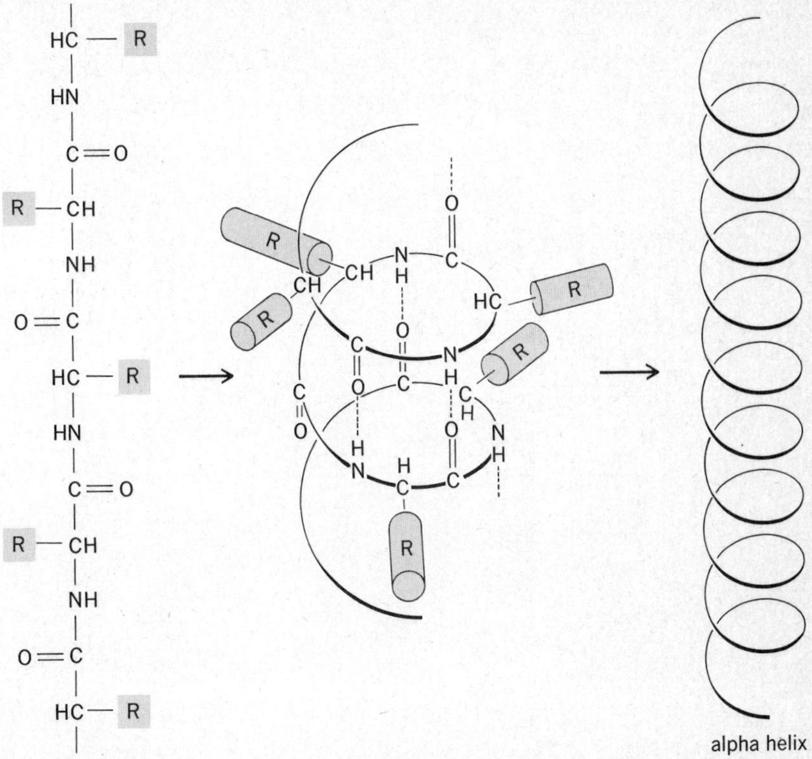

FIGURE 1-19
The alpha-helix configuration of proteins results from the electric attraction

$$\begin{array}{cc} H & O \\ | & | \\ \end{array}$$

between polarized —N— and —C— segments of the peptide bonds which occur at regular intervals along the protein chain.

alpha helix

lar repeating pattern of attraction occurs along the protein chain, which produces a coiled configuration known as an *alpha helix* (Fig. 1-19).

In addition to the electric interactions between peptide bonds, many of the side chains of the amino acids contain charged groups that interact with other charged groups in the molecule, tending to distort the regular coiled pattern of the alpha helix. Thus, part of a protein molecule may have the alpha-helix configuration and another part may be in the form of a random coil. The alpha helix can even form larger coils because of electric interactions between the side chains of amino acids. Besides the electric interactions, which influence the shape of the protein molecule, true chemical bonds can be formed between side chains of certain amino acids. The amino acid cysteine, which contains a sulfur atom, is able to form a chemical bond with another cysteine molecule, giving rise to a disulfide bond between the cysteine amino acids. Such a bond may be formed between widely separated cysteine molecules within the protein chain and thus hold widely separated segments

of the molecule in close proximity to each other. Figure 1-20 illustrates a few of the many three-dimensional patterns that can be formed by protein molecules.

This description of the general properties of protein structure is based upon many investigations of many different proteins. Actually we know relatively little about the specific structure of most protein molecules in the body, and only a few have been analyzed in detail. The problems involved in determining the structure of a specific protein molecule are enormous. In 1958, Frederick Sanger was awarded the Nobel Prize in chemistry for determining the amino acid sequence in the protein insulin, which required years of careful chemical analysis. Insulin, $C_{257}H_{380}N_{64}O_{76}S_6$, a very small protein, with only 51 amino acids, contains 17 of the 20 known amino acids and is composed of 783 atoms. Since Sanger's work, several other proteins have been studied in detail, but we still know the complete structure of only a few protein molecules. In some cases we know the three-dimensional shape of the molecule without knowing its amino acid sequence, whereas in others we know the sequence but

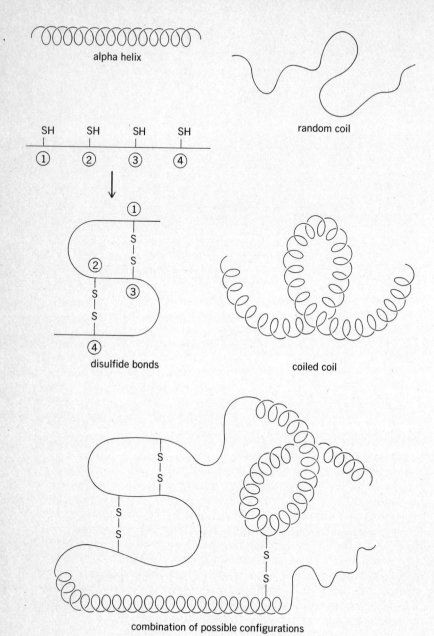

alpha helix

random coil

SH   SH   SH   SH

① ② ③ ④

disulfide bonds

coiled coil

combination of possible configurations

FIGURE 1-20
Some possible three-dimensional configurations of protein molecules.

not the shape. For the chemist, determination of the shape of the molecule and its amino acid sequence are two separate operations; for the cell, however, separate sets of information for the shape and sequence do not appear to be necessary, since once the sequence of amino acids is constructed, the shape of the molecule follows automatically in most cases from the distribution of charges along the protein chain.

Synthesizing a protein molecule in a test tube is enormously difficult; yet it can be performed by a living

$$CH_3-CH_2-CH_2-(CH_2)_{12}-CH_2-CH_2-COOH$$
saturated fatty acid

FIGURE 1-21
Hydrocarbon structure of saturated and polyunsaturated fatty acids.

$$CH_3-(CH_2)_3-CH_2-CH=CH-CH_2-CH=CH-CH_2-(CH_2)_5-CH_2-COOH$$
polyunsaturated fatty acid

cell in a matter of seconds. To synthesize a protein a cell must have information about the sequence of amino acids in the molecule; then it must have a mechanism for transcribing this information from its blueprint into the finished product so that each amino acid will be placed in the proper sequence to form a specific protein. How cells synthesize proteins will be discussed in Chap. 4. Since 1963 several laboratories in China, Germany, and the United States have reported the chemical synthesis of insulin in a test tube. In terms of the overall complexity of the product, this accomplishment represents one of the most remarkable achievements in the history of chemistry.

Lipids The old saying that oil and water do not mix helps define the class of organic molecules known as *lipids* (Greek *lipos,* fat). A lipid is defined as a molecule which is relatively insoluble in water but is soluble in organic solvents such as acetone, chloroform, ether, or benzene. The members of this class thus have a physical property in common rather than similar molecular structures. The fats belong to this class, and the term fat is often used interchangeably with the term lipid since the majority of the lipids in the body are fats. Strictly speaking, however, the fats constitute only one of several subclasses of lipid molecules.

Lipid molecules contain very few polar or ionic chemical groups, which accounts for their insolubility in water. Lipids are composed largely of hydrogen and carbon. The simplest lipids are the hydrocarbons, like the example of *n*-octane shown here, which consists solely of these two elements:

$$CH_3-CH_2-CH_2-CH_2-CH_2-CH_2-CH_2-CH_3$$

Oils and gasoline are mixtures of hydrocarbons, the longer carbon chains being found in the more viscous oils and the shorter chains found in the more volatile gasoline. (Prehistoric plant life buried for millions of years is the source of the crude oil from which petroleum products are refined. These hydrocarbons are derived from all classes of biological organic molecules but few simple hydrocarbons are present in living organisms.)

The lipids found in the body can be divided into three subclasses on the basis of their chemical structures: the *neutral fats*, the *phospholipids*, and the *steroids*. All three subclasses have the common property that their molecules are relatively insoluble in water but are soluble in organic solvents. Taken together, the lipids compose 15 percent of the total body weight and about 40 percent of the organic matter in the body.

Neutral fats The neutral fats constitute the majority of the lipids in the body, and it is these molecules which are generally referred to simply as fat. The basic component of the neutral fat molecule is a straight-chain hydrocarbon containing a carboxyl group at one end, known as a *fatty acid* (Fig. 1-21). Sixteen- and eighteen-carbon fatty acids are the most common in the body although shorter and longer fatty acids are also present. Because fatty acids are synthesized in the body by linking together two-carbon fragments, most fatty acids have an even number of carbon atoms. Another variable is the presence or absence of double bonds. A carbon atom is able to form four chemical bonds with other atoms, and two carbon atoms are sometimes joined by two bonds rather than a single one.

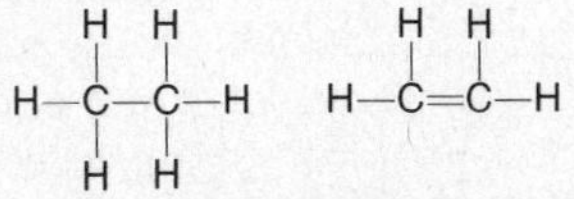

$$H-\overset{\overset{\displaystyle H}{|}}{\underset{\underset{\displaystyle H}{|}}{C}}-\overset{\overset{\displaystyle H}{|}}{\underset{\underset{\displaystyle H}{|}}{C}}-H \qquad H-\overset{\overset{\displaystyle H}{|}}{C}=\overset{\overset{\displaystyle H}{|}}{C}-H$$

When a fatty acid molecule contains a double bond, it is said to be *unsaturated.* The carbon atoms joined by the double bond cannot be associated with as many hydrogen atoms as singly bonded carbon atoms, and thus these atoms are not saturated with hydrogen. If more than one double bond is present, the fatty acid is said to be *polyunsaturated* (Fig. 1-21). Animal fats generally contain a high proportion of saturated fatty acids, whereas vegetable fats contain more polyunsaturated fatty acids. There is some evidence that fatty deposits in blood vessels, which can block blood flow, are more readily formed when the diet contains a large proportion of saturated fatty acids, thus the current emphasis in advertising on the proportion of polyunsaturated fats in salad and cooking oils.

In addition to fatty acids, neutral fat molecules contain a three-carbon molecule, *glycerol,* to which the

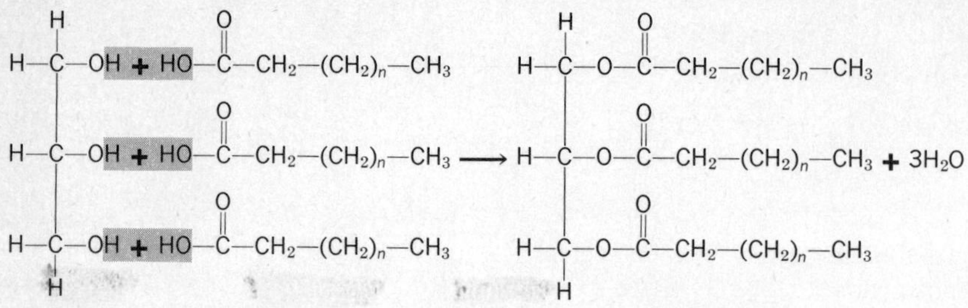

$$
\begin{array}{l}
\text{H}-\overset{\text{H}}{\underset{}{\text{C}}}-\text{OH} \; + \; \text{HO}-\overset{\text{O}}{\underset{}{\text{C}}}-\text{CH}_2-(\text{CH}_2)_n-\text{CH}_3 \\[4pt]
\text{H}-\overset{}{\underset{}{\text{C}}}-\text{OH} \; + \; \text{HO}-\overset{\text{O}}{\underset{}{\text{C}}}-\text{CH}_2-(\text{CH}_2)_n-\text{CH}_3 \\[4pt]
\text{H}-\overset{}{\underset{\text{H}}{\text{C}}}-\text{OH} \; + \; \text{HO}-\overset{\text{O}}{\underset{}{\text{C}}}-\text{CH}_2-(\text{CH}_2)_n-\text{CH}_3
\end{array}
\longrightarrow
\begin{array}{l}
\text{H}-\overset{\text{H}}{\underset{}{\text{C}}}-\text{O}-\overset{\text{O}}{\underset{}{\text{C}}}-\text{CH}_2-(\text{CH}_2)_n-\text{CH}_3 \\[4pt]
\text{H}-\overset{}{\underset{}{\text{C}}}-\text{O}-\overset{\text{O}}{\underset{}{\text{C}}}-\text{CH}_2-(\text{CH}_2)_n-\text{CH}_3 \; + \; 3\text{H}_2\text{O} \\[4pt]
\text{H}-\overset{}{\underset{\text{H}}{\text{C}}}-\text{O}-\overset{\text{O}}{\underset{}{\text{C}}}-\text{CH}_2-(\text{CH}_2)_n-\text{CH}_3
\end{array}
$$

glycerol   3 fatty acids   neutral fat (triglyceride)

FIGURE 1-22
Neutral fat (triglyceride) consists of three fatty acids attached to the three hydroxyl groups of glycerol.

fatty acids are attached. Although present in the neutral fat molecule, glycerol is not itself a lipid since it is readily soluble in water due to the three hydroxyl, —OH, groups. Thus a neutral fat molecule consists of three fatty acid chains joined through their carboxyl group to the three hydroxyl groups of glycerol (Fig. 1-22) and is often referred to as a *triglyceride* because of the three fatty acids attached to the backbone of the glycerol molecule. Since the three acids need not be identical, a variety of neutral fats can be formed with fatty acids of different chain lengths and degrees of saturation.

Most people who eat too much tend to become fat. If the body receives more food than it can utilize in carrying out its normal functions, it converts the excess into molecules of neutral fat rather than discarding the unused material. One can reduce the amount of fat in the body by going on a diet and lowering the food intake. If the body is receiving less food than it requires to carry out its functions, it calls upon its reserve of fat, which can be broken down and in the process release potential energy. Thus, one of the major functions of neutral fat is to act as a reservoir of potential energy.

In addition to storing energy, the fat deposits of the body serve a number of other functions. The layers of fat below the surface of the skin provide thermal insulation to protect the organism from cold; the blubber (fat) of a whale is an extreme example. Human females normally have more fat underlying their skin than males, and these fatty deposits are responsible for the curves of the female figure.

Phospholipids   The phospholipids are similar in structure to the neutral fats since they are usually composed of glycerol and fatty acids, but they contain only two fatty acids, the third hydroxyl group of the glycerol being linked to a phosphate group (Fig. 1-23). In addition, a small nitrogen-containing group is usually attached to the phosphate, and both are usually ionized; thus, portions of the molecule are electrically charged, unlike the neutral fats. In some phospholipids, molecules other than glycerol provide the site of attachment for the two fatty acid chains and phosphate. In spite of this polar portion of the molecule, phospholipids are not very soluble in water because of the large hydrocarbon chains of their fatty acids.

The charged portion of a phospholipid is at the glycerol end of the molecule, as shown in Fig. 1-23. This end of the molecule associates with polar water molecules, and the neutral hydrocarbon chains associate with other lipids. Because of this division between the charged and neutral portion of the molecule, phospholipids tend to become oriented at water-lipid interfaces, as shown in Fig. 1-24.

The functions of phospholipids in the body are related to this property. Phospholipids form one of the major components of cell membranes, the hydrocarbon portion making it possible for the membrane to act as a barrier between two water compartments while the charged portion provides the structural orientation for the phopholipid within the membrane. We shall have more to say about the role of phospholipids in the cell membrane in Chap. 2.

Steroids   The third class of lipids, the steroids (Greek: solid), differs structurally from the other two. Four interconnected rings of carbon atoms form the basic structure

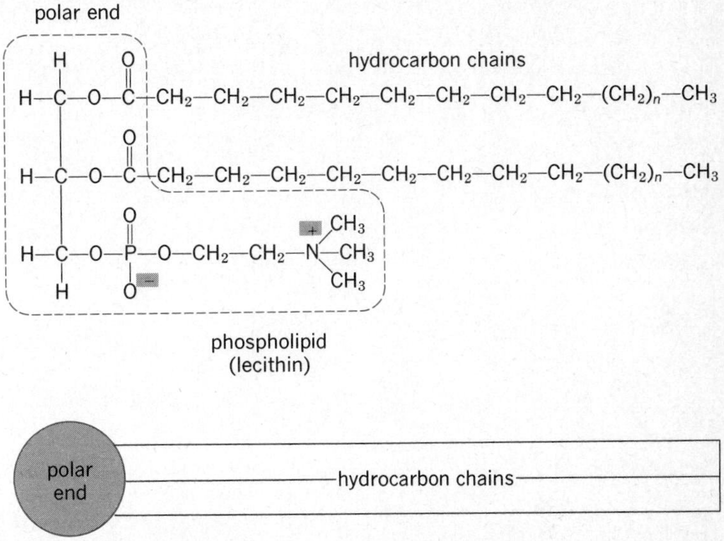

polar end

hydrocarbon chains

$$H-C-O-C-CH_2-CH_2-CH_2-CH_2-CH_2-CH_2-CH_2-CH_2-(CH_2)_n-CH_3$$

$$H-C-O-C-CH_2-CH_2-CH_2-CH_2-CH_2-CH_2-CH_2-CH_2-(CH_2)_n-CH_3$$

$$H-C-O-P-O-CH_2-CH_2-N-CH_3$$

phospholipid
(lecithin)

FIGURE 1-23
Polarized structure of phospholipid molecules.

polar end

hydrocarbon chains

water phase

lipid phase

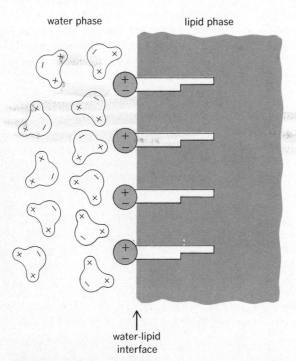

water-lipid interface

FIGURE 1-24
Orientation of phospholipid molecules at the boundary between a water and lipid phase.

of all steroid molecules (Fig. 1-25). Few polar groups are attached to the steroid rings, which are thus insoluble in water and qualify as lipids.

Steroids include such molecules as cholesterol, a structural component in cell membranes, and many of the hormones in the body, such as the male and female sex hormones, testosterone and estrogen (Fig. 1-25).

Carbohydrates Although carbohydrates account for only 1 percent of the total body weight, they play a central role in the chemistry of the body. It is the chemical breakdown of carbohydrate molecules into carbon dioxide and water which provides most of the chemical energy utilized by cells. Although carbohydrates do not provide the only source of chemical energy (we have already seen that lipids can be used this way, and we shall see later that even proteins can provide chemical energy), they remain the most readily available source of chemical energy and are utilized by many cells in preference to other types of molecules. Some tissues, such as the brain, rely solely on carbohydrates for energy.

The term carbohydrate is derived from the general formula for most of these molecules, $C_n(H_2O)_n$, where n is any whole number. As the formula indicates, for every atom of carbon in the molecule there is the equivalent of one molecule of water. Thus, carbohydrates are hydrated (containing water) carbon chains.

The simplest carbohydrates are the sugars, and the

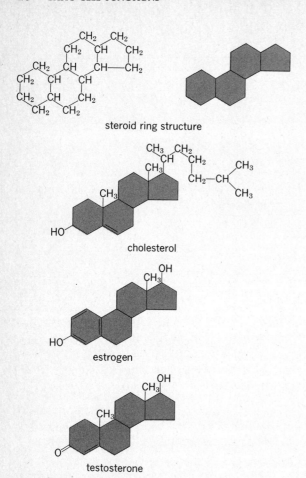

steroid ring structure

cholesterol

estrogen

testosterone

FIGURE 1-25
Carbon-ring structure of various steroids, including the female and male sex hormones estrogen and testosterone.

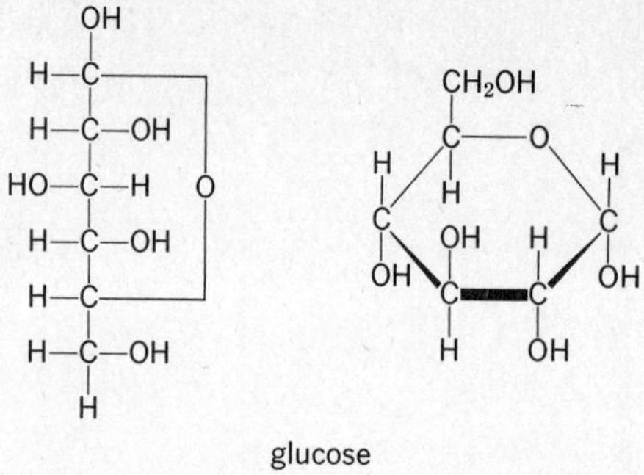

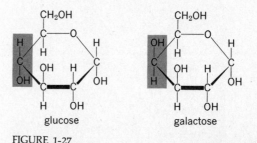

glucose

FIGURE 1-26
Two ways of diagraming the structure of the carbohydrate glucose.

1-27, a different sugar results. Most sugars in the body contain five or six carbon atoms.

Larger carbohydrate molecules can be formed by linking a number of sugars together, much as amino acids are linked to form proteins. Table sugar, sucrose (Fig. 1-28), is composed of two sugars, glucose and fructose, linked together by a chemical bond formed when a molecule of water is removed from the two sugars. A molecule containing two sugars is known as a *disaccharide* (Greek *sakcharon,* sugar). When many sugars are linked together, the molecule is known as a *polysaccharide.*

The most important polysaccharides in living organisms are starch, glycogen, and cellulose, all three of which are composed of thousands of repeating units of the monosaccharide glucose. The difference between these three structures depends upon how the glucose units are joined. Cellulose is a structural polysaccharide

most important sugar in the body is *glucose,* often called blood sugar (Fig. 1-26). Glucose is a six-carbon sugar having the formula $C_6H_{12}O_6$, which agrees with the general formula for carbohydrates, $C_6(H_2O)_6$. Sugars are quite soluble in water because of their numerous polar —OH groups. Two ways of representing the structure of a sugar molecule are shown in Fig. 1-26; the first is the conventional representation for organic molecules, but the second gives a better idea of the three-dimensional structure of glucose. Five carbon atoms are linked with an oxygen atom in a ring, which lies in an essentially flat plane with the hydrogen and hydroxyl groups lying above and below it. If one of the hydroxyl groups below the ring is shifted to a position above the ring, as shown in Fig.

glucose   galactose

FIGURE 1-27
The difference between the sugars glucose and galactose depends upon the position of the hydroxyl group on the fourth carbon atom.

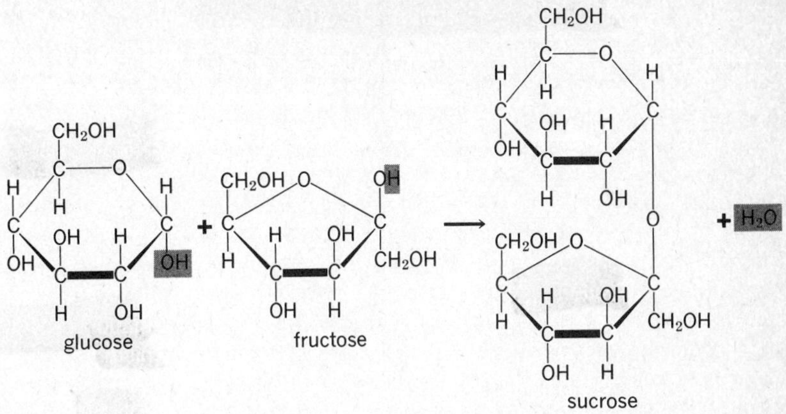

FIGURE 1-28
Sucrose (table sugar) is a disaccharide; it is formed by linking together two monosaccharides, glucose and fructose.

found in plants, where it forms the rigid walls of plant cells and is the major component of wood. It is composed of straight chains of glucose units strung end to end to form long fibers. Man cannot utilize the cellulose in the vegetables he eats because the intestinal tract cannot break up these giant molecules into individual glucose units, which the body could use; but starch, which is also of plant origin, is readily digested by man. Glycogen is of animal origin and is often called animal starch. By combining a large number of glucose units into the polysaccharide glycogen, the body can store limited amounts of carbohydrate, which can be made available when sudden increases in energy production are required. Plants can store much larger quantities of carbohydrate as starch than animals can store as glycogen. Plants store chemical energy primarily in the form of starch,

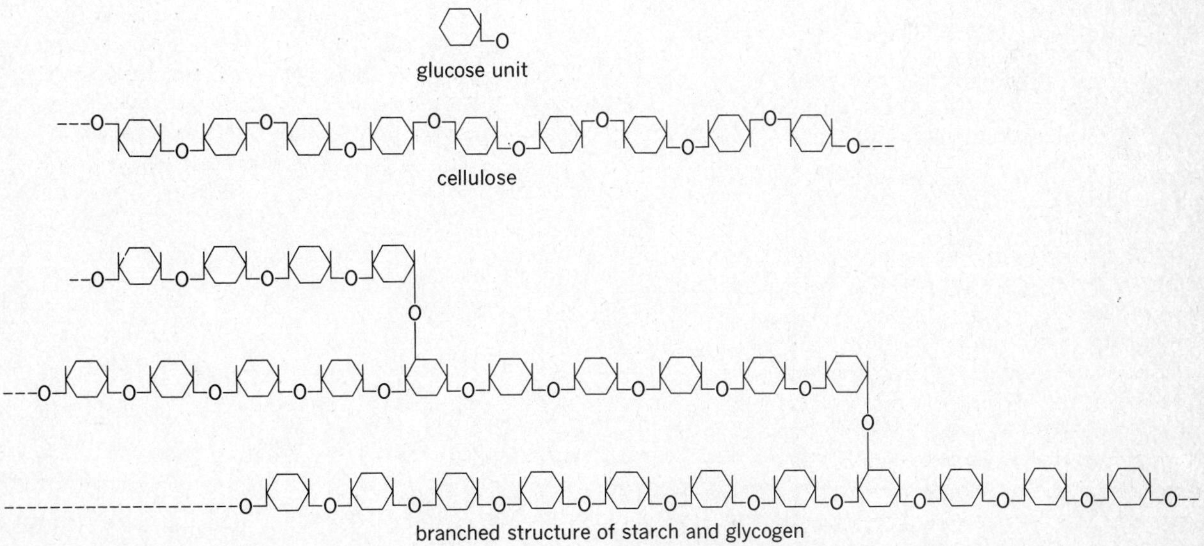

glucose unit

cellulose

branched structure of starch and glycogen

FIGURE 1-29
The polysaccharides cellulose, starch, and glycogen consist of numerous molecules of glucose linked together to form straight and branched chains.

whereas animals store energy in the form of neutral fat. Unlike cellulose, glycogen and starch are composed of branched chains of glucose (Fig. 1-29).

Intermediates   Within a cell, molecules of carbohydrate, lipid, and protein are continually undergoing chemical reactions which break these structures down into smaller molecular units, and new molecules are simultaneously being constructed out of small molecular units. These chemical reactions occurring within the cell are known collectively as *metabolism* (Greek: change). In cellular metabolism, special molecules are synthesized which perform specific functions within the cell while other molecules are broken down in order to release the potential energy stored in their chemical bonds. The carbohydrates, lipids, and proteins lie either at the end or at the beginning of a chain of chemical reactions. The *intermediates* represent the many types of molecules formed during the synthesis and degradation of the molecules of the body. As intermediates between the raw materials and the finished products of the body, such molecules often have no other function in the cell than to form a link in the chain of chemical reactions which leads ultimately to the end product. The intermediates have no one physical or chemical property in common since they comprise all the intermediate molecular structures de-

rived from carbohydrates, lipids, and proteins. We shall encounter some intermediates in Chap. 3, when we discuss the metabolic reactions which provide energy for the cell.

Nucleic acids   Although, as a class, nucleic acids contribute very little to the body's weight, these are the largest and most specialized molecules in the body. It is the nucleic acids which determine whether one is a man or a mouse, whether a cell is a muscle cell or a liver cell. These are the molecules which contain the genetic information providing the blueprint for constructing an organism.

Nucleic acids are of two types: *deoxyribonucleic acid* (DNA) and *ribonucleic acid* (RNA). The DNA molecule has the primary genetic information coded within its molecular structure; RNA molecules function chiefly in transcribing the information contained in the DNA molecule into a form that can be utilized by the cell to build specific structures to carry out specific functions. The shape of these molecules is so intimately related to their function within the cell that we shall postpone a discussion of nucleic acid structure until Chap. 4, where we consider how information is passed from cell to cell and how it is decoded and used to guide the formation of cell structures.

Movement of molecules across cell membranes

The development of the microscope in the seventeenth century revealed for the first time the existence of living organisms so small that they could not be seen with the naked eye. Robert Hooke (1635–1703) introduced the microscope into England and on April 13, 1663, demonstrated its remarkable powers before the Royal Society in London. As one of his demonstrations he placed a slice of cork (plant tissue) under a microscope, describing his observations as follows: "Our microscope informs us that the substance of cork is altogether filled with air and that the air is perfectly enclosed in little boxes or cells distinct from one another." It is from this description of Hooke's that the term cell was introduced into the biological literature. Hooke calculated that there must be over 1 billion little boxes in 1 in.^{3} of cork tissue, but he failed to realize that each of these air-filled boxes had once been occupied by a living cell. All that remained was the rigid outermost cellulose cell wall which surrounded the living plant cell.

Careful examination of many different organs and tissues has revealed that all living things are composed of microscopic compartments, or cells. Single-celled organisms carry out all their living functions within the confines of a single cell. In man, the total functioning of the body involves the interactions of an estimated 40

2

trillion cells, each of which has a life of its own. The common denominator of all living systems, the functional unit of life, is the cell.

General cell structure

Biologists once referred to the material found within a cell as *protoplasm* and studied the properties of protoplasm as if it were a substance having specific properties of its own and within which resided that most illusive quality—life. Protoplasm was considered to be a granular, gel-like, colloidal mixture of chemical compounds which performed all the chemical transformations associated with the living process. As techniques were developed for examining smaller and smaller portions of the cell, the concept of a cell as a semihomogeneous protoplasmic matrix, much like a bag of thick soup, slowly disappeared as the diversity of structure and function within a single cell became apparent. The interior of the cell, far from being a semiuniform medium, is structurally divided into a number of compartments, known as *cell organelles,* which have different chemical compositions and carry out specific functions. The combined interactions of these organelles within the cell contribute to the total quality we call life.

If a living cell is examined under a microscope, numerous floating granules and particles, some of which appear to be long filamentous structures, can be seen (Fig. 2-1). The largest visible structure is the centrally located nucleus. Most of the granular organelles of the cell are found in the area outside of the nucleus, known as the *cytoplasm*. When a living cell is closely observed for several minutes, the organelles within the cytoplasm are seen to move about slowly, suggesting a fluid medium. This medium within the cell is mostly water, which accounts for about 80 percent of the cell's weight. In addition to the intracellular water, all cells are surrounded by an extracellular watery medium. Such an abundance of water, which is highly transparent to light, makes a living cell almost invisible through an ordinary light microscope, and many devices are used to make the cellular structure more visible. The photograph in Fig. 2-1 was made with a special phase-contrast microscope, which takes advantage of differences in the velocity of light when it passes through materials of slightly different composition or thickness to produce an image of the cell. Dyes are often applied to the cell; they combine with molecules within the structural elements of the cell, staining them so that they become visible with an ordinary light microscope.

Even with special microscopes and staining techniques, only a portion of the cell's structure can be seen through a light microscope. Considerable magnification is necessary to make even these structures visible to the eye. The smallest object the naked eye can resolve is about 0.1 mm (0.0039 in.) in diameter. A light microscope can magnify an object about 2,000 times, producing an image of a 10-μm cell that is about 0.8 in. in diameter. But even at this magnification many cell structures are too small to be seen. The cell membrane and the membranes surrounding many of the organelles are about 75 Å thick. After a magnification of 2,000 times, they are only 0.0006 in. thick and still invisible under the light microscope. The great diversity of structures within the cell began to be discovered only after the development of the *electron microscope* in the late 1940s.

A beam of electrons can be focused by an electron microscope to form an image of a specimen just as a beam of light can be focused by a light microscope. The magnifying power of the light microscope is limited by the wavelength of light. The wavelength of a high-velocity electron may be of the order of 0.05 Å, which is about 100,000 times smaller than the wavelength of green light. The smaller wavelength of electrons increases the magnifying power of the electron microscope some 200-fold over that of the light microscope, giving a total magnification of 400,000. At the present time electron microscopes can resolve structures as small as 5 to 10 Å. Some large molecules, such as DNA and proteins, can actually be seen with an electron microscope.

To interpret the image of cell structures obtained with the electron microscope correctly we must know something of how this instrument works. Within the electron microscope, electrons are accelerated to very high velocities by high-voltage electric fields and focused upon the specimen by magnetic lenses. If an electron does not strike any of the atoms in the specimen, it passes through it and is focused on a photographic plate, where it reacts with silver atoms in the plate, resulting in a white spot in the final photograph. If, on the other hand, the electron collides with a structure in the specimen and is either captured or deflected to one side, it will not reach the photographic plate and a black spot or shadow is produced in that area.

Since electrons are very easily scattered by atoms in the air, the air inside the microscope must be pumped out, leaving a vacuum surrounding the specimen. Be-

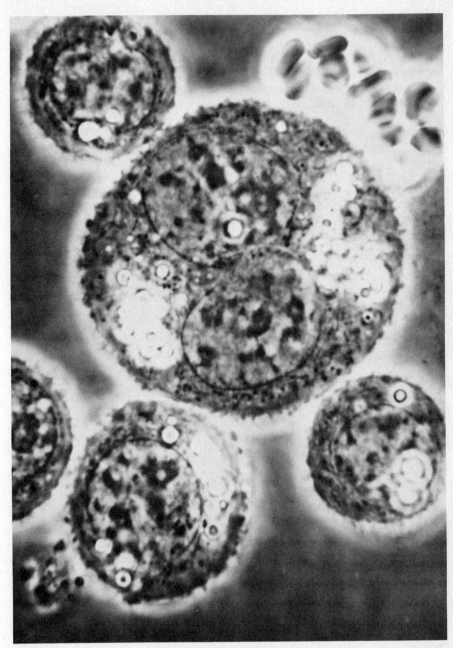

FIGURE 2-1
Living cells as seen with a phase-contrast microscope.

cause cells cannot survive in a vacuum, living cells cannot be examined in the electron microscope. Cells are so large compared with electrons that few electrons would manage to pass all the way through an entire cell. Therefore the cell must be cut into many thin sections, some as thin as 200 Å, and these sections, rather than the whole cell, are examined. In order to maintain the cell structure during sectioning, it is embedded in plastic and treated with a heavy metal, which increases the contrast of the image by blocking electrons more effectively, much as colored dyes increase contrast. Thus, the specimen placed in the electron microscope is far from being a living cell. It is perhaps surprising that the image obtained from a cell in this manner has any resemblance to the real world of living cells. Almost every structure seen with an electron microscope has been suspected

at some time or another of being an artifact resulting from the method of preparing the cell for examination. Although this possibility must always be considered when examining biological structures with the electron microscope, independent methods for studying cell structures have confirmed the presence of most of the structures observed with the aid of the electron microscope.

Because very thin sections of cells are used in electron microscopy, the final image represents only a very small portion of the whole cell, and the total three-dimensional image of a cell and its organelles must be reconstructed from a number of sections taken through different portions of the cell and at different angles. Figure 2-2 illustrates some of the problems of interpreting the overall shape of an object from thin sections made through the object at various angles. When examining an

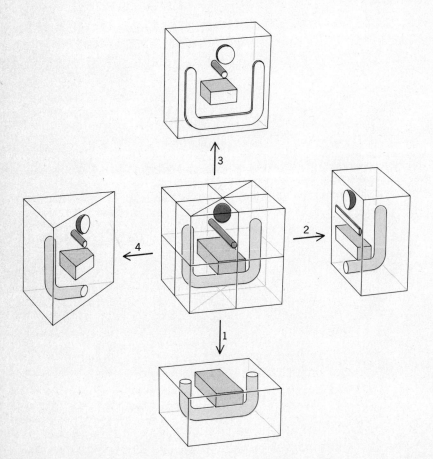

FIGURE 2-2
Four thin sections cut through a cell at different angles, showing the different shapes of the cell structures apparent when only a portion of the structure is included.

electron micrograph, one must always remember that many structures which appear independent may actually be connected together through portions of the cell not included in the thin section. Thus, a section through a ball of string would appear as a series of lines and dots even though it was originally one continuous piece of string.

An electron micrograph of a thin section through a liver cell (Fig. 2-3) shows the cytoplasm filled with organelles and particles of various sizes and shapes. Comparing this photograph with Fig. 2-1, which depicts a living cell as seen with a light microscope, we see that many of the particles which appear as barely visible granules in Fig. 2-1 have considerable internal structure in Fig. 2-3. Some of the organelles and structures found in a typical cell are diagramed in Fig. 2-4. Each of these structures plays a specific role in cell function. We postpone discussion of the properties and functions of organelles until later chapters, where each one will be discussed in relation to its particular role in cell function.

Rather than being a single homogeneous fluid, the substance of the cell is divided into a large number of separate compartments, each having a different chemical composition and being capable of carrying out different sets of chemical reactions. The cell is like a chemical laboratory with different chemical reactions occurring in various test tubes (cell organelles). Not only does a membrane separate the entire cell from its external environment, but membranes cover each of the cell organelles and separate their contents from other parts of the cell. In fact, there are very few structures in the cell that are not composed of membranes, which are the basic structural units of the cell. As a first approximation, a cell can be considered as a collection of chemical compartments separated by membranes (Fig. 2-5), whose primary function is to act as barriers separating the different chemical substances in the various organelles. However, membranes cannot be completely impenetrable if the compartments in the cell are to receive raw materials and release finished products. Some molecules must be able to cross these barriers; conversely, other molecules must be restrained. Thus, we say that these membranes are *semipermeable* in that they allow some substances to pass but not others. Before we consider the movements of molecules across membranes, we must first consider some of the ways in which molecules are able to move from one area to another, specifically bulk flow and diffusion.

Bulk flow and diffusion

Bulk flow

When a ball is thrown, all the molecules in the ball travel through the air as a unit. The same is true of the flow of liquids when a river flows or water pours from a faucet. In both these cases the water molecules are moving together as a unit in one direction, and this type of movement is known as *bulk flow*. Bulk flow results from forces which push the molecules from one point to another. The driving force for bulk fluid flow is pressure, pressure P being defined as the amount of force acting on a unit area of surface. Figure 2-6 illustrates some of the basic principles of bulk flow. Whenever there is a difference, Δ, in pressure, $\Delta P = P_1 - P_2$, between two regions of a liquid, fluid flows from the high-pressure region toward the low-pressure region. The magnitude of the flow F_B in liters per minute is directly proportional to the difference in pressure, ΔP; the greater the pressure difference, the greater is the amount of flow. Thus, we can write an equation for the bulk flow of a fluid as

$$F_B = k_B \Delta P \qquad \qquad \textbf{2-1}$$

where k_B is the bulk-flow constant, which is the proportionality constant between the pressure difference ΔP and flow F_B. The numeral value of k_B depends upon the type of fluid and geometry of the container through which the fluid is flowing. Equation **2-1** applies to the bulk flow of gas as well as a liquid.

Blood flow and the movement of gases between the lungs and the atmosphere are examples of bulk flow resulting from pressures created by the contraction of the heart muscle and respiratory muscles, respectively. Bulk flow must be clearly distinguished from the random movements of individual molecules known as *diffusion,* which leads to most of the exchange of molecules between a cell and its immediate environment, the movements of molecules throughout the cell, and their exchange with various cellular compartments.

Diffusion

All molecules undergo continuous motion. The warmer an object is, the faster its molecules move. This thermal motion enables molecules to move from one region of a solution to another, the velocity at which they travel depending upon the temperature and the mass of the molecule. At body temperature, an average molecule of

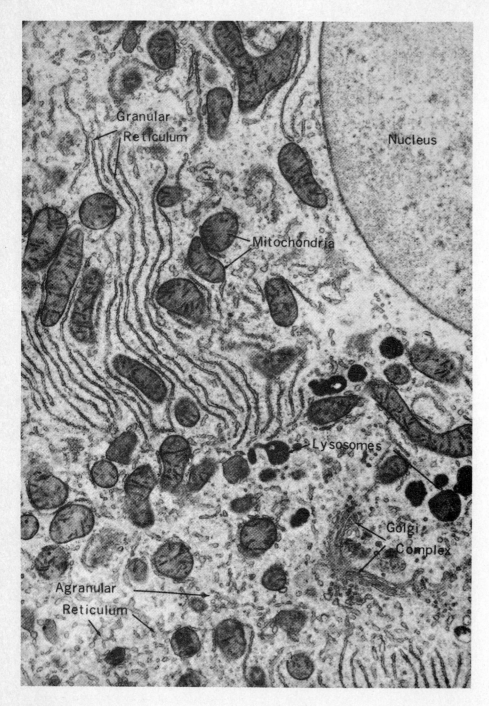

FIGURE 2-3
Electron micrograph of a thin section through a rat liver cell.
(From K. R. Porter, in T. W. Goodwin and O. Lindberg (eds.),
"Biological Structure and Function," vol. 1, Academic Press, Inc., New
York, 1961.)

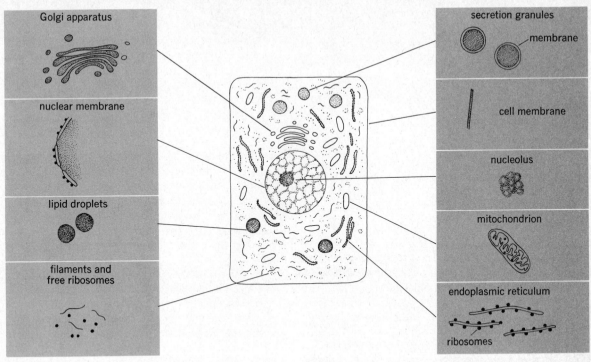

FIGURE 2-4
Diagram of the major organelles and structures in cells.

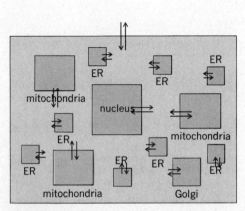

FIGURE 2-5
Diagram of a cell as a collection of chemical compartments
separated by membranes. (ER is endoplasmic reticulum.)

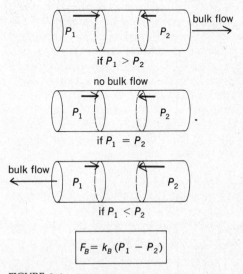

$$F_B = k_B (P_1 - P_2)$$

FIGURE 2-6
Bulk flow of a liquid or a gas results from a difference in
pressure ΔP between two regions of the fluid.

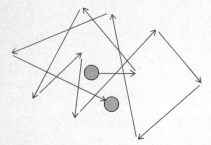

FIGURE 2-7
Path traveled by a single water molecule in solution. Changes in direction result from collisions with other molecules in the solution.

water is moving about 1,500 mi/hr, whereas a molecule of glucose, which is 10 times heavier, is moving about 500 mi/hr. In a liquid like water, where individual molecules are separated from each other by about 3 Å, such rapidly moving molecules cannot travel very far before colliding with other molecules. They bounce off each other like rubber balls, undergoing millions of collisions every second. Each collision alters the direction of the molecule's movement, so that the path of any one molecule through a solution may look something like that shown in Fig. 2-7. Such movement is said to be *random*, since a molecule moving in one direction may in the next instant be moving in the opposite direction. There is no way of predicting the direction of a molecule's motion; all directions are assumed to be equally probable.

The watery medium surrounding cells and within cells contains billions of molecules, all undergoing random movements (Fig. 2-8A). Because the movements are random and because billions of molecules are involved, at any instant there are about as many molecules moving to the right within a given volume as there are molecules moving to the left, as many moving up as down, etc. If the molecules cross imaginary boundaries surrounding a small volume of solution, a certain number of molecules cross each surface in a given period of time. The term *flux* is used to describe the amount of material (in this case the number of molecules) which crosses a unit area moving from one side to the other in a unit of time. Any volume of solution can be looked upon as being composed of a number of small volume elements joined together at their surfaces (Fig. 2-8B). Thus, the flux of molecules across the surfaces of these volume elements represents the exchange of molecules between the different portions of the total solution. If the number of

molecules in a unit of volume is doubled, the flux of molecules at each surface is doubled since twice as many molecules are moving in any one direction at a given time. Thus, the concentration of molecules (number of molecules in a unit of volume) in any region of a solution determines the magnitude of the flux across the surfaces of this region. We can formulate this simple relationship in an equation

$$f = k_D C \qquad \qquad \text{2-2}$$

where f is the flux, or number of molecules crossing a unit area in a unit time; C is the concentration, or number of molecules per unit of volume; and k_D is a proportionality constant, known as the *diffusion coefficient*. The value of the diffusion coefficient depends upon the molecular weight and chemical structure of the molecule and the chemical composition and temperature of the solution.

The diffusion of glucose between two adjacent volume elements is illustrated in Fig. 2-9. Glucose is present in compartment 1 at an initial concentration of 20 mmoles/liter, and initially there is no glucose in compartment 2. The random movements of the glucose molecules in compartment 1 carry some of them into compartment 2. The flux of glucose from 1 to 2 is, by Eq. 2-2, $f_{1-2} = k_D C_1$, where C_1 is the initial concentration of glucose in compartment 1. As the number of glucose molecules in compartment 2 increases, some of them move at random back into compartment 1. The magnitude of the glucose flux from 2 to 1 depends upon the concentration of glucose in compartment 2, C_2, at any given time

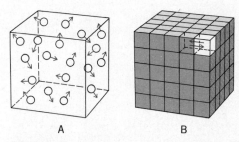

A   B

FIGURE 2-8
Collection of molecules undergoing random motion A in a small volume element results in approximately equal numbers of molecules moving up and down, right and left, etc. A large volume of solution is equivalent to a number of small volume elements B in which the fluxes of the molecules between the volume elements depend upon the concentration of molecules in each volume element.

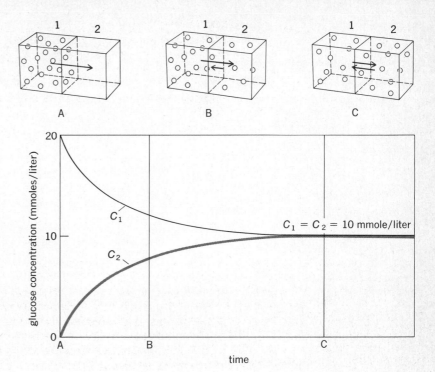

FIGURE 2-9
Diffusion of glucose between two compartments of equal volume. A Initial conditions: No glucose is present in compartment 2. B Some glucose molecules have moved into compartment 2, and some of them are moving at random back into compartment 1. C Diffusion equilibrium has been reached, the flux of glucose between the two compartments being equal in the two directions.

and is, by Eq. **2-2**, $f_{2-1} = k_D C_2$. Therefore, as the concentration of glucose in 2 increases, the flux of glucose from 2 to 1 also increases. The net increase in the number of glucose molecules in compartment 2 is given by the difference between the flux of glucose into 2 from 1 and the flux of glucose out of 2 back into 1. This difference between the two one-way fluxes is known as the *net flux F*. The net movement of glucose from compartment 1 to 2 causes the concentration in 1 to decrease and the concentration in 2 to increase. Eventually the concentrations in the two compartments become equal, the glucose fluxes in the two directions become equal, and the net transfer of glucose becomes zero. The system has now reached *diffusion equilibrium,* and no further change in the concentration of the two compartments occurs.

Several important properties of the diffusion process should be noted from this example. In the overall diffusion process three fluxes can be identified, the two one-way fluxes occurring in opposite directions from one compartment to the other and the net flux, which is the difference between them. It is the net flux *F* which is the most important component in the diffusion process since it is the net amount of material moving from one location

to another. In spite of the fact that there is no force acting to direct the movements of individual molecules, the *net diffusion of molecules always proceeds from a high to a low concentration.* The magnitude of the net flux is determined by the difference in concentration between two compartments. Thus, the basic diffusion equation relating the net flux to the concentration difference between two compartments becomes

$$F = k_D (C_1 - C_2) \qquad \text{2-3}$$

Just as heat never flows spontaneously from a cold object to a warm object, making the cold object colder and the warm object warmer, diffusion can never lead to the spontaneous net movement of molecules from a region of low concentration to a region of high concentration.

Many properties of living organisms are closely associated with the process of diffusion. Oxygen enters the blood from the lungs by diffusion; molecules leave the blood and enter the extracellular compartments of the tissues by diffusion; and the exchange of molecules between compartments within cells and between the cell and the extracellular environment occurs largely by diffusion. The rate of diffusion is thus an important factor in

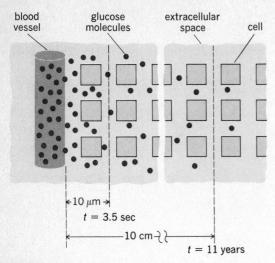

blood
vessel   glucose   extracellular
            molecules   space   cell

←10 μm →

t = 3.5 sec

←——— 10 cm ─⌇}——→

t = 11 years

FIGURE 2-10

The time required for diffusion to raise the concentration of glucose at a point 10 μm (about one cell diameter) away from a blood vessel to 90 percent of the blood glucose concentration is about 3.5 sec, whereas it would take over 11 years for glucose to reach the same concentration at a point 10 cm (3.9 in.) away.

determining the rate at which nutrients can reach a cell, and it can also be a factor in determining how rapidly a cell, or a system of cells, is able to respond when a chemical signal is transmitted from one compartment to another by diffusion.

Although individual molecules travel at very high velocities, the number of collisions they undergo prevents them from traveling very far in a straight line. In order to measure the overall rate of a diffusion process, we must determine how long it takes the concentration of molecules in a particular region to reach a given level when molecules are diffusing into this region from a source some distance away. Figure 2-10 illustrates the diffusion of glucose from the blood into the extracellular space surrounding the cells of a tissue. The question we wish to answer is: How long will it take the concentration of glucose to reach 90 percent of the glucose concentration in the blood at a point 10 μm and at a point 10 cm away from the blood vessel? The answer is 3.5 sec versus 11 years. Thus although diffusion can distribute molecules rapidly over short distances, i.e., inside cells or within the extracellular regions surrounding a few layers of cells, diffusion becomes an extremely slow process when distances of a few inches or more are involved. For an organism as large as a man, the diffusion of oxygen and nutrients from the body surface to the interior

tissues of the body would be far too slow to provide adequate nourishment. The circulatory system, operating by bulk flow, provides the mechanism for rapidly transporting materials over large distances, whereas diffusion provides the means of entry and exit from the blood. Because cells depend upon the diffusion of materials from the blood, all the cells in the body must be close to a blood vessel.

The rate at which diffusion is able to move molecules into and out of a cell is one of the major factors limiting the size to which a cell can grow and function. The nutrients entering the cell at its surface reach the volume under this surface by diffusion. The larger the cell, the greater is the volume of cytoplasm that must be supplied by diffusion across a given unit of surface area. A cell does not have to be very large before diffusion becomes too slow to provide sufficient nutrients to the volume of cytoplasm underneath the surface. The center of a 20-μm cell would become equilibrated with oxygen in about 15 msec whereas it would require 256 days for oxygen to reach the center of a cell the size of a basketball. Some cells which have a fairly large cell volume have solved this problem by assuming a shape that still provides a small volume/surface area ratio. Thus, muscle cells are often more than 10 cm long but only 10- to 100-μm thick. Thus, a molecule has to diffuse only a short distance from the surface to reach the center of the muscle cell.

Membrane permeability

The surface of the cell is the only portion coming in direct contact with the fluid immediately surrounding the cell, the *extracellular fluid*. Although no structure at the cell surface shows up under the light microscope, the rates at which molecules enter and leave the cell led biologists more than a century ago to postulate the existence of a highly selective barrier at the cell surface, called the *cell membrane*.[1] All nutrients must enter the cell across this membrane, and the waste products of cellular metabolism, as well as substances synthesized by the cell, must leave the cell through this barrier. The cell membrane is involved in several other important cell functions. The electric activity associated with the excitation of nerve

[1] Technically the membrane surrounding the surface of the cell is known as the plasma membrane. The membranes surrounding the various organelles within the cell are referred to as mitochondrial membranes, nuclear membranes, etc. Collectively these membranes are called cell membranes.

and muscle depends on the properties of the cell membrane surrounding these cells, and the membrane is often the site at which a chemical messenger released by one cell acts upon another cell to alter its specific activity, e.g., the rate of secreting a chemical substance or the force with which a muscle cell contracts. The variety of properties associated with cell membranes will be discussed later as they relate to specific functions carried out by the various organs in the body. In this chapter we restrict our discussion to the permeability and structural properties of cell membranes.

Diffusion across cell membranes

The rate at which a substance crosses the cell membrane can be measured by determining the rate at which the intracellular concentration C_i increases when the cell is placed in a solution containing the substance. If the volume of solution outside the cell is very large, the concentration of the substance in this solution, C_o, undergoes very little change as the material diffuses into the small intracellular volume (see Fig. 2-11). As with all diffusion processes, the net flux of material across the membrane is from the region of high concentration outside the cell, C_o, to the region of low concentration inside the cell, C_i. The net flux into the cell is, by Eq. **2-3**, $F = k_p(C_o - C_i)$, where k_D has become k_p, the *permeability constant* of the cell membrane. The numerical value of k_p depends upon

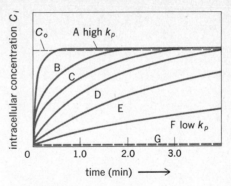

FIGURE 2-12
Comparison of the different rates at which various molecules A, B, . . . , G are able to diffuse across the cell membrane from an extracellular medium of constant concentration C_o. Molecule A enters the cell rapidly and has a high permeability constant k_p. Molecule F enters the cell very slowly, and molecule G is unable to cross the membrane at all.

the thickness of the cell membrane, its chemical composition, the temperature, and the particular type of molecule diffusing across the membrane. Net diffusion of material into the cell continues until the intracellular and extracellular concentrations equal each other, $C_i = C_o$. The rates at which a number of different molecules diffuse into a cell are shown in Fig. 2-12. The more permeable the membrane, the faster a substance enters the cell and the larger the value of the membrane permeability constant k_p. Such measurements reveal several properties of the cell membrane. The permeability constants of the cell membrane are found to be a thousand to a million times smaller than the diffusion coefficients of the same molecules in water, indicating that the molecules are entering the cell at a much slower rate than if there were no barrier and they were diffusing through a continuous water medium. Not only is the membrane a barrier which slows down the entry of molecules, it is also a highly selective barrier which allows some molecules to enter much more rapidly than others. In fact, some molecules, particularly proteins and other highly charged large molecules, diffuse across the membrane so slowly that the membrane can be considered to be impermeable to them.

Molecular structure and permeability

When the permeability constants of a number of different organic molecules are examined in relation to their molecular structure, a correlation between the two emerges. Molecules with few polar or ionic groups are found to enter cells rapidly and thus have large permeability con-

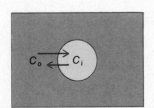

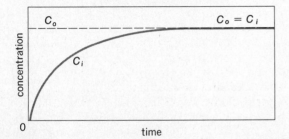

FIGURE 2-11
Change in intracellular concentration C_i as material diffuses from a constant extracellular concentration C_o across the cell membrane into the cell.

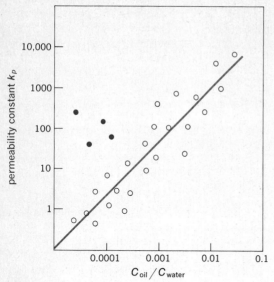

FIGURE 2-13
Relationship between the lipid solubility (C_{oil}/C_{water}, partition coefficient) and the rate at which molecules diffuse across a membrane measured in terms of their permeability constants k_p. Some molecules, e.g., the ions Na^+, K^+, and Cl^-, have a large k_p but are not lipid-soluble, as indicated by the colored dots in the graph. Each dot represents a different type of molecule.

solubility, the larger is its permeability constant, which suggests that the membrane material is composed primarily of nonpolar lipids. Molecules with a high lipid solubility will dissolve in the lipid portions of the membrane, giving rise to a high concentration of the molecule in the membrane. Thus the magnitude of the flux of the molecule through the membrane, which depends on its concentration in the membrane, will be greater than for molecules with low lipid solubility. If a molecule cannot dissolve in the lipid barrier, it will be excluded from the cell unless some other route of entry exists.

Although most molecules cross the cell membrane at a rate proportional to their lipid solubility, some molecules penetrate the membrane much more rapidly than would be predicted on this basis. Sodium, potassium, and chloride ions, the sugars such as glucose, and the amino acids—all these enter the cell very rapidly, yet they are almost insoluble in lipids. According to size these molecules fall into two classes: the ions which are very small charged particles, and the sugars and amino acids which are much larger charged molecules. When the membrane permeability constant calculated from the passive diffusion of these charged molecules into the cell is compared with their molecular diameter, the smaller molecules are found to enter more rapidly than the large molecules (Fig. 2-14). Charged molecules larger than about 8 Å in diameter enter cells very slowly by passive diffusion. This relation between molecular size and rate of entry suggests that the small charged molecules may be passing

stants, whereas polar or ionized molecules enter very slowly or not at all. This suggests that the relatively nonpolar molecules are passing through a membrane composed of nonpolar molecules in which other nonpolar molecules can readily dissolve. The solubility of a substance in a nonpolar medium, such as oil, can be measured by shaking an aqueous solution of the substance with a layer of oil. Some of the molecules dissolve in the oil phase and some in the water phase. The ratio of the two concentrations, C_{oil}/C_{water}, known as a *partition coefficient,* is a measure of the solubility of the molecule in oil relative to water. If the ratio is 1.0, the substance is equally soluble in oil and in water; a ratio greater than 1.0 means that the molecule is more soluble in oil than in water and vice versa. Most biological molecules are less soluble in oil than in water, since they usually contain at least a few polar groups. When the permeability constants of a number of different molecules are plotted against their oil/water partition coefficients, a broad linear relationship is found (Fig. 2-13). The larger the partition coefficient of a molecule, i.e., the greater its lipid

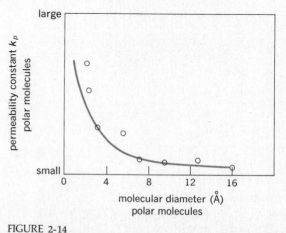

FIGURE 2-14
Relationship between the diameter of polar molecules and the rate at which the molecules cross the cell membrane.

through small holes, or pores, in the membrane. Molecules larger than the pore would be unable to cross the membrane except by passing through the lipid portions of the membrane, where their rate of entry would be low because of their low lipid solubility. The rates at which molecules cross the cell membrane by diffusion is consistent with a membrane structure composed primarily of lipids and having pores with an average diameter of about 8 Å. It has been calculated that pores occupying less than 1 percent of the total membrane surface area would account for observed diffusion rates.

Thus far we have considered the physical factors that influence the movement of molecules through the membrane resulting from the random diffusion of molecules. Such movements do not involve any chemical interactions between membrane and molecule. Although pores may account for the rapid entry of small polar molecules, such as water and ions, they cannot be responsible for the observed rapid entry of large polar molecules such as the sugars and amino acids, which appear to cross the membrane by a series of specific chemical reactions with the membrane structure.

Mediated-transport systems

Amino acids and sugars are polar molecules generally larger than 8 Å in diameter. Passive diffusion of such molecules through a lipid membrane with 8-Å pores would occur slowly if at all. The cell, however, depends upon the entry of large quantities of these molecules to provide the raw materials for protein synthesis and a source of chemical energy. Moving such molecules across the cell membrane requires special mechanisms, which are built into the membrane structure. Experimental evidence suggests that these molecules bind to specific sites on the membrane surface, which in some manner facilitate their movement through the membrane. This process is called *mediated transport*. The properties of chemical specificity, competition, and saturation distinguish molecules which cross by mediated transport from those which cross by simple diffusion.

Properties distinguishing mediated transport from diffusion

Specificity One of the characteristics of mediated-transport systems is their ability to handle a specific group of chemical substances. Although both amino acids and sugars undergo mediated transport, the system which

transports amino acids does not transport sugars. Some mediated-transport systems are so selective that they can distinguish between molecules which have only slightly different geometrical shapes, even though these molecules contain the same number of atoms and chemical groups. There are a number of different mediated-transport systems in the cell membrane, and each is able to react with only a very select group of chemical substances. Therefore, the membrane, in addition to restricting the entry of molecules on the basis of lipid solubility and size, also provides a mechanism whereby very specific chemical substances can enter the cell by these mediated-transport systems.

Saturation In simple diffusion the one-way flux of molecules entering a cell at any instant is proportional to the concentration of the substance outside the cell, C_o, and is given by $f = k_p C_o$. If C_o is doubled, the flux into the cell is doubled (see Fig. 2-15A). No matter how large C_o, the flux into the cell is always proportional to the concentration. In mediated-transport systems, however, the flux increases with extracellular concentration only up to a certain point (Fig. 2-15B). Thus, there is a maximum rate at which material can enter a cell through a mediated-transport system. An analogous situation is a building with a limited number of doors. If the number of people trying to enter the building is small, the rate at which they can enter is proportional to the number of people outside the building. As the number of people increases, the number of available doors limits the entry rate. Only a certain number of people can pass through a single door in a given period of time. When this maximum is reached, increasing the number of people outside the building will not increase the entry rate. When the maximum velocity of entry is reached, the system is said to be *saturated*. A transport system is saturated when all the specific sites on the membrane are fully occupied and operating at their maximum capacity. Some transport systems become saturated at very low concentrations, whereas others become saturated only at concentrations far above those normally found in the body. The limited capacity of mediated-transport systems to move molecules through the membrane has important consequences in various organs in the body, particularly in the kidney, which utilizes mediated-transport systems to regulate the excretion of solutes in the urine.

Competition A third characteristic of mediated-transport systems is the competition that occurs between similar molecules which enter the cell by means of the same transport site. If two molecules A and B, such as the

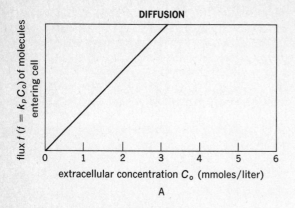

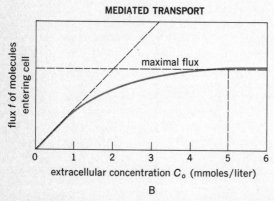

FIGURE 2-15
Relationship between the one-way flux of molecules into a cell and the extracellular concentration for A a molecule entering by simple diffusion and B a molecule entering by a mediated-transport system, showing a saturation of the transport flux when the extracellular concentration C_o is 5 mmoles/liter or greater. The dotted line in B represents the flux that would be expected if the molecule entered by simple diffusion.

amino acids glycine and alanine, which both enter the cell by the same transport system, are present simultaneously outside the cell, they must compete with each other for the available transport sites on the membrane (Fig. 2-16). The presence of B decreases the rate at which A enters the cell if B occupies some of the binding sites that would otherwise be available to A. Molecules entering by simple diffusion do not compete with each other, since their entry is not restricted to a limited number of sites.

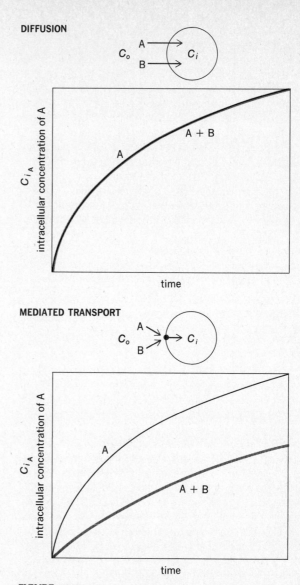

FIGURE 2-16
Competition between two molecules A and B for a limited number of binding sites in the membrane of a mediated-transport system decreases the rate at which A can enter the cell. If A can cross by diffusion, the rate at which A enters the cell is not affected by the presence of B.

The carrier hypothesis

The properties of specificity, saturation, and competition suggest that a specific chemical site within the membrane is able to combine with the transported molecule S. The

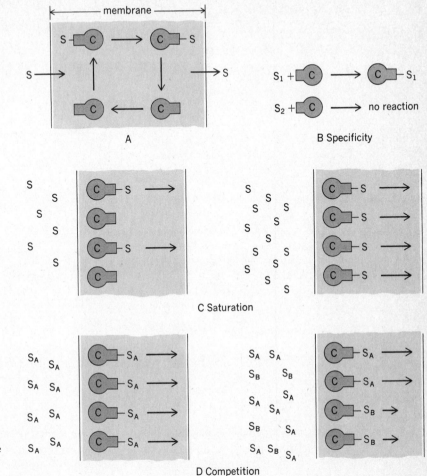

FIGURE 2-17
A **Carrier model for the mediated transport of molecules across the cell membrane.** S, transported molecule; C, carrier molecule; and C-S, carrier-solute complex. Examples of specificity B, saturation C, and competition D.

sites are called carriers C, and the carrier hypothesis is a model which attempts to explain the properties of mediated-transport systems in terms of the movements of the carrier and the carrier-solute complex (C–S) through the membrane (Fig. 2-17A). The movements of the carrier molecule are restricted to the membrane, and the simplest assumption is that carriers move by simple thermal diffusion so that the net diffusion of carrier through the membrane is always from a region of high carrier concentration to a region of low carrier concentration. A free carrier molecule is able to combine with a solute molecule at one surface of the membrane to form a carrier-solute complex, which then moves through the membrane to the opposite surface, where the solute

dissociates from the carrier and leaves the membrane. The carrier molecule then returns to the other surface, where it can repeat the cycle.

Specificity is determined by the chemical nature of the carrier molecule, which is able to react with only certain specific molecules (Fig. 2-17B). Because the number of carrier molecules in the membrane is finite, saturation occurs when all the carriers are combined with solute (Fig. 2-17C). Increasing the concentration of solute outside the membrane does not increase the rate of entry since no more carrier molecules are available to carry solute across the membrane. Competition between solutes decreases their rate of entry, as some of the carrier sites are occupied by the competing molecules, making

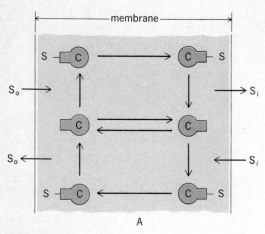

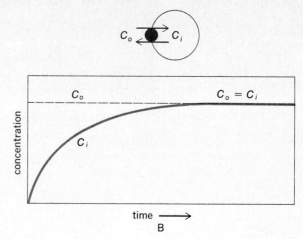

FIGURE 2-18
Facilitated diffusion. A Carrier model of the facilitated diffusion of a solute S through a cell membrane in combination with carrier molecule C. B Changes in intracellular concentration C_i with time, resulting from the entry of molecules by facilitated diffusion.

less total carrier available for each solute (Fig. 2-17D).

Although the carrier model offers an explanation of how mediated-transport systems may operate, the existence of actual carrier molecules in the membrane has yet to be proved. Calculations based upon the carrier-model assumptions have shown that only a small number of carrier molecules need be present in the membrane to provide the observed rates of solute movement across the membrane. For example, there may be only 1,000 carrier molecules involved in sugar transport in the entire membrane of a red blood cell compared with the several hundred million protein molecules and more than 2 trillion total molecules found in a single cell. Such small quantities of material make the problem of chemical isolation and identification very difficult. Educated guesses at the moment predict that the carrier molecules, if they exist, are probably proteins, since proteins are the only molecules known which possess the high degree of chemical specificity shown by mediated-transport systems. Just how the combination of a molecule with a carrier leads to an increase in the rate of passage through the membrane is unknown. Possibly the carrier-solute complex is more lipid-soluble than the solute molecule alone, thus allowing the molecule to enter the lipid portion of the membrane.

The behavior of carrier-mediated transport systems can be further divided into two classes: *facilitated-diffusion*

systems, which lead to an equal distribution of molecules across the membrane, and *active-transport systems,* which can move solutes from a low-concentration region on one side of a membrane to a high-concentration region on the other side.

Facilitated diffusion

When the change in the intracellular concentration C_i is measured as a function of time in a facilitated-diffusion system in a single experiment, the results appear very much like a simple diffusion process, since the final intracellular concentration equals the extracellular concentration, $C_i = C_o$ (Fig. 2-18B). It is only when the concentration of the medium is varied and competing molecules are added to the system that specificity, saturation, and competition appear.

A proposed carrier model for facilitated diffusion is shown in Fig. 2-18. The carrier molecules can undergo identical reactions at both membrane surfaces and can move molecules in either direction across the membrane. As the concentration of solute increases within the cell, some of the solute molecules react with free carriers at the inner surface of the membrane and are moved out of the cell. When the intracellular concentration becomes equal to the extracellular concentration, molecules move out of the cell as rapidly as they enter and no further change in the intracellular concentration occurs. If extracellular concentration exceeds intracellular concentrations, there is net movement into the cell. In facilitated diffusion all net movement of molecules across the membrane is from a region of high concentration to a region of low concentration.

The most important example of facilitated diffusion in the body is the movement of glucose (blood sugar) across the membranes of most cells. This relatively large polar molecule would not be expected to diffuse readily through the lipid or pore regions of the membrane. Facilitated diffusion through the membrane provides a mechanism for supplying cells with this essential compound.

Active transport

Not only do some molecules cross membranes more rapidly than expected, but some are even moved from a region of low concentration into a region of higher concentration. Such movement is quite contrary to that expected for a diffusing molecule since diffusion alone can never lead to the net movement of molecules uphill against a concentration gradient. In order for molecules to move against a concentration gradient, energy must

be supplied. The movement of molecules across a cell membrane from a region of low to high concentration at the expense of energy provided by chemical reactions within a cell is known as *active transport*.

A carrier model of active transport is shown in Fig. 2-19. If a higher concentration of molecules on one side of a membrane than on the other is to be achieved, molecules must be able to move more readily in one direction through the membrane than in the other. In a facilitated-diffusion system we saw that the reactions between solute molecule and carrier were symmetrical on both sides of the membrane. In contrast, for active transport to occur, these reactions must be asymmetric, allowing the solute molecule to combine more readily with a carrier at one surface than at the other. Such an asymmetry can be achieved if the carrier molecule undergoes a reaction at one surface of the membrane which alters its structure in such a way that it now reacts differently with a solute molecule. If the carrier combines more readily with the solute molecule, it is said to be a *high-affinity carrier;* if it reacts less readily, it is a *low-affinity carrier.* When the concentration of molecules on both sides of the membrane is equal, more molecules combine

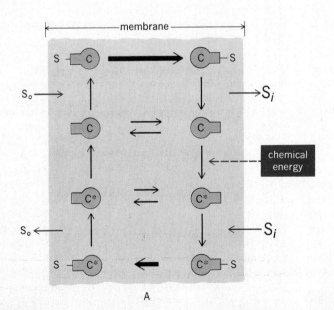

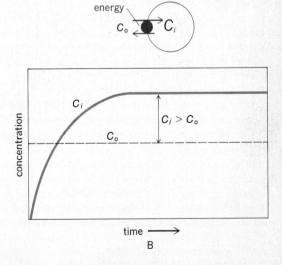

FIGURE 2-19
Active transport. A Carrier model of the active transport of solute S through the cell membrane in combination with carrier molecule C. Chemical energy modifies the carrier molecule C to form C*, which has a lower affinity for solute at

the inner surface of the membrane. B Time course of changes in intracellular concentration C_i during active transport, resulting in an intracellular concentration of solute higher than the extracellular concentration C_o.

with the high-affinity carriers on one side, than with low-affinity carriers on the other side. Thus, there is a net flux of molecules through the membrane, and the concentration on the side with the low-affinity carriers continues to increase. When the concentration on the low-affinity side of the membrane has increased to the point where the number of molecules moving in both directions through the membrane becomes equal, a steady state is reached and no further change in the concentrations occurs. Such active-transport systems can achieve concentrations 30 to 50 times higher on one side of the membrane than on the other.

Energy provided by cellular metabolism is required to change the carrier structure at one surface of the membrane. At the opposite surface, the carrier may spontaneously revert to its original shape as it releases its solute. Inhibiting the source of energy provided by the cell prevents an active-transport system from moving molecules against a concentration gradient. Since no carrier molecule has yet been isolated and identified, the mechanism by which energy is linked to the carrier system is still a matter of speculation, and the carrier model presented above is only a hypothetical mechanism.

Perhaps the most important active-transport system in the body involves pumping sodium and potassium ions across cell membranes. The extracellular fluids of the body normally have a sodium concentration of 144 mmoles/liter and a potassium concentration of 4.4 mmoles/liter. Inside the cell the respective concentrations are 15 and 150 mmoles/liter. Thus, the concentration of sodium outside the cell is about 10 times greater than inside, and the potassium concentration inside the cell is about 35 times greater than outside. These concentration gradients are maintained by the active transport of potassium into and sodium out of the cell. Experimental evidence suggests that these two ions use the same carrier molecule. It has been estimated that as much as 10 to 40 percent of the total energy produced by the cell may be utilized for the transport of sodium and potassium ions across the membrane. Since small ions can also cross the membrane by diffusion through pores, two routes are available to ions crossing the membrane. The electric activity of nerve and muscle cells depends upon ion movements across their cell membranes. The importance of passive ion movement and active ion transport in producing electric activity will be described in a later chapter. All molecules which cross membranes through a mediated-transport system generally also cross by passive diffusion to some extent; the contribution of passive diffusion to the total flux of the

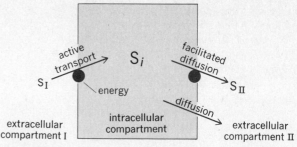

FIGURE 2-20
Transcellular active transport between two extracellular compartments I and II. Solute S enters the cell by active transport from compartment I and leaves the cell by diffusion or facilitated diffusion from the high intracellular concentration S_I into the lower concentration S_{II} in compartment II.

molecule across the membrane is generally very small owing to the restricted diffusion of large polar molecules through the lipid regions of the membrane.

Amino acids are actively transported into cells, attaining intracellular concentrations 2 to 20 times higher than the extracellular concentrations. Sugars, which enter most cells by facilitated diffusion, are actively transported across the walls of the intestine and in the kidneys. In this case the active-transport systems move molecules from an extracellular compartment, into a cell, and out of the cell into a second extracellular compartment (Fig. 2-20). Active transport may be operating at one end of the cell to move molecules up a concentration gradient, while at the other end of the cell the molecules leave by diffusion or facilitated diffusion down a concentration gradient into the second extracellular compartment. The net result is the active movement of molecules from one extracellular compartment having a low concentration to a second extracellular compartment containing the molecules at a higher concentration. Such transcellular transport systems are involved in the absorption of nutrients from the intestinal tract into the blood and the transport of solutes between the tubules of the kidney and the blood during the formation of urine.

Figure 2-21 summarizes the permeability properties exhibited by cell membranes. The ability of the membrane to act as a selective barrier depends upon (1) the lipid composition of the membrane, which allows nonpolar molecules of all sizes to cross by dissolving in the lipid matrix; (2) the pores, which allow small molecules, particularly ions, to cross without passing through the lipid

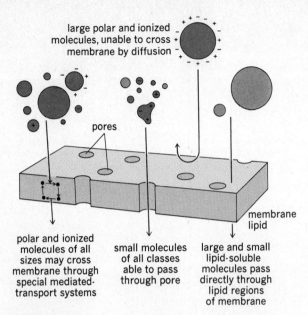

large polar and ionized molecules, unable to cross membrane by diffusion

pores

membrane lipid

polar and ionized molecules of all sizes may cross membrane through special mediated-transport systems

small molecules of all classes able to pass through pore

large and small lipid-soluble molecules pass directly through lipid regions of membrane

FIGURE 2-21
Summary of the various pathways by which molecules can cross cell membranes.

matrix; and (3) the special mediated-transport systems, which facilitate the movement of polar molecules and ions through the lipid matrix. Unable to cross the membrane are the large polar and ionized molecules which cannot dissolve in the lipid matrix, are too large to pass through the pores, and do not interact with any of the mediated-transport systems; but even these molecules can enter the cell very slowly through pinocytosis and phagocytosis.

Pinocytosis and phagocytosis

One more mechanism for moving molecules across cell membranes remains to be discussed. If certain types of cells are closely observed under a microscope, small vesicles can be seen to form at the surface of the cell and move into its central regions. A portion of the cell membrane invaginates and then detaches itself from the surface, forming an isolated vesicle within the cell (Fig. 2-22). This process is known as *pinocytosis* (Greek: cell drinking) if the contents of the vesicle are fluid and *phagocytosis* (Greek: cell eating) if the contents are a solid. Although the vesicle has brought a portion of the extracellular medium inside the cell, the contents of the vesicle are still separated from the intracellular medium

by a membrane. Once within the cell, chemical transformations of the membrane or the contents of the vesicle allow the material within the vesicle to leak out into the intracellular medium. Proteins and various ions appear to increase the rate of vesicle formation in certain cells, but little is known of the specific mechanisms which cause these dynamic changes in the cell membrane. Since the quantity of material these processes move is small and the rates slow, pinocytosis and phagocytosis do not provide a major route for the entry of molecules into the cell; however, they do provide a mechanism whereby very large molecules, such as proteins and nucleic acids, can reach the interior of a cell in small quantities. Because of these mechanisms no molecule is ever completely excluded from a cell. The ability of white blood cells to engulf bacteria by phagocytosis provides an important defense mechanism for protecting the body against bacterial infections and will be described more fully in Chap. 15.

Osmosis

A net movement of water across a cell membrane occurs whenever there is a difference between the water concentrations on the two sides of the membrane. If water happens to be the only molecule able to move through the membrane, the net movement of water into or out of the cell wall leads to a change in cell volume. *Osmosis* is the net movement of water between two compartments as a result of a difference in water concentration when the movement of solute is restricted by a barrier (membrane).

How can a difference in water concentration be established across a membrane? One usually thinks of the concentration of a substance in a solution as referring to the amount of solute dissolved in a given volume of solvent; e.g., a 1 molar (1 M) solution of glucose contains 1 g molecular weight of glucose (1 mole) dissolved in enough water to form 1 liter of solution. However, the water in this same solution also has a concentration. A liter of pure water weighs about 1,000 g. The molecular weight of water is 18; thus, the concentration of pure water is $1,000/18 = 55.5$ moles/liter. However, if a solute molecule such as glucose is dissolved in water, the concentration of water in the resulting solution is less than that of pure water. Each molecule of solute added to a solution occupies an element of volume formerly occupied by water molecules. The more solute molecules added, the greater the number of water molecules dis-

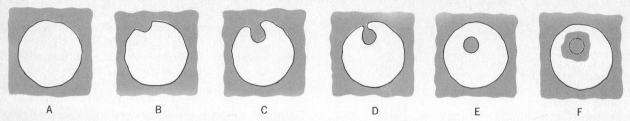

FIGURE 2-22
Sequence of membrane changes during the entry of material into a cell by pinocytosis or phagocytosis.

NaCl$_2 \simeq$ ●●
Glucose $\simeq$ ●

placed. Figure 2-23 illustrates this decrease in water concentration that results from the addition of solute. The degree to which the concentration of water is decreased by the addition of solute depends upon the *number* of particles of solute in solution and not upon the chemical nature of the solute molecule. In reality all molecules do not have an identical effect upon lowering the water concentration of a solution. This is because molecules differ in size and electric charge which produce slight differences in their interaction with water molecules. These effects, however, are small, and the major fact remains that it is the total number of solute molecules in solution and not their chemical nature that determines the concentration of water. Thus, 1 mole of glucose added to 1 liter of water decreases the water concentration to approximately the same extent as 1 mole of an amino acid, or 1 mole of a protein, or 1 mole of any molecule which exists as a single particle in solution. A molecule which ionizes in solution, such as sodium chloride, which forms the two ions Na^+ and CL^-, decreases the water concentration in proportion to the number of

ions formed. Hence, 1 mole of sodium chloride lowers the water concentration twice as much as 1 mole of glucose because sodium chloride forms two ions in solution for every molecule added.

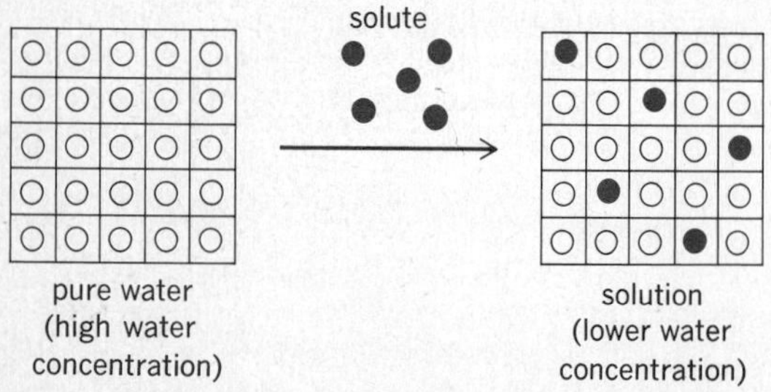

Since the concentration of water depends upon the number of solute particles in solution rather than their chemical properties, it is useful to have a concentration term which refers to the total concentration of all solute particles in a solution. The total solute concentration of a solution is known as the *osmolarity.* One osmole is equal to one mole of an ideal nonionizing molecule. Thus, a $1\,M$ solution of glucose has a concentration of 1 osmole/liter, but a $1\,M$ solution of sodium chloride contains 2 osmoles of particles per liter of solution. A liter of solution containing 1 mole of glucose and 1 mole of sodium chloride has an osmolarity of 3 osmoles/liter since it would contain 3 moles of solute particles. The concentration of water in any two solutions having the same osmolarity is always the same, since the total number of solute particles is the same. A 3 osmolar solution may contain 1 mole of glucose and 1 mole of sodium chloride, or 3 moles of glucose, or $1\frac{1}{2}$ moles of sodium chloride, or any other combination of solutes so long as the total number of solute particles is equal to the number of solute particles in a $3\,M$ solution of an ideal non-ionizing molecule. Although osmolarity refers to the concentration of solute particles in solution, it is indirectly a measure of the water concentration in the solution; the *higher* the osmolarity of a solution, the *lower* is the water concentration.

Figure 2-24 shows two compartments separated by a membrane. The concentration of solute in compartment 2 is higher (4 osmoles/liter) than the concentration of solute in compartment 1 (2 osmoles/liter). Thus, there is a solute concentration gradient that could lead to a net diffusion of solute from 2 to 1. This difference in solute

FIGURE 2-23
Decrease in water concentration resulting from the addition of solute molecules.

solute

pure water
(high water
concentration)

solution
(lower water
concentration)

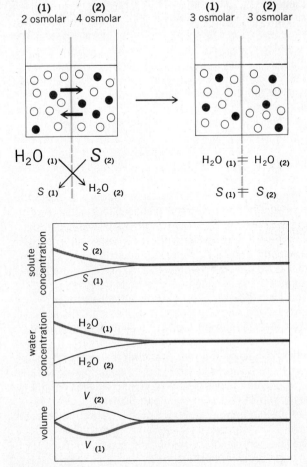

FIGURE 2-24
Changes in water concentration, solute concentration, and compartment volume when both water and solute are able to diffuse across a membrane separating two compartments that initially contained different concentrations of solute.

concentration also means a difference in water concentration across the membrane, with the water in compartment 1 having a higher concentration than the water in compartment 2. Therefore, water undergoes a net movement from compartment 1 to 2, in the opposite direction from the net solute movement. If both water and solute can cross the membrane, movement proceeds in both directions until the concentrations of both water and solute are equal in the two compartments. The movement of water out of compartment 1 and into compartment 2 lowers the concentration of water in 1 and raises the

concentration of water in 2. This also has the effect of concentrating the solute in compartment 1 and diluting the solute in compartment 2. While the water is moving between the two compartments, solute is diffusing in the opposite direction and also tends to bring the water and solute concentrations in the two compartments into equilibrium. The net result is the redistribution of water and solute between the two compartments so that the final concentration of solute in both compartments is 3 osmoles/liter. However, the final volumes of the two compartments remain unchanged since the water and solute molecules have merely exchanged positions between the two compartments. Transitory changes in volume may occur because of differences in the rates at which solute and water move between the two compartments.

If we make one alteration in this system, namely, replacing the permeable membrane with a membrane permeable to water molecules but impermeable to the solute, the final result is different (Fig. 2-25). Initially, the same concentration gradients are found across the membrane as in the previous example. Water moves from its high concentration in compartment 1 into compartment 2, which has a lower water concentration, but since there can be no movement of solute in the opposite direction to compensate for the water movement, the volume of compartment 2 must increase. Since solute molecules cannot leave compartment 1, the loss of water from this compartment has the effect of increasing the concentration of solute there, and the water entering compartment 2 dilutes the solute concentration. This system also comes to equilibrium when the concentrations of water and solute in the two compartments become equal, only now the equilibrium has resulted from the transfer of water alone and has led to a change in the final volume of the two compartments. This example illustrates the following generalization: In order for a volume change to occur the movement of solute molecules must be restricted by a semipermeable membrane. During osmosis the water always moves from a region of high water concentration (low solute, i.e., osmolar concentration) to a region of low water concentration (high solute, i.e., osmolar concentration).

The cell behaves like an osmotic system; it is surrounded by a semipermeable membrane highly permeable to water, and its cytoplasm contains many molecules which are unable to cross the cell membrane. The cell membrane is highly elastic, allowing the cell to shrink and expand when the cell is placed in media of differing water concentrations. The permeability constant of the cell

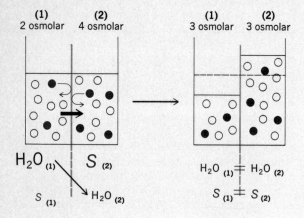

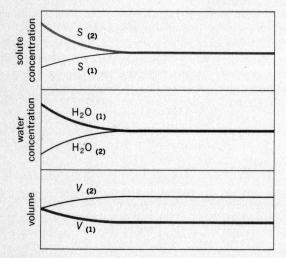

FIGURE 2-25
Changes in water concentration, solute concentration, and compartment volume when only water is able to diffuse across a membrane separating two compartments that initially contained different concentrations of solute.

membrane to water is larger than to almost any other substance. This high membrane permeability to water means that any change in the water concentration surrounding a cell rapidly leads to a change in cell volume. Since changes in cell volume are completed so rapidly, for all practical purposes, the concentration of water within a cell is always equal to the concentration of water in the extracellular fluids.

When a cell is in osmotic equilibrium, the water concentration is the same on the two sides of the membrane. Thus, the total solute (osmolar) concentration on the two sides of the membrane must also be the same. Any change in the solute concentration also alters the water concentration. If the solute cannot cross the cell membrane, any change in its concentration produces a water concentration gradient across the membrane, leading to a net flux of water into or out of the cell and a change in cell volume. Thus, the volume of a cell ultimately depends upon the concentration of nonpenetrating solutes on the two sides of the cell membrane.

About 90 percent of the solutes in the extracellular fluids of the body are the sodium and chloride ions, which can diffuse into the cell through the pores in the cell membrane. However, the membrane contains an active-transport system for pumping sodium ions out of the cell, the result being that there is no net movement of sodium into the cell. Sodium appears as if it were unable to cross the membrane since, for every sodium ion entering the cell, another sodium ion is returned to the outside by the active-transport system. Since chloride ions follow the sodium because of the electric attraction between the positive and negative charges of the two ions, both sodium and chloride ions act as if they were nonpenetrating solutes and any changes in the concentration of sodium chloride outside the cell alter the concentration of extracellular water and lead to a change in cell volume.

Inside the cell the major solute particles are potassium ions and a number of organic solutes, most of which are unable to cross the membrane because they contain polar or ionized groups, which limit their diffusion through the lipid portions of the membrane, and they are too large to pass through the pores in the membrane. Since potassium ions can leak out of the cell but are actively transported into the cell, the net effect is the same as if potassium could not cross the membrane. Thus, sodium chloride outside the cell and potassium and organic solutes inside the cell represent the major effective nonpenetrating solutes which determine the water concentrations on the two sides of the membrane. Solutes which are not actively transported across the membrane but which can cross it by simple diffusion have no net effect upon the cell volume since they eventually reach the same concentration on both sides of the membrane and thus have the same effect on water concentration of the two sides of the membrane. It is only the nonpenetrating or effectively nonpenetrating solutes which can alter cell volume by producing a change in the water concentration across the membrane.

Red blood cells removed from the body and placed

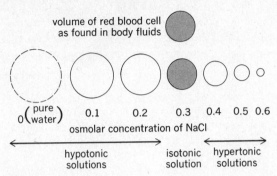

FIGURE 2-26

Comparison of the volume of a red blood cell in solutions of different sodium chloride concentrations with the volume of a red blood cell surrounded by body fluids.

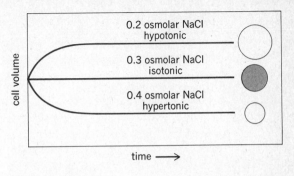

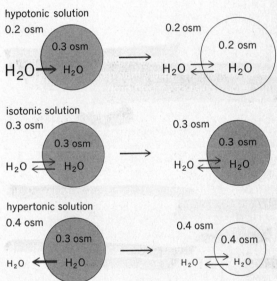

FIGURE 2-27

Changes in cell volume that result from the diffusion of water when a cell is placed in a hypotonic, isotonic, or hypertonic solution of sodium chloride.

in solutions containing different concentrations of sodium chloride change their volume in each solution according to the water concentration inside relative to the concentration outside the cell (Fig. 2-26). If the outside water concentration is greater, water moves into the cell, which swells. The cell shrinks if the concentration of water outside is less than inside. By comparing the volume of the cell as it exists in the body with the volume in each of the sodium chloride solutions, one can determine the concentration of the sodium chloride solution in which the cell volume remains the same as when surrounded by body fluids. This solution must have the same number of nonpenetrating solute particles as the red cell. Such a solution is said to be *isotonic* to those cells. An isotonic solution is any solution in which the cell volume remains the same as when surrounded by body fluids. Solutions in which the cell swells are called *hypotonic*, and solutions in which the cell shrinks are *hypertonic* solutions (Fig. 2-27). The terms isotonic, hypotonic, and hypertonic express the effects of solutions on the volume of any cell as compared with its initial volume.

In the example above, a solution of 0.15 M sodium chloride was found to be isotonic for the human red blood cell; its osmolar concentration is 0.3 osmole/liter. Any other solution containing 0.3 osmole/liter of *nonpenetrating* solutes will be isotonic for the same cell. But if a cell is placed in a 0.3 osmolar solution of *penetrating* solute, the cell will swell up and burst just as if it had been placed in pure water, since the solute is able to enter the cell and will have the same effect on lowering the intracellular water concentration as it does on the extracellular water.

The difference in water concentration that leads to the cell swelling is due to the additional nonpenetrating solutes within the cell.

A red cell placed in pure water, a hypotonic solution, swells until it bursts. Such a cell can never reach a stable volume because there are no solute molecules in the extracellular medium; thus, the intracellular water containing solute molecules always has a lower concentration than the extracellular water. As the cell volume becomes larger and larger, the membrane becomes more and more leaky until even the hemoglobin molecules escape, a loss known as *osmotic hemolysis*. Since red cells would be hemolized if water were injected into a blood vessel, solutions to be injected into the bloodstream must be nearly isotonic with the blood. Most drugs to be injected into the body are dissolved in isotonic saline (sodium chloride solution).

As we shall see in later chapters, the water concentration of the body fluids is kept nearly constant. The kidneys, by regulating the excretion of water and solutes in the urine, are primarily responsible for maintaining the osmolarity of the body fluids.

Membrane structure

We have noted that nonpolar (lipid-soluble) molecules diffuse into cells more rapidly than molecules which contain numerous polar or ionized groups. This observation strongly suggests that the barrier at the surface of the cell is composed of lipids, the solubility of a molecule in this lipid surface determining the ease with which the molecule can enter the cell. Chemical analysis of plasma membranes isolated from a variety of cells has shown that lipids are indeed a major component of the membrane structure, constituting about 50 percent of the membrane dry weight.

The majority of the lipids present in the membrane are phospholipids (see Chap. 1), which have a bimodal structure in that one end of the molecule is electrically charged whereas the fatty acid end is nonpolar. The polar end of the phospholipid will interact electrically with other polar molecules, including water, and the fatty acid chains tend to associate with each other and other nonpolar molecules. Therefore, when phospholipids are mixed with water they tend to associate in a regular pattern (Fig. 2-28) based on the physical properties of the polar and nonpolar ends of the molecules. A series of layers two phospholipid molecules thick is formed, in which the polar ends of the molecules are oriented toward the polar water molecules and the nonpolar fatty acid chains are oriented to the interior of the bilayer.

The phospholipids in cell membranes appear to be organized into a single bimolecular layer similar to those which form spontaneously in phospholipid-water mixtures. Thus structure formation does not necessarily imply a mechanical placing of each molecule into a fixed position within the structure. Cells need only be able to synthesize individual molecules which then spontaneously aggregate according to their individual chemical and physical properties to form multimolecular structures.

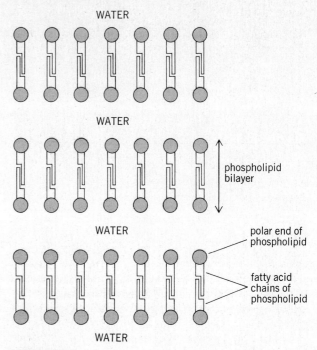

FIGURE 2-28
Bilayer organization of phospholipid molecules in water.

In addition to phospholipid, the surface membranes surrounding cells also contain cholesterol, a steroid. Cholesterol is an uncharged, nonpolar lipid which is thought to reside between the fatty acid chains of the phospholipids, maintaining a looser, more flexible packing of the fatty acid chains.

The physical properties of phospholipid bilayers have been extensively studied and found to have permeability characteristics very similar to those of intact cell membranes, including a very high permeability to water. The water molecule, although polar and not soluble in lipid, appears to be small enough, about 3 Å in diameter, to pass through the spaces between the fatty acid side chains in the membrane. The high permeability of membranes to certain ions and the processes of carrier-mediated transport cannot, however, be explained by the properties of the phospholipid bilayer alone.

The second major component found in isolated cell membranes is protein, which accounts for most of the remaining nonlipid dry weight of the membrane. Since proteins contain many charged and polar amino acid side

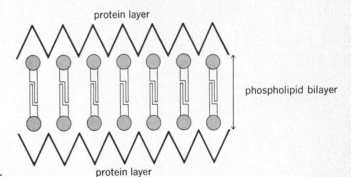

protein layer

phospholipid bilayer

protein layer

FIGURE 2-29
Danielli model of membrane structure.

chains it is logical to assume that these charged groups would associate with the polar water molecules and the polar ends of the phospholipids. Thus, Danielli, in 1935, proposed a model for the structure of cell membranes (Fig. 2-29), which consisted of a bimolecular layer of phospholipids covered on its two surfaces by a layer of protein.

Viewed with the electron microscope, the cell membrane appears as a very thin line at the surface of the cell, about 75 Å thick. At very high magnification the line can be resolved into two dark lines separated by a light interspace (Fig. 2-30). The heavy metals used to increase the contrast of electron micrographs are believed to react strongly with charged proteins and the polar ends of phospholipid molecules but do not combine in large quantities with the nonpolar hydrocarbon lipid chains. This suggests that the two dark lines represent the protein layers and the polar ends of the phospholipids; the light middle layer corresponds to the lipid fatty acid chains. The membranes surrounding organelles in the cytoplasm of the cell show the same three-layered structure. The similar appearance of all membranes in the cell has led to the *unit-membrane hypothesis,* which states that all membranes are composed of protein and lipid layers arranged in a pattern similar to that of the Danielli model.

Although all membranes have been found to contain phospholipid and protein, they do not all contain the same types of lipid and protein molecules or the same relative proportions of these two classes of molecules. These differences in chemical composition reflect the different functional properties of membranes found in different areas of the cell and surrounding different types of cells.

In recent years new information about the proper-

ties of the lipids and proteins in cell membranes has necessitated some modifications in the Danielli model of membrane structure. These modifications emphasize the very dynamic nature of cell membranes. The Danielli lipid bilayer model (Fig. 2-29) gives the impression of a very rigid, orderly arrangement of parallel fatty acid chains perpendicular to the membrane surface. It now appears that the fatty acid chains are not in a rigid configuration but are actually very mobile, so that the interior of the membrane is in an almost liquid state, with the fatty acid chains wiggling back and forth but with the polar ends of the phospholipids retaining a crystalline organization at each surface of the membrane. In fact, the entire phospholipid molecule appears to be able to move laterally within the surface planes of the membrane. Thus the phospholipid bilayer of the membrane is actually a highly mobile, fluid structure which has the characteristic of a liquid crystal. Evidence is accumulating which indicates that this fluid state of the membrane lipids is essential for the operation of various mediated-transport systems which involve the movement of carriers within the membrane.

The development of more sophisticated methods of isolating and studying the structure of protein molecules has also led to a considerable modification of our concept of the organization of protein molecules in the membrane. Just as phospholipids have a bimodal structure with polar and nonpolar ends, many of the membrane proteins appear to be constructed so that most of the nonpolar amino acid side chains are located in one area of the molecule and the polar side chains in another, producing a bimodal distribution of charged and uncharged regions within the molecule. Rather than lying flat on the surfaces of the phospholipids, these proteins are now thought to be inserted into the membrane so that the nonpolar portion

FIGURE 2-30
Electron micrograph of a human red cell membrane. (*From
J. D. Robertson, in Michael Locke (ed.), "Cell Membranes in
Development," Academic Press, Inc., New York, 1964.*)

of the molecule interacts with the nonpolar fatty acid chains and the polar end of the protein is exposed to the aqueous environment at the two surfaces of the membrane (Fig. 2-31). Some of the proteins extend all the way through the membrane and have part of their structure exposed on both surfaces; others are located on only the outer or inner surface. Most of these proteins are not anchored to the membrane in a fixed position but are free to move along the surface of the membrane just as the phospholipid molecules are able to undergo lateral movement. Thus, the membrane proteins appear to float in a sea of lipid and form a mosaic pattern on the surface of the membrane.

Although holes or pores in the membrane have not been identified in the electron microscope, some of the proteins which extend all the way through the membrane may provide the channels which allow ions such as sodium, potassium, and chloride to readily cross cell membranes. The presence of positively charged side chains on the proteins lining these channels may provide a type of polar pore that would be selective for the negatively charged chloride ions but exclude the positive ions, and negatively charged channels of various shapes would distinguish between sodium and potassium ions on the basis of their ionic size.

The membrane of the red blood cell has been the

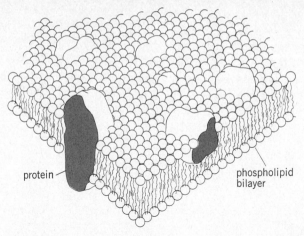

FIGURE 2-31
Fluid mosaic model of the cell membrane. (*Redrawn from Singer and Nicolson, Science,* **175:***723. Copyright 1972 by the American Association for the Advancement of Science.*)

protein

phospholipid bilayer

most extensively studied because of the ease with which the membrane can be isolated by osmotic hemolysis. At least 20 different proteins have been identified in the membranes of red cells, and there are probably additional proteins present in very small amounts. The distribution of the protein in these membranes is quite asymmetric, with different types of proteins being located on the outer surface than are present on the inner surface. In addition it appears that there is considerably more protein on the inner surface than on the outer surface. This asymmetric distribution of protein is consistent with the known asymmetric functioning of cell membranes such as the active transport of various substances across the membrane in a specific direction.

In addition to the lipids and proteins found in all membranes, the cell membrane generally contains small amounts of carbohydrate. This carbohydrate is attached to proteins (glycoproteins) and is located only at the outer surface of the membrane. These carbohydrates are associated with the binding sites on the membrane surface for viruses and various chemical substances, such as hormones, which modify cell function. In addition these carbohydrates are involved in the aggregation of cells to form tissues and in the recognition of foreign cells which enter the body. The surfaces of cancer cells show chemical differences from normal cells, which are related to their ability to break away from surrounding cells, migrate through the body, and invade normal tissues.

Membrane junctions

In addition to providing a barrier to the movements of molecules between the intracellular and extracellular media, cell membranes are also involved in interactions between cells to form organized tissues. Some cells, particularly those of the blood, do not associate with other cells but remain as independent cells suspended in a fluid, the blood plasma. Most cells, however, are closely packaged to form tissues and organs and are not free to float around the body. But even the cells in an organized tissue are not packaged so tightly that the adjacent cell surfaces are in direct contact with each other. There usually exists a gap of at least 200 Å between the opposing membranes of adjacent cells; this gap is filled with extracellular fluid and provides the pathway for the extracellular diffusion of substances within a tissue.

The forces that bring about the organization of cells into tissues and organs are poorly understood but appear to depend in part on the ability of the cell surface to recognize and associate with similar types of cells. Thus, if the cells of the kidney are gently dissociated from each other so that they form a suspension of free cells and are placed in a sterile glass chamber with appropriate nutrients, they continue to live and grow and, most important, organize themselves into tubular structures that resemble the organized structure of the kidney. If a suspension of dissociated liver and kidney cells are mixed together in tissue culture, not only do both types of cells continue to grow but the two types of cells begin to associate with cells of a similar type and become organized into separate clusters of kidney and liver tissue.

Some tissues can be separated into individual cells by fairly gentle procedures such as shaking the tissue in a medium from which calcium ions have been removed. The absence of calcium ions in the extracellular medium decreases the ability of some cells to stick together, suggesting that the doubly charged calcium ion, Ca^{2+}, may act as a link between cells by combining with negatively charged groups on adjacent cell surfaces. Many types of cells, however, require more drastic chemical procedures to separate them from their neighbors.

The electron microscope has revealed that a variety of cells are physically joined together by specialized types of junctions. One type of special junction, known as a *desmosome* (Greek *desmos,* binding), is illustrated in Fig. 2-32. This type of junction is found between the cells in the skin and in heart muscle as well as in a variety of other tissues in the body. The desmosome consists of two opposed membranes that remain separated by about

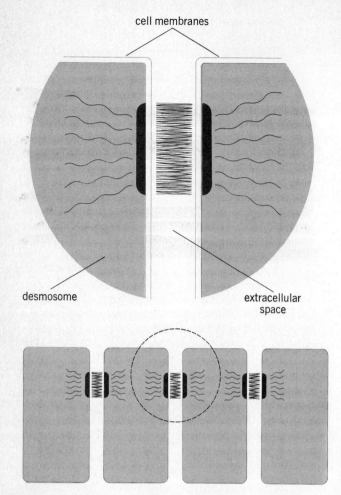

cell membranes

desmosome

extracellular space

FIGURE 2-32
Schematic diagram of desmosome structure linking cells together. Membranes of adjacent cells are not in contact in the region of the desmosome.

200 Å but show a dense accumulation of matter at each membrane surface and some dense material between the two membranes. In addition, fibers extend from the inner surface of the membrane into the cytoplasm and appear to add structural stability to the desmosome. The function of the desmosome appears to be that of holding adjacent cells together in areas that are subject to considerable stretching. such as in the skin and heart muscle. Desmosomes are usually disk-shaped and thus could be likened to rivets or spot-welds as a means of linking cells together.

Another type of membrane junction is found in epithelial cells that line the surfaces of internal cavities in the body, such as the lining of the intestinal tract. Such an epithelial layer of cells usually separates two compart-

ments having differing chemical compositions; thus, the intestinal epithelium lies between the lumen of the intestinal tract containing the products of food digestion and the blood vessels that pass beneath the epithelial layer. These epithelial layers generally mediate the passage of molecules between the two compartments. It was noted earlier that most cells are separated from each other by a gap of at least 200 Å which is filled with extracellular fluid. However, such a layer of cells would make a very leaky barrier that would allow even large protein molecules to diffuse between the two compartments by passing between adjacent cells. In fact, however, the intestinal epithelium in an adult is impermeable to protein, and the reason is the presence of a special type of junction joining these epithelial cells near their luminal border. These

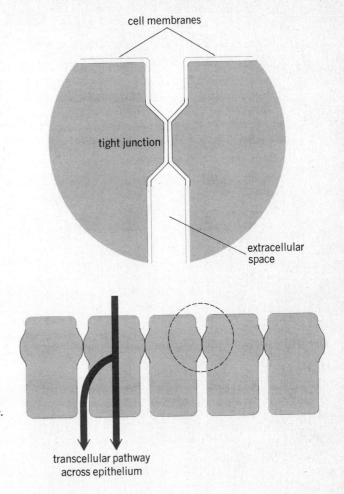

FIGURE 2-33
Schematic diagram of structure of tight
junctions linking epithelial cells together.
The membranes of adjacent cells are in
contact, forming an impermeable barrier
across the epithelial-cell layer. All
molecules crossing this layer must pass
through the cells (arrow).

junctions, known as *tight junctions* (Fig. 2-33), are an
actual fusing of the two adjacent cell membranes so that
there is no gap between the adjacent cells in the region
of the tight junction. This type of junction extends around
the circumference of the cell and effectively closes off
the extracellular route for the passage of molecules between the epithelial cells. Therefore, in order to cross the
epithelium, a molecule must first cross the cell membrane
of an epithelial cell, pass through the cytoplasm, and exit
through the membrane on the opposite side of the cell.
Thus, the tight junction, in addition to helping to hold cells
together, also seals the passageway between adjacent
cells.

When a molecule to which the cell membrane is
impermeable is injected into a cell through a micropipette

inserted into the cell, it will remain within the cell and
cannot pass into adjacent cells. When this experiment is
performed on certain types of cells, however, the marker
molecule is found to appear in the cell adjacent to the
injected cell but not in the extracellular medium, suggesting that there is a direct channel linking the cytoplasms
of the two cells. The electron microscope has revealed
a structure in these types of cells known as a *gap junction*
(Fig. 2-34) which is intermediate in structure between a
desmosome and a tight junction. In the region of the gap
junction the two opposing cell membranes come within
20 to 40 Å of each other but do not appear to fuse as
in the case of the tight junction. Small channels about
15 Å in diameter extend across this small gap and directly
link the cytoplasm of the two cells. The small diameter

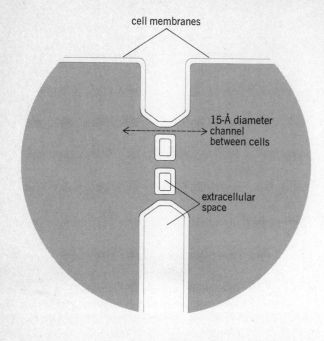

cell membranes

15-Å diameter channel between cells

extracellular space

of these channels limits the size of the molecules that can pass between the connected cells to small molecules and ions such as sodium and potassium but exclude the exchange of large protein molecules. A variety of cell types possess gap junctions, including the muscle cells of the heart and smooth muscle cells which surround various hollow cavities in the body, such as blood vessels and the intestinal tract. As we shall see in Chap. 8, the gap junctions in these cells play a very important role in the transmission of electrical activity between adjacent smooth and cardiac muscle cells. Gap junctions are not present between skeletal muscle cells.

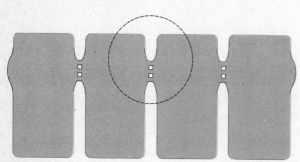

FIGURE 2-34
Schematic diagram of structure of gap junctions between cells. The membranes of adjacent cells become joined, forming channels 15 Å in diameter between the cytoplasms of the connected cells.

Energy and cellular metabolism

Metabolism (Greek: change) refers to the total collection of chemical reactions that occur within a living organism. The overall process of metabolism consists, to a large extent, of two generalized classes of chemical reactions, those which result in the fragmentation of a molecule into smaller and smaller parts, known as *degradative* or *catabolic reactions,* and those which put small molecular fragments together to form larger molecules, known as *synthetic* or *anabolic reactions.*

The molecular composition of a cell undergoes a continuous transformation as some molecules are broken down while others of the same type are being synthesized. Virtually all the organic components of a cell are broken down and replaced with new molecules within a relatively short period of time compared with the lifetime of the cell. Some molecules are replaced every few minutes whereas others may take years to be completely replaced by newly synthesized molecules. Chemically no person is the same at noon as at 8 o'clock in the morning since during even this short period much of the body's structure has been torn apart and replaced with newly synthesized molecules. In adults the composition of the body is in a state of *dynamic equilibrium* in which the rates of chemical synthesis and degradation are in balance. Because of this continuous molecular turnover, any factor

3

which disturbs the balance between catabolism and anabolism can lead to the destruction of the cell or to abnormal growth.

There is also a continuous breakdown of those organic molecules which release chemical energy. This energy is used by the cell to maintain its structure and function, i.e., to synthesize molecules, for muscle contraction and the active transport of molecules across cell membranes, and to maintain body temperature.

Although each type of cell has a slightly different chemical composition and metabolism, all cells are similar in that they are able to release and utilize chemical energy. Most cells are also able to synthesize and degrade carbohydrates, lipids, proteins, and nucleic acids. In this chapter we focus attention on three basic aspects of metabolism: (1) The mechanisms for achieving highly specific types of chemical transformation within a cell. Most organic molecules do not readily undergo chemical reactions in a test tube, yet within a cell hundreds of different chemical reactions proceed simultaneously in an orderly fashion. How are these transformations achieved? (2) The mechanisms for releasing energy from molecules and making it available for the performance of cellular functions. (3) The mechanisms for regulating the chemical reactions within a cell. How are the rates of anabolism and catabolism kept in balance? How are the rates of energy production increased or decreased to meet the varying energy requirements of the cell during different states of activity?

Before considering the specific types of chemical reactions occurring in a cell, we must examine some basic properties common to all chemical reactions whether they take place in a living cell or in a test tube. In the reaction between the elements hydrogen and oxygen to form a molecule of water

$$H—H \quad + \quad O \longrightarrow H—O—H$$
Hydrogen   Oxygen   Water

the chemical bond joining the two hydrogen atoms has been broken, and two new chemical bonds between the atoms of hydrogen and oxygen have been formed. This process of breaking chemical bonds between atoms and forming new sets of linkages is the primary characteristic of a chemical reaction. A molecule of water can also be formed by a reaction between two molecules which contain hydrogen and oxygen, e.g., the reaction between an alcohol and acetic acid:

$$CH_3O—H + HO—\overset{\displaystyle O}{C}CH_3 \longrightarrow CH_3O—\overset{\displaystyle O}{C}CH_3 + HO—H$$
Methanol   Acetic acid   Methyl acetate   Water

Here again chemical bonds between atoms are broken and new linkages established. The H— from alcohol combines with the HO— from acetic acid to form water. The molecules taking part in this reaction are composed of many atoms with many different sets of chemical bonds. Several ways of breaking these bonds and rearranging them to form new combinations are possible. Which bonds will be broken and what types of new linkages will be formed are determined by the geometry of the molecules and the distribution of energy in the molecules at the time of their reaction. The concept of energy is the key to understanding the properties of chemical reactions and ultimately the properties of living cells.

Energy

All physical and chemical change involves a redistribution of energy. But just what is energy? It is not a "thing" that can be described in the terms, size, shape, or mass, used to describe the static physical properties of matter. It is definable only in dynamic terms, specifically, the ability to produce change. Indeed, the presence of energy is revealed only when change is occurring. Accordingly, energy is defined as the ability to produce change or, more precisely, as the ability to perform work. Work is therefore a measure of energy, and the magnitude of work can be measured in terms of the forces acting upon matter and the resulting displacements of the matter. Whenever a change occurs in a system, energy must be transferred from one part of the system to another, but although energy is transferred between different parts of a system during physical or chemical change, the total energy content of the system remains constant. That energy is neither created nor destroyed during any physical or chemical process is one of the most important axioms of science, known as the *first law of thermodynamics* or, more generally, as the *law of the conservation of energy.*

The total energy content of any physical object consists of two components, the energy associated with the object because of its motion, called *kinetic energy,* and the energy associated with the object because of its position or internal structure, called *potential energy.* The amount of kinetic energy a moving object has is determined by its mass and its velocity. The faster an object moves, the greater its kinetic energy, and the larger the mass of an object, the greater is its kinetic energy at any given velocity. Movement is, therefore, a form of energy, and energy must be transferred to an object in order to produce movement. Kinetic energy is also associated with the motion of individual molecules. It is this kinetic energy of molecular motion that we recognize as heat. The hotter an object is, the faster its molecules move and the greater their kinetic energy. Thus, heat is a form of energy—the kinetic energy of molecular motion.

Potential energy is the energy associated with an object due to its structure or position. It is energy which has the potential of becoming kinetic energy when it is released. Thus, the energy transferred to a book by lifting it onto a table is present in the book as potential energy, as can be demonstrated by allowing the book to fall to the floor; the potential energy of the book due to its position above the floor is converted into kinetic energy of motion as the book falls. When the book strikes the floor, the kinetic energy of its motion is converted into sound and heat energy. Both sound and heat are phenomena associated with increased molecular motion and are thus forms of kinetic energy.

Chemical energy is also a form of potential energy locked within the structure of molecules. It can be released during a chemical reaction, which alters the structural arrangement of the atoms in a molecule. The chemical potential energy that can be stored in the structure of a molecule is analogous to the potential energy that can be stored in a stretched spring. Work is required to stretch a spring, the amount of work done representing the amount of potential energy stored in the stretched spring. If the spring is released, the stored potential energy is released and appears as motion and heat as the spring returns to its original position. A chemical bond between two atoms is the result of electric forces of attraction between the charged subatomic portions of the atoms. The balance of these forces in a chemical bond stores potential energy. If the chemical bond is broken, the potential energy is released. Conversely, in some chemical reactions energy must be added to the molecules to form new chemical bonds. This is analogous to stretching a spring to increase its potential energy.

Since energy can neither be created nor destroyed during a chemical reaction, the difference in potential energy between the reactant molecules and the product molecules must be the amount of energy added or released during the reaction. For example, the reaction between hydrogen and oxygen to form water proceeds with the release of a considerable amount of heat energy. The amount of heat is measured in units of *calories,* where a calorie is the amount of heat energy required to raise the temperature of one gram of water one degree Celsius. Energies associated with chemical reactions are generally of the order of several thousand calories and are reported as kilocalories (1 kcal = 1000 calories). The formation of 1 mole of water from hydrogen and oxygen releases 68 kcal of heat energy, the full reaction being

$$H_2 + O \longrightarrow H_2O + 68 \text{ kcal/mole}$$

The chemical potential energy stored in 1 mole of water molecules is less by 68 kcal than that originally present in the hydrogen and oxygen molecules.

This process can be reversed; i.e., the chemical bonds of a water molecule can be broken and rearranged to give the original hydrogen and oxygen molecules. However, to accomplish this, 68 kcal of energy must be added to 1 mole of water to provide the higher energy content of the hydrogen and oxygen molecules. The reverse reaction can be written

$$68 \text{ kcal/mole} + H_2O \longrightarrow H_2 + O$$

How can a molecule of water obtain the energy required to convert it into hydrogen and oxygen? During a collision between two moving molecules, the kinetic energy of one molecule can be transferred to the other molecule, just as in the collision of two billiard balls. The energy gained by the molecule in the collision may disrupt the electron structure of the molecules in such a way that the kinetic energy is transferred into increased potential energy within the molecular structure. Through collisions with other molecules a water molecule may obtain a potential-energy content equal to that of the free hydrogen and oxygen molecules and thus be able to break down into free hydrogen and oxygen atoms.

Any chemical reaction can be reversed if sufficient energy is made available to the system. The reactions between hydrogen, oxygen, and water can therefore be represented by

$$H_2 + O \rightleftharpoons H_2O + 68 \text{ kcal/mole}$$

where the double arrow indicates the forward and reverse reactions.

Determinants of chemical reaction rates

Chemical concentration: law of mass action

If a chemical bond is to be formed between two molecules, they must come close enough for their electrons to interact, as happens if the random motion of the molecules brings them together in a collision. If one molecule of A and one molecule of B are present in 1 liter of solution, the probability of their colliding with each other due to their random movements is extremely slight, and thus the rate at which product molecules can be formed from the interaction of A and B is also very small. But if several million molecules of A and B are present in the same volume, the probability of collisions between A and B is greatly increased, and the rate of product formation is increased. Thus, the concentration of molecules in a solution is an important determinant of their overall rate of interaction.

In a reversible chemical reaction such as

$$\text{A} + \text{B} \underset{(2)}{\overset{(1)}{\rightleftharpoons}} \text{C} + \text{D}$$
Reactants   Products

no further net change in the concentration of reactants or products will occur when the rates of the forward and reverse reactions become equal. Under these conditions of chemical equilibrium, molecules of A and B still react to form C and D, but since molecules of C and D also react to form A and B at the same rate, no net change in the concentration of A, B, C, or D occurs. If, after chemical equilibrium has been reached, the concentration of B is increased by adding more B to the solution, the rate of the forward reaction (1) increases since the larger number of B molecules increases the probability of a collision with an A molecule. The higher rate of the forward reaction leads to greater formation of C and D. As the concentration of C and D rises, the rate of the reverse reaction (2) also rises, until a new equilibrium is achieved. Note that the effect of increasing the concentration of B is to increase the concentration of the product molecules C and D. Since no new molecules of A have been added to the system, and since C and D are formed from A and B, any increase in the concentration of C and D must be accompanied by a decrease in the concentration of A. If the concentration of B molecules is made very large, almost all the A molecules present can be converted into product molecules.

The conversion of A molecules into product molecules C and D can also be influenced by changes in the concentration of product molecules. If the concentration of C is lowered by removing it from the reaction mixture, the rate of the reverse reaction (2) is lowered. Therefore, A and B molecules are not replaced as rapidly as they react to form C and D. The net effect is a decrease in the concentration of A and B. Thus, the conversion of molecule A into product molecules can be increased either by raising the concentration of B or by lowering the concentration of one of the product molecules.

This effect of molecular concentration on chemical reactions, known as the *law of mass action*, follows directly from the effect of concentration on the frequency of

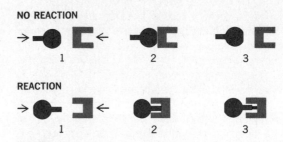

NO REACTION

1   2   3

REACTION

1   2   3

FIGURE 3-1
Two molecules must collide in the correct orientation in order to form a chemical bond.

molecular collisions. Raising the concentration of any one molecular species in a chemical reaction increases the rate at which that molecule undergoes a reaction with other molecules in the system and thus alters the concentration of both product and reactant molecules.

In metabolism, the products formed in one reaction may become the reactant molecules necessary for another reaction, which in turn forms products taking part in further reactions, so that many reactions are interconnected by having certain molecules in common. In such a situation, altering the concentration of one of the molecules participating in a large set of interconnected reactions can affect a great many reactions within the cell, leading to increases or decreases in the concentrations of certain key molecules and shifting the balance between the rates of degradative and synthetic reactions. Because of their interactions through the mass-action principle a large number of reactions can be controlled by regulating the concentration of certain key molecules in the system.

In order to react, the specific regions of the molecules where the bonds are to form must come within reach of each other. If these areas are on opposite sides of the molecules from the region of contact during a collision, no bond formation can occur (Fig. 3-1). Thus, for a reaction to take place not only must two molecules collide but the appropriate regions on the surfaces of the molecules must interact with each other.

Activation energy

If hydrogen and oxygen are mixed together at room temperature, water is formed at an extremely slow rate, but if the mixture is heated, water is formed rapidly. The addition of heat energy accelerates the reaction. At room temperature few of the collisions between hydrogen and oxygen result in the formation of new chemical bonds because the electric forces holding a molecule together in a stable configuration must be disrupted before a new arrangement of the electrons in a chemical bond can occur. To upset the balance of forces in the molecule, a quantity of energy known as *activation energy* must be added. The activation energy can be acquired through collisions with other molecules, in which kinetic energy transferred to the molecule increases the potential energy stored in its structure. Heating the mixture of hydrogen and oxygen increases the kinetic energy of the molecules, more of which now have sufficient kinetic energy to react. Heat also increases the rate of molecular motion and thus increases the frequency of molecular collisions.

As a mechanical analogy to a chemical reaction, consider the potential energy of a ball resting in a depression on top of a hill as representing the chemical potential energy stored in a molecule (Fig. 3-2). If the ball rolls down the hill, it converts potential energy into kinetic energy, just as chemical potential energy can be released as the kinetic energy of heat during a reaction. The ball

will not roll down the hill spontaneously; it must be given a slight push. This push represents the activation energy necessary to initiate a chemical reaction. The magnitude of the activation energy can be represented by the size of the hump which the ball must roll over before it can roll down the hill. The magnitude of the activation energy determines the rate of the reaction since only molecules which have acquired this amount of energy are able to react. At any given temperature the larger the activation energy, the fewer are the number of molecules in the population that have this amount of energy and thus the slower the reaction.

The cells in the human body are maintained at a nearly constant temperature of 37°C (98.6°F). The body temperature provides an environment in which molecules can acquire the energy required for their activation, but since it remains nearly constant, *changes* in chemical reaction rates within cells cannot be due to *changes* in temperature. The rates of chemical reactions in cells are accelerated by special protein molecules known as *enzymes*.

Enzymes

During the nineteenth century both biologists and chemists were fascinated and frustrated by this ability of cells to achieve high rates of organic reactions without high temperatures. During this period chemists began to understand the chemical properties of organic molecules found in living organisms. Although chemists could make some of the same organic molecules in a test tube that a living cell makes in the body, they generally had to use high temperatures and reagents which could not possibly exist within a cell.

The fermentation of grape juice to produce wine received considerable attention from biologists and chemists when it was discovered that it was carried out by microscopic unicellular plant cells known as yeast. These cells metabolize the sugar in grape juice, converting it into alcohol and carbon dioxide.

$$C_6H_{12}O_6 \xrightarrow{\text{yeast}} 2CH_3CH_2OH + 2 CO_2$$

Glucose   Ethyl alcohol   Carbon dioxide

Louis Pasteur in the 1850s and 60s undertook a study of the fermentation process by yeast cells. All his attempts to make an extract from the yeast cells that would convert sugar into alcohol were unsuccessful. He therefore concluded that the ability of living cells to carry out organic reactions depended upon some ''vital process'' associated with the living cell, a process that could

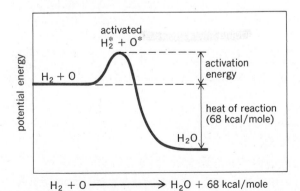

FIGURE 3-2
Changes in the potential energy of a ball moved from the top to the bottom of a hill compared with changes in chemical potential energy during a chemical reaction.

not be associated into nonliving chemical fragments. The negative results of Pasteur's experiments appeared to support the vitalistic concept of life as a process dependent upon forces unique to living systems. Pasteur's conclusion was, of course, incorrect; unfortunately the methods he used to prepare the extracts of the yeast cells destroyed many of the molecules, particularly the protein molecules, present in the living cells. In 1897, Buchner succeeded in preparing a cell-free extract of yeast that could convert glucose into alcohol and carbon dioxide. He was successful because his extraction method did not damage the protein molecules. Further experiments over the next 30 years led to the conclusion that the proteins in Buchner's extract were responsible for accelerating chemical reactions in cells. These proteins were called *enzymes*, meaning ''in yeast.''

During this same period, chemists working with inorganic molecules were beginning to discover ways of accelerating the rates of chemical reactions which did not involve raising the temperature. Adding small amounts of certain substances, such as powdered platinum, was found to accelerate some reactions. Such substances are called *catalysts*. A catalyst is not chemically altered by the reaction it catalyzes and can be used over and over again; thus only small amounts of a catalyst are required to transform large amounts of reactants into products. A catalyst cannot make a reaction occur which is energetically impossible according to the laws of thermodynamics; it only accelerates the rates of reactions that would still occur spontaneously in its absence, although at a much slower rate. Thus, a catalyst is similar in its effects to temperature and concentration, both of which accelerate chemical reactions.

The behavior of enzymes fits this description of a catalyst. Since all enzymes are proteins, an enzyme can be defined as a protein catalyst. (Although all enzymes are proteins, not all proteins are enzymes.) In order to accelerate a reaction an enzyme must come into contact with the reactant molecules, known as the *substrates* of the enzyme. The reaction between enzyme and substrate can be written

$$S + E \longrightarrow ES \longrightarrow P + E$$

| Substrate | Enzyme | Enzyme-substrate complex | Product | Enzyme |

The enzyme combines with substrate to form an enzyme-substrate complex, which breaks down to release product molecules and free enzyme. At the end of the reaction, the enzyme molecule is free to undergo further reactions with additional substrate molecules. The overall effect has been to accelerate the conversion of substrate molecules into product molecules with the enzyme acting as a catalyst.

$$Substrates \xrightarrow{enzyme} products$$

The number of substrate molecules which can react with a single enzyme molecule in 1 sec varies from 1 to over 20,000, depending upon the type of enzyme and reaction. Because enzymes have a high molecular activity and are not used up in the reactions they catalyze, they can be effective at very low concentrations in cells. In some cases only a few molecules of enzyme per cell are sufficient to provide the cell with the products of reactions catalyzed by the enzyme.

Although enzymes are not destroyed during the reactions they catalyze, like all biological molecules, they are in a state of dynamic equilibrium. The lifetime of a given enzyme may be weeks or months, but eventually it is broken down and replaced with a newly synthesized enzyme. One of the mechanisms available to the cell for regulating its metabolism is the ability to control the rate of enzyme synthesis and breakdown, thus controlling the total amount of enzyme present in the cell at a given time and thereby determining the overall rate at which the product catalyzed by the enzyme is formed.

Besides accelerating the rates of chemical reactions in cells, enzymes are highly specific for the type of substrate molecule they act upon. Thousands of different reactions occur within a cell, and almost all of them are catalyzed by different enzymes. Some enzymes are so specific that they interact with only one particular type of substrate molecule and no other. At the other extreme, some enzymes interact with a wide range of different substrate molecules that contain a particular type of chemical bond or grouping. Enzymes are generally named by adding the suffix *-ase* to the name of the substrate or to the type of reaction catalyzed by the enzyme. For example, the enzyme lactic dehydrogenase catalyzes the removal of hydrogen atoms from lactic acid.

Mechanisms of enzyme action

The structure of proteins holds the key to the mechanism of enzyme action. In Chap. 1, the general structure of these giant molecules, composed of hundreds of amino acids, was described along with the problems of determining their complete amino acid sequences and three-dimensional structures. The complete protein structure of only a few enzymes has been determined, but after years of research a number of general statements can be made about the mechanisms of enzyme activity.

Since a substrate molecule is usually much smaller than a large protein molecule, the substrate can come into contact with only a small portion of the enzyme surface. The specific area of the enzyme molecule which interacts with the substrate is known as its *active site*. The particular amino acids that come into contact with the substrate at the active site need not be located in sequence along the protein chain but may be at widely separated points along the chain that are brought together by coiling of the protein (Fig. 3-3).

The high specificity of an enzyme for a particular substrate results from the three-dimensional protein structure. The structure of the substrate molecule must fit the three-dimensional structure of the enzyme active site for the two to interact, the interaction being analogous to a lock and key, where the shape of the key must fit the shape of the lock. Figure 3-4 illustrates the relations between enzyme structure and substrate structure. Once the enzyme and substrate become associated with each other at the active site, the enzyme can increase the probability of a chemical reaction involving the substrate by a number of mechanisms, all explainable in terms of the principles of mass action and activation energy described earlier.

When the substrate is bound to the enzyme, random movements of the substrate are prevented and the molecule is oriented in a fixed position on the surface of the enzyme, thereby increasing the probability of reaction at specific sites on the substrate.

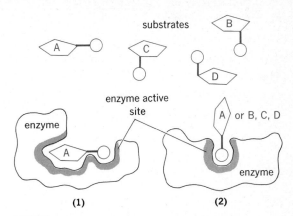

FIGURE 3-4
Relation between enzyme structure and substrate structure. (1) **Enzyme whose active-site geometry matches that of only a single substrate, A.** (2) **Enzyme whose active site is able to react with a particular grouping of atoms attached to a number of different substrates.**

Positioning the substrate on the enzyme surface may also distort the substrate structure, thereby producing a strain on some of its chemical bonds—like bending a stick to the point where it almost breaks—thus increasing the probability of their reaction. Instead of a one-step reaction, in which a reactant molecule is directly converted into product, enzyme-mediated reactions have an intermediate step in which the substrate first reacts with the enzyme to form an enzyme-substrate product, which in turn undergoes a second reaction to form the final product and return the enzyme to its original state (Fig. 3-5). By converting a one-step reaction with a large activation energy into a two-step reaction, each step of which has a smaller activation energy, the overall rate of product formation is increased since at any given temperature the probability of the system's acquiring the large amount of energy required to activate the single-step reaction is much less than the probability of its acquiring the smaller amount necessary for each step of the two-step reaction.

Coenzymes

Many enzymes are totally inactive in the absence of additional low-molecular-weight substances known as cofactors or *coenzymes*. Many enzymes require small amounts of certain metal ions such as copper, zinc, iron, cobalt, or magnesium, which bind the substrate molecule to the enzyme active site. For example (Fig. 3-6), if both the substrate and the enzyme active site have negatively

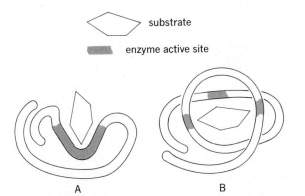

substrate

enzyme active site

A   B

FIGURE 3-3
Location of the active-site region of enzyme molecules. A Active site consisting of a number of adjoining amino acids within the protein chain. B Active site consisting of three separate regions of the protein chain brought into relation with each other by the shape of the protein and shape of the substrate.

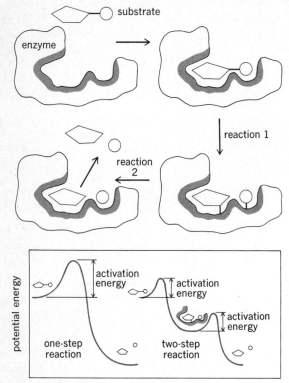

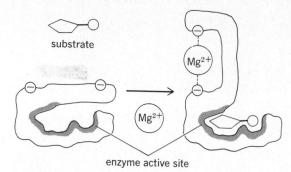

FIGURE 3-7
Effect of magnesium ion on the shape of an enzyme molecule assures the accessibility of the substrate to the enzyme active site.

FIGURE 3-5
Reaction of enzyme with substrate to form an intermediate enzyme-substrate product, which then undergoes a second reaction to give the final product and return the enzyme to its original state.

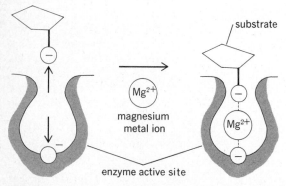

FIGURE 3-6
Inorganic ions such as magnesium may activate enzymes by binding a charged substrate to the charged surface of the enzyme.

charged groups, they repel each other, but a metal ion having a double positive charge can link substrate and enzyme together. In other cases the metal ion may be required to maintain the proper configuration of the enzyme molecule so that the substrate has access to the active site (Fig. 3-7). The function of many of the trace elements in the body (Chap. 1) appears to be related to their role as coenzymes for specific enzymes.

In addition some enzymes are active only in the presence of specific low-molecular-weight organic molecules, which usually participate chemically in the reaction catalyzed by the enzyme. As an example, the enzyme lactic dehydrogenase requires the coenzyme nicotinamide adenine dinucleotide (NAD) in a reaction whereby two atoms of hydrogen are transferred from lactic acid to the coenzyme molecule NAD:

$$\underset{\text{Lactic acid}}{CH_3-\overset{\displaystyle OH}{\underset{\displaystyle }{C}H}-COOH} + NAD \xrightarrow{\text{lactic dehydrogenase}}$$

$$\underset{\text{Pyruvic acid}}{CH_3-\overset{\displaystyle O}{\underset{\displaystyle }{C}}-COOH} + NADH_2 \quad \textbf{3-1}$$

In one sense the NAD molecule acts as a substrate for the enzyme along with lactic acid, but the chemical fate of the $NADH_2$ is quite different from that of pyruvic acid. Although pyruvic acid can be metabolized through a number of reactions into various other products, including carbon dioxide and water, the $NADH_2$ is converted back into NAD through a second reaction, in which it donates its two hydrogen atoms to another molecule which at this time we shall call X:

$$X + NADH_2 \longrightarrow XH_2 + NAD$$

Thus, the NAD molecule has acted as a temporary carrier for the hydrogen atoms as they are transferred from one molecule to another. Only small amounts of the coenzyme NAD are necessary to maintain the enzyme's activity because of the continuous regeneration of NAD from $NADH_2$.

Vitamins Around the turn of the century, biologists studying the dietary requirements of animals found that trace amounts of certain organic molecules must be present in the diet to maintain normal health and growth. The Polish biochemist Funk, in 1912, believing these substances to belong to a group of organic molecules known as amines, gave them the name vitamins (Latin *vita*, life). A young animal placed on a diet lacking vitamins ceases to grow and eventually sickens and dies. Very small amounts of these vitamins will restore the animal to health. Over 20 different substances have been shown to be essential in small amounts for the normal growth and health of various animal species. The vitamins as a class have no particular chemical structure in common. They are often identified by letters of the alphabet, such as vitamin A, B, C, D, E, and K, since their effects were often discovered before their chemical structures were determined. Most, but not all, vitamins function in the body as coenzymes. The essential ingredient in the coenzyme NAD, for example, is the vitamin niacin.

Vitamins are required in the diet because the body is unable to synthesize them, or is unable to synthesize adequate amounts of them. However, plants and bacteria have the enzyme machinery necessary for vitamin synthesis, and it is from these sources that we ultimately obtain vitamins. Since the coenzyme molecule can be used over and over again in a cycle of reactions, only small quantities of vitamins are required in the diet to replace those destroyed or excreted from the body. By themselves vitamins do not provide chemical energy for the body, although they may participate in chemical reactions which release energy from other molecules. Increasing the amount of vitamins in the diet does not necessarily increase the activity of those enzymes for which the vitamin functions as a coenzyme. Only very small quantities of coenzymes are necessary to produce fully active enzymes. Increasing the coenzyme concentration above this level does not increase the enzyme's activity. Obviously, a lack of vitamins in the diet causes ill effects since they are essential for the activity of many enzymes, especially those involved in providing chemical energy for cellular functions.

Since different vitamins act as coenzymes for different enzymes, the symptoms resulting from a deficiency of a particular vitamin differ, depending upon which vitamin is deficient. In general, quantities of vitamins in the diet that exceed the small quantities required for the functioning of enzymes probably have no beneficial effects, although, in some cases, large quantities of a vitamin may have an effect upon the body that is unrelated to its normal role as a coenzyme. Moreover, in the case of the fat-soluble vitamins, such as vitamin A and D, large quantities are known to have quite harmful effects.

Regulation of enzyme-mediated reactions

The rate of an enzyme-mediated reaction depends on three major factors: concentration of substrates, concentration of enzyme, and activity of individual enzyme molecules. Increasing the concentration of substrate will, by mass action, increase the rate of the reaction. The concentration of a given substrate in a cell may vary for a variety of reasons. It may increase because the circulation delivers more substrate to the cell or because of changes in membrane transport which allow more substrate to enter a cell or because other reactions in the cell are producing the substrate at a faster rate. When the reaction is mediated by an enzyme, there is a limit to the extent to which mass action will increase the rate of the reaction. Since the substrate combines with the enzyme during the reaction and there are a finite number of a particular type of enzyme molecules in the cell, as the substrate concentration increases, a point is reached at which all the enzyme molecules are reacting with substrate. At this point the enzyme is said to be *saturated* with substrate, and further increases in substrate concentration will not increase the rate of the enzyme-catalyzed reaction.

The second way of altering a reaction rate is to change the concentration of enzyme; this has the effect of providing more or less sites with which substrate can react. Thus, for any given substrate concentration, the rate of a reaction will depend on the amount of enzyme present. Certain reactions proceed faster in some cells than in others because more enzyme is present. In a given cell the amount of enzyme depends on the rate at which enzyme is synthesized and its rate of degradation. (Enzymes are proteins, and their synthesis will be discussed in the next chapter dealing with protein synthesis.) Most enzymes in the cell are synthesized and broken down at a fairly constant rate and thus their concentrations do not change appreciably during different states of cell activity. Certain enzymes, however, increase in concentration when their substrate is present or increased in the cell, a process known as *enzyme induction*. A cell need not waste energy and nutrients synthesizing such enzymes unless their substrates are present, and if large amounts of substrate are present, the cell can increase its synthesis of the enzyme and thereby its capacity to metabolize the substrate. In contrast to enzyme induction, the products of certain reactions inhibit the synthesis of the enzyme mediating the reaction, a process known as *enzyme repression;* by this means a buildup of product turns off its own production by repressing the synthesis of the enzymes responsible for its formation. The mechanisms of enzyme induction and repression will be described in the next chapter.

The third factor determining reaction rates is the activity of the individual enzyme molecules. As we have seen, the molecular activity of different enzymes may vary from a rate of one substrate reaction per enzyme molecule per second to rates as high as 20,000 substrate reactions per enzyme molecule per second. Not only do different enzymes differ in their molecular activity but the same enzyme molecule may have different molecular activities under different conditions.

The activity of an enzyme ultimately depends on the properties of the active site on the enzyme where the combination with substrate occurs. If the shape of the active site is altered, the molecular activity of the enzyme will also be altered. This occurs when the enzyme combines with another molecule known as a *modulator*. The combination of the modulator molecule with the enzyme alters the shape of the protein and thus the shape of the active site. The modulator molecule is not a substrate of the enzyme and usually has a quite different chemical structure from that of the substrate. It binds to the enzyme at a distinctly separate site from the active site which binds substrate. Enzymes whose activity is altered by modulator molecules are known as *allosteric enzymes* (Greek *allos,* other; *steric,* shape). A diagrammatic example of allosteric inhibition of enzyme activity is given in Fig. 3-8. Only certain key enzymes in the cell are subject to allosteric regulation, but through this process a wide variety of reactions can be coordinated in the cell; the products of certain metabolic processes are modulator molecules which alter the activity of key allosteric enzymes in other metabolic processes.

We have already discussed a second determinant of enzyme activity: the availability of coenzymes. If the concentration of the coenzyme for a specific enzyme is absent, that enzyme will not be active; therefore, an increase in coenzyme concentration will increase enzyme activity.

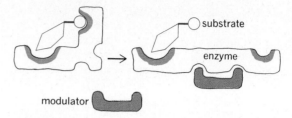

FIGURE 3-8
Allosteric inhibition of enzyme activity. The modulator molecule alters the shape of the enzyme's active site by combining with some other region of the enzyme molecule.

Figure 3-9 summarizes the factors which determine the rates of enzyme-mediated reactions; these include the concentration of substrate, coenzyme, and enzyme as well as the molecular activity of the enzyme.

Regulation of multienzyme metabolic pathways

The sequence of enzyme-mediated reactions leading to the formation of a particular product is known as a *metabolic pathway*. For example, the 19 reactions which convert glucose to carbon dioxide and water are the metabolic pathway for glucose degradation. Another set of enzymes is used to synthesize fat; this is the metabolic pathway for lipid synthesis. Moreover, since glucose can be converted into fat, there must be some metabolic connections between these two pathways and some mechanism for determining whether glucose will be degraded to carbon dioxide and water or converted into fat. Ultimately the regulation of these pathways depends upon the interaction of the factors discussed in the previous section which determine the rates of individual reactions. In this section we apply these general principles to the regulation of the flow of material through multienzyme metabolic pathways.

Consider a general metabolic pathway consisting of four enzymes (e), leading from an initial substrate A to the end product E, through a series of intermediates, B, C, and D.

$$A \xrightarrow{e_1} B \xrightarrow{e_2} C \xrightarrow{e_3} D \xrightarrow{e_4} E$$

By mass action, increasing the concentration of A will lead to an increase in the concentration of B and so on until eventually there is an increase in the concentration of the end product E. If, as a result of the increasing concentration of A and the intermediates, one of the enzymes in the pathway becomes saturated with substrate, the rate of the overall pathway will no longer increase with increasing concentration of A. Such a reac-

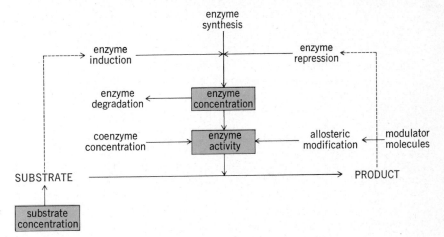

FIGURE 3-9
Factors which affect the rate of
enzyme-mediated reactions.

tion becomes rate-limiting for the flow of material through the entire pathway. Therefore, by regulating the activity of this one rate-limiting enzyme, the rate of flow through the whole pathway can be increased or decreased. Thus it is not necessary to alter the activities of all the enzymes in a metabolic pathway to be able to control the overall rate at which the final end product is produced. These rate-limiting enzymes are usually the sites of allosteric regulation in metabolic pathways. Let us assume that enzyme e_2 is the rate-limiting enzyme in this pathway. The end product E may interact with enzyme e_2 to produce an allosteric inhibition of its activity, and thus E becomes the modulator molecule for this pathway. This form of *end-product inhibition* is quite common in many synthetic pathways. It effectively prevents an excessive accumulation of end product when the end product is not being utilized and there is a considerable amount of the initial substrate A available to the cell.

Let us now add another level of complexity. Suppose the reaction

$$A \xrightarrow{e_1} B$$

is spontaneously and easily reversible:

$$A \underset{e_1}{\overset{e_1}{\rightleftharpoons}} B$$

Note that the same enzyme (e_1) catalyzes the reaction in both directions. This is a very important point: An enzyme speeds up a reaction but does not determine the *direction* in which the reaction *spontaneously* proceeds. Therefore, in an easily reversible reaction, increasing the concentration or activity of the mediating enzyme will speed the reaction in both directions.

In theory, all chemical reactions are reversible. However, whether any reaction is spontaneously reversi-

ble to any extent, given a finite amount of time, depends upon the amount of chemical energy exchanged during the reaction. If, in a given reaction, there has been little change in the energy content of the substrate and product molecules, the reaction is readily reversible. However, if large exchanges of energy have occurred, this energy must be made available if the reaction is to be reversed. Therefore, in practice, reactions can be classified into two categories, those that are spontaneously reversible because they involve only small exchanges of energy and those which are not spontaneously reversible because of the release of large quantities of energy.

Reversible reaction: $A \underset{e_1}{\overset{e_1}{\rightleftharpoons}} B$
  $+$ small amount of energy

Irreversible reaction: $C \xrightarrow{e_3} D$
  $+$ large amount of energy

It should be emphasized that it is the quantity of energy released in the reaction and not the property of the enzymes which determines the spontaneous reversibility of a given reaction; enzymes speed up a reaction but do not determine the direction in which the reaction spontaneously proceeds. Thus far we have been discussing spontaneous reversibility. However, the product D in the spontaneously irreversible reaction above can be converted into C, provided a source of sufficient energy is added to the system. Often this energy can be provided by coupling the reaction to the simultaneous breakdown of another substrate which releases large quantities of energy. Thus, if the conversion of W to X occurs with the release of a large amount of energy,

$$W \xrightarrow{e} X + \text{large amount of energy}$$

then coupling this reaction with the former essentially irreversible reaction leads to

$$D + W \xrightarrow{e_5} C + X$$

Note that we have not truly reversed the reaction $C \longrightarrow D$ but rather have substituted a quite different reaction, catalyzed by a quite different enzyme.

Thus, an irreversible step can be "reversed" through an alternative route, using a second enzyme and an additional substrate to provide the required energy. Any step in a metabolic sequence may consist of one of the following three types of reactions: (1) a low-energy reversible reaction catalyzed by a single enzyme,

$$P \underset{}{\overset{e}{\rightleftharpoons}} Q$$

(2) a high-energy irreversible reaction catalyzed by a single enzyme,

$$R \xrightarrow{e} S$$

and (3) two high-energy irreversible reactions catalyzed by two separate enzymes. (The bowed arrows emphasize that two quite distinct enzymes are involved in the two directions.)

$$T \underset{e_b}{\overset{e_a}{\rightleftharpoons}} U$$

In the third case, the reaction rate in either direction can be altered selectively by regulating the activities or concentrations of the specific enzymes.

In summary, consider the following modification of our general metabolic pathway:

$$A \underset{}{\overset{e_1}{\rightleftharpoons}} B \underset{}{\overset{e_2}{\rightleftharpoons}} C \underset{e_5}{\overset{e_3}{\rightleftharpoons}} D \xrightarrow{e_4} E$$

The first two reactions are low-energy reversible reactions, each catalyzed by a single enzyme. The third reaction is a high-energy reaction having separate enzymes catalyzing the reactions in each direction. The fourth reaction is a high-energy irreversible reaction catalyzed by a single enzyme. In this metabolic pathway there is no route by which E can be converted back into A because of the irreversible last step. However, D and other intermediates can all lead to the formation of A by reversal of the pathway. The third step in the pathway is mediated by two separate enzymes and provides an example of how the direction of flow in a metabolic pathway can be regulated. If the two separate enzymes e_3 and e_5 are subject to separate allosteric control, it is possible to inhibit e_3, allowing the reactions to proceed in the direction of product E or alternatively to inhibit e_3 and allow D to be converted into A.

The combination of reversible and irreversible reactions, together with the variety of mechanisms for controlling enzyme activities, provides the cell with a wide range of controls for regulating metabolic activity. When one considers the thousands of reactions that occur in the body, the permutations and combinations of possible control points, the overall result is indeed staggering, and the details of metabolic regulation at the enzymatic level are beyond the scope of this book. In the remainder of the chapter we shall consider the general pathways by which cells obtain energy and metabolize carbohydrates, fats, and proteins, and indicate some of the key control points in these pathways.

ATP and cellular energy transfer

The role of ATP

The functioning of a cell depends upon its ability to extract and use the chemical potential energy locked within the structure of organic molecules. When a mole of glucose is broken down into carbon dioxide and water, 686 kcal of energy is released. In a test tube all this energy is released as heat which can be used to perform work, as in a steam engine. However, a cell cannot transform heat energy into work. In contrast to the test-tube reaction, some of the chemical energy released within a cell is captured by directly transferring chemical potential energy from one molecule to another; i.e., energy lost by one molecule is transferred to the chemical structure of another molecule and thus does not appear as heat. Consider a reaction in which a molecule AX breaks down into a molecule of A and X and releases chemical energy as heat:

$$AX \longrightarrow A + X + energy \qquad 3\text{-}2$$

Then consider a second reaction, in which molecules B and X plus energy react to form BX, where BX has a higher energy content than the sum of the energies contained in B and X:

$$B + X + energy \longrightarrow BX \qquad 3\text{-}3$$

If these two reactions can be coupled together, the energy released by the first can be used to provide the energy needed in the second. This can be accomplished if AX reacts directly with B to form BX:

$$AX + B \longrightarrow A + BX \qquad 3\text{-}4$$

Here X no longer appears as a separate molecule since it has been transferred directly from AX to B, and in the

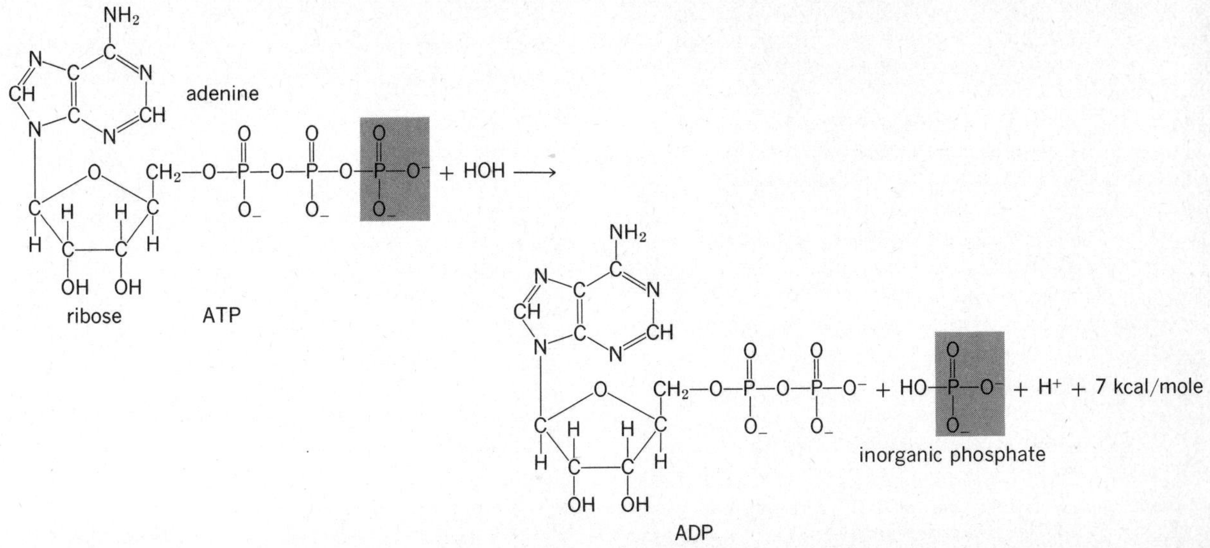

FIGURE 3-10
Breaking the chemical bond that joins the terminal phosphorus atom to the ATP molecule releases 7 kcal/mole of energy and forms ADP and inorganic phosphate.

process energy has been transferred from AX to BX and no longer appears as heat. The energy transferred to BX can be transferred to yet another molecule, C, in a reaction similar to **3-4**.

$$C + BX \longrightarrow CX + B \qquad \text{3-5}$$

Energy was added to BX in **3-4** and transferred to CX in **3-5**. In this series of reactions BX has acted as an energy carrier, transferring energy from AX to CX in the process of transferring X between the two molecules.

The cell obtains useful energy from degradative reactions which release energy by coupling them to energy-carrier molecules. In all living cells, from bacteria to man, the major energy-carrier molecule corresponding to BX is *adenosine triphosphate* (ATP). This molecule consists of three components: a ring structure known as *adenine;* a five-carbon sugar molecule, *ribose;* and three phosphate groups linked in series to the sugar. When the bond between the terminal phosphorus atom and oxygen is broken in the presence of water, 7 kcal/mole is released and the resulting product is adenosine diphosphate (ADP) and inorganic phosphate (P_i) (Fig. 3-10):

$$ATP + H_2O \longrightarrow ADP + P_i + 7 \text{ kcal/mole}$$

Conversely, in order to synthesize ATP from ADP and P_i, 7 kcal/mole must be added to the molecule:

$$ADP + P_i + 7 \text{ kcal/mole} \longrightarrow ATP + H_2O$$

Energy is added to ADP and released from ATP in the process of transferring phosphate from one molecule to another, giving a series of reactions comparable to **3-4** and **3-5**.

$$AP + ADP \longrightarrow A + ATP$$
$$\underline{C + ATP \longrightarrow CP + ADP}$$
Net reaction: $AP + C \longrightarrow A + CP$

The energy stored in ATP is used to perform work by the cell, i.e., muscle contraction, the active transport of molecules across cell membranes, and the synthesis of organic molecules.

Energy is constantly being cycled through ATP molecules in the cell. A typical ATP molecule may exist for only a few seconds before its energy is transferred to another molecule, and the ADP then formed is rapidly reconverted into ATP through coupling to energy-releasing reactions, namely, the breakdown of carbohydrates, lipids, or proteins. Although energy is stored in the structure of the ATP molecule, its function is not to store energy but to transfer energy from one molecule to another. Thus the total energy stored in all the ATP molecules of the cell can supply the cell's energy requirements for only a fraction of a minute. The actual storage of energy in the body is performed by the molecules of carbohydrate, lipid, and protein which are continuously transferring energy to ATP. Most of the transfer of energy to ADP to form ATP takes place in the specialized cell organelles known as mitochondria.

Mitochondria

The Greek words from which the mitochondria get their name (*mitos,* thread, and *chondros,* granule) characterize the appearance of these organelles under a light microscope. Viewed with the higher magnification of an electron microscope, a mitochondrion is found to have a highly organized structure (Fig. 3-11). Two membranes are clearly discernible, an outer smooth membrane surrounding the mitochondrion and an inner membrane, which at intervals folds into the central portion of the mitochondrion forming *cristae* (Latin: crests or ridges). The central space of the mitochondrion is known as the *matrix* (Fig. 3-12). Most mitochondria are cylindrical in shape, although some are spherical. The total number of mitochondria in a cell varies with the size and the energy expenditure of the cell. Cells using large amounts of ATP, such as heart muscle cells, have large numbers of mitochondria, whereas cells with much smaller energy requirements have fewer. A liver cell has about 800 mitochondria, the much smaller sperm cell about 20.

During the 1940s methods were developed for isolating the various subcellular organelles from cells; they consist of first breaking cells by subjecting a tissue to shearing forces in a glass homogenizer (Fig. 3-13); the resulting homogenate consists of a suspension of cell organelles plus all the cytoplasm compounds not associated with cell structures. The homogenate is then placed in a high-speed centrifuge, which causes the particles to settle out as a result of centrifugal force. The particles with a high density and large size are the first to accumulate at the bottom of the centrifuge tube; smaller particles sedimenting more slowly form several layers on top of the denser particles. By varying the speed and time of centrifugation, it is possible to collect various particulate fractions of the cell as well as the supernatant fraction, which contains the nonorganelle substances of the cell. To obtain some of the lighter elements, forces as high as 100,000 times that of gravity must be created. With the isolation of mitochondria from the cell, it became possible to study their chemical components and biochemical activity directly.

The primary function of the mitochondria is to couple to the synthesis of ATP some of the energy released during the breakdown of organic molecules. The mitochondrion has often been called the powerhouse of the cell since 95 percent of the ATP molecules synthesized from glucose breakdown and 100 percent of that from fatty acid breakdown are synthesized within this organelle. This ATP synthesis occurs through a highly specialized coupling process known as *oxidative phosphorylation.*

Oxidative phosphorylation

In the breakdown of carbohydrates, fats, and proteins many of the chemical reactions involve the removal of hydrogen atoms from various intermediates formed during degradation. These reactions usually require a coenzyme, such as NAD, to which the hydrogen atoms are transferred during the reaction, e.g., the conversion **3-1** of lactic acid to pyruvic acid. A generalized equation for the removal of hydrogen from a molecule can be written

$$BH_2 + NAD \longrightarrow B + NADH_2$$

Here some of the chemical potential energy stored in the BH_2 molecule has been transferred to the $NADH_2$ along with the two hydrogen atoms.

As a result of such hydrogen transfers, much of the chemical potential energy in molecules of carbohydrate, fat, and protein can be transferred to coenzyme molecules. The process of oxidative phosphorylation uses this energy to synthesize ATP. Since the hydrogen atoms can be obtained from a wide variety of organic molecules in the cell, the process illustrates how energy obtained from many sources can be funneled into the common energy carrier, ATP. How does the energy transferred to the hydrogen-carrier molecules get passed on to ATP?

The transfer of energy from $NADH_2$ to ATP involves a complex series of reactions which take place in the inner membranes of the mitochondria. These membranes contain a group of iron-containing proteins known as cytochromes (because they are brightly colored) which accept a pair of electrons from the hydrogens in $NADH_2$ (Fig. 3-14) and, after passing these electrons through a sequence of different cytochromes, eventually donate them to molecular oxygen, which then reacts with the hydrogen ions formed at the beginning of the sequence of reactions to form water. The overall reaction can be written

$$NADH_2 + \tfrac{1}{2}O_2 \xrightarrow{\text{cytochromes}}$$
$$NAD + H_2O + 52 \text{ kcal/mole}$$

The total result is energetically similar to the reaction between hydrogen and oxygen to form water with the

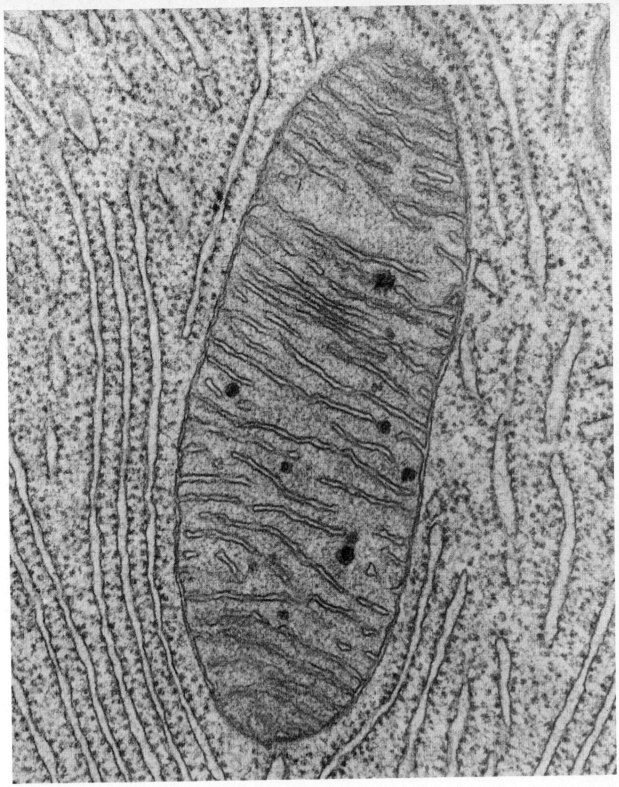

FIGURE 3-11
Electron micrograph of a mitochondrion and surrounding cytoplasm. (*Micrograph courtesy of Dr. K. R. Porter.*)

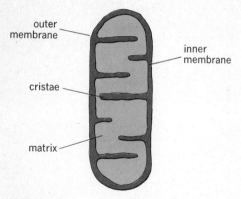

FIGURE 3-12
Mitochondrial structure.

release of energy. Why must the cell proceed through such a complex series of reactions to achieve what is essentially the formation of water from hydrogen and oxygen? If $NADH_2$ reacted directly with oxygen, all 52 kcal of energy would be released. However, only 7 kcal of energy would form ATP from ADP and inorganic phosphate, and the remaining 45 kcal of energy would appear as wasted heat. In contrast, by passing through the cytochrome chain, the same 52 kcal of energy is released but in small steps along with each electron transfer between successive cytochromes. The energy released at

a number of these steps can be coupled to ATP synthesis, resulting in the formation of three molecules of ATP from each reaction between $NADH_2$ and oxygen rather than only one. Since each ATP formed requires 7 kcal of energy and three molecules of ATP are formed, a total of $3 \times 7 = 21$ kcal of energy is transferred to ATP molecules out of a total of 52 kcal released from the reaction of $NADH_2$ with oxygen. The efficiency of the process is thus $\frac{21}{52} \times 100 = 40$ percent, the remaining 60 percent of the energy appearing as heat. The overall process of oxidative phosphorylation is summarized in Fig. 3-15.

Several general properties of oxidative phosphorylation should be noted. $NADH_2$ is not the only coenzyme that can donate hydrogen to the cytochrome system. In some cases the hydrogen donor enters the cytochrome chain at a point beyond the first site of ATP formation and as a result only two molecules of ATP may be formed from each pair of hydrogen atoms rather than three. One of the overall results of oxidative phosphorylation is the regeneration of the coenzyme NAD which is now available to pick up another pair of hydrogen atoms and transfer them to the cytochrome chain.

If oxygen is not available to the cytochrome system, ATP will not be formed by the mitochondria. Since the mitochondria provide about 95 percent of the ATP formed in most cells, the absence of oxygen means the absence of the energy provided by ATP that is required to maintain cell structure and function, and the cell will die. The

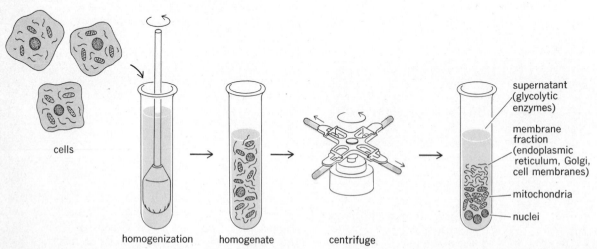

FIGURE 3-13
Isolation of cellular components by homogenization and centrifugation.

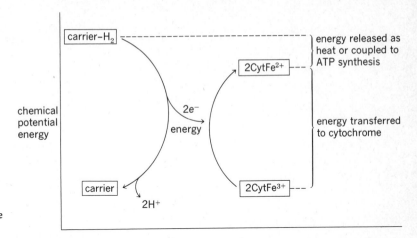

FIGURE 3-14
Changes in chemical potential energy accompanying the transfer of electrons from carrier–H_2 to the iron atoms in the cytochromes.

cause of death from cyanide poisoning is the same as that from a lack of oxygen. Cyanide poisoning is a direct consequence of the chemical reaction between cyanide and the final cytochrome in the chain, preventing electron transfer and blocking the production of ATP by the mitochondria. The requirement of all higher forms of life for oxygen is thus related to the use of oxygen by the mitochondria to transfer chemical energy to ATP. The oxygen combines with hydrogen and ends up in a molecule of water. Nearly 99 percent of all molecular oxygen used by a cell is used in this process of oxidative phosphorylation to form ATP.

Metabolic pathways

The total metabolism of a cell can be divided into a number of metabolic pathways involving the major classes of organic molecules: the carbohydrates, lipids, and proteins. The pathways for the breakdown of these substances are coupled to the formation of ATP to provide the energy necessary for maintaining cell structure and function. There are also pathways for the synthesis of these substances. In addition there are a number of pathways for synthesizing special end products, such as hormones, that may be secreted from cells. The metabolism of a cell can be further subdivided according to the location of the various enzymes that participate in a given metabolic pathway. Many of the enzymes in a particular pathway are located in one of the cell organelles, just as the enzymes for oxidative phosphorylation are located in the mitochondria. We shall not attempt to describe all

the metabolic pathways in a cell but shall focus on the major pathways through which carbohydrates, lipids, and proteins are broken down or synthesized.

Carbohydrate metabolism

The primary function of carbohydrates in the body is to provide a readily available source of energy that can be coupled to the synthesis of ATP. The major carbohydrate in the body is glucose, which can be metabolized in the presence of oxygen to carbon dioxide and water with the release of 686 kcal/mole of energy:

$$C_6H_{12}O_6 + 6O_2 \longrightarrow 6CO_2 + 6H_2O + 686 \text{ kcal/mole}$$

Since 7 kcal/mole of energy is required to synthesize ATP from ADP and P_i, it would be theoretically possible to synthesize $\frac{686}{7} = 98$ moles of ATP per mole of glucose by coupling all the energy released from the breakdown of glucose to the synthesis of ATP. However, only 38 moles of ATP are formed, the remainder of the energy appearing as heat. The total reaction as it occurs in the cell can be written

$$C_6H_{12}O_6 + 6O_2 + 38ADP + 38P_i \longrightarrow$$
$$6CO_2 + 6H_2O + 38ATP + 38H_2O + 420 \text{ kcal/mole}$$
$$\text{(heat)}$$

The total 686 kcal of energy released by glucose is released not in any one reaction but in small amounts in each of the 19 reactions breaking down glucose to carbon dioxide and water. Some of these reactions are coupled to the synthesis of ATP; others simply release energy directly as heat.

The metabolic pathway for glucose degradation to

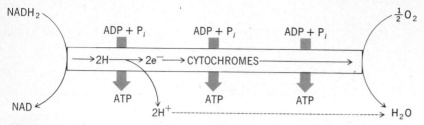

FIGURE 3-15
Energy is coupled to the formation of ATP at three points in
the cytochrome chain during oxidative phosphorylation.

carbon dioxide and water can be divided into two parts, the first involving the breakdown of glucose to pyruvic acid (and in some cases to lactic acid) and the second the conversion of pyruvic acid to carbon dioxide and water in the presence of oxygen. The enzymes mediating these two sets of reactions are found in different parts of the cell. The mechanism of ATP synthesis is also different. The significance of this division of glucose metabolism will be become apparent as we discuss the chemical events that occur in the two pathways.

Glycolysis   The sequence of enzyme-mediated reactions leading from a molecule of glucose to two molecules of either pyruvic acid or lactic acid is known as *glycolysis*. The glycolytic enzymes are found in the cytoplasm of the cell and are not associated with any particular organelle.

The reactions leading to pyruvic acid are shown in detail in Fig. 3-16 in order to illustrate some general principles about the transformation of chemical structure during metabolism. Each step in the sequence modifies the chemical structure of the preceding molecule only slightly, but the total sequence of reactions alters the structure of the glucose molecule markedly. These reactions illustrate the general principle that chemical transformations in the cell occur through a number of small alterations in chemical structure rather than by radical changes in a single reaction. Note that all the intermediates between glucose and pyruvic acid contain an ionized phosphate group. In discussing the permeability of the cell membrane, we found that ionized molecules were generally unable to diffuse across the lipid barrier of the membrane. Thus, once glucose has been phosphorylated in reaction (1), the intermediates of glycolysis are trapped within the cell and cannot leak out across the cell membrane.

One of the major functions of glycolysis is to trans-

fer some of the chemical energy of glucose to ATP in the absence of oxygen. It is somewhat surprising, therefore, to find that the first reaction is the transfer of phosphate from ATP to glucose to form glucose 6-phosphate. In reaction (3), a second molecule of ATP is used in the transfer of phosphate to fructose 6-phosphate. Thus, two molecules of ATP have been used in getting to fructose 1,6-diphosphate, but no new ATP has yet been synthesized. It would appear uneconomical to use ATP when the objective is to synthesize ATP; however, the ATP used during the initial stages of glycolysis is recovered in reaction (7). In the early stages of glycolysis ATP provides the ionized phosphate groups that trap the intermediates within the cell and forms the intermediates required for the later stages of glycolysis.

Reaction (4) splits a six-carbon molecule into two different three-carbon molecules, one of which is converted into the other in reaction (5), the overall result being to convert a six-carbon molecule into two identical three-carbon molecules. Reaction (7) leads to the first direct synthesis of ATP in the glycolytic pathway as one of the phosphate groups in 1,3-diphosphoglycerate is transferred directly to ADP. This type of direct energy transfer of phosphate from a substrate to ADP is known as *substrate phosphorylation*. (Note that this substrate phosphorylation does not involve molecular oxygen.) In this reaction some of the energy in the glucose intermediate is transferred directly to ATP, in contrast to the process of oxidative phosphorylation, in which energy is first transferred to a hydrogen-carrier molecule which in turn transfers its energy to ATP via the cytochromes.

Since the six-carbon glucose molecule was converted into two three-carbon molecules during reactions (4) and (5), two molecules of ATP can be synthesized from glucose during reaction (7), one from each of the three-carbon molecules. This reaction replaces the two

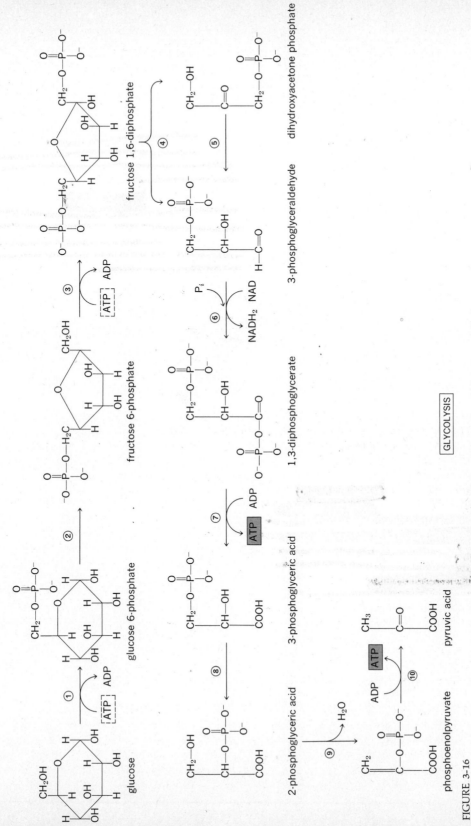

FIGURE 3-16
The glycolytic pathway by which glucose is converted into two molecules of pyruvic acid.

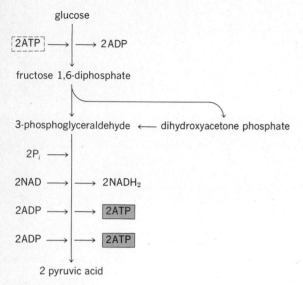

FIGURE 3-17
Major steps leading from glucose to pyruvic acid during glycolysis.

molecules of ATP used in reactions (1) and (3). During reaction (10), two additional molecules of ATP are synthesized by the direct process of substrate phosphorylation. These two molecules represent a net gain of ATP by the cell.

The important steps in the aerobic glycolytic pathway are summarized in Fig. 3-17. Two molecules of ATP are used in the initial states of glycolysis, but four molecules of ATP are eventually formed, giving a net gain by the cell of two ATP. In addition, ATP can also be synthesized from the $NADH_2$ formed during reaction (6) by the process of oxidative phosphorylation. This leads to the synthesis of an additional six molecules of ATP, three from each of the $NADH_2$ molecules formed in reaction (6). The process of oxidative phosphorylation leads to the eventual transfer of the two hydrogen atoms to oxygen and thus releases NAD, which can again proceed through reaction (6).

In the absence of oxygen $NADH_2$ cannot be converted to NAD by transfer of the hydrogen atoms and electrons to the cytochrome system. If there were no other mechanism for converting $NADH_2$ back into NAD, the entire glycolytic process would come to a halt after the small amount of NAD in the cell had been converted to $NADH_2$ during reaction (6) because reaction (6) cannot occur without NAD. Thus, lack of oxygen would stop glycolysis. However, there is another means of generating NAD from $NADH_2$, as shown in Fig. 3-18. The addition of two hydrogen atoms to pyruvic acid results in the formation of lactic acid; $NADH_2$ is the hydrogen donor for this reaction and in the process is converted to NAD. Thus, even in the absence of oxygen (anaerobic conditions), $NADH_2$ can be converted to NAD, and glycolysis can continue, the end product being lactic acid rather than pyruvic acid. This breakdown of glucose to lactic acid is known as *anaerobic glycolysis*. Note that the amount of ATP synthesized per glucose molecule during anaerobic glycolysis is less than during aerobic glycolysis (two versus eight) because of the absence of the additional six molecules of ATP synthesized from the two $NADH_2$ by oxidative phosphorylation.

Some organisms, e.g., certain species of bacteria, can function and multiply under completely anaerobic conditions, obtaining their energy by anaerobic processes very similar to glycolysis. Yeast cells survive and multiply under anaerobic conditions by using energy derived from the conversion of glucose to alcohol by fermentation, which is identical to glycolysis except that in the terminal stages pyruvic acid is converted into alcohol. All higher organisms, including man, are aerobic organisms and require molecular oxygen to survive and function, although certain processes in the human body can, and do, function in the absence of oxygen for short periods of time since cells can synthesize some ATP by anaerobic glycolysis.

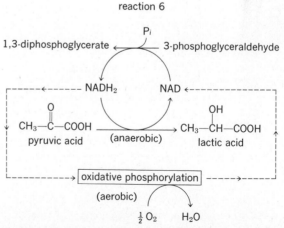

FIGURE 3-18
Pathways for the conversion of $NADH_2$ to NAD.

The Krebs cycle   The second portion of the pathway of glucose degradation involves the conversion of pyruvic acid into carbon dioxide and water in the presence of molecular oxygen and the coupling of these reactions to the synthesis of ATP. The enzymes mediating these reactions are localized in the mitochondria, in contrast to the free distribution of the glycolytic enzymes in the cytoplasm of the cell. Figure 3-19 shows the nine reactions in the breakdown of pyruvic acid. The overall reaction can be written

$$2CH_3-\overset{\overset{\displaystyle O}{\|}}{C}-COOH + 5O_2 \longrightarrow 6CO_2 + 4H_2O$$

The three carbon atoms in the pyruvic acid molecule eventually end up in carbon dioxide. The first molecule of carbon dioxide is formed in reaction (1) and is derived from the carboxyl group of pyruvic acid. Note that the oxygen atoms in carbon dioxide are derived from the oxygen in pyruvic acid, not from molecular oxygen. In this same reaction the remaining two-carbon fragment of pyruvic acid, known as acetate, is transferred to coenzyme A, forming acetyl CoA. The coenzyme A molecule acts as a carrier for acetate, transferring it from one molecule to another, just as NAD is a carrier for hydrogen and ATP is a carrier for energy. Coenzyme A is a derivative of pantothenic acid (one of the B vitamins) and participates in numerous reactions in the cell in which an acetate fragment is involved. There is no net consumption of coenzyme A since it is returned to its original state in reaction (2), when it transfers acetate to the oxaloacetic acid molecule.

In reaction (2), the two-carbon acetate fragment of acetyl CoA is transferred to the four-carbon oxaloacetic acid to form the six-carbon citric acid. The remaining seven reactions operate in a cycle leading from the six-carbon citric acid to the four-carbon oxaloacetic acid, which is then converted back to citric acid through the addition of acetate. This cycle is known as the *Krebs cycle* in honor of Hans Krebs, who established the sequence of reactions in the early 1940s and for this work received the Nobel Prize in 1952. For obvious reasons the cycle is also referred to as the *citric acid* or *tricarboxylic acid cycle.*

A summary of the essential reactions in the degradation of pyruvic acid is given in Fig. 3-20. The flavine adenine dinucleotide (FAD) shown in the figure is a hydrogen-carrier coenzyme molecule similar to NAD, but it is derived from the vitamin riboflavin rather than niacin. The Krebs-cycle reactions have converted the three carbon atoms in pyruvic acid into three molecules of carbon

dioxide and transferred hydrogen atoms to hydrogen carriers. Thus far, no molecular oxygen has been used, and no ATP has been synthesized. Most of the chemical energy in pyruvic acid has been transferred to the hydrogen carriers during this sequence of reactions.

The hydrogen atoms then proceed through the process of oxidative phosphorylation, during which ATP is synthesized. A total of five pairs of hydrogen atoms is transferred to hydrogen carriers during each reaction of the Krebs cycle. Since each pair of hydrogen atoms transferred to oxygen leads to the synthesis of three molecules of ATP,[1] a total of 15 molecules of ATP is synthesized during the degradation of each pyruvic acid molecule.

Figure 3-21 summarizes the pathways for ATP synthesis utilizing the energy released during the breakdown of glucose through the glycolytic and Krebs-cycle reactions. In the presence of oxygen, 2 molecules of ATP are formed by direct substrate phosphorylation during glycolysis, and 36 molecules of ATP are formed by oxidative phosphorylation utilizing the 10 pairs of hydrogen atoms from the Krebs cycle and the 2 pairs from glycolysis. Thus, a total of 38 molecules of ATP can be formed from each molecule of glucose. Since a total of 686 kcal of energy is released for each mole of glucose metabolized to carbon dioxide and water and 7 kcal of energy has been transferred to each mole of ATP formed, $38 \times \frac{7}{686} \times 100 = 39$ percent of the total potential energy in glucose has been transferred to ATP. The remaining 61 percent of the energy appears as heat.

In the absence of oxygen, oxidative phosphorylation cannot occur, and the only ATP synthesized is the two molecules formed by direct substrate phosphorylation during anaerobic glycolysis. The energy transferred to ATP during anaerobic glycolysis represents only $2 \times \frac{7}{686} \times 100 = 2$ percent of the total potential energy in glucose. Thus $\frac{36}{38} \times 100 = 95$ percent of the ATP that can be synthesized from the energy released from glucose depends upon oxygen and the oxidative phosphorylation occurring in the mitochondria.

Glucose synthesis and storage   The body can synthesize glucose. Is the metabolic pathway for glucose synthesis the reversal of that for glucose breakdown, so that intermediates in the pathway can be used to synthesize glu-

---

[1] Actually $FADH_2$ leads to the synthesis of only two ATP because of the stage at which it enters the cytochrome chain, but a third molecule of ATP is formed by a substrate-level phosphorylation during the formation of succinic acid from α-ketoglutaric acid. The net result is equivalent to three ATPs being formed from each carrier–H_2 molecule.

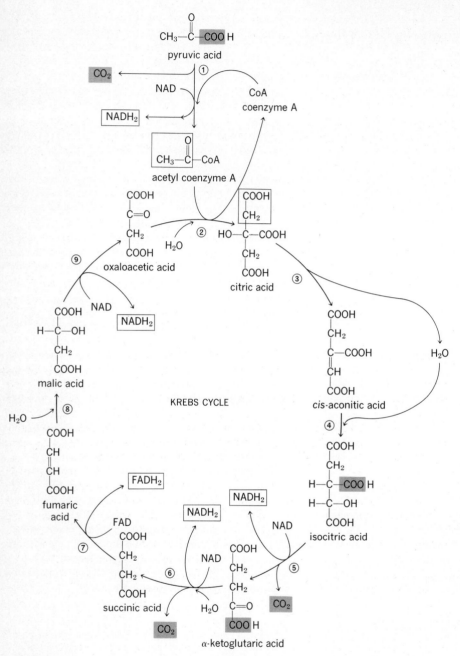

FIGURE 3-19
Krebs-cycle reaction occurring in the mitochondria.

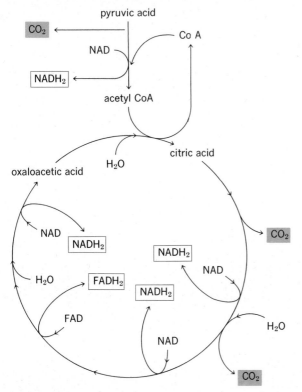

FIGURE 3-20
Summary diagram of the Krebs-cycle reactions.

cose? The answer is both *yes* and *no,* depending on which portion of the pathway is considered. The first step in glycolysis,

$$\text{Glucose} + \text{ATP} \xrightarrow{\text{hexokinase}} \text{glucose 6-phosphate} + \text{ADP}$$

is an irreversible reaction. However, glucose can be formed from glucose 6-phosphate via a different reaction mediated by a second enzyme, glucose phosphatase:

$$\text{Glucose 6-phosphate} \xrightarrow{\substack{\text{glucose} \\ \text{phosphatase}}} \text{glucose} + \text{P}_i$$

Most cells have only the hexokinase enzyme but do not contain glucose phosphatase and thus cannot form glucose from glucose 6-phosphate. On the other hand, liver cells contain both enzymes and, as we shall see, the liver is the major organ that maintains fairly constant the glucose levels in the blood by releasing glucose which it has synthesized from various intermediates.

The remainder of the glycolytic pathway from glucose 6-phosphate to pyruvic acid is reversible. Thus, any of the intermediates in this portion of the pathway can be used to synthesize glucose 6-phosphate. Some of the individual reactions in this pathway are spontaneously reversible and are catalyzed by a single enzyme in both directions. At other points in the pathway, there are two irreversible reactions catalyzed by separate enzymes in the two directions. These enzymes provide control points for regulating the direction of flow in the glycolytic pathway.

At the junction between glycolysis and the Krebs cycle, pyruvic acid is converted to acetyl CoA. This reaction is irreversible and is mediated by a single enzyme such that the flow of material can occur only from glycolysis to the Krebs cycle but not in the reverse direction. The consequence of this irreversible reaction is that acetyl CoA cannot be used to form glucose. As we shall see shortly, the end product of lipid breakdown, acetyl CoA, feeds into the Krebs cycle at this point and thus cannot be used to form glucose. The reversible and irreversible pathways of glucose metabolism are illustrated in Fig. 3-22.

In contrast to plants, animals and man have a limited capacity for the storage of glucose. Most of the glucose in food is either broken down to synthesize ATP or is converted into fat for storage. The small amount that is stored, primarily in the muscles and liver, is in the form of the polysaccharide *glycogen,* which is similar to the plant polysaccharide, starch. The synthesis of glycogen begins with glucose 6-phosphate, the first intermediate in glycolysis. After a few steps, the glucose units are linked together to form the long, branched-chain polysaccharide glycogen. The final reaction which links the glucose units to glycogen is an irreversible reaction. Therefore, the release of glucose from glycogen requires a second enzyme and is also an irreversible reaction. The two enzymes catalyzing these opposed reactions provide a very important control point for regulating the breakdown and synthesis of glycogen in cells.

Lipid metabolism

The neutral fats, consisting of three fatty acids attached to the three-carbon molecule glycerol, constitute the majority of the lipids in the body. The breakdown and synthesis of neutral fats are closely associated with the metabolism of glucose because of the formation of intermediates common to both pathways. The initial step in the breakdown of neutral fat involves splitting off the three fatty acids from glycerol (Fig. 3-23). The fatty acids

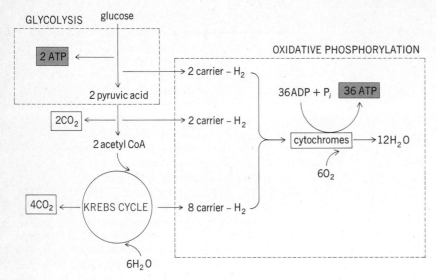

FIGURE 3-21
Summary diagram of the formation of ATP during the aerobic breakdown of glucose to carbon dioxide and water.

and glycerol then proceed through two separate pathways. Glycerol, being a three-carbon carbohydrate, enters the glycolytic pathway (by forming dihydroxyacetone phosphate, Fig. 3-16), from which point it can go through the rest of the carbohydrate pathway. Conversely, glucose can provide the glycerol required for the synthesis of neutral fats.

The breakdown of fatty acids requires coenzyme A and hydrogen carriers such as NAD. The coenzyme A molecule attaches to the carboxyl group at the end of the fatty acid molecule. One molecule of ATP is utilized in this initial step, just as in the initial reactions occurring in the breakdown of glucose. A series of reactions then occurs during which a molecule of water is added and four hydrogen atoms are transferred to two hydrogen-carrier molecules, one each of NAD and FAD:

$$R—CH_2—CH_2—CH_2—\overset{\overset{O}{\|}}{C}—CoA \xrightarrow[\text{2 carrier 2 carrier–}H_2]{H_2O}$$

$$R—CH_2—\overset{\overset{O}{\|}}{C}—CH_2—\overset{\overset{O}{\|}}{C}—CoA$$

These reactions involve only the first three carbon atoms in the fatty acid molecule. The bond between the second and third carbon atoms is now broken, releasing a molecule of acetyl CoA and at the same time adding a molecule of coenzyme A to the end of the remaining fatty acid fragment.

$$R—CH_2—\overset{\overset{O}{\|}}{C}—CH_2—\overset{\overset{O}{\|}}{C}—CoA + CoA \longrightarrow$$

$$R—CH_2—\overset{\overset{O}{\|}}{C}—CoA + CH_3—\overset{\overset{O}{\|}}{C}—CoA$$

The net result is the transfer of the two terminal carbon atoms in the fatty acid to acetyl CoA and the simultaneous formation of two carrier–H_2 molecules. The new fatty acid–CoA molecule can also proceed through the same series of reactions, transferring two more carbons to acetyl CoA. By repetition of this process, the entire fatty acid molecule is broken down to acetyl CoA, two carbons at a time. The overall reaction for a fatty acid containing 18 carbon atoms can be written

$$C_{18} \text{ fatty acid} + ATP + 9CoA + 16NAD + 7H_2O \longrightarrow$$
$$9 \text{ acetyl CoA} + 16 \text{ carrier–}H_2 + ADP + P_i$$

Thus far, no ATP has been formed directly from the breakdown of fatty acids, but a number of molecules of carrier–H_2 and acetyl CoA have been formed. Recall that acetyl CoA is also formed from pyruvic acid at its entry into the Krebs cycle. Acetyl CoA derived from the fatty acids can enter the Krebs cycle at this point and be broken down to carbon dioxide and water and the energy released coupled to ATP synthesis. The carrier–H_2 molecules resulting from fatty acid breakdown can transfer their hydrogens to the cytochromes in the mitochondria and thus provide an additional source of energy for ATP synthesis. Because of the close association between the products of fatty acid breakdown, carrier–H_2 and acetyl CoA, and the metabolism of these products in the mitochondria, it is not surprising to find that the enzymes which mediate the breakdown of fatty acids are located in the mitochondria. Figure 3-24 shows the close relationship between neutral fat and glucose metabolism.

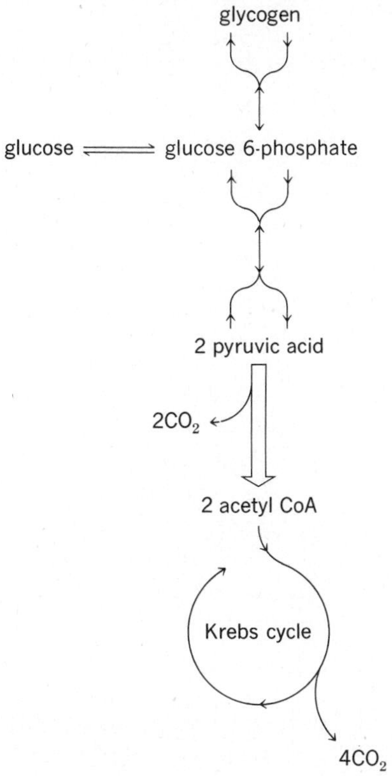

The energy released during the breakdown of neutral fat is coupled to the synthesis of ATP at a number of points. Glycerol passing through the glycolytic pathway leads to direct substrate phosphorylation of ATP, as well as the formation of $NADH_2$ and eventually acetyl CoA. The $NADH_2$ and acetyl CoA are coupled to ATP synthesis in the mitochondria, the net result being the formation of 22 molecules of ATP from each molecule of glycerol.[1] From each 18-carbon fatty acid, 147 molecules of ATP can be synthesized from the $NADH_2$, $FADH_2$, and acetyl CoA formed during fatty acid breakdown. Since there are three fatty acids in each molecule of neutral fat, a total of 441 ATP can be formed. Adding to this the 22 ATP formed from glycerol breakdown gives a total of 463 molecules of ATP formed from the breakdown of each molecule of neutral fat. Table 3-1 compares the molecular weight of glucose and neutral fat and the amount of ATP that can be formed from each. Note that, per gram, the

[1] In the process of getting from glycerol to dehydroxyacetone phosphate one molecule of ATP is used and one molecule of $NADH_2$ is formed. Thus the net ATP that is formed after $NADH_2$ has donated its hydrogen to the cytochromes is only 2ATP rather than 3.

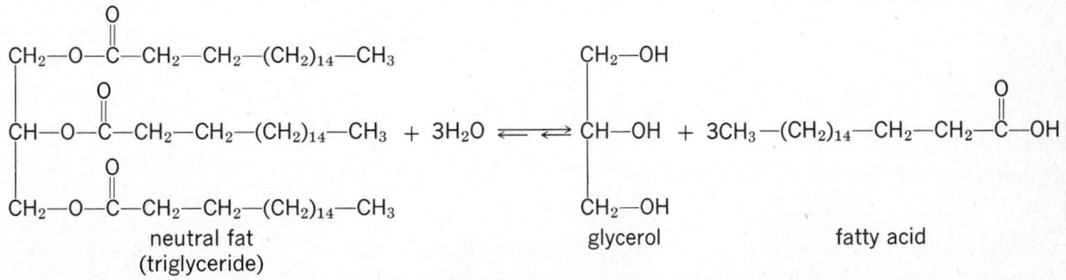

FIGURE 3-22
Reversible and irreversible reactions in the pathway leading to the breakdown and synthesis of glucose and glycogen. Double-headed arrows indicate reversible reactions. Single-headed arrows indicate irreversible reactions.

neutral fat
(triglyceride)

glycerol   fatty acid

FIGURE 3-23
Breakdown of a neutral fat molecule into a molecule of glycerol and three fatty acids.

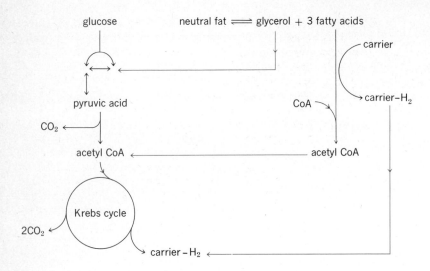

FIGURE 3-24
Pathways linking fat metabolism to the metabolism of carbohydrates.

breakdown of neutral fat leads to the synthesis of about three times as much ATP as the breakdown of a gram of glucose. Fat is thus a more efficient means of storing chemical energy in a cell than is carbohydrate.

The synthesis of fatty acids is mediated by an entirely separate set of enzymes. They are located in the cytoplasm rather than in the mitochondria, in contrast to the enzymes for fatty acid breakdown. The starting point for fatty acid synthesis is the two-carbon fragment, acetyl CoA, and the fatty acid chain is built up two carbons at a time, just as the breakdown proceeded two carbons at a time. Since a fatty acid has a higher energy content than an equivalent number of acetyl CoA molecules, energy in the form of ATP is required for fatty acid synthesis. In addition, hydrogen, in the form of coenzyme hydrogen carriers, must be provided.

We can now identify the route by which carbohydrate is converted into fat. The breakdown of glucose produces acetyl CoA, the main ingredient for fatty acid

synthesis. It also provides the ATP and hydrogen-carrier molecules which are required. Thus, in the presence of excess carbohydrate, since there is a limited capacity to store glycogen or ATP, the accumulation of acetyl CoA will, by mass action, lead to the synthesis and storage of fat. When the metabolic demands of the cell for ATP are increased, the stored lipid can be broken down to acetyl CoA and metabolized in the Krebs cycle to produce ATP. It should be emphasized, however, that once glucose is converted to fatty acids, the fat cannot be used to form glucose because of the irreversible reaction between pyruvic acid and acetyl CoA. Thus the acetyl CoA produced by lipid breakdown cannot be used to synthesize glucose.

Protein metabolism

The metabolism of protein is far more complicated than lipid or carbohydrate metabolism. Proteins are formed from 20 different amino acids, all of which have different chemical structures and require different pathways for their synthesis and degradation. Several enzymes can break down proteins into amino acids, some acting on the terminal amino acids in the chain and others breaking the bonds between a specific set of amino acids within the chain. The net result of this enzyme action is the production of a pool of amino acids in the cell.

Proteins $\longrightarrow$ amino acids

We shall not consider the specific routes by which each

TABLE 3-1
ATP formed per gram of glucose or triglyceride (fat)

	Grams per mole	Moles ATP per mole	Moles ATP per gram
Glucose	180	38	$\frac{38}{180} = 0.21$
Triglyceride (C_{18})	842	463	$\frac{463}{842} = 0.55$

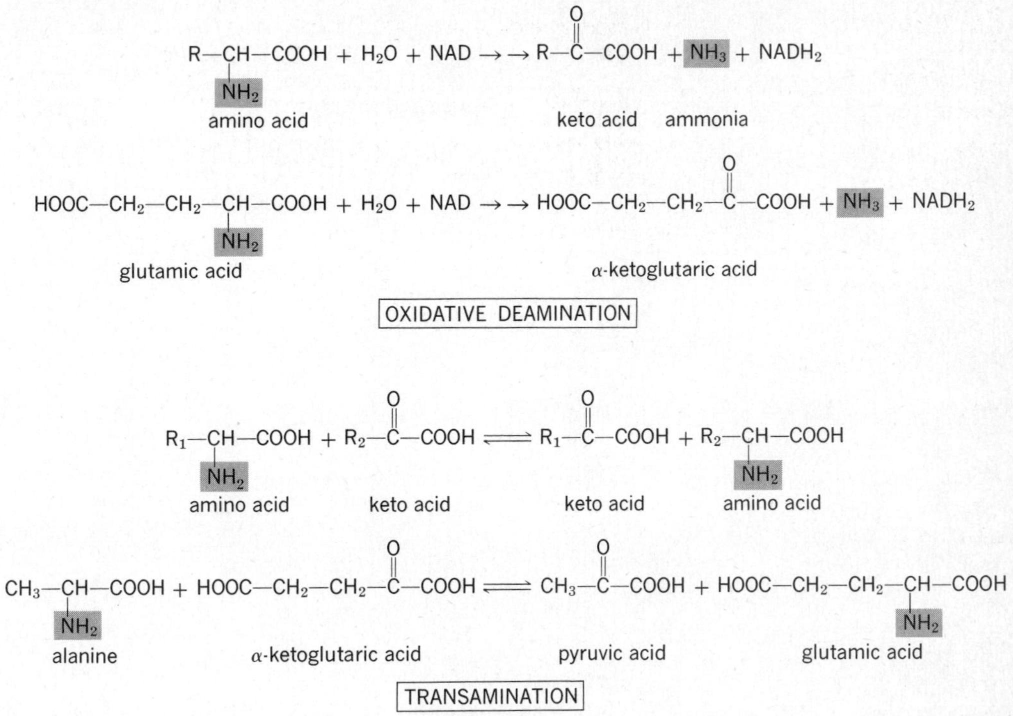

$$R-\underset{\underset{NH_2}{|}}{CH}-COOH + H_2O + NAD \rightarrow \rightarrow R-\underset{\underset{\|}{O}}{C}-COOH + NH_3 + NADH_2$$

amino acid   keto acid   ammonia

$$HOOC-CH_2-CH_2-\underset{\underset{NH_2}{|}}{CH}-COOH + H_2O + NAD \rightarrow \rightarrow HOOC-CH_2-CH_2-\underset{\overset{O}{\|}}{C}-COOH + NH_3 + NADH_2$$

glutamic acid   α-ketoglutaric acid

OXIDATIVE DEAMINATION

$$R_1-\underset{\underset{NH_2}{|}}{CH}-COOH + R_2-\underset{\overset{O}{\|}}{C}-COOH \rightleftharpoons R_1-\underset{\overset{O}{\|}}{C}-COOH + R_2-\underset{\underset{NH_2}{|}}{CH}-COOH$$

amino acid   keto acid   keto acid   amino acid

$$CH_3-\underset{\underset{NH_2}{|}}{CH}-COOH + HOOC-CH_2-CH_2-\underset{\overset{O}{\|}}{C}-COOH \rightleftharpoons CH_3-\underset{\overset{O}{\|}}{C}-COOH + HOOC-CH_2-CH_2-\underset{\underset{NH_2}{|}}{CH}-COOH$$

alanine   α-ketoglutaric acid   pyruvic acid   glutamic acid

TRANSAMINATION

FIGURE 3-25
Oxidative deamination and transamination of amino acids.

amino acid is synthesized and degraded by the cell since a few examples will suffice to indicate the general pattern.

Unlike carbohydrates and lipids, amino acids contain nitrogen in addition to carbon, hydrogen, and oxygen. Once the nitrogen in the amino group is removed, the remainder of the molecule is converted into intermediates that can enter the pathways of glucose metabolism. Cells have several ways of removing nitrogen from amino acids. The amino acid may react with water and a hydrogen carrier to form a keto acid and ammonia in a reaction known as *oxidative deamination* (Fig. 3-25). The second possibility involves the transfer of the amino group to a keto acid (Fig. 3-25). This type of reaction is known as *transamination.*

The examples given in Fig. 3-25 also illustrate the link between amino acid metabolism and carbohydrate metabolism. α-Ketoglutaric acid, formed from the amino acid glutamic acid, is one of the intermediates in the Krebs cycle (Fig. 3-19). In the second example, the transamination of alanine gives pyruvic acid, which can be metabolized to carbon dioxide and water or used to synthesize glucose or lipid. Although we have given only two examples of amino acid metabolism, the other 18 amino acids can be transformed into intermediates that at some point can enter into the glycolytic or Krebs-cycle reactions. Conversely carbohydrates can be used to synthesize amino acids provided there is a source of nitrogen to supply the amino group. Ammonia acts as a source of nitrogen for the amino group in some cases, but transamination predominates. All 20 amino acids cannot be synthesized by the cells in the body. A group of about eight, known as *essential amino acids,* must be provided in the human diet. Amino acids in the diet also provide the amino groups used for transamination reactions during the synthesis of new amino acids.

The complete degradation of carbohydrates and lipids leads to carbon dioxide and water, which can be considered the waste products of metabolism. The car-

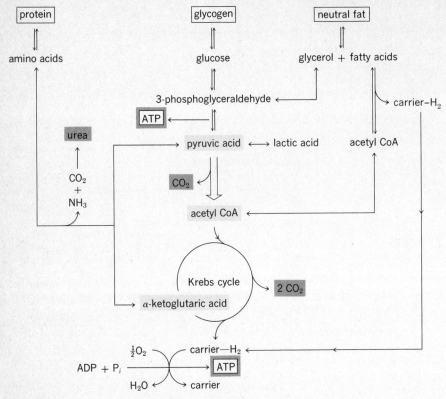

FIGURE 3-26
Interrelations between the metabolism of carbohydrates, fats, and proteins.

bon dioxide is eliminated by the lungs, and excess water is excreted primarily in the urine. In protein metabolism, the end products also include some nitrogen-containing compounds, e.g., ammonia. Ammonia is very reactive and highly toxic to cells in large quantities. The cells of the liver are able to detoxify ammonia rapidly by converting it into the inert molecule urea. The net reaction for urea synthesis is

$$Energy + CO_2 + 2NH_3 \longrightarrow NH_2\!-\!\overset{\displaystyle O}{\overset{\|}{C}}\!-\!NH_2 + H_2O$$

Actually about eight different reactions, involving a number of intermediates, are required for urea synthesis. Once formed, urea is excreted from the body in the urine. In addition to nitrogen, a few of the amino acids contain atoms of sulfur, which eventually end up as sulfate, SO_4^{2-}, which can also be excreted in the urine.

Interconversion of carbohydrate, lipid, and protein

Having discussed the metabolism of the three major classes of organic molecules found in cells, we can now examine how each class is related to the others and to the process of synthesizing ATP. Figure 3-26 shows the major pathways we have discussed and the relation of the common intermediates. All three classes of molecules can enter the Krebs cycle through some intermediate, and thus all three can be used as a source of chemical potential energy for the synthesis of ATP. Hydrogen atoms are the substrate required for the oxidative phosphorylation of ATP. Some of these hydrogens are derived from the Krebs-cycle reaction directly; others may come from glycolysis or the breakdown of fatty acids. Glucose can be converted into fat or into amino acids by way of the common intermediates such as pyruvic acid, α-keto-

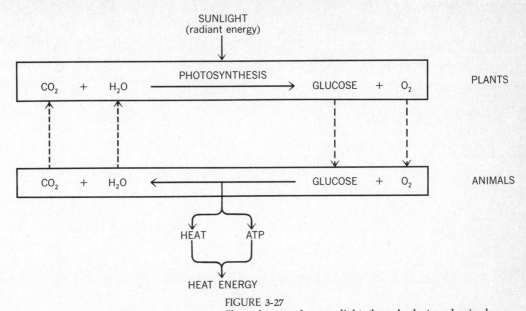

FIGURE 3-27
Flow of energy from sunlight, through plants and animals.

glutaric acid, and acetyl CoA. Similarly amino acids can be converted into lipids through common intermediates. Some amino acids can also be converted into glucose provided they can be transformed into the appropriate intermediates in the glycolytic pathway. Fatty acids cannot be converted into glucose because of the irreversibility of the pyruvic acid–to–acetyl CoA reaction, but glycerol can be converted into glucose. Metabolism is thus a highly integrated process in which all classes of molecules can be used, if necessary, to provide energy for the cell through ATP synthesis and in which each class of molecule can to a large extent provide the raw materials required to synthesize members of other classes.

Photosynthesis

Animals cannot convert carbon dioxide and water into glucose and oxygen by a reversal of the pathway for glucose breakdown:

$$\text{Glucose} + 6O_2 \rightleftharpoons 6CO_2 + 6H_2O + 686 \text{ kcal/mole}$$

In contrast, plants are able to capture the energy in sunlight and use this energy to synthesize glucose by the process of *photosynthesis*. This process of photosynthesis in plants is the origin of the oxygen in the earth's atmosphere that is used by animals in the release of energy from glucose. Figure 3-27 illustrates the interdependence between plants and animals. All life is ultimately dependent upon the energy derived from sunlight. Plants trap some of this radiant energy by photosynthesis, forming glucose and oxygen. Animals release this energy stored in glucose by photosynthesis and use it to perform cellular functions. This energy ultimately appears as heat which cannot be used by either plants or animals. Thus there is a continuous flow of energy from the sun through all living organisms on the earth.

Protein synthesis, heredity, and cell development

4

One of the outstanding accomplishments of twentieth-century biology has been the elucidation of the chemical basis of heredity and its relationship to protein synthesis. Whether an organism is a man or a mouse, has blue eyes or black, has light skin or dark is determined by the type of hereditary information passed on from parents to offspring. The transmission of hereditary characteristics from cell to cell and from parents to offspring depends upon molecules which contain coded information built into their molecular structure. These molecules are the giant deoxyribonucleic acid (DNA) molecules found in the nucleus of the cell. This coded information is used by a cell to synthesize proteins, both structural proteins and enzymes. In Chap. 3 we described the importance of enzymes in catalyzing the wide variety of chemical reactions which occur in cells. The total enzyme composition of a cell determines its metabolic activity, which in turn determines its chemical composition and functional activity. Hereditary information is translated from DNA into cell structure and function indirectly through the process of protein synthesis.

A single DNA molecule may carry the information required for the synthesis of many different protein molecules. This information can be divided into separate units, known as *genes*. Most genes contain the information corresponding to a single protein, but some carry information used to regulate the activity of other genes or to synthesize molecules of ribonucleic acid (RNA), required for the assembly of protein molecules. The total DNA present in a human cell has been estimated to carry information corresponding to 2 to 3 million genes.

As an example of the expression of genetic information, consider eye color in man. The color is due to the presence of particular kinds of molecules (pigments) within cells of the eye. A sequence of enzyme-mediated reactions is required to synthesize these pigments. The genes determining eye color control the synthesis of specific enzymes in this synthetic pathway. The type of colored pigment synthesized depends upon the type of enzymes, which in turn depends upon the type of genes. Thus, the genes for eye color do not contain information about the chemical structure of the eye pigments themselves but the information required to synthesize the enzymes which mediate the formation of eye pigment.

Although DNA molecules contain the information necessary for synthesis of specific proteins, the DNA molecule does not itself participate directly in the assembly of a protein molecule. The DNA molecules of a cell are located in the nucleus, where as most protein synthesis occurs in the cytoplasm. The transfer of information from the DNA molecules in the nucleus to the site of protein synthesis in the cytoplasm is effected by molecules of RNA, which are assembled on the surface of the DNA molecules. The molecular arrangement of the DNA molecule determines the arrangement of molecular subunits in RNA, and by this process information is passed from DNA to RNA. The RNA molecules are much smaller than the DNA molecules, and each RNA molecule carries the information content of a single gene (or at most a few genes) into the cytoplasm, where the actual assembly occurs. Figure 4-1 shows the general pathway by which the information stored in DNA is able to influence the metabolism of a cell through the process of protein synthesis.

With this general orientation we can now examine the detailed mechanisms by which information coded into the structure of DNA is used by a cell to assemble a protein from individual amino acids.

The cell nucleus

The structure of the nucleus as revealed by the electron microscope is shown in Fig. 4-2, where it is immediately seen that the nucleus does not have the wide variety of organelles found in the cytoplasm. The only prominent structural feature present is the irregularly shaped, densely staining, threadlike region known as the *nucleolus* (little nucleus). The remainder of the nucleus appears to consist of granules of varying density.

Chemical analysis of isolated nuclei indicates that practically all the cell's DNA is located in the nucleus; however, a small amount is associated with the mito-

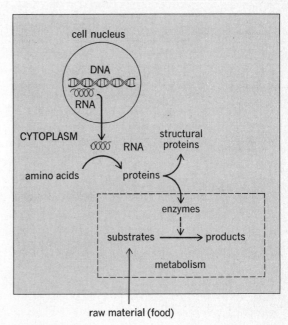

FIGURE 4-1
The relations between metabolism, the synthesis of protein enzymes, and the transfer of the information required for protein synthesis from the DNA molecules in the nucleus to the site of protein synthesis in the cytoplasm by RNA molecules.

chondria in the cytoplasm. DNA is a long, threadlike molecule about 20 Å in diameter. These molecules form a loose network of thin filaments responsible for the granular appearance of the nuclear contents. Some regions of the DNA threads are moderately coiled, giving rise to areas of variable density in the nucleus. This description applies to the nucleus when the cell is not dividing. When division is occurring, the loose network of DNA filaments becomes highly coiled and condenses to form short rod-like bodies. Because these structures have an affinity for the dyes used to make cell structures visible under a light microscope, they were named *chromosomes* (colored bodies).

In addition to DNA, the nucleus contains about 10 percent of the total cellular RNA, most of which is associated with the nucleolus. Biochemical studies indicate that RNA is synthesized in the nucleus and passes into the cytoplasm. The nucleolus appears to be the site at which special types of RNA molecules are synthesized.

Surrounding the nucleus is a double nuclear membrane, which is traversed at intervals with porelike struc-

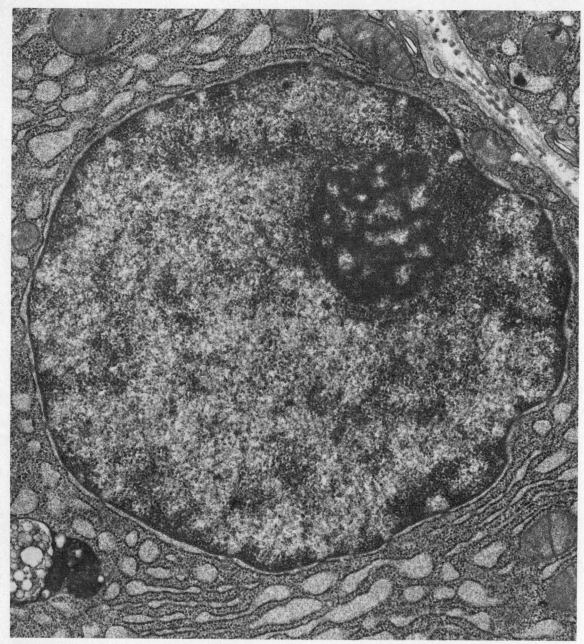

FIGURE 4-2
Electron micrograph of the nucleus in a pancreas cell. (*Courtesy of Keith R. Porter.*)

tures about 500 Å in diameter. These pores should not be confused with the cell-membrane pores, which are estimated to be only 7 to 8 Å in diameter. Differences in the nuclear and cytoplasmic composition of various small-molecular-weight molecules suggest that, like other membranes, the nuclear membrane acts as a selective barrier allowing the entry and exit of only certain types of molecules.

Nucleic acid structure

DNA structure

DNA molecules are the largest molecules in the cell. Like a protein molecule, a DNA molecule consists of a sequence of molecular subunits linked together to form long chains. A single cell may contain a number of DNA molecules of different lengths and thus different molecular weights. The molecular weights of isolated DNA molecules have been found to fall within the range of 1 million to several billion. (Thus, DNA molecules are 10 to 100,000 times larger than most proteins.)

The chemical subunits of nucleic acids are known as *nucleotides.* Each nucleotide consists of three component parts: phosphate, a sugar, and a ring-shaped organic molecule, known as a *base,* which contains several atoms of nitrogen (Fig. 4-3). The nucleotides in DNA contain the sugar deoxyribose, thus, the name deoxyribonucleic acid. Any one of four different bases may be linked to the sugar to form the nucleotide unit. According to their structure, the four bases can be divided into two classes: the purine bases, adenine (A) and guanine (G), which have two rings, and the pyrimidine bases, cytosine (C) and thymine (T), which have only a single ring. Nucleotides are linked together to form the long chains which make up the DNA molecule. The linkage holding two adjacent nucleotides together occurs between the phosphate group of one nucleotide and the sugar of the next (Fig. 4-4), a pattern that produces a repeating sugar-phosphate chain along the DNA molecule. Attached to each sugar is a base located to the side of the sugar-phosphate chain. This general arrangement is similar to protein structure, in which the R groups of the amino acids project to the side of the peptide chain (Fig. 1-19).

In 1953, James D. Watson and Francis H. C. Crick proposed a model for the three-dimensional structure of the DNA molecule which has led to our present knowledge of how genetic information is coded in the DNA molecule and translated ultimately into protein synthesis.

For this work Watson and Crick were awarded the 1962 Nobel Prize. They found that a DNA molecule consists of not one but two chains of nucleotides, which are coiled about each other to form a double helix (Fig. 4-5). It was found that the total number of purine bases and pyrimidine bases in DNA is the same, that the amounts of the purine adenine and pyrimidine thymine are equal, and that the amounts of the purine guanine and pyrimidine cytosine are similarly equal. These data suggested that a purine base is paired with a pyrimidine base, more specifically, that adenine is paired with thymine and guanine with cytosine. Such a pairing between the purine and pyrimidine bases produces a unit about 11 Å in diameter, which is the distance found to separate the two sugar-phosphate chains in the DNA double helix. Watson and Crick proposed that these purine-pyrimidine pairs, A–T and G–C, form bridges between the two sugar-phosphate chains.

The helical configuration of the DNA molecule is formed by hydrogen bonds between the repeating nucleotide units. Watson and Crick further suggested that the sequence of bases along the DNA molecule might provide the mechanism for storing coded genetic information. The specific pairing between A and T and between G and C has important implications both for the replication of the DNA molecules and the mechanism by which information is transferred from DNA to RNA during protein synthesis.

RNA structure

The structure of RNA is similar in many respects to the structure of DNA. Nucleotides composed of phosphate, sugar, and base form the repeating subunits of RNA, but the sugar in the nucleotides is *ribose* rather than deoxyribose, thus the name ribonucleic acid. The nucleotides in both DNA and RNA contain adenine, guanine, and cytosine bases. The fourth base is different. In DNA it is the pyrimidine base thymine; in RNA it is the pyrimidine base uracil (U). A comparison of the nucleotide compositions of DNA and RNA is shown in Fig. 4-6.

Like DNA, RNA consists of a chain of nucleotides connected by phosphate-sugar linkages, but unlike DNA, RNA has only a single chain of nucleotides rather than a double chain. Thus, in RNA there is no equality between the numbers of purine and pyrimidine bases. However, some base pairings do occur between A and U and between G and C in the same single-stranded chain when the chain folds back upon itself, so that portions of the RNA molecule have a double-helical structure (Fig. 4-7). As we shall see, RNA performs several different functions

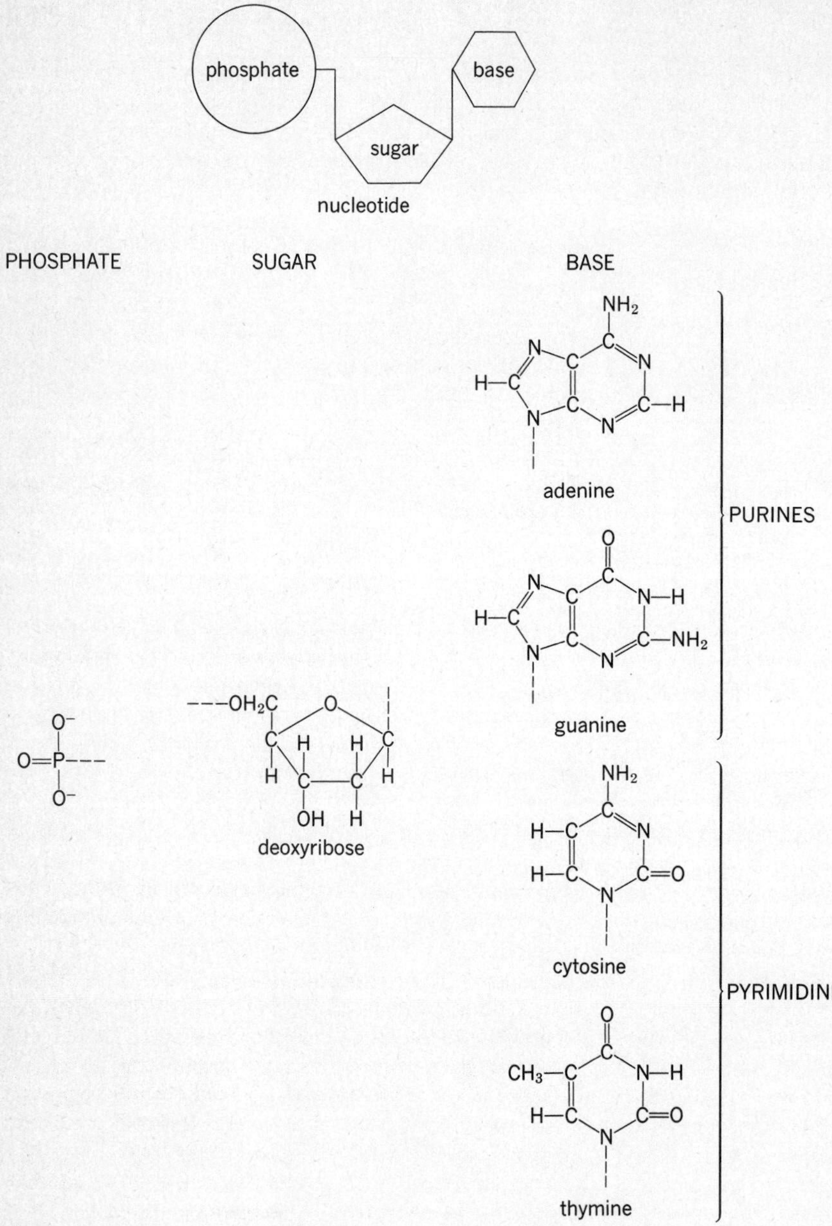

FIGURE 4-3
Chemical structures of the four nucleotides found in DNA.

during protein synthesis. Corresponding to the different functions of RNA, there are several classes of RNA molecules, ranging in molecular weight from about 25,000 to several million.

In summary, RNA, although similar to DNA, has the following important differences: It is a single-stranded molecule rather than double-stranded; it has a slightly different sugar and base composition; and different

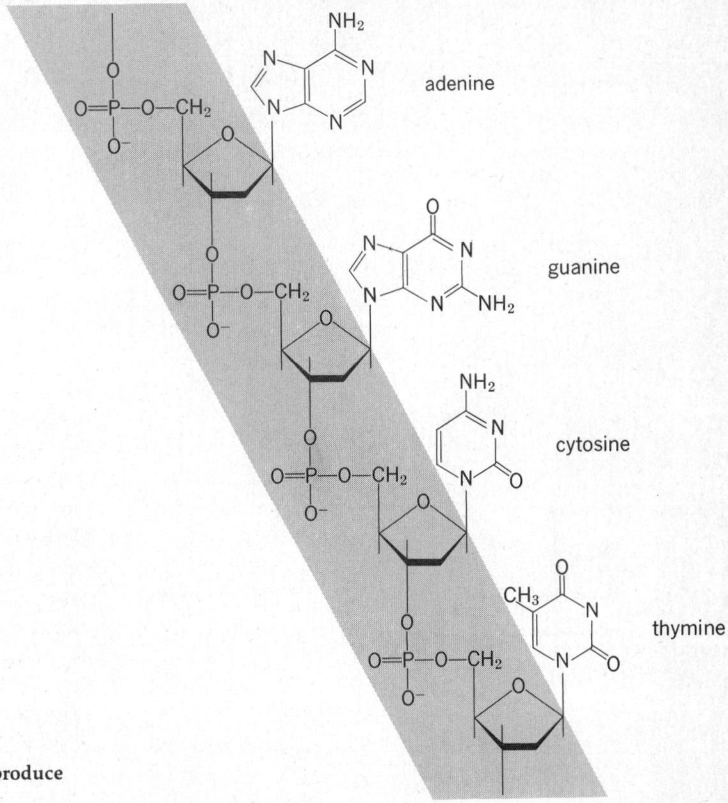

adenine

guanine

cytosine

thymine

FIGURE 4-4
**Linkages between nucleotides produce
the nucleotide chains of DNA.**

classes of RNA molecules perform different functions in the cell. In addition RNA molecules are continually being synthesized and broken down, and thus any one molecule of RNA exists for only a relatively short period of time before it is destroyed and replaced with newly synthesized RNA. In contrast, the DNA molecule is highly resistant to metabolic degradation and is not synthesized until the time of cell division, when it makes one copy of itself to pass on to the newly formed cell.

The genetic code

In 1943, the first experimental evidence was presented which showed conclusively that DNA (rather than protein or some other substance) is the molecule which transmits genetic information from cell to cell. The structure of DNA indicated that this information must be coded in terms of the linear sequence of the four bases along the DNA chain; i.e., the sequence of bases must somehow determine the sequence of amino acids in proteins. Since there

are 20 different amino acids in proteins but only 4 different bases in DNA, a single base cannot be the code word for a single amino acid. The problem of cracking the genetic code thus became the problem of determining what specific sequence of bases in DNA corresponds to a specific amino acid in a protein. In 1961, the first code words in the genetic language were deciphered, and since then most of the others have been determined.

The genetic language is very similar in principle to a written language, like English, which consists of a set of symbols forming the letters of an alphabet. The letters are arranged in specific sequences to form specific words, and the words are arranged in linear sequences to form sentences. The 26 letters of the English language can be arranged in sequence to form words which vary in length from single-letter words like *a* and *I* to very large words like *antidisestablishmentarianism.* Words can then be assembled into sentences of varying lengths. Unlike English, the genetic language contains only four letters, corresponding to the four bases A, G, C, and T. The words in the genetic language are the base sequences which

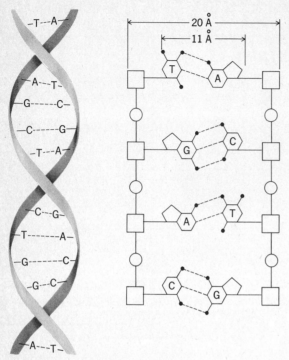

FIGURE 4-5
Base pairings between the two nucleotide chains account for the double-helical structure of DNA.

each codon is a three-letter word, a total of $4 \times 4 \times 4 = 64$ different code words can be formed (Table 4-1). With a total of 64 possible code words in the genetic language a new problem arises. Since there are only 20 different amino acids, what is the meaning of the remaining 44 words?

They might be nonsense words and not correspond to any amino acid. Thus, in English the three-letter words *cat, tag,* and *gag* all specify particular things, but the same four letters can also form three-letter "words" that are nonsense, such as *cta, atg,* and *gga.* When the genetic code was first being deciphered, it appeared as if many of the three-letter combinations were indeed nonsense, but as more is learned, we find that few if any of the 64 possible code words are nonsense words. The evidence now indicates that several different codons may specify the same amino acid. Thus, the codons CGA, CGG, CGT, and CGC all specify the same amino acid, alanine. The same situation is found in English, where several different words can be used to specify the same object, such as *house, home, dwelling, residence.* Codes in which several different code words describe the same subject are called *degenerate codes.* The genetic code does not have a unique set of code words and thus is degenerate. Later we shall consider some of the implications of a degenerate genetic code when we discuss genetic mutation.

Although most of the 64 codons now appear to specify particular amino acids, a few codons carry a

are the words for specific acids. Each word in the genetic language is only three letters long, i.e., consists of a sequence of three bases. Thus, word size is constant rather than variable. The three-letter code words are arranged in a specific sequence in DNA, and this sequence specifies the amino acid sequence in a single protein molecule. The total sequence of bases which specify the sequence of amino acids in a single protein chain forms the gene corresponding to that specific protein. Thus, a gene is equivalent to a sentence in English. The entire collection of genes in a single cell is equivalent to a book composed of many sentences.

The code word (three-base sequence) for a single amino acid is known as a *codon,* and the codon sequence in a gene therefore determines the sequence of amino acids in a protein. Since there are 20 different amino acids, the genetic code must have at least 20 different codons. How can the four letters of the DNA alphabet A, G, C, and T be arranged to form at least 20 different words, or codons? If each codon were a two-letter word (two-base sequence), it would be possible to form only $4 \times 4 = 16$ different words (Table 4-1). However, since

TABLE 4-1
Codons formed from the four bases of DNA arranged in sequences of one, two, and three bases

Singlet code (4 words)	Doublet code (16 words)				Triplet code (64 words)			
					AAA	AAG	AAC	AAT
					AGA	AGG	AGC	AGT
					ACA	ACG	ACC	ACT
					ATA	ATG	ATC	ATT
					GAA	GAG	GAC	GAT
					GGA	GGG	GGC	GGT
A	AA	AG	AC	AT	GCA	GCG	GCC	GCT
G	GA	GG	GC	GT	GTA	GTG	GTC	GTT
C	CA	CG	CC	CT	CAA	CAG	CAC	CAT
T	TA	TG	TC	TT	CGA	CGG	CGC	CGT
					CCA	CCG	CCC	CCT
					CTA	CTG	CTC	CTT
					TAA	TAG	TAC	TAT
					TGA	TGG	TGC	TGT
					TCA	TCG	TCC	TCT
					TTA	TTG	TTC	TTT

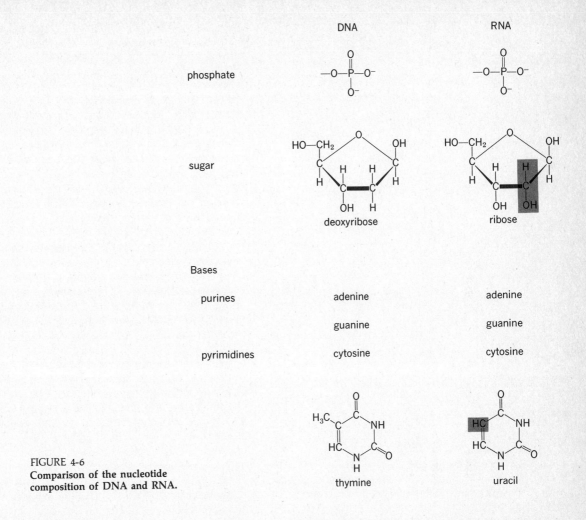

FIGURE 4-6
Comparison of the nucleotide composition of DNA and RNA.

different type of information. A single DNA molecule contains a number of genes specifying a number of different proteins, but the molecule consists of a continuous unbroken sequence of the four different bases. How does a cell determine where a particular gene begins and ends? Some codons appear to perform the function of punctuation marks in the genetic message, designating the beginnings and endings of genes.

Protein synthesis

Messenger RNA

Although the DNA molecules are found within the cell nucleus, almost all protein synthesis occurs in the cytoplasm. Coded information is transferred from DNA to the protein-synthesizing sites in the cytoplasm by means of RNA molecules, known as *messenger* RNA (mRNA), synthesized on the surface of DNA, where the sequence of nucleotides in DNA determines the sequence of nucleotides in mRNA.

Just as proteins are assembled from individual amino acids, the nucleic acids are assembled from individual nucleotides. The nucleotides are synthesized by a series of enzyme-mediated reactions. Separate biochemical pathways are followed for the synthesis of the sugar and the different bases, which are then assembled to form nucleotide triphosphates. We have already encountered one type of nucleotide triphosphate, namely, the energy-carrying molecule adenosine triphosphate (ATP), which consists of the base adenine, the sugar ribose, and three phosphate groups. Thus, in addition to

During the synthesis of mRNA, the bonds between these pairs in DNA are broken, and the two chains partially unwind and separate. Bonds are then formed between the bases of the free nucleotide triphosphates and the bases in one of the separated DNA chains. Thus, DNA acts as a mold, or template, to order the sequences of bases in RNA. The base adenine in the free nucleotide pairs with the base thymine in DNA, and the base uracil in the free nucleotide pairs with the base adenine in DNA. In a similar fashion, free nucleotide C pairs with G in DNA, and free nucleotide G pairs with C in DNA, the result being a sequence of bases in RNA which is a mirror image of the base sequence in DNA. Thus, if the DNA codon contains the base sequence C–G–T, the corresponding codon sequence in messenger RNA is G–C–A.

Once the appropriate free nucleotide triphosphates have been base-paired to the corresponding bases in DNA, the nucleotides are joined to each other by the enzyme RNA-polymerase, which causes pyrophosphate (P–P) to be split off from the nucleotide triphosphate in the process of linking one nucleotide to the next, to form the sugar-phosphate backbone of mRNA (Fig. 4-8). This enzyme is active only in the presence of DNA and does not link the free nucleotide triphosphates together (in what would be a random sequence) in its absence. The enzyme appears to move along the DNA strand, linking one nucleotide at a time into the growing mRNA chain.

Since DNA contains two chains of nucleotides, one of which is the mirror image of the other, but mRNA consists of only a single chain of nucleotides, are two mRNA molecules formed, corresponding to the two chains in DNA? Present evidence indicates that only a single strand of DNA is used to form a single mirror-image strand of mRNA. Synthesis is initiated on one of the two strands of DNA in which there are exposed clusters of pyrimidine nucleotides; this appears to act as the code specifying the point at which mRNA synthesis begins.

The formation of mRNA does not involve making a mirror image of the entire DNA molecule. Only selected portions of the DNA molecule are copied at any one time; the portion copied by mRNA contains the genetic information of a single gene or a small number of genes in sequence, because only the portion of the DNA double helix at the region where mRNA is to be synthesized becomes unwound. Later we shall discuss some of the mechanisms controlling initiation of mRNA synthesis and how, at any one time, only certain genes in DNA lead to protein synthesis while others are inactive.

In summary, the synthesis of mRNA on the surface of DNA leads to the formation of a molecule having a

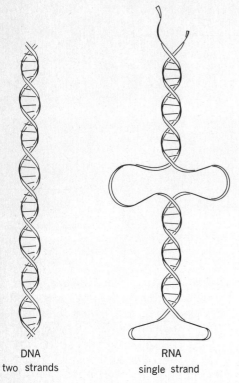

DNA
two strands

RNA
single strand

FIGURE 4-7
Double-stranded structure of DNA and single-stranded structure of RNA.

being an energy carrier involved in many reactions in the cell, ATP is also one of the nucleotide building blocks used in the synthesis of RNA, the others being GTP, CTP, and UTP. These nucleotides are usually synthesized by the transfer of energy from ATP to the diphosphate forms of the nucleotides in reactions of the general type

$$ATP + \begin{cases} UDP \\ GDP \\ CDP \end{cases} \longrightarrow ADP + \begin{cases} UTP \\ GTP \\ CTP \end{cases}$$

Thus, within the cell there is a general pool of nucleotide triphosphates, which are the building blocks for RNA synthesis. How are these free nucleotides assembled into a linear sequence so that the RNA molecule contains some of the coded information that is present in DNA? Recall that in DNA two chains of nucleotides are joined together by base pairing in such a way that the base adenine (A) is paired with the base thymine (T) and the base guanine (G) is paired with the base cytosine (C).

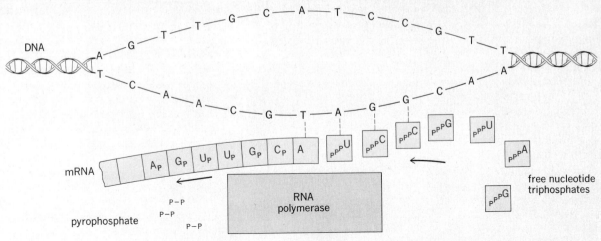

FIGURE 4-8
Synthesis of mRNA on the surface of one of the two strands
of a DNA molecule through base pairing between free
ribonucleotides and the bases in DNA.

linear sequence of bases which is the mirror image of the base sequence in DNA. Since a three-base sequence in DNA forms a codon for one amino acid, the corresponding mirror-image three-base sequence in RNA also forms a codon for one specific amino acid; e.g., the DNA codon, C–G–T, becomes the RNA codon, G–C–A. There has been a one-to-one transfer of information from DNA to mRNA.

Ribosomes

Once a molecule of mRNA has been synthesized on a DNA molecule, it leaves the nucleus, possibly by way of the pores in the nuclear membrane, and enters the cytoplasm. There the mRNA becomes attached to small particles known as *ribosomes* (because they were found to contain ribonucleic acid), which are the sites where the physical process of protein assembly occurs. In Fig. 4-9, the ribosomal particles can be seen scattered throughout the cytoplasm of the cell. Some ribosomes are attached to the surface of membranes, whereas others are free in the cytoplasm. The membranes to which the ribosomes are attached form a network of interconnected sheets within the cytoplasm and are known collectively as the *endoplasmic reticulum*. When seen under the electron microscope, thin cross sections of these membranes look like long tubules or flattened sacs. Their actual structure in three dimensions, however, is that of a loose interconnected network of folded membranes (Fig. 4-10).

There are two types of endoplasmic reticulum: the rough or *granular endoplasmic reticulum,* which has ribosomes attached to its surface, and the smooth or *agranular endoplasmic reticulum,* which does not have ribosomes on its surface. Both types can be seen in Fig. 4-9. From observations of a number of different types of cells which have different protein-synthesizing capacities, it is possible to make the following general statements about the distribution of ribosomes in a cell: (*1*) Cells that synthesize large amounts of protein have numerous ribosomes, whereas cells that synthesize small amounts have relatively few. (*2*) Cells that synthesize proteins eventually to be secreted from the cell have most of their ribosomes attached to the membranes of the endoplasmic reticulum. (*3*) Rapidly growing cells synthesizing large amounts of protein to be used for the internal functioning of the cell rather than for export have large numbers of free ribosomes and relatively few membrane-bound ribosomes.

The ribosomes, not the membranes of the endoplasmic reticulum, are the sites of protein synthesis. Proteins synthesized by ribosomes attached to the surface of the endoplasmic reticulum are released into the lumen of the reticulum and are eventually secreted from the cell. The smooth endoplasmic reticulum is not involved in the process of protein synthesis. The enzymes associated with a number of biochemical pathways have been found in the agranular reticulum. The enzymes for fatty acid and steroid synthesis appear to be located in these mem-

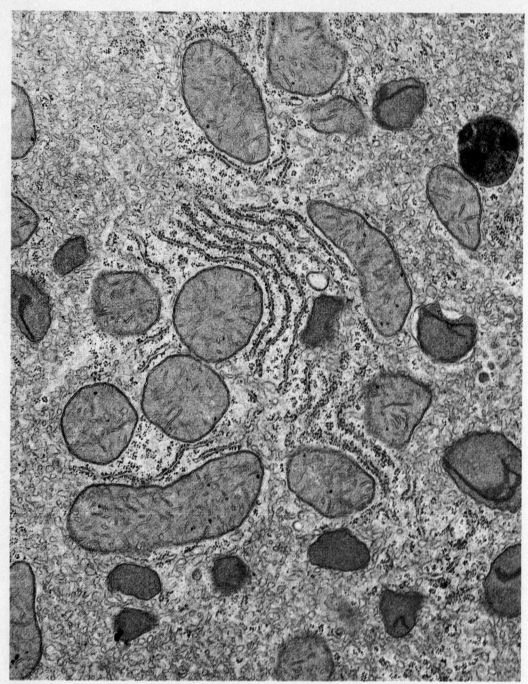

FIGURE 4-9
Electron micrograph of a portion of the cytoplasm of a
hamster liver cell, showing free ribosomes and ribosomes
attached to the surface of the endoplasmic reticulum. (*From*

*Keith R. Porter, in T. W. Goodwin and O. Lindberg (eds.), "Biological
Structure and Function," vol. 1, Academic Press, Inc., New York,
1961.)*

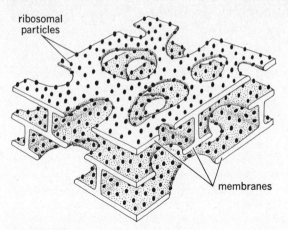

FIGURE 4-10
Three-dimensional structure of endoplasmic reticulum.

branes, and cells which produce large quantities of ste-roid hormones are found to have an extensive agranular reticulum. The enzymes responsible for metabolizing various foreign organic molecules, such as drugs, also appear to be located in the agranular endoplasmic reticulum.

The ribosomal particles are about 230 Å in diameter, are composed of RNA and protein, and are sometimes referred to as ribonucleoprotein particles (RNP). The ribosomal RNA (rRNA) accounts for about 80 percent of the total RNA in the cell and has a different function from mRNA. Each ribosomal particle actually consists of two subunit particles of unequal size. The larger subunit, known as the 60S particle, accounts for about two-thirds of the weight of the total ribosome; the small subunit particle is known as the 40S particle. Both contain RNA and protein.

After leaving the nucleus the mRNA becomes attached to the 40S subunit of the ribosome (Fig. 4-11). While the amino acids are being joined to make a protein chain, the strand of mRNA is moved along the ribosome. Some of the proteins within the ribosomal structure act as enzymes which catalyze the reactions linking the amino acids together as the ribosome moves along. Other ribosomal proteins are involved in the physical events which cause the movement of mRNA across the ribosome, as well as in the supply of energy for these processes.

The exact function of rRNA remains one of the major problems in protein synthesis. Ribosomes isolated from one species of animal can be mixed, under appropriate conditions, with mRNA isolated from another spe-

cies of animal and the system is able to synthesize proteins in a test tube. The proteins synthesized by such a system are always characteristic of the animal from which the mRNA was obtained. Thus, rRNA does not appear to carry coded information for the synthesis of specific types of proteins. rRNA appears to be necessary for the organization of the ribosomal proteins in a pattern of association within the ribosome essential for their interaction during protein synthesis.

Transfer RNA

How do individual amino acids become arranged in the proper sequence to form a protein molecule? By themselves free amino acids do not have the ability to bind to the bases in RNA. Orientation of the amino acids on the mRNA molecule involves yet a third class of RNA known as *transfer RNA* (tRNA).

Rather than reacting directly with mRNA, a free amino acid first becomes chemically bonded to one end of a molecule of tRNA. There are a number of different tRNA molecules, each of which is specific for a different amino acid. Each tRNA molecule contains within its nucleotide sequence a specific three-base codon which is able to pair with a specific codon in mRNA. The binding of a specific tRNA containing one specific amino acid to its corresponding codon in mRNA provides the mechanism for arranging the amino acids in sequence. The three-base sequence in tRNA which pairs with a codon in mRNA is known as an anticodon.

The combination of an amino acid and a molecule of tRNA occurs in the cytoplasm and is mediated by a specific enzyme known as amino acyl synthetase. This reaction utilizes energy released in the splitting of ATP to form the chemical bond between the amino acid and tRNA:

$$\text{tRNA} + \text{amino acid} + \text{ATP} \longrightarrow$$
$$[\text{tRNA}]\text{—amino acid} + \text{AMP} + \text{PP}$$

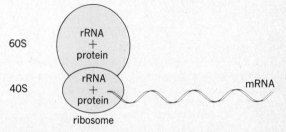

FIGURE 4-11
mRNA attaches to the 40S portion of the ribosomal particle.

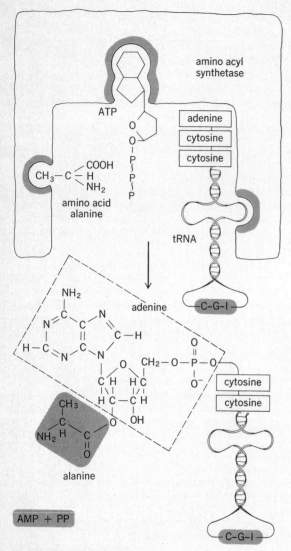

FIGURE 4-12
Diagrammatic representation of the binding sites on amino acyl synthetase, the enzyme responsible for linking specific amino acids to specific tRNA molecules.

There are as many different amino acyl synthetase enzymes as there are amino acids. Each enzyme has three binding sites, one for a specific amino acid, one for a specific type of tRNA, and one for ATP (Fig. 4-12). Joining a given amino acid to the correct tRNA molecule depends upon the specificity of the amino acyl synthetase enzyme; this in turn depends upon the configuration of

the active sites on the enzyme's surface, which must correspond to the shape of the specific amino acids and tRNA molecule.

The tRNA molecules have a molecular weight of about 25,000 and contain only about 80 nucleotides. Thus, tRNA is a relatively small molecule compared with the other types of nucleic acids. About 10 percent of the bases in tRNA are unusual in that they are different from the standard four bases A, G, C, and U. The unusual bases are thought to prevent regions of the tRNA molecule from forming intramolecular base pairings; therefore, sections of the molecule are exposed where extramolecular base pairing can occur. Such exposed regions may provide the sites of attachment to the amino acyl synthetase enzyme, mRNA, and rRNA.

The complete base sequence in some of the tRNA molecules has been established. In these molecules the base sequence corresponding to the anticodon is located in the middle of the molecule in one of the exposed loops. The anticodon for the amino acid alanine, C–G–I, actually includes one of the unusual bases, inosine (I). The region of the tRNA molecule containing the anticodon is able to undergo base pairing with the corresponding codon in mRNA (Fig. 4-13). For alanine, the mRNA codon is G–C–C, which pairs with the anticodon C–G–I in alanine tRNA. Thus, specific tRNAs which base-pair with specific codons in mRNA provide the mechanism for placing amino acids in the proper sequence along the mRNA chain.

Protein assembly

All the different component parts involved in the assembly of protein molecules have now been described. The final interactions between mRNA, ribosomes, and tRNA take place on the surface of the ribosomal particles. Synthesis is initiated by the attachment of one end of the mRNA to a ribosomal particle. This initial reaction appears to be quite specific since the coded message in mRNA is read in one direction, always beginning at the same end of the mRNA molecule, just as in English one reads a sentence from left to right, never from right to left. Protein synthesis begins at the free-amino end of the protein chain and proceeds toward the free-carboxyl end.

The various tRNAs, with their specific amino acids, base-pair through their anticodons with the corresponding codons in mRNA. Enzymes within the ribosomal particle then catalyze the formation of a peptide bond between two adjacent amino acids while they are still attached to their tRNA molecules. Thus, one end of the growing protein chain is always attached to a molecule

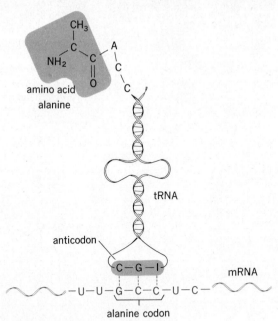

FIGURE 4-13
Base pairing between the anticodon region of a tRNA molecule with the corresponding codon region of an mRNA molecule.

of tRNA. The large 60S subunit of the ribosome has two binding sites for tRNA; one holds the tRNA that is also attached to the growing peptide chain, and the second holds the tRNA containing the next amino acid to be added to the peptide chain. With the formation of a peptide bond between the protein chain and the next amino acid, the initial tRNA is released from the mRNA and the peptide chain is transferred to the next tRNA. The ribosome now moves one codon space down the mRNA, making room for the binding of the next amino acid–tRNA molecule (Fig. 4-14). This process is repeated over and over as each amino acid is added in succession to the growing peptide chain. When the ribosome reaches the end of the mRNA or the end of a coded sequence for a single protein, the completed protein is released from the ribosome. It has been estimated that a single protein molecule containing 150 amino acids can be synthesized in about 60 sec.

The same strand of mRNA can be used more than once to synthesize several molecules of the same protein because the message in mRNA is not destroyed during protein assembly. In fact, the same strand of mRNA can

simultaneously synthesize several molecules of the same protein. As mRNA is moved across the ribosome during the process of assembling a protein, it may become attached to a second ribosome and begin the synthesis of a second molecule of the same protein. Thus, a number of ribosomes may be attached to the same strand of mRNA forming what is known as a *polyribosome*. Polyribosomes containing as many as 70 ribosomes attached to a single strand of mRNA have been observed with the electron microscope. Each ribosome has a growing peptide chain attached to it. Ribosomes near the beginning of the chain are associated with very short peptide chains representing the first few amino acids in the protein; ribosomes near the end are associated with a protein chain which is almost completed (Fig. 4-15). The steps leading from DNA to a completed protein are summarized in Table 4-2.

TABLE 4-2
Sequence of events leading from DNA to protein synthesis

Transcription of DNA to mRNA†
1 The two strands of the DNA double helix separate in the region of the gene to be transcribed.
2 Free nucleotide triphosphates base-pair with the nucleotide bases in one strand of DNA.
3 The nucleotide triphosphates are linked together by RNA polymerase to form a strand of mRNA containing a sequence of bases corresponding to the DNA base sequence.

Initiation of amino acid assembly
4 mRNA passes from the nucleus to the cytoplasm where one end of the mRNA molecule binds to a ribosome.
5 Free amino acids combine with their corresponding tRNAs in the presence of specific amino acyl synthetase enzymes in the cytoplasm.
6 Two amino acid–tRNA complexes bind to sites on the ribosome.
7 The three base anticodons in tRNA pair with the corresponding codons in mRNA attached to the ribosome.

Elongation of protein chain
8 One amino acid is transferred to the adjacent amino acid–tRNA complex.
9 The amino acid free tRNA is released from the ribosome.
10 A new amino acid–tRNA complex is attached to the vacated site on the ribosome.
11 mRNA moves one codon step along the ribosome.
12 Steps 8 to 11 are repeated over and over.

Termination of protein chain assembly
13 The completed protein chain is released from the ribosome when the termination codon in mRNA is reached.

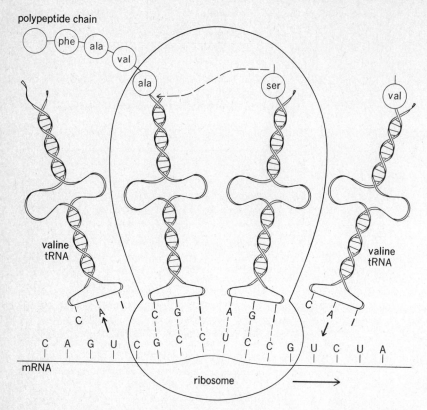

FIGURE 4-14
Sequence of events during the movement of a ribosome along a strand of mRNA.

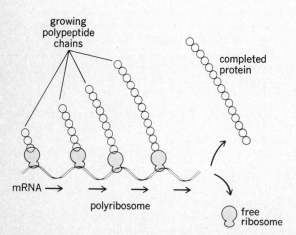

FIGURE 4-15
A polyribosome, showing the different stages of protein assembly as each ribosome moves along the strand of messenger RNA.

Protein secretion

Cells which secrete proteins are faced with the problem not only of synthesizing the protein but of transferring these large molecules, which cannot diffuse across cell membranes, to the outside of the cell. One of the characteristics of these cells is the presence of dense granules in the cytoplasm known as *secretory* or *zymogen (enzyme) granules*. These granules are surrounded by a membrane and contain high concentrations of protein (Fig. 4-16). Examination of a cell after it has been stimulated to secrete shows very few granules in the cytoplasm, suggesting that these granules are released during the process of secretion.

The Golgi apparatus In addition to the ribosomes and endoplasmic reticulum, a second membrane organelle, known as the *Golgi apparatus*, appears to be directly involved in the secretion of protein from the cell. In 1898 the Italian cytologist Camillo Golgi first observed this structure in nerve cells. It has since been found to be present in most cells. The Golgi apparatus consists of

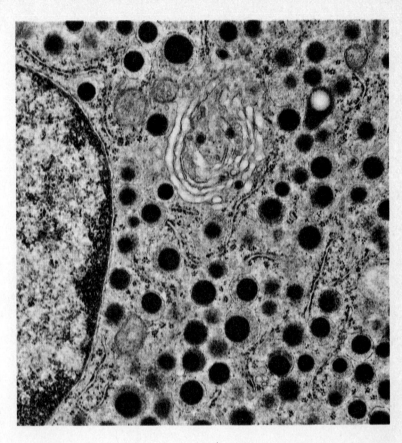

FIGURE 4-16
Secretory granules in the cytoplasm of a pancreatic cell. (*Courtesy of A. Like.*)

a series of flattened sacs and vesicles generally located near the nucleus of the cell (Fig. 4-17). The Golgi apparatus in secretory cells becomes larger during periods of intense secretory activity, suggesting that it plays a role in the secretory process.

Examination of numerous electron micrographs of gland cells at various stages of the secretory process has uncovered the following relationship between protein synthesis, the endoplasmic reticulum, Golgi apparatus, and protein secretion. Proteins that are to be secreted from the cell are synthesized on the surface of the ribosomes attached to the endoplasmic reticulum. The mRNAs for these proteins are coded so that they bind only to the ribosomes that are attached to the endoplasmic reticulum. The completed protein then passes into the lumen of the endoplasmic reticulum. As the protein accumulates in the endoplasmic reticulum, portions of the reticulum break away, forming small membrane-bound vesicles containing the synthesized protein. These

vesicles accumulate in the region of the Golgi apparatus where they fuse with other vesicles, forming the stacks of Golgi membranes. Within the Golgi apparatus, the protein solution within the vesicles becomes progressively more concentrated. This may occur by the removal of salt and water, leaving only a concentrated solution of proteins within the vesicles. One of the characteristics of secreted proteins, in contrast to intracellular proteins, is that they are usually glycoproteins containing small amounts of carbohydrate incorporated into their structure. The Golgi apparatus contains the enzymes responsible for the addition of these carbohydrate units to the protein. Thus the Golgi apparatus both concentrates and modifies the structure of the secretory products. The final product, as it leaves the Golgi apparatus, is neatly packaged in concentrated form and now becomes a secretory granule.

The mechanism of the release of these secretory granules varies in different types of gland cells. Usually

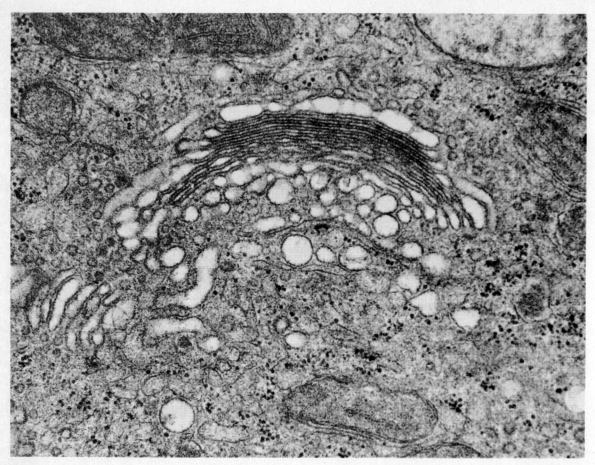

FIGURE 4-17
The Golgi apparatus. (*From W. Bloom and D. W. Fawcett,*
"Textbook of Histology," 9th ed., W. B. Saunders Company,
Philadelphia, 1968.)

the secretory granule migrates from the region of the Golgi apparatus to the cell membrane, where the membrane of the secretory granule fuses with the cell membrane. This may also be part of the pathway by which new membrane material is formed and added to the cell membrane. Figure 4-18 diagrams the sequence of steps from the synthesis of the secretory protein to its release from the cell.

The Golgi apparatus is also the site of formation of the cell organelles known as *lysosomes*. These vesicular bodies contain a variety of very potent digestive enzymes which are capable of chopping up almost any molecule with which they come into contact. These enzymes are normally prevented from digesting the cell itself because they are surrounded by a membrane which is impermeable to them. Once a cell has died, these potent enzymes are released locally, digesting cell debris. It now appears that some diseases, for example gout (an inflammation caused by the accumulation of uric acid in the joints), may be related to the release of lysosomal enzymes due to the selective disruption of the lysosomal membranes by certain chemical agents. As we shall see in Chap. 15, these lysosomes play an important role in the destruction of bacteria in the body. Bacteria, which are engulfed by special cells known as phagocytes, become fused with the lysosomes and are digested.

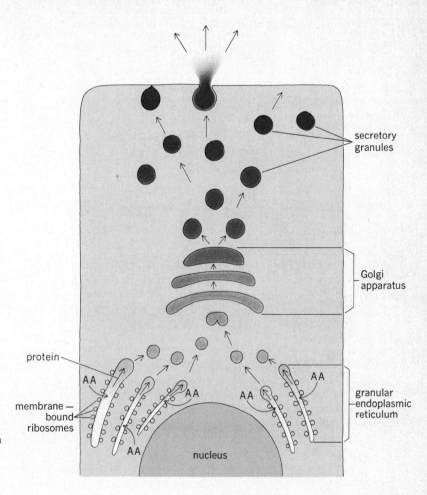

secretory
granules

Golgi
apparatus

granular
endoplasmic
reticulum

protein

AA

AA

AA

AA

AA

membrane —
bound
ribosomes

AA

nucleus

FIGURE 4-18
Pathway leading to the release of protein from a secretory cell. (AA; free amino acids assembled into proteins on ribosomes.)

Regulation of protein synthesis

Possible sites of regulation

The different types of cells in the body synthesize different proteins, and no one type of cell synthesizes all its proteins at the same rate. Cells thus have mechanisms which regulate both the types and rates of protein synthesis.

From our discussion of the mechanism of protein synthesis we can identify a number of different stages in the process at which protein synthesis might be regulated (Fig. 4-19):

1 If the ability to synthesize a specific mRNA on the surface of DNA were blocked, the protein corresponding to that particular gene would not be synthesized by the cell.

2 Regulation can also occur at the ribosome level by modifying the ability of mRNA to combine with the ribosome, the movement of the ribosome along mRNA, or the activity of the various ribosomal enzymes. Drugs are known which block protein synthesis at the ribosomal level or which disrupt the structure of the ribosome so that the message in mRNA is read incorrectly, giving rise to abnormal inactive proteins.

3 Although the mRNA molecule can be used over and over again to synthesize many molecules of the same protein, it is eventually destroyed and must be replaced with newly synthesized mRNA. Thus, by controlling the rate at which mRNA is destroyed relative to the rate at which it is synthesized, it is possible to regulate the amount of a particular protein being synthesized.

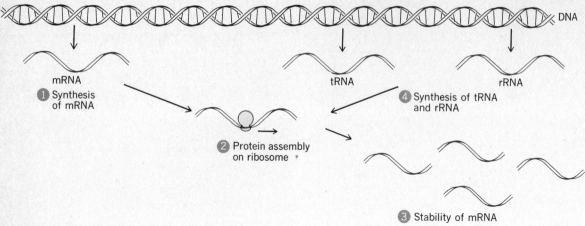

FIGURE 4-19
Possible sites at which the rate of protein synthesis may be regulated.

4 In addition to coding the sequence of amino acids in proteins, the DNA molecule also codes the base sequences in tRNA and rRNA. These molecules are synthesized on the surface of DNA by base-pairing in a manner similar to the synthesis of mRNA. Specific regions of the DNA molecule code for the synthesis of tRNA and rRNA, and these regions are associated with the nucleolus. Thus, a mechanism which regulates the rate at which tRNA and rRNA is synthesized might also affect the process of protein synthesis.

We have listed the sites at which the regulation of protein synthesis *may* occur. A brief discussion of enzyme induction and repression will serve to illustrate one class of these regulatory mechanisms. Further information can be found by consulting the references to Chap. 4 at the back of the book.

Induction and repression

The functions of DNA and its relation to protein synthesis are better understood in bacteria than in any other type of cell because they are a relatively simple type of cell than can be subjected to a wide variety of experimental techniques more easily than cells from higher organisms.

The aspect of protein synthesis that has been most extensively studied is the mechanism by which the synthesis of particular proteins can be turned on and off. If the type of bacterium known as *Escherichia coli* is grown in a nutrient medium containing the disaccharide lactose, an enzyme, galactosidase, can be isolated which cata-

lyzes the splitting of lactose into a molecule of glucose and a molecule of galactose:

$$\text{Lactose} \xrightarrow{\text{galactosidase}} \text{glucose} + \text{galactose}$$

Glucose and galactose are metabolized by the cell to provide energy and a source of carbon for the synthesis of other types of molecules, but in the absence of galactosidase the molecules of lactose are not broken down and cannot be used. If *E. coli* are grown in a nutrient medium that does not contain lactose, the cells are found to have very few molecules of galactosidase, about three molecules of enzyme per cell. If lactose is then added to the medium, a rapid synthesis of galactosidase takes place. In the presence of lactose, each cell has about 3,000 molecules of the enzyme. The presence of lactose has *induced* the synthesis of the enzyme required for its metabolism. Such a system has many advantages since the cell need not expend energy synthesizing an enzyme unless there is a substrate present for the enzyme to act upon.

Inhibition of enzyme synthesis in the presence of substrate is also observed in bacteria. The synthesis of the amino acid histidine, which is required for protein synthesis, involves a series of reactions catalyzed by 10 different enzymes. If *E. coli* are grown in a medium to which histidine has been added, the cells do not synthesize the 10 enzymes needed for histidine synthesis but rather utilize the histidine in the medium. When histidine is removed from the medium, the 10 enzymes are rapidly synthesized to provide the histidine needed for growth. In this example, it appears that the presence of histidine has *repressed* the synthesis of the enzymes utilized in its synthesis. This type of system is useful to a cell because the end product of a biochemical pathway may be able to regulate the synthesis of the enzymes in that pathway. As the concentration of the end product is increased, the synthesis of the enzymes in the pathway is reduced and the rate of product formation is decreased. This process is similar in effect to the end-product inhibition described in the preceding chapter, in which the end product of a biochemical pathway decreases the activity of a particular enzyme in the pathway by combining with the enzymes and altering the shape of the active site. The difference is that, in repression, the synthesis of a series of enzymes is prevented, whereas in end-product inhibition the activity of an enzyme that has already been synthesized is reduced. Figure 4-20 is a diagrammatic summary of induction, repression, and end-product inhibition.

In general, both induction and repression involve a number of enzymes simultaneously. Thus, in histidine repression, 10 different enzymes are simultaneously repressed by histidine. Genetic mapping of the genes which code for each of these different enzymes has shown that most of the genes are located next to each other in the DNA molecule and that one strand of mRNA is formed at this site which contains the information for the synthesis of all 10 enzymes. Repression of enzyme synthesis involves blocking the synthesis of this mRNA molecule. A collection of genes which can be induced or repressed as a unit is known as an *operon*.

Figure 4-21 diagrams the mechanism of enzyme induction found in bacterial cells. A repressor gene contains the coded information for the synthesis of a repressor protein molecule. The repressor protein is synthesized in the standard manner on the surface of a ribosome from mRNA synthesized on the DNA molecule at the site of the repressor gene. Once formed, the repressor protein can combine with the initial gene of an operon, known as the *operator gene*. The combination of the repressor protein with the operator gene prevents synthesis of the operon's mRNA. As long as the repressor is combined with the operator gene, enzyme synthesis is prevented since no mRNA can be formed.

The induction of enzyme synthesis is believed to involve the combination of the inducer molecule, such as lactose in the example above, with the repressor protein, thereby leading to inactivation of the repressor so that it is unable to combine with the operator gene. Since

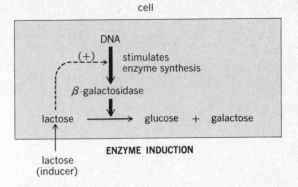

ENZYME INDUCTION

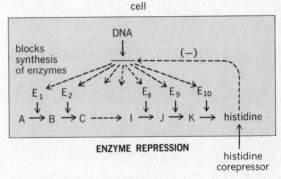

ENZYME REPRESSION

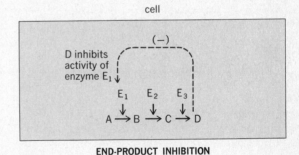

END-PRODUCT INHIBITION

FIGURE 4-20
Three mechanisms regulating enzyme activity. A plus sign indicates stimulation, and a minus sign indicates inhibition.

the operator gene is not inhibited by the presence of an active repressor molecule, it can initiate the synthesis of the mRNA. This mRNA then can initiate enzyme synthesis on the ribosomes. Thus, induction of enzyme synthesis really involves the inhibition of an inhibitor (the repressor molecule) usually present in the cell.

Enzyme repression by the end product of a biochemical pathway involves a similar set of interactions

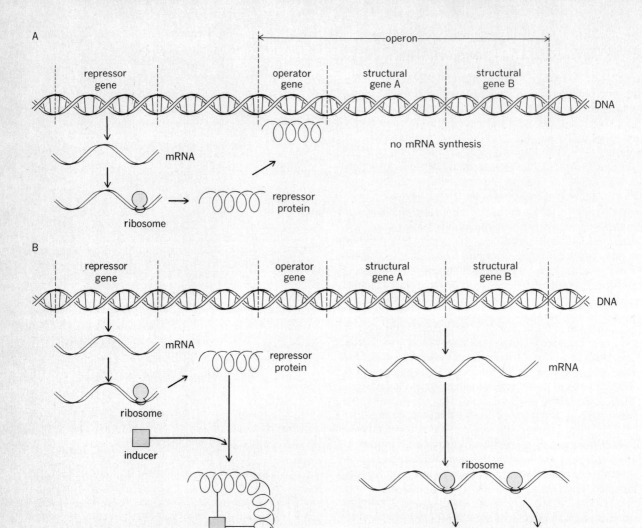

FIGURE 4-21
Enzyme induction. A No enzyme is formed in the absence of
an inducer molecule. B Enzyme is formed in the presence of
an inducer molecule due to the inactivation of the repressor
protein.

except that the repressor molecule synthesized by the
repressor gene is normally inactive. The end-product
molecule, histidine in the example above, combines with
the inactive repressor molecule, forming an active repressor molecule which is able to block the synthesis of
mRNA at the site of the operator gene and thus block
the synthesis of the corresponding enzymes.

Not all genes in DNA are subject to induction and
repression; many appear to lack operator genes and are
thus continuously active in the synthesis of mRNA independent of the presence or absence of inducers or repressors. However, the proteins synthesized by these
genes may be subject to other types of control, such as
end-product inhibition.

Many of the enzymes in human cells have been shown to undergo induction and repression in the presence or absence of certain substrate molecules, and there is some indication that one of the mechanisms by which hormones in the body influence the metabolism of cells is by interactions with DNA that alter protein synthesis by the cell. Thus, the mechanism of action of some hormones may be similar to that of inducers and repressors.

Cell development

The development of the human body from a single fertilized egg cell involves cell growth, cell replication, and cell differentiation. A cell takes in raw materials from its environment and synthesizes the various components of its structure. If the rate at which these components are synthesized exceeds the rate of degradation, the cell grows. In Chap. 2 we discussed the limitation placed upon cell size by the rate at which raw materials can diffuse from the external environment into the cell. The slow rate of diffusion limits the size to which a single cell can grow and still obtain adequate amounts of the raw materials required to maintain its structure and function. Therefore, growth of cells can continue only in association with repeated cell divisions, during which a parent cell splits into two daughter cells, each initially about half as large as the parent.

Starting with a single fertilized egg cell, the first division produces two cells. When these daughter cells divide, they produce two cells, giving a total of four cells. When these four cells divide, they produce a total of eight cells, and these eight produce sixteen cells, ... Thus, starting from a single cell, four division cycles will produce 16 cells (2^4), 10 division cycles will produce $2^{10} = 1,024$ cells, and 20 division cycles will produce $2^{20} = 1,048,576$ cells. If the development of the human body proceeded at a constant rate by the repeated cycle of cell division and growth, it would require only about 46 division cycles to produce the 40 trillion cells in the adult body, starting from a single cell.

There is more to development, however, than just cell growth and cell division. Not all the cells in the body have identical chemical compositions or functions, in spite of the fact that all are descended from a single cell. Somewhere along the sequence from a single cell to the fully developed organism, the process of cell differentiation occurs, during which certain cells develop the contractile ability of muscle cells while other cells differentiate into specialized nerve cells or bone cells, etc. How does the information present in the DNA molecules of a single fertilized egg cell manage to regulate the development of such a wide variety of cell types? We shall return to this question in a subsequent section.

Replication of DNA

When a cell divides, forming two new independent cells, the genetic information stored in its DNA must be duplicated and a copy passed on to each of the daughter cells. DNA is the only molecule in a cell which is able to form a duplicate copy of itself without requiring a set of instructions from some other component in the cell. Thus, mRNA molecules can be formed only in the presence of DNA, which provides the information for the ordering of its base sequence. Likewise, protein can only be formed if mRNA is present to provide the information necessary for ordering the sequence of amino acids in the protein, and the other molecules of a cell are formed by the action of enzymes resulting from protein synthesis.

The replication of DNA is, in principle, very similar to the process whereby mRNA is synthesized on the surface of DNA through base pairings between the bases in DNA and a collection of free nucleotides. Figure 4-22 illustrates the replication of DNA. Each of the two strands of nucleotides in the DNA double helix is joined together by base pairings between a purine and pyrimidine base. During DNA replication, the two strands of DNA separate, and the exposed bases in each strand can act as a template with which free nucleotides can base-pair, forming a complementary strand. The free nucleotides used to form DNA are triphosphates containing one of the four bases, A, G, C, or T, and the sugar deoxyribose. Once the free nucleotides have base-paired with the bases in each of the two strands of DNA, the enzyme DNA polymerase joins the nucleotides together in a reaction which splits off a molecule of pyrophosphate (P–P). In contrast to the synthesis of mRNA, however, both strands of DNA act as templates for the synthesis of new DNA molecules. The end result is two identical molecules of DNA, each containing one strand of nucleotides present in the DNA molecule before duplication and one strand newly synthesized using the old strand as a template. When two identical molecules of DNA have been formed, one copy is passed on to each of the two daughter cells during cell division. Thus, each daughter cell receives the same set of genetic instructions as was originally present in the parent cell.

Cell division

The ability of a cell to divide and thereby reproduce itself is a major characteristic of most (but not all) living cells.

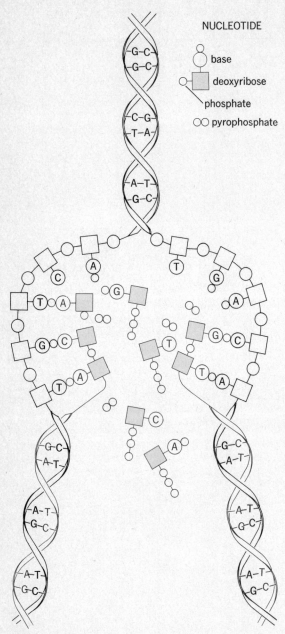

NUCLEOTIDE

○ base
□ deoxyribose
| phosphate
○○ pyrophosphate

FIGURE 4-22
Replication of DNA involves the pairing of free nucleotides with the bases of each DNA strand, giving rise to two new DNA molecules, each containing one old and one new nucleotide strand.

We have described above how a molecule of DNA is able to reproduce itself. A cell, however, is much more than a single molecule; it is a complex organized collection of membrane structures and molecules. The division of such a complex structure into two roughly equivalent parts involves marked alterations in the structure and metabolism of the cell. Aside from the morphological description of the various stages of the division process as visualized with a light or electron microscope, we know surprisingly little about the underlying molecular events which occur during division or how the process is initiated and regulated. We therefore limit our discussion of cell division to a brief description of the various stages of the process and focus our attention on the net result, which is the duplication, packaging, and distribution of DNA molecules to the daughter cells.

Although the time between cell divisions varies considerably in different types of cells in the body, the more rapidly growing cells divide about once every 24 hr. During most of this period there is no visible evidence that the cell will undergo a division at some later time. In a "24-hr" cell, visual changes in cell structure begin to appear 23 hr after the last division, and 1 hr later the cell has divided into two cells. The period between the end of one division and the beginning of the next division is known as *interphase*. Since the actual process of cell division lasts only about 1 hr, the cell spends most of its time in the interphase, and most of the cell properties described in this book are properties of interphase cells.

One very important event related to the subsequent cell division does occur during the interphase, namely, the replication of DNA. A few hours after cell division, in the early stages of the interphase, the process of DNA replication begins and proceeds over a period which may last 10 to 12 hr. At the end of this period each DNA thread has been duplicated, but the duplicate threads remain joined together over a small segment of their length in the region known as the *centromere* (Fig. 4-23B).

Just prior to cell division the duplicated DNA threads become highly coiled and condense to form rod-shaped bodies known as *chromosomes*. Isolated chromosomes consist of DNA, protein, and small amounts of RNA. The combination of nucleic acid and protein forms a nucleoprotein known as *chromatin*. The exact function of the protein associated with DNA is still unclear; it has been implicated as a possible factor in controlling gene activity by blocking the synthesis of mRNA at various sites along the DNA molecule, and it may also play a role in the coiling of chromatin threads to form chromosomes. The total amount of DNA found in a single human cell during interphase, when the chromatin threads are in the

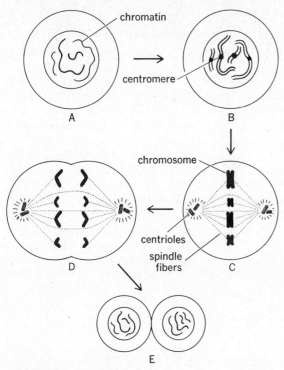

FIGURE 4-23
Chromosome condensation and separation during mitotic cell division.

As division proceeds, each chromosome separates at the centromere, and the two identical segments move toward the opposed centrioles (Fig. 4-23D). The spindle fibers act as if they were pulling the chromosome segments toward the poles although the actual mechanism of chromosome movement is still unknown. As the chromosome segments move toward opposite poles of the cell, the cell begins to constrict along a plane perpendicular to the spindle apparatus and constriction continues until the cell has been completely separated into two halves. Following division the spindle elements dissolve, the chromosomes become uncoiled, and a new nuclear membrane is formed in each daughter cell (Fig. 4-23E).

This process of cell division is known as *mitosis* (Greek *mitos*, thread) and is the mechanism for distributing equal amounts of DNA from cell to cell during most cell divisions in the body. The reproductive cells, which give rise to the egg cells and the sperm cells, undergo a slightly altered form of division, known as *meiosis* (Greek: diminution). In man, meiosis in the reproductive cell reduces the number of chromosomes distributed to the daughter cells from 46 to 23. Meiosis consists of two cell divisions in succession. Before the first division there are 46 chromatin threads, each of which undergoes duplication. During the condensation of the chromatin threads into chromosomes, the duplicated maternal and paternal chromatin threads pair with each other, forming a four-stranded chromosome (Fig. 4-24). Thus, there are 23 four-stranded chromosomes instead of 46 two-stranded chromosomes. Although paired together, the two strands of a maternal chromosome are not attached to the two strands of the paternal chromosome by a centromere. When the cell divides, the two maternal strands pass into one of the daughter cells and the paternal strands into the other. Since each of the 23 four-stranded chromosomes becomes attached to the spindle fibers at random, one chromosome may have its maternal strands oriented toward one pole of the cell and the next chromosome has its maternal strands oriented toward the opposite pole (Fig. 4-24C). Thus, when the maternal and paternal strands separate, the daughter cells, in general, receive a mixture of maternal and paternal chromosomes. It would be extremely improbable that all 23 maternal chromosomes would end up in one cell and all 23 paternal chromosomes in the other.

Since each chromosome in the daughter cells still has two strands, the next cell division proceeds without a DNA replication stage and is very similar to a normal mitotic division (except that only 23 chromosomes are involved) in which one of the duplicate strands is passed

uncoiled state, could form a thread about 70 in. long, a distance which is about 100,000 times the diameter of a typical cell; thus the coiling and condensation of chromatin to form chromosomes prior to cell division provide a convenient way of transferring these long threads to daughter cells.

The first sign that a cell is going to divide is the appearance of the chromosomes in the nucleus, formed by the coiling of the extended chromatin threads. Each chromosome consists of two identical chromatin threads attached at the centromere. Human cells except the male and female reproductive cells contain 46 chromosomes.

As the duplicated chromatin threads begin to coil, forming the chromosome, the nuclear membrane breaks down and a new structure appears in the cell, the *spindle apparatus,* which consists of a number of thin fibers. Some of these fibers pass unbroken from one side of the cell to the other between two small cylindrical bodies known as *centrioles* (Fig. 4-23C). Other fibers pass from the centrioles and are attached to the centromere region of a chromosome.

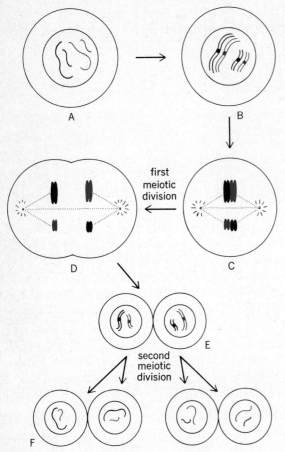

FIGURE 4-24
Separation of chromosomes during meiotic cell division.

the same type of function. If one of the genes, either the maternal or paternal, dominates over the other corresponding gene, said to be *recessive,* only a single type of protein is synthesized, corresponding to the dominant gene.

The random distribution of maternal and paternal chromosomes during meiosis provides the basis for much of the variability in the genetic constitution of the offspring from a single set of parents. Over 8 million (2^{23}) different combinations of maternal and paternal chromosomes can result from the distribution of the 23 chromosomes during meiosis. Since each parent can produce any one of 8 million different sex cells, the random combination of a sperm cell with an egg cell may produce a fertilized egg cell having any one of over 70 trillion possible combinations of maternal and paternal chromosomes. This is only the minimum number of possible combinations of genetic material since segments of the maternal and paternal chromosomes may exchange with each other in a process known as *crossing over,* which occurs during chromosomal pairing before the first meiotic division. This forms chromosomes containing both maternal and paternal genes in the same chromosome. Thus all the genes initially present in a paternal or maternal chromosome do not necessarily pass into the same cell during meiosis. Crossing over greatly increases the number of possible combinations of maternal and paternal genes which end up in a given egg or sperm cell.

Cell differentiation

Mitotic cell division results in the distribution of identical sets of chromosomes to each of the daughter cells. Thus, each cell in the body, with the exception of the reproductive cells, contains an identical set of DNA molecules and, thereby, genetic information. How, then, is it possible for one cell to become a muscle cell and synthesize muscle proteins while another cell containing the same molecules of DNA differentiates into a nerve cell and synthesizes a different set of proteins?

Cell differentiation in a multicellular organism is an extremely complex process which involves not only the development of special properties in individual cells but also the organization and distribution of various types of cells to form tissues and organs such as the stomach, the heart, an ear, or a finger. All the information required to form these structures appears to be present in the fertilized egg cell, but we have only the barest hints as to how these transformations are brought about.

That different types of cells in the same body have different properties and different chemical compositions suggests that different combinations of genes are active

on to each of the daughter cells. The end result of the two meiotic divisions is the production of four cells having half the number of chromosomes as the original parent cell. Each of the four daughter cells contains a mixture of maternal and paternal chromosomes. When the egg is fertilized by the sperm, the two sets of 23 chromosomes are brought together in a single cell and the combined total of 46 are transferred to each successive cell during cell division. Thus, the nonreproductive cells in the body contain two sets of chromosomes, maternal and paternal, which means that each cell contains two sets of corresponding genes. Thus, a cell contains a maternal and a paternal gene both carrying the information required for the synthesis of a particular type of protein. If both genes are able to initiate protein synthesis, then two different proteins may be formed by the cell, both of which have

in the different cells. Thus, the genes which contain the information responsible for the synthesis of muscle proteins are able to synthesize mRNA in muscle cells and, thereby, muscle proteins. These same genes are also present in nerve cells but are unable to form the corresponding mRNA and thus do not synthesize muscle proteins. Other genes are active in the nerve cell which are not active in muscle cells. These observations suggest that the difference between a nerve and a muscle cell depends upon which genes in the cell are able to form mRNA and, thereby, the corresponding proteins. The problem of cell differentiation is thus related to the general problem of regulating protein synthesis. Certain genes appear to be "turned on" or "turned off" during cell differentiation. Mechanisms similar to those employed in enzyme induction and repression may be involved. The proteins associated with DNA in the chromatin threads of the nucleus may inhibit various genes by preventing access to the DNA surface, where mRNA can be formed. To understand the complex process of cell differentiation a great deal more must be learned about the processes which regulate protein synthesis in the cells of higher organisms.

Mutation

In order to form the 40 trillion cells of the human body a minimum of 40 trillion individual cell divisions must occur, and the DNA molecules in the original fertilized egg cell must be replicated at least 40 trillion times. If a secretary typed the same letter 40 trillion times, one would expect to find some typing errors. It is not surprising, then, to find that errors also occur during the duplication of the DNA molecule, which result in an altered sequence of bases and a change in the genetic message. What is surprising is that DNA can be replicated so many times with relatively few errors. Any alteration in the genetic message carried by DNA is known as a *mutation.*

Mutations may result from a multitude of factors, all of which ultimately lead to an alteration in the base sequence in DNA and thus a change in the genetic information. The simplest type of mutation occurs when a single base is incorrectly inserted at the wrong position in DNA. For example, the base sequence C–G–T forms the DNA codon for the amino acid alanine; if the guanine (G) is replaced by adenine (A) in this sequence, it becomes C–A–T, which is the codon for valine.

It is during the replication of DNA, when free nucleotides are being incorporated into new strands of DNA,

that incorrect base substitution is most likely to occur. Even in the absence of specific agents which increase the likelihood of mutations, mistakes in copying DNA do occur, so that the mutation rate is never zero. The major factors in the environment which increase the mutation rate are certain chemicals and various forms of ionizing radiation, such as x-rays, cosmic rays, and atomic radiation. Most of these factors cause the breakage of chemical bonds, so that incorrect pairings between bases occur or the wrong base is incorporated when the broken bonds are reformed.

Cells also possess mechanisms for protecting themselves against certain types of mutation. For example, if an abnormal base pairing occurs in DNA, such as C with T (C normally pairs with G), a special set of enzymes will cut out the segment of one strand of DNA containing the abnormal base T, allowing the normal strand to resynthesize the deleted segment by normal base pairing. Certain diseases in which the cells of the skin tend to become cancerous when exposed to the ultraviolet radiation in sunlight appear to result from a lack of the DNA repair enzymes in these cells, leading to a high rate of mutation.

Mutations can also occur in which large sections of DNA are deleted from the molecule or in which single bases are added to, or deleted from, the molecule. Such mutations may cause the loss of an entire gene or group of genes or may cause the misreading of a large sequence of bases. Figure 4-25 shows the effect of removing a single base in a sequence on the reading of the genetic code. Since the code is read in sequences of three bases, the removal of one base alters not only the codon containing that base but causes a misreading of all subsequent bases by shifting the reading sequence. Addition of an extra base would also cause a similar phase-shift misreading.

When a single base is incorrectly substituted in a codon, only a single amino acid in the protein coded by the gene containing the mutation is affected, but because the genetic code is degenerate, having several different codons representing the same amino acid, a mutation in a single codon does not always alter the type of amino acid coded. We saw that substituting A for G in the alanine codon C–G–T produced the valine codon C–A–T. If, however, the mutation caused cytosine (C) to be substituted for thymine (T), the new codon C–G–C is one of several that correspond to alanine, and the protein formed by the mutant gene will not be altered in amino acid sequence in spite of the change in base sequence.

Assume that a mutation has altered the codon so that it now codes for a different amino acid, say the

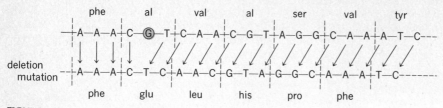

FIGURE 4-25
A deletion mutation caused by the loss of a single base, G, in the DNA sequence and the resulting misreading of a large number of codons.

alanine C–G–T to valine C–A–T mutation. What effect does such a mutation have upon the cell? The effect depends upon both the type of gene and where in the gene the mutation has occurred. Although proteins are composed of many amino acids, the properties of a protein often depend upon a very small region of the total molecule, such as the active-site region of an enzyme. If the mutation alters an amino acid not in the region of an active site, there may be little or no change in the properties of the protein. On the other hand, if the mutation alters an amino acid in the active-site region of the protein, a marked change in the properties of the protein may occur. If the protein is an enzyme, a mutation may render it totally inactive or change its specificity for substrate. The mutated enzyme may even catalyze an entirely different type of reaction. Thus, the alteration of a single amino acid in a protein may lead to a protein whose activity is unchanged, increased, or decreased or a protein with entirely new properties.

Let us assume that the mutation leads to a protein enzyme that is totally inactive. If the enzyme is in the Krebs cycle or glycolytic pathway, the loss of the enzyme may lead to the death of the cell since the cell is unable to catalyze the reactions leading to the production of ATP. On the other hand, the enzyme may be involved in the synthesis of an amino acid. If the cell exists in a medium containing preformed amino acids, the cell's functioning will not be impaired by the loss of the ability to synthesize its own amino acids from raw materials. In this case the mutation has led to a totally inactive enzyme, but since even the normal enzyme carries out no critical function in the cell, its loss does not affect the life of the cell. The effect of such a mutation, however, becomes apparent if the cell is deprived of preformed amino acids. There the enzyme performs a critical function, and the loss of its activity by mutation may lead to the death of the cell. All these effects may result from altering one base in the DNA molecule, which leads to an alteration in one amino acid in one specific type of protein.

Mutation may have any one of three effects upon a cell: (*1*) It may cause no noticeable change in the cell's functioning; (*2*) it may modify cell function but still be compatible with cell growth and reproduction; and (*3*) it may lead to the death of the cell. Any one of these three types of mutation can occur in any of the 40 trillion cells in the human body, but the consequences of mutation depend upon the type of cell in which it occurs. If a liver cell in an adult undergoes a mutation that causes it to function abnormally or even die, the effect upon the total organism is usually negligible since there are thousands of similar liver cells performing similar functions and the loss of one cell does not affect the overall functioning of the liver or the total organism. In contrast, since all the cells in the body are descended from a fertilized egg cell through repeated cell division, any mutation occurring during the early stages of development which does not lead to the immediate death of the cell is passed on to most of the cells in the developing organism. If the mutation is in an egg or sperm cell, all the cells descended from them inherit the mutation. Thus, mutations in the reproductive cells of the body do not affect the individual in which they occur but do affect the children produced by the individual.

Many types of abnormal functions in the body are the result of genetic mutations which are passed on from generation to generation. Inherited diseases of this type are often due to a single gene, which either fails to produce an active protein or produces an abnormally active protein. Such inherited diseases have been termed *inborn errors in metabolism.* An example is phenylketonuria, a disorder that can lead to one form of mental retardation in children. Because of a single abnormal enzyme, the victim is unable to convert the amino acid phenylalanine into the amino acid tyrosine at a normal rate. Phenylalanine is therefore diverted into other biochemical pathways in large amounts, giving rise to products which apparently interfere with the normal activity of the nervous system. These products, excreted in large amounts in the urine, account for the name of the disease. The symptoms of the disease are prevented if the content of phenylalanine in the diet is restricted during childhood, thus preventing the accumulation of the products formed from phenylalanine. As our knowledge of biochemistry, genetics, and disease increases, more and more diseases are being found that are passed on from generation to generation and are the result of abnormally functioning enzymes.

Although mutations may alter the properties of the protein coded by the mutant gene, all such mutations are not necessarily harmful to the cell in which they occur In fact, mutation is the mechanism by which organic evolution occurs. Mutations may alter the activity of an enzyme in such a way that it is more rather than less active, or they may introduce an entirely new type of enzyme activity into a cell. If an organism carrying such a mutant gene is able to perform some function more effectively than an organism lacking the mutant gene, it has a better chance of surviving and passing the mutant gene on to its descendants. If the mutation produces an organism which functions less effectively than organisms lacking the mutation, the organism is less likely to survive and pass on the mutant gene. This is the principle underlying *natural selection*. Although any one mutation, if it is able to survive in the population, may cause only a very slight alteration in the properties of a cell, given enough time, a large number of small changes can accumulate to produce very large changes in the structure and function of an organism. The evolution of life on earth from the first cell to man has proceeded over a period of some 3 billion years. During this long period of time the environment has selected those mutations which best enable the cells carrying them to survive and propagate.

Viruses

Our discussion of DNA, protein synthesis, and cell growth would not be complete without a brief mention of viruses and their relation to cell function and disease. The structure of the most elementary virus particles consists of nothing more than a molecule of nucleic acid surrounded by several molecules of protein. Isolated virus particles are inert; they do not move, metabolize, reproduce themselves, or show any of the properties attributed to living systems, but when a virus particle enters a living cell, it is able to reproduce thousands of copies of itself. A virus is thus a replicating molecular unit which is dependent upon living cells to provide the energy and molecular machinery necessary for its own duplication.

Viral nucleic acid carries the coded information for a limited number of proteins. Some of these genes code for the structural proteins in the virus particle; others code for protein enzymes involved in the synthesis of the viral nucleic acid. The nucleic acid in some virus particles is RNA rather than DNA, thus providing an exception to the rule that genetic information is always carried by DNA molecules.

The study of virus multiplication has also led to a revision of our concepts of the flow of genetic information from DNA to RNA to protein. An enzyme has been found in certain RNA-containing viruses which will catalyze the synthesis of DNA, using the virus RNA as a template. This is known as *reverse transcription*. The DNA formed can then be incorporated into the cell's pool of DNA and passed from cell to cell during DNA replication accompanying cell division. Transcription of this viral DNA that was formed from viral RNA leads to the formation of new virus particles by the transcription process similar to that of forming messenger RNA from normal DNA.

When a virus infects a cell, it is generally only the nucleic acid portion of the virus which actually enters the cell. Once there, the viral nucleic acid uses the biochemical machinery of the cell to synthesize virus-type proteins and to replicate itself, using nucleotides and energy sources provided by the cell. The synthesis of a complete virus particle proceeds in several steps. Shortly after the virus has entered a cell, the cell begins to synthesize protein enzymes that are involved in the replication of viral nucleic acid, the necessary information coming from the genes in the viral nucleic acid. Thus, the virus provides the source of the mRNA, but the cell's own ribosomes and raw materials are used to synthesize viral proteins. Shortly after the viral enzymes appear in the cell, the viral nucleic acid begins to replicate. Accompanying the increase in viral nucleic acid is the synthesis of virus structural proteins, which have also been coded by information provided by the viral nucleic acid. Finally, the viral proteins and viral nucleic acid combine to form complete new virus particles.

The effect of virus replication on cell function depends upon the type of virus. Some virus particles, upon entering a cell, multiply very rapidly, eventually killing the cell. During the multiplication of the virus, the cell's ability to synthesize its own proteins is decreased as the virus takes over control of the protein-synthesizing machinery of the cell and directs it toward the synthesis of virus-related proteins. In contrast, other types of virus particles may replicate very slowly. The viral nucleic acid may even become associated with the cell's own DNA molecules, replicate along with the cell DNA, and be passed on to the daughter cells during cell division. Such a virus may propagate along with the cell over many cell generations without having any obviously deleterious effect upon the cell. A change in the cell's physical or chemical environment may cause the virus to suddenly cease to be dormant and enter a stage of rapid multiplication which kills the cell.

Much of our information about the properties of nucleic acids and the control of protein synthesis has come from studying the multiplication of these simple virus particles. It is hoped that understanding the principles of virus multiplication will enable scientists to develop better methods of preventing and curing virus infection in man, which causes such diseases as measles, mumps, chicken pox, polio, the common cold, and perhaps even certain forms of cancer.

Cancer

The interrelationships of all the topics discussed in this chapter are perhaps best illustrated by the abnormal type of cell growth known as *cancer,* which is an uncontrolled multiplication of cells in the body.

Different types of body cells normally divide and grow at different rates: the cells lining the intestines divide about once every 24 hr, whereas adult nerve and muscle cells are unable to divide at all. Even the same type of cell may be able to multiply rapidly under certain conditions but slowly under others. Liver cells normally divide infrequently and at a rate just sufficient to replace those which have been damaged and die; but if a section of the liver is surgically removed, the remaining cells divide rapidly and multiply until the amount of tissue that was removed has been replaced. Many types of cells which do not divide or divide infrequently in the body multiply rapidly when removed and cultured in an artificial medium outside the body. For example, cells removed from embryonic chicken heart tissue have continued to grow and divide in a test tube for over 30 years, whereas if they had remained in the chicken, they would have stopped dividing when the heart was fully formed. Cell differentiation is often accompanied by loss of the ability to divide, as shown most clearly in the case of adult nerve and muscle cells. Although the body provides many examples of the regulation of cell division and growth, we know almost nothing about the mechanisms which control cell division.

A cell becomes cancerous when its growth is no longer regulated by whatever processes control normal cell growth. The major abnormality of cancer is not that the cells grow rapidly but that they do not stop growing. Any cell in the body has the potential of becoming a cancer cell. Although different factors can induce cancerous growth in normal tissues, the underlying mechanism that transforms a normal cell into a cancerous cell is still unknown. Ionizing radiation and various chemical agents can lead to the production of cancer cells; for example, some of the chemical components in cigarette smoke have been implicated in lung cancer. Viruses have been isolated that induce cancer in some species of animals, although no virus specifically related to human cancer has yet been isolated. It has been suggested that the initial event leading to the production of a cancer cell may be a mutation or some alteration in the cell's mechanisms which control the translation of genetic information into protein synthesis. The variety of agents implicated in the production of cancer may all be operating at a common point within the cell.

Whatever the mechanisms are which regulate the growth of normal cells in the body, they cannot control the growth of a cancer cell. As the cancer cell continues to grow and divide, it forms a mass of growing cells known as a *malignant tumor.* As the tumor grows, it invades the surrounding tissues, disrupts the structure and function of organs, and eventually leads to the death of the organism. If cancer is detected in the early stages of its growth, the cancer cells can often be removed by surgery. One of the properties of cancer cells is their lack of adhesiveness to other cells, which allows them to break away from the parent tumor and spread by way of the circulatory system to other parts of the body (metastasis), where they continue to grow, forming multiple tumor sites. After the cancer cells have spread to different parts of the body, surgery is impossible as a means of removing all the cancer cells, thus, the importance of detecting cancer in its early stages, when it can be effectively treated.

The problem of finding a cure for cancer is the problem of finding some agent which will stop the growth of cancer cells without damaging normal cells. Various drugs and ionizing radiation, which tend to damage rapidly dividing cells in preference to nondividing cells, have been used in the treatment of cancer, but these agents also damage normal cells, particularly the rapidly dividing cells of the blood-forming tissues and the intestinal tract. Such treatments can prolong the life of the patient but seldom result in a permanent cure of the disease. Perhaps as we learn more about the molecular mechanisms which regulate normal cell growth and the translation of genetic information into protein synthesis, we shall be able to control the growth of cancer cells.

BIOLOGICAL CONTROL SYSTEMS PART

2

The internal environment and homeostasis

5

The first section of this book described certain basic processes common to individual cells. Indeed, these processes, taken together, define the term *living*. The emergence of cell aggregations, i.e., multicellular organisms, during the course of evolution created new problems of a kind never encountered by single-celled organisms. This chapter attempts to define these problems and outlines the general mechanisms by which they are overcome. Our analysis rests heavily upon a concept known as the *internal environment,* which cannot really be understood without at least a brief description of that initial biological event—the origin of life.

The origin of life

In 1924 the Russian biochemist A. I. Oparin proposed the idea that life began by a progressive series of reactions leading to the formation of organic molecules and their organization into living cells. Crucial to his hypothesis was the belief that physical and chemical conditions of the earth when life arose were very different from what they are now. He concluded that given certain chemical and physical properties of the earth, and enough time, it was inevitable that the inanimate world would give rise to living systems through the operation of physical and chemical laws. Oparin's hypothesis now appears valid and is supported by most recent findings from many different scientific disciplines.

The earth began approximately 5 billion years ago, probably as an extraordinarily hot mass of atoms, hydrogen being the most abundant. It is likely that the atoms gradually became sorted out on the basis of their differing weights; heavier elements sinking to the center of the gaseous ball, lighter elements, such as silicon, forming the next shell, which was surrounded by the lightest elements, hydrogen, oxygen, nitrogen, and carbon.

In time, the lighter elements began to combine to form molecules; the most abundant were probably H_2, O_2, H_2O, N_2, NH_3, and methane, CH_4. Oxygen also reacted readily with elements in the middle layer to give oxides such as silicon dioxide, SiO_2, and salts such as calcium phosphate, $CaHPO_4$. In this manner, the free oxygen, O_2, was rapidly removed from the atmosphere.

As the gas ball began to cool, the middle shell congealed to form the earth's crust whereas the outer, lighter layers remained in the gaseous state. Thus, the earth's crust consisted of a collection of metallic oxides, phosphates, sulfates, and mineral salts such as sodium chloride and potassium chloride, and was covered with a gaseous atmosphere of H_2, N_2, H_2O, CH_4, and NH_3.

As the earth cooled still more, the water vapor in the atmosphere condensed and the rains came—millions upon millions of years of rain, forming the oceans, lakes, and rivers. Dissolved in these oceans were large quantities of the atmospheric gases. Slowly the chemical composition of the oceans changed as wind and water eroded the earth's surface and rivers fed by the rains carried dissolved mineral elements down to the sea. The oceans became salty, and their salt concentration continued to increase as the water evaporated from them and then returned to the earth as rain, washing still more minerals into the sea.

The stage was now set for the formation of simple organic molecules. Water was the solvent, the basic elements of nitrogen, hydrogen, oxygen, and carbon were supplied by the gases, and the energy sources were the sun's radiation and discharges of lightning. The formation of organic molecules occurred solely by chance, but given the environmental conditions described above, it was virtually impossible for it not to happen. This has been brilliantly demonstrated in experiments in which methane, ammonia, hydrogen, and water were placed in test tubes and subjected to various forms of radiation and electric discharges. After several days the tubes were found to contain a wide variety of organic molecules, including acetic and lactic acids, sugars, and amino acids (indeed the amino acids synthesized in these experiments are the very ones most abundant in most proteins today).

Furthermore, under appropriate conditions, single molecules can form stable larger molecules; e.g., heat alone forms stable polypeptides from a mixture of single amino acids. Many such polypeptides must have been formed, but only the most stable remained intact. The step to ATP was also probably relatively easy; phosphates were present in the sea and the organic components, ribose and adenine, have both been synthesized in the electric-discharge experiments described above. With the development of self-replicating molecules, DNA and RNA, all the major organic components were present.

The story up to this point is referred to as *chemical evolution*. The next events are far less clear, namely, how these organic components, dissolved in the seas, came together to form a cell. The thought that enough of each critical molecular species came together by random movements and in some haphazard manner were bound together into a self-reproducing structural entity is enough to make the mind boggle. But time is the key—millions upon millions of years, probably 2 billion years from the earth's birth to the first cell. One important hypothesis concerning the possible emergence of discrete particles from the seawater mixture of chemicals depends upon a recently demonstrated characteristic of the polypeptides synthesized in test tubes; when treated with hot water, these *proteinoids* separate out of the solution as membrane-bounded microspheres, which manifest differential permeability to different solutes and actually can split ATP (Fig. 5-1). Moreover, electron micrographs reveal double-walled membranes not too unlike the cell membranes of living cells. This is only one of several ingenious hypotheses for the development of cells, and the question remains highly controversial.

The development of discrete unicellular organisms is the first stage of *biological evolution*. These first cells must have been anaerobic since there was no free oxygen in the atmosphere. Only after evolution of chlorophyll and photosynthesis did the green plants liberate enough oxygen into the atmosphere to permit the survival of aerobic animal organisms.

For all these early unicellular organisms, the sea was home and sustenance. Organic molecules were taken up from the sea as nutrients, and waste products were eliminated into the vast oceans where they were so diluted that their toxicity was eliminated. Less obvious, but just as important, was the total dependence of the cell upon the inorganic composition of the sea surrounding it. The interposition of a membrane between the cell interior and the sea made it possible for the cell to evolve a chemical composition differing greatly from that of sea-

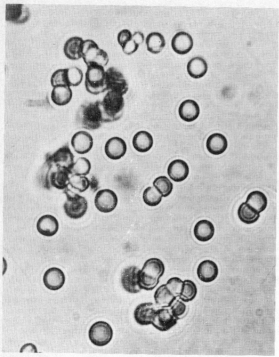

FIGURE 5-1

Phase-contrast micrograph of proteinoid, membrane-bonded microspheres. (*Courtesy of S. W. Fox and R. McCauley, Institute of Molecular Evolution, University of Miami.*)

water. Cytoplasmic composition thus depended upon the membrane's permeability characteristics and carrier-mediated transport systems. For example, the internal sodium concentration was made lower than that of sea-water by development of the membrane active-transport system for sodium; it must have had immense survival value for the cell because the process is found almost universally in animal cells.

Thus, the cell membrane became the link between the sea and the cytoplasm of the unicellular organism. Since the properties of the membrane had evolved in relationship to the specific composition of the sea surrounding the cells, its capacity to maintain the cytoplasmic composition depended upon the stable physico-chemical composition of its seawater environment. Moreover, the molecular structure of the cell membrane is itself highly sensitive to changes in the chemical composition of the fluid bathing it; e.g., a decrease in ex-tracellular calcium causes marked alterations in cell-membrane permeability. For these reasons, a stable inorganic extracellular environment is an absolute necessity for cell function and survival.

The internal environment

In the course of evolution, multicellular organisms made their appearance. At first, they were probably no more than a collection of unicellular organisms adhering to each other. But with time, the aggregates developed cell specialization and true complex multicellular organisms emerged. As they became larger, the interior cells became physically separated from the external environment, the sea. How could these cells obtain oxygen and nutrients? How could they eliminate their waste products? How could their cell membranes function in the absence of a stable fluid surrounding them? If the cell could not reach the sea, the sea would have to be brought to the cell. The animal would have to internalize the sea and create an inner environment so that each cell, regardless of its location, would be bathed by a fluid with which it could exchange oxygen, nutrients, and wastes and which had the correct physicochemical composition for stable membrane and cell function.

Each cell of a complex multicellular organism, such as man, is surrounded by a fluid known as *extracellular fluid*, i.e., fluid outside the cell. Claude Bernard, the great nineteenth-century French physiologist, first clearly enunciated the concept that extracellular fluid constitutes the immediate environment for the cells bathed by it. In other words, this fluid, the body's *internal environment*, acts as the medium for exchange of nutrients and wastes and provides the stable physicochemical environment required for membrane and cell function.

In the earliest and simplest multicellular organisms, the extracellular fluid was doubtless the sea itself, perco-lating through the intercellular spaces. With further complexity, mechanisms evolved which allowed the animal to produce its own extracellular fluid instead of using sea-water per se. Evolution, acting over millions of years, has resulted in the existence today of extracellular fluids which differ considerably from seawater and from each other, but in virtually all animals, the evolution of ex-tracellular fluid from seawater is evident from its composition. The most striking parallel is the high sodium concentration in the two solutions. Indeed, it was this similarity that was largely responsible for the early conclusion that life evolved in the sea.

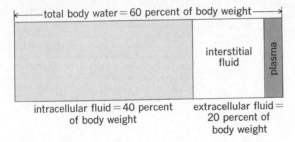

FIGURE 5-2
Body-fluid compartments.

Body-fluid compartments

As mentioned in Chap. 1, water is by far the most abundant component of animals and in man constitutes approximately 60 percent of his total body weight (Fig. 5-2). There is considerable variation between individuals due primarily to the amount of fat in the body. Since fat tissue contains little water, the percentage of total body weight which is water is lower in a fat person (as low as 40 percent) and higher in a thin person (as high as 80 percent).

Body water can be classified by location as inside cells (*intracellular fluid*) and outside cells (*extracellular fluid*). The extracellular fluid is the internal environment of the body and constitutes approximately 20 percent of body weight. Usually about twice as much water is found inside cells; i.e., 40 percent of body weight is intracellular fluid.

The extracellular fluid is further divided between two so-called *compartments,* or general locations. Approximately 80 percent of the extracellular fluid surrounds all the body's cells, and because it lies between cells and tissues, it is known as intercellular or, more often, *interstitial fluid.* The remaining 20 percent of the extracellular fluid constitutes the fluid portion of the blood, or *plasma.* The plasma is continuously circulated by the action of the heart to all parts of the body. Thus, the plasma is the dynamic component of the extracellular fluid. As will be seen in Chap. 9, the plasma exchanges oxygen, nutrients, wastes, and other metabolic products with the interstitial fluid as the blood passes through the capillaries of the body. In this manner, the interstitial fluid bathing the cells is continuously refreshed.

Because of the capillary exchanges between plasma and interstitial fluid, solute concentrations are virtually identical in the two fluids, except for protein. With this major exception, the entire extracellular fluid may be considered to have a homogeneous composition.

In contrast to the extracellular fluid, the composition of intracellular fluid varies considerably from cell to cell. However, certain aspects of cell composition are at least qualitatively similar between cells and, just as importantly, differ greatly from extracellular fluid. The major positive ion of intracellular fluid is potassium, in contrast to the dominant position of sodium in the extracellular fluid. Magnesium is also present in much higher concentration in cells. The chloride concentration of cell fluid is much lower than that of extracellular fluid. The major reasons for these marked differences in ionic concentrations between intracellular and extracellular fluid are the active-transport systems and transmembrane potentials (Chaps. 2 and 7).

Homeostasis

Bernard not only recognized the existence of the internal environment but appreciated its vital role. "It is the fixity of the internal environment which is the condition of free and independent life. . . . All the vital mechanisms, however varied they may be, have only one object, that of preserving constant the conditions of life in the internal environment." Led to these conclusions, in large part, by his studies of glucose and the liver, Bernard found that the liver releases glucose into the blood when the blood glucose concentration is low. He reasoned, correctly, that the glucose concentration of the blood, and therefore of the entire extracellular fluid, was being kept constant by the activity of the liver. He then went on to formulate the general concept of an internal environment.

The ideas embodied in the quotation are convenient handles by which to grasp a multitude of specialized activities carried out by the tissues and organs of complex multicellular organisms. The concentrations of oxygen and carbon dioxide, of organic nutrients and wastes, of the inorganic ions, the temperature—all these must remain relatively unchanged in the body fluids. Virtually every activity of the body in some way contributes to the maintenance of this stability: the liver acts as a metabolic factory, adding or removing organic molecules, as needed; the lungs take in oxygen and eliminate carbon dioxide at rates required to keep the pressures of these gases constant in the arterial blood; the gastrointestinal tract takes into the body ingested water, nutrients, and salts; the kidney eliminates just the right amount of wastes, water, and salts; and so on down the long list of bodily activities. Each cell of the body contributes, in its own way, to the survival of the total organism by helping to maintain the stable conditions required for life in the fluid bathing them all.

This concept of the maintenance of a stable internal environment was further elaborated and supported by the American physiologist W. B. Cannon, who emphasized that such stability could be achieved only through the operation of carefully coordinated physiological processes which he termed *homeostatic*. The activities of tissues and organs must be regulated and integrated with each other in such a way that any change in the internal environment automatically initiates a reaction to minimize the change. *Homeostasis* denotes the stable conditions which result from these compensating regulatory responses. Some changes in the composition of the internal environment do occur, of course, but the fluctuations are minimal and are kept within narrow limits through the multiple coordinated homeostatic processes, descriptions of which constitute the bulk of the remaining chapters.

Concepts of regulation and relative constancy have already been introduced in the context of a single cell. In Chap. 3, we described how metabolic pathways within a cell are regulated by the principle of mass action and by changes in enzyme activity so as to maintain the concentrations of the various metabolites within the cell. Chapter 4 described the control of protein synthesis and the mechanisms by which the genetic apparatus maintains a constant species line. In man, these basic intracellular primitive regulators still remain, so that each individual cell exhibits some degree of self-regulation, but the existence of a multitude of different cells organized into specialized tissues, which are further combined to form organs, obviously imposes the need for overall regulatory mechanisms to coordinate and integrate the activities of all cells. For this, intercellular communication over relatively long distances is essential. Such communication is accomplished by means of nerves and the blood-borne chemical messengers known as *hormones*.

The mechanisms by which these two communications systems operate are the subject of the next two chapters, but their overall role and the basic characteristics of homeostatic processes can be appreciated only in terms of *control systems*, to which we now turn.

General characteristics of control systems

We shall define a homeostatic, or control, system as a collection of interconnected components which functions to keep a physical or chemical parameter of the body relatively constant.

Let us first analyze a nonbiological control system

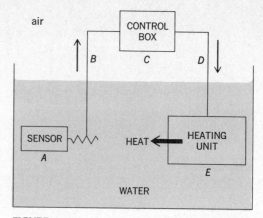

FIGURE 5-3
Components of a control system for regulating the temperature of a water bath.

(Fig. 5-3) designed to maintain the temperature of a water bath at approximately 30°C despite fluctuations in room temperature from 25 to 10°C. Since the water temperature is always to be higher than the room temperature, there is a continuous loss of heat to the room from the water. Moreover, the lower the room temperature, the greater this heat loss. Accordingly, the water must be continuously heated in order to offset the loss, and the degree of heating must be altered whenever the room temperature changes. This adjustment of heat input to heat loss so that water temperature remains approximately 30°C is the job of the control system.

The first system component, known as a *sensor*, is the temperature-sensitive instrument A. It generates an electric current the magnitude of which is inversely proportional to the water temperature; in other words, the higher the temperature, the less the current flow. (This current is always so small that its magnitude per se does not affect the water temperature.) The current flows through wire B into the control box C constructed in such a way that the amount of current flowing out of the box in wire D is directly proportional to that entering along B. The current from the box is fed to the heating unit E within the water bath. Its activity and therefore the amount of heat it produces per unit time is directly proportional to the strength of the signal to it, i.e., the current flow in D.

We fill the bath with water at room temperature, close the switches, and allow the system to operate (Fig. 5-4). Current generated in the sensor A controls the

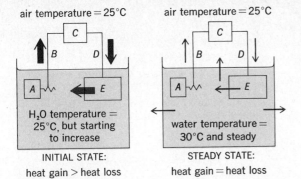

INITIAL STATE:
heat gain > heat loss

STEADY STATE:
heat gain = heat loss

FIGURE 5-4
Effects of filling a water bath with water at room temperature and allowing the system to operate. Shown are the initial state, i.e., just after the water was added, and the ultimate steady state achieved. Note the differences in output of the sensor and the heating unit in the two states.

output of the heating unit by way of the control box. Because of the initially low water temperature, the magnitude of current flow from A is large and the heating unit is running full blast. This heats the water rapidly, but as the water temperature rises, two opposing events occur: (*1*) More heat is lost from the bath to the room and (*2*) the signal from A decreases and results in a decreased input to the heating unit, thereby decreasing the amount of heat it produces. The system ultimately stabilizes at a particular water temperature when heat loss to the room exactly equals heat gain from the unit. At this point, the system is said to be in a *steady state;* input equals output, and the temperature remains steady. Actually, as shown in Fig. 5-5, there is always some oscillation around the steady-state temperature because of the time required for the heating unit to heat up or cool down.

The steady-state temperature is determined, in large part, by the characteristics of the sensor and the control box C, since these components are what determine the output of the heating unit. If we had chosen a heat-sensitive sensor which generates only half as much current at any given temperature as A does, the input to the control box would always be less, the output from the control box would be less, and the heating unit would always be generating less heat. Therefore, the steady-state temperature of the bath would be lower. Similarly, by altering the transforming function, i.e., the relationship between current in and current out, of the control box, we alter the steady-state temperature ultimately reached. In any control system, the actual steady

state, or so-called set point of the system, depends upon the characteristics of the individual components of the system. Our components were chosen to achieve a set point of 30°C.

Most important is that this type of system resists any changes from the set point. In our example the control system automatically prevents any significant deviation of the water temperature from the steady state established. Suppose that after steady state (30°C) has been reached, the room temperature is suddenly lowered (and kept low) so that the loss of heat from water to room is increased. This loss unbalances heat loss and heat gain, and the water temperature falls. But the decrease in water temperature immediately increases the current generated in A; therefore, more current flows out of the control box C and the heating unit increases its activity, thereby raising the water temperature back toward its original value. A new steady state will be reached when heat loss once again equals heat gain, both having been increased by a proportionate amount. What is the new steady-state temperature at this point? If the system is extremely sensitive to change, the temperature will be only very slightly below what it was before the room temperature was lowered *but it cannot be precisely the same.* Compensation is incomplete because the new steady state depends upon the maintenance of an increased heat production to balance the increased heat loss, and this increased heat production is due to the increased signal coming from A. The reason A generates more current is the slightly lower water temperature. If the temperature actually returned completely to normal, the signal from A would return to normal, heat production

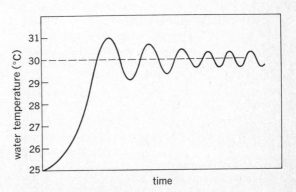

FIGURE 5-5
Change in water temperature over time during operation of the system shown in Fig. 5-4.

would return to normal, and the water temperature would immediately decrease as heat loss again became greater than heat gain (recall that the room temperature is being maintained lower than normal). Thus, control systems of this type cannot absolutely prevent changes from occurring in the physical or chemical variable being regulated, but they do keep such changes within very narrow limits, dependent upon the sensitivity of the system.

We can now summarize the basic characteristics of a control system in general terms. There must be a component which is sensitive to the variable being regulated and which changes its output as the variable changes. There must be a continuous flow of information from this sensor to an integrating control box, from which, in turn, the so-called command signal flows to an apparatus which responds to the signal by altering its rate of output (heat, in our example).

There remains one more concept which we have described but not named: *feedback*. The ultimate effects of the change in output by the system must somehow be made known (fed back) to the sensor which initiated the sequence of events. Our example illustrated perhaps the commonest form of feedback: When the water temperature is reduced, the sensor *A* detects this change and relays the information to the control box, which in turn signals the heating unit to increase its output; sensor *A* is "informed" of this change in output by the resulting rise in water temperature; accordingly, its current generation is again altered, which, in turn, results in an alteration of input to the control box and thereby the heating unit. The water temperature acts as a continuous link between the sensor and the heating unit (without feedback, the sensor's signal would be unrelated to the heating-unit output and the system would be unable to maintain constancy). This type of feedback, in which an increase in the output of the system results in a decrease in the input, is known as *negative feedback*. It clearly leads to stability of a system and is crucial to the efficient operation of homeostatic mechanisms.

Note that a major characteristic of negative-feedback control systems is that they *restore* the regulated variable toward normal after its initial displacement, but they cannot *prevent* the initial displacement. Suppose, however, that we add another component to our temperature-control system of Fig. 5-4, namely, a thermosensor on the *outside* of the water bath which can detect changes in room temperature. Now, when the room temperature is lowered, as in our example, this information is immediately relayed to the control box which causes the heating unit to increase its output. In this manner, additional

heat can be supplied to the water bath *before* the water temperature begins to fall. Thus, the use of the external sensor permits the system to *anticipate* a pending fall in water temperature and begin to take action to counteract the change before it occurs. This provides *feedforward* information which has the net effect of minimizing fluctuations in the level of the parameter being regulated. We shall see that the body frequently makes use of feedforward control in conjunction with negative-feedback systems.

There is, however, another type of feedback known as *positive feedback* in which an initial disturbance in a system sets off a train of events which increases the disturbance even further. Generally, such cycles occur explosively; they usually lead to instability and are quite uncommon. However, several important positive-feedback relationships occur in the body, blood clotting being an example.

Components of living control systems

The reflex arc

Homeostatic control systems in living organisms manifest virtually the same characteristics as those just described, but some of the terminology is different. The components are shown in Fig. 5-6 and are completely analogous to those of Fig. 5-3.

A *stimulus* is defined as a detectable change in the

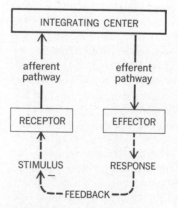

FIGURE 5-6
General components of a biological control system, the reflex arc. The response of the system has the effect of counteracting or eliminating the original stimulus. This phenomenon of negative feedback is emphasized by the minus sign in the feedback loop.

environment, such as a change in temperature, potassium concentration, pressure, etc. A *receptor* is the component which receives the stimulus, i.e., detects the environmental changes (it is identical to the sensor of the previous section). The stimulus acts upon the receptor to alter the signal emitted by the receptor, and this signal is the information relayed to the control box, or *integrating center*. The pathway between the receptor and the integrating center is known as the *afferent pathway*.

The integrating center usually receives input from many receptors, some of which may be responding to quite different types of stimuli. Thus, the output of the integrating center reflects the net effect of the total afferent input; i.e., it represents an integration of numerous and frequently conflicting bits of information.

The output of the integrating center is then relayed to the last component of the system, the device whose change in activity constitutes the overall response of the system. This component, known as an *effector,* is analogous to the heating unit of our previous example. The information going from the integrating center to the effector is like a command directing the effector to alter its activity. The pathway along which this information travels is known as the *efferent pathway*.

As a result of the effector's response, the original stimulus (environmental change) which triggered this entire sequence of events may be counteracted (at least in part), just as in our example the increased heat production by the heating unit caused the lowered water temperature to return toward normal. As the stimulus is diminished by the effector's response, the activity of the receptor is diminished, so that the flow of information from receptor to integrating center returns toward the original level and, in turn, the effector's activity is turned toward its previous rate. Thus, the counteracting of the stimulus by the effector's response constitutes the *negative feedback*.

Most biological control systems belong to the general category of stimulus-response sequences known as reflexes. A *reflex* is the sequence of events elicited by a stimulus, though we may be aware only of the final event in the sequence, the *reflex response* (pulling one's hand away from a hot stove, for example). The entire reflex, including the response, often occurs without any awareness on the part of the person. The pathway mediating the reflex is known as the *reflex arc*, and its components are, as described above:

1 Receptor
2 Afferent pathway
3 Integrating center

4 Efferent pathway
5 Effector

Sometimes the term reflex is restricted to situations in which the first four components are all parts of the nervous system. However, the afferent and efferent information can be carried by nervous or hormonal pathways. In any case, two different components must serve as afferent and efferent pathways; thus the input to, or output from, the integrating center may both be neural, but they must be two different nerve fibers. Of course, they can also both be hormonal, or one neural and the other hormonal. Depending on the specific nature of the reflex, the integrating center may reside in either the nervous system or an endocrine gland.

Finally, we must identify the effectors, the cells whose outputs constitute the ultimate responses of the reflexes. Actually, most cells of the body act as effectors in that their activity is subject to control by nerves or hormones. There are, however, two specialized tissues, muscle and gland, which comprise the major effectors of biological control systems. Muscle cells are specialized for the generation of force and movement. Glands consist of epithelial cells specialized for the function of *secretion,* an overall process which usually involves (*1*) selective uptake of substances by the gland cell, (*2*) chemical modification of these substances in the cell, and (*3*) selective release of the finished products as the gland's secretion.

To summarize, most biological control systems function to keep a physical or chemical parameter of the body relatively constant. One may analyze any such system by answering a series of questions: (*1*) What is the parameter (blood glucose, body temperature, blood pressure, etc.) which is being maintained constant in the face of changing conditions? (*2*) Where are the receptors which detect changes in the state of this parameter? (*3*) Where is the integrating center to which these receptors send information and from which information is sent out to the effectors, and what is the nature of these afferent and efferent pathways? (*4*) What are the effectors and how do they alter their activities so as to restore the regulated parameter toward normal (i.e., the set point of the system)?

Some semantic problems   There are problems which arise when one attempts to categorize the components of some complex reflex arcs according to the five terms listed above. For example, in the reflex arc shown in Fig. 5-7, the stimulus, receptor, and final effector (muscle) are quite clear, as is the designation of the first nerve (*A*)

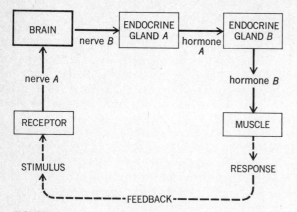

FIGURE 5-7
Complex reflex arc composed of multiple hormones and nerve fibers.

in the chain and the last hormone (*B*) as afferent and efferent pathways. But what about the brain, nerve *B*, hormone *A*, and the two endocrine glands? Actually, one is dealing here with a chain of small reflex arcs all subserving the overall reflex arc. Thus, endocrine gland *A* may be considered to be an effector whose response is the secretion of hormone *A*, or it may be viewed as an integrating center; in the latter case, nerve *B* becomes its afferent input and hormone *A* becomes its efferent output. It is fruitless to assign rigid terms to the interior components of such complex reflex chains. It is far more important for the reader to appreciate the sequence of events rather than worry about the labels.

Another problem is that some reflex arcs lack the usual afferent pathway. This may seem puzzling, but the following example may help (Fig. 5-8). Parathormone is a hormone which is secreted by the parathyroid glands and which acts upon bone, causing it to increase its release of calcium into the blood. When the blood calcium concentration is decreased for any reason, an increased amount of parathormone is secreted into the blood; the blood-borne parathormone reaches bone throughout the body and induces it to release more calcium into the blood. Clearly, in this reflex, blood-borne parathormone acts as the efferent pathway and bone is the effector. But we have made no mention yet of the receptors or the afferent pathway involved. Actually, the parathyroid gland cells are themselves sensitive to the calcium concentration of the blood supplying them; thus, the same cells which produce the hormone act as the receptors

for this reflex. Clearly, there is no afferent arc in such a situation since the receptors and integrating center (i.e., hormone-producing cells) are one and the same.

Finally, we must point out a problem to which we will return in several subsequent chapters. In the most narrow sense of the word, a reflex is an involuntary, unpremeditated, unlearned response to a stimulus; the pathway over which this chain of events occurs is "built in" to all members of a species. Examples of such basic reflexes would be pulling one's hand away from a hot stove or shutting one's eyes as an object rapidly approaches the face. However, there are also many responses which appear to be automatic and stereotyped but which actually are the result of learning and practice. For example, an experienced driver performs many complicated acts in operating his car; to him these motions are, in large part, automatic, stereotyped, and unpremeditated, but they occur only because a great deal of conscious effort was spent to learn them. We shall refer to such acts as *learned* or *acquired.* In general, most reflexes, no matter how basic they may appear to be, are subject to alteration by learning; i.e., there is often no clear distinction between a basic reflex and one with a learned component.

Local homeostatic responses

Besides reflexes, another group of biological responses is of immense importance for homeostasis. We shall call them *local responses.* Local responses are initiated by a change in the external or internal environment, i.e., a stimulus, which acts upon cells in the immediate vicinity

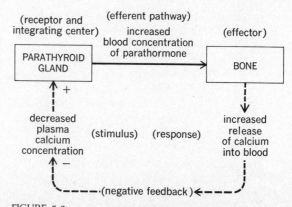

FIGURE 5-8
Homeostatic reflex by which plasma calcium concentration is controlled. Note the absence of an afferent pathway.

of the stimulus inducing an alteration of cell activity with the net effect of counteracting the stimulus. Thus, a local response is, like a reflex, a sequence of events proceeding from stimulus to response, but, unlike a reflex, the entire sequence occurs only in the area of the stimulus, no hormones or nerves being involved.

Two examples should help clarify the nature and significance of local responses: (1) Damage to an area of skin causes the release of certain chemicals from cells in the damaged area which help the local defense against further damage; (2) an exercising muscle liberates chemicals into the extracellular fluid which act locally to dilate the blood vessels in the area, thereby permitting the required inflow of additional blood to the muscle. The great significance of such local responses is that they provide individual areas of the body with mechanisms for local self-regulation.

Chemical mediators

It should be evident that the *sine qua non* of reflexes is the ability of cells to communicate with one another, i.e., the capacity of one cell to alter the activity of another. When a hormone is involved in a reflex, it is clear that the communication between cells, i.e., between endocrine gland cell and effector, is accomplished by a chemical agent, the hormone (the blood, of course, acting as the delivery service). What has not been said, however, is that virtually all nerve fibers in the body also communicate with each other or with effectors by means of chemical agents or mediators. Thus, one neuron alters the activity of the next neuron in a reflex chain by releasing a substance from its ending; this *chemical transmitter* diffuses across the very narrow space separating the two neurons and acts upon the second, altering its activity. Similarly, chemical transmitters released from the ends of the neurons going to effectors constitute the immediate signal, or input, to the effector cells.

The detailed physiology of these chemical transmitters and of neuron-neuron or neuron-effector communication will be described in Chaps. 6 and 8. We mention them here to emphasize that chemical mediators, whether they are secreted by endocrine gland cells or released from neuronal endings, constitute the ultimate messages by which one cell signals another to alter its activity. This is true not only for reflexes but for local responses, as the examples illustrate.

In future chapters, we shall describe the roles of a considerable number of chemical transmitters, but one aspect of their physiology is quite striking (and confusing), namely, the fact that the same chemical may act as a transmitter in many different sites with widely differing effects. For example, *acetylcholine* is one of the transmitters for communication between many neurons, neurons and skeletal muscle, neurons and heart muscle, neurons and gland cells, neurons and smooth muscles. *Epinephrine* and *norepinephrine* also have a wide spectrum of functions. These three have received the most study in the past but others (*dopamine, histamine, serotonin,* etc.) are rapidly gaining on them. Recently, an entirely new set of substances has gained the limelight—the family of fatty acids called *prostaglandins.* They were originally discovered in semen but are now known to be present in most tissues. Their precise physiological roles are yet to be established, but they may well be involved in the function of smooth muscle, nerves, the liver, adipose tissue, the circulation, and the reproductive organs. We shall discuss these possibilities in subsequent chapters where relevant.

Receptor sites

The mechanisms by which chemical mediators act to alter cellular activity represent one of the most intensively studied areas in physiology today. Present thinking centers around the concept of *membrane receptor sites.*[1] It is believed that the first step in the action of most, if not all, chemical mediators is their combination with certain specific molecules, i.e., receptor sites, within the cell-membrane structure. This combination of chemical mediator and membrane component somehow alters the membrane structure. In some cases it leads to changes in membrane permeability and the rate at which a particular substance is transported across the membrane; in the main, however, the precise mechanisms by which this combination induces an alteration in the cell's activity, i.e., a response, are unknown.

One extremely important characteristic of chemical mediation that is understandable in terms of membrane receptor sites is *specificity.* A chemical mediator—hormone, neural transmitter, or locally released chemical agent—influences only certain cells and not others, although there is considerable variation in the degrees of specificity manifested by the different mediators. The likeliest explanation is that membranes of different cell types all differ; accordingly, only certain cells types, frequently just one, possess the precise membrane receptor site required for combination with a given chemical mediator.

---

[1] This term is frequently shortened in scientific literature to receptor, a usage that unfortunately can cause confusion because receptor sites are totally different from afferent receptors, described earlier. The reader must carefully distinguish between these terms.

Conversely, the different molecular structures of the mediators explain why only certain mediators, but not others, can influence any given cell type.

The balance concept and chemical homeostasis

One of the most important concepts in the physiology of control systems is that of balance. Our example of the water bath was really a study of heat balance within the water bath, i.e., the control system functioned to maintain a precise balance between the rates at which heat was added to and left the bath. Almost every homeostatic system in the body can be studied in terms of balance; some regulate the balance of a physical parameter (heat, pressure, flow, etc.), but most are concerned with the balance of a chemical component of the body. This section is intended to provide a foundation of general principles upon which can be built the study of any specific chemical substance.

Figure 5-9 is a generalized schema of the possible pathways involved in the balance of a chemical substance. The *pool* occupies a position of central importance in the balance sheet; it is the body's readily available quantity of the particular substance and is frequently identical to the amount present in the extracellular fluid. The pool functions as "middleman," receiving from and contributing to all the other pathways.

The pathways on the left of the figure are sources of *net gain* to the body. A substance may be ingested and then absorbed from the gastrointestinal (GI) tract. It is important to realize that all that is ingested may not be absorbed; some may either fail to be absorbed or be consumed by the bacteria residing in the gut. The lungs offer another site of entry to the body for gases (O_2) and airborne chemicals. Finally, the substance may be synthesized by cells within the body itself. (Bacteria in the GI tract may also synthesize nutrients which can be used by the body.)

The pathways to the right of the figure are sources of *net loss* from the body. A substance may be excreted in the urine, feces, expired air, and menstrual flow, as well as from the surface of the body (skin, hair, nails, sweat, tears, etc.). The substance may be catabolized or transformed within the body to some other chemical; this fate—the opposite of synthesis—represents net loss of the substance.

The central portion of the figure illustrates the *distribution* of the substance within the body. From the readily available pool, it may be taken up by storage depots; conversely, material may leave the storage depots to reenter the pool. Finally, the chemical may be incorporated into some other molecular structure (fatty acids into membranes, iodine into thyroxine, etc.). This process is reversible in that the substance is liberated again whenever the more complex molecule is broken down. For this reason, this pathway differs from catabolism or transformation by which the substance is irretrievably lost. This pathway is also distinguished from storage in that the latter has no function other than the passive one of storage, whereas the incorporation of the substance into

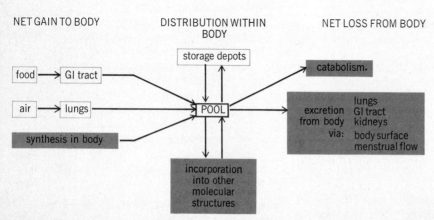

FIGURE 5-9
Balance diagram for a chemical substance.

other molecules is done to fulfill an active function of the substance (e.g., the function of iodine in the body is to provide an essential component of the thyroxine molecule).

Of course, it should be recognized that every pathway of this generalized schema is not applicable to every substance; for example, the mineral electrolytes cannot be synthesized or catabolized by the body.

The orientation of the figure illustrates two important generalizations concerning the balance concept: (1) The total body balance depends upon the rates of total body net gain and net loss; (2) the pool concentration depends not only upon total body losses and gains but upon exchanges of the substance within the body.

It should be apparent that, with regard to total body balance, three states are possible: (1) Loss exceeds gain, the total amount of the substance in the body decreases, and the person is said to be in *negative balance;* (2) gain exceeds loss, the total amount of the substance in the body increases, and the person is said to be in *positive balance;* (3) gain equals loss, and balance is stable. Physiology is, in large part, concerned with the homeostatic mechanisms which match gain with loss to achieve a stable balance. Pool size, too, tends to be maintained relatively constant as a result of these homeostatic mechanisms as well as those operating on the pathways for internal exchange. Clearly a stable balance can be upset by alteration of the magnitude of any single pathway in the schema, e.g., severe negative water balance can occur in the presence of increased sweating. Conversely, stable balance can be restored by homeostatic control of the pathways. Therefore, much research has been concerned with determining which are the key homeostatically controlled pathways for each substance, what are the specific mechanisms involved, and what are the limits of intake and output beyond which balance cannot be achieved.

Essential nutrients: an example

There are many substances which are required for normal or optimal body function but which are synthesized by the body either not at all or in amounts inadequate to achieve balance. They are known as *essential nutrients.* Because they are all excreted or catabolized at some finite rate, a continuous new supply must be provided by the diet. Approximately 50 in number, they are water, 8 amino acids, several unsaturated fatty acids, approximately 20 vitamins, and a similar number of inorganic minerals.

It should be reemphasized that the term essential nutrient is reserved for substances that fulfill *two* criteria: They not only must be essential for good health but must not be synthesized by the body in adequate amounts. Thus, glucose, although "essential" for normal metabolism, is not classified as an essential nutrient because the body normally can synthesize all it needs.

The physiology of each essential nutrient can be studied in terms of the schema of Fig. 5-9, i.e., by analyzing each pathway relevant for that nutrient. There is considerable variation in the relative importances of the pathways for the homeostatic regulation of the different nutrients. Thus, in Chap. 11 we shall see that ingestion and urinary excretion are the main controlled variables for water. In contrast, control of iron balance, as described in Chap. 10, is dependent largely upon the control of iron absorption by the gastrointestinal tract. Balance of essential amino acids is achieved in still another way; the rate of catabolism of these amino acids (specifically the loss of the NH_2 group from the amino acid) is reduced in the presence of amino acid deficiency (and increased in the presence of excess).

The reason for placing so much emphasis on the pathways which are homeostatically controlled is that alteration of the nutrient flow via these pathways constitutes the mechanism by which balance is achieved wherever a primary change occurs in any of the other pathways. For example, iron balance can be upset either by a primary change in intake or excretion; in either case, balance can be reestablished by a compensating change in the rate of absorption.

The key event in triggering off the homeostatic response is a change in the body content of the nutrient, manifested frequently as a change in pool content, i.e., a change in the internal environment. Therefore, it is important to realize that nothing in the body is maintained *absolutely* constant; relatively small deflections from normal are continuously occurring and constitute the signals for the control systems which then operate to limit the extent of deviation.

Let us look more closely at the spectrum of conditions possible as we alter primary intake of an essential nutrient, assuming all the other pathways are functioning normally. At zero intake, balance is impossible since excretion or catabolism, or both, cannot be reduced to zero; accordingly negative balance persists as the body stores of the nutrient are progressively depleted. Thus, the minimal combined rate of excretion and catabolism sets the lower limit for achievement of balance. At the other end of the spectrum, the maximal combined rate of excretion and catabolism sets the upper limit for achievement of balance, i.e., the maximal intake of nutrient compatible with balance. If intake is greater than this, body stores will continuously increase, with the potential for toxic effects. This occurs for certain of the fat-soluble vitamins (A, D, and K) and is a problem for other nutrients as well.

Between the extremes there obviously exists a range of intakes at which balance can be maintained without overt manifestations of either deficiency or toxicity. How wide this range is depends upon how much the rates of excretion or catabolism can be altered. (For example, sodium balance can be achieved readily at intakes between 0.3 and 25 g/day.) It must be reemphasized that, despite the fact that balance is stable at all intakes in this range, there are small differences in body content and pool size over the range, since these differences are required to drive the homeostatic control systems.

The ultimate aim of nutrition is to determine which intake in this range is the *optimal* intake for each nutrient. The critical question may be stated as follows: Does ingestion of more than the minimal amount of a nutrient produce greater health, growth, intelligence, etc., i.e., is there an amount of nutrient which is *optimal* for health rather than merely adequate for avoiding disease? But this question raises the further question: Optimal for what? For example, most Americans ingest very large quantities of protein (which supplies essential amino acids), and this almost certainly has contributed to our increased body height. This would seem desirable, but in experimental animals it has been found that protein intakes which are optimal for maximal total body growth enhance the development of cancers and arteriosclerosis as well. Clearly, "optimal for what?" cannot be answered without a clearly definable "what" and a careful consideration of many factors, including existing environmental influences and life style.

Biological rhythms

A striking characteristic of many bodily functions is the rhythmical changes they manifest. Body temperature, for example, fluctuates considerably during a normal 24-hr period, as do the concentrations of many hormones. Such daily rhythms are called *circadian* (literally, around the day). Most likely, they occur as a result of changes in the set points of the control systems regulating them, but we generally are uncertain as to their precise mechanisms or significance. We do know that modern life, with its stepped-up pace, rapid changes, and creation of artificial environments, may frequently disrupt them with, as yet, unknown results. Many of the body's rhythms are much longer than 24 hr. The menstrual cycle is the best-known longer cycle, but there may well be others with even greater time spans.

Central to this field is the problem of "biological clocks" within the body which set the rhythms. This question has received particular attention by physiologists of reproduction, but it is now recognized as a critical area in the biology of growth, aging, and many other fields relevant to human physiology.

Neural control mechanisms

One of the greatest achievements of biological evolution has been the development of the human nervous system. It is made up of the brain and spinal cord as well as the many nerve processes which pass between these two structures and the muscles, glands, or receptors which they innervate.

In Chaps. 16 to 18 we shall consider specific aspects of neural activity concerned with the special senses (sight, hearing, etc.), the coordination of muscle activity in posture and movement, and the relation between brain activity and consciousness, memory, emotions, etc. In this chapter we are concerned with the basic components common to all neural mechanisms. We shall review the principles of electricity and describe the electric signal (*action potential*) generated in and propagated along nerve cells, the functional anatomy of the individual nerve cell (*neuron*), the processes by which action potentials are normally initiated within the body, the passage of information from one neuron to another, and some basic patterns of neural interaction. Finally, we shall introduce the basic organization and major divisions of the nervous system.

electrical signal - action potential
individual nerve - neuron cell

6

SECTION A
MEMBRANE POTENTIALS

Basic principles of electricity

All chemical reactions are basically electric in nature since they involve exchanging or sharing negatively charged electrons between atoms to form ions or bonds. Most chemical reactions result in neutral molecules, containing equal numbers of electrons and protons, but in some cases ions are formed, which have a net electric charge, e.g., the ionized groups in organic molecules, such as the negative carboxyl group $RCOO^-$ and the positive amino group RNH_3^+, or the inorganic ions such as sodium, potassium, and chloride (Na^+, K^+, and Cl^-). With the exception of water, the major chemical components of the extracellular fluid are the sodium and chloride ions, whereas the intracellular fluid contains high concentrations of potassium ions and organic molecules containing ionized groups, particularly proteins and phosphate compounds, RPO_4^{2-}. Since the environment of the cell contains many charged particles, it is not too surprising to discover that electrical phenomena resulting from the interaction of these charged particles play a significant role in cell function.

According to the laws of physics, all physical and chemical phenomena can be described using only five fundamental units of measurement, length, time, mass, temperature, and electric charge. Each of these units is independent in the sense that none can be defined in terms of a combination of the others. For example, velocity is not a fundamental unit since it can be defined in terms of a length moved in a given period of time. Force is not a fundamental unit since it is defined by Newton's law as $F = ma$, force equals mass times acceleration, and thus is defined in terms of mass, length, and time. In contrast, electric charge is a fundamental unit of measurement and cannot be defined in terms of length, time, mass, or temperature.

When labels are applied to identify parts of a complex event, the choice is often quite arbitrary. For instance, a man walking in a straight line from point A to point B moves in a specific direction determined by the location of points A and B. What happens if he walks from point B to point A? Obviously he must walk the same distance as before but in the opposite direction. The two events, walking from A to B and walking from B to A, are identical except for direction. To describe these events, one must be able to distinguish between the two directions. One can call movement from A to B forward

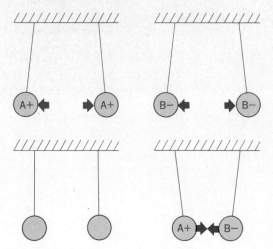

FIGURE 6-1
Two cork balls having the same type of electric charge (both positive or both negative) repel each other. Two cork balls having opposite electric charges, A+ and B−, attract each other.

and from B to A backward, or one can label movement from A to B positive and movement in the opposite direction negative. Which label one chooses to describe the directions is arbitrary. When describing electrical phenomena, labels are needed for certain symmetrical events. The initial choice of the label is arbitrary, but once chosen, it must be used consistently throughout the discussion.

There are two types of electric charge in the universe; they behave symmetrically; how we label them is arbitrary. If we take four small balls of cork suspended by silk threads and apply one type of electric charge to two of the balls labeled A and the other type of electric charge to the other two balls labeled B, we can observe some of the fundamental properties of electric charge. When the two balls labeled A are brought close together, they repel each other; the same thing happens when the two balls labeled B are brought together (Fig. 6-1). This illustrates one of the fundamental principles of electricity, namely *like electric charges repel each other*. There is an electric force operating between two like charges which moves them apart. In contrast, if we bring two unlike electric charges near each other, they attract each other; there is an electric force pulling unlike charges together (Fig. 6-1).

If we decide to call A charges positive ($+$) and B charges negative ($-$), we can summarize our observations by saying that *positive charge repels positive charge and negative charge repels negative charge, whereas positive and negative charges attract each other.* The labeling of positive and negative charges of electricity occurred historically before it was known that atoms consisted of protons and electrons. When electrons were discovered, they behaved like the charged particles which had been arbitrarily labeled negative. Protons, on the other hand, behave like the electric charges labeled positive. If the original labeling had been the opposite, we would call electrons positive and protons negative.

All electrical phenomena result from the interaction of the two types of electric charge in atoms, the negative electrons and the positive protons. The total numbers of positive and negative particles in the universe are believed to be equal; thus, the universe is electrically neutral, as are atoms, which contain an equal number of protons and electrons. But in order to observe electrical phenomena, we must separate positive and negative electric charges. The cork balls labeled A are positive because they contain more protons than electrons, and the B ones are negative because they contain more electrons than protons.

When positive and negative charges are separated, an electric force draws the opposite charges together. Why there is such a force cannot be answered in terms of other physical properties of matter since electric charge is a fundamental property of matter. However, the force can be measured, and the relation between the amount of force, the quantity of charge, and the distance separating the charges can be studied. It is found that the amount of force acting between electric charges increases when the charged particles are moved closer together and with increasing quantity of charge (Fig. 6-2).

In Chap. 3, energy E was defined as the ability to do work, and work W was defined as the product of force F and distance X; $W = FX$. If we allow oppositely charged particles to come together as a result of the attracting force between them, work will be done by these moving particles since a force will be exerted over a distance (Fig. 6-3). Likewise, to separate oppositely charged particles we must exert a force on them opposite to that of their electric attracting force. Energy, in the form of work, must be added to the system in order to separate the charges. If a number of positive charges are placed at point A and a similar number of negative charges at point B, there will be an electric force between points A and B tending to draw them together. If a positive charge is allowed to move from A to B, it does work, and the amount of work done depends upon both the

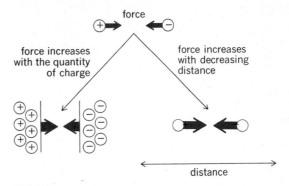

FIGURE 6-2
The electric force of attraction between positive and negative charges increases with the quantity of charge and with decreasing distance between charges.

total number of charges at A and B and the distance between them. Thus, when electric charges are separated, they have the "potential" of doing work if they are allowed to come together again. *Voltage is a measure of the potential of separated electric charge to do work and is defined*

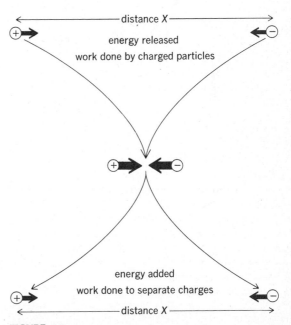

FIGURE 6-3
When opposite charges come together, energy is released and can be used to perform work. Work must be done (energy added) in order to separate opposite charges.

as the amount of work done by an electric charge when moving from one point in a system to another. Voltage is always measured with respect to two points in a system; thus, one refers to the potential difference between two points, and the units of measurement are known as volts. The terms voltage and potential difference are synonymous. Since the total amount of charge that can be separated in most biological systems is very small, the potential differences are small. The voltage measured across a nerve cell membrane is approximately 70 millivolts (mV), or about $\frac{1}{10}$ V. A millivolt is $\frac{1}{1,000}$ volt.

The movement of electric charge is known as *current.* If electric charge is separated between two points, there is a potential difference between these points, and the electric force of attraction between the opposite charges tends to make charges flow, producing a current. The amount of charge that does move, the current, depends upon the nature of the material lying between the separated charges. The space between the two points may be occupied by copper wire, a solution of water and ions, glass, or rubber, or it may be entirely empty, a vacuum. The ability of an electric charge to move through these different media varies, the flow depending upon the number of charged particles in the material that are able to move and thus carry current. The amount of current also depends upon the interactions between the moving charges and the material, such as frictional interactions if the charges collide with other molecules while moving through the material. The hindrance of the movement of electric charge through a particular material is known as *resistance.* Thus, for a given amount of charge separation (a given voltage) the amount of current flow depends upon the resistance of the material between the separated charges. The higher the resistance of the material, the lower the amount of current flow for any given voltage. Some materials, like glass and rubber, have such high electric resistance that the amount of current flow through them, even when high voltages are applied, is very small. Such materials, known as insulators, are used to prevent the flow of current. Thus, the rubber insulation around electric wires prevents the flow of current from the wire to areas outside it. Materials having low resistance to current flow are known as conductors.

Pure water is a relatively poor conductor because it contains very few charged particles, but when sodium chloride is added, the solution becomes a relatively good conductor with a low resistance because the sodium and chloride ions provide charges that can carry the current. The water compartments inside and outside the cells in the body contain numerous charged particles (ions)

which are able to move between areas of charge separation. Lipids contain very few charged groups and thus have a high electric resistance. The lipid components of the cell membrane provide a region of high electric resistance separating two water compartments of low resistance.

The electric and chemical properties of a resting cell

One can determine the presence of an electric-potential (voltage) difference across cell membranes by inserting a very fine electrode into the cell, another into the extracellular fluid surrounding the cell, and connecting the two to a voltmeter (Fig. 6-4). In this manner it has been found that all cells of the body exhibit a membrane potential oriented so that the inside of the cell is negatively charged with respect to the outside. (As we shall see, this is true for nerve and muscle cells only when they are not being stimulated.) This potential is called the *resting membrane potential;* its magnitude varies from 5 to 100 mV, depending upon the type of cell and its chemical environment.

The normal ionic composition of the fluid bathing the cells (the *extracellular fluid*) is approximately that listed in Table 6-1. (Note that the chemical composition of the

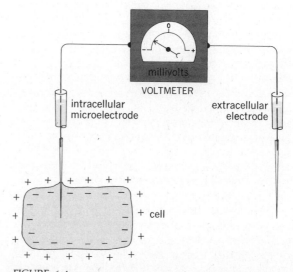

FIGURE 6-4
Intracellular microelectrode used to measure electric-potential difference across the cell membrane.

TABLE 6-1
Distribution of ions across the cell membrane of a cat nerve cell, concentration in millimoles per liter of water

Ion	Extracellular concentration	Intracellular concentration
Na	150	15
Cl	125	10
K	5	150

intracellular fluid is entirely different.) There are many other substances in the extracellular fluid, such as Mg^{2+}, Ca^{2+}, HCO_3^-, PO_4^{2-}, SO_4^{2-}, glucose, urea, amino acids, and hormones, but sodium and potassium play the most important roles in the generation of the resting membrane potential.

Diffusion potentials

Given that we are dealing with two solutions of different ionic composition and, for the moment, ignoring the nature of the membrane which separates them, consider how the solutions interact. Figure 6-5 depicts two dilute solutions of sodium chloride, the solution on side 1 at a concentration of $0.1\ M$ and that on side 2 at $0.01\ M$. The barrier separating the solutions is extremely permeable to all ion species. Both sodium and chloride are more concentrated on side 1, and they will diffuse down their concentration gradients, moving from side 1 to side 2. However, the mobility of chloride, i.e., the ease with which it can move through the solution, is about 50 percent greater than that of sodium; thus, chloride will move to side 2 more rapidly than sodium and side 2 will, at least transiently, become slightly negatively charged with respect to side 1. This electric gradient is due to the differential diffusion of charged particles in solution and is called a *diffusion potential*. The potential developed by a system such as that in Fig. 6-5 will disappear over time as the concentrations of sodium and chloride in sides 1 and 2 become equal.

What would happen at the junction between sides 1 and 2 if side 1 contained a solution of $0.15\ M$ NaCl (similar to extracellular fluid) and side 2 contained $0.15\ M$ KCl (similar to intracellular fluid) (Fig. 6-6)? Again, we ignore the membrane that separates the two compartments. Chloride concentrations on the two sides are equal, but those of sodium and potassium are not. The mobility of potassium, like that of chloride, is about 50 percent greater than that of sodium; thus, potassium will diffuse down its concentration gradient faster than so-

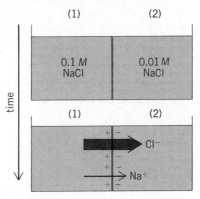

FIGURE 6-5
Generation of a diffusion potential by differential ion movement through a completely permeable membrane.

dium, i.e., positive charge will initially leave side 2 faster than it will enter, and side 2 will become electronegative with respect to side 1. Again, with time, equilibrium will be reached and the potential difference will disappear.

Equilibrium potentials   Now consider the situation of Fig. 6-7 but assume that a selectively permeable membrane separates the two compartments, such that potassium can pass through but sodium and chloride cannot. In this case, all the sodium will remain on side 1, and some of the potassium, diffusing down its concentration gradient, will be added to it; side 1 will become relatively positive.

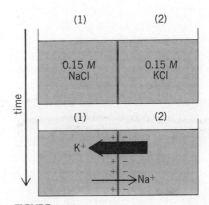

FIGURE 6-6
Generation of a diffusion potential by potassium movement through a completely permeable membrane.

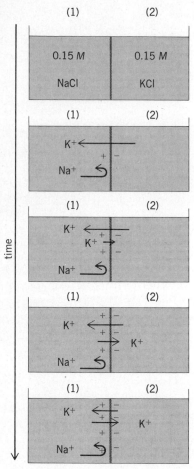

FIGURE 6-7
Generation of a diffusion potential across a membrane permeable only to potassium.

In contrast to the previous examples, this diffusion potential will not disappear with time and will be of greater magnitude, the actual magnitude depending upon the concentration gradient for potassium.

The concentration gradient, which causes net diffusion of particles from a region of higher to a region of lower concentration, is called the *concentration force.* As this force moves potassium from side 2 to side 1 and side 1 becomes increasingly positive, the electric-potential difference itself begins to influence the movement of the positively charged potassium particles; they are attracted by the relatively negative charge of side 2 and repulsed by the positive charge of side 1. This attraction because

of a difference in electric charge (or repulsion because of a similarity of charge) is the *electric force.* As long as the concentration force driving potassium from side 2 to 1 is greater than the electric force driving in the opposite direction, there will be net movement of potassium from side 2 to 1 and the potential difference will increase. Side 1 will become more and more positive until the electric force opposing the entry of potassium equals the concentration force favoring entry. The membrane potential at which the electric force equals in magnitude and opposes in direction the concentration force is called the *equilibrium potential.* At the equilibrium potential there is no net movement of the ion because the forces acting upon it are exactly balanced.

It can be seen that the value of the equilibrium potential for any ion depends upon the concentration gradient for that ion across the membrane; if the concentrations on the two sides were equal, the concentration force would be zero and the electric potential required to oppose it would also be zero. The larger the concentration gradient, the larger is the equilibrium potential. Using potassium concentrations typical for neurons and extracellular fluid, the equilibrium potential for potassium is close to 90 mV, the inside of the cell being negative with respect to the outside.

If the membrane separating sides 1 and 2 is replaced with one permeable only to sodium, the initial net flow of the positively charged sodium will be from side 1 to 2 and side 2 will become positive (Fig. 6-8). This movement of sodium down its concentration gradient is opposed by the electric force generated by that movement. A sodium equilibrium potential will be established with side 2 positive with respect to side 1, at which point net movement will cease. For neurons, the sodium equilibrium potential is 60 mV, inside positive. *Thus, the direction of the diffusion potential across the membrane is determined both by the permeability properties of the membrane and the orientation of the concentration gradients.* The diffusion potential for each ion species is different from those for other ion species since the concentration gradients are different.

The resting cell membrane potential It is not difficult to move from these hypothetical experiments to a nerve cell at rest where (1) the potassium concentration is much greater inside the cell than out and the sodium concentration gradient is the opposite and (2) the cell membrane is some 50 to 75 times more permeable to potassium than to sodium. Given these characteristics, it should be evident that a diffusion potential will be generated across the membrane largely because of the movement of po-

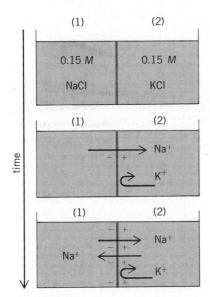

FIGURE 6-8
Generation of a diffusion potential across a membrane permeable only to sodium.

tassium down its concentration gradient so that the inside of the cell is negative with respect to the outside. The experimentally measured membrane potential is not, however, equal to the potassium equilibrium potential because the membrane is not perfectly impermeable to sodium and some sodium continually diffuses down its electric and concentration gradients, adding a small amount to the inside of the cell. Thus the measured membrane potential, of a neuron at least, is closer to −70 mV than to the potassium equilibrium potential of −90 mV. An important result of this fact is that, since the membrane is not at the potassium equilibrium potential, there is a continual net diffusion of potassium out of the cell.[1]

The membrane pump   If there is net movement of sodium into and potassium out of the cell, why do the concentration gradients not run down? The reason is that active-transport mechanisms in the membrane utilize forces derived from cellular metabolism to pump the sodium back out of the cell and the potassium back in. For each sodium pumped from the cell, a potassium is moved in; one positive ion is exchanged for another. Under these circumstances the pump itself does not contribute directly

[1]In contrast, chloride is at its equilibrium potential (−70 mV); accordingly, there is no net chloride flux in the resting neuron.

to the membrane potential.[2] The pump does make an essential indirect contribution, however, because it maintains the concentration gradients down which the ions diffuse. The membrane potential is then due directly to the diffusion of these ions.

Summary   Some potassium ions diffuse out of the cell down their concentration gradient and some move in down the electric gradient, but because the membrane potential is not as negative as the potassium equilibrium potential (−70 mV rather than −90 mV), less potassium enters passively than leaves. The difference is relatively small and is made up by active transport via the membrane pump; therefore the *total* potassium entering equals that leaving, and the resting cell neither gains nor loses potassium (Fig. 6-9). Sodium is driven passively into the cell by both electric and concentration forces but, because the membrane permeability of a resting cell to sodium is so low, the amount entering is small. There is no passive force to remove sodium from the cell, and that

[2]Whenever a pump does not exchange the ions on a one-to-one basis, it directly produces charge separation; such a pump is called an electrogenic pump. In those cases in which an electrogenic pump has been identified in excitable tissues its overall contribution to the generation of the membrane potential is small relative to that contributed by the diffusion potentials.

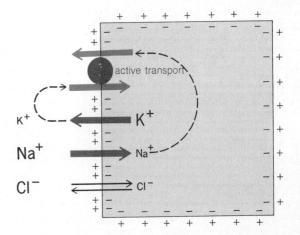

FIGURE 6-9
Steady-state total fluxes of sodium, potassium, and chloride ions across the cell membrane. The net flux of sodium and potassium ions by diffusion is balanced by the active transport of these ions in the opposite direction across the membrane (i.e., potassium in and sodium out). There is no net flux of chloride ions because they are in electrochemical equilibrium across the membrane.

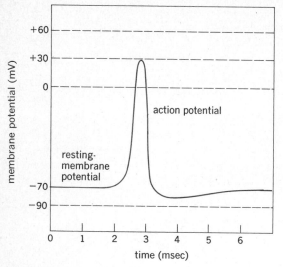

FIGURE 6-10
Changes in membrane potential during an action potential.
(The sodium equilibrium potential is +60 mV; the potassium
equilibrium potential is −90 mV.)

−70 to +30 mV and then rapidly returning to its original value. This rapid change of membrane potential, which may last only $\frac{1}{1000}$ sec, is called an *action potential*. Of all the types of cells in the body, only nerve and muscle cells are capable of producing action potentials; this property is known as *excitability*. Excitable membranes, besides generating action potentials, are able to transmit them along their surfaces. Thus, the action potential is the signal which is transmitted from one part of the nerve or muscle cell to another.

How is an excitable membrane able to make rapid changes in its membrane potential? How does a change in the environment (a stimulus) interact with an excitable membrane to bring about an action-potential response? How is an action potential propagated along the surface of an excitable membrane? These questions will be discussed in the following sections. The terms *depolarize, hyperpolarize,* and *repolarize* will be used frequently. The membrane is said to be *depolarized* when the membrane potential is less negative than the resting membrane potential, i.e., closer to zero, and *hyperpolarized* when it is more negative than the resting level. When the membrane potential is changing so that it moves toward or even above zero, it is *depolarizing,* and when it moves away from zero back toward its resting level, it is *repolarizing* (Fig. 6-11). When it moves beyond its resting level it is *hyperpolarizing.*

which enters must be actively transported out by the membrane pump. The amount of sodium pumped out equals, in most cases, the amount of potassium pumped in (Fig. 6-9).

When more than one ion species can diffuse across the membrane, the membrane permeability properties as well as the concentration gradient of each species must be considered when accounting for the membrane potential. If the membrane is impermeable to a given ion species, no ion of that species can cross the membrane and contribute to a diffusion potential, regardless of the electric and concentration gradients that may exist. And, for a given concentration gradient, the greater the membrane permeability to an ion species is, the greater the influence that ion species will have on the diffusion potential. Since the resting membrane is much more permeable to potassium than to sodium (and the concentration differences are approximately similar), the resting membrane potential is much closer to the potassium equilibrium potential than to that of the sodium.

Action potentials

During periods when nerve and muscle cells appear to be physiologically active the membrane potential undergoes rapid alteration (Fig. 6-10), suddenly changing from

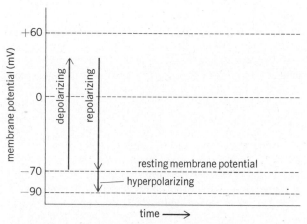

FIGURE 6-11
As potentials become more positive inside than the resting potential they are said to be *depolarizing;* those returning toward the resting membrane potential are said to be *repolarizing,* and those becoming more negative inside than the resting membrane potential are *hyperpolarizing.*

Ionic basis of the action potential

Action potentials can be explained by the concepts already developed for the origins of resting membrane potentials. This explanation, known as the *ionic hypothesis,* was developed mainly by the English scientists A. L. Hodgkin and A. F. Huxley, who received the Nobel Prize in 1963. We have seen that the magnitude of the resting membrane potential depends upon the concentration gradients of and membrane permeabilities to ions, particularly sodium and potassium. This situation is true for the period of the action potential as well. Obviously, then, the action potential must result from a transient change in either the concentration gradients or the membrane permeabilities. The latter is the case. In the resting state the membrane is 50 to 75 times more permeable to potassium than to sodium ions. Thus, the magnitude and polarity of the resting potential are due almost entirely to the movement of potassium ions out of the cell. During an action potential, however, the permeability of the membrane to sodium and potassium ions is markedly altered. In the rising phase of the action potential the membrane permeability to sodium ions undergoes a 600-fold increase and sodium ions rush into the cell, whereas there is little change in the potassium permeability of the membrane. During this period more positive charge is entering the cell in the form of sodium ions than is leaving in the form of potassium ions, and thus the membrane potential decreases and eventually reverses its polarity, becoming positive on the inside and negative on the outside of the membrane. In this phase the membrane potential approaches but does not quite reach the sodium equilibrium potential.

Action potentials in neurons last about 1 msec (0.001 sec). What causes the membrane to return so rapidly to its resting level? The answer to this question is twofold: (*1*) The increased sodium permeability (*sodium activation*) is rapidly turned off (*sodium inactivation*), and (*2*) the membrane permeability to potassium increases over its resting level. The timing of these two events can be seen in Fig. 6-12. As the membrane becomes more positive inside and the drive for sodium entry is reduced, sodium permeability decreases toward its resting value, and sodium entry rapidly decreases. This alone would restore the potential to its resting level. However, the entire process is speeded up by a simultaneous increase in potassium permeability which causes more potassium ions to move out of the cell down their concentration gradient. These two events, sodium inactivation and increased potassium permeability, allow potassium diffu-

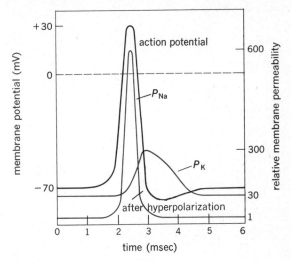

FIGURE 6-12
Changes in membrane permeability to sodium and potassium ions during an action potential.

sion to regain predominance over sodium diffusion, and the membrane potential rapidly returns to its resting level. In fact, while the potassium permeability is greater than normal, there is generally a small hyperpolarizing overshoot of the membrane potential (*after hyperpolarization,* Fig. 6-12).

In our description it may have seemed as though the sodium and potassium fluxes across the membrane involved large numbers of ions. Actually, only one of every 100,000 potassium ions within the cell need diffuse out to charge the membrane potential to its resting value, and very few sodium ions need enter the cell to cause the depolarization during an action potential. Thus, there is virtually no change in the concentration gradients during an action potential. Yet if the tiny number of additional ions crossing the membrane with each action potential were not eventually restored, the concentration gradients for sodium and potassium across the cell membrane would gradually disappear. As might be expected, an accumulation of sodium and loss of potassium are prevented by the continuous action of the membrane active-transport system for sodium and potassium. This restoration occurs mainly after the action potential is over. The number of ions that cross the membrane during an action potential is so small, however, that the pump need not keep up with the action-potential fluxes, and hundreds of action potentials can occur even if the pump is stopped experimentally. Note, however, that the pump plays no direct role in the generation of the action potential itself.

Mechanism of permeability changes

The cause of the permeability changes which underlie action potentials has been elucidated by experiments in which membrane permeability and ion fluxes are monitored as action potentials are triggered by means of electrical stimulation applied to the nerve cell membrane. The stimulation is accomplished through use of two electrodes, one placed in the cell and the other in the extracellular fluid surrounding it. With appropriate settings, the electrodes can be made to add positive charge to the outside of the membrane while simultaneously removing positive charge from the inside of the cell, thereby causing the membrane potential to increase, i.e., become hyperpolarized. If the settings are reversed, the electrodes remove positive charge from the outside while adding it to the inside, thereby reducing the membrane potential, i.e., depolarizing the cell membrane.

Measurement of membrane permeability changes during these experiments revealed that the permeability to sodium was altered whenever the membrane potential was changed; specifically, hyperpolarization of the membrane caused a decrease in sodium permeability whereas depolarization caused an increase in sodium permeability. In light of our previous discussion of the ionic basis of membrane potentials, it is very easy to confuse the cause-and-effect relationships of the statement just made. Earlier we pointed out that an increase in sodium permeability *causes* membrane depolarization; now we are saying that depolarization *causes* an increase in sodium permeability. Combining these two distinct causal relationships yields the positive-feedback cycle (Fig. 6-13) responsible for the rising phase of the action-potential spike: Depolarization alters the cell membrane structure so that its permeability to sodium increases; because of increased sodium permeability, sodium diffuses into the cell; this addition of net positive charge to the cell further depolarizes the membrane, which, in turn, produces a still greater increase in sodium permeability, which, in turn, causes. . . . We still have no physicochemical explanation of how the sodium permeability of the membrane is altered when the level of the membrane potential changes. Whatever mechanisms may be responsible, they appear to be present only in the cell membranes of nerve and muscle cells. In other types of cells, changing the membrane potential does not cause a change in membrane permeability.

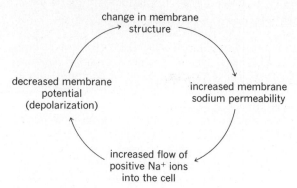

FIGURE 6-13
Positive-feedback relationship between membrane depolarization and increased sodium permeability which leads to the rapid rising phase of the action potential. This relationship is called *sodium activation.*

Finally, as described above, the rising phase of the action potential is terminated and repolarization is brought about mainly by a rapid shutting off of the increased sodium permeability. The cause of this sodium inactivation is unknown. Similarly, the cause of the changes in potassium permeability which contribute to repolarization is unknown.

Threshold

Given the positive-feedback cycle just described, one might conclude (wrongly) that an action potential will be triggered whenever the cell membrane is even slightly depolarized, either by electric input, as in our experiment, or by the physiological means to be described later. But such is not the case, as demonstrated in Fig. 6-14. In this experiment, a series of depolarizing stimuli is delivered electrically to the cell membrane. The first stimulus, which is fairly weak, causes a transient, small membrane depolarization *but no action potential.* The next stimulus is doubled in strength and causes twice the amount of depolarization but again the change is transient and does not trigger an action potential. Only when the stimulus strength has been increased even more does an action potential occur. Thus, partially depolarizing an excitable membrane initiates an action potential only when the strength of the stimulus is sufficient to depolarize the membrane potential to a critical level, known as the *threshold potential.* Such a stimulus is known as a *threshold stimulus.* Stimuli weaker than this are known as *subthreshold stimuli* and do not initiate an action potential. Stimuli of more than threshold magnitude (*suprathreshold stimuli*) elicit action potentials but, as can be seen in Fig. 6-14, the action-potential response is no different from that following a threshold stimulus. The threshold potential of most excitable membranes is 5 to 15 mV more depolarized than the resting membrane potential. Thus, if the resting potential of a neuron is −70 mV, the threshold potential may be −60 mV; in order to initiate an action potential in such a membrane, the potential must be decreased by at least 10 mV.

What is the explanation for the phenomenon of threshold; i.e., why does not any slight amount of initial depolarization trigger an action potential? The answer lies in the fact that, at any potential between resting and threshold, sodium movement into the cell is less than potassium movement out, despite the increased sodium permeability.[1] This prevents any further depolarization beyond that induced directly by the stimulus and drives the potential rapidly back toward the resting level as soon as the stimulus is removed. In contrast, at potentials just barely above threshold, the sodium permeability has been increased so much (as a result of the stimulus-induced depolarization) that its inflow exceeds potassium outflow. This net inflow of positively charged ions depolarizes the membrane still further, which increases sodium permeability even more. Thus the positive-feedback relationship can be effective in causing an action potential only after the membrane has been initially depolarized to the critical threshold value by a stimulus. The nature of the physiologically occurring stimuli will be described in the subsequent sections on receptors and synapses.

It should be reemphasized that, once threshold is reached, the process is no longer dependent upon stimulus strength. The depolarization continues on to become an action potential solely because the membrane permeability changes allow sodium ions to diffuse down the electric and concentration gradients which exist across the membrane. Thus, action potentials triggered either by stimuli just strong enough to depolarize the membrane a bit above threshold or by very strong stimuli are identical. Action potentials occur either maximally as determined by the electrochemical conditions across the

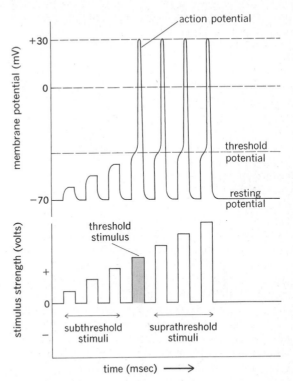

FIGURE 6-14
Decreasing membrane potential with increasing strength of depolarizing stimulus. When the membrane potential reaches the threshold potential, action potentials are generated. Increasing the stimulus strength above threshold levels does not alter the action-potential response. (The after-hyperpolarization has been omitted from this figure and from Figs. 6-15 and 6-16.)

membrane or they do not occur at all. Another way of saying this is that action potentials are *all or none*.

Because of the all-or-none nature of the action-potential response, a single action potential cannot convey any information about the magnitude of the stimulus which initiated it, since a threshold-strength stimulus and one of twice threshold strength give the same response. Since the function of nerve cells in the body is to transmit information by propagating action potentials, one may ask how a system operating according to an all-or-none principle can convey information about the strength of a stimulus. How can one distinguish between a loud noise and a whisper, a light touch and a pinch? The answer, as we shall see later, depends upon the number of action potentials transmitted per unit time, i.e., the frequency of action potentials, and not upon their size.

[1] Recall that the flux of an ion across a membrane depends not only upon membrane permeability but also upon the electric potential. Depolarizing the membrane increases the flux of potassium out of the cell even though there is no change in the potassium permeability of the membrane at this time. Thus, although the sodium flux into the cell increases during depolarization because of the potential-dependent changes in membrane sodium permeability, the flux of potassium out remains slightly greater than that of sodium in until threshold is reached. Beyond threshold, sodium flux is greater than potassium flux.

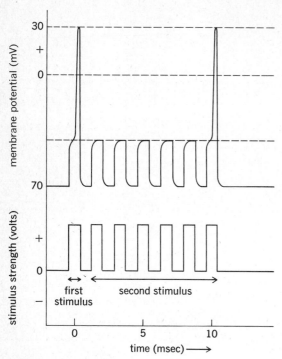

FIGURE 6-15
Following generation of an action potential, the membrane remains refractory to a second threshold-level stimulus for several milliseconds.

Refractory periods

How soon after firing an action potential can an excitable membrane be stimulated to fire a second one? If we apply a threshold-strength stimulus to a membrane and then stimulate the membrane with a second threshold-strength stimulus at various time intervals following the first, the membrane does not always respond to the second stimulus (Fig. 6-15). Even though identical stimuli are applied to the membrane, it appears unresponsive for a certain time. The membrane during this period is said to be *refractory* to a second stimulus.

Instead of applying the second stimulus at threshold strength, if we increase it to suprathreshold levels, we can distinguish two separate refractory periods associated with an action potential (Fig. 6-16). During the 1-msec period of the action-potential spike a second stimulus will not produce a second action-potential response no matter how strong it is. The membrane is said to be in its *absolute refractory period*. Following the absolute

refractory period there is an interval during which a second action-potential response can be produced but only if the stimulus strength is considerably greater than threshold level. This is known as the *relative refractory period* and can last some 10 to 15 msec or longer.

The mechanisms responsible for the refractory periods are related to the membrane mechanisms that alter the sodium and potassium permeability. The absolute refractory period corresponds with the period of sodium permeability changes, and the relative refractory period corresponds roughly with the period of increased potassium permeability. Following an action potential, time is required to return the membrane structure to its original resting state. The system may be likened to the cocking of a spring; until the spring (sodium permeability mechanism) has been reset, it cannot be released again.

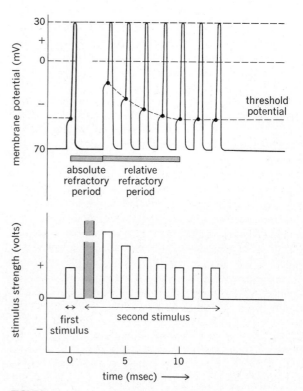

FIGURE 6-16
The magnitude of a single second stimulus necessary to generate a second action potential during the refractory period is greater than the initial stimulus and decreases in magnitude as the time between the first and second stimulus increases. Immediately following an action potential, the membrane is absolutely refractory to all stimulus strengths.

The refractory periods limit the number of action potentials that can be produced by an excitable membrane in a given period of time. Recordings made from nerve cells in the intact organism that are responding to physiological stimuli indicate that most nerve cells in the body respond at frequencies between 0 and 100 action potentials per second although some nerve cells may produce much higher frequencies for brief periods of time.

Action-potential propagation

Everything we have discussed so far concerns a signal, in the form of an electrochemical change across the membrane, brought about by a stimulus, but for this signal to serve as a means of communication there must be a way for it to travel from one part of the cell to another, and this brings us to the topic of action-potential propagation.

One particular action potential does not itself travel along the membrane; rather, each action potential triggers, by local current flow, a new one at an adjacent area of membrane. The old action potential provides the electric stimulus that depolarizes the new membrane site to just past its threshold potential. Once this has happened, the sodium activation cycle at the new membrane site takes over and an action potential occurs there. Once the new site is depolarized to threshold, the action potential generated there is solely dependent upon the electrochemical gradients and membrane permeability properties at the new site. Since these factors are identical to those involved in the generation of the old one, the new action potential is virtually identical to the old. This is an important point because it means that no distortion occurs as the signal passes along the membrane; the signal (action potential) arriving at the end of the membrane is precisely identical to the initial one.

Follow the steps in action-potential propagation, concentrating initially on the shaded area in Fig. 6-17. Ignore for the present the question of how the first action potential was initiated and start with the fact that sodium permeability has changed so that sodium rushes in across the membrane, making the inside of the cell relatively more positive and leaving the outside more negative (Fig. 6-17A and B). Looking at the phenomenon in tiny fragments of time as we are, we can assume that at this first instant the remainder of the cell membrane is at its normal resting potential; accordingly, the shaded area of membrane has an electric potential different from that of the adjacent areas. Like charges attract and unlike charges repel; thus, current (defined as the flow of positive

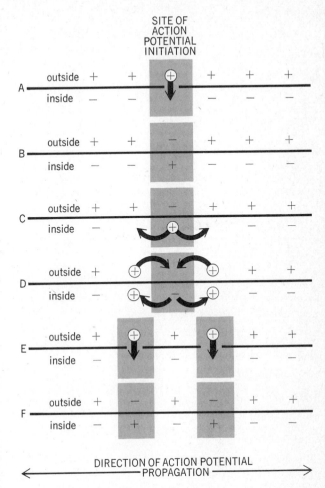

FIGURE 6-17
Mechanisms of action-potential propagation.

charge) flows *away* from the activated membrane region through the cytoplasm (Fig. 6-17C) and *toward* the activated region through the extracellular fluid (Fig. 6-17C and D).

What is the effect of these local current flows on the as-yet-unexcited regions of the membrane adjacent to the shaded area? The addition of positive charge to the inside of the cell and the removal of positive charge from the outside decrease the potential difference across the membrane; this initial depolarization, due to local current flow, acts as the stimulus to trigger an action potential. Meanwhile, at the original membrane site, sodium inactivation is occurring and potassium permeability is increasing so that the membrane is repolarizing. These processes repeat themselves until the end of the membrane is reached.

The steps illustrated in Fig. 6-17C and D are examples of local current flow. Let us look at this process in more detail. Note that, as in all biological systems, the current is carried by ions such as K^+, Na^+, Cl^-, and HCO_3^-. By convention the direction of movement of the positive ions is designated the direction of current flow, but negatively charged particles can and do move in the opposite direction.

The change in electric charge or potential, however, passes from one point to another much faster than the ions themselves move between these same two points. This can best be understood by considering water movement through a water-filled pipe. If the pressure is changed at one end of the pipe, e.g., by putting in more water, the water flow at the opposite end is changed long before those molecules that were put in could have traveled to the outflow end. This occurs because the influence of the molecules (in this example the increase in water pressure) is transmitted much faster than the molecules themselves.

In much the same way, the influence of electric charge is transmitted from one area to another much faster than the ions could move between the same two points. But here the analogy between water flow in the fluid-filled pipe and current flow in the cytoplasm or extracellular fluid begins to break down. In the case of the pipe, the flow out one end equals the input. But current flow in the cytoplasm is more like water flow through a leaky hose; charge is lost across the membrane, with the result that the flow out the other end is less than the input. In fact, when conducting current, cells are so leaky that the current almost completely dies out within a few millimeters of its point of origin. For this reason, local current flow is *decremental;* i.e., its amplitude decreases with increasing distance.

But it does not matter that cytoplasm is such a poor conductor of electric current and that local current flow is decremental because membranes of nerves and muscles depend upon local current flow only over very short distances. Only that amount of local current flow which is necessary to depolarize adjacent membrane areas to threshold is required. Once threshold is reached, an action potential occurs at the new site. The action potentials are transmitted *without decrement,* because new action potentials are continually generated along the membrane.

Direction of action-potential propagation In the example discussed above (Fig. 6-17) the membrane is stimulated in the middle; in this case the action potential spreads in both directions away from the site of stimulation. Local current flow occurs in all directions in which there is an electric gradient (meaning, of course, that it also flows toward the original site of stimulation); however, the membrane areas which have just undergone an action potential are refractory and cannot undergo another; thus, the only direction of action-potential propagation is away from the stimulation site (Fig. 6-17G).

The action potentials in skeletal muscle membrane are initiated near the middle of the cell and propagate from this region toward the two ends, but in most nerve-cell membranes action potentials are initiated at one end of the cell and propagate in only one direction toward the other end of the cell. Realize, though, that this unidirectional propagation of action potentials is determined by the stimulus location rather than an intrinsic inability to conduct in the opposite direction.

Velocity of action-potential propagation The velocity with which an action potential is transmitted down the membrane depends upon fiber diameter and whether or not the fiber is myelinated. The larger the fiber diameter, the faster is the action-potential propagation, because a large fiber offers less resistance to local current flow; therefore adjacent regions of the membrane are brought to threshold faster.

Myelinization is the second factor influencing propagation velocity. *Myelin* is a fatty covering present around most nerve membranes. Myelin electrically insulates the membrane, making it more difficult for current to flow between intra- and extracellular fluid compartments. In effect, it "reduces the leak in the hose"; less current passes out through the myelin-covered section of the membrane during local current flow so that there is a lesser change in the voltage gradient along the fiber (Fig. 6-18). Action potentials do not occur along the sections of membrane protected by myelin; they occur only where the myelin coating is interrupted (called the nodes of Ranvier) and the membrane is exposed to the extracellular fluid (Figs. 6-19 and 6-20). Thus the action potential appears to jump from one node to the next as it propagates along myelinated fiber, and for this reason this method of propagation is called *saltatory conduction,* from the Latin *saltare,* to leap. The membrane of nodes adjacent to the active node is brought to threshold faster and undergoes an action potential sooner than if myelin were not present. The velocity of action-potential propagation in large myelinated fibers can exceed 250 mi/hr.

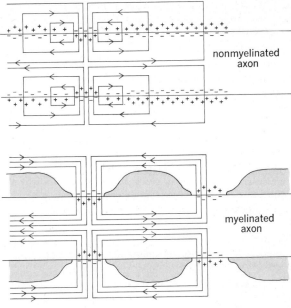

FIGURE 6-18
Ion current flows during an action potential in a
nonmyelinated and a myelinated axon.

SECTION B
PHYSIOLOGICAL ACTIVITIES OF NEURONS

Functional anatomy of neurons

The basic unit of the nervous system is the individual
nerve cell, or *neuron*. (However, only 10 percent or so of
the cells in the nervous system are neurons; the remainder are glial cells, which probably sustain the neurons
metabolically and support them physically.) The human
nervous system is thus composed of the neurons and glial
cells which make up the brain and spinal cord as well
as the many nerve processes which pass between these
two structures and the receptors, muscles, or glands
which they innervate. The brain and spinal cord, together
forming the *central nervous system,* are protected by being
housed within the bony skull and vertebral column
(backbone). All nerve cells or parts of nerve cells which
lie outside the skull or vertebral column are known collectively as the *peripheral nervous system.*

The neuron can be divided structurally into three
parts, each associated with a particular function (Fig.
6-21): (*1*) the dendrites and cell body, (*2*) the axon, and
(*3*) the axon terminals. The *dendrites* form a series of highly

branched cell outgrowths connected to the cell body and
may be looked upon as an extension of the cell membrane of the neuron cell body. The dendrites and cell
body are the site of most of the specialized junctions with
other neurons through which signals are passed to the
cell. Moreover, the cell body contains the nucleus and
many of the organelles involved in metabolic processes
and is responsible for maintaining the metabolism of the
neuron and for its growth and repair.

The *axon,* or *nerve fiber,* is a single long process
extending from the cell body, usually considerably longer
than the dendrites. The first portion of the axon plus the
part of the cell body where the axon is joined is known
as the *initial segment.* The axon can give off branches
called *collaterals* along its course, and near the end it
undergoes considerable branching into numerous *axon
terminals,* the last part of which is enlarged and is responsible for transmitting a signal from the neuron to the cell
contacted by the axon terminal. Neurons assume many
different shapes, depending upon their role (Fig. 6-22),
and sometimes the axons and dendrites are hard to distinguish; for example, the long process between the receptors and cell body of an afferent neuron (Fig. 6-22)
is technically a dendrite since it conducts action potentials toward the cell body, yet it looks like an axon and
is, in fact, often called an axon or nerve fiber.

Yet, regardless of their shape, neurons can be
divided into three classes: afferent neurons, efferent neurons, and interneurons (Fig. 6-23). Afferent and efferent
neurons lie largely outside the skull or vertebral column,
and interneurons lie within the central nervous system.

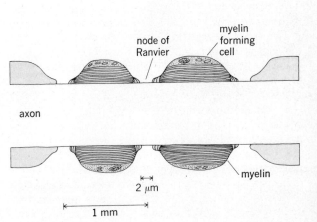

FIGURE 6-19
Structure of a myelinated axon. The surface of the axon is
exposed to the extracellular fluid at the nodes of Ranvier.

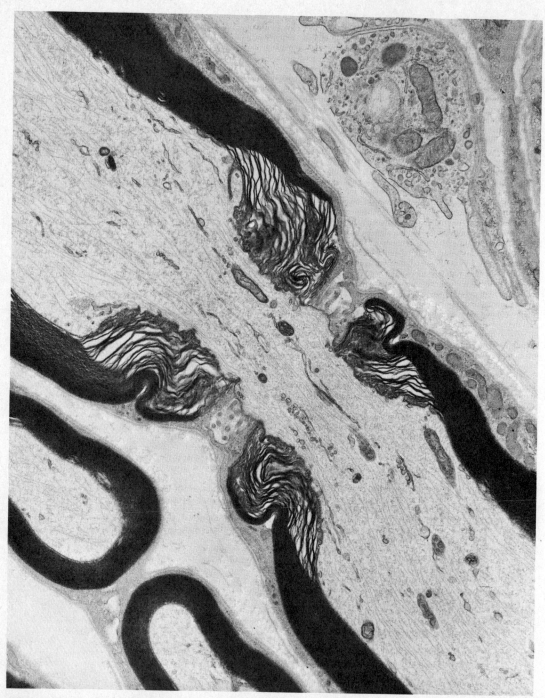

FIGURE 6-20
Electron micrograph of a myelinated axon (mouse sciatic nerve) in the region of a node of Ranvier. The axon, bordered by dark areas of myelin, is oriented diagonally in this micrograph (top left and bottom right). The myelin appears to fray on both sides of the node (cf. Fig. 6-19). *(From Keith R. Porter and Mary A. Bonneville, "An Introduction to the Fine Structure of Cells and Tissues," 2d ed., Lea & Febiger, Philadelphia, 1964.)*

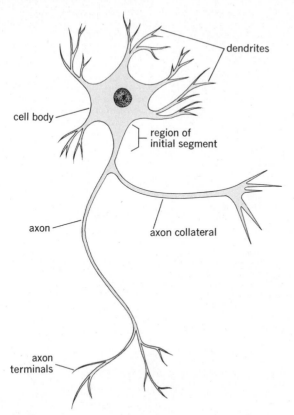

FIGURE 6-21
Highly diagrammatic representation of a neuron.

Labels in figure: dendrites; cell body; region of initial segment; axon; axon collateral; axon terminals

At their peripheral endings afferent neurons have *receptors,* which, in response to various physical or chemical changes in their environment, cause action potentials to be generated in the afferent neuron. The afferent neurons carry information from the receptors *into* the brain or spinal cord. After transmission to the central nervous system, some of this afferent information may be perceived as a conscious sensation. Efferent neurons transmit the final integrated information *from* the central nervous system out to the effector organs (muscles or glands). Efferent neurons which innervate skeletal muscle are also called *motor neurons.* The third group of nerve cells, the *interneurons,* both originate and terminate within the central nervous system, and 99 percent of all nerve cells belong to this group. The interneurons and their connections in large part account for thoughts, feelings, learning, language, etc. The number of interneurons in the pathway between afferent and efferent neurons varies

according to the complexity of the action. One type of basic reflex, the stretch reflex, has no interneurons, the afferent neuron synapsing directly upon the efferent neuron, whereas stimuli invoking memory or language may involve thousands of interneurons.

Given this brief description of neural structure and function, we may now proceed to describe how neural activity, i.e., action potentials, is initiated in the body and how its rate can be altered to relay different types of information. Action potentials are normally initiated in three ways: by the activation of receptors, by synaptic input from other neurons, or, in some neurons, by spontaneous activity.

Receptors

Information about the external world and internal environment exists in different energy forms—pressure, temperature gradients, light, sound waves, etc—but only receptors can deal with these energy forms. The rest of the nervous system can extract meaning only from action potentials (or, over very short distances, from small, subthreshold changes in membrane potential). Thus, regardless of its original energy form, information must be translated into the language of action potentials.

The devices which do this are *receptors.* They are either specialized peripheral endings of neurons or separate cells intimately connected to them. Not all neurons have specialized receptor regions, only that small class of neurons called *afferent* neurons. There are several types of receptors, each of which is specific; i.e., it responds more readily to one form of energy than to others, although virtually all receptors can be activated by several different forms of energy. For example, the receptors of the eye normally respond to light, but they *can* be activated by intense mechanical stimuli like a poke in the eye. Usually much more energy is required to excite a receptor by energy forms to which it is not specific. On the other hand, most receptors are exquisitely sensitive to their specific energy form. Olfactory receptors can respond to a concentration as low as three or four odorous molecules, and visual receptors can respond to the smallest known quantity of light.

Receptor activation:
the generator potential

Here we describe only the general mechanisms for receptor activation and use the simple case in which the receptor is the peripheral ending of the afferent neuron

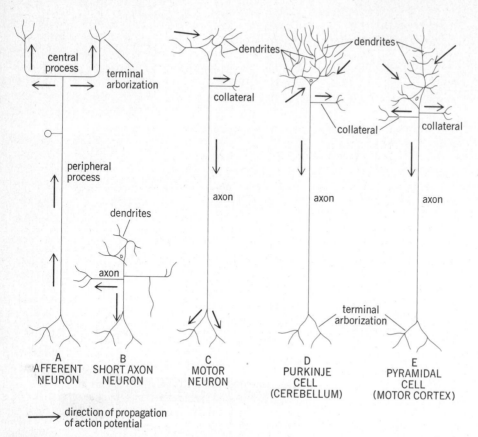

central
process

terminal
arborization

peripheral
process

dendrites

axon

A
AFFERENT
NEURON

dendrites

axon

B
SHORT AXON
NEURON

dendrites

collateral

axon

C
MOTOR
NEURON

dendrites

collateral

axon

D
PURKINJE
CELL
(CEREBELLUM)

dendrites

collateral

axon

terminal
arborization

E
PYRAMIDAL
CELL
(MOTOR CORTEX)

→ direction of propagation
of action potential

FIGURE 6-22
Examples of the different shapes of neurons. Arrows indicate
direction of impulse propagation.

itself and responds to pressure or mechanical deforma-
tion. In subsequent chapters, we shall describe the many
other kinds of receptors which respond to such stimuli
as light, sound, and heat. The receptor action (in this
case the transformation of mechanical energy to the
electrochemical energy of action potentials) occurs at the
very end of the afferent neuron in the portion which lacks
a myelin sheath (Fig. 6-24).

Mechanical stimuli, such as pressure, bend,
stretch, or press upon the receptor membrane, and
somehow, perhaps by opening up pores in the membrane, increase membrane permeability. With the increased permeability, ions move across the membrane
down their electric and concentration gradients. Although
these ion movements have not been worked out as clearly
as those associated with action potentials, the permeability increases appear to be nonselective and to apply
to all small ions. The effects of the membrane permeability

changes to be described below for the mechanoreceptor
are possibly similar in every receptor, even when it is
distinct from the neuron.

Remember that the intracellular fluid of all nerve
cells has a higher concentration of potassium and a lower
concentration of sodium than the extracellular fluid and
that the inside of a resting neuron is about 70 mV negative with respect to the outside. The effect of a nonselective increase in membrane permeability at a stimulated
receptor is a net outward diffusion of a small number of
potassium ions and the simultaneous movement of a
larger number of sodium ions in. The result is net movement of positive charge into the cell, leading to a decrease in membrane potential (depolarization). The
movement of sodium into the nerve fiber thus plays the
major role in depolarizing the nerve ending, potassium
and other ions such as chloride being less important. This
initial depolarization of the receptor is known as the *gener-*

ator potential. It occurs at the unmyelinated nerve ending where the cell membrane has a very high threshold; thus, the generator potential cannot, itself, depolarize the nerve ending enough to cause an action potential there. Rather, the depolarization, i.e., the generator potential, is conducted by local current flow a short distance from the nerve ending to the first node in the myelin sheath where the membrane's threshold is lower. The magnitude of the generator potential decreases with distance from its site of origin (this is true of all potentials due to local current flow) but, if the amount of depolarization which reaches the first node is large enough to bring the membrane there to threshold, an action potential is initiated. The action potential (not the generator potential) then propagates along the nerve fiber.

If, after one action potential has been fired, the depolarization at the first node is still above threshold, another action potential will occur; as long as the first node is depolarized to threshold, action potentials continue to fire and propagate along the membrane of the afferent neuron. In fact, it is much more common for stimuli to cause trains, or bursts, of action potentials than single ones.

Generator potentials are not all or none; their amplitude and duration vary with stimulus strength and other parameters to be discussed shortly. A change in the

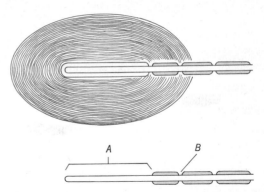

FIGURE 6-24
(*Top*) **A mechanoreceptor (pacinian corpuscle) in which the nerve ending is modified by cellular structures.** (*Bottom*) **The naked nerve ending of the same mechanoreceptor. The generator potential arises at the nerve ending** *A,* **and the action potential arises at the first node in the myelin sheath** *B.* (*Adapted from Loewenstein.*)

generator potential is reflected via local current flow in a similar change in the degree of depolarization at the first node. Because of this relationship, the amplitude and duration of the generator potential determine action-potential frequency in the afferent neuron.

Before discussing the factors that can alter the amplitude of the generator potential, we want to emphasize the fact that action potentials and generator potentials represent quite separate events. It is the action potential which travels along the nerve fiber to its terminations, not the generator potential. The latter is a local response whose only function is to trigger the action potential. The differences between them are summarized in Table 6-2.

Amplitude of the generator potential

Since the amplitude and duration of the generator potential determine the number of action potentials initiated at the first node, the factors controlling these parameters are important. They vary with the stimulus intensity, rate of change of stimulus application, summation of successive generator potentials, and adaptation. Note that generator-potential amplitude determines action-potential *frequency,* i.e., number of action potentials fired per unit time; it does *not* determine action-potential *size.* The action potential is all or none; its amplitude is always the same regardless of the size of the stimulus.

Intensity and velocity of stimulus application   Generator potentials become larger with greater intensity of the stimu-

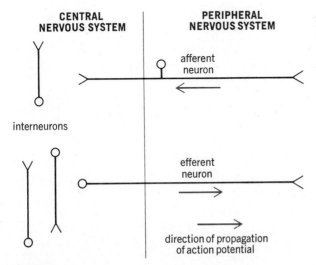

FIGURE 6-23
Three classes of neurons. Note that interneurons are entirely within the central nervous system.

TABLE 6-2
The differences between generator potentials and action potentials

Generator potentials	Action potentials
1 Graded response; amplitude increases with increasing stimulus strength or increasing velocity of stimulus application.	1 All-or-none response; once stimulus strength is great enough to bring the membrane to threshold, further increases in stimulus strength do *not* cause an increase in amplitude.
2 Can be added together; if a second stimulus arrives before the generator potential of the first stimulus is over, the generator potential from the second stimulus is added to the depolarization from the first.	2 Cannot be added together.
3 Has no refractory period.	3 Has a refractory period of about 1 msec.
4 Is conducted passively and decreases in magnitude with increasing distance along the nerve fiber.	4 Is propagated without loss of amplitude along the length of the nerve fiber.
5 Duration is greater than 1 to 2 msec and varies.	5 Duration is 1 to 2 msec.

lus (Fig. 6-25) because the permeability changes increase and the transmembrane ion movements are greater. The explanation for this phenomenon may be an increase in the number or size of pores which permit transmembrane ion flow. The amplitude of the generator potential also rises with a greater rate of change of stimulus application. This applies also to the rate of removal of the stimulus. Thus, although one would expect the amplitude of generator potentials and the rate of firing of action potentials in the afferent neuron to decrease as the stimulus is removed, some receptors give rise to an *off response,* which is a further depolarization of the receptor membrane leading to a burst of action potentials.

Summation of generator potentials   Another way of varying the amplitude of the generator potential is by adding two or more together. This is possible because generator potentials are graded phenomena and because they last 5 to 10 msec; indeed, they may last as long as the stimu-

lus is applied. If the nerve ending is stimulated again before the generator potential from a preceding stimulus has died away, the two potentials sum and make a larger single generator potential.

Adaptation   *Adaptation* is a decrease in frequency of action potentials in the afferent neuron despite a constant stimulus energy. Adaptation is not due to nerve fatigue. Fatigue (in the sense of wearing down from overuse) does not occur in nerve membrane. Adaptation occurs for three reasons: (*1*) The stimulus energy can be dissipated in the tissues as it passes through them to reach the receptor. As the energy loss gradually increases, the amount of energy reaching the receptor decreases. Thus, the membrane permeability changes and generator-potential amplitudes decrease with time. (*2*) The responsiveness of the receptor membrane can decrease with time so that the generator-potential amplitude drops even though the energy reaching the receptor remains unchanged. These two reasons for adaptation are probably the most frequent. In both the lower action-potential frequency is due to lower amplitude of the generator poten-

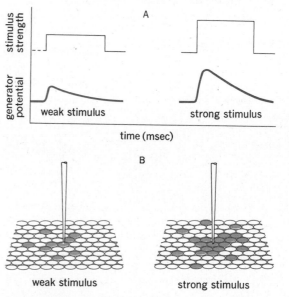

FIGURE 6-25
A weak stimulus produces a small generator potential. The amplitude of the generator potential increases as stimulus intensity increases. B This increase in amplitude possibly occurs because more pores open in the membrane.

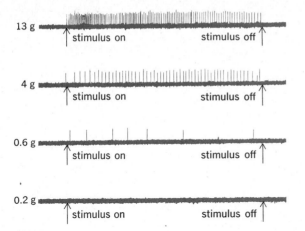

FIGURE 6-26
Action potentials in a single afferent nerve fiber in response to the application of various constant-pressure stimuli to the mechanosensitive receptor ending. (*Adapted from Hensel and Bowman.*)

tial. (3) Even if the generator-potential amplitude remains unchanged, there can be a decreased frequency of action potentials in response to a constant stimulus because of membrane changes at the first node. The actual reasons for adaptation vary in different receptor types. Adaptation of an afferent neuron in response to the constant stimulation of its mechanoreceptor ending can be seen in the first line of Fig. 6-26. Some receptors adapt completely so that in spite of a constantly maintained stimulus, the transmission of action potentials stops. In some extreme cases, the receptors fire only once at the stimulus application or release. In contrast to these rapidly adapting receptors, slowly adapting types merely drop from an initial high action-potential frequency to a lower level, which is then maintained for the duration of the stimulus.

Summary The magnitude of a given generator potential can vary with stimulus intensity, rate of change of stimulus application, summation, adaptation, and cessation of the stimulus, but in spite of these, the amplitude of the generator potential in no way determines the amplitude of the action potentials in the afferent neuron. It does determine how many, if any, action potentials will occur.

Intensity coding

We are certainly aware of different stimulus intensities. How is information about stimulus strength relayed by action potentials of constant amplitude? One way is related to the frequency of action potentials; increased stimulus strength means a larger generator potential and higher frequency of firing of action potentials. A record of an experiment in which increased stimulus intensity is reflected in increased action-potential frequency in a single afferent nerve fiber is shown in Fig. 6-26.

There is an upper limit to this positive correlation between stimulus intensity and action-potential frequency. When stimulus strength becomes very great, the generator potential reaches a maximum and a further increase in rate of firing action potentials by that receptor cannot occur. However, even though that particular receptor cannot generate a higher frequency of action potentials in the afferent neuron, receptors at other branches of the same neuron can be stimulated. Most afferent neurons have many branches, each with a receptor at its ending, but the receptors at different branches do not respond with equal ease to a given stimulus; some are less easily excited and respond only to stronger stimuli. Thus, as stimulus strength increases, more and more receptors begin to respond. Action potentials generated by these receptors propagate along the branch to the main afferent nerve fiber, and if the membrane is not refractory, they increase the frequency of action potentials there.

In addition to the increased frequency of firing in a single neuron, similar receptors on the nerve endings of other afferent neurons are also activated as stimulus strength increases, because stronger stimuli usually affect a larger area. For example, when one touches a surface lightly with a finger, the area of skin in contact with the surface is small and only receptors in that area of skin are stimulated. But pressing the finger down firmly upon the surface increases the area of skin stimulated. This "calling in" of receptors on additional nerve cells is known as *recruitment*. These generalizations are true of virtually all afferent systems: Increased stimulus intensity is signaled both by an increased firing rate of action potentials in a single nerve fiber and by recruitment of receptors on other afferent neurons in the surrounding area.

By the mechanisms discussed in this section different energy forms activate specific receptors to supply information about the kind of stimulus and its duration and intensity. This information is then transmitted in the form of action potentials along the afferent neuron. Next to be considered is the mechanism by which the activity is transferred from one neuron to another.

Synapses

A synapse is an anatomically specialized junction between two neurons where the electric activity in one neuron influences the excitability of the second.

Most synapses occur between the axon terminals of one neuron and the cell body or dendrites of a second. The neurons conducting information toward synapses are called *presynaptic neurons,* and those conducting information away are *postsynaptic neurons.* Figure 6-27 shows how, in a multineuronal pathway, a single neuron can be postsynaptic to one group of cells and, at the same time, presynaptic to another.

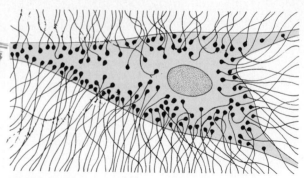

FIGURE 6-28
Synaptic endings on a neuron.

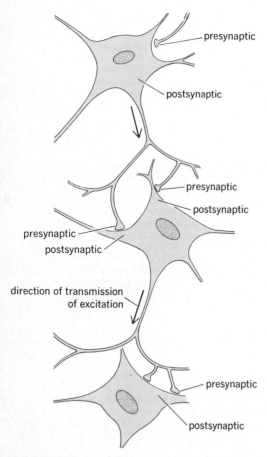

FIGURE 6-27
A single neuron postsynaptic to one group of cells and presynaptic to another.

Every postsynaptic neuron has thousands of synaptic junctions on the surface of its dendrites or cell body so that information from hundreds of presynaptic nerve cells converges upon it; the illustration drawn to demonstrate this fact (Fig. 6-28) has far too few. A single motor neuron in the spinal cord probably receives some 15,000 synaptic endings, and it has been calculated that certain neurons in the brain receive even more. Each activated synapse produces a small electric signal, either excitatory or inhibitory, in the postsynaptic cell for a brief time. The picture we are left with is one of thousands of synapses from many different presynaptic cells *converging* upon a single postsynaptic cell. The level of excitability of this cell at any moment, i.e., how close the membrane potential is to threshold, depends upon the number of synapses active at any one time and how many are excitatory or inhibitory. If the postsynaptic neuron reaches threshold and generates a response, action potentials are transmitted out along its axon to the terminal branches, which *diverge* to influence the excitability of many other cells. Figure 6-29 demonstrates the neuronal relationships of convergence and divergence.

In this manner, postsynaptic neurons function as neural *integrators;* i.e., their output reflects the sum of all the incoming bits of information arriving in the form of excitatory and inhibitory synaptic inputs.

Functional anatomy

Figure 6-30 shows the anatomy of a single synaptic junction. The axon terminal of the presynaptic neuron ends in a slight swelling, the *synaptic knob.* A narrow extracellular space, the *synaptic cleft,* separating the pre- and postsynaptic neurons prevents direct propagation of the action

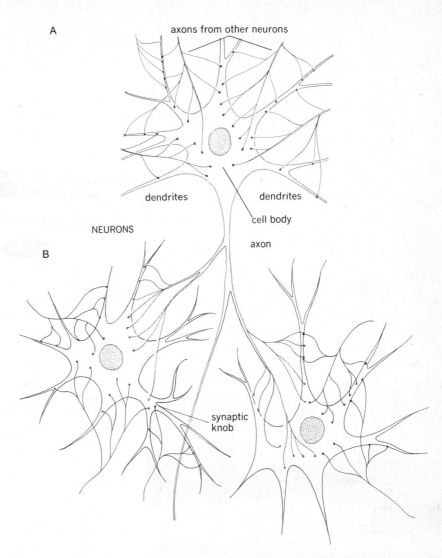

FIGURE 6-29
A Convergence of neural input. B Divergence of neural output.

potential from the presynaptic neuron to the postsynaptic cell. Information is transmitted across the synaptic cleft by means of a chemical agent (in most cases its exact chemical nature is not known) stored in small, membrane-enclosed *vesicles* in the synaptic knob. When an action potential in the presynaptic neuron reaches the

axon terminal and depolarizes the synaptic knob, small quantities of the chemical transmitter are released from the synaptic knob into the synaptic cleft. Little is known about the events which couple action potentials of the presynaptic-cell membrane with the secretion of the transmitter substance. Once released from the vesicles,

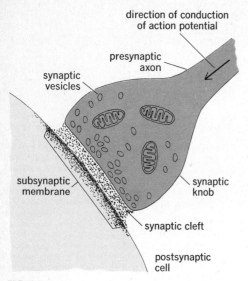

FIGURE 6-30
A synapse.

the transmitter diffuses across the synaptic cleft and combines with receptor sites[1] on the part of the postsynaptic cell lying right under the synaptic knob (*subsynaptic membrane*). The combination of the transmitter with the receptor sites causes changes in the permeability properties of the subsynaptic membrane and in the membrane potential of the postsynaptic cell. There is a delay between excitation of the nerve terminal of the presynaptic neuron and membrane-potential changes in the postsynaptic cell; it lasts less than $\frac{1}{1000}$ sec and is called the *synaptic delay*. Its duration is a function of the release mechanism which frees transmitter substance from the synaptic knob, since the time required for the transmitter to diffuse across the synaptic cleft is negligible. Synaptic activity is terminated when the transmitter is chemically transformed into an ineffective substance, simply diffuses away from the receptor sites, or is taken back up by the synaptic knob.

Excitatory synapse The two different kinds of synapses, excitatory and inhibitory, are classified by their effect on the postsynaptic cells. An excitatory synapse, when activated, increases the likelihood that the membrane poten-

[1]Again, the reader is cautioned not to confuse the term "receptor site" (signifying a membrane molecular configuration with which chemical transmitters combine) with the afferent receptors described in the previous section.

tial of the postsynaptic cell will reach threshold and that the cell will undergo an action potential. Here the effect of the chemical-transmitter receptor site combination is to increase the permeability of the subsynaptic membrane to positively charged ions so that they are free to move according to the electric and chemical forces acting upon them. The mechanisms responsible for the increased permeability are not known.

At the subsynaptic membrane of excitatory synapses there occurs the simultaneous movement of a relatively small number of potassium ions out of the cell and a larger number of sodium ions in. The *net* movement of positive ions is into the neuron, which slightly depolarizes the postsynaptic cell. This potential change, called the *excitatory postsynaptic potential* (EPSP), brings the membrane closer to threshold (Fig. 6-31). The EPSP, like the generator potential of the receptor, is a local, passively propagated potential; its only function is to help trigger an action potential.

Inhibitory synapse Activation of an inhibitory synapse produces changes in the postsynaptic cell which lessen the likelihood that the cell will undergo an action potential. At inhibitory synapses the combination of the chemical transmitter with the receptor sites on the subsynaptic membrane also changes the permeability of the membrane, but only the permeabilities to potassium and chloride ions are increased; sodium permeability is not. The greater permeability to potassium is responsible for the changes in membrane potential associated with the acti-

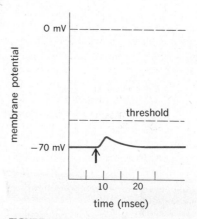

FIGURE 6-31
Excitatory postsynaptic potential (EPSP). Stimulation of the presynaptic neuron is marked by the arrow. Note the short synaptic delay before the postsynaptic cell responds.

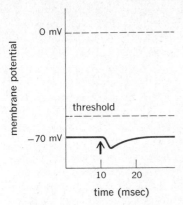

FIGURE 6-32
Inhibitory postsynaptic potential (IPSP). Stimulation of the presynaptic neuron is marked by the arrow. Note the short synaptic delay.

vation of an inhibitory synapse. Earlier it was noted that if the membrane were permeable only to potassium ions the resting membrane potential would equal the potassium equilibrium potential; i.e., the resting membrane potential would be −90 mV instead of −70 mV. The increased potassium permeability at an activated inhibitory synapse makes the postsynaptic cell more like the hypothetical cell that is permeable only to potassium ions. Consequently, the membrane potential becomes closer to a true potassium equilibrium potential. This increased negativity (hyperpolarization) is an *inhibitory postsynaptic potential* (IPSP) (Fig. 6-32). Thus, when a neuron is acted upon by an inhibitory synapse, its membrane potential is moved farther away from the threshold level.

The rise in chloride-ion permeability lessens the likelihood that the cell will reach threshold, the reason being that it increases the tendency of the membrane to stay at the resting potential because the equilibrium potential of chloride is very close to the resting membrane potential. The greater chloride permeability is important when EPSPs and IPSPs arrive at the postsynaptic cell simultaneously because stabilization of the membrane at its resting potential makes it less likely that it will change toward threshold.

Activation of the postsynaptic cell

A feature that makes postsynaptic integration possible is that, in most neurons, one excitatory synaptic event is not enough by itself to change the membrane potential of the postsynaptic neuron from its resting level to thresh-

old; e.g., a single EPSP in a motor neuron is estimated to be only 0.5 mV whereas changes of up to 25 mV are necessary to depolarize the membrane from its resting level to threshold. Since a single synaptic event does not bring the postsynaptic membrane to its threshold level, an action potential can be initiated only by the combined effects of many synapses. Of the thousands of synapses on any one neuron, probably hundreds are active simultaneously (or at least close enough in time that the effects of later synaptic events occur before the potential changes caused by the first disappear), and the membrane potential of the postsynaptic neuron at any one moment is the resultant of all the synaptic activity affecting it at that time. There is a general depolarization of the membrane toward threshold when excitatory synaptic activity predominates (this is known as *facilitation*) and a hyperpolarization when inhibition predominates (Fig. 6-33).

Let us perform a simplified experiment to see how two EPSPs, two IPSPs, or an EPSP plus an IPSP interact (Fig. 6-34). Let us assume that there are three synaptic inputs to the postsynaptic cell; *A* and *B* are excitatory and *C* is an inhibitory synapse. There are stimulators on the axons to *A*, *B*, and *C* so that each of the three inputs can be activated individually. A very fine electrode is placed in the postsynaptic neuron and wired to record the membrane potential. In part I of the experiment we shall test the interaction of two EPSPs by stimulating *A* and then, a short while later, stimulating *A* again. Part I

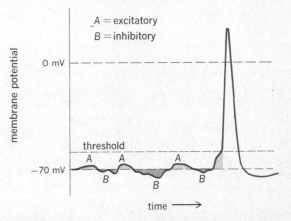

FIGURE 6-33
Intracellular recording from a postsynaptic cell during episodes when (A) excitatory synaptic activity predominates and the cell is facilitated and (B) inhibitory synaptic activity dominates.

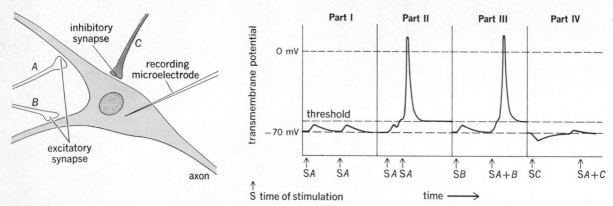

FIGURE 6-34
Interaction of EPSPs and IPSPs at the postsynaptic neuron.

of Fig 6-34 shows that no interaction occurs between the two EPSPs. The reason is that the change in membrane potential associated with an EPSP is fairly short-lived. Within 100 msec the permeability properties of the subsynaptic membrane return to normal, and the excess positive charge in the cell moves back out into the extracellular fluid. By the time axon A is restimulated, the postsynaptic cell has returned to its resting condition. In part II of the experiment, axon A is stimulated again before the effect of the first stimulus has died away, and the potential change from the second stimulation of axon A can be added to the EPSP caused by the first. The two synaptic potentials summate, and because this summation is due to successive stimulation of the same presynaptic fiber, it is called *temporal summation*. In part III, axons A and B are stimulated simultaneously. The potential change attributable to the activity of synapse B can be added to the EPSP caused by the activity of synapse A. The two EPSPs summate, and, because they originate at different places on the postsynaptic neuron, this is called *spatial summation*. The summation of EPSPs can bring the membrane to its threshold so that an action potential is initiated. So far we have tested only the patterns of interaction of excitatory synapses. What happens if an excitatory and inhibitory synapse are activated so that their effects occur at the postsynaptic cell simultaneously? Since the EPSP and IPSP are due to the movement of different ions, they do not exactly cancel each other, and there is a slight depolarization (Fig. 6-34, part IV). Inhibitory potentials can also show temporal and spatial summation.

In the above examples we referred to the threshold

of the postsynaptic neuron. However, the fact is that different parts of a neuron have different thresholds. The neuronal cell body and larger dendritic branches reach threshold when their membrane is depolarized about 25 mV from the resting level, but in many cells the *initial segment* [that part of the neuron cell body from which the axon leaves plus the first, unmyelinated portion of the axon itself (Fig. 6-21)] has a threshold which is less than half that.

As described above, the subsynaptic membrane is depolarized at an activated excitatory synapse and hyperpolarized at an activated inhibitory synapse. By the mechanisms of local current flow described earlier, current flows through the cytoplasm from an excitatory synapse and toward an inhibitory synapse (Fig. 6-35). Thus, the entire cell body, including the initial segment, becomes slightly depolarized during activation of an excitatory synapse and slightly hyperpolarized during activation of an inhibitory synapse. In cells whose initial-segment threshold is lower than that of their dendrites and cell body, the initial segment is activated first whenever enough EPSPs summate, and the action potential originating there is propagated both down the axon and back over the cell body.

Synaptic events last more than 10 times as long as action potentials do. In the event that the initial segment is still depolarized above threshold after an action potential has been fired and the refractory period is over, a second action potential will occur. In fact, the greater the depolarization due to synaptic events, the greater is the number of action potentials fired (up to the limit imposed by the duration of the absolute refractory period).

Neuronal responses are almost always in the form of so-called bursts or trains of action potentials.

The significance of the lower threshold of the initial segment can best be demonstrated if we suppose for a moment that the threshold were the same over the entire neuron so that an action potential could be initiated with equal ease at any point of the cell body or dendrites. In Fig. 6-36 the greatest input to the cell is clearly inhibitory, but in the upper left corner three active excitatory synapses are clustered together. At this one point there is sodium-ion movement into a relatively small portion of the cell. If this small region had a low threshold, an action potential could be initiated at this site and conducted over the entire cell membrane despite the fact that most of the input of the cell is inhibitory. This possible bias of the cell's activity by synapse grouping is greatly lessened

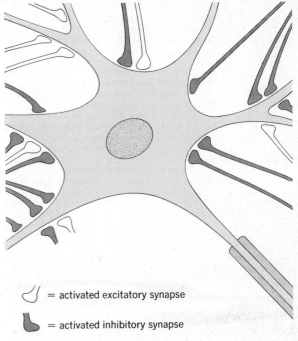

= activated excitatory synapse

= activated inhibitory synapse

FIGURE 6-36
Synaptic input to a neuron.

by the fact that the initial segment acts to average all the synaptic input.

On the other side of the coin, those synapses right next to the initial segment have a greater influence upon cell activity than those at the ends of the dendrites, and thus synaptic placement provides a mechanism for giving different inputs a greater or lesser influence on the postsynaptic cell's output. Finally, it should be noted that, in some cells, action potentials can be initiated in regions other than the initial segment.

Chemical transmitters

As described above, presynaptic neurons influence postsynaptic neurons only by means of chemical transmitters. This section lists some of the most important facts relating to them.

1 There appear to be many different synaptic transmitters. However, all synaptic endings from a single presynaptic cell probably liberate the same one. Some of the substances thought to be transmitters are acetylcholine, norepinephrine, γ-aminobutyric acid (GABA), serotonin, and dopamine. Their specific roles will be described in more detail in subsequent chapters.

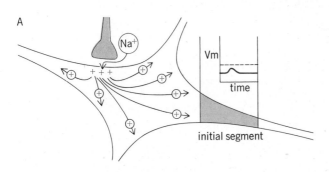

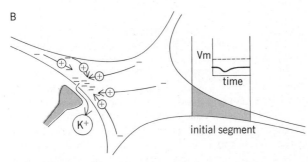

FIGURE 6-35
Comparison of excitatory (A) and inhibitory (B) synapses showing direction of current flow through the postsynaptic cell following synaptic activation. A Current flows through the cytoplasm of the postsynaptic cell away from the excitatory synapse, depolarizing the cell. B Current flows through the cytoplasm of the postsynaptic cell toward an inhibitory synapse, hyperpolarizing the cell. Arrows show direction of positive ion flow.

2 Synapses can operate in only one direction because all the transmitter is stored on one side of the synaptic cleft and all the receptor sites are on the other side. This is in contrast to action potentials, which can travel along a nerve fiber in either direction. Because of the one-way conduction across synapses, action potentials pass through the nervous system in only one direction.

3 The chemical transmitter continues to combine with the receptor sites on the subsynaptic membrane until it is inactivated by combining chemically with another substance, is taken up again by the presynaptic endings, or simply diffuses away from the synaptic region. In some cases, the transmitter substance is not removed immediately from the receptor site, and a single volley of impulses in the presynaptic fibers causes prolonged firing in the postsynaptic neuron.

4 Synapses are vulnerable to many drugs and toxins which can modify the synthesis, storage, release, inactivation, or uptake of the transmitter substance, or block the receptor sites on the subsynaptic membrane to prevent combination with the transmitter. For example, the toxin produced by the tetanus bacillus acts at the inhibitory synapses upon the motor neuron, presumably by blocking the receptor sites. This eliminates inhibitory input to the motor neuron and permits unchecked influence of the excitatory inputs, leading to muscle spasticity and seizures. The spasms of the jaw muscles appearing early in the disease are responsible for the common name of lockjaw.

Presynaptic inhibition

One way an inhibitory influence can be exerted upon the postsynaptic neuron has been discussed, i.e., the activation of an inhibitory synapse. This type of inhibition hyperpolarizes the postsynaptic cell, thereby depressing *all* information fed into the cell. A second type of inhibitory influence, called *presynaptic inhibition,* provides a means by which certain inputs to the postsynaptic cell can be selectively altered. In presynaptic inhibition the postsynaptic cell is not necessarily depressed; rather, its membrane potential is determined by a different group of synaptic inputs and, therefore, by different bits of information.

Presynaptic inhibition works by affecting the transmission at a single excitatory synapse. At excitatory synapses, the amount of chemical transmitter released from the synaptic knob is directly related to the amplitude of the action potential in the axon terminal, and the magnitude of the resulting EPSP in the postsynaptic cell, in turn, is directly related to the amount of chemical transmitter released. In other words, an excitatory synapse is less effective if the amplitude of the action potential in the presynaptic terminal is depressed. [Action potentials usu-

ally have an amplitude of about 100 mV (-70 to $+30$ mV).] This is brought about, in presynaptic inhibition, by means of a synaptic junction between one neuron (neuron A, Fig. 6-37) and the synaptic knob of a second (neuron B). When activated, this synapse slightly depolarizes the membrane of the synaptic knob of neuron B. The depolarization is not great enough to cause an action potential in neuron B; it only brings the membrane potential of the synaptic knob closer to threshold and reduces the amplitude of the action potentials when neuron B is fired. Thus, the amount of chemical transmitter released from the synaptic knob of neuron B is decreased, the permeability changes in the postsynaptic cell (neuron C) are smaller, the size of the EPSP is decreased, and neuron C will be less influenced by inputs from neuron B. Note that this phenomenon provides an exception to the rule that action potentials are always the same along the entire length of the nerve fiber.

Neuroeffector communication

To complete our analysis of the steps involved in neural control, we must at least mention neuroeffector communication. Efferent neurons innervate muscle or gland cells. Information is transmitted from axons to these effector cells by means of chemical transmitters. When an action potential reaches the terminal portions of an axon, it causes the release of transmitter which diffuses to the effector cell and alters its activity. The structures of these axon terminals and their anatomical relationships to the effector cells vary, depending upon the effector cell type; they will be described in detail in subsequent chapters. The neuroeffector transmitters are well characterized (in contrast to the synaptic transmitters); they are either acetylcholine or norepinephrine.

Patterns of neural activity

The purpose of this section is to illustrate through a few specific examples the varying degrees of complexity exhibited by neural control systems and to review the neural mechanisms presented in the previous sections of this chapter. We start with one of the simpler reflexes—the flexion reflex.

The flexion reflex

The sequence of events elicited by a painful stimulus to the toe leads to withdrawal of the foot from the source

chemical energy of action potentials in afferent neurons. Information about the intensity of the stimulus is coded both by the frequency of action potentials in single afferent fibers and by the number of different afferent fibers activated. The afferent fibers branch after entering the spinal cord, each branch terminating at a synaptic junction with another neuron. In the flexion reflex, the second neurons in the pathway are interneurons. Because interneurons are interspersed between the afferent and efferent limbs of the reflex arc, the flexion reflex is one of the very large class of *polysynaptic* reflexes.

The afferent-neuron branches serve different functions. (*1*) Some branches synapse with interneurons whose processes carry the information to higher brain centers; it is only after the information transmitted in these pathways reaches the brain that the conscious correlate of the stimulus, i.e., the sensation of pain, is experienced. (*2*) Other branches synapse with different interneurons which, in turn, synapse upon the efferent neurons innervating flexor muscles. These muscles, when activated, cause flexion (bending) of the ankle and withdrawal of the foot from the stimulus. If the stimulus is very intense and the afferent discharge of very high frequency, the number of motor neurons in the discharge zone is large, and muscles of other joints are activated so that in response to a strong stimulus, the knee and thigh also flex. As the injured leg is flexed away from the stimulus, the opposite leg is extended more strongly to support the added share of the body's weight. This involvement of the opposite limb is called the *crossed-extensor reflex.*

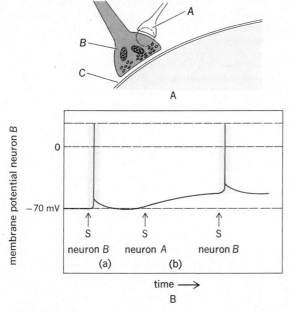

FIGURE 6-37
A **Presynaptic inhibition.** B **Amplitude of the action potential in the synaptic knob: (a) normally and (b) during presynaptic inhibition.**

of injury: this is an example of the flexion, or withdrawal, reflex (Fig. 6-38). Receptors in the toe are stimulated and transform the energy of the stimulus into the electro-

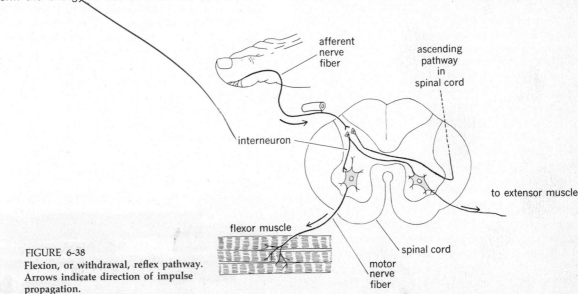

FIGURE 6-38
Flexion, or withdrawal, reflex pathway. Arrows indicate direction of impulse propagation.

Still other branches of the afferent neurons activate interneurons which inhibit the motor neurons of antagonistic muscles whose activity would oppose the reflex flexion (reciprocal innervation). In this case, the motor neurons innervating the foot and leg extensor muscles (which straighten the ankle and leg) are inhibited.

The neural networks activated in the flexion reflex are among the simplest in man, yet the reflex response is purposeful and coordinated. The foot flexion away from the stimulus removes the receptors from the stimulus source, and further damage is prevented (an example of negative feedback). The purposefulness of the reflex response is independent of the sensation of pain, for in normal man the withdrawal occurs before the sensation of pain is experienced. In fact, the reflex response can still be elicited after the spinal cord pathways, which would normally transmit the afferent information up to the brain, have been severed. An important characteristic of this type of control system is that the observed response varies with stimulus strength (between threshold and maximal stimulus strength). In general, low levels of stimulation lead to responses localized to the area of stimulus application whereas higher levels of stimulation cause more widespread (and even opposite) responses. Thus, to take another example, the hotter an object, the more intense is the response to touching it; first one simply withdraws the finger; if it is hotter, the wrist and arm are pulled away; if hotter still, one jumps away with appropriate exclamations.

The flexion reflex demonstrates the coordinated, functional operation of a reflex arc, but it does not represent all parts of the nervous system which can be included in neural control systems.

Swallowing

A swallow is brought about by the coordinated interaction of some 20 muscles whose efferent neurons' cell bodies lie in the lower part of the brain. If food is to be moved from the mouth to the stomach, the muscles of the throat and esophagus (Fig. 12-7) must contract in a precise sequence over time, the upper muscles contracting first to push the food downward while the lower regions of the esophagus are relaxed to allow the food to enter. After the food enters these lower regions, the muscles there must contract to push the food still farther down. The contractions of the upper regions must be maintained during this time to prevent the food from moving back up. And, during a swallow, the respiratory passages must be protected so that food does not enter them. In order for each of these muscles to contract, they must be activated by the motor neurons specific to them. Accord-

ingly, swallowing requires an orderly sequence of excitation of many neurons.

Thus, a single brief stimulus triggers off a chain of responses which may last more than 10 sec. Moreover, once initiated, the course of the swallow is largely unaffected by further sensory input. Contrast this to the flexion reflex in which the stimulus induces a very brief and nearly simultaneous response. Clearly, there must be a far more complex network of interacting neurons built into the swallowing reflex to ensure the timing of the response. However, in one sense, swallowing is a less complex reflex than the flexion reflex in that it elicits a stereotyped all-or-none response whose pattern and magnitude are not altered by the stimulus strength.

Respiratory movements

The control system which mediates respiratory movements has one component which is qualitatively different from any present in the flexion reflex and swallowing, namely, a group of neurons whose membranes manifest rhythmic spontaneous depolarization; i.e., the membrane, in the absence of any external input, rhythmically depolarizes to threshold and fires action potentials. This probably occurs because of a gradual change in membrane permeabilities, either an increasing sodium or decreasing potassium permeability. These neurons, which are the key neurons in controlling respiration, are strongly influenced by afferent input, but this input only plays upon a basic rhythm generated spontaneously. Spontaneously depolarizing neurons such as these provide, as it were, "biological clocks" for the nervous system and may be of considerable importance in other rhythmic activities, including the sleep-wake cycle, the reproductive cycle, daily variations in hormone secretion, etc. Above all, the question of spontaneous activity of neurons may be basic to one's view of man and will be discussed again in Chap. 18.

This completes our analysis of the functional characteristics of the nervous system, and we now turn to a description of its anatomy.

SECTION C
DIVISIONS OF THE NERVOUS SYSTEM

Recall that the brain and spinal cord constitute the central nervous system and the nerve processes connecting these structures with the receptors, muscles, and glands form the peripheral nervous system. These nerve processes can be further classified as the afferent or efferent division of the peripheral nervous system.

TABLE 6-3
Differences between somatic and autonomic nervous systems

Somatic nervous system

1 Fibers do not synapse once they have left the central nervous system.
2 Ends at skeletal muscle.
3 Always leads to excitation of the effector organ.

Autonomic nervous system

1 Fibers synapse once in ganglia after they have left the central nervous system.
2 Ends at smooth or cardiac muscle or gland.
3 Can lead to excitation or inhibition of the effector organ (see text).

Peripheral nervous system: afferent division

Afferent neurons convey information from receptors in the periphery to the central nervous system. Their cell bodies are outside but close to the brain or spinal cord. From the region of the cell body, one long process extends away from the central nervous system to innervate the receptors; commonly, the process branches several times as it nears its destination, each branch innervating one receptor. A second process passes from the cell body into the central nervous system where it branches; these branches terminate in synaptic junctions with other neurons.

The neurons described in the receptor section of this chapter are typical afferent neurons, and so are the neurons which constitute the afferent pathways of the flexion and swallowing reflexes. Regardless of whether the initial stimulus is a prick of the skin, stretch of skeletal muscle, distension of the intestine, or a loud sound, the action potentials relaying information to the central nervous system travel along neurons which are structurally very similar. These make up the afferent division of the peripheral nervous system.

Afferent neurons are sometimes called *primary afferents* or *first-order neurons* because they are the first cells activated in the synaptically linked chains of neurons which handle incoming information. They are also frequently called sensory neurons, but we hesitate to use this term because it implies that the information transmitted by these neurons is destined to reach consciousness, and this is not always true. For example, we have no conscious awareness of our blood pressure even though we have receptors sensitive to this variable.

Peripheral nervous system: efferent division

The efferent division is more complicated than the afferent, being subdivided into a *somatic nervous system* and an *autonomic nervous system*. Although this separation is justified by many anatomical and physiological differences, the simplest distinction between the two is that the somatic nervous system innervates skeletal muscle and the autonomic nervous system innervates smooth and cardiac muscle and glands. Other differences are listed in Table 6-3.

Somatic nervous system

The somatic division of the peripheral nervous system is made up of all the fibers going from the central nervous system to skeletal muscle cells. The cell bodies of these neurons are located in groups within the brain or spinal cord; their large-diameter, myelinated axons leave the central nervous system and pass directly, i.e., without any synapses, to skeletal muscle cells. The transmitter substance released by these neurons is *acetylcholine*. Because activity of somatic efferent neurons causes contraction of the innervated skeletal muscle cells, these neurons are often called *motor neurons*. Motor neurons can be activated by local reflex mechanisms, as in the flexion reflex, or they can be activated by pathways which descend from higher brain centers, but in either case, their excitation always leads to *contraction* of skeletal muscle cells; there are no *inhibitory* somatic motor neurons.

Autonomic nervous system

Fibers of the autonomic division of the peripheral nervous system innervate cardiac and smooth muscle cells or glands. Anatomic and physiologic differences within the autonomic nervous system are the basis for its further subdivision into *sympathetic* and *parasympathetic* components (Table 6-4). The two divisions leave the central nervous system at different levels, the sympathetic from the thoracic and lumbar regions of the spinal cord and the parasympathetic from brain and sacral portion of the

TABLE 6-4
Divisions of the peripheral nervous system

I Afferent system
II Efferent systems
  a Somatic nervous system
  b Autonomic nervous system
    1 Sympathetic nervous system
    2 Parasympathetic nervous system

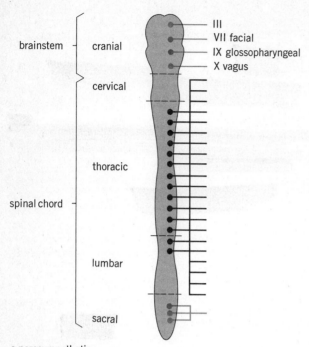

brainstem — cranial
- III
- VII facial
- IX glossopharyngeal
- X vagus

cervical

spinal chord — thoracic

lumbar

sacral

● parasympathetic
● sympathetic

FIGURE 6-39

Origins of the sympathetic and parasympathetic branches of the autonomic nervous system.

spinal cord (Fig. 6-39). Thus, the sympathetic division is also called the *thoracolumbar division,* and the parasympathetic is called the *craniosacral division.* Although the two divisions leave at different levels, the heart and many glands and smooth muscles are innervated by both sympathetic and parasympathetic nerve fibers; i.e., they receive *dual innervation.* (Do not confuse this with reciprocal innervation, which is the inhibition of an active muscle's antagonist.)

The fibers of the autonomic nervous system synapse once after they have left the central nervous system and before they arrive at the neuroeffector junctions (Fig. 6-40). These synapses outside the central nervous system occur in cell clusters called *ganglia.* The fibers passing between the central nervous system and the ganglia are the *preganglionic* autonomic fibers; those passing between the ganglia and the effector organ are the *postganglionic* fibers. The two divisions of the autonomic nervous system differ with respect to the locations of their ganglia. The sympathetic ganglia lie close to the spinal cord in a

well-defined chain called the *sympathetic trunk,* or they lie halfway between the spinal cord and innervated organ. Thus, the sympathetic preganglionic fiber is short and the postganglionic fiber is long. In contrast, the parasympathetic ganglia lie within the walls of the effector organ, and the preganglionic fiber is much longer than the postganglionic fiber.

In both sympathetic and parasympathetic divisions, the chemical transmitter for the ganglionic synapse between pre- and postganglionic fibers is *acetylcholine.* The chemical transmitter at the junction between the *parasympathetic* postganglionic fiber and the effector organ is also acetylcholine. Fibers that release acetylcholine are called *cholinergic* fibers. The transmitter between the *sympathetic* postganglionic fiber and the effector organ is *norepinephrine,* a member of the chemical family of catecholamines (Fig. 6-40).[1] However, early experiments on the function of the sympathetic nervous system were done using epinephrine rather than the closely related norepinephrine. Moreover, the epinephrine was called by its British name, adrenaline, so that fibers which release norepinephrine came to be called *adrenergic* fibers.

Many drugs stimulate or inhibit the synaptic functions of the autonomic nervous system. Those whose actions "mimic" the actions of the sympathetic nervous system are called *sympathomimetic* drugs. Amphetamine and phenylephrine are drugs of this class. Conversely, several choline compounds and the mushroom poison muscarine mimic parasympathetic actions and are *parasympathomimetic* drugs. Drugs which block the actions of the autonomic nervous system are *sympatholytic* or *parasympatholytic.* The plant belladonna yields atropine, which is a parasympatholytic drug. In fact, the plant received its common name because extracts of the leaf caused dilation of the pupil of the eye (by blocking the parasympathetic constriction of the iris sphincter muscle) and transformed the users into "beautiful women." (The plant's other name, deadly nightshade, arose because a lethal overdose of the extract was a favorite poison in the Middle Ages.)

The actions of the autonomic nervous system depend upon the nature of not only the chemical released by the postganglionic cell but also the effector cell's receptor sites and intracellular machinery. Four classes of receptor sites have been identified, and their differences explain seemingly contradictory functions of the divisions of the autonomic nervous system.

[1] There are a few exceptions to this statement; they will be identified where appropriate.

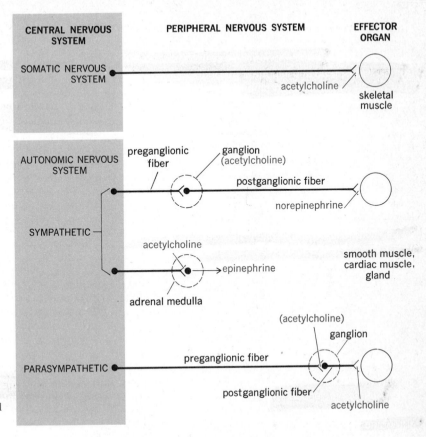

| CENTRAL NERVOUS SYSTEM | PERIPHERAL NERVOUS SYSTEM | EFFECTOR ORGAN |

SOMATIC NERVOUS SYSTEM

acetylcholine

skeletal muscle

AUTONOMIC NERVOUS SYSTEM

preganglionic fiber

ganglion (acetylcholine)

postganglionic fiber

norepinephrine

SYMPATHETIC

acetylcholine

epinephrine

smooth muscle, cardiac muscle, gland

adrenal medulla

(acetylcholine)

ganglion

preganglionic fiber

PARASYMPATHETIC

postganglionic fiber

acetylcholine

FIGURE 6-40
Efferent divisions of the peripheral nervous system.

One "ganglion" in the sympathetic nervous system never developed long postganglionic fibers; instead, upon activation of its preganglionic nerves, the cells of this "ganglion" discharge their transmitters into the blood stream. This "ganglion," called the *adrenal medulla,* is therefore an endocrine gland. It releases a mixture of about 80 percent epinephrine and 20 percent norepinephrine. These substances, properly called hormones rather than neurotransmitters, are transported via the blood and interstitial fluid to receptor sites on effector cells sensitive to them. The receptor sites may be the same ones that sit beneath the terminals of the sympathetic postganglionic neurons and are normally activated by the transmitter delivered directly to them, or they may be other, noninnervated sites.

An important characteristic of autonomic nerves is *dual innervation* of most effector organs, i.e., innervation of the same organ by both divisions. Whatever one division does to the effector organ, the other division frequently does just the opposite. For example, action potentials over the sympathetic nerves to the heart increase the heart rate; action potentials over the parasympathetic fibers decrease it. In the intestine, activation of the sympathetic fibers reduces contraction of the smooth muscle in the intestinal wall; the parasympathetics increase contraction. Dual innervation with fibers of opposite action provides for a very fine degree of control over the effector organ. It is like equipping a car with both an accelerator and a brake. With only an accelerator, one could slow the car simply by decreasing the pressure on the accelerator, but the combined effects of releasing the accelerator and applying the brake provide faster and more accurate control. Skeletal muscle cells, in contrast, do not receive dual innervation; they are activated by somatic motor neurons which are only excitatory. To prevent the sympathetic and parasympathetic systems' opposing effects from conflicting with each other, the two systems are usually activated reciprocally; i.e., as the activity of one system is enhanced, the activity of the other is depressed.

Because glands, smooth muscles, and the heart participate as effectors in almost all bodily functions, it follows that the autonomic nervous system has an extremely widespread and important role in the homeostatic control of the internal environment. It exerts a wide array of effects which can be very difficult to remember. Some common denominators are that, in general, the sympathetic system helps the body to cope with challenges from the outside environment, and the parasympathetic seems to be more responsible for internal housekeeping, such as digestion, defecation, and urination. The sympathetic system is utilized in situations involving stress or strong emotions such as fear or rage, whereas the parasympathetic system is most active during recovery or at rest. The sympathetic nervous system provides the responses to a situation leading to "fight or flight." For example, the sympathetic system increases blood flow to exercising muscles and sustains blood pressure in case of severe blood loss; it decreases activity of the gastrointestinal tract, increases the metabolic production of energy, and increases sweating, changes which provide energy utilization most appropriate to the emergency. These and many other functions of the sympathetic nervous system (and parasympathetic nervous system) will be discussed in relevant places throughout the book.

Autonomic responses usually occur without conscious control or awareness as though they were indeed autonomous (in fact, the autonomic nervous system has been called the involuntary nervous system). However, it is wrong to assume that this need always be the case, for recent experiments have indicated that discrete visceral and glandular responses can be learned. For example, to avoid an electric shock, a rat can learn to selectively increase or decrease its heart rate and a rabbit can learn to constrict the vessels in one ear while dilating those in the other. The implications of such voluntary control of autonomic functions in human medicine are enormous.

These experiments are also important in showing that small segments of the autonomic response can be regulated independently; thus, overall autonomic responses, made up of many small components, are quite variable. Rather than being gross, undiscerning discharges, they are finely tailored to the specific demands of any given situation.

For purposes of general orientation we now turn to a brief description of the anatomy of the central nervous system; the physiology of each region will be described in future chapters.

Central nervous system

Spinal cord

The central nervous system is divided into two main parts, the spinal cord and the brain (Fig. 6-41). The spinal cord is a slender cylinder about as big around as the little finger. Figure 6-42 shows the basic division of the spinal cord into central gray and peripheral white regions. The central area (*gray matter*) is largely filled with the cell bodies and dendrites of interneurons and efferent neurons and the entering fibers of afferent neurons. The cell bodies of neurons in the gray matter are often clustered together with other neurons having similar functions into groups called *nuclei*. The peripheral region of the spinal cord (*white matter*) is made up of nerve axons which transmit action potentials between different levels of the spinal cord or between the spinal cord and brain. The fatty myelin coating of these fibers gives the region its whitish appearance and hence its name. The fibers can be divided into bundles called *pathways* or *tracts*, which are organized according to their function and are made up of nerve-cell processes which transmit the same *types* of information. For example, those nerve fibers which transmit information about light pressure to the brain are all in the same pathway. The spinal white matter contains many tracts, some descending to convey information from brain to spinal cord, others ascending to transmit in the opposite direction (Fig. 6-43). The tracts and even the nuclei can be visualized as columns extending the length of the spinal cord.

Groups of afferent fibers enter the spinal cord on the side toward the back of the body: these groups form the *dorsal roots*. Efferent fibers leave the spinal cord on the opposite side via the *ventral roots*. Shortly after leaving the cord, the ventral and dorsal roots from the same level combine to form a *spinal nerve*. There are 31 symmetrically arranged pairs of spinal nerves in the peripheral nervous system (Fig. 6-41). Spinal nerves (and some cranial nerves) are mixed, containing afferent and efferent, somatic and autonomic fibers.

Brain

The three primary divisions of the brain are the brainstem, cerebellum, and cerebrum (Figs. 6-41 and 6-44 and Table 6-5). The *brainstem* is composed of medulla, pons, and midbrain. It is literally the stalk of the brain, through which pass all the nerve fibers relaying signals of afferent input and somatic or autonomic output between the spinal cord and the higher brain centers. The brainstem contains the cell bodies of motor neurons which control the skeletal

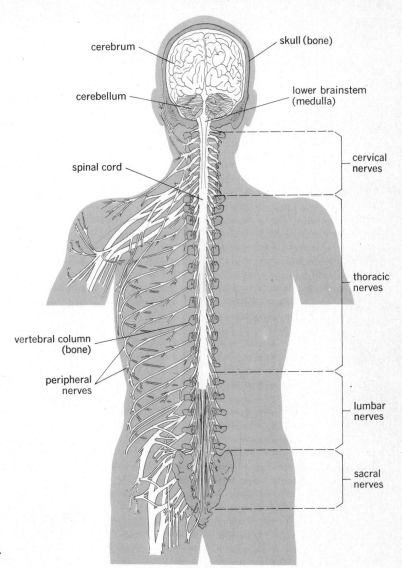

cerebrum

skull (bone)

cerebellum

lower brainstem
(medulla)

cervical
nerves

spinal cord

thoracic
nerves

vertebral column
(bone)

peripheral
nerves

lumbar
nerves

sacral
nerves

FIGURE 6-41
Nervous system viewed from behind.
(*Adapted from Woodburne.*)

muscle of the head. It also gives rise to one of the major parasympathetic nerves, the *vagus,* which innervates the heart and smooth muscle and glands of most other thoracic and abdominal organs, and to the parasympathetic fibers innervating the head. The brainstem also receives many afferent fibers from the head and visceral cavities. Running through the entire brainstem is a core of tissue called the *reticular formation* which is composed of a diffuse collection of small, many-branched neurons. The neurons of the reticular formation receive and integrate information from many afferent pathways as well as from many other regions of the brain. Some reticular-formation neurons are clustered together, forming certain of the brainstem nuclei and "centers." We have already discussed the neurons which dictate the stereotyped pattern of swallowing; they make up the swallowing center, which is located in the reticular formation, and there are cardiovascular, respiratory, and vomiting centers, too.

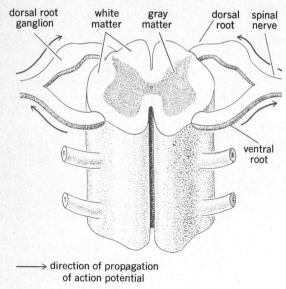

FIGURE 6-42
Section of the spinal cord (ventral view, i.e., from the front).

TABLE 6-5
Divisions of the central nervous system

I Spinal cord
II Brain
  a Brainstem
    1 Medulla
    2 Pons
    3 Midbrain
  b Cerebellum
  c Cerebrum
    1 Cortex
    2 Basal ganglia
    3 Thalamus
    4 Hypothalamus
    5 Other areas

The output of the reticular formation can be divided functionally into descending and ascending systems. The descending components influence the function of both somatic and autonomic efferent neurons, and the ascending components affect such things as wakefulness and the direction of attention to specific events.

The *cerebellum* plays only a minor part in the interaction of autonomic and somatic function. Its chief role, the unconscious coordination of muscle movements, is discussed in Chap. 17.

The large part of brain remaining when the brainstem and cerebellum are excluded is called the *cerebrum.* Its outer portion, the *cortex,* is a cellular shell about $\frac{1}{4}$ in.

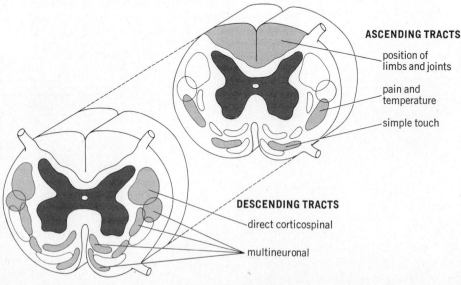

FIGURE 6-43
Some principal tracts of the spinal cord.

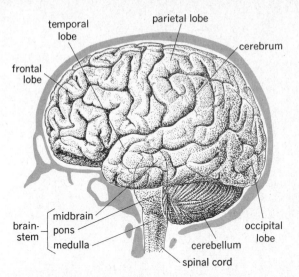

FIGURE 6-44
The three major divisions of the brain: the brainstem, cerebellum, and cerebrum. The outer layer of the cerebrum consists of four lobes, as shown.

thick, containing about 14 billion neurons and covering the entire surface of the cerebrum. It forms the outer rim of the brain cross section in Fig. 6-45. Popular opinion calls the cortex the "site of the mind and the intellect." Scientific opinion considers it to be an integrating area necessary for the bringing together of basic afferent information into complex perceptual images and ultimate refinement of control over both the autonomic and somatic systems, although somatic reactions are more severely affected by cortical damage than are autonomic activities. The cortex is divided into several parts, or *lobes,* frontal, parietal, occipital, and temporal (Fig. 6-44), the functions of which will be discussed in later chapters. The cortex is an area of gray matter, so called because of the predominance of cell bodies. In other parts of the cerebrum, nerve-fiber tracts predominate, their whitish myelin coating distinguishing them as white matter. The *subcortical nuclei* form an area of gray matter lying, as the name suggests, beneath the surface of the cortex (Fig. 6-45) and contribute to the coordination of muscle movements. The *thalamus* (Fig. 6-45) is a way station and important integrating center for all sensory input (except smell) on its way to the cortex. It also contains a significant portion of the reticular system. The *hypothalamus,*

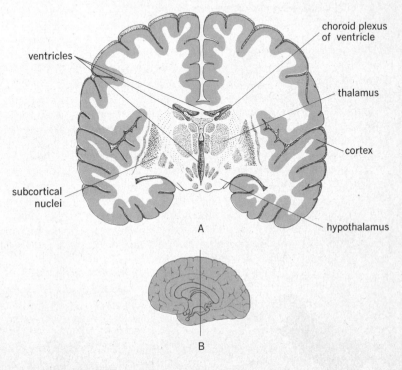

FIGURE 6-45
A Cross section of brain at the level indicated in B.

which lies below the thalamus, is a tiny region whose volume is about 5 to 6 cm³; it is responsible for the integration of many basic behavioral patterns which involve correlation of autonomic, endocrine, and somatic functions (Fig. 6-45). Indeed, the hypothalamus appears to be the single most important control area for regulation of the internal environment. It is also one of the brain areas associated with emotions; stimulation of some hypothalamic areas leads to behavior interpreted as rewarding or pleasurable, and stimulation of other areas is associated with unpleasant feelings. Neurons of the hypothalamus are also affected by a variety of hormones and other circulating chemicals.

Blood supply, blood-brain barrier phenomena, and cerebrospinal fluid Glucose is the only substrate that can be metabolized sufficiently rapidly by the brain to supply its energy requirements, and most of the energy from glucose is transferred to high-energy ATP molecules during its oxidative breakdown. The glycogen stores of brain are negligible; thus the brain is completely dependent upon a continuous blood supply of glucose and oxygen. Although the adult brain is only 2 percent of the body weight, it receives 15 percent of the total blood supply at rest to support the high oxygen utilization. If the oxygen supply is cut off for 4 to 5 min or if the glucose supply is cut off for 10 to 15 min, brain damage will occur. In fact, the most common cause of brain damage is stoppage of the blood supply (a *stroke*). The cells in the region deprived of nutrients cease to function and die.

The central nervous system receives a rich blood supply, but the exchange of substances between the blood and brain or spinal cord is handled differently from the somewhat unrestricted movement of substances from capillaries in other organs. Dye injected into a vein stains all the organs of the body except the brain and spinal cord. A complex group of *blood-brain barrier* mechanisms closely controls both the kinds of substances which enter the extracellular space of brain and the rate at which they enter. The barrier probably comprises both anatomical structures and physiological transport systems which handle different classes of substances in different ways. The blood-brain barrier mechanisms precisely regulate the chemical composition of the extracellular space of the brain and prevent harmful substances from reaching neural tissue.

In addition to its blood supply, the central nervous system is perfused by a second fluid, the cerebrospinal fluid. This clear fluid fills the four cavities (*ventricles*) (Fig. 6-46) within the brain and surrounds the outer surfaces

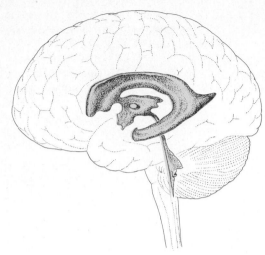

FIGURE 6-46
Cerebral ventricles.

of the brain and spinal cord. The brain literally floats in a cushion of cerebrospinal fluid. Since it is a soft and delicate tissue about the consistency of jelly, the brain is thus protected from sudden and jarring movements of the head. Cerebrospinal fluid is formed within the ventricles by highly vascular tissues, the *choroid plexes*. A blood-brain barrier is present here, too, between the capillaries of the choroid plexus and the cerebrospinal fluid which flows next to the brain. Consistent with the barrier mechanisms, cerebrospinal fluid is a selective secretion, not a simple filtrate of plasma. For example, protein, potassium, and calcium concentrations are lower in cerebrospinal fluid than in a filtrate of plasma, whereas the sodium and chloride are higher. The mechanisms responsible for this selective formation of spinal fluid are not known.

As the cerebrospinal fluid flows from its origin at the choroid plexes within the ventricles, substances diffuse between it and the extracellular space of brain tissue since the walls of the ventricles are permeable to most substances. This exchange across the ventricular walls allows nutrients to enter brain tissue from cerebrospinal fluid and end products of brain metabolism to leave. However, the major sites of nutrient and end-product exchange are the cerebral capillaries. The cerebrospinal fluid moves from its origin, back through the interconnected ventricular system to the brainstem, where it passes through small holes from the ventricle out to the surface of the brain and spinal cord. Aided by circulatory, respiratory, and postural pressure changes, the cerebrospinal fluid finally flows to the top of the outer surface of the brain, where it enters large veins through one-way valves. If the path of flow is obstructed at any point between its site of formation and its final reabsorption into the vascular system, cerebrospinal fluid builds up, causing hydrocephalus, or "water on the brain." The pressure against which cerebrospinal fluid continues to be secreted is quite high—high enough to damage the brain; and mental retardation accompanies severe, untreated cases.

Hormonal control mechanisms

7

Glands are classified into two groups on the basis of the site into which the secretion is released. *Endocrine glands* have no ducts and secrete material into the bloodstream—more precisely, into the extracellular space around the gland cell, from which the secretion then diffuses into capillaries (or lymphatics). *Exocrine glands* secrete into ducts leading to a specific compartment or surface, such as the lumen of the gastrointestinal tract or the skin surface. Certain organs, for example, the pancreas, have both exocrine and endocrine portions. The term endocrine is usually used synonymously with "hormone secreting"; this usage, which we shall also adopt, is not, however, formally correct since there are glands, notably the liver, which secrete nonhormonal materials into the blood.

The endocrine system constitutes the second great communications system of the body, the hormones serving as blood-borne messengers which regulate cell function. We shall define a *hormone* as a chemical substance synthesized by a specific organ or tissue and secreted into the blood, which carries it to other sites in the body, where its actions are exerted. The word specific in the definition is important to distinguish true hormones from another class of substances, the so-called *parahormones*, which are metabolic end products produced by *many* organs of the body and exert effects on distant sites, e.g., the hydrogen ion. In terms of chemical structure, hormones generally fall into two categories: steroids and amino acid derivatives, the latter ranging in size from small molecules containing single amine groups to very large proteins.

Hormones serve to control and integrate many

bodily functions: reproduction (Chap. 14), organic metabolism and energy balance (Chap. 13), and mineral metabolism (Chap 11). The ability to reproduce is absolutely dependent upon a normally functioning endocrine system; in contrast, no other bodily function absolutely requires hormonal control, and, for this reason, the endocrine system is not strictly essential for life. However, having made this statement, we must quickly point out that such a life would be extremely precarious and abnormal; individuals would be unable to adapt to environmental alteration or stress, and their physical and mental abilities would be drastically impaired. They would require the constant attention bestowed upon a hothouse plant.

In recent years, it has become clear that the nervous and endocrine systems actually function as a single interrelated system. The central nervous system, particularly the hypothalamus, plays a crucial role in controlling hormone secretion; and conversely, hormones markedly alter neural function and strongly influence many types of behavior. These interrelationships form the area of study known as *neuroendocrinology* and will be described in the section dealing with control of hormone secretion.

Despite the many ways in which the hormones differ from each other, several general characteristics and principles apply to virtually all of them; these are described in this chapter and form the foundation for the detailed specific descriptions to follow in relevant chapters. This chapter is full of illustratory examples, but these scattered bits of information are given here *only* to explain general principles. Factual information concerning each hormone will be repeated more fully in subsequent chapters.

Hormone–target-organ cell specificity

Hormones travel in the blood and are therefore able to reach virtually all tissues. This is obviously very different from the efferent nervous system, which can send messages selectively to specific organs. Yet, the body's response to hormones is not all-inclusive but highly specific, in some cases involving only one organ or group of cells. In other words, despite the ubiquitous distribution of a hormone via the blood, only certain cells are capable of responding to the hormone; they are known as *target-organ cells*. By unknown evolutionary mechanisms, cells have become differentiated so as to respond in a highly characteristic manner only to certain hormones. As described in Chap. 5, this ability to respond depends upon specific receptor sites on cell components. Specialization

of target-organ receptor sites explains the specificity of action of hormones; e.g., thyroid-stimulating hormone is produced by the anterior pituitary and affects significantly only the thyroid gland and no other tissue, and insulin causes increased glucose uptake by many cells but not by all, the brain cells being one of the important exceptions.

General factors which determine the blood concentrations of hormones

Rate of secretion

With few exceptions, hormones are not secreted at constant rates. As emphasized previously, a regulatory system must be capable of altering its output. Normally, some secretion is always occurring; therefore, the rate can be increased or decreased. This pattern is completely analogous to the phenomena of tonic activity, facilitation, and inhibition manifested by the nervous system.

As was described in Chap. 4, secretion encompasses two processes, intracellular synthesis and release into the blood. For very short periods of time, release can occur in the absence of synthesis because all endocrine cells store some finished product, but over any prolonged period synthesis must obviously keep pace with release. Too little information is available to permit dissociation of the inputs which control these distinct processes, and so we generally incorporate the two into the general term secretion. Hormone deficiencies can be caused by failure of synthesis due to lack of an essential chemical needed for production of the particular hormone. Thus, hormone production requires a supply of chemical precursors either from the diet or other body cells (Fig. 7-1), and the manner in which the rest of the body metabolizes these precursors becomes of great importance.

Rates of inactivation and excretion

The concentration of a hormone in the plasma depends not only upon the rate of secretion but also upon the rate of removal from the blood. Sometimes the hormone is inactivated by the cells upon which it acts, but for most hormones, the pathway of removal from the blood is the liver or the kidneys. Accordingly, patients with kidney or liver disease may suffer from excess of certain hormones solely as a result of reduced hormone inactivation. In any case, it is essential to realize that all hormones are continuously removed by excretion or inactivation, so that maintenance of blood concentrations requires continuous secretion.

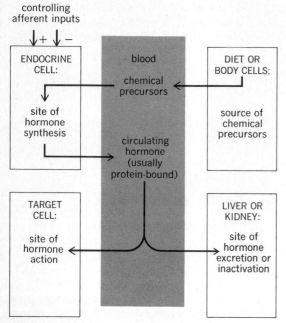

controlling
afferent inputs

↓ + ↓ −

FIGURE 7-1
General factors determining blood concentrations of hormones.

Because, for many hormones, the rate of urinary excretion is directly proportional to the rate of glandular secretion, physiologists often use the rate of excretion as an indicator of secretory rate.

Transport in the blood

Many of the hormone molecules which circulate in the blood are bound to various plasma proteins; the free moiety is usually quite small and is in equilibrium with the bound fraction:

"Free" hormone + protein ⇌ hormone − protein

It is important to realize that only the free hormone can exert effects on the target-organ cells.

Summary example: thyroxine

In subsequent sections, we shall devote considerable attention to the afferent inputs which control the secretion of various hormones but very little to the other factors described above, which also help determine the concentration of circulating hormone. Therefore in order to illustrate the general principles, we here describe these factors in some detail as they apply to one endocrine gland, the thyroid (Fig. 7-2).

The thyroid, located in the neck, secretes several hormones, but the principal one is *thyroxine,* an iodine-containing amino acid. Thyroid physiology received its greatest stimulus when it was discovered that the enlarged thyroids (goiters) so common among inland populations were completely preventable by the administration of small quantities of iodine, as little as 4 g/year. The other ingredient for thyroxine synthesis is the amino acid *tyrosine,* which can be produced from a wide variety of substances within the body and therefore offers no supply problem.

Much of the ingested iodine absorbed by the gastrointestinal tract (which converts it to iodide, the ionized form of iodine) is removed from the blood by the thyroid cells, which manifest a remarkably powerful active-transport mechanism for iodide. Once in the gland, the iodide is reconverted to an active form of iodine, which immediately combines with tyrosine. Two molecules of iodinated tyrosine then combine to give thyroxine, which becomes bound to a polysaccharide-protein material known as *thyroglobulin.* The normal gland may store several weeks' supply of thyroxine in this bound form. Hormone release into the blood occurs by enzymatic splitting of the thyroxine from the thyroglobulin and the entry of this freed thyroxine into the blood. Finally, the overall process is controlled by a pituitary *thyroid-stimulating hormone,* which stimulates certain key rate-limiting steps and thereby alters the rate of thyroxine secretion. A variety of defects—dietary, hereditary, or disease-induced—may decrease the amount of thyroxine released into the blood. One such defect results from dietary iodine deficiency. However, this deficiency need not lead to permanent reduction of thyroxine secretion because the thyroid gland enlarges (goiter) so as to provide greater utilization of whatever iodine is available. This response is mediated by thyroid-stimulating hormone, which induces thyroid enlargement whenever the blood concentration of thyroxine decreases, regardless of cause (the mechanism is described in a subsequent section).

Once in the blood, most of the thyroxine becomes bound to certain plasma proteins and circulates in this form. This protein-bound thyroxine is in equilibrium with a much smaller amount of free thyroxine, the latter being the effective hormone.

Finally, thyroxine may be removed from the blood and excreted by the liver and kidneys or inactivated by catabolism in many tissues. These processes may assume considerable importance in certain disease states, particularly those involving the kidneys or liver.

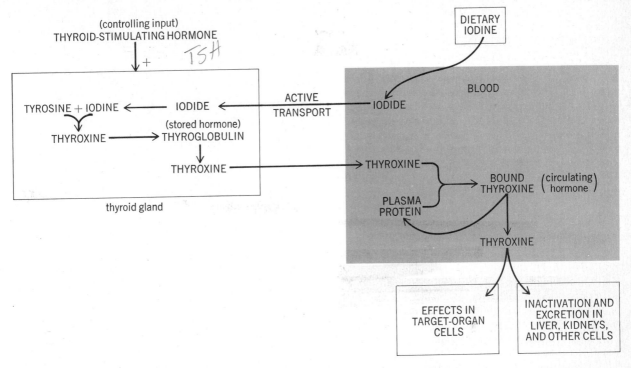

FIGURE 7-2
Summary of thyroxine pathways. Besides iodine, the diet must also supply the amino acids which are used for the synthesis of tyrosine, colloid globulin, and plasma protein.

Thyroid-stimulating hormone is produced by the anterior pituitary. Iodine is converted to iodide in the process of absorption by the gastrointestinal tract.

Mechanisms of hormone action

Common denominators of hormonal effects

Hormones exert their effects by altering the rates at which specific cellular processes proceed. It must be emphasized that hormones never initiate a process[1]; they merely alter its rate. For example, the absence of insulin results in markedly reduced glucose uptake by cells but not

[1] This statement refers to the primary biochemical effect of the hormone (see below) rather than to the biological events ultimately resulting from these effects. For example, the pituitary hormone LH induces ovulation, and ovulation will not occur in its absence. Accordingly, LH "initiates" ovulation. However, the biochemical reactions (RNA and protein synthesis) which ultimately lead to ovulation are not "all-or-none" events, and their rates are altered, not initiated, by LH. There is no contradiction since it is clear that a certain level of change in the biochemical reaction rates is required to trigger off an essentially "all-or-none" event—ovulation.

absolute cessation. The specific cell processes accelerated or decelerated by hormones are numerous and varied, but most of them fit into one of two general categories (both of which require the combination of hormone with a specific receptor site), namely, alteration of the activity of a crucial enzyme, and alteration of the rate of membrane transport of a substance.

Alteration of enzyme activity What is meant by a "crucial enzyme" in this context? The reader should review the section on "regulation of multienzyme pathways" in Chap. 3. Recall that most metabolic reactions are truly reversible whereas others proceed generally in one direction under the influence of one set of enzymes and in the reverse direction under the influence of a second set of enzymes. The enzymes which catalyze these "one-way" reactions are the primary ones regulated by various hormones. Let us consider, as an example, the relationship between glucose and glycogen in the liver:

$$\text{Glucose} \underset{\text{enzyme B}}{\overset{\text{enzyme A}}{\rightleftharpoons}} \text{glycogen} \qquad \textbf{7-1}$$

Although there are actually multiple steps in both pathways, each catalyzed by a different enzyme, two enzymes (A + B) are particularly critical since they catalyze the major irreversible reactions in opposing directions. The hormone insulin increases the activity of enzyme A and thereby stimulates the formation of glycogen from glucose. In contrast, the hormone epinephrine increases the activity of enzyme B and thereby facilitates the catabolism of glycogen to glucose.

How is the activity of a particular type of enzyme increased? One way is for the cell to produce more of the enzyme, a prominent effect of certain hormones. In Chap. 4 we described the mechanics of protein synthesis and the control of these processes. Hormones may exert effects on the genetic apparatus of their target-organ cells to induce (or repress) the synthesis of RNA and, in turn, the proteins (enzymes) whose synthesis is directed by the particular RNA. This may well be the major biochemical action of most steroid hormones (although it is by no means limited to steroid hormones).

There are, however, other ways by which enzyme activity may be altered without a change in the total number of enzyme molecules in the cells, i.e., with no change in enzyme synthesis. Many enzymes exist within a cell in both active and inactive forms; thus, the number of active enzymes can be increased by converting some of the inactive molecules into active ones. This appears to be the common denominator for a number of hormones.

It is of great interest that certain hormones induce both the synthesis of new enzyme molecules and an increased activity of the enzyme molecules already present in the cell. The advantages of this dual effect are considerable: Induction of new enzyme synthesis requires hours to days, whereas the activation of molecules already present can occur within minutes. Thus, the hormone simultaneously exerts a rapid effect and sets into motion a long-term adaptation.

Alteration of membrane transport The effect of many hormones is to facilitate or inhibit the transport of substances into the cell. For example, glucose enters most cells by carrier-mediated, facilitated diffusion, and insulin somehow affects cell membranes so as to increase the rate of glucose transport (this is quite distinct from the insulin effect on enzyme activity described above). Other hormones inhibit glucose transport. A similar type of pattern involving hormonally mediated inhibition and stimulation also operates for the membrane transport of amino acids and other organic metabolites. Finally, the transport of ions and water by kidney cells and others is also influenced by hormones. The inhibition or stimulation of membrane transport is without question one of the major modes of action of many hormones, but we do not know the specific chemical or physical mechanism by which any of these actions is exerted.

Direct and indirect effects

Another important consideration is that of direct versus indirect hormone effects. Because intracellular chemical reactions are so closely interrelated, and because all cells of the body are interconnected by the blood, it should be evident that a single effect may set into motion an almost endless chain of subsequent events, which may well be more important than the initial event for both the single cell and the total body.

Indirect effects within a cell Again we take insulin as our example. As just described, one of its major effects is to increase the transport of glucose into cells, increasing the cellular concentration of glucose and, by mass action, the rates of the many intracellular chemical reactions in which glucose participates. For example, looking again at the synthesis of glycogen (Eq. **7-1**), it should be evident that an increased glucose concentration drives this reaction to the right, resulting in synthesis of more glycogen. The important generalization to be derived from this example is that the direct effect of a hormone may initiate multiple indirect effects within the cell.

Indirect effects on other cells When insulin is injected into a person, the blood concentration of glucose rapidly decreases because glucose is leaving the blood and being taken up by cells all over the body. Conversely, insulin deficiency causes the blood glucose to rise because of deficient uptake. When the blood glucose becomes very high, large quantities of glucose, sodium, and water appear in the urine (the mechanism will be described in Chap. 13). These urinary losses are not due to any direct effect of insulin on the kidney but result indirectly from the high blood glucose. Despite its indirect nature, this urinary loss is one of the major causes of sickness and death in patients with insulin deficiency (diabetes mellitus).

Cyclic AMP A major goal of endocrinologists is to identify the precise initial biochemical actions of hormones on

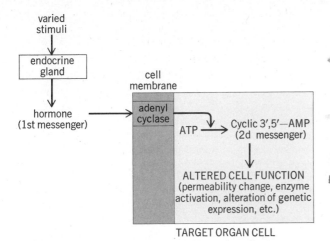

FIGURE 7-3
Second-messenger or cyclic AMP mechanism of hormone action.

their target-organ cells. For example, we have pointed out that an effect of epinephrine is to activate the crucial enzyme (enzyme B in Eq. **7-1**) leading to glycogen breakdown. But exactly how does it do this? Does it act directly on the enzyme itself or does it act several steps removed from the enzyme? To take another example, just what does insulin do to cells that leads to the increased transport of glucose? In recent years it has become apparent that a large number of hormones actually have *identical* initial biochemical actions, namely, activation of the enzyme adenyl cyclase, which is found in cell membranes throughout the body. This has become known as the *second-messenger* or *cyclic AMP* system which is summarized in Fig. 7-3. Adenyl cyclase is a membrane-bound enzyme which, when activated, catalyzes the transformation of cell ATP (on the inner side of the cell membrane) to another molecule known as cyclic AMP. The sole action of the hormone is to interact with a receptor site on the cell membrane so as to activate adenyl cyclase. The cyclic AMP generated as a result then acts within the cell as a "second messenger" to produce the alteration of cell function associated with that hormone. Note that the hormone itself does not gain entry to the cell; this explains how those protein hormones which cannot penetrate cell membranes can still alter cell function.

At least 12 hormones have been shown to exert their biochemical actions by stimulating the intracellular synthesis of cyclic AMP. Several others may reduce cell concentrations of cyclic AMP by inhibiting adenyl cyclase. Let us take as an example the action of epinephrine on glucose formation from glycogen in the liver (Fig. 7-4). Epinephrine activates adenyl cyclase which then catalyzes the formation of cyclic AMP. In turn, cyclic AMP stimulates the conversion of an inactive enzyme to an active form (enzyme B in Eq. **7-1**) which catalyzes the critical reaction leading to the breakdown of glycogen. The figure is highly simplified in that cyclic AMP actually triggers off a complex sequence of multiple reactions leading to the ultimate generation of the final active enzyme.

For his elucidation of this beautiful unifying principle, Earl W. Sutherland was awarded the Nobel Prize for medicine and physiology in 1971. Moreover, cyclic AMP plays important roles in nonendocrine regulatory mechanisms as well. Some examples are the control of antibody production and the regulation of vision, and the list will surely grow rapidly as further experiments are done. However not all hormones act via cyclic AMP, the most notable exceptions being the various steroid hormones. It is also unlikely that all the actions of all nonsteroid hormones can be explained by cyclic AMP.

The perceptive reader may have detected an apparent inconsistency in this description of cyclic AMP; if the generation of cyclic AMP is the common biochemical action of many hormones, why do not all these hormones produce identical effects in the body? The answer is that the adenyl cyclase systems in different target-organ cells differ in their abilities to be activated by different hormones. This is best explained by qualitative differences in membrane receptor sites in different tissues.

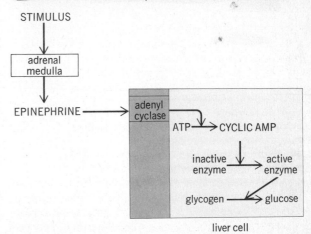

FIGURE 7-4
Effect of epinephrine on the liver via cyclic AMP.

Thus the adenyl cyclase receptor sites in liver cells are able to interact with epinephrine but not parathyroid hormone, whereas just the reverse holds true for bone.

There is a second problem relating to cyclic AMP and specificity: How is it that several hormones, all of which influence cyclic AMP, can have different effects on the same cell? The most likely explanation is that there are qualitatively different receptor sites in the cell, some of which respond to one hormone, some to another. This hypothesis goes on to postulate a compartmentalization of cyclic AMP within the cell, i.e., cyclic AMP generated as a result of the interaction between hormone A and its specific receptor site does not gain access to the same intracellular site on which cyclic AMP generated from the interaction of hormone B and its receptor site does (and vice versa).

Mechanism of steroid-hormone action   As mentioned above, the common denominator of most (if not all) of the effects of the steroid hormones is an increased synthesis of proteins (enzymes, structural proteins, etc.) by their specific target-organ cells. This increased protein synthesis is the result of hormone-induced stimulation of the synthesis of RNA. The actual mechanism by which steroid hormones influence nuclear constituents so as to alter the rate of RNA synthesis is presently the subject of intense study. The first step is the passage of the steroid across the outer cell membrane into the cell cytoplasm (steroid hormones, unlike protein hormones and amines, are quite lipid soluble and can readily cross cell membranes). Once in the cell, the hormone binds to a hormone-specific soluble cytoplasmic protein. Accordingly, this protein, which must contain specific molecular receptor sites for the particular steroid hormone, is known as the ''receptor.'' The hormone-protein complex then moves into the nucleus and combines with a chromatin protein (i.e. a protein associated with DNA) specific for it (note that this system involves a cascade of receptors, for the nuclear protein functions as a receptor for the original hormone-protein complex). It is this molecular interaction that triggers off the specific RNA synthesis, i.e., the transcription of the relevant DNA sequences, although how it does so is quite unknown. This system has come to be known as the ''mobile-receptor model'' since it involves the movement of the hormone-receptor complex from the cytoplasm into the nucleus. In contrast, the cyclic AMP system described earlier utilizes fixed receptor sites on the outer cell membrane of target-organ cells.

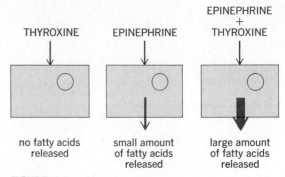

THYROXINE	EPINEPHRINE	EPINEPHRINE + THYROXINE
no fatty acids released	small amount of fatty acids released	large amount of fatty acids released

FIGURE 7-5
Ability of thyroxine to permit epinephrine-induced liberation of fatty acids from adipose tissue cells.

Hormone interactions on target-organ cells

Because virtually all hormones are always being secreted at some rate, finite blood concentrations of all hormones exist at all times. These concentrations may vary over wide ranges in response to stimuli, but since the blood contains *some* of each hormone, cells are constantly exposed to the simultaneous effects of many hormones. This allows for complex hormone-hormone interactions on the target-organ cells, the most important phenomenon being that of *permissiveness*. In general terms, frequently hormone A must be present for the full exertion of hormone B's effect. In essence, A is ''permitting'' B to exert its action. Generally, only a very small quantity of the permissive hormone is required. For example (Fig. 7-5), the hormone epinephrine causes marked release of fatty acids from adipose tissue only in the presence of thyroid hormone. Many of the defects seen when an endocrine gland is removed or ceases to function because of disease actually result from loss of the permissive powers of the hormone secreted by that gland.

Pharmacological effects

Administration of very large quantities of a hormone may have results which are never seen in a normal person, although these so-called *pharmacological effects* sometimes occur in endocrine diseases when excessive amounts of hormone are secreted. These effects are of great importance in medicine since hormones in pharmacological doses are used as therapeutic agents. Perhaps the most famous example is that of the adrenal hormone cortisol, which is highly useful in suppressing allergic and inflammatory reactions. Mental changes, including outright psychosis, may also be induced by large quantities of cortisol and are frequently a striking symptom of patients suffering from hyperactive adrenal glands.

When one considers the complexities of hormone interrelationships, it is not surprising that removal of an endocrine gland may produce a bewildering array of abnormalities, just as marked hypersecretion may induce many effects. Until fairly recently, endocrine physiology consisted largely of studying these effects of excess and deficit. This kind of study yields important data but does not really clarify the basic reflex control of the hormones, i.e., what environmental change induces increased or decreased secretion. Perhaps the major problem was (and still is to some extent) the lack of sensitive methods for measuring hormone concentrations in the blood. The development of such methods has led to some striking changes in both knowledge and viewpoint. For example, it has recently been demonstrated that blood concentrations of growth hormone are not increased during periods of rapid growth; this does not mean that growth hormone is unimportant for growth but only that other factors are the primary regulators of growth, growth hormone itself performing a more permissive type of role. In contrast, blood concentrations of growth hormone change markedly in response to a variety of stimuli, and these observations have forced reevaluation of the entire physiology of this hormone.

Control of hormone secretion

Later chapters will describe the detailed physiology of the hormones and the reflexes in which they participate, which are analogous to the completely neural reflexes described in Chap. 6, the blood-borne hormone serving as a pathway in the reflex arc. We now describe types of direct input which act upon the endocrine cells to cause production and release of the hormones into the blood. If an endocrine gland cell is viewed as being analogous to a neuron, the subject of this section is analogous to the facilitating and inhibiting synaptic input to the neuron.

Table 7-1 summarizes the major hormones. It is only too evident that there are a large number of hormones, each with its own unique controlling input. Memorizing the table at this point would serve no purpose. Each hormone will be described in detail later. The group is presented here partly for reference but more importantly to illustrate the several common denominators of

TABLE 7-1
Summary of the major hormones and their most important immediate controls

Gland	Hormone	Secretion directly controlled by
Anterior pituitary	Growth hormone	Hypothalamic GH-releasing factor
	Thyroid-stimulating hormone (TSH)	Hypothalamic TSH-releasing factor and thyroxine
	ACTH	Hypothalamic ACTH-releasing factor and cortisol
	Gonadotropic hormones: FSH, LH	Hypothalamic FSH- and LH-releasing factors and Female: estrogen and progesterone Male: testosterone
	Prolactin	Hypothalamic prolactin-inhibiting factor
Thyroid	Thyroxine	TSH
	Calcitonin	Plasma calcium concentration
Adrenal cortex	Cortisol	ACTH
	Aldosterone	Angiotensin and plasma K^+ concentration
Gonads		
Female: ovaries	Estrogen and progesterone	FSH and LH
Male: testes	Testosterone	LH
Posterior pituitary	Oxytocin and antidiuretic hormone (ADH, vasopressin)	Action potentials in hypothalamic secretory neurons
Adrenal medulla	Epinephrine and norepinephrine	Preganglionic sympathetic neurons
Parathyroids	Parathyroid hormone (PH)	Plasma calcium concentration
Pancreas	Insulin and glucagon	Plasma glucose concentration†
Kidneys	Renin-angiotensin and erythropoietin	See Chaps. 10, 11
Gastrointestinal tract	See Chap. 12	See Chap. 12

†Other important inputs which control insulin and glucagon secretion will be described in Chap. 13.

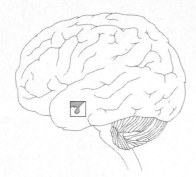

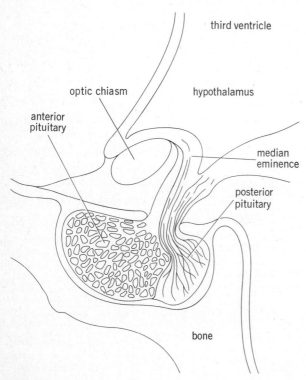

FIGURE 7-6
Relationship of the pituitary to the brain and hypothalamus.
(*Adapted from Guillemin and Burgus.*)

hormonal control; an appreciation of these basic patterns should greatly facilitate the understanding of each specific hormone as it is discussed in future chapters. (Table 7-1 lists only the hormones considered most important; future chapters will mention additional hormones, known or suspected to exist; their omission from this table does not imply that they are not true hormones.)

It is evident from the table that the *pituitary gland* (or hypophysis) is of major importance in hormone secretion. This gland lies in a pocket of bone just below the hypothalamus (Fig. 7-6) to which it is connected by a stalk containing neurons and small blood vessels, the functions of which will be described below. It is composed of three lobes, each of which is a more or less distinct gland. They are the *anterior, intermediate,* and *posterior lobes.* Synonyms for the anterior and posterior lobes are *adenohypophysis* and *neurohypophysis,* respectively. In man, the intermediate lobe is rudimentary, and its function is unclear. It contains two substances known as melanocyte-stimulating hormones (MSH) which are known to cause skin darkening in lower vertebrates; however, their function in man is unknown.

The anterior pituitary hormones

Despite its close proximity to the brain, the anterior pituitary is not neural but is composed of true glandular tissue, which produces at least six different protein hormones. A variety of evidence has established that each hormone is secreted by a different cell type. Secretion of each of the six hormones occurs independently of the others; i.e., the anterior pituitary comprises, in effect, six endocrine glands anatomically associated in a single structure.

The major function of two of the anterior pituitary hormones is to stimulate the secretion of other hormones: (*1*) *Thyroid-stimulating hormone* (TSH) induces secretion of thyroid hormone[1] from the thyroid. (*2*) *Adrenocorticotropic hormone* (ACTH), meaning "hormone which stimulates the adrenal cortex," is responsible for stimulating the secretion of *cortisol.* Thus, the important target organs for TSH and ACTH are the thyroid and adrenal cortex, respectively.

Two other anterior pituitary hormones, the *gonadotropic hormones, follicle-stimulating hormone* (FSH) and *luteinizing hormone* (LH), primarily control the secretion of sex hormones (*estrogen, progesterone,* and *testosterone*) by the gonads. The gonadotropins differ from TSH and ACTH in that, besides controlling the secretion of other hormones, they have a second major role, the growth and development of the reproductive cells, the sperm and ova. The gonads are the sole target organs for the anterior pituitary gonadotropins.

It should now be clear why the anterior pituitary

[1]Thyroid hormone is a term which includes two closely related hormones (thyroxine and triiodothyronine) secreted by the thyroid gland. For simplicity, we shall refer in this book only to thyroxine which is produced in greater quantity.

ANTERIOR PITUITARY

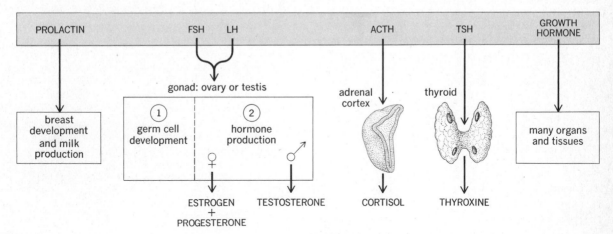

FIGURE 7-7
Target organs and functions of the six anterior pituitary hormones.

is frequently called the master gland; it secretes six hormones itself and controls the secretion of three or four (depending upon the person's sex) other hormones.

What about the two remaining anterior pituitary hormones; do they also control the secretion of some other hormones? The answer is *no. Prolactin's* major target organs are the breasts, and *growth hormone* exerts multiple metabolic effects upon many organs and tissues. The target organs and functions of the anterior pituitary hormones are summarized in Fig. 7-7.

Let us now return to the target organs of the tropic hormones. We have said that the secretion rates of thyroid hormone, cortisol, and the gonadal sex hormones are stimulated by anterior pituitary tropic hormones. Are they also controlled by other types of input? The answer is *no* in normal (and nonpregnant) persons. The sole control of these hormones is via the pituitary. This can easily be proved by observing them after surgical removal of the anterior pituitary; the secretion of thyroid hormone, cortisol, and sex hormones ceases almost completely. The thyroid gland, most of the adrenal cortex, and the gonads greatly decrease in size (Fig. 7-8) and take on a nonfunctioning appearance; these observations clearly show that the tropic hormones control not only the secretion of their target-gland hormones but the growth and development of the target glands themselves. The reason the adrenal cortex does not atrophy completely is that,

as shown in Table 7-1, the cells which secrete the second major cortical hormone, *aldosterone,* are primarily controlled not by ACTH but by other inputs to be described in Chap. 11.

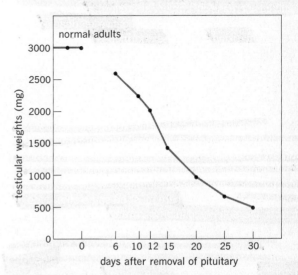

FIGURE 7-8
Decrease in weight of testes after removal of the pituitary from adult rats. (*Adapted from Steinberger and Nelson.*)

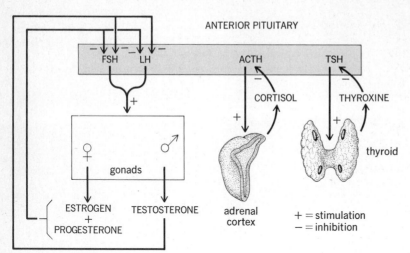

ANTERIOR PITUITARY

+ = stimulation
− = inhibition

FIGURE 7-9
Negative feedback of target-organ hormones on their respective anterior pituitary tropic hormones.

Control of anterior pituitary hormone secretion

What direct inputs control secretion of the anterior pituitary hormones? One important type of input for the tropic hormones is the target-gland hormone itself, a beautiful example of negative feedback (Chap. 5). Thus, ACTH stimulates cortisol secretion and increases the blood concentration of cortisol, which acts upon the anterior pituitary to inhibit ACTH release. In this manner, any increase in ACTH secretion is partially prevented by the resultant increase in cortisol secretion, as illustrated in Fig. 7-9. This same pattern of negative feedback is exerted by thyroxine and the sex hormones on their respective pituitary tropic hormones (the sex hormone effects are actually more complex than those shown in the figure, as will be discussed in Chap. 14). It is evident that such a system is highly effective in damping hormonal responses, i.e., limiting the extremes of hormone secretory rates. However, if this negative-feedback relationship were the sole source of anterior pituitary control, there would be no way of altering anterior pituitary output; some unchanging equilibrium blood concentrations of pituitary and target-gland hormone would always be maintained. Obviously, there must be some other type of input to the anterior pituitary. In reality, this other input is the major controller of anterior pituitary function.

The major inputs controlling release of anterior pituitary hormones are a group of so-called releasing factors produced in the hypothalamus. An appreciation of the anatomical relationships between the hypothala-

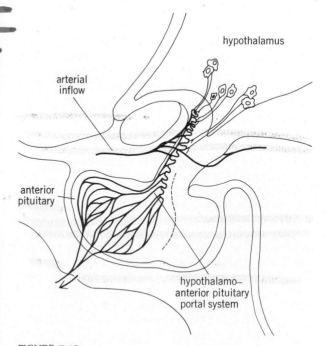

FIGURE 7-10
Schematized summary of hypothalamus–anterior pituitary vascular connections. The hypothalamic neurons, which secrete releasing factors, end on the capillary loops which are the beginning of the portal system carrying blood from the hypothalamus to the anterior pituitary.

mus and anterior pituitary is essential for understanding this process. Although the anterior pituitary lies just below the hypothalamus, there are no important neural connections between the two, but there is an unusual capillary-to-capillary connection (Fig. 7-10); a group of capillaries in the base (*median eminence*) of the hypothalamus recombine into small vessels which pass down through the connecting stalk into the anterior pituitary, where they branch and form most of the anterior pituitary capillaries. Thus they offer a local route for flow of capillary blood from hypothalamus to anterior pituitary. It has been termed the *hypothalamopituitary portal system*, one of the shortest and most critical vascular connections in the entire body. The axons of neurons which originate in diverse areas of the hypothalamus are believed to terminate in the median eminence around the hypothalamic capillary origins of the portal vessels. These neurons secrete into the capillaries substances which are carried by the portal vessels to the anterior pituitary, where they act upon the various pituitary cells to control hormone secretion (Fig. 7-11). We are dealing with multiple discrete substances, each controlling the release of only one (or, at most, two) type of pituitary hormone. Most of these substances stimulate release of their relevant hormones and are therefore termed *hypothalamic releasing factors* (TSH-releasing factor, ACTH-releasing factor, etc.). At least one, that which controls prolactin secretion, inhibits rather than stimulates prolactin release and is termed *prolactin-inhibiting factor* (PIF). Moreover, the system may be even more complex than this in that prolactin, growth hormone, and perhaps other hormones may be controlled by dual systems of hypothalamic substances, one inhibitory and the other stimulatory.

Each factor appears to be secreted only by neurons in a discrete portion of the hypothalamus, i.e., one group of neurons secretes PIF, a different group secretes TSH-releasing factor, etc. Regardless of the origin of the hypothalamic neurons, the releasing factors are all secreted into the hypothalamopituitary portal vessels. The alert reader will have recognized that these substances fulfill all the criteria for our definition of a hormone and really should be called that instead of factor. This change in nomenclature is presently under way. It should also be noted that the "releasing" factors control not only the release of their respective pituitary hormones, but their synthesis as well.

These relationships form much of the foundation of neuroendocrinology. We stated earlier that the nervous and endocrine systems actually function as a single inter-

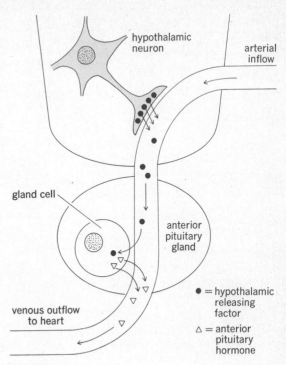

FIGURE 7-11
Control of anterior pituitary secretion by a hypothalamic releasing factor.

related system; the anterior pituitary may be the master gland, but its function is primarily controlled by the hypothalamus via the releasing factors. It now appears that many diseases characterized by inadequate secretion of one or more pituitary hormones really are due to hypothalamic malfunction rather than primary pituitary disease.

Our analysis has pushed the critical question one step further: The hypothalamic releasing factors control anterior pituitary function, but what controls secretion of the releasing factors? The answer is neural and hormonal input to the hypothalamic neurons which secrete the releasing factors. The hypothalamus receives neural input, both facilitory and inhibitory, from virtually all areas of the body; the specific type of input which controls the secretion rate of the individual releasing factors will be described in future chapters when we discuss the relevant anterior pituitary or target-gland hormone. It suffices for now to point out that endocrine disorders may be generated, via alteration of hypothalamic activity, by all

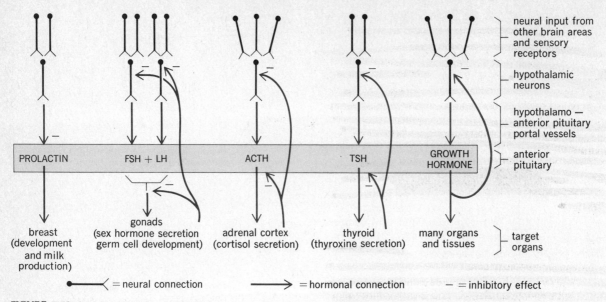

| | | | | | neural input from other brain areas and sensory receptors |
| PROLACTIN | FSH + LH | ACTH | TSH | GROWTH HORMONE | hypothalamic neurons |

hypothalamo — anterior pituitary portal vessels

anterior pituitary

breast (development and milk production) — gonads (sex hormone secretion germ cell development) — adrenal cortex (cortisol secretion) — thyroid (thyroxine secretion) — many organs and tissues — target organs

●———< = neural connection   ———> = hormonal connection   − − = inhibitory effect

FIGURE 7-12
Summary of neural–anterior pituitary–target-organ relationships. This model will almost certainly become more complicated as research progresses. For example, current evidence suggests the possible existence of dual systems of hypothalamic substances, one inhibitory and the other stimulatory, for both growth hormone and prolactin.

manner of neural activity, such as stress, anxiety, etc. An example is sterility caused by severe emotional upsets.

Hormonal influences upon the hypothalamus are also important. Some of the negative-feedback effect of thyroxine, cortisol, and sex hormones upon pituitary tropic-hormone secretion is actually mediated via the hypothalamus, i.e., by inhibition of releasing-factor secretion. For example, cortisol acts not only directly upon the anterior pituitary to inhibit ACTH secretion but also upon the hypothalamus to inhibit ACTH-releasing-factor secretion, an event which also reduces ACTH secretion. Presently, there is still controversy over the quantitative importance of these two negative-feedback sites for the various hormones, but the generalization that both sites are involved, albeit to varying extents for each hormone, seems likely. In addition, it is quite likely that growth hormone, one of the two anterior pituitary hormones which have no target-organ hormones, exerts a negative-feedback control, via the hypothalamus, over its own secretion. The situation for prolactin is presently too unclear to permit speculation. Our description of the interrelationships of the hypothalamus, anterior pituitary, and target glands is now complete, as shown in Fig. 7-12.

Before leaving this subject, we present a few of the many types of experimental findings which support the picture just given, partly because of their inherent interest but also because they may facilitate understanding.

1 Electric stimulation of discrete areas of the hypothalamus elicits secretion of anterior pituitary hormones.

2 Conversely, destruction of these same areas results in marked reduction of anterior pituitary secretion and prevents the increases which normally occur in response to physiological stimuli.

3 Transplantation of the pituitary to a distant site in the body produces the same effects as item 2, and retransplantation to an area just under the hypothalamus results in restoration of function after the regeneration of hypothalamo-pituitary portal vessels.

4 Destruction of the portal vessels and preventing regeneration produces the same effects as item 2.

5 Substances have been extracted from the hypothalamus which cause selective release of anterior pituitary hormones. Recently, two of the releasing factors (those for TSH and LH) have been isolated, identified, and synthesized. The feat of isolation and identification of TSH-releasing factor required 7 tons of hypothalamic tissue obtained from 5 million sheep (the tissue was obtained

from slaughterhouses); the final yield of pure releasing factor was 1 mg!

6 Anterior pituitary tissue can be removed from the body and grown in tissue culture, but the cells rapidly decrease in size and become nonfunctional. Addition of hypothalamic extract to the tissue culture causes redevelopment of the cells, which actually begin to secrete hormones in the test tube.

7 In experiments studying prolactin, the findings have been just the opposite of all those above; e.g., the pituitary in tissue culture normally secretes large quantities of prolactin, and secretion decreases when hypothalamic extract is added.

Hypothalamus–posterior-pituitary function

The posterior pituitary lies just behind the anterior pituitary in the same bony pocket at the base of the hypothalamus, but its structure is totally different from its neighbor's. The posterior pituitary, or neurohypophysis, is actually an outgrowth of the hypothalamus and is true neural tissue. Two well-defined clusters of hypothalamic neurons send out nerve fibers which pass by way of the connecting stalk to end within the posterior pituitary in close proximity to capillaries (Fig. 7-13). The two hormones, oxytocin and antidiuretic hormone (ADH, vasopressin), released from the posterior pituitary are actually synthesized in the hypothalamic cells; enclosed in small vesicles, they move slowly (3 mm/day) down the cytoplasm of the neuron axons to accumulate at the nerve endings. Release into the capillaries occurs in response to generation of an action potential within the nerve. Thus, these hypothalamic neurons secrete hormones in a manner quite analogous to that described previously for hypothalamic releasing factors, the essential difference being that the releasing factors are secreted into capillaries which empty directly into the anterior pituitary whereas the posterior pituitary capillaries drain primarily into the general body circulation.

It is evident, therefore, that the term posterior pituitary hormones is somewhat of a misnomer since the hormones are actually synthesized in the hypothalamus, of which the posterior pituitary is merely an extension. However, when oxytocin and antidiuretic hormone were discovered, this fact was not known, and both were believed to be synthesized in small posterior pituitary cell bodies now known to be connective-tissue cells. Indeed, the entire posterior pituitary can be surgically removed with only a temporary loss of oxytocin and ADH secretion; capillaries soon grow up along the connecting stalk, and hormone release returns to normal levels. We must there-

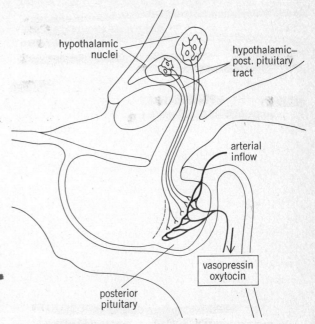

FIGURE 7-13
Relationship between the hypothalamus and posterior pituitary. (*Adapted from Guillemin and Burgus.*)

fore add two more names to our growing list of hormones under direct or indirect control of the hypothalamus.

Epinephrine and the adrenal medulla

As described in Chap. 6, the adrenal medulla is really an overgrown sympathetic ganglion whose cell bodies do not send out nerve fibers but release their active substances directly into the blood, thereby fulfilling the criteria for an endocrine gland. In man, the hormone released by the medulla is for the most part epinephrine (a smaller amount of norepinephrine is also secreted), a substance closely related to norepinephrine but with several distinctly different functional properties. In controlling epinephrine secretion, the adrenal medulla behaves just like any sympathetic ganglion and is dependent upon stimulation by sympathetic preganglionic fibers. Destruction of these incoming nerves causes marked reduction of epinephrine release and failure to increase secretion in response to the usual physiological stimuli.

The adrenal medulla is best viewed as a general reinforcer of sympathetic activity. Its secretion of epinephrine into the blood serves to increase the overall sympathetic functions of the body. We shall discuss in later chapters the specific reflexes which cause enhanced sympathetic activity and elicit epinephrine secretion; suffice it to say that these reflexes are under the strong control of higher brain centers, particularly the hypothalamus. Here is another hormone which is literally a product of the nervous system.

The reader must carefully distinguish the adrenal medulla from the adrenal cortex, which surrounds it. They are completely distinct structures, and the adaptive value of their anatomical proximity is presently unclear.

Hormones not directly or indirectly controlled by the hypothalamus or pituitary

The remaining hormones of Table 7-1 are controlled by mechanisms other than the hypothalamopituitary system. The control mechanisms which govern aldosterone and the hormones secreted by the kidneys will be discussed in Chaps. 10 and 11. The various hormones produced by the gastrointestinal tract are released into the blood in response to events occurring within the tract itself and will be described in Chap. 12.

This leaves only the control mechanisms for the hormones of the pancreas and parathyroid glands; the cells of these glands respond directly to the glucose[1] and calcium concentrations, respectively, of the blood supplying them. As will be described in later chapters, the major function of *insulin* and *glucagon* is regulation of blood glucose concentration, whereas the major function of *parathyroid hormone* is regulation of blood calcium concentration. The control mechanisms are therefore quite appropriate and remarkably simple; the hormone-secreting cells are themselves sensitive to the plasma substance their hormones regulate. It is likely that such regulatory systems represent the earliest type of control mechanism evolved by complex organisms.

Summary

Table 7-1 should now appear less formidable. The control mechanisms for most of the hormones involve the direct or indirect participation of the hypothalamus and pituitary. The anterior pituitary hormones are controlled primarily by releasing factors secreted into the hypothalamopituitary portal vessels by neurons in the hypothalamus. In turn, the four anterior pituitary tropic hormones control hormone secretion by their target-organ glands, the thyroid, adrenal cortex, and gonads. These glands exert a negative-feedback control over their own secretion via

[1]Other important inputs which control insulin and glucagon secretion will be described in Chap. 13.

the effects of their hormones on both the hypothalamus and anterior pituitary. The hypothalamus itself also produces two hormones, oxytocin and ADH, which are released from nerve endings in the posterior pituitary. And finally, the hypothalamus exerts profound control over the autonomic nervous system, including the adrenal medulla. Thus, this small area of brain, which weighs 4 g in the adult human being, acts as a compact integrating center, receiving messages, both neural and hormonal, from all areas of the body and sending out efferent messages via both the nerves and hormones. As we shall see, it also regulates body temperature, food intake, water balance, and a host of other autonomic, endocrine, and behavioral activities. Despite the central role of the hypothalamopituitary system, there are important hormones whose secretion is controlled, at least in part, by completely distinct mechanisms, including aldosterone from the adrenal cortex, insulin and glucagon from the pancreas, parathyroid hormone, several kidney hormones, and a group of gastrointestinal hormones.

With this basic foundation laid, future chapters will describe the relevant functions of each major hormone and the specific environmental changes and afferent inputs which induce reflex alteration of their secretion rates.

The problem of multiple hormone secretion

The phenomenon of multiple hormone secretion by a single gland is clearly evident in Table 7-1. In certain glands, such as the pancreas, the hormones are secreted by completely distinct cells; in other glands, it is likely that, although some separation of function exists, a single cell may secrete more than one hormone. Even in such cases, it is essential to realize that each hormone has its own unique control mechanism; i.e., there is no massive undifferentiated release of the multiple hormones.

The adrenal cortex and gonads offer a particularly vexing problem of multiple hormone secretion. The hormones secreted by these organs all have a common ringlike type of lipid structure known as a steroid. Subtle changes in the steroid molecule produce great alterations of physiological activity (compare, for example, the structures of the male sex hormone, testosterone, and one of the female sex hormones, progesterone, in Fig. 7-14). The biochemical pathways leading to steroid synthesis are complex and overlapping so that the adrenals and gonads secrete similar steroids. Only the major ones are

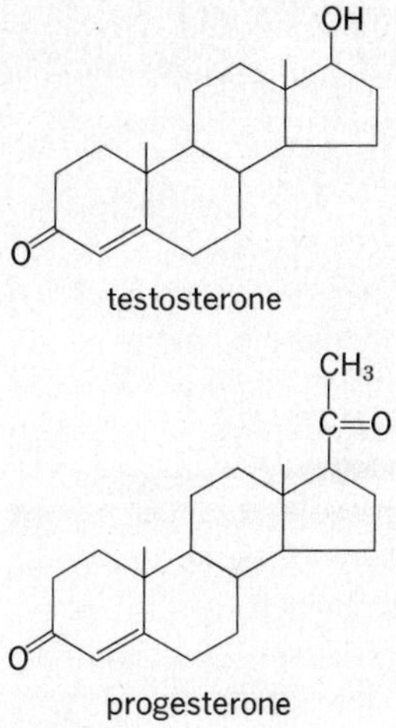

testosterone

progesterone

FIGURE 7-14
Structures of testosterone and progesterone.

presented in Table 7-1, but the reader should also be aware of several facts:

1 There are several closely related male sex hormones, all termed *androgens*. Testosterone is the dominant hormone of the group, and we shall usually use this name rather than the more general term, androgens, when referring to the male gonadal hormones.

2 The name estrogen actually is a general term used to include several closely related, normally secreted female sex hormones.

3 The adrenal glands normally secrete significant quantities of androgen and estrogen.

4 The male and female gonads normally secrete very small quantities of estrogen and androgen, respectively, but the amounts are probably too small to be of significance (except as a reminder that there are no chemicals unique to males or females).

This crossover of secretory capacities can lead to impor-

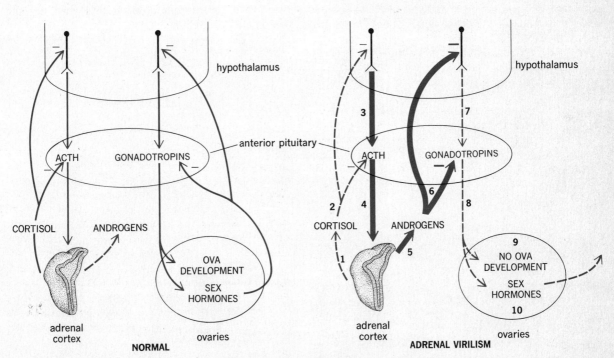

FIGURE 7-15
Sequence of events leading to masculinization, ovarian atrophy, and sterility in a woman with a defect in the adrenal synthesis of cortisol.

tant disease-induced abnormalities. We shall describe one of these, *adrenal virilism,* as an example of how a single defect can play havoc with many control mechanisms (Fig. 7-15). A pathway for synthesis of androgen branches from the normal synthetic pathway for cortisol. A very small quantity of this androgen is released from the adrenal cells into the blood. The basic defect in adrenal virilism is congenital lack of the enzyme which catalyzes a step in cortisol formation below this branch pathway; therefore, the synthetic pathway is blocked at stage 1. The result is decreased secretion of cortisol, leading to subnormal blood concentration of this hormone (2). The hypothalamopituitary system is therefore not subjected to the normal negative-feedback effect of cortisol, and the secretion of ACTH-releasing factor and ACTH markedly increase (3 and 4). The tropic hormone profoundly stimulates the adrenal. However, because of the existence of the enzymatic block preventing cortisol formation, the stimulating effects of ACTH result in greater production and release of the adrenal androgen (5) formed via the branch pathway. The effects of the male sex hormone are most striking in the female, in whom they induce a masculine appearance. Finally, the negative-feedback effect (6) of these large quantities of androgen in the blood shuts off the secretion of the gonadotropin-releasing factor (7) and thereby the gonadotropins (8), resulting in gonadal atrophy and sterility (9 and 10). All these defects can be rapidly reversed merely by giving cortisol to the patient.

A frequent source of confusion concerning the endocrine system concerns not multiple hormone secretion by a single gland but the mixed general functions exhibited by the gonads, pancreas, kidneys, and gastrointestinal tract. All these organs contain endocrine gland cells but also perform other completely distinct non-endocrine functions. For example, most of the pancreas is concerned with the production of digestive enzymes; these are produced by exocrine glands; i.e. the secretory products are transported not into the blood but into ducts leading, in this case, into the intestinal tract. The endocrine function of the pancreas is performed by completely distinct nests of endocrine cells scattered throughout the pancreas. This pattern is true for all the organs of mixed function; the endocrine function is always subserved by gland cells distinct from the other cells which constitute the organ.

Muscle

A variety of movements occur in the body, including the movements of chromosomes to the poles of a cell during cell division, the movements of hairlike processes (cilia) on the surfaces of certain cells, the movements of sperm cells propelled by tail-like processes, and the movement of white blood cells which permits them to migrate out of the blood and into a tissue. Even blood clotting involves a contractile process. Thus, the property of movement is not unique to muscle. In other cells, however, movement is secondary to performing some other function of the cell. In contrast, the primary function of muscle is to generate force and produce movement.

Three different types of muscle cells can be identified on the basis of structure and contractile properties: (1) *skeletal muscle*, (2) *smooth muscle*, and (3) *cardiac muscle*. Most skeletal muscle, as the name implies, is attached to the bones of the body, and its contraction is responsible for the movements of parts of the skeleton. Skeletal muscle movement is also involved in other activities of the body such as the voluntary release of urine and feces. Skeletal muscle is under the control of the somatic nervous system. The movements produced by skeletal muscle are primarily involved with interactions between the body and the external environment.

Smooth muscle surrounds such hollow chambers

8

in the body as the stomach and intestinal tract, the urinary bladder, blood vessels, and uterus. The contraction of smooth muscle is controlled in large part by the autonomic nervous system and thus is not normally under direct conscious control. The contraction of smooth muscle is associated with processes which regulate the internal environment of the body. The third type of muscle, cardiac muscle, is the muscle of the heart and, like smooth muscle, is primarily under the control of the autonomic nervous system. Although there are significant differences in the structure, contractile properties, and control of these three types of muscle, the physicochemical principles underlying their contractile activity are similar.

Structure of skeletal muscle

Skeletal muscle is the largest tissue in the body, accounting for 40 to 45 percent of the total body weight. Each muscle cell is cylindrical, having a diameter of 10 to 100 μm and may be up to 1 ft long. A single muscle cell is known as a *muscle fiber*. The term *muscle* refers to a number of muscle fibers bound together by connective tissue. Thus, the relation between a single muscle fiber (cell) and a muscle is similar to that between a single nerve fiber (axon) and a nerve composed of many axons.

Over 600 muscles can be identified in the human body. Some of them are very small, consisting of only a few hundred fibers; larger muscles may contain several hundred thousand fibers. Surrounding the individual fibers is a network of connective tissue through which blood vessels and nerve fibers pass to the muscle fibers. In some muscles the individual fibers are as long as the muscle itself; but most fibers are shorter than the total muscle, and their ends are attached to the connective-tissue network interlacing the muscle fibers.

Generally each end of the whole muscle is attached to a bone by bundles of collagen fibers known as *tendons*. Collagen is a fibrous protein produced by cells known as fibrobasts and is the major element of connective tissue throughout the body. It has great strength but no active contractile properties. The collagen fibers in the connective tissue and the tendons act as a structural framework to which the muscle fibers and bone are attached. The forces generated by the contracting muscles are transmitted by the connective tissue and tendons to the bones. The transmission of force from muscle to bone is like a number of men pulling on a rope, each man corresponding to a single muscle fiber and the rope to

the connective tissue and tendons. Some tendons are very long, and the site of attachment of the tendon to the bone is far removed from the muscle. Thus, some of the muscles which move the fingers are found in the lower portion of the arm between the elbow and the wrist, as one can observe by wiggling one's fingers and feeling the movements of the muscles in the lower arm; they are connected to the finger bones by long tendons. If the muscles which moved the fingers were located in the fingers themselves, we would have very fat fingers.

Figure 8-1 shows a section through a skeletal muscle as seen with a light microscope. In between the individual muscle fibers are capillary blood vessels containing red blood cells. The most striking feature of the muscle fibers is the series of transverse light and dark bands forming a regular pattern along the fiber. Both skeletal and cardiac muscle fibers have this characteristic banding and are known as *striated muscles;* smooth muscle cells show no banding pattern. Although the pattern appears to be continuous across a single fiber, the fiber is actually composed of a number of independent cylindrical elements in the cytoplasm of the fiber known as *myofibrils* (Fig. 8-2). Each myofibril is about 1 to 2 μm in diameter and continues through the length of the muscle fiber. Myofibrils occupy about 80 percent of the fiber volume and vary in number from several hundred to several thousand, depending on the fiber diameter. The myofibrils form a longitudinal striation of the muscle; the banding patterns in the myofibrils form transverse striations.

Viewed with the electron microscope, the structures responsible for the banding patterns become evident. The myofibrils consist of smaller *myofilaments* (Fig. 8-2), which form a regular repeating pattern along the length of the fibril. One unit of this repeating pattern is known as a *sarcomere*. It is the functional unit of the contractile system in muscle, and the events occurring in a sarcomere are duplicated in the other sarcomeres along the myofibrils.

Each sarcomere contains two types of myofilaments: thick and thin. The thick myofilaments, 150 Å in diameter, are located in the central region of the sarcomere where their orderly parallel arrangement gives rise to the dark bands, known as *A bands*, that are seen in striated muscles. These thick filaments contain the protein known as *myosin*. The thin myofilaments, 50 Å in diameter, contain the protein *actin* and are attached at either end of the sarcomere to a structure known as the *Z line*. Two successive Z lines define the limits of one sarcomere. The Z lines consist of short elements which interconnect the thin filaments from two adjoining sar-

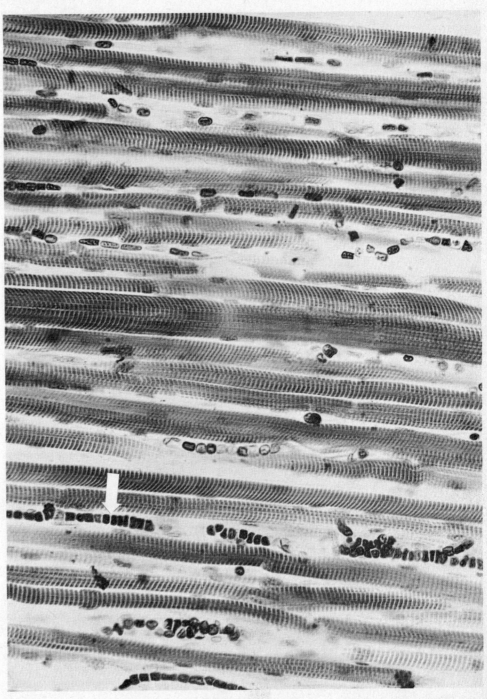

FIGURE 8-1
**Photomicrograph of striated skeletal muscle fibers. Arrow
indicates capillary blood vessel.** (*From Edward K. Reith and
Michael H. Ross, "Atlas of Descriptive Histology," Harper & Row,
Publishers, Incorporated, New York,1968.*)

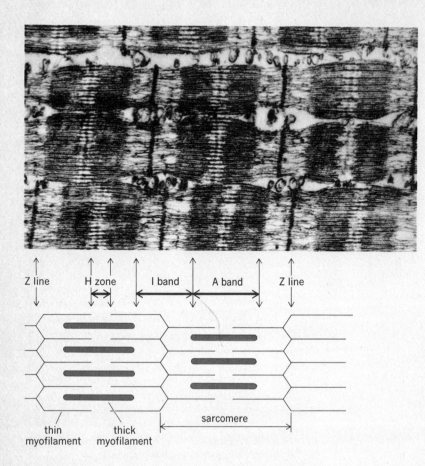

Z line   H zone   I band   A band   Z line

thin myofilament   thick myofilament   sarcomere

FIGURE 8-2
Electron micrograph showing three myofibrils in a single muscle fiber. Diagramed below is the organization of thick and thin myofilaments in a myofibril which gives rise to the banding pattern seen in striated muscles. (*From H. E. Huxley and J. Hanson, in G. H. Bourne (ed.), "The Structure and Function of Muscle," vol. 1, Academic Press, Inc., New York, 1960.*)

comeres and thus provide an anchoring point for the thin filaments. The thin filaments extend from the Z lines toward the center of the sarcomere where they overlap with the thick filaments.

In addition to the A band and Z line already mentioned, two other bands will be identified so that changes in the banding patterns occurring during contraction can be related to the relative positions of the thick and thin filaments in the sarcomere. The *I band* (Fig. 8-2) represents the region between the ends of the A bands of two adjoining sarcomeres. This band contains that portion of the thin filaments which do not overlap with the thick filaments and is bisected by the Z line. Because it contains only thin filaments it usually appears as a light band separating the dark A bands. Finally, the *H zone* appears as a thin, lighter band in the center of the A band which corresponds to the space between the ends of the thin filaments. Only thick filaments are found in the H-zone region.[1]

A cross section through myofibrils in the region of the A band where both thick and thin filaments overlap (Fig. 8-3) shows their hexagonal arrangement. Each thick filament is surrounded by six thin filaments, and each thin filament is surrounded by three thick. Thus there are twice as many thin as thick filaments in the region of overlap. An average muscle fiber contains about 16 billion thick and 64 billion thin filaments.

At higher magnification, in the region of overlap in

[1]A thin dark band can be seen in the center of the H zone. This is known as the M line and is produced by processes which link the thick myofilaments together, maintaining their orderly, parallel arrangement.

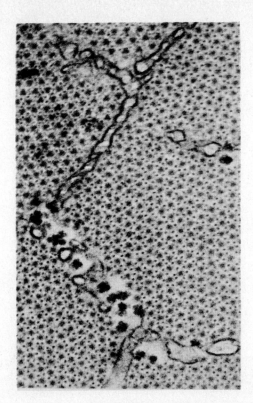

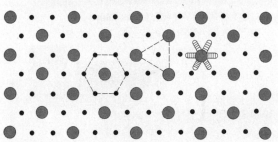

FIGURE 8-3
Electron micrograph of a cross section through six myofibrils in a single skeletal muscle fiber. Diagramed below are the hexagonal arrangements of the thick and thin filaments and the bridges extending from a thick filament to each of the surrounding thin filaments. [*From H. E. Huxley, J. Mol. Biol.,* **37:**507–520 (1968).]

the A band (Fig. 8-4), the gap between thick and thin filaments appears to be bridged by projections at intervals along the filaments. Although it is not evident from this electron micrograph, these projections, or *cross bridges,* are part of the structure of the myosin molecules in the thick

filaments. These cross bridges are arranged in a spiral around the thick filament. One turn of the spiral gives rise to six bridges which are able to interact with the actin in the six thin filaments surrounding each thick filament (Fig. 8-3). Figure 8-5 summarizes the levels of filamentous structure within a muscle fiber.

Molecular basis of contraction

Sliding-filament theory

H. E. Huxley used the electron microscope to examine muscle in a resting, relaxed state and after different degrees of shortening. The changes in sarcomere structure found at the different muscle lengths is shown in Fig. 8-6. His crucial observation was that, as the muscle becomes shorter and shorter, the thick and thin filaments slide past each other, but the lengths of the individual thick and thin filaments do not change. Thus, the width of the A band remains constant, corresponding to the constant length of the thick filaments. The I band narrows as the thick filaments approach the Z line. As the thin filaments move past the thick filaments, the width of the H zone between the ends of the thin filaments becomes smaller and may disappear altogether when the thin filaments meet at the center of the sarcomere. With further shortening, new banding patterns appear as thin filaments from opposite ends of the sarcomere begin to overlap. These observations of the changes in banding pattern during contraction led to the *sliding-filament theory of muscle contraction,* which states that muscle shortening results from the relative movement of the thick and thin filaments past each other.

The structures which actually produce the sliding of the filaments are the myosin cross bridges which swivel in an arc around their fixed positions on the surface of the thick filaments, much like the oars of a boat (Fig. 8-7). The movement of the cross bridges in contact with the actin thin filaments produces the sliding of the thick and thin filaments past each other. Since one movement of a cross bridge will produce only a small displacement of the thin filament relative to the thick, the cross bridges must undergo many repeated cycles of movement during a contraction. During contraction each cross bridge undergoes its own independent cycle of movement so that at any one instant during contraction only about 50 percent of the bridges are attached to the actin thin filaments; the others are at intermediate stages of the cycle.

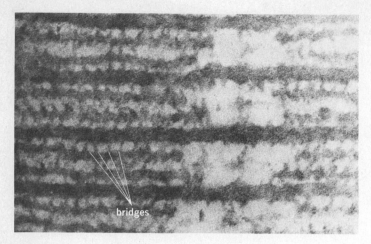

bridges

FIGURE 8-4
High-magnification electron micrograph in the region of the A band in a myofibril. The H zone can be seen at the right. Bridges between the thick and thin filaments can be seen at regular intervals along the filaments. Note the bridges in the A-band region of Fig. 8-2. (*From H. E. Huxley and J. Hanson, in G. H. Bourne (ed.), "The Structure and Function of Muscle," vol.1, Academic Press, Inc., New York, 1960.*)

Actin, myosin, and ATP

What are the properties of actin and myosin which produce this cyclic activity of the cross bridges? Actin is a globular-shaped molecule about 55 Å in diameter, which has a reactive site on its surface that is able to combine with myosin. These globular molecules of actin are arranged in two chains which are helically intertwined to form the thin myofilaments (Fig. 8-8). Myosin is a much larger molecule that is shaped like a lollypop (Fig. 8-9) with a large globular end attached to a long tail. The myosin molecules are arranged within a thick filament so that the molecules are oriented tail to tail in the two halves of the filament and the globular ends extend to the sides, forming the cross bridges.

The globular end of myosin contains a reactive site that is able to bind to the reactive site on the actin molecule. In addition, the globular end contains a separate active site that is able to split ATP. Thus myosin is an enzyme (myosin ATPase) whose substrate is ATP. (Magnesium is a cofactor required to bind ATP to the active site of myosin ATPase.) However, myosin alone has a very low ATPase activity (slow rate of splitting ATP), but when a myosin cross bridge combines with actin in the thin filaments, the activity of myosin ATPase is increased considerably. [Thus the binding of actin appears to have an allosteric effect (see Chap. 3) upon the active site of myosin ATPase, converting it from a relatively inactive to a very active enzyme.] The energy released from the splitting of ATP produces cross-bridge movement although we still do not understand, in molecular terms, how this occurs. Presumably this involves a change in

the shape of the globular end of the myosin molecule while it is attached to the actin thin filament.

As described above, since a single movement of the myosin bridge produces only a small displacement of the thin filament with respect to the thick, these bridges must undergo repeated cycles of activity to produce the degree of shortening observed during muscle contraction. This means that the myosin bridge must be able to detach itself from actin, rebind to a new actin site, and repeat the cycle of bridge movement. What causes the dissociation of myosin bridges from actin at the end of a bridge movement?

The dissociation is achieved by the binding (not splitting) of a molecule of ATP to myosin. The process of binding ATP appears to break the linkage between actin and myosin.

$$A \cdot M + ATP \longrightarrow A + M \cdot ATP$$

The reaction returns the bridge to its initial state so that it can now undergo binding to a new actin site and repeat the cycle of cross-bridge movement. The three basic reactions in the cross-bridge cycle are summarized in Fig. 8-10.

The importance of ATP in dissociating actin and myosin at the end of a bridge cycle is illustrated by the phenomenon of *rigor mortis* (death rigor), in which the muscles of the body become very stiff and rigid shortly after death. This results directly from the loss of ATP in the dead muscle cells. In the absence of ATP the myosin cross bridges are able to combine with actin but the bond between them is not broken. The thick and thin filaments become cross-linked to each other and cannot be pas-

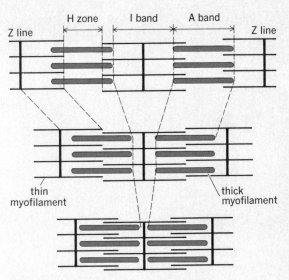

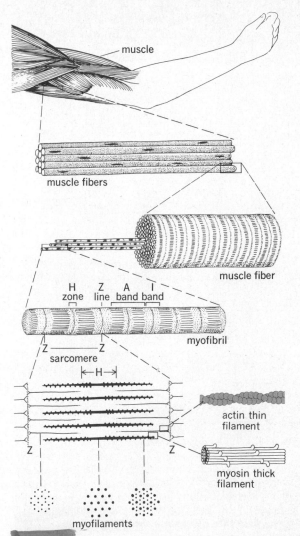

FIGURE 8-5
Levels of fibrillar organization within a skeletal muscle. The three diagrams at bottom show the pattern of myofilaments in cross section. (*Redrawn from Bloom and Fawcett.*)

FIGURE 8-6
Changes in banding pattern resulting from the movements of thick and thin filaments past each other during contraction.

Regulator proteins

Since a muscle cell contains all the ingredients necessary for cross-bridge activity—actin, myosin, ATP, and magnesium ions—the question arises: Why are muscles not in a continuous state of contractile activity? The reason is that the myosin bridges can be prevented from combining with actin by the two regulator proteins *troponin* and *tropomyosin*. These proteins are associated with the thin filaments in muscle. The interactions between these proteins and actin prevent actin from combining with myosin in a resting muscle, perhaps by blocking or changing the shape of the actin reactive site. These regulator proteins thus act as natural inhibitors of the contractile process. The problem now becomes, not what prevents contractile activity from occurring continuously, but whatever starts the process in the first place? The role of turning contractile activity on and off falls to the calcium ion.

Calcium inhibits the inhibitory effects of troponin and tropomyosin. For example, if calcium ions are injected into a muscle fiber, it immediately contracts. The site of action of calcium ions appears to be the troponin molecule. The binding of calcium to the troponin molecule produces a change which is transmitted through the tropomyosin molecule to the actin molecules, the overall effect of which is to restore the ability of actin to combine with myosin and initiate cross-bridge activity (Fig. 8-11).

sively pulled apart by stretch, thus the rigid condition of the dead muscle. At the molecular level of actin and myosin, we can identify two very specific roles for ATP: (1) The splitting of bound ATP by myosin ATPase provides the energy for the movement of the cross bridges and (2) the binding (not splitting) of ATP to myosin dissociates actin from the myosin cross bridges during the contraction cycle of the bridges.

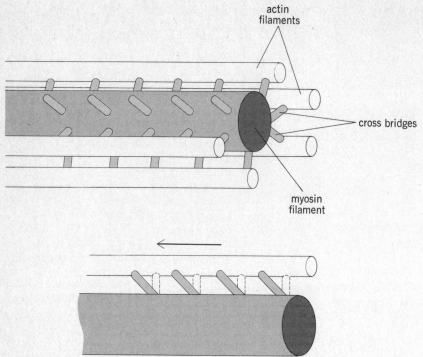

FIGURE 8-7
**Oscillating movements of myosin cross bridges produce the
relative movement of thick and thin filaments. For clarity, one
actin filament has been omitted.** (*Adapted from Anthony and
Kolthoff.*)

Thus, muscle contraction is initiated when calcium is
made available to troponin and ceases when calcium is
removed. The mechanisms which regulate the availability
of calcium ions to the contractile machinery of muscle
are coupled to electrical events that occur in the muscle
membrane.

Excitation-contraction coupling

The cell membranes of muscle are excitable membranes
capable of generating and propagating action potentials
by mechanisms very similar to those discussed in nerve
cells (Chap. 6). An action potential in the muscle cell
membrane provides the signal for the initiation of contractile activity within the muscle cell. The mechanism by
which an electric signal in the membrane triggers off the
chemical events of contraction is the process known as
excitation-contraction coupling and is mediated by calcium
ions.

Sarcoplasmic reticulum   In a resting muscle the concentration of free calcium ions is very low, and the regulatory
proteins, in the presence of ATP, are able to maintain
actomyosin in its dissociated state. An action potential
in the cell membrane leads to an increase in the intracellular calcium ion concentration. These calcium ions
which react with the regulator proteins, blocking their
inhibitory action, initiate the cycle of cross-bridge activity which leads to shortening of the muscle. The source

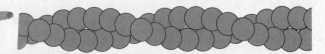

FIGURE 8-8
**Structure of thin myofilament composed of two helical chains
of globular actin monomers.**

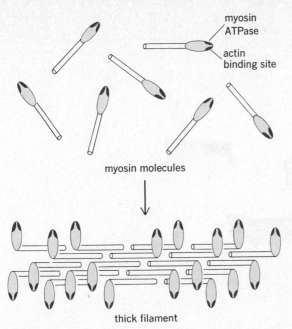

myosin
ATPase

actin
binding site

myosin molecules

↓

thick filament

FIGURE 8-9
Aggregation of myosin molecules to form thick filaments, with the globular heads of the myosin molecules forming the cross bridges.

of these calcium ions is the *sarcoplasmic reticulum.*

The sarcoplasmic reticulum in muscle is homologous in its general structure to the sheets of intracellular membranes (endoplasmic reticulum) found in most cells. In muscle the sarcoplasmic reticulum forms a sleevelike

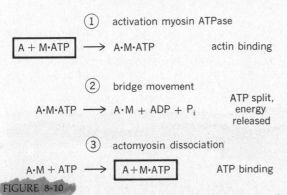

① activation myosin ATPase

$$A + M \cdot ATP \longrightarrow A \cdot M \cdot ATP \qquad \text{actin binding}$$

② bridge movement

$$A \cdot M \cdot ATP \longrightarrow A \cdot M + ADP + P_i \qquad \begin{array}{l}\text{ATP split,}\\ \text{energy}\\ \text{released}\end{array}$$

③ actomyosin dissociation

$$A \cdot M + ATP \longrightarrow A + M \cdot ATP \qquad \text{ATP binding}$$

FIGURE 8-10
The sequential steps in the interaction of actin, myosin, and ATP leading to cross-bridge movement.

structure which surrounds each of the myofibrils (Fig. 8-12). Its structure is closely associated with the repeating pattern of the sarcomeres in the myofibrils. At regular intervals, associated with each sarcomere, the reticulum enlarges to form what are known as *lateral sacs*. These lateral sacs contain the calcium ions that are released following membrane excitation. In between the lateral sacs is found a small tubule that runs transversely around the myofibril, known as the *transverse tubule* (t tubule). This t tubule interconnects with other t tubules surrounding myofibrils and eventually joins the surface membrane of the muscle fiber. The lumen of the t tubule is thus continuous with the extracellular medium surrounding the muscle fiber. The repeating organization of the sarcoplasmic reticulum and t tubules with the sarcomere structure varies in different species of animals and in different types of muscle in the same animal. In frog skeletal muscle the combination of the lateral sacs and t tubule (which is known as a triad because of its appearance in sections seen with the electron microscope) surrounds the region of the Z lines in the myofibrils, as seen in Fig. 8-13. In human skeletal muscle these triads are located opposite the junctions of the A and I bands, but in human cardiac muscle, which is very similar to skeletal muscle in its organization of thick and thin filaments into myofibrils, the t tubules are found associated with the Z lines rather than at the A-I junctions.

Frog muscle can be stimulated with a microelectrode so that only a small region of the muscle membrane is excited. When the membrane is stimulated in the region of the A bands, no contraction occurs even though an action potential has been generated. When the membrane is stimulated in the region of the Z lines, however, which are the location of the junctions of the t tubules with the surface membrane, the underlying myofibrils immediately contract. This suggests that the t tubules are involved in transmitting the electric signal in the membrane into the muscle fiber where it can trigger contraction. The details of this process are still unclear, but it appears that the action potential in the membrane produces an electric signal that is passed along the t tubules to the lateral sacs of the sarcoplasmic reticulum, causing them to release calcium into the muscle cytoplasm in the immediate vicinity of the regulator proteins, thereby initiating a cycle of cross-bridge activity by the mechanisms described above.

How is contraction, once initiated by the release of calcium from the sarcoplasmic reticulum, turned off?

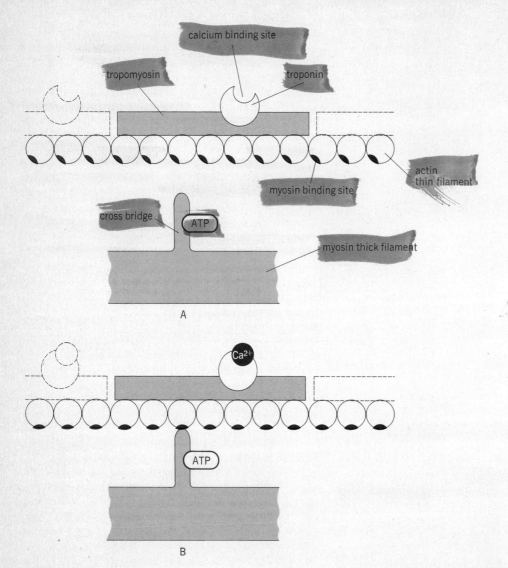

FIGURE 8-11

Diagrammatic representation of the role of regulatory proteins, troponin and tropomyosin, in cross-bridge activity. A Relaxed muscle: Troponin and tropomyosin interact with the actin monomers in the thin filament, preventing the attachment of myosin cross bridges to actin. **B Contracting muscle: The binding of calcium ions to troponin produces a change that is transmitted to the actin monomers through tropomyosin, allowing the myosin cross bridges to combine with actin.**

It is turned off by lowering the intracellular calcium concentration, thereby removing calcium from the regulator proteins. The membranes of the sarcoplasmic reticulum have the ability to concentrate calcium in the lumen of the lateral sacs, using energy derived from the splitting of ATP. This is the third important role of ATP in the process of muscle contraction. The action potential and its release of calcium from the lateral sacs last only a few milliseconds. Immediately following this electrical

activity, the calcium pumps in the membranes of the reticulum begin pumping the released calcium back into the lateral sacs. This process of reaccumulating the released calcium takes much longer than the initial release, and contractile activity of the cross bridges proceeds for several hundred milliseconds after the release of calcium until the concentration of free calcium becomes so low that the regulator proteins are no longer inhibited and they again begin to block actin and myosin interactions.

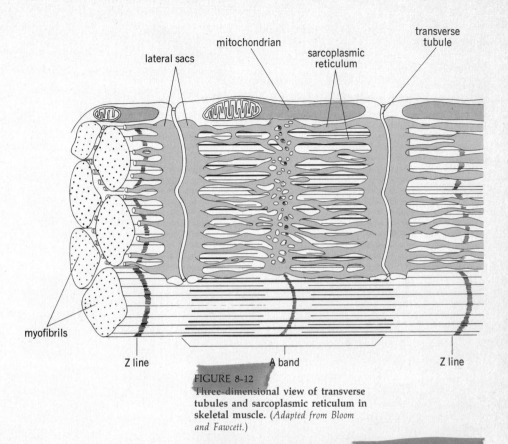

lateral sacs mitochondrian sarcoplasmic reticulum transverse tubule

myofibrils

Z line   A band   Z line

FIGURE 8-12
Three-dimensional view of transverse tubules and sarcoplasmic reticulum in skeletal muscle. (*Adapted from Bloom and Fawcett.*)

Thus, if contractile activity is to last for more than a few hundred milliseconds, repeated action potentials must occur to maintain the free-calcium concentration surrounding the myofibrils at a high enough level to inhibit troponin and tropomyosin inhibition of actomyosin. Figure 8-14 summarizes the role of calcium in excitation-contraction coupling.

Drugs are known which will interfere with various stages in this process of excitation-contraction coupling. Thus some drugs can block the release of calcium from the reticulum and maintain the muscle in a relaxed state even in the presence of action potentials in the muscle membrane. Other drugs, such as caffeine, in higher concentrations than are found in coffee or tea, can cause the release of calcium and produce contractures of muscle in the absence of action potentials. Some of the drugs which are used to increase the force of cardiac muscle contraction in heart disease act by increasing the release of calcium from the sarcoplasmic reticulum.

Membrane excitation   The whole process of contraction begins at the cell surface where the muscle interacts with

its environment. What are the mechanisms that lead to the initiation of action potentials in muscle cell membranes? There are three answers to this question, depending on the type of muscle that is being considered: (*1*) stimulation by a nerve fiber, (*2*) stimulation by hormones and chemical agents, and (*3*) spontaneous electrical activity within the membrane itself. Stimulation by nerve fibers is the only one of the three mechanisms by which skeletal muscles are normally excited in the body. We shall find that the other two mechanisms as well as nerve fibers are involved in initiating excitation in smooth and cardiac muscle.

The axonal process of a nerve fiber forms a junction with a skeletal muscle membrane which resembles in general structure and function the synaptic junctions between two nerve fibers described in Chap. 6. These junctions between nerve and skeletal muscle are known as *neuromuscular* or *myoneural* junctions. The nerve cells which form myoneural junctions with skeletal muscles are known as *motor neurons* (somatic efferent), and the cell bodies of these neurons are located in the brain and spinal cord. The axons of these motor neurons are mye-

FIGURE 8-13

Electron micrograph of frog skeletal muscle, showing the sarcoplasmic reticulum between the myofibrils and the transverse tubule opposite the Z lines. (In human skeletal muscle the transverse tubule is located opposite the junction of the A and I bands, Fig. 8-12.) (*Courtesy of Clara Franzini-Armstrong.*)

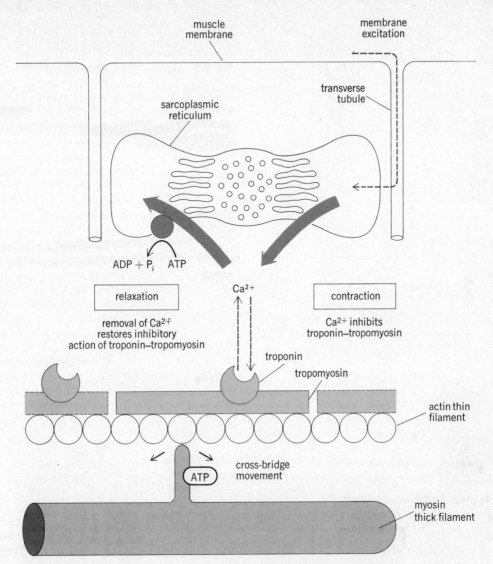

muscle
membrane

membrane
excitation

transverse
tubule

sarcoplasmic
reticulum

ADP + P$_i$ ATP

relaxation

Ca^{2+}

contraction

removal of Ca^{2+}
restores inhibitory
action of troponin–tropomyosin

Ca^{2+} inhibits
troponin–tropomyosin

troponin

tropomyosin

actin thin
filament

cross-bridge
movement

ATP

myosin
thick filament

FIGURE 8-14
Summary of the role of calcium in muscle
excitation-contraction coupling.

linated and are generally the largest-diameter axons in
the body. They are thus able to propagate action potentials at high velocities, sending signals to the muscle that
can rapidly initiate muscle activity.

As the motor axon approaches the muscle, it divides into many branches, each of which forms a single
myoneural junction with a muscle fiber (Fig. 8-15). Thus,
each motor neuron is connected through its branching
axon to several muscle fibers. The combination of the
motor neuron and the muscle fibers it innervates is known
as a *motor unit.* Although each motor neuron innervates
many muscle fibers, each muscle fiber is innervated by
only a single motor neuron.

As a branch of the motor axon approaches the
muscle surface, it loses its myelin sheath and further
divides into a fine terminal arborization which lies in

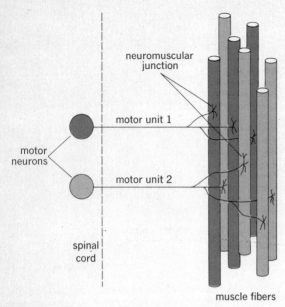

neuromuscular
junction

motor unit 1

motor
neurons

motor unit 2

spinal
cord

muscle fibers

FIGURE 8-15
Muscle fibers associated with two motor neurons, forming two
motor units within a muscle.

grooves on the muscle membrane. The region of the muscle membrane which lies directly under the terminal portion of the axon has special properties and is known as the *motor end plate.*

The terminal ends of the motor axon contain membrane-bound vesicles resembling the synaptic vesicles found at synaptic junctions (Fig. 8-16). These vesicles contain the chemical transmitter *acetylcholine* (abbreviated ACh). When an action potential in the motor axon arrives at the myoneural junction, it depolarizes the nerve membrane, initiating the reactions which lead to a fusion of the transmitter vesicles with the nerve membrane, allowing them to release their contents of acetylcholine into the space separating the nerve and muscle membranes.

Once acetylcholine is released from the nerve ending, it diffuses across the extracellular cleft between the nerve and muscle membrane and combines with receptor sites on the motor-end-plate membrane. This combination causes the permeability of this membrane to sodium and potassium ions to increase, leading to a depolarization of the muscle end plate. This depolarization of the end plate is known as the *end-plate potential* (EPP). The mechanism leading to the formation of the EPP in muscle is analogous to that of the EPSP (excitatory postsynaptic

potential) produced at synaptic junctions. The magnitude of a muscle EPP, however, is much larger than a single EPSP because the surface area between the nerve and muscle is larger and much larger amounts of transmitter agent (ACh) are released. The magnitude of a single muscle EPP is sufficiently large to exceed the threshold potential of the muscle membrane and initiate an action potential which is propagated over the surface of the muscle membrane by the same mechanism described for the propagation of action potentials in nerves.

Since, at synaptic junctions a single EPSP is not sufficient to depolarize the postsynaptic membrane of a nerve fiber to threshold, several EPSPs must occur (temporal and spatial summation) in order to elicit a single action potential in the postsynaptic membrane. In contrast, there is a one-to-one transmission of nerve action potentials to muscle action potentials at the myoneural junction. A further difference between synaptic and myoneural junctions should be noted. At a synaptic junction it is possible to produce an inhibitory postsynaptic potential (IPSP) which hyperpolarizes the postsynaptic membrane and decreases the probability of firing an action potential. No such inhibitory potentials are found in human skeletal muscle; all myoneural junctions are excitatory. Thus the only way to inhibit the electrical activity in the muscle membrane is to inhibit the initiation of action potentials in the motor neuron by synaptic activity (or lack of activity) at the level of the motor neurons in the spinal cord and brain.

In addition to the receptor sites for acetylcholine in the motor-end-plate membrane, these membranes also contain the enzyme acetylcholine-esterase which destroys ACh. The molecules of ACh released from the motor-neuron endings have a lifetime of only about 5 msec before they are destroyed by this enzyme. Once ACh is destroyed, the muscle membrane permeability to sodium and potassium ions returns to its initial state and the depolarized end plate returns to its resting potential.

There are many ways in which events at the neuromuscular junction can be modified by disease or drugs. For example, the deadly South American Indian arrowhead poison, curare, is strongly bound to the acetylcholine receptor site, but it does not change membrane permeability, nor is it destroyed by acetylcholineesterase. When a receptor site is occupied by curare, acetylcholine released from an axon cannot interact with the motor end plate. Thus, the motor nerves conduct normal action potentials and release acetylcholine, but there is no resulting muscle action potential or contraction. Since the skeletal muscles responsible for breathing

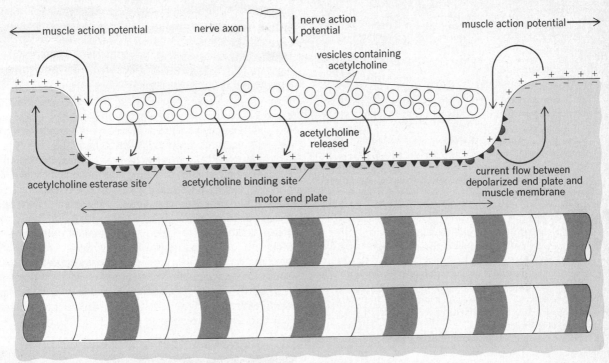

FIGURE 8-16
Events occurring at a neuromuscular junction which lead to an action potential in the muscle membrane.

movements depend upon neuromuscular transmission to initiate their contraction, death comes from asphyxiation.

Neuromuscular transmission can also be altered by inhibition of acetylcholine-esterase, e.g., by some organophosphates, which are the main ingredients in many pesticides and some nerve gases developed for biological warfare. When the enzyme is inhibited, acetylcholine is not destroyed and its prolonged action maintains depolarization of the muscle cell. The failure of repolarization prevents new action potentials from being initiated, so that the muscle does not contract in response to nerve stimulation. The result is paralysis of the muscle and death from asphyxiation.

A third group of substances affects the release of acetylcholine from the nerve terminals, thereby interfering with normal action at the neuromuscular junction. Botulinus toxin, produced by the bacterium *Clostridium botulinum*, blocks the release of acetylcholine in response to an action potential and thus prevents excitation of the muscle membrane. Botulinus toxin is responsible for a

type of food poisoning and is one of the most deadly poisons known. Less than 0.0001 mg is sufficient to kill a man, and half of a pound could kill the entire human population.

The molecular events leading to muscle contraction are summarized in Table 8-1.

Mechanics of muscle contraction

Contraction refers to the active process of generating a force in a muscle. This force, generated by the contractile proteins, is exerted parallel to the muscle fiber. The force exerted by a contracting muscle on an object is known as the muscle *tension*, and the force exerted on a muscle by the weight of an object is known as the *load*. Thus, muscle tension and load are opposing forces. To lift a load the muscle tension must be greater than the load.

When a muscle shortens and lifts a load, the muscle contraction is said to be *isotonic* (constant tension) since

TABLE 8-1
Sequence of events between nerve action potential and contraction and relaxation of a muscle fiber

1 An action potential is initiated and propagated in a motor axon as a result of synaptic events on the cell body and dendrites of the motor neuron in the central nervous system.
2 The action potential in the motor axon causes the release of acetylcholine from the axon terminals at the neuromuscular junction.
3 Acetylcholine is bound to receptor sites on the motor-end-plate membrane.
4 Acetylcholine increases the permeability of the motor end plate to sodium and potassium ions, producing an end-plate potential (EPP).
5 EPP depolarizes the muscle membrane to its threshold potential, generating a muscle action potential which is propagated over the surface of the muscle membrane.
6 Acetylcholine is rapidly destroyed by acetylcholine-esterase on the end-plate membrane.
7 Muscle action potential depolarizes transverse tubules at the A-I junction of sarcomeres.
8 Depolarization of transverse tubules leads to the release of calcium ions from the lateral sacs of the sarcoplasmic reticulum surrounding the myofibrils.
9 Calcium ions bind to troponin-tropomyosin in the thin actin myofilaments, releasing the inhibition that prevented actin from combining with myosin.
10 Actin combines with myosin-ATP:

$$A + M \cdot ATP \longrightarrow A \cdot M \cdot ATP$$

11 Actin activates the myosin ATPase, which splits ATP, releasing energy used to produce a movement of the myosin cross bridge:

$$A \cdot M \cdot ATP \longrightarrow A \cdot M + ADP + P_i$$

12 ATP binds to the myosin bridge, breaking the actin-myosin bond and allowing the cross bridge to dissociate from actin:

$$A \cdot M + ATP \longrightarrow A + M \cdot ATP$$

13 Movements of the cross bridges lead to relative movement of the thick and thin filaments past each other.
14 Cycles of cross-bridge contraction and relaxation continue as long as the concentration of calcium remains high enough to inhibit the action of the troponin-tropomyosin system.
15 Concentration of calcium ions falls as they are moved into the lateral sacs of the sarcoplasmic reticulum by an energy-requiring process which splits ATP.
16 Removal of calcium ions restores the inhibitory action of troponin-tropomyosin, and in the presence of ATP, actin and myosin remain in the disassociated relaxed state.

the load remains constant throughout the period of shortening. When shortening is prevented by a load that is greater than muscle tension, or when a load is supported in a fixed position by the tension of the muscle, the development of tension occurs at constant muscle length and is said to be an *isometric* contraction (constant length). The internal physicochemical events are the same in both isotonic and isometric contractions. Supporting a weight in a fixed position involves isometric contractions whereas body movements involve isotonic contractions.

Figure 8-17 illustrates the general method of recording isotonic and isometric contractions of isolated muscles. During an isotonic contraction the distance the muscle shortens and the time of the contraction are recorded. To measure an isometric contraction the muscle is attached at one end to a rigid support and at the other to a force transducer, which controls the movement of a recording pen in proportion to the force exerted. Thus, the two parameters recorded during an isotonic contraction are distance shortened and time, and the two parameters recorded during an isometric contraction are muscle tension and time.

Single twitch

The mechanical response of a muscle to a single action potential in a muscle cell is known as a *twitch*. Figure 8-18 shows the main features of an isometric and an isotonic twitch. Following excitation there is an interval of a few milliseconds known as the *latent period*, before the tension begins to increase in an isometric twitch. The time from the start of tension development to the peak of tension is the *contraction time*. Not all skeletal muscles contract at the same rate. Some fast fibers have contraction times as short as 10 msec, whereas slow fibers may take 100 msec or longer. The time from peak tension until the tension has decreased to zero is known as the *relaxation time*. In the example of a typical isometric twitch, shown in Fig. 8-18, the entire sequence of contraction and relaxation lasts about 150 msec.

Comparing an isometric twitch with an isotonic twitch in the same muscle, one can see from Fig. 8-18 that the duration of the isotonic twitch is considerably shorter. The latent period, on the other hand, is considerably longer than in an isometric twitch. Once shortening begins, it proceeds at a constant velocity over about 70 percent of the total distance shortened by the muscle, as can be seen from the straight-line relation between distance shortened and time. The slope of this line gives the velocity of shortening. Both the velocity of shortening and the duration of an isotonic twitch depend upon the

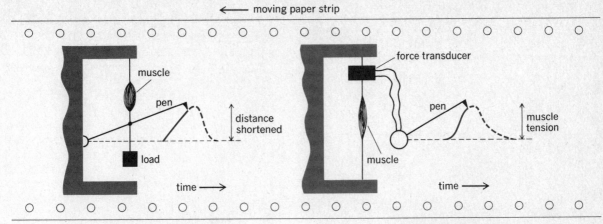

← moving paper strip

muscle

pen

distance
shortened

load

time →

force transducer

pen

muscle

muscle
tension

time →

ISOTONIC CONTRACTION

ISOMETRIC CONTRACTION

FIGURE 8-17
**Methods of recording isotonic and isometric muscle
contractions.**

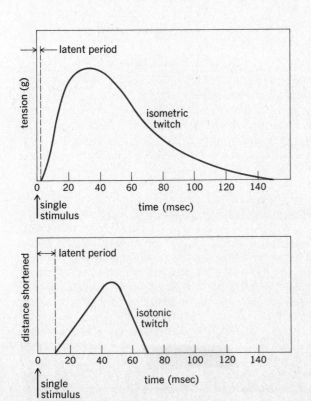

FIGURE 8-18
**Isometric and isotonic skeletal muscle twitches as a function of
time.**

magnitude of the load being lifted by the muscle (Fig.
8-19). At heavier loads the latent period lasts longer but
the velocity of shortening, the duration of the isotonic
twitch, and the distance shortened all decrease. Eventually a load is reached that the muscle is unable to lift.
At this maximum load, the velocity of shortening is zero,
and the contraction of the muscle becomes isometric.
The maximum velocity of shortening occurs when there
is zero load on the muscle. The relation between the load
on a muscle and the velocity at which the muscle lifts
the load is shown in Fig. 8-20.

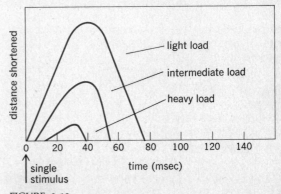

FIGURE 8-19
**Change in the isotonic twitch response of a muscle fiber with
different loads.**

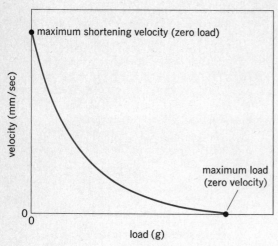

FIGURE 8-20
Shortening velocity as a function of load.

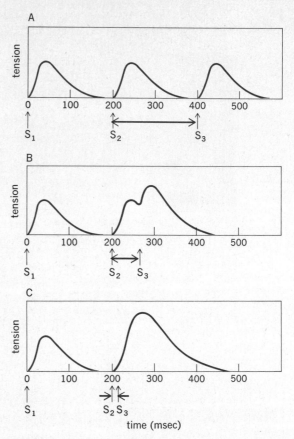

FIGURE 8-21
Summation of isometric contractions produced by shortening the time between stimulus S_2 and S_3.

Summation of contractions

The muscle action potential lasts about 1 to 2 msec and is over before muscle tension has even begun to increase; however, the mechanical response (twitch) which follows may last several hundred milliseconds. Thus it is possible for a second action potential to be initiated during this period of mechanical activity. Figure 8-21 illustrates the isometric contractions of a muscle in response to three successive stimuli. In Fig. 8-21A the isometric twitch following the first stimulus S_1 lasts 150 msec. The second stimulus S_2, applied to the muscle 200 msec after S_1 when the muscle has completely relaxed, causes a second identical isometric twitch. In Fig. 8-21B the interval between S_1 and S_2 remains 200 msec, but a third stimulus is applied 60 msec after S_2, when the mechanical response resulting from S_2 is beginning to decrease. Stimulus S_3 induces a contractile response and the resulting peak tension is greater than that of a single isometric twitch. In Fig. 8-21C the interval between S_2 and S_3 is further reduced to 10 msec. The resulting peak tension is even greater than in Fig. 8-21B, and the rise in tension forms a smooth curve. Here the mechanical response to S_3 appears as a continuation of the mechanical response already induced by S_2.

The property of skeletal muscle contraction in which the mechanical response to one or more successive stimuli is added on to the first is known as *summation*. *Tetanus* is the mechanical response of a muscle resulting from repetitive stimulation at a frequency that is suffi-

ciently rapid to produce a sustained maximal summation. The greater the frequency of stimulation, the greater is the tension produced until a maximal frequency is reached beyond which the tension no longer increases (Fig. 8-22). This is the greatest tension the muscle can develop and is generally about three to four times greater than the isometric twitch tension produced by a single stimulus.

Just as the contraction time of different muscle fibers varies considerably, so does the stimulus frequency that will give a maximal tetanus. The slower the contraction of the fiber, the lower is the frequency of stimulation needed. Frequencies of about 30 per second may produce a tetanus in slow fibers, but frequencies of 100 per second or more are necessary in very rapidly contracting fibers.

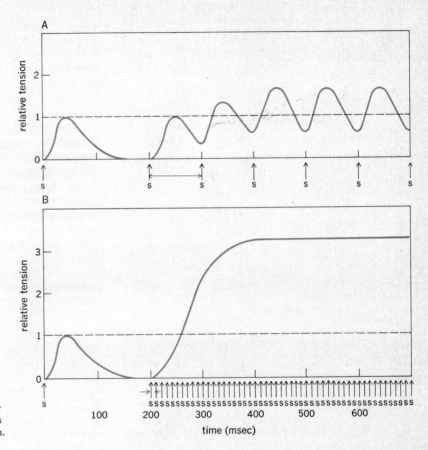

FIGURE 8-22
Isometric tetanic contractions produced
by multiple stimuli of A 10 stimuli per
second and B 100 stimuli per second as
compared with a single isometric twitch.

Summation and tetanus also occur when a muscle contracts isotonically, repetitive stimulation leading in this case to greater shortening. During a maximal isotonic tetanus, a lightly loaded muscle fiber shortens about 40 percent of its resting length. At heavier loads the maximum degree of shortening during tetanic stimulation is less.

How long the muscle tension can be maintained in a tetanically contracted state depends upon the muscle's ability to supply ATP to the contractile proteins. As the energy supply of the muscle is depleted, the force of contraction lessens and eventually falls to zero. This drop in tension with prolonged stimulation is known as *muscle fatigue.* The time of onset of fatigue varies considerably for different types of muscle fibers. Some of the factors contributing to fatigue will be discussed later in the chapter.

What is the underlying explanation for this phe-

nomenon of summation? A logical (but incorrect) explanation of this phenomenon, based on the role of calcium in excitation-contraction coupling, might be to propose that the amount of calcium released from the sarcoplasmic reticulum during a single action potential was only sufficient to inhibit some of the troponin-tropomyosin present in the muscle and thus turn on only a portion of the total contractile apparatus. Multiple stimuli would release more calcium, leading to the activation of a greater number of cross bridges, until a frequency was reached at which sufficient calcium would be present to allow activation of all the cross bridges. However, the fact is that more than sufficient calcium is released by a single action potential to inhibit all the troponin-tropomyosin in the muscle. Thus, following a single action potential the muscle proteins are producing the maximal amount of tension that they are capable of producing.

The explanation of summation involves the passive

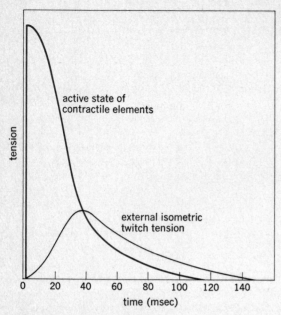

FIGURE 8-23
Active-state tension in the muscle in response to a single stimulus as a function of time.

elastic properties of the muscle. The tension produced *internally* by the muscle proteins is referred to as the *active state* and follows the time course of the calcium release and reuptake. The time course of the internal active state, as compared with the *external* tension exerted by a muscle on an external object, is illustrated in Fig. 8-23. Note the differences in magnitude and time course of these two tensions.

The discrepancy between the external tension and the internal tension of the active state is due to the structure of the muscle and the element of time. Tension is transmitted from the cross bridges through the thick and thin filaments, across the Z lines, and eventually through the extracellular connective tissue and tendons to the bone. All these structures have a certain amount of elasticity and are collectively known as the *series elastic element*. The series elastic element is equivalent to a spring that is placed between the contractile components of the muscle (the cross bridges) and the external object. When the cross bridges contract, the tension produced stretches the spring, which in turn transmits the tension to the external object. To illustrate the consequence of this linkage, consider what would happen if a strong man attempted to lift a brick that was attached to a very weak

spring. Although a great deal of force may be exerted by the man on the spring, he would at first succeed only in stretching the spring, not in lifting the brick, until a point is reached at which the tension in the stretched spring becomes equal to the weight of the brick. Note that it will take a finite amount of time to stretch the spring to this extent.

In the muscle the contractile components in their fully active state begin to stretch the series elastic element immediately following the release of calcium from the sarcoplasmic reticulum. While the contractile elements are stretching the series elastic element, the tension of the active state is decreasing as calcium is being pumped back into the sarcoplasmic reticulum. In a single twitch, the active state decreases before it is able to stretch the series elastic element to a tension equal to the maximal active state and thus less than the full internal tension is transmitted to the external object. When tetanus occurs, the active state is maintained long enough to completely stretch the series elastic element to a tension equal to that of the active state. Thus the increase in external tension that accompanies an increased frequency of stimulation is the result of the increased amount of time during which the contractile elements are maintained in their active state, enabling them to fully stretch the series elastic component. In an isotonic contraction the increased length of the latent period with increasing load is also explained by this same mechanism. The long latent period represents a period during which the series elastic element is being stretched to a tension equivalent to the load before the muscle will begin to lift the load. The heavier the load, the longer it takes to develop the amount of stretching of the series elastic element required to lift the load, and thus the longer the latent period.

Length-tension relationship

One of the classic observations in muscle physiology is the relationship between muscle length and the tension the muscle can develop at that length. A relaxed muscle has properties similar to a passive rubber band; when an outside force is applied to the muscle, it is stretched, and the greater the external force stretching the muscle, the longer it becomes. If the maximal tetanic tension developed by a muscle is plotted against the muscle length at which the isometric tension is measured, the length-tension relationship shown in Fig. 8-24 is obtained. If the length at which the muscle develops maximum tension, l_0, is used as a reference point, the different muscle lengths at which isometric tension is measured

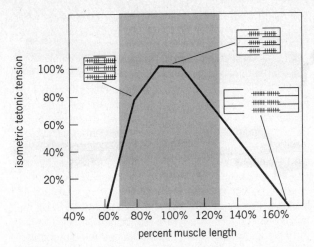

FIGURE 8-24
Variation in isometric tetanus tension with muscle length. The shaded band represents the range of length changes (from 70 to 130 percent) that can occur in the body while the muscles are still attached to bones. (*Adapted from Gordon, Huxley, and Julian.*)

can be expressed in terms of the percent of the l_0 muscle length. If the muscle is set at a length equal to about 60 percent of l_0, it develops no tension when stimulated. As the length of the muscle is increased, the isometric tension rises to a maximum at l_0 and further lengthening of the muscle causes a drop in active tension. When the muscle is stretched to about 175 percent of l_0 or beyond, it no longer develops tension. Thus, the length of a muscle determines the amount of isometric tension it can develop.

This relationship is explained by the sliding-filament model. Passively stretching a muscle changes the amount of overlap between the thick and thin filaments in the myofibrils. Passively stretching a muscle to the point (175 percent of l_0) where it can no longer develop active tension pulls the thick and thin filaments so far apart that there is no overlap between the two. Since active tension is developed by the interaction of the myosin bridges with the actin molecules in the thin filaments, when there is no overlap there can be no bridge interaction with the thin filaments and no tension is developed. More and more bridges overlap with thin filaments at shorter muscle lengths, increasing the active tension in proportion to the number of active cross bridges. At l_0 there is a maximal overlap of thick and thin filaments, and tension is maximal. At lengths less than l_0 two factors lead to decreasing

tension: (*1*) The thin filaments in the two halves of the sarcomere begin to overlap, interfering with cross-bridge interaction, thus decreasing the total number of active cross bridges, and (*2*) the thick filaments become compressed against the two Z lines.

In the body, where muscles are attached to bones, the relaxed length of the muscle is very nearly l_0 and thus at the optimal length for force generation. The total range of length changes that a skeletal muscle can undergo while still attached to bone is limited to a maximum of about 30 percent of the resting length and is often much less. Thus, even at maximal extension or flexion of a limb, its muscles are still able to develop more than 50 percent of their maximum tension. However, in a muscular organ such as the heart, which is not attached to bones, the range over which muscle fiber length can be varied during the filling of the heart with blood is considerably greater.

Control of muscle tension

Since a muscle is composed of many muscle fibers, the total tension a muscle can develop depends upon two factors: (*1*) the number of muscle fibers in the muscle that are contracting at any given time and (*2*) the amount of tension developed by each contracting fiber.

The number of fibers in a muscle that are contracting at any given time depends upon the number of motor neurons to the muscle that are being stimulated. Recall that each motor neuron innervates several muscle fibers, forming a motor unit (Fig. 8-15). Stimulation of a motor neuron produces a contraction in all the muscle fibers in the motor unit. The total tension a muscle develops can thus be varied by varying the number of motor units that are activated. This is determined by the activity of the synaptic inputs to the motor neurons in the brain and spinal cord. A single motor neuron may receive as many as 15,000 synaptic endings, which converge from many different sources, the balance between excitatory and inhibitory synaptic input determining whether a given motor neuron will fire or not fire. The process of increasing the number of active motor neurons (and thus the number of active motor units) is known as *recruitment*.

The number of muscle fibers associated with a single motor axon varies considerably in different types of muscles. In muscles which are able to produce very delicate movements, such as those in the hand and eye, the size of the individual motor units is small (e.g., in a muscle of the eye one motor neuron innervates only about 13 muscle fibers). In the more coarsely controlled

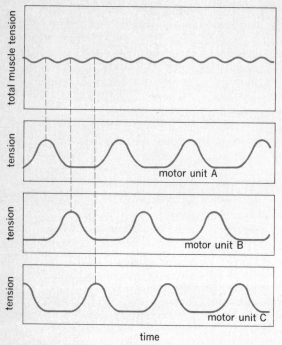

FIGURE 8-25
Asynchronous motor-unit activity maintains a nearly constant tension in the total muscle.

ever were to fire simultaneously, the resulting movement would be a jerky series of contractions and relaxations. The asynchronous activity of motor units is one of the factors responsible for the smooth movements produced by contracting muscles in the body.

In addition to being able to vary the number of active motor units, the tension produced by individual fibers can be varied by the mechanisms discussed previously. Thus increasing the frequency of action potentials delivered to a motor unit produces increases in motor-unit tension through the process of summation and tetanus. In general, when a motor neuron is stimulated by synaptic input, it discharges a burst of action potentials rather than a single action potential. Thus the contractions of the muscle fibers in the body are not usually single twitches but brief summations or more prolonged tetanic contractions.

The tension produced by individual muscle fibers also varies with their length, which determines the amount of thick- and thin-filament overlap. The duration

muscles of the back and legs, each motor unit contains hundreds of muscle fibers (e.g., one motor unit in the large calf muscle of the leg contains about 1,730 muscle fibers). The smaller the size of the motor units, the more precisely the tension of the muscle can be controlled by the recruitment of additional motor units.

The motor neurons to a given muscle fire in an asynchronous pattern determined by their individual synaptic inputs. Thus, the muscle fibers in some motor units may be contracting while other motor units are relaxing. This asynchronous activity in the pool of motor units has several consequences for the development of muscle tension. In muscles which are active for long periods of time, such as the postural muscles which support the weight of the body, the asynchronous activity of their motor units tends to prevent fatigue that might otherwise result from prolonged continuous activity. Some units are active while others rest briefly, only to return to activity as others rest. This pattern of asychronous activity is able to maintain a nearly constant tension in the muscle, as illustrated in Fig. 8-25. Furthermore, if the motor units

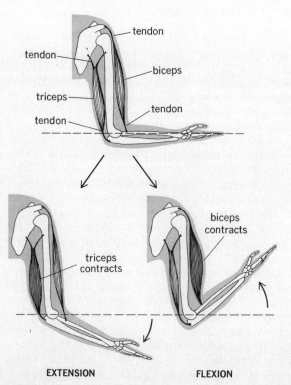

FIGURE 8-26
Antagonistic muscle for flexion and extension of the forearm.

TABLE 8-2
Factors determining total muscle tension

I Number of muscle fibers that are contracting
   a Recruitment of motor units (synaptic input to motor neurons)
   b Size of motor units (number of muscle fibers per axon)
   c Asynchronous activity of motor units
II Tension produced by each contracting muscle fiber
   a Frequency of action potentials in motor neuron (summation and tetanus)
   b Muscle fiber length (degree of overlap of thick and thin myofilaments)
   c Duration of activity (fatigue)
   d Chemical composition and characteristics of individual fibers

of stimulation also becomes a factor in determining fiber tension since prolonged stimulation may result in fatigue. Finally, as we shall discuss in more detail later, individual fibers show distinct chemical differences which result in marked differences in their rates of contraction, tension development, and ease of fatigue. The multiple factors involved in the control of total muscle tension are summarized in Table 8-2.

Lever action of muscles

A contracting muscle exerts a force on bones through its connecting tendons. When the force is great enough, the bone moves as the muscle shortens. A contracting muscle exerts only a pulling force; as the muscle shortens, the bones attached to it are pulled toward each other. *Flexion* of a limb is its bending or movement toward the body, and *extension* is straightening or movement away from the body. These motions require at least two separate muscles, one to cause flexion and the other extension. From Fig. 8-26 it can be seen how contraction of the biceps causes flexion of the forearm and contraction of the triceps causes its extension. Both muscles exert a pulling force upon the forearm when they contract. Groups of muscles which produce oppositely directed movements of a limb are known as *antagonists*. Other sets of antagonistic muscles are required to cause side-to-side movements or rotation of a limb. In some muscles contraction leads to two types of limb movement. In Fig. 8-27 contraction of the gastrocnemius muscle in the calf causes foot extension and flexion of the lower leg, as in walking. Contraction of the gastrocnemius at the same time as that of the quadriceps femoris (which causes extension of the lower leg) prevents the knee joint from bending, leaving only the ankle joint capable of moving; the foot is extended, and the body rises on tiptoe.

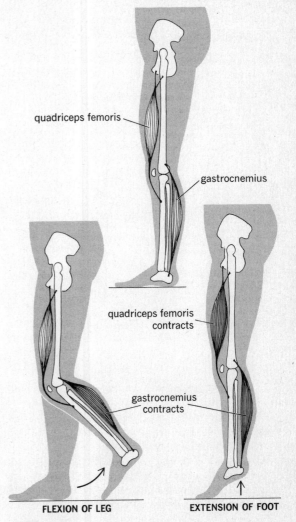

FIGURE 8-27
Flexion of the leg or extension of the foot follows contraction of the gastrocnemius muscle, depending on the activity of the quadriceps femoris muscle.

The arrangement of the muscles, bones, and joints in the body forms lever systems. The basic principle of a lever can be illustrated by the flexion of the forearm by the biceps muscle (Fig. 8-28). The biceps exerts an upward pulling force on the forearm about 2 in. away from the elbow. A 25-lb weight held in the hand exerts a downward force of 25 lb about 14 in. from the elbow. It can be demonstrated according to the laws of physics that a rigid body (the forearm) is in mechanical equilib-

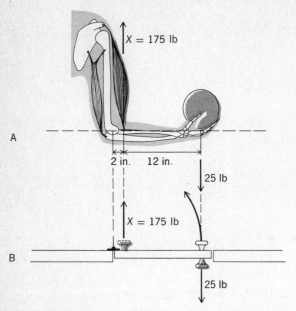

FIGURE 8-28

A **Mechanical equilibrium of forces acting on the forearm while supporting a 25-lb load.** B **Equivalent mechanical system operating on a hinged door.**

The lever system of the arm amplifies the movements of the muscle. Short, relatively slow movements of the muscle produce longer and faster movements of the hand. Thus, a pitcher can throw a basball at 100 mi/hr even though his muscles shorten at only a fraction of this velocity. Skeletal muscles shorten at the rate of about 5 to 10 muscle lengths per second. Thus, the longer the muscle, the faster is its velocity of shortening, even though each sarcomere in a long and short muscle may be shortening at the same velocity.

Energy metabolism of muscle

If a muscle is to contract and relax, ATP must be available to perform three major functions: (*1*) The energy released from ATP splitting is directly coupled to the movement of the cross bridges; (*2*) binding ATP to myosin without splitting is necessary to break the actomyosin bond and allow the cross bridge to operate cyclically; (*3*) energy released from ATP splitting is utilized by the sarcoplasmic reticulum to accumulate calcium ions, producing relaxation.

rium (not accelerating) if the product of the downward force (25 lb) and its distance from the elbow (14 in.) is equal to the product of the upward force exerted by the muscle (X) and its distance from the elbow (2 in.). From this relationship one can calculate the force the biceps must exert on the forearm to support a 25-lb load: $25 \times 14 = 2X$, thus $X = 175$ lb. Such a system is working at a mechanical disadvantage since the force exerted by the muscle is considerably greater than the load it is supporting. Figure 8-28B illustrates the same situation in terms of a hinged door. If the doorknob were placed near the hinge instead of at the other side, a much greater force would be required to open the door. In athletes with very powerful muscles, the large forces exerted by contracting muscles under conditions of maximum exertion sometimes tear the tendon away from the muscle or bone, or in rare cases break the bone.

The mechanical disadvantage under which the muscles operate is offset by increased maneuverability. In Fig. 8-29, when the biceps shortens 1 in., the hand moves through a distance of 7 in. Since the muscle shortens 1 in. in the same amount of time that the hand moves 7 in., the velocity at which the hand moves is seven times faster than the rate of muscle contraction.

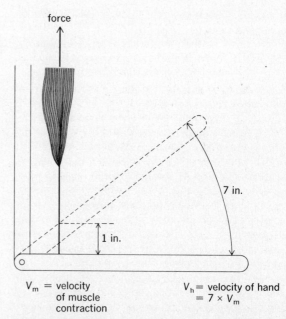

V_m = velocity of muscle contraction

V_h = velocity of hand = $7 \times V_m$

FIGURE 8-29

Small movements of the biceps muscle are amplified by the lever system of the arm, producing large movements of the hand.

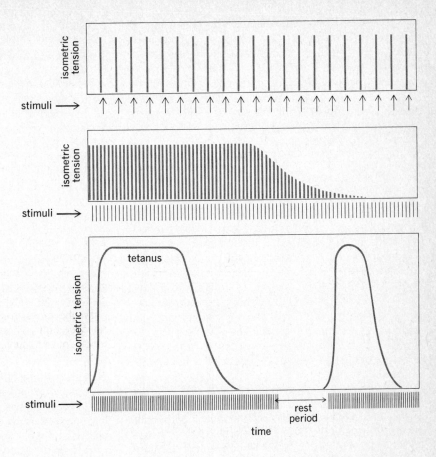

FIGURE 8-30
Muscle fatigue resulting from prolonged stimulation and recovery of ability to contract after a period of rest.

If an isolated muscle is given an adequate supply of oxygen and nutrients which can be broken down to provide ATP, it can continue to give a series of twitch responses to low-frequency stimulation for long periods of time. Under these conditions the muscle is able to synthesize ATP at a rate sufficient to keep up with the rate of ATP breakdown. If the rate of stimulation is increased, the twitch responses soon begin to grow weaker and eventually fall to zero (Fig. 8-30). This drop in tension following prolonged stimulation is muscle fatigue. If rates of stimulation produce tetanic contractions, fatigue occurs even sooner. If the stimulation is stopped and a period of rest is allowed before resuming stimulation, the muscle briefly recovers its ability to contract before again undergoing fatigue. When the muscle is completely fatigued and is unable to develop tension, the concentration of ATP in the muscle is very low. During recovery the ATP concentration rises as metabolism replaces the ATP broken down during contraction.

Recall from Chap. 3 that, during breakdown of carbohydrates, fats, and proteins, energy is continually being transferred to molecules of ATP which then transfer this energy to various processes in the cell, such as the movement of the myosin cross bridges and the accumulation of calcium by the sarcoplasmic reticulum. If a muscle had to rely on its supply of previously synthesized molecules of ATP for contraction, it would be completely fatigued within a few twitches. Therefore, if a muscle is to maintain its contractile activity, molecules of ATP must be synthesized as rapidly as they are broken down. There are three sources for supplying this ATP: (*1*) *creatine phosphate,* (*2*) substrate phosphorylation during glycolysis, and (*3*) oxidative phosphorylation in the mitochondria.

When contraction is initiated by the release of calcium, the myosin ATPase begins to break down ATP at a very rapid rate. The increase in ADP and P_i concentrations resulting from this breakdown of ATP leads, ultimately, to increased rates of oxidative phosphorylation and glycolysis by the mechanisms described in Chap. 3. However, a short period of time elapses before these multienzyme pathways begin to deliver newly formed ATP at a high rate. It is the role of creatine phosphate to provide the energy for ATP formation during this interval.

Creatine phosphate (CP) provides the most rapid means of forming ATP in the muscle cell. This molecule contains energy and phosphate, both of which can be transferred to a molecule of ADP to form ATP and creatine (C):

$$CP + ADP \rightleftharpoons C + ATP \qquad \text{8-1}$$

A single enzyme catalyzes this reversible reaction. Energy is stored in creatine phosphate in resting muscle by the reversal of reaction 8-1. The high levels of ATP in a resting muscle favor, by mass action, the formation of creatine phosphate, and during periods of rest the muscle builds up a concentration of creatine phosphate that is about five times that of ATP. When the ATP level rapidly falls at the beginning of contraction, mass action favors the rapid formation of ATP from creatine phosphate mediated by this single enzymatic reaction. The creatine phosphate system is so efficient that the actual concentration of ATP in the cell changes very little at the start of contraction but the concentration of creatine phosphate falls rapidly.

If contractile activity is to be continued for more than a few seconds, the muscle cell must be able to derive ATP from sources other than creatine phosphate. At moderate levels of muscle activity (moderate rates of ATP breakdown) most of this ATP can be formed by the process of oxidative phosphorylation. Carbohydrates, fats, and proteins can all provide sources of energy for this process.

During very intense exercise, when the breakdown of ATP is very rapid, a number of factors begin to limit the cell's ability to replace ATP by oxidative phosphorylation: (1) the delivery of oxygen to the muscle, (2) the availability of substrates such as glucose, and (3) the rates at which the enzymes in the metabolic pathways can process these substrates. Any of these may become rate limiting under various conditions. Since oxidative phosphorylation depends upon the utilization of oxygen, the continued formation of ATP by this process depends upon an adequate delivery of oxygen to the muscle by the circulatory system. It is this delivery of oxygen to the cell which eventually becomes rate limiting for most forms of prolonged intense exercise, such as running.

Even when adequate oxygen is delivered to the muscle, the rate at which oxidative phosphorylation can produce ATP may be inadequate to keep pace with the rapid rate of ATP breakdown during very intense exercise. When the level of exercise exceeds about 50 percent of maximum (50 percent of the maximal rate of ATP breakdown) anaerobic glycolysis begins to contribute an increasingly significant fraction of the total ATP produced by the muscle.

Although the aerobic process of oxidative phosphorylation produces large quantities of ATP (36 of the 38 ATP formed from each molecule of glucose) the enzymatic machinery of this pathway is relatively slow. The glycolytic pathway, although producing only small quantities of ATP from the breakdown of glucose, can operate at a much higher rate. Thus, in the same amount of time that oxidative phosphorylation can produce 36 molecules of ATP from 1 glucose molecule, about 64 molecules of ATP can be formed by glycolysis through the breakdown of 32 molecules of glucose to lactic acid. Not only is glycolysis faster than oxidative phosphorylation, but it can proceed in the absence of oxygen, leading to the formation of lactic acid as its end product. Thus, during intense exercise, even if adequate oxygen is available, anaerobic glycolysis becomes an additional source for rapidly supplying the muscle with ATP, and lactic acid, the end product of this process, begins to diffuse out of the muscle tissue and accumulate in the blood.

Although anaerobic glycolysis can produce ATP very rapidly, it has the disadvantage of requiring very large quantities of glucose to produce relatively small amounts of ATP. The ability of muscle to store glucose in the form of glycogen provides the muscle with a certain degree of independence from externally supplied glucose. During intense exercise, the glycogen content of the muscle falls progressively, the rate of fall depending upon the intensity of the exercise. The onset of fatigue from exercise lasting more than a few minutes correlates closely with the depletion of the muscle glycogen stores. Finally, in very intense exercise, myosin ATPase may break down ATP faster than even glycolysis can replace it from existing glycogen stores, and fatigue occurs rapidly as the cells' ATP is depleted. The pathways providing ATP for cross-bridge activity are illustrated in Fig. 8-31.

In contrast to true muscle fatigue, psychological fatigue may cause an individual to stop exercising even though his muscles are not depleted of ATP and are still able to contract. An athlete's performance depends not only on the physical state of his muscles but also upon his "will to win."

Following an intense period of exercise a number of changes have occurred in the muscle cell; creatine phosphate levels have decreased, and much of the mus-

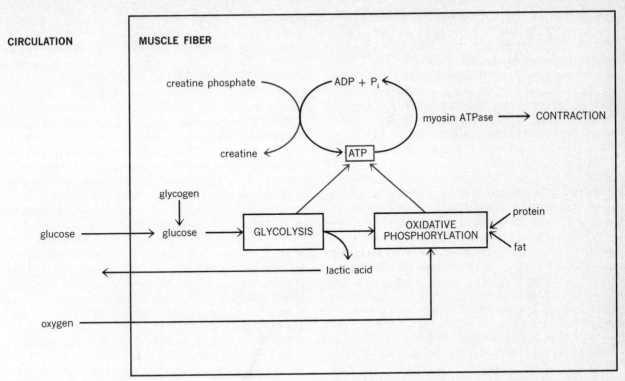

FIGURE 8-31
Biochemical pathways producing ATP utilized during muscle contraction.

cle glycogen may have been converted to lactic acid. To return the cell to its original state, the glycogen stores must be replaced and creatine phosphate resynthesized; both processes require energy. Thus, even though the muscle has stopped contracting, it continues to consume oxygen at a high rate to provide the energy necessary for these synthetic processes. Increased oxygen uptake may proceed for quite some time after the end of exercise, as seen by the fact that one continues to breathe deeply and rapidly after a period of intense exercise. The longer and more intense the exercise, the longer it takes to restore the muscle to its original state. The efficiency with which muscle converts chemical energy to work (movement) is about 40 percent. However, if the energy necessary to return glycogen and creatine phosphate to their original levels in the muscle is included, the overall efficiency is only about 20 percent, the remaining 80 percent of the energy released appearing as heat. The more intense the exercise, the greater is the amount of

heat produced. This increased heat production may place a severe stress upon the body's ability to maintain a constant body temperature, especially on a hot day. On the other hand, the process of shivering reflects the body's use of this same source of heat energy to maintain body temperature in a cold environment. A more detailed discussion of temperature regulation will be found in Chap. 13.

Muscle growth and differentiation

All skeletal muscle fibers are not identical in such properties as their speed of contraction or their enzymatic capacity to produce ATP. Furthermore, it is well known that the capacity of an individual for exercise, both in terms of strength and endurance, can be increased by regular exercise (training). These variations in the contractile properties of muscle, its growth, and development are strongly dependent upon the activity of the motor neurons to the muscle.

Muscle fibers are formed during embryological development by the fusion of a number of small, mononucleated myoblast cells to form long, cylindrical, multinucleated muscle fibers. It is only after the fusion of the myoblasts that the muscle fibers begin to form actin and myosin filaments and are capable of contraction. Once the myoblasts have fused, the resulting muscle fibers no longer have the capacity for cell division and any increase in the size of the muscle results from a growth in the length and diameter of these preformed fibers. This stage of muscle differentiation appears to be completed by about the time of birth, and a newborn child already has all the skeletal muscle fibers it will ever have.

Following the fusion of the myoblasts to form muscle fibers, the motor neurons begin to send axon processes into the muscle, forming myoneural junctions and bringing the muscle under the control of the central nervous system. From this point on there is a very critical dependence of the muscle fiber on its motor neuron, not only to provide a means of initiating contraction but also for the continued survival and development of the muscle fiber. If the nerve fibers to a muscle are severed or the motor neurons destroyed, the denervated muscle fibers become progressively smaller, their content of actin and myosin decreases, and there is a proliferation of connective tissue around the muscle fibers. This decrease in muscle mass following denervation is known as *denervation atrophy*. A muscle can also atrophy with its nerve supply intact if it is not used for a long period of time, as when a broken arm or leg is immobilized in a cast. This is known as *disuse atrophy*. Since atrophy can be prevented by electrically initiating action potentials in the muscle membrane in the absence of nerve fibers, some degree of electrical activity in the muscle membrane is necessary for the maintenance of the functional state of the muscle. In contrast to the decrease in muscle mass that results from a lack of regular neural stimulation of the muscle, increased amounts of neural activity, which accompany repeated exercise, produce changes in the chemical composition of muscle fibers, leading in some cases to a considerable increase in the size (hypertrophy) of the individual muscle fibers. Action potentials in nerve fibers appear to release chemical substances which influence the biochemical activities of the muscle fiber. The chemical identity of these tropic agents and their site of action in the muscle are unknown.

Because of this strong influence of the motor neurons on the chemical properties of muscle fibers, it is not surprising to find distinct differences in the enzymatic composition of the muscle fibers in the body which result from long-term (weeks or months) differences in the amount of neural activity to the muscles. Three classes of skeletal muscle fibers can be identified on the basis of their rate of utilizing ATP (speed of contraction) and their capacity to generate ATP. These three classes are (*1*) high-oxidative slow-twitch fibers, (*2*) high-oxidative fast-twitch fibers, and (*3*) low-oxidative fast-twitch fibers.

The speed with which a muscle fiber contracts is determined primarily by the rate at which myosin ATPase splits ATP and thus the rate at which cross bridges can undergo repeated cycles of activity. Although differences in the myosin ATPase activity produce different speeds of contraction in the different muscle types (Fig. 8-32), the total amount of tension generated depends on the number of actin and myosin filaments in the cross-sectional area of the muscle. Thus large-diameter fibers contain more actin and myosin and produce more tension, regardless of the myosin ATPase.

The second major difference between the various types of muscle fibers is the type of enzymatic machinery available for synthesizing ATP. The high-oxidative fibers have a high rate of oxidative phosphorylation and contain numerous mitochondria. The activity of the glycolytic enzymes in these cells is relatively low. Thus, most of the ATP produced by these cells must be derived from oxidative phosphorylation and is dependent upon an adequate supply of oxygen to these cells. These fibers are found to be surrounded by numerous capillary blood vessels which deliver oxygen and nutrients to the muscle fiber (Fig. 8-33). These high-oxidative fibers also contain a protein known as *myoglobin* which is red in color. Myoglobin is very similar to the protein, hemoglobin, found in the red cells of the blood. Myoglobin binds oxygen and increases the rate of oxygen diffusion into the muscle cell as well as providing a mechanism for storing small amounts of oxygen within the muscle. Muscle fibers containing myoglobin are often referred to as red muscle.

Low-oxidative fibers have an enzymic composition that is geared to the utilization of glucose and the glycolytic pathway for the production of most of its ATP. These fibers contain few mitochondria and have a high content of glycolytic enzymes. Because they contain little myoglobin they are often referred to as white muscle. Very few capillaries are found in the vicinity of these fibers (Fig. 8-33). These fibers also contain large quantities of glycogen to provide an immediately available source of glucose for the anaerobic glycolytic pathway. In contrast, the high-oxidative fibers contain relatively little glycogen.

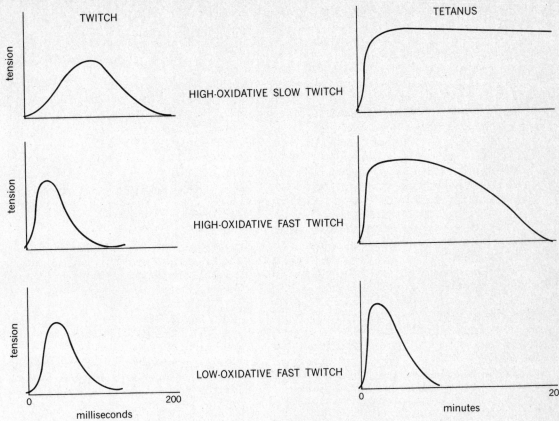

FIGURE 8-32
Twitch and fatigue characteristics of the three types of skeletal muscle fibers.

Recalling the various factors associated with ATP production and utilization which may become rate limiting for muscle contraction, we can predict some of the properties of a muscle fiber that we would expect to result from placing a given type of myosin ATPase in an environment which has a particular capacity for ATP production. The low-oxidative fast-twitch white fibers split ATP very rapidly and are able to produce ATP rapidly by anaerobic glycolysis. However, these fibers also fatigue very rapidly (Fig. 8-32), as their high rate of ATP splitting quickly depletes their glycogen stores. These fibers are generally found to have a large diameter (Fig. 8-33) and are thus able to produce large amounts of tension but only for short periods of time before they fatigue. At the other extreme there are the high-oxidative, slow-twitch red fibers which have a high rate of oxidative phospho-

rylation and are able to keep pace with the relatively slow rate of ATP breakdown. These fibers are very difficult to fatigue (Fig. 8-32) since the high rate of blood flow to these fibers delivers oxygen and nutrients at a sufficient rate to keep up with the relatively slow rate of ATP breakdown by myosin ATPase. These fibers are relatively small in diameter (Fig. 8-33) and thus do not produce the large amounts of tension developed by white fibers.

The third class of fibers, the high-oxidative fast twitch, has properties intermediate between the other two types. These fibers can maintain their contractile activity for longer periods than the fast white fibers because of their capacity to utilize oxidative phosphorylation for some of their ATP requirements and their well-developed blood supply which can deliver oxygen and nutrients to the fibers. However, at high rates of activity,

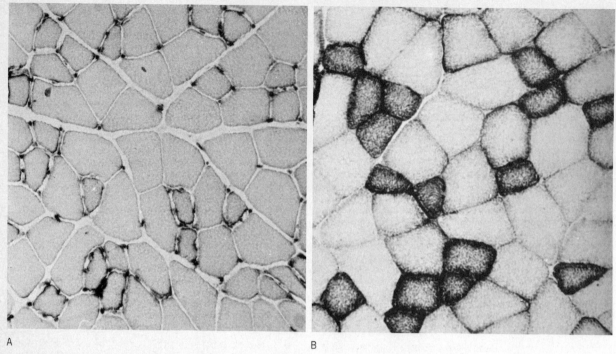

A   B

FIGURE 8-33
Cross sections of skeletal muscle, showing individual muscle fibers which have been stained according to their chemical composition. A The capillaries surrounding the muscle fibers have been stained. Note the large number of capillaries surrounding the small-diameter, high-oxidative fibers. B Darkly stained fibers reveal the presence of high concentrations of oxidative enzymes in the high-oxidative, small-diameter fibers. (*Courtesy of John A. Faulkner.*)

the high rate of ATP splitting by the fast myosin ATPase exceeds the capacity of oxidative phosphorylation to supply ATP and these fibers eventually fatigue (Fig. 8-32). Table 8-3 provides a summary of the characteristics of these three types of skeletal muscle fibers.

The contractile activity that skeletal muscles are called upon to perform varies in different locations in the body. The muscles which support the weight of the body, the postural muscles of the back and legs, must be able to maintain their activity for long periods of time without fatigue, whereas the muscles in the arms are intermittently called upon to rapidly produce large amounts of tension associated with the lifting of objects. Thus, there is a spectrum of activities that a skeletal muscle may be called upon to perform, ranging from long-duration, low-intensity endurance-type activity, to short-duration, high-intensity strength activities. Some muscles which perform predominantly one form of contractile activity are often composed of only one type of muscle

fiber, usually of the high-oxidative type. More commonly, however, a muscle is required to perform endurance-type activity under some circumstances and high-intensity strength activity under others. These muscles generally contain a mixture of the three types of muscle fibers.

Different patterns of neural activity to a muscle occur during different types of exercise. At low strengths, just the high-oxidative slow-twitch fibers are recruited. Stronger contractions result from the added recruitment of the high-oxidative fast-twitch and eventually the low-oxidative fast-twitch fibers. In endurance exercise the frequency of action potential to the muscle is lower, and most of the high-oxidative motor units are recruited in an asynchronous pattern. Thus, the pattern of neural activity varies from high-frequency, short-duration discharges to most motor units in high-intensity, short-duration exercise, to low-frequency, long-duration discharges to high-oxidative motor units during endurance exercise. Changes in the type or amount of activity a

TABLE 8-3
Properties of three types of skeletal muscle fibers

	High-oxidative slow-twitch	High-oxidative fast-twitch	Low-oxidative fast-twitch
Speed of contraction	Slow	Fast	Fast
Myosin ATPase activity	Low	High	High
Primary source of ATP production	Oxidative phosphorylation	Oxidative phosphorylation	Anaerobic glycolysis
Glycolytic enzyme activity	Low	Intermediate	High
Number of mitochondria	Many	Many	Few
Capillaries	Many	Many	Few
Myoglobin content	High	High	Low
Muscle color	Red	Red	White
Glycogen content	Low	Intermediate	High
Fiber diameter	Small	Intermediate	Large
Rate of fatigue	Slow	Intermediate	Fast

muscle is called upon to perform alter the pattern of neural activity to the muscle and gradually produce changes in the chemical composition of the muscle.

The capacity of a muscle for activity can be altered in two ways: (1) by transformation of one biochemical type of fiber into another and (2) by the growth in size (hypertrophy) of the muscle fibers. Endurance types of exercise (running and swimming) are associated with a transformation of low-oxidative fast-twitch fibers into high-oxidative fast-twitch fibers. Endurance exercise also leads to an increase in the number of mitochondria in the high-oxidative fibers and an increase in the number of capillaries surrounding these fibers. These changes are accompanied by relatively small increases in the mass (strength) of the muscle and result in a muscle which has an increased capacity for long-duration, relatively low-intensity activity.

High-intensity, short-duration exercise, such as weight lifting, produces quite a different pattern of change in the muscle. The short-duration, high-frequency discharges to the low-oxidative fast-twitch fibers induces hypertrophy in these fibers, with increased synthesis of actin and myosin filaments and a large increase in the mass and strength of the muscle. The extreme result of this type of exercise is the bulging muscles of a professional weight lifter. Because different types of exercise produce quite different chemical changes in skeletal muscle, an individual performing regular exercises to improve his muscle performance must be careful to chose a type of exercise that is compatible with the type of activity he ultimately wishes to perform. Thus lifting weights will not improve the endurance of a long-distance runner and jogging will not produce bulging biceps in a weight lifter. As we shall see in later chapters, endurance exercise not only produces changes in the skeletal muscles of the body but also produces changes in the respiratory and circulatory systems which improve the delivery of oxygen and nutrients to the muscle fibers.

Smooth muscle

Smooth muscle is found in the walls of hollow organs such as the intestinal tract, blood vessels, the air passages to the lungs, the urinary bladder, and the uterus. It may also be found as single cells distributed throughout an organ such as the spleen or in small groups of cells attached to the hairs in the skin. Because of the diversity of smooth muscle function we shall not attempt to describe the specific properties of any one type of smooth muscle but shall identify their general properties. The reader must keep in mind that any one specific smooth muscle may not exhibit all these properties. In later chapters, as we discuss each organ system, the specific factors affecting the activity of the smooth muscle in that specific organ will be described.

Smooth muscle structure

Smooth muscle fibers are considerably smaller than skeletal muscle fibers, being only 2 to 20 μm in diameter and are spindle-shaped rather than cylindrical. Each fiber has a single nucleus located in the central portion of the cell. The most noticeable morphological factor distinguishing smooth from either skeletal or cardiac muscle is the

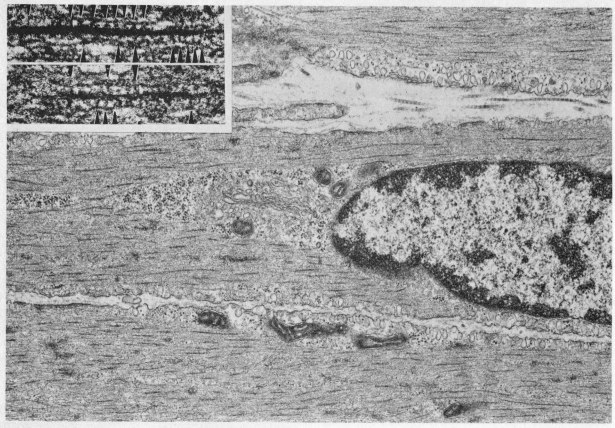

FIGURE 8-34

Electron micrograph of portions of three smooth muscle fibers. (Insert) Higher magnification of thick filaments with projections (arrows) suggestive of cross bridges connecting adjacent thin filaments. [*From A. P. Somlyo, C. E. Devine, Avril V. Somlyo, and R. V. Rice, Phil. Trans. R. Soc. Lond. B.* **265**:223–229 (1973).]

absence of striated banding patterns in the cytoplasm (thus the name smooth muscle). The banding patterns of striated muscle arise from the regular arrangement of thick and thin myofilaments in the myofibrils of these muscles. As can be seen from Fig. 8-34, smooth muscle does not contain myofibrils. However, myosin thick filaments and actin thin filaments can be seen distributed throughout the cytoplasm, oriented parallel to the muscle fiber but not organized into regular units of filaments as in striated muscle. At high magnification the thick filaments can be seen to possess cross bridges, just as in skeletal muscle. Since troponin and tropomyosin can be isolated from smooth muscle, there is every reason to believe that the molecular events of force generation in

these cells, following the release of calcium ions, are similar to those in skeletal muscle. It is still unclear exactly how the thick and thin filaments are anchored in smooth muscle so that the relative sliding of the filaments past each other leads to a shortening of the cell.

The random overlap of thick and thin filaments in smooth muscle may provide part of the explanation for the observation that smooth muscle can develop active tension over a wide range of muscle lengths (Fig. 8-35). The less ordered arrangement of the thick and thin filaments is such that there is almost always some overlap between the two sets of filaments over wide ranges of muscle lengths. This property allows the smooth muscles surrounding hollow organs, such as the stomach and

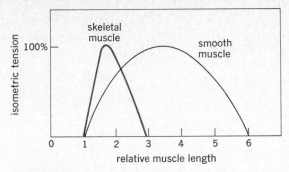

FIGURE 8-35
Variation in isometric tension with muscle length for skeletal and smooth muscle.

bladder, whose walls become stretched as the contents are filled with food or urine, to actively exert a force on the contents of these cavities. If such organs were surrounded by skeletal muscle, a moderate amount of stretching would pull the thick and thin filaments beyond the range of overlap and no active tension could be developed.

A gram of smooth muscle contains only about 10 percent of the amount of actin and myosin found in a gram of striated muscle. The concentrations of ATP and creatine phosphate are also lower in smooth muscle. Furthermore, the actin-activated myosin-ATPase activity of most smooth muscle is relatively low. Thus, both the rate of contraction and the total tension developed by smooth muscles are generally much less than in striated muscle.

A further morphological difference between smooth muscle and skeletal muscle is the lack of a well-developed sarcoplasmic reticulum and transverse tubule system. A series of vesicles can be seen just below the cell membrane (Fig. 8-34) and may be involved in the storage and release of calcium, but smooth muscles do not have the well-developed system of intracellular membranes that is associated with excitation-contraction coupling in striated muscle. The calcium ions for the initiation of contraction in skeletal muscle come from the lateral sacs within the muscle, and skeletal muscle is able to contract in the complete absence of extracellular calcium ions. Many types of smooth muscle, however, will not contract when stimulated unless calcium is present in the extracellular medium. In these cells the surface membrane appears to have properties similar to the membranes of the sarcoplasmic reticulum in striated muscles; they allow calcium to enter the cell when the membrane is stimulated and pump calcium out of the cell to produce relaxation. Because of the large diameter of skeletal muscle fibers, if they depended upon only their surface membrane to regulate the access of calcium to the cell, there would be a long delay between the electrical activity in the membrane and the contraction (because of the slow rate of calcium diffusion over long distances). This does not present a problem for smooth muscle because of its small diameter.

Classification of smooth muscle

When the total range of smooth muscle properties is examined there emerge two general classes of smooth muscle, known as *single-unit* and *multiunit smooth muscle*. A given smooth muscle generally exhibits properties which are characteristic of one of these two classes. However, these classifications are not absolute, and some smooth muscles may show some characteristics common to both classes. Multiunit smooth muscle has functional properties which resemble those of skeletal muscle, whereas, as we shall see, single-unit smooth muscle more closely resembles cardiac muscle in its functional activity.

Multiunit smooth muscle Smooth muscles showing multiunit type of activity are found in the larger arteries and in some areas of the intestinal and reproductive systems and make up the pilomotor muscles that are attached to the hairs of the skin. A major characteristic of multiunit smooth muscles is that contractile activity is initiated in them by electrical activity in the nerve fibers to the muscle, just as in the case of skeletal muscle. The innervation of the smooth muscle is by way of the sympathetic and parasympathetic branches of the autonomic nervous system rather than the somatic nerves which innervate skeletal muscle. However, the autonomic nerve fibers do not form discrete junctions with the smooth muscle membrane as the somatic neurons do on skeletal muscle.

Distributed along the terminal branches of the autonomic nerves are a series of regions where the axon appears swollen. These regions are filled with membrane-bound vesicles which are presumably the location of the chemical transmitters. Upon stimulation of the nerve, these vesicles fuse with the nerve membrane and release their transmitter into the extracellular space. Since a single axon releases transmitter from several such regions along its length, it affects numerous surrounding smooth muscle cells. Moreover, the released transmitter is able to diffuse to a number of smooth muscle cells in the vicinity.

Because of the relatively large distances between the nerve terminals and the smooth muscle membrane, there is a considerably longer period of time between nerve stimulation and contractile response in smooth muscle than is found in skeletal muscle where only a few hundred angstroms separate the nerve ending from the motor end plate. Thus, a single autonomic nerve fiber may affect the activity of a number of smooth muscle fibers even though it does not form an anatomical junction with any given cell.

The total surface of the smooth muscle membrane appears to contain receptor sites that can combine with the nerve transmitter agents; this contrasts with the localization of receptors in one region under the nerve terminal in the motor end plate of skeletal muscle. Binding of the transmitter to the membrane receptor alters the permeability of the smooth muscle membrane to ions, which leads to changes in the membrane potential resembling those seen at synaptic junctions between nerve fibers. Either excitatory or inhibitory potentials may be produced, depending on the nature of the chemical transmitter released. Just as with synaptic junctions, a single action potential in the nerve fiber produces only a small subthreshold change in the smooth muscle membrane potential. Multiple action potentials in the nerve are required to depolarize the smooth muscle membrane to threshold and initiate an action potential. Once an action potential is produced, it releases calcium into the cell which initiates contraction.

In addition to being stimulated by autonomic nerves, multiunit smooth muscle may also be induced to contract by hormones, such as epinephrine, which reach the cell by way of the circulatory system. These hormones apparently react with receptor sites on the membrane surface, producing changes in the membrane potential by mechanisms similar to those produced by neurotransmitter agents at synaptic junctions.

Single-unit smooth muscle   Single-unit smooth muscle resembles cardiac muscle since both types of muscle can undergo spontaneous, rhythmical contractions in the absence of nerve or hormonal input. Spontaneous electrical and mechanical activity occurs in single-unit smooth muscles that have been removed from the body and are no longer attached to nerves. Thus, this spontaneous activity appears to be an inherent property of the muscle cell itself. The molecular properties of the cells which produce this spontaneous activity are not understood. The membrane potentials in these cells show spontaneous fluctuations in potential which are responsible for

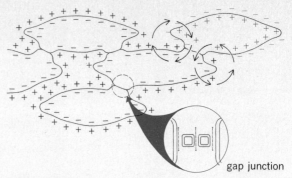

gap junction

FIGURE 8-36
Action potentials from one smooth muscle cell can be conducted to others through gap junctions between adjacent cells.

initiating the cells' mechanical activity. Examples of smooth muscle showing single-unit-type activity are the vascular smooth muscle in small arteries and veins, intestinal smooth muscle, and the smooth muscle of the uterus.

The term single-unit smooth muscle is applied to these cells because large numbers of them show a synchronized electrical and mechanical activity and thus respond as if they were a single unit. What is responsible for coordinating the simultaneous activity of so many separate cells? The electron microscope has revealed that the membranes of adjacent single-unit smooth muscle cells are joined together to form gap junctions (Chap. 2). In the regions of the gap junctions the membranes of the adjacent cells join to form small channels directly linking the cytoplasms of the joined cells (Fig. 8-36). These gap junctions provide a low-resistance pathway for the conduction of electrical activity from cell to cell. Thus an action potential in one cell is conducted from cell to cell through the gap junctions and the cells so joined together electrically respond as a single unit. As we shall see later, cardiac muscle cells are also linked together by gap junctions.

The basic electrical event in the membranes of single-unit smooth muscle appears to be a spontaneous depolarization of the membrane which occurs at regular intervals and is known as a *pacemaker potential*. This pacemaker potential appears as a relatively slow depolarization of the membrane potential which, when it reaches the threshold potential, triggers off an action potential. Following repolarization of the action potential, the mem-

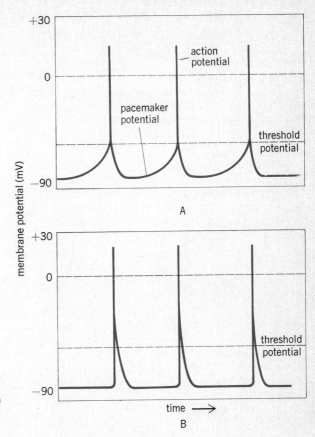

FIGURE 8-37
A Spontaneous action potentials in pacemaker cells. B Action potentials in nonpacemaker cells connected to a pacemaker cell by gap junctions.

brane again begins to undergo depolarization leading to a second action potential, and so on (Fig. 8-37). In a population of cells joined together by gap junctions, all the cells do not show spontaneous pacemaker potentials. Cells which do not have pacemaker activity, however, fire action potentials at the same rate as the pacemaker cells because the action potential from the pacemaker cell has been propagated throughout the population of cells through the gap junctions (Fig. 8-37). Thus the pacemaker cell sets the pace at which all the cells in the population fire action potentials.

In addition to the pacemaker potentials, much slower waves of depolarization and repolarization are often observed upon which are superimposed bursts of action potentials (Fig. 8-38). The origin of these slow waves of activity is still unclear. They may reflect the electrical summation of a number of interconnected pacemaker cells which fail to trigger propagated action potentials. When the mechanical activity of the muscle is recorded simultaneously with the electrical activity (Fig.

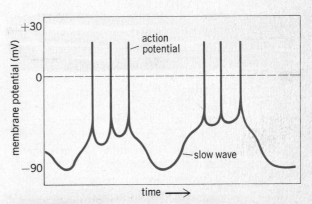

FIGURE 8-38
Slow-wave oscillations in a membrane potential triggering bursts of action potentials.

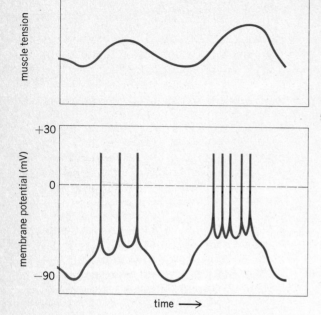

FIGURE 8-39
Mechanical activity of smooth muscle corresponding to the frequency of slow-wave potential changes and the frequency of generated action potentials.

sulting mechanical activity of the muscle increases. Stretching the muscle leads to depolarization of the membrane, which increases the frequency of action potentials, which produces a contraction which tends to oppose further stretch of the muscle. This provides a form of negative feedback which tends to keep the length of the muscle constant. In contrast, stretching a multiunit innervated smooth muscle usually does not elicit a contraction. Thus, this response to stretch is a property of single-unit smooth muscle and probably reflects the instability inherent in their membranes. Agents which lead

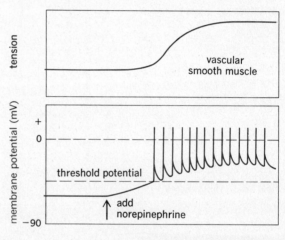

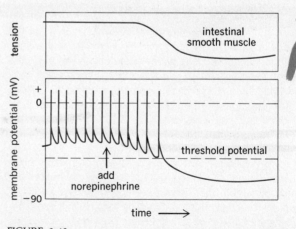

FIGURE 8-40
Different responses of vascular smooth muscle and intestinal smooth muscle to norepinephrine released from a sympathetic nerve ending.

8-39), the contractions of the muscle are seen to correspond to the periods when action potentials are occurring in the muscle and waves of contraction and relaxation may occur which correspond to the slow-wave frequency. Furthermore, the intensity of the mechanical response increases with the frequency of the action potentials. This is equivalent to the mechanical summation of activity in skeletal muscle that results from multiple action potentials.

Although the responses we have discussed thus far result from spontaneous electrical activity within the muscle itself, this activity can be modified by external agents such as nerves, hormones, metabolic intermediates, mechanical stretch, and a variety of drugs. All these agents appear to act upon the smooth muscle membrane to produce either a depolarization or a hyperpolarization of the membrane potential. If the membrane becomes depolarized, so that it is nearer to threshold, the frequency of spontaneous action potentials increases and the re-to a hyperpolarization of the membrane, on the other hand, tend to decrease the frequency of spontaneous action potentials, inhibiting contraction and leading to smooth muscle relaxation.

As was indicated earlier, the specific properties of smooth muscles show great diversity between different tissues. One form of this diversity is illustrated in Fig. 8-40, which shows the opposite response of vascular and intestinal single-unit smooth muscle to sympathetic stimulation. The same neurotransmitter, norepinephrine, is released from the sympathetic ending in each tissue, yet the vascular smooth muscle becomes depolarized and contracts whereas the intestinal smooth muscle becomes hyperpolarized and relaxes. How can the same chemical agent produce such opposite effects? The molecular mechanism is unknown but it seems clear that the different responses are not produced by the chemical effects of norepinephrine directly. The neurotransmitter merely acts as a trigger at the receptor site which sets in motion a set of permeability changes which were initially built into the structure of the cell membranes of these two different cells. The mechanism necessary to produce a given type of permeability change must already be built into the membrane and norepinephrine acts only to release that built-in mechanism, not to produce the permeability change itself.

Cardiac muscle

Cardiac (heart) muscle has properties similar to those of both skeletal and single-unit smooth muscle. It is a striated muscle having myofibrils with thick and thin myofilaments. The sliding-filament type of contraction is found in cardiac muscle, which has a length-tension relationship similar to that shown by skeletal muscle.

Cardiac muscle also has a well-developed sarcoplasmic reticulum. Action potentials in the cardiac muscle membrane lead to the release of calcium from the sarcoplasmic reticulum and thereby to the activation of the actomyosin contractile system.

The metabolism of cardiac muscle is designed for endurance rather than speed or strength. A continuous supply of oxygen must be maintained to the heart muscle if it is to continue to supply ATP to the contractile machinery. Cardiac cells deprived of oxygen for as little as 30 sec cease to contract, and heart failure ensues.

Cardiac muscle most resembles single-unit smooth muscle in its spontaneous activity and the presence of gap junctions. The properties of cardiac muscle will be developed more fully as they apply to the functioning of the cardiovascular system discussed in the next chapter.

COORDINATED BODY FUNCTIONS PART

3

Circulation

9

Since the cells of higher animals are organized into tissues and organs which perform specialized functions, there must be a system to transport materials between them. This function is served by the *circulation,* which comprises the *blood* and the apparatus moving it, the *cardiovascular system.*

Because of its transport function, the blood participates directly or indirectly in virtually all body functions. These discrete blood functions will be described in subsequent relevant chapters. We accordingly present here only the summary of blood composition necessary for our description of the cardiovascular system.

Blood is composed of specialized cells and a liquid, *plasma,* in which they are suspended. The cells are the *red blood cells,* or *erythrocytes,* the *white blood cells,* and the *platelets.* Ordinarily, the constant motion of the blood keeps the cells well dispersed throughout the plasma, but if a sample of blood is allowed to stand (clotting prevented), the cells slowly sink to the bottom. This process can be speeded up by centrifuging. By this means, the percentage of total blood volume which is cells, known as the *hematocrit,* can be determined. The normal hematocrit is approximately 45 percent. The total blood volume of an average man is approximately 8 percent of his total body weight. Accordingly, for a 70-kg man

ucts / Bulk flow across the capillary wall: distribution of the extracellular fluid

Veins

Determinants of venous pressure / Effects of venous constriction on resistance to flow / The venous valves

Lymphatics

Functions of the lymphatic system / *Return of excess filtered fluid / Return of protein to the blood / Specific transport functions / Lymph nodes* / Mechanism of lymph flow

SECTION C   INTEGRATION OF CARDIOVASCULAR FUNCTION: REGULATION OF SYSTEMIC ARTERIAL PRESSURE

Arterial pressure, cardiac output, and arteriolar resistance

Cardiovascular control centers in the brain

Receptors and afferent pathways

Arterial   baroreceptors / Other   baroreceptors / Chemoreceptors / Summary

SECTION D   CARDIOVASCULAR PATTERNS IN HEALTH AND DISEASE

Hemorrhage and hypotension

The upright posture

Exercise

Hypertension

Congestive heart failure

"Heart attacks" and arteriosclerosis

Total blood weight = 0.08 × 70 kg = 5.6 kg

One kilogram of blood occupies approximately 1 liter; therefore

Total blood volume = 5.6 liters

The hematocrit is 45 percent; therefore

Total cell volume[1] = 2.5 liters
Plasma volume = 5.6 − 2.5 liters = 3.1 liters

Plasma is an extremely complex liquid. It consists of a large number of organic and inorganic substances dissolved in water. The most abundant solutes by weight are the proteins, which together compose approximately 7 percent of the total plasma weight. The *plasma proteins* vary greatly in their structure and function, but they can be classified, according to certain physical and chemical reactions, into two broad groups, the *albumins* and the *globulins*. The albumins are three to four times more abundant than the globulins and usually are of smaller molecular weights. The plasma proteins, with notable exceptions, are synthesized by the liver, the major exception being the group known as *gamma globulins,* which are formed in the lymph nodes and other lymphoid tissues

(Chap. 15). The plasma proteins serve a host of important functions which will be described in relevant chapters, but it must be emphasized that normally they are *not* taken up by cells and utilized as metabolic fuel. Accordingly, they must be viewed quite differently from most other organic constituents of plasma, such as glucose, which use the plasma as a vehicle for transport but function in cells. The plasma proteins function in the plasma itself or, under certain circumstances, in the interstitial fluid.

In addition to the organic solutes—proteins, nutrients, and metabolic end products—plasma contains a large variety of mineral electrolytes, the concentrations of which are shown in Table 9-1, along with that of protein. The value in millimoles per liter for protein may seem puzzling in view of the statement that protein is the most abundant plasma solute by *weight*. Remember, however, that molarity is a measure not of the weight but of the *number* of molecules or ions per unit volume. Protein molecules are so large in comparison with sodium ions that a very small number of them greatly outweighs a much larger number of sodium ions. The osmolarity (and, therefore, water concentration) of a solution depends upon the *number,* not the weight, of the solute particles present. Accordingly, sodium is the single most important determinant of total plasma osmolarity.

---

[1]Since the vast majority of all blood cells are erythrocytes, the total cell volume is approximately equal to the erythrocyte volume.

TABLE 9-1
Plasma concentrations of electrolytes and protein

Constituent	Grams per liter	Millimoles per liter
Sodium, Na^+	3.39	144
Chloride, Cl^-	3.55	100
Bicarbonate, HCO_3^-	1.50	25
Potassium, K^+	0.17	4.4
Calcium, Ca^{2+}	0.10	2.5
Phosphate, HPO_4^{2-} or $H_2PO_4^-$	0.10	1.0
Magnesium, Mg^{2+}	0.04	1.5
Protein	70	2.5

Overall design of the cardiovascular system

The cardiovascular system (Fig. 9-1) comprises a set of tubes, *blood vessels,* through which blood flows and a pump, the *heart,* which produces this flow. Physiology as an experimental science began in 1628, when William Harvey demonstrated that the entire system forms a circle, so that blood is continuously being pumped out of the heart through one set of vessels and returning to the heart via a different set. In man, as in all mammals, there are actually two circuits, both originating and terminating in the heart, which is divided longitudinally into two functional halves. Blood is pumped via one circuit (the *pulmonary circulation*) from the right half of the heart through the lungs and back to the left half of the heart. It is pumped via the second circuit (the *systemic circulation*) from the left half of the heart through all the tissues of the body, except, of course, the lungs, and back to the right half of the heart. In both circuits, the vessels carrying blood away from the heart are called *arteries,* and the vessels carrying blood from the lungs and tissues back to the heart are called *veins.* In the systemic circuit, blood leaves the left half of the heart via a single large artery, the *aorta.* From the aorta branching arteries conduct blood to the various organs and tissues. These arteries divide in a highly characteristic manner into progressively smaller branches, much of the branching occurring within the specific organ or tissue supplied. The smallest branches, called *arterioles,* differ structurally and functionally from the arteries. Ultimately the arterioles branch into a huge number of very small, thin vessels, termed *capillaries.* The capillaries unite to form larger vessels (*venules*) which, in turn, unite to form fewer and still larger vessels, termed *veins.* The veins from different organs and tissues unite to form two large veins, the *inferior vena cava* (from

the lower portion of the body) and the *superior vena cava* (from the upper half of the body). By these two veins blood is returned to the right half of the heart. The entire systemic circuit can be visualized, therefore, as two trees, one arterial and the other venous, having the same origin (the heart) and being connected by fine twigs (capillaries) which unite the smallest branches of each tree (Fig. 9-2).

The pulmonary circulation is composed of a similar circuit. Blood leaves the right half of the heart via a single large artery, the *pulmonary artery,* which divides into two arteries, one supplying each lung. Within the lungs, the arteries continue to branch, ultimately forming arterioles, which then divide into capillaries. These capillaries unite to form small venules, which unite to form larger and larger veins. The blood leaves the lungs via the largest of these, the *pulmonary veins,* which empty into the left half of the heart. The blood flowing through the systemic veins, right half of the heart, and pulmonary arteries has a low oxygen content. As this blood flows through the

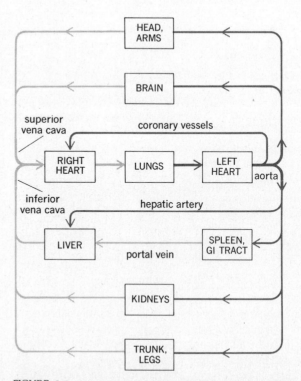

FIGURE 9-1
Diagrammatic representation of the cardiovascular system in the adult human being. Darker shading indicates blood with high oxygen content.

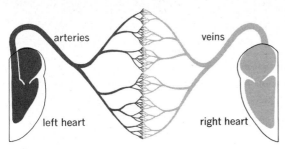

FIGURE 9-2
Systemic circulation as two trees connected by capillaries. As indicated by the color change, oxygen leaves the blood during passage through the capillaries. (*Adapted from Rushmer.*)

lung capillaries, it picks up large quantities of oxygen; therefore, the blood in the pulmonary veins, left heart, and systemic arteries has a high oxygen content. As this blood flows through the capillaries of tissues and organs throughout the body, much of this oxygen leaves the blood, resulting in the low oxygen content of systemic venous blood.

In a normal person, blood can pass from the systemic veins to the systemic arteries only by first being pumped through the pulmonary circuit, thus oxygenating all the blood returning from the body tissues before it is pumped back to them. Normally the total volumes of blood pumped through the pulmonary and systemic circuits during a given period of time are equal. In other words, the right heart pumps the same amount of blood as the left heart. Only under unusual circumstances, such as malfunction of one half to the heart, do these volumes differ from each other, and then only transiently.

It should be evident that all the blood pumped by the right heart flows through the lungs; in contrast, only a fraction of the total left ventricular output flows through any single organ or tissue. In other words, the systemic circulation comprises numerous different pathways "in parallel." They all originate as large arteries branching off from the aorta. The only significant deviation from this pattern is the blood supply to the liver, much of which is not arterial but venous blood, which has just left the spleen and gastrointestinal tract.

The dominant feature of the cardiovascular system is the pumping of blood by the heart. In a resting normal man, the amount of blood pumped simultaneously by each half of the heart is approximately 5 liters/min. During heavy work or exercise, the volume may increase as much as fivefold to 25 liters/min. We shall discuss first

the basic mechanisms by which the heart pumps blood, then the functions of the vascular system, and finally, the mechanisms by which the entire system is controlled according to the requirements of the body. Despite the preeminent position of the heart, the blood vessels play critical roles in the circulation of the blood. These vessels are not merely inert plumbing. Each type has a highly characteristic structure and function. Cardiovascular mechanisms will have meaning for the reader only as he is able to perceive how they contribute to the overall function, an adequate flow of blood to the tissues.

SECTION A
THE HEART

Anatomy

The heart is a muscular organ located in the chest (*thoracic*) cavity and covered by a fibrous sac, the *pericardium*. Its walls are composed primarily of muscle (*myocardium*), the structure of which is different from either skeletal or smooth muscle. The inner surface of the myocardium, i.e., the surface in contact with the blood within the heart chambers, is lined by a thin layer of cells (endothelium).

The human heart is divided longitudinally into right and left halves (Fig. 9-3), each consisting of two chambers, an *atrium* and a *ventricle*. The cavities of the atrium and ventricle on each side of the heart communicate with each other, but the right chambers do not communicate directly with those on the left. Thus, right and left atria and right and left ventricles are distinct.

Perhaps the easiest way to picture the architecture of the heart is to begin with its fibrous skeleton, which comprises four rings of dense connective tissue joined together (Fig. 9-4). To the tops of these rings are anchored the muscle masses of the atria, pulmonary artery, and aorta. To the bottom are attached the muscle masses of the ventricles. The connective-tissue rings form the openings between the atria and ventricles and between the great arteries and ventricles. To these rings are attached four sets of valves (Figs. 9-4 and 9-5).

Between the cavities of the atrium and ventricle in each half of the heart are the *atrioventricular valves* (*AV valves*), which permit blood to flow from atrium to ventricle but not from ventricle to atrium. The right and left AV valves are called, respectively, the *tricuspid* and *mitral* valves. When the blood is moving from atrium to ventricle, the valves lie open against the ventricular wall, but when the ventricles contract, the valves are brought together

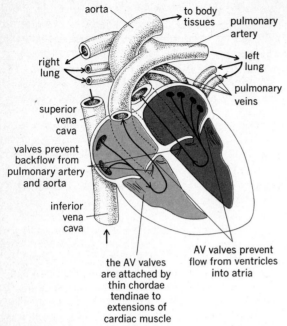

FIGURE 9-3
Diagrammatic section of the heart. The arrows indicate the direction of blood flow. The intracardiac pulmonary artery and aorta have been cut away; their locations are indicated by the dashed lines. (*Adapted from McNaught and Callender.*)

venules, veins; left atrium; left ventricle; aorta. The driving force for this flow of blood, as we shall see, comes solely from the active contraction of the cardiac muscle. The valves play no part at all in initiating flow and only prevent the blood from flowing in the opposite direction.

The cells constituting the heart walls do not exchange nutrients and metabolic end products with the blood within the heart chambers. The heart, like all other organs, receives its blood supply via arterial branches (*coronary arteries*) which arise from the aorta.

The walls of the atria and ventricles are composed of layers of cardiac muscle which are tightly bound together and completely encircle the blood-filled chambers. Thus, when the walls of a chamber contract, they come together like a squeezing fist, thereby exerting pressure on the blood they enclose. Cardiac muscle cells combine certain of the properties of smooth and skeletal muscle. The individual cell is striated (Fig. 9-6), containing both the thick myosin and thin actin filaments described for skeletal muscle. Cardiac cells are considerably shorter than the long, cylindrical skeletal fibers and have several branching processes. The processes of adjacent cells are joined end to end at structures known as intercalated disks, within which are points of membrane fusion, i.e. "tight junctions," which allow action potentials to be

by the increasing pressure of the ventricular blood and the atrioventricular opening is closed. Blood is therefore forced into the pulmonary artery (from the right ventricle) and into the aorta (from the left ventricle) instead of back into the atria. To prevent the valves themselves from being forced upward into the atrium, they are fastened by fibrous strands to muscular projections of the ventricular walls. These muscular projections do *not* open or close the valves: they act only to limit the valves' movements and prevent them from being everted.

The openings of the ventricles into the pulmonary artery and aorta are also guarded by valves, which permit blood to flow into these arteries but close immediately preventing reflux of blood in the opposite direction. There are no true valves at the entrances of the venae cavae and pulmonary veins into the right and left atrium, respectively.

We can now list the structures through which blood flows in passing from the systemic veins to the systemic arteries: superior or inferior venae cavae; right atrium; right ventricle; pulmonary arteries, arterioles, capillaries,

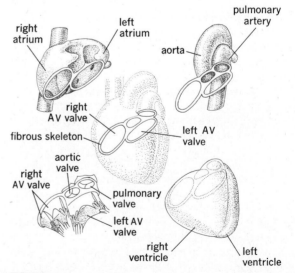

FIGURE 9-4
Anatomic components of the heart. Four connective-tissue rings compose the fibrous skeleton of the heart. To these are attached the valves, the trunks of the aorta and the pulmonary artery, and the muscle masses of the cardiac chambers. (*Adapted from Rushmer.*)

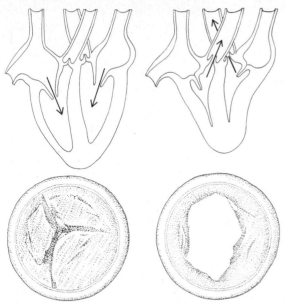

FIGURE 9-5
(*Above*) **Position of the heart valves and direction of blood flows during ventricular relaxation** (*left*) **and ventricular contraction** (*right*). (*Below*) **Aortic valve closed** (*left*) **and open** (*right*). (*Adapted from Carlson, Johnson, and Cavert.*)

sloshed back and forth within the ventricular cavities instead of being ejected into the aorta and pulmonary arteries. In other words, the complex muscle masses which form the ventricular pumps must contract more or less simultaneously for efficient pumping.

Such coordination is made possible by two factors, already mentioned: (*1*) The tight junctions allow spread of an action potential from one fiber to the next so that excitation in one muscle fiber spreads throughout the heart; (*2*) the specialized conducting system within the heart facilitates the rapid and coordinated spread of excitation. Where and how does the action potential first arise, and what are the path and sequence of excitation?

Origin of the heart beat

Cardiac muscle cells, like certain forms of smooth muscle, are autorhythmic; i.e., they are capable of spontaneous, rhythmical self-excitation. When the individual cells of a salamander embryo heart are separated and placed in salt solution, the individual cells are seen beat-

transmitted from one cardiac cell to another, in a manner similar to that in smooth muscle.

Besides the usual type of cardiac muscle shown in Fig. 9-6, certain areas of the heart contain specialized muscle fibers which have a different appearance and are essential for normal excitation of the heart. They constitute a network known as the *conducting system* of the heart and also are in contact with fibers of the usual cardiac muscle to form gap junctions which permit passage of action potentials from one cell to another.

Heart-beat coordination

Contraction of cardiac muscle, like all other types, is triggered by depolarization of the muscle membrane. In dealing with the mechanisms by which membrane excitation is initiated and spread within the heart we may wonder what would happen if all the many muscle fibers in the heart were to contract in a random manner. One result would be lack of coordination between pumping by each corresponding atrium and ventricle, but this defect is dwarfed by the more serious lack of muscle coordination within the ventricles. The blood would be

FIGURE 9-6
Electon micrograph of cardiac muscle. Note the striations similar to those of skeletal muscle. The dark bands indicate interdigitating areas called intercalated disks. (*Courtesy of D. W. Fawcett.*)

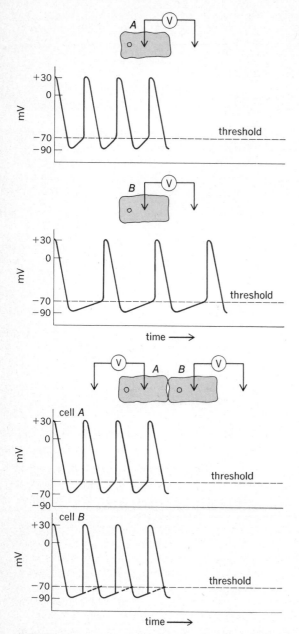

FIGURE 9-7

Transmembrane potential recordings from cardiac muscle cells grown in tissue culture. The dashed lines in the bottom recording of cell *B* indicate the course depolarization would have followed if the cells had not been joined.

ing spontaneously. But they are beating at different rates. Figure 9-7 shows recordings of membrane potentials from two such cells, the most important feature of which is the gradual depolarization causing the membrane potential to reach threshold, at which point an action potential occurs. Following the action potential, the membrane potential returns to the initial resting value, and the gradual depolarization begins. It should be evident that the slope of this depolarization, i.e., the rate of membrane potential change per unit time, determines how quickly threshold is reached and the next action potential elicited. Accordingly, cell *A* has a faster rate of firing than cell *B*. This capacity for autonomous depolarization toward threshold makes the rhythmical self-excitation of the muscle cells possible. It is due to a decreasing membrane permeability to potassium, but just how this change is generated ''spontaneously'' remains obscure.

In the course of the salamander experiment above, many individual cells form gap junctions. When such a gap junction is formed between two cells previously contracting autonomously at different rates, both the joined cells contract at the faster rate (Fig. 9-7). In other words, the faster cell sets the pace, causing the initially slower cell to contract at the faster rate. The mechanism is straightforward: The action potential generated by the faster cell causes depolarization, via the gap junction, of the second cell's membrane to threshold, at which point an action potential occurs in this second cell. The important generalization emerges that, because of the gap junctions between cardiac muscle cells, all the cells are excited at the rate set by the cell with the fastest autonomous rhythm.

Precisely the same explanation holds for the origination of the heart beat in the intact heart; several areas of the adult mammalian heart demonstrate these same characteristics of autorhythmicity and pacemaking, the one with the fastest inherent rhythm being a small mass of specialized myocardial cells embedded in the right atrial wall near the entrance of the superior vena cava (Fig. 9-8). Called the *sinoatrial (SA) node*, it is the normal pacemaker for the entire heart. Figure 9-9 is an intracellular recording from an SA node cell; note the slow depolarization toward threshold which initiates the action potential. Compare this SA nodal action potential to that of unspecialized nonautorhythmical atrial cells, which fail to show the pacemaker potential. In unusual circumstances, if some other area of the heart becomes more excitable and develops a faster spontaneous rhythm than the SA node, the new area begins to determine the rhythm for the entire heart.

Sequence of excitation

The cells of the SA node make contact with the surrounding atrial myocardial fibers (Figs. 9-8 and 9-10). From the SA node, the wave of excitation spreads throughout the right atrium along ordinary atrial myocardial cells, passing from cell to cell by way of the gap junctions. There is also a specialized bundle of fibers which conducts the impulse from the SA node directly to the left atrium, thereby ensuring the virtually simultaneous contraction of both atria. How does the excitation spread to the ventricles? At the base of the right atrium very near the wall between the ventricles (*interventricular septum*), the wave of excitation encounters a second small mass of specialized cells, the *atrioventricular* (AV) *node*. This node and the bundle of fibers leaving it constitute the only myocardial link between the atria and ventricles, all other areas being separated by nonconducting connective tissue. This anatomical pattern ensures that excitation will travel from atria to ventricles only through the AV node, but it also means that malfunction of the AV node may completely dissociate atrial and ventricular contraction. The AV node manifests one particularly important characteristic; the propagation of action potentials through the node is delayed for approximately 0.1 sec, allowing the atria to contract and empty their contents into the ventricle before ventricular contraction. The wave of excitation travels between the SA and AV nodes,

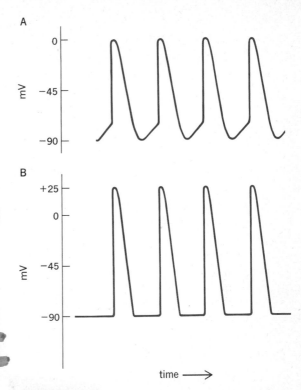

FIGURE 9-9
Transmembrane potential recordings from A SA node cell and B atrial muscle fiber.

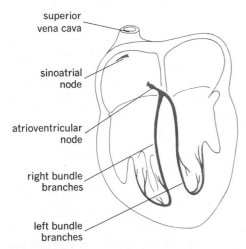

FIGURE 9-8
Conducting system of the heart. The interatrial and SA-node–AV-node conducting bundles are not shown in the figure. (*Adapted from Rushmer.*)

in large part, along ordinary atrial myocardial fibers. However, there may also be some contribution by several more or less specialized fiber bundles which conduct directly between these two nodes.

After leaving the AV node, the impulse travels rapidly along specialized myocardial fibers which run down the interventricular septum as bundles which then spread throughout much of the right and left ventricular myocardium. Finally, these fibers make contact with unspecialized myocardial fibers through which the impulse spreads from cell to cell in the remaining myocardium. The rapid conduction along these fibers and highly diffuse distribution cause depolarization of all right and left ventricular cells more or less simultaneously and ensure a single coordinated contraction.

Refractory period of the heart

The pumping of blood requires alternate periods of contraction and relaxation. Imagine the result of a prolonged

atrial excitation

begins   complete

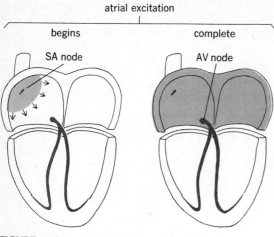

ventricular excitation

begins   complete

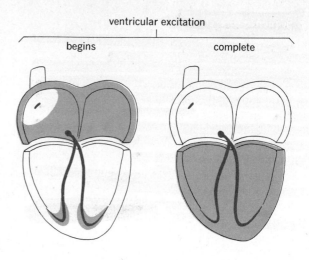

FIGURE 9-10
Sequence of cardiac excitation. Atrial excitation is complete
before ventricular excitation begins because of the delay at the
AV node. (*Adapted from Rushmer.*)

tetanic contraction of cardiac muscle like that described
for skeletal muscle. Obviously, pumping would cease and
death would ensue. In reality such contractions never
occur in the heart because of the long *refractory period*
of cardiac muscle. Recall that in any excitable membrane
an action potential is accompanied by a period during
which the membrane is completely insensitive to a stimulus regardless of intensity. Following this absolute refractory period comes a second period during which the
membrane can be depolarized again but only by a more
intense stimulus. In skeletal muscle, the absolute refractory periods are very short (1 to 2 msec) compared with
the duration of contraction, and a second contraction can
be elicited before the first is over. In contrast, the absolute
refractory period of cardiac muscle lasts almost as long
as the contraction (250 msec), and the muscle cannot
be excited in time to produce summation (Fig. 9-11).

A common situation explainable in terms of the
refractory period is shown in Fig. 9-12. In many people,
drinking several cups of coffee causes increased excitability of certain areas of the atria or ventricle due to the
action of caffeine. When one of these areas (*ectopic foci*)
fires just after completion of a normal contraction but
before the next SA nodal impulse, a premature wave of
excitation causes a contraction. As a result, the next
normal SA nodal impulse occurs during the refractory
period of the premature beat and is not propagated since
the myocardial cells are not excitable (the SA node still

fires because it has a shorter refractory period). The
second SA nodal impulse after the premature contraction
is propagated normally. The net result is an unusually
long delay between beats. The contraction after the delay
is unusually strong, and the person is aware of his heart

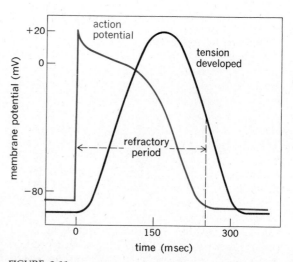

FIGURE 9-11
Relationship between membrane potential changes and
contraction in a single cardiac muscle cell. The refractory
period lasts almost as long as the contraction.

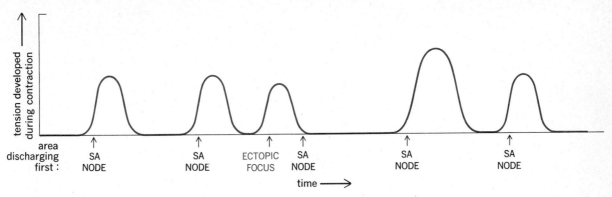

FIGURE 9-12

Effect of an ectopic discharge on ventricular contraction. The arrows indicate the times at which the SA node or ectopic focus fires. The premature beat induced by the ectopic discharge makes the ventricular muscle refractory at the time of the next SA-node impulse. The failure of this impulse to induce a contraction results in a longer period than normal before the next beat.

pounding. Moreover, if the hyperexcited area continued to discharge at a rate higher than that of the SA node, it might then capture the role of pacemaker and drive the heart at rates as high as 200 to 300 beats per minute, compared with a normal rate of 70.

Many similar examples are important clinically and help to clarify the normal physiological process. For example, disease commonly damages human cardiac tissue and hampers conduction through the AV node. Frequently, only a fraction of the atrial impulses are transmitted into the ventricles; thus, the atria may have a rate of 80 beats per minute and the ventricles only 60. If there is complete block at the AV node, none of the atrial impulses get through but a portion of the ventricle just below the AV node usually begins to initiate excitation at its own spontaneous rate. This rate is quite slow, generally 25 to 40 beats per minute, and completely out of synchrony with the atrial contractions, which continue at the normal higher rate. Under such conditions, the atria are totally ineffective as pumps since they are usually contracting against closed AV valves, but atrial pumping, as we shall see, is relatively unimportant for cardiac functioning (except during relatively strenuous exercise). Some patients have transient recurrent episodes of complete AV block signaled by fainting spells (due to decreased brain blood flow). These spells result because the ventricles do not begin their own impulse generation immediately and cardiac pumping ceases temporarily.

Partial AV block need not be caused by disease and may frequently represent a normal life-saving adap-

tation. Imagine a patient with an ectopic area driving his atria at 300 beats per minute. Ventricular rates this high are very inefficient because of inadequate filling time. Fortunately, the long refractory period of the AV node may prevent passage of a significant fraction of the impulses and the ventricles beat at a slower rate.

Another group of abnormalities apparently is characterized by a prolonged or unusual conduction route so that the impulse constantly meets an area which is no longer refractory and keeps traveling around the heart in a so-called "circus" movement. This may lead to continuous, completely disorganized contractions (*fibrillation*), which can cause death if they occur in the ventricles. Indeed, ventricular fibrillation is the immediate cause of death from electrocution.

The electrocardiogram

The *electrocardiogram* (ECG)[1] is primarily a tool for evaluating the electric events within the heart. The action potentials of cardiac muscle can be viewed as batteries which cause current flow throughout the body fluids. These currents produce voltage differences at the body surface which can be detected by attaching small metal plates at different places on the body. Figure 9-13 illustrates a typical normal ECG recorded as the potential difference between the right and left wrists. The first wave P represents atrial depolarization. The second complex

[1] Traditionally, electrocardiogram has been abbreviated EKG, the K being derived from the Greek word for heart. ECG is now the preferred term.

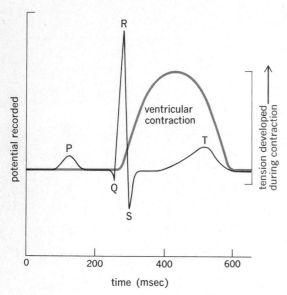

FIGURE 9-13
Typical electrocardiogram. P: atrial depolarization; QRS: ventricular depolarization; T: ventricular repolarization.

QRS, occurring approximately 0.1 to 0.2 sec later, represents ventricular depolarization. The final wave T represents ventricular repolarization. No manifestation of atrial repolarization is evident because it occurs during ventricular depolarization and is masked by the QRS complex. Figure 9-14A gives one example of the clinical usefulness of the ECG. This patient is suffering from partial AV nodal block so that only one-half the atrial impulses are being transmitted. Note that every second P wave is not followed by a QRS and T. Because many myocardial defects alter normal impulse propagation, and thereby the shapes of the waves, the ECG is a powerful tool for diagnosing heart disease.

Mechanical events of the cardiac cycle

Fluid always flows from a region of higher pressure to one of lower pressure. (This important concept will be developed formally in a later section; at present, we need deal with it only as an intuitively obvious phenomenon.) The sole function of the heart is to pump blood through the various organs of the body; the fluid pressure generated by cardiac contraction accomplishes the task, the heart valves serving only to direct the flow. The orderly

process of depolarization described in the previous section triggers contraction of the atria followed rapidly by ventricular contraction. The flow and pressure changes induced by these contractions are summarized in Fig. 9-15. The reader should follow our analysis of the heart picture and the pressure profile in each phase carefully since an understanding of the *cardiac cycle* is essential. *Systole* is the name of the period of ventricular contraction; *diastole* is ventricular relaxation. We start our analysis with the events of late diastole, considering first only the left heart, events on the right being qualitatively identical.

Late diastole

The left atrium and ventricle are both relaxed; left atrial pressure is very slightly higher than left ventricular (because blood is entering the atrium from the pulmonary veins); therefore, the AV valves are open, and blood is passing from atrium to ventricle. This is an important point: The ventricle receives blood from the atrium throughout most of diastole, not just when the atrium contracts. Indeed, at rest, approximately 80 percent of ventricular filling occurs before atrial contraction.[1] Note that the aortic valve (between aorta and left ventricle) is closed because the aortic pressure is higher than the ventricular pressure. The aortic pressure is slowly falling

---

[1] It is for this reason that the conduction defects discussed above, which eliminate the atria as efficient pumps, do not seriously impair cardiac function, at least at rest. Many persons lead relatively normal lives for many years despite atrial fibrillation. Thus, in many respects, the atrium may be conveniently viewed as merely a continuation of the large veins.

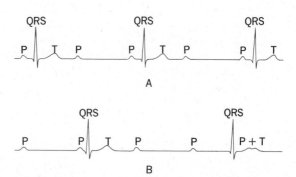

FIGURE 9-14
Electrocardiograms from two persons suffering from atrioventricular block. A Partial block; one-half of the atrial impulses are transmitted to the ventricles. B Complete block; there is absolutely no synchrony between atrial and ventricular electric activities.

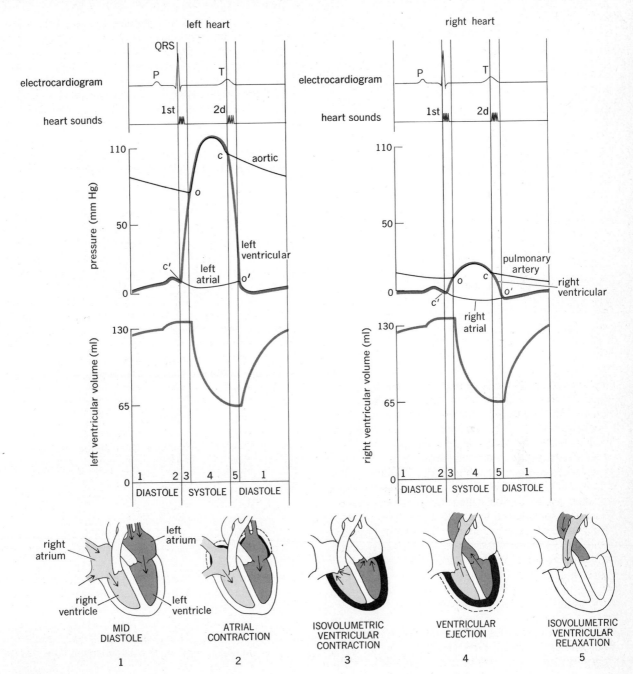

FIGURE 9-15

(*Left*) Summary of events in the left heart and aorta during the cardiac cycle. At *c′* the AV valve closes; at *o′* it opens. At *o* the aortic valve opens; at *c* it closes. The contracting portions of the heart are shown in black. (*Right*) Summary of events in the right heart and pulmonary arteries during the cardiac cycle. At *c′* the AV valve closes; at *o′* it opens. At *o* the pulmonary valve opens; at *c* it closes. (*Adapted from Ganong.*)

because blood is moving out of the arteries and through the vascular tree; in contrast, ventricular pressure is rising slightly because blood is entering from the atrium, thereby expanding the ventricular volume. At the very end of diastole, the SA node discharges, the atrium depolarizes (as shown by the P wave of the ECG), the atrium contracts (note the small rise in atrial pressure), and a small volume of blood is added to the ventricle. The amount of blood in the ventricle just prior to systole is called the *end-diastolic volume*.

Systole

The wave of depolarization passes through the ventricle (QRS complex) and triggers ventricular contraction. As the ventricle contracts, it squeezes the blood contained in it and ventricular pressure rises steeply. Almost immediately, this pressure exceeds the atrial pressure and closes the AV valve, thus preventing backflow into the atrium. Since for a brief period the aortic pressure still exceeds the ventricular, the aortic valve remains closed and the ventricle does not empty despite contraction. This early phase of systole is called *isovolumetric ventricular contraction* because ventricular volume is constant; i.e., the lengths of the muscle fibers remain approximately constant as in an isometric skeletal-muscle contraction. This brief phase ends when ventricular pressure exceeds aortic, the aortic valve opens, and *ventricular ejection* occurs. The ventricular-volume curve shows that ejection is rapid at first and then tapers off. *The ventricle does not empty completely;* the amount remaining after ejection is called the *end-systolic volume*. As blood flows into the aorta, the aortic pressure rises with ventricular pressure. Atrial pressure also rises slowly throughout the entire period of ventricular ejection because of continued flow of blood from the veins. Note that peak aortic pressure is reached before the end of ventricular ejection; i.e., the pressure actually is beginning to fall during the last part of systole despite continued ventricular ejection. This phenomenon is explained by the fact that the rate of blood ejection during this last part of systole is quite small (as shown by the ventricular volume curve) and is less than the rate at which blood is leaving the aorta (and other large arteries) via the arterioles; accordingly the volume and, therefore, the pressure within the aorta begin to decrease.

Early diastole

When contraction stops, the ventricular muscle relaxes rapidly owing to release of tension created during contraction. Ventricular pressure therefore falls almost imme-

diately below aortic pressure, and the aortic valve closes. However, ventricular pressure still exceeds atrial so that the AV valve remains closed. This phase of early diastole, obviously the mirror image of early systole, is called *isovolumetric ventricular relaxation*. It ends as ventricular pressure falls below atrial, the AV valves open, and ventricular filling begins. Filling occurs rapidly at first and then slows down as atrial pressure decreases. The fact that ventricular filling is almost complete during early diastole is of the greatest importance; it ensures that filling is not seriously impaired during periods of rapid heart rate, e.g., exercise, emotional stress, fever, despite a marked reduction in the duration of diastole. When rates of approximately 200 beats per minute or more are reached, however, filling time is inadequate and cardiac pumping is impaired. Significantly, the AV node in normal adults does not conduct at rates greater than 200 to 250 per minute.

Pulmonary circulation pressures

Figure 9-15 summarizes the simultaneously occurring events in the right heart and pulmonary arteries, the patterns being virtually identical to those just described for the left heart. There is one striking quantitative difference: The ventricular and arterial pressures are considerably lower during systole. The pulmonary circulation is a low-pressure system (for reasons to be described in a later section). This difference is clearly reflected in the ventricular architecture, the right ventricular wall being much thinner than the left. Note, however, that despite the lower pressure, the right ventricle ejects the same amount of blood as the left.

Heart sounds

Two heart sounds are normally heard through a stethoscope placed on the chest wall (see Fig. 9-15). The first sound, a low pitched *lub*, is associated with closure of the AV valves at the onset of systole; the second, a high-pitched *dub*, is associated with closure of the pulmonary and aortic valves at the onset of diastole. These sounds, which result from vibrations caused by valvular closure, are perfectly normal, but heart murmurs are frequently (although not always) a sign of heart disease. When blood flows smoothly in a streamline manner, i.e., layers of fluid sliding evenly over one another, it makes no sound, but turbulent flow produced by unusually high velocities makes a noise. This noise is heard as a murmur or sloshing sound. Turbulence can be produced by blood flowing rapidly in the usual direction through an abnormally narrowed valve, backward through a damaged leaky valve, or between the two atria or two ventricles via a small hole in the septum. The exact timing and location of the murmur provide the physician with a powerful diagnostic clue. For example, a murmur heard throughout systole suggests a narrowed pulmonary or aortic valve or a hole in the interventricular septum. The diagnosis can then be completed by the use of specialized techniques.

The cardiac output

The volume of blood pumped *by each ventricle* per minute is called the *cardiac output*, usually expressed as liters per minute. It must be remembered that the cardiac output is the amount of blood pumped by *each* ventricle, *not* the total amount pumped by both ventricles. The cardiac output is determined by multiplying the *heart rate* and the volume of blood ejected by each ventricle during each beat (*stroke volume*):

cardiac output = heart rate × stroke volume
liters/min   beats/min   liters/beat

For example, if each ventricle has a rate of 72 beats per minute and ejects 70 ml with each beat, what is the cardiac output?

CO = 72 beats/min × 0.07 liter/beat = 5.0 liters/min

These values are approximately normal for a resting adult. During periods of exercise, the cardiac output may reach 20 to 25 liters/min. Obviously, heart rate or stroke volume or both must have increased. Physical exercise is but one of many situations in which various tissues and organs require a greater flow of blood; e.g., flow through skin vessels increases when heat loss is required, and flow through intestinal vessels increases during digestion. Some of the increased flow can be obtained merely by decreasing blood flow to some other organ, i.e., by redistributing the cardiac output, but most of the supply must come from a greater total cardiac output. The following description of the factors which alter the two determinants of cardiac output, heart rate and stroke volume, applies in all respects to both the right and left heart since stroke volume and heart rate are the same for both the right and left ventricles.

Control of heart rate

The rhythmic discharge of the SA node occurs spontaneously in the complete absence of any nervous or hormonal influences. However, it is under the constant influence of both nerves and hormones. A large number of parasympathetic and sympathetic fibers end on the SA node as well as on other areas of the conducting system. The parasympathetics to the heart are contained in the vagus nerves. As shown in Fig. 9-16, stimulation of these nerves (or local application of acetylcholine) causes slowing of the heart and, if strong enough, may stop the heart completely for some time. The effects of the sympathetic nerves are just the reverse; nerve stimulation (or local application of norepinephrine) increases the heart rate, whereas cutting the sympathetics slows the heart. Finally, cutting the parasympathetics causes the heart rate to increase. The experiments in which the nerve is sectioned demonstrate that both the sympathetic and parasympathetic nerves normally are discharging at some finite rate. Apparently, in the resting state, the parasympathetic influence is dominant since simultaneous removal of all nerves causes the heart rate to increase to approximately 100 beats per minute. This is the inherent autonomous discharge rate of the SA node.

Figure 9-17 illustrates the nature of the sympathetic and parasympathetic influence on SA node function. Sympathetic stimulation increases the slope of the pacemaker potential, which causes the cell to reach threshold more rapidly and is responsible for the rate change. Stimulation of the parasympathetics has the opposite effect; the slope of the pacemaker potential decreases, threshold is reached more slowly, and heart rate decreases. The underlying alterations of membrane permeability induced by the sympathetic mediator, norepinephrine, and the parasympathetic mediator, acetylcholine, are still uncertain.

Factors other than the cardiac nerves also can alter heart rate. Epinephrine, the hormone liberated from the adrenal medulla, speeds the heart; this is not surprising since epinephrine is a blood-borne sympathetic mediator similar in structure to norepinephrine (Chaps. 6 and 7).

The heart rate is also sensitive to many other factors, including temperature, plasma electrolyte concentrations, and hormones other than epinephrine. However, these are generally of lesser importance, and the heart rate is primarily regulated very precisely by balancing the slowing effects of parasympathetic discharge against the accelerating effects of sympathetic discharge, both operating on the SA node (Fig. 9-18).

Control of stroke volume

There are always considerable amounts of blood remaining in the ventricles after contraction. The volume of blood ejected during each ventricular contraction is obvi-

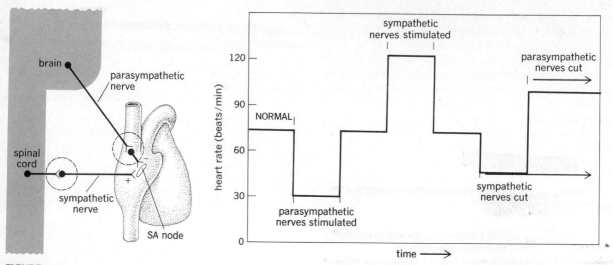

FIGURE 9-16
Effects of autonomic nerves on heart rate. Sympathetic nerves stimulate the SA node whereas parasympathetics inhibit it. The parasympathetic fibers to the heart are contained in the vagus nerves. The values shown during nerve stimulation do not represent the maximal changes obtainable.

ously the difference between the volume of blood contained in the ventricle at the end of diastole (*end-diastolic volume*) and the volume remaining at the end of systole:

Stroke volume

= end-diastolic volume − end-systolic volume

An increased end-diastolic volume usually leads to an increased stroke volume. We shall refer to this as *intrinsic control* because it represents an inherent property of cardiac muscle. The stroke volume can also be increased by the nerves to the heart and several circulating hormones; we shall call this *extrinsic control*.

Intrinsic control: the relationship between stroke volume and end-diastolic volume It is possible to study the completely intrinsic adaptability of the heart by means of the so-called heart-lung preparation (Fig. 9-19). Tubes are placed in the heart and vessels of an anesthetized animal so that blood flows from the very first part of the aorta (just above the exit of the coronary arteries) into a blood-filled reservoir and from there into the right atrium. The blood then is pumped by the right heart as usual via the lungs into the left heart. The net effect is to nourish the heart and lungs normally and to deprive the rest of the animal's body of blood and cause death, thereby abolishing all nervous and hormonal activity. A key feature of this preparation is that the pressure causing blood flow into the heart can be altered simply by raising or lowering the reservoir. This is analogous to altering the venous and atrial pressures and causes changes in the quantity of blood entering the ventricles during diastole. When the reservoir is raised, the following sequence of events is observed:

1 Diastolic filling increases, thereby increasing end-diastolic volume.

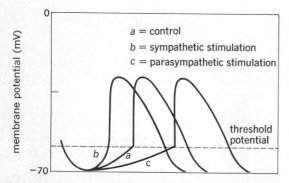

FIGURE 9-17
Effects of sympathetic and parasympathetic nerve stimulation on the slope of the pacemaker potential of an SA node cell. (*Adapted from Hoffman and Cranefield.*)

2 Stroke volume increases as a result of a more forceful contraction but not enough to eject all the blood which entered during the previous diastole, and end-systolic volume increases.

3 For several more beats diastolic filling slightly exceeds systolic ejection despite progressively more forceful contractions, and the end-diastolic volume becomes progressively larger.

4 Ultimately the distended heart contracts forcefully enough so that the stroke output becomes equal to the diastolic filling.

The net result is a new steady state, in which the ventricle is distended and diastolic filling and stroke volume are both increased, but equal. The mechanism underlying this completely intrinsic adaptation is that cardiac muscle, like other muscle, increases its strength of contraction when it is stretched. Thus, in the experiment above, the increased diastolic volume stretches the ventricular muscle fibers and causes them to contract more forcefully. This relationship was expounded by the British physiologist Starling, who observed that there was a direct proportion between the diastolic volume of the heart, i.e., the length

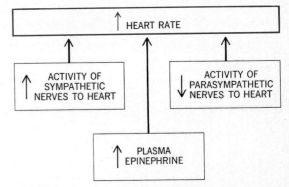

FIGURE 9-18
Summary of major factors which influence heart rate. All effects are exerted upon the SA node. The figure, as drawn, shows how heart rate is increased; conversely, heart rate is slowed when sympathetic activity and epinephrine are decreased and when parasympathetic activity is increased.

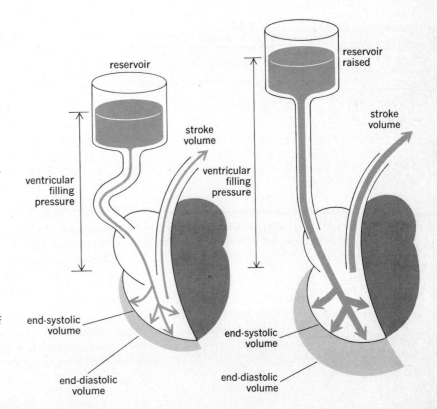

FIGURE 9-19
Demonstration of the intrinsic control of stroke volume (Starling's law of the heart). By raising the reservoir, the pressure causing ventricular filling is increased. The increased filling distends the ventricle, which responds with an increased strength of contraction.

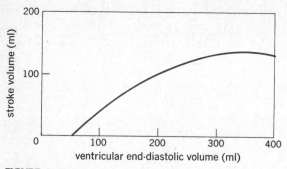

FIGURE 9-20

Relationship between ventricular end-diastolic volume and stroke volume (Starling's law of the heart). The data were obtained by progressively increasing ventricular filling pressure, as in Fig. 9-19.

of its muscle fibers, and the force of contraction of the following systole. It is now referred to as *Starling's law of the heart.* A typical response curve obtained by progressively increasing end-diastolic volume is shown in Fig. 9-20. Note that marked overstretching causes the force of contraction, and thereby the stroke volume, to fall off. Thus, heart muscle manifests a length-tension relationship very similar to that described earlier for skeletal muscle (Fig. 8-27) and explainable in terms of the sliding-filament mechanism of muscle contraction. However, unlike skeletal muscle, cardiac muscle length in the resting state is less than that which yields maximal tension during contraction so that an increase in length produces an increase an contractile tension.

This intrinsic relationship between end-diastolic volume and stroke volume, originally demonstrated in the heart-lung preparation, applies equally to the intact human being. End-diastolic volume, therefore, becomes a crucial determinant of cardiac output.

What then are the factors which determine end-diastolic volume, i.e., the degree of ventricular distension just before systole? The simplest way to approach this question is to view the ventricle as an elastic chamber, much like a balloon. A balloon enlarges when one blows into it because the internal pressure acting upon the wall becomes greater than the external pressure. The more air blown in, the higher the internal pressure becomes. The degree of distension therefore depends upon the pressure difference across the wall and the distensibility of the wall. This is precisely the situation for the ventricle.

We shall ignore the problem of ventricular distensibility (for the reason that it does not appear to be under physiological control) and concentrate only on the *transmural,* or across-the-wall, difference in pressure.

What is the ventricular transmural pressure at the end of diastole? The internal pressure, of course, is the fluid pressure exerted by the blood against the walls. The external pressure surrounding the heart is the pressure within the chest cavity (thorax), or *intrathoracic pressure.* A typical end-diastolic ventricular blood pressure is 4 mm Hg. Physiologists state pressures with the atmospheric pressure, i.e., the pressure of the atmospheric air surrounding the body, given as zero. Thus, when we say the end-diastolic pressure is 4 mm Hg, we really mean 4 mm Hg greater than atmospheric pressure. If we assume the atmospheric pressure to be 760 mm Hg, the true internal pressure is 764 mm Hg. This distinction must be remembered, especially when one is discussing events within the thoracic cage, because, for reasons to be described in Chap. 10, the intrathoracic pressure of the fluid surrounding the heart, lungs, and all other intrathoracic structures is *less* than atmospheric. This subatmospheric intrathoracic pressure is frequently termed a "negative" pressure, but this terminology should be avoided since there is no such thing in nature as a negative pressure. The intrathoracic pressure averages approximately 5 mm Hg *less* than atmospheric pressure (or a true pressure of 755 mm Hg); accordingly, the pressure difference acting to distend the ventricles at the end of diastole is $4 + 5 = 9$ mm Hg, that is, $764 - 755 = 9$ mm Hg.

We can now answer our original question. End-diastolic ventricular volume, i.e., distension, can be increased either by increasing the intraventricular blood pressure or by decreasing the intrathoracic pressure, or both. The latter always occurs during inspiration (see Chap. 10) and accounts, in part, for the increased stroke volume which characteristically occurs during inspiration. However, it is primarily by changes in the end-diastolic intraventricular pressure that end-diastolic volume is controlled. As can be seen from Fig. 9-15, the pressure within the ventricles during diastole closely approximates the atrial pressure, since the valves are open and the chambers are connected by the wide AV orifice through which blood is flowing into the ventricle. Accordingly, the diastolic ventricular blood pressure is determined by the atrial pressure. As we shall see, the atrial pressure, in turn, is determined by the rate of blood flow from the veins into the atria.

The significance of this mechanism should now be apparent; an increased flow of blood from the veins into the heart automatically forces an equivalent increase in cardiac output by distending the ventricle and increasing stroke volume, just like the heart-lung apparatus when we increased ''venous return'' by elevating the reservoir. This is probably the single most important mechanism for maintaining equality of right and left output. Should the right heart, for example, suddenly begin to pump more blood than the left, the increased blood flow to the left ventricle would automatically produce an equivalent increase in left ventricular output and blood would not be allowed to accumulate in the lungs. Another example of Starling's law has already been described above, namely, the pounding that occurs after a premature contraction. Recall that an unusually long period elapses between the premature contraction and the next contraction; the period for diastolic filling is increased, end-diastolic volume increases, and the force of contraction is increased (Fig. 9-12). It is this strong contraction, which may actually lift the heart against the chest wall, that the person is aware of. In summary, end-diastolic volume and stroke volume are generally increased whenever atrial pressure increases; the factors which determine atrial pressure will be described in a subsequent section.

Extrinsic control: the sympathetic nerves   Until recently, it was generally assumed that the mechanism described by Starling's law could explain almost all changes in stroke volume observed under physiological conditions. During exercise, for example, it was believed that the increased venous return produced by various factors led to cardiac distension and increased stroke output. However, experiments have not borne out this hypothesis; x-ray photographs of exercising men (and other mammals) have clearly shown that the normal heart does not usually distend during exercise and may even decrease in size, despite the fact that stroke volume usually increases.

It must be stressed that the relationship described by Starling's law is not invalid but simply is not the sole determinant of ventricular strength of contraction. The other major factor is the sympathetic nerves, which are distributed not only to the SA node and conducting system but to all myocardial cells. The effect of the sympathetic mediator, norepinephrine, is to increase ventricular (and atrial) *contractility*, defined as the strength of contraction at any given initial muscle-fiber length, i.e., end-diastolic volume. Not only is the contraction more powerful but both it and relaxation occur more rapidly. These

latter effects are quite important since, as described earlier, increased sympathetic activity to the heart also increases heart rate. As heart rate increases, the time available for diastolic filling decreases, but the more rapid contraction and relaxation induced simultaneously by the sympathetic neurons partially compensate for this problem by permitting a larger fraction of the cardiac cycle to be available for filling. Moreover, because the ventricles relax so rapidly after a contraction the intraventicular pressure falls rapidly, thereby creating an enhanced pressure gradient for flow of blood into the ventricles. This is, in essence, a ''sucking'' effect which facilitates ventricular filling. The ability of these effects to maintain diastolic filling is, of course, not unlimited, and diastolic filling is significantly reduced at very high heart rates. The significance of this interplay between diastolic filling time, heart rate, and contractility will be analyzed further in the later section on exercise.

Circulating epinephrine produces changes in contractility similar to those induced by the sympathetic nerves to the heart. Moreover, a decreased contractility can be obtained by reducing the rate of sympathetic discharge below the usual tonic level. The mechanism by which norepinephrine and epinephrine increase contractility probably involves an increased liberation of calcium during excitation. In contrast to the sympathetic nerves, the parasympathetic nerves to the heart have relatively little effect on ventricular contractility.

The interrelationship between the intrinsic (Starling's law) and extrinsic (cardiac nerves) mechanisms as measured in a heart-lung preparation is illustrated in Fig. 9-21. The dashed line is the same as the line shown in Fig. 9-20 and was obtained by slowly raising ventricular pressure while measuring end-diastolic volume and stroke volume; the solid line was obtained similarly for the same heart but during sympathetic-nerve stimulation. Starling's law still applies, but during nerve stimulation the stroke volume is greater at any given end-diastolic volume. In other words, the increased contractility leads to *a more complete ejection* of the end-diastolic ventricular volume.

In summary (Fig. 9-22) stroke volume is controlled both by an intrinsic cardiac mechanism dependent only upon changes in end-diastolic volume and by an extrinsic mechanism mediated by the cardiac sympathetic nerves (and circulating epinephrine). The contribution of each mechanism in specific physiological situations and the reflexes controlling the nerves are described in a later section.

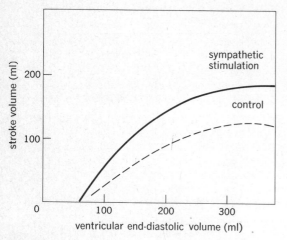

FIGURE 9-21
Effects on stroke volume of stimulating the sympathetic nerves to the heart.

Summary of cardiac output control

A summary of the major factors which determine cardiac output is presented in Fig. 9-23, which combines the information of Fig. 9-18 (factors influencing heart rate) and Fig. 9-22 (factors influencing stroke volume).

SECTION B
THE VASCULAR SYSTEM

To comprehend the functional characteristics of the various blood-vessel types, one must be familiar with the basic physical principles underlying the flow of fluids. These principles were introduced in the discussion of *bulk flow* in Chap. 2 and are amplified below.

Basic principles of pressure, flow, and resistance

Fluid flows through a tube in response to a difference (*gradient*) in pressure between the two ends of the tube. The total volume flowing per unit time is directly proportional to the pressure ΔP. It is not the absolute pressure in the tube which determines flow but the difference in pressure between the two ends (Fig. 9-24). This direct proportionality of flow F and pressure difference ΔP can be written

$$F = k\,\Delta P$$

where k is the proportionality constant describing how much flow occurs for a given pressure difference. Knowing only the pressure difference between two ends of a tube is not enough to determine how much fluid will flow; one must also know the numerical value of k. This constant is simply a measure of the ease with which fluid will flow through a tube. Usually we do not deal directly with k but with the reciprocal of k, known as *resistance R:*

$$R = \frac{1}{k}$$

Thus, the larger k is, the smaller R is. R answers the question: How difficult is it for fluid to flow through a tube at any given pressure? Its name, resistance, is therefore quite appropriate. Substituting $1/R$ for k, the basic equation becomes

$$F = \frac{\Delta P}{R}$$

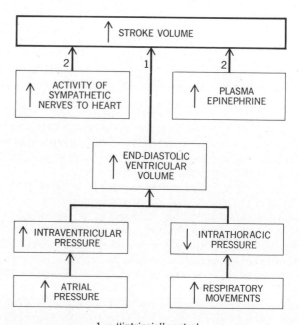

1 = "intrinsic" control
2 = "extrinsic" control

FIGURE 9-22
Major factors which influence stroke volume. The figure as drawn shows how stroke volume is increased; a reversal of all arrows in the boxes would illustrate how stroke volume is decreased. Refer to the text for details.

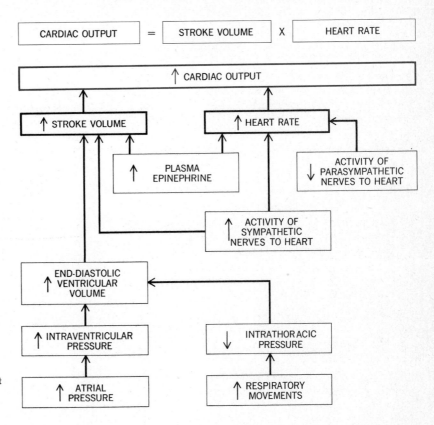

FIGURE 9-23
Major factors determining cardiac output
(an amalgamation of Figs. 9-18 and
9-22).

In words, flow is directly proportional to the pressure difference and inversely proportional to the resistance. This basic equation applies to the flow of fluids in general, not just in living systems.

Determinants of resistance

Resistance (see Table 9-2) is essentially a measure of friction since it is basically the friction between tube wall and fluid and between the molecules of the fluid themselves which opposes flow. Resistance depends upon the nature of the fluid and the geometry of the tube.

Nature of the fluid Maple syrup flows less readily than water because the molecules of maple syrup slide over each other only with great difficulty. This property of fluids is called *viscosity*. As we shall see, changes in blood viscosity can have important effects on blood flow because they change the resistance to flow. Generally, however, the viscosity of blood is relatively constant and only a minor factor in determining resistance.

Geometry of the tube Both the length and radius of a tube affect its resistance. Resistance to flow is directly proportional to the length of the tube, but since the lengths of the blood vessels remain constant in the body, we shall not be concerned with this relationship. Because of complex relationships between the tube wall and the fluid, the resistance increases markedly as tube radius decreases. The exact relationship is given by the following formula, which states that the resistance is inversely proportional to the fourth power of the radius (the radius multiplied by itself four times):

$$R \propto \frac{1}{r^4}$$

The extraordinary dependence of resistance upon radius can be appreciated by the fact (shown in Fig. 9-25) that doubling the radius increases the flow sixteenfold. As we shall see, the radius of blood vessels can be changed significantly and constitutes the most important factor in the control of resistance to blood flow.

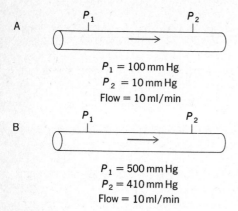

FIGURE 9-24
Flow between two points within a tube is proportional to the pressure *difference* between the points.

Arteries

The aorta and other arteries have thick walls containing large quantities of elastic tissue. Although they also have smooth muscle, there is apparently little fluctuation in the state of muscle activity, and the arteries can be viewed most conveniently simply as elastic tubes. Because the arteries have large radii, they serve as low-resistance pipes conducting blood to the various organs. Their second major function is to act as a pressure reservoir driving blood through the tissues.

Arterial blood pressure

Recall once more the factors determining the pressure within an elastic container, e.g., a balloon filled with water. The pressure inside the balloon depends upon the volume of water within it and the distensibility of the balloon, i.e., how easily its walls can be stretched. If the walls are very stretchable, large quantities of water can go in with only a small rise in pressure; conversely, a small quantity of water causes a large pressure rise in a balloon which strongly resists stretching.

TABLE 9-2
Factors which determine resistance to flow

$$\text{Resistance} \propto \begin{cases} \text{viscosity of fluid} \\ \text{tube length} \\ 1/(\text{tube radius})^4 \end{cases}$$

These principles can now be applied to an analysis of arterial function. The contraction of the ventricles ejects blood into the pulmonary and systemic arteries during systole. If a precisely equal quantity of blood were to flow simultaneously out of the arteries via arterioles, the total volume of blood in the arteries would remain constant and arterial pressure would not change. Such, however, is not the case. As shown in Fig. 9-26, a volume of blood equal to only about one-third the stroke volume leaves the arteries during systole. The excess volume distends the arteries and raises the arterial pressure. When ventricular contraction ends, the stretched arterial walls recoil passively (like a stretched rubber band upon release) and the arterial pressure continues to drive blood through the arterioles. As blood leaves the arteries, the pressure slowly falls, but the next ventricular contraction occurs while there is still adequate blood in the arteries to stretch them partially, so that the arterial pressure does

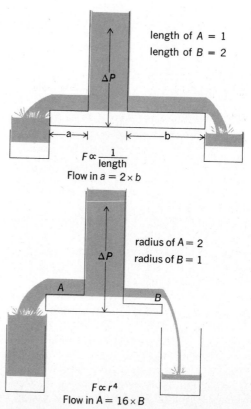

FIGURE 9-25
Effects of tube length and radius on resistance to flow.

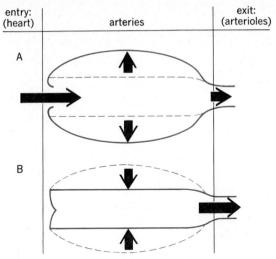

entry:
(heart)

arteries

exit:
(arterioles)

A

B

FIGURE 9-26
Movement of blood into and out of the arteries during the
cardiac cycle. The lengths of the arrows denote relative
quantities. During systole A, less blood leaves the arteries than
enters and the arterial walls stretch. During diastole B, the
walls recoil passively, driving blood out of the arteries.

not fall to zero. In this manner, the arterial pressure
provides the immediate driving force for tissue blood flow.
The aortic pressure pattern shown in Fig. 9-27 is typical
of the pressure changes which occur in all the large
systemic arteries. The pulmonary-artery pressure profile
is similar but with all pressures smaller. The maximum
pressure is reached during peak ventricular ejection and
is called *systolic pressure*. The minimum pressure obviously
occurs just before ventricular contraction and is called
diastolic pressure. They are generally recorded as systolic/
diastolic, that is, 125/75 mm Hg in our example. The
pulse, which can be felt in an artery, is due to the difference between systolic and diastolic pressure. This difference (125 − 75 = 50) is called the *pulse pressure*. The
factors which alter pulse pressure are the following: (*1*)
An increased stroke volume tends to elevate systolic
pressure because of greater arterial stretching by the
additional blood. (*2*) Decreased arterial distensibility, as
in arteriosclerosis, may cause a marked increase in systolic pressure because the wall is stiffer; i.e., any given
volume of blood produces a greater pressure rise.

It is evident from the figure that arterial pressure
is constantly changing throughout the cardiac cycle and
the average pressure (*mean pressure*) throughout the cycle
is not merely the value halfway between systolic and

diastolic pressure, because diastole usually lasts longer
than systole. Actually, the true mean arterial pressure can
be obtained only by complex methods, but for most purposes it is reasonably accurate to assume the mean
pressure is the diastolic pressure plus one-third of the
pulse pressure. Thus, in our example,

$$\text{Mean pressure} = 75 + (\tfrac{1}{3} \times 50) = 92 \text{ mm Hg}$$

The mean arterial pressure is actually the most important
of the pressures described because it is the *average* pressure driving blood into the tissues throughout the cardiac
cycle. In other words, if the pulsatile pressure changes
were eliminated and the pressure throughout the cardiac
cycle were always equal to the mean pressure, the total
flow would be unchanged. It is a closely regulated quantity; indeed, the reflexes which accomplish this regulation
constitute the basic cardiovascular control mechanisms
and will be described in detail in a later section.

One last point: We can refer to "arterial" pressure
without specifying to which artery we are referring because the aorta and other arteries have such large diameters that they offer only negligible resistance to flow and
the pressures are therefore similar everywhere in the
arterial tree.

Measurement of arterial pressure

Both systolic and diastolic blood pressure are readily
measured in human beings with the use of a sphygmomanometer (Fig. 9-28). A hollow cuff is wrapped around
the arm and inflated with air to a pressure greater than
systolic blood pressure (Fig. 9-28A). The high pressure
in the cuff is transmitted through the tissues of the arm
and completely collapses the arteries under the cuff,
thereby preventing blood flow to the lower arm. The air

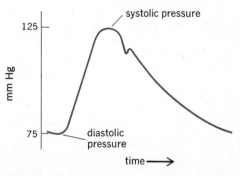

FIGURE 9-27
Typical aortic pressure fluctuations during the cardiac cycle.

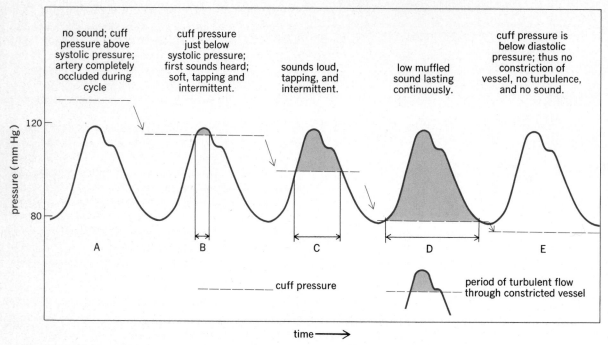

FIGURE 9-28
Sounds heard through a stethoscope while the cuff pressure of
a sphygmomanometer is gradually lowered. Systolic pressure is
recorded at B and diastolic pressure at the point of sound
disappearance.

in the cuff is now slowly released, causing the pressure
in the cuff and arm to drop. When cuff pressure has fallen
to a point just below the systolic pressure (Fig. 9-28B),
the arterial blood pressure at the peak of systole is
greater than the cuff pressure, causing the artery to
expand and allow blood flow for this brief time. During
this interval, the blood flow through the partially occluded
artery occurs at a very high velocity because of the small
opening for blood passage and the large pressure gradient. The high-velocity blood flow produces turbulence
and vibration, which can be heard through a stethoscope
placed over the artery just below the cuff. The pressure
measured on the manometer attached to the cuff at which
sounds are first heard as the cuff pressure is lowered
is identified as the systolic blood pressure. These first
sounds are soft tapping sounds, corresponding to the
peak systolic pressure reached during ejection of blood
from the heart. As the pressure in the cuff is lowered
further, the time of blood flow through the artery during
each cycle becomes longer (Fig. 9-28C). The tapping

sound becomes louder as the pressure is lowered. When
the cuff pressure reaches the diastolic blood pressure,
the sounds become dull and muffled, as the artery remains open throughout the cycle, allowing continuous
turbulent flow (Fig. 9-28D). Just below diastolic pressure
all sound stops as flow is now continuous and nonturbu-
lent through the completely open artery. Thus, systolic
pressure is measured as the cuff pressure at which
sounds first appear and diastolic pressure as the cuff
pressure at which sounds disappear.

Arterioles

Each organ or tissue obviously receives only a fraction
of the total left ventricular cardiac output. The typical
distribution for a resting normal adult is given in Fig. 9-29.
The digestive tract (including the liver), the kidneys, and
the brain receive the largest supplies. Perhaps the most
striking aspect of brain blood flow is its remarkable con-

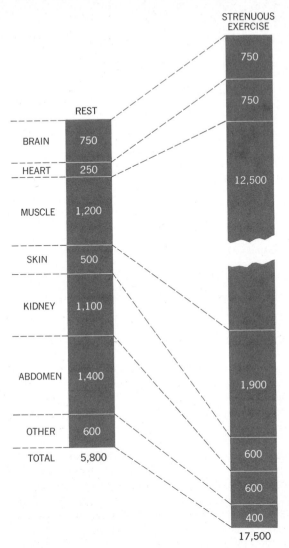

REST

BRAIN	750
HEART	250
MUSCLE	1,200
SKIN	500
KIDNEY	1,100
ABDOMEN	1,400
OTHER	600
TOTAL	5,800

STRENUOUS EXERCISE

750
750
12,500
1,900
600
600
400
17,500

FIGURE 9-29
Distribution of blood flow to the various organs and tissues of the body at rest and during strenuous exercise. The numbers show blood flow in milliliters per minute. (*Adapted from Chapman and Mitchell.*)

stancy. It really requires little, if any, additional energy to think; i.e., whether we are staring blankly into space or contemplating the theory of relativity, the total energy consumption of the brain remains virtually unchanged. In contrast, the energy consumption of muscular tissues of the body (heart, skeletal muscle, uterus, etc.) varies directly with the degree of muscle activity. Now compare the values at rest with those for exercise in Fig. 9-29. There is a large increase in blood flow to the exercising skeletal muscle and heart. Skin blood flow also has increased, kidney flow has decreased, and brain flow is unchanged. Obviously, the total cardiac output has increased; but, more important for our present purposes, the distribution of flow has greatly changed. Cardiac output is distributed to the various organs and tissues according to their functions and needs at any given moment. The remainder of this section describes the physiology of the arterioles, that segment of the vascular tree primarily responsible for control of blood-flow distribution.

Figure 9-30 illustrates the major principles in terms of a simple model, a fluid-filled tank with a series of compressible outflow tubes. What determines the rate of flow through each exit tube? As always,

$$\text{Flow} = \frac{\Delta P}{R}$$

Since the driving pressure is identical for each tube, differences in flow are completely determined by differences in the resistance to flow offered by each tube. The lengths of the tubes are approximately the same, and the viscosity of the fluid is a constant; therefore, differences in resistance offered by the tubes are due solely to differences in their radii. Obviously, the widest tube has the greatest flow. If we equip each outflow tube with an adjustable cuff, we can obtain any combination of *relative* flows we wish.

This analysis can now be applied to the cardiovascular system. The tank is analogous to the arteries, which serve as a pressure reservoir, the major arteries themselves being so large that they contribute little resistance to flow. The smaller terminal arteries begin to offer some resistance, but it is relatively slight. Therefore, all the arteries of the body can be considered a single pressure reservoir. The arteries branch within each organ into the next series of smaller vessels, the arterioles, which are now narrow enough to offer considerable resistance. The arterioles are the major site of resistance in the vascular tree and are therefore analogous to the outflow tubes in the model. Their radii are subject to precise physiological control; their walls consist of relatively little elastic tissue but much smooth muscle, which can relax or constrict, thereby changing the radius of the inside (*lumen*) of the arteriole. *The pattern of blood-flow distribution depends primarily upon the degree of arteriolar smooth-muscle constriction within each organ and tissue.*

The mechanisms controlling arteriolar constriction

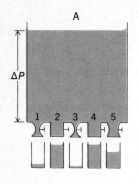

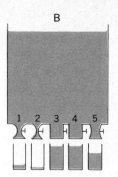

pressure reservoir
("arteries")

variable-resistance
outflow tubes
("arterioles")

flow to
different "organs"
$F = \Delta P / R$

FIGURE 9-30
Physical model of the relationship between arterial pressure, arteriolar radius in different organs, and blood-flow distribution. Blood has been shifted from organ 2 to organ 3 (in going from A to B) by constricting the "arterioles" of 2 and dilating those of 3.

and dilation fall into two general categories: (*1*) local controls, which serve the metabolic needs of the specific tissue in which they occur, and (*2*) reflex controls, which integrate and coordinate the needs of the whole body.

Local controls

Active hyperemia Certain organs and tissues, particularly the heart, skeletal muscle, and other muscular organs, manifest an increased blood flow (*hyperemia*) any time their metabolic activity is increased. For example, the blood flow to exercising skeletal muscle increases in direct proportion to the increased activity of the muscle. This phenomenon, known as *active hyperemia,* is the direct result of arteriolar dilation within the more active organ. This vasodilation does not depend upon the presence of nerves or hormones but is a locally mediated response. The adaptive value of the phenomenon should be readily apparent; an increased rate of activity in an organ automatically produces an increased blood flow to that organ by relaxing its arterioles.

What causes the arterioles to dilate? It is evident that the mechanism must involve local chemical changes which result from the increased cellular activity, but the relevant chemical changes have not yet been precisely identified. Moreover, the relative contributions of the various local factors thus far implicated seem to vary, depending upon the organs involved, e.g., cardiac versus skeletal muscle. At present, therefore, we can only name some of the factors which appear to be involved. A *decreased oxygen concentration* occurs locally as a result of increased utilization of oxygen by the more active cells. Conversely, local concentrations of *carbon dioxide* and *hydrogen ion* increase. Other *metabolites* also increase in concentration as a result of the greater metabolic activity. The concentrations of certain ions, particularly *potassium,*

frequently increase (perhaps as a result of enhanced movement out of muscle cells during the more frequent action potentials). Local osmolarity also increases (i.e., water concentration decreases) as a result of the increased breakdown of high-molecular-weight substances. Changes in all these variables—decreased oxygen, increased carbon dioxide, ions, osmolarity, and metabolites—have been shown to cause arteriolar dilation under controlled experimental conditions, and they probably all contribute, more or less, to the active hyperemia response. It must be emphasized that all these chemical changes act *directly* upon the arteriolar smooth muscle, causing it to relax (dilate), no nerves or hormones being involved.

It should not be too surprising that the phenomenon of active hyperemia is most highly developed in heart and skeletal muscle, which show the widest range of normal metabolic activities of any organs or tissues in the body. It is highly efficient, therefore, that their supply of blood be primarily determined *locally* by their rates of activity. The gastrointestinal tract also manifests a great capacity for active hyperemia in keeping with its relatively wide range of metabolic activity.

Reactive hyperemia If a tourniquet is placed around the upper arm and tightened to shut off the arterial inflow, the tissues of the arm are deprived of blood (*ischemia*). When the tourniquet is loosened, the arm becomes red and very warm, signs of a greatly increased blood flow. Increased flow following a period of ischemia, called *reactive hyperemia,* occurs because the ischemia has produced local arteriolar dilation. The explanation appears to be quite similar to that for active hyperemia. While the blood flow was reduced, the supply of oxygen to the tissue was diminished and the local oxygen concentration decreased. Simultaneously, the concentrations of carbon dioxide, hydrogen ion, and metabolites all increased because they were not removed by the blood as fast as they were produced. In other words, the events of active and reactive hyperemia are similar because both reflect an imbalance of blood supply and level of cellular metabolic activity. The adaptive value of reactive hyperemia is that a tissue which suffers ischemia, say as a result of partial occlusion of the artery supplying it, automatically tends to maintain its blood supply because of local arteriolar dilation.

Histamine and the response to injury   Injury to the skin (and probably other tissues as well) causes local release from cells of a chemical substance known as *histamine.* It makes arteriolar smooth muscle relax and is probably a major cause of vasodilation in an injured area. This phenomenon, a part of the general process known as inflammation, will be described in detail in Chap. 15 and is mentioned here only to point out that histamine is *not* the vasodilator substance responsible for active or reactive hyperemia. In addition to histamine, several other locally released chemicals alter arteriolar tone in response to tissue injury or blood-vessel damage and will also be described in Chap. 15.

Reflex controls

Sympathetic nerves   Most arterioles in the body receive a rich supply of sympathetic postganglionic nerve fibers. These nerves (with one major exception) release norepinephrine, which acts upon vascular smooth muscle to cause vasoconstriction. The only organs whose arterioles are not significantly influenced by these constrictor fibers are the brain and the heart.[1] (The adaptive significance for these exceptions will be clear after the role of these sympathetic nerves has been explained.) If almost all the nerves to arterioles are constrictor in action, how can reflex arteriolar dilation be achieved? Since the sympathetic nerves are seldom completely quiescent but discharge at some finite rate, which varies from organ to organ, the nerves always cause some degree of tonic constriction; from this basal position, further constriction is produced by increased sympathetic activity, whereas dilation can be achieved by decreasing the rate of sympathetic activity below the basal level. The skin offers an

excellent example of these processes (Fig. 9-31). Skin arterioles of a normal unexcited person at room temperature are already under the influence of a high rate of sympathetic discharge; an appropriate stimulus (fear, loss of blood, etc.) causes reflex enhancement of this activity; the arterioles constrict further, and the skin pales. In contrast, an increased body temperature reflexly inhibits the sympathetic nerves to the skin, the arterioles dilate, and the skin flushes. This generalization cannot be stressed too strongly: *Control of the sympathetic constrictor nerves to arteriolar smooth muscle can accomplish either dilation or constriction.*

In contrast to the processes of active and reactive hyperemia, the primary functions of these nerves are concerned *not* with the coordination of *local* metabolic needs and blood flow but with reflexes that help maintain an adequate blood supply at all times to vital organs such as the brain and heart. As their common denominator these reflexes have the regulation of arterial blood pressure; they will be described in detail in a subsequent section.

There is, as we said, one exception to the generalization that sympathetic nerves to arterioles release norepinephrine. A group of sympathetic (*not* parasympathetic) nerves to the arterioles in skeletal muscle instead releases acetylcholine, which causes arteriolar dilation and increased blood flow. It still must be emphasized that *most* sympathetic nerves to skeletal muscle arterioles release norepinephrine; thus skeletal muscle arterioles receive a dual set of sympathetic nerves. The only known function of the vasodilator fibers is in the response to exercise or stress and will be described in a later section; the vasoconstrictor fibers mediate all other situations involving neural control of skeletal muscle arterioles.

Parasympathetic nerves   With but one major exception (the blood vessels of certain areas in the genital tract), there is no significant parasympathetic innervation of arterioles. It is true that stimulation of the parasympathetic nerves to certain glands is associated with an increased blood flow, but this may be secondary to the increased metabolic activity induced in the gland by the nerves, with resultant local active hyperemia.

Hormones   Several hormones cause constriction or dilation of arteriolar smooth muscle, one of the most important being epinephrine, the hormone released from the adrenal medulla. In most vascular beds, epinephrine, like the sympathetic nerves, causes vasoconstriction; surprisingly, in other vascular beds, epinephrine may induce vasodilation. However, it is likely that the effects of cir-

---

[1] The vasculature of the brain and heart receives sympathetic neurons the activity of which may be altered under certain circumstances. However, present evidence indicates that these neurons are usually of negligible importance when compared with the local control of these vascular beds.

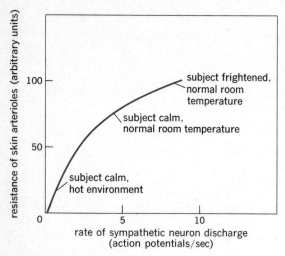

FIGURE 9-31
Dependence of skin arteriolar resistance upon activity of sympathetic nerves to the arterioles.

culating epinephrine on arterioles are quantitatively of little significance when compared with those exerted by norepinephrine released from sympathetic-nerve endings. Angiotensin, a hormone to be discussed in Chap. 12, may also strongly constrict arterioles under certain conditions.

Figure 9-32 summarizes the factors which determine arteriolar radius.

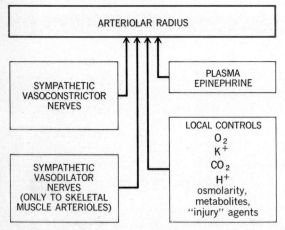

FIGURE 9-32
Major factors affecting arteriolar radius.

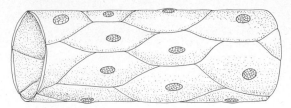

FIGURE 9-33
Typical capillary.

Capillaries

At any given moment, approximately 5 percent of the total circulating blood is flowing through the capillaries. Yet this 5 percent is really the only blood in the cardiovascular system which is performing the ultimate function of the entire system, namely, the exchange of nutrients and metabolic end products. All other segments of the vascular tree subserve the overall aim of getting adequate blood flow through the capillaries. The capillaries permeate every tissue of the body; no cell is more than 0.005 in. from a capillary. Therefore, diffusion distances are very small, and exchange is highly efficient. There are thousands of miles of capillaries in an adult person, each individual capillary being only about 1 mm long.

Capillaries throughout the body vary somewhat in structure, but the typical capillary (Fig. 9-33) is a thin-walled tube of endothelial cells without elastic tissue, connective tissue, or smooth muscle to impede transfer of water and solutes. The flat cells which constitute the endothelial lining interlock like pieces of a jigsaw puzzle. This thin capillary membrane behaves as though it were perforated by small pores through which water and solute particles smaller than proteins readily move. The permeability of capillaries varies throughout the body, liver capillaries being the "leakiest" and brain capillaries the "tightest." No capillary is so leaky, however, that it will allow erythrocytes to escape.

Anatomy of the capillary network

Figure 9-34 illustrates diagrammatically the general anatomy of the small vessels which constitute the so-called microcirculation. Blood enters the capillary network from the arterioles. Most tissues appear to have two distinct types of capillaries: "true" capillaries and thoroughfare channels. The thoroughfare channels connect arterioles and venules directly. From these channels exit and re-enter the network of true capillaries across which materials actually exchange. The site at which a true capillary

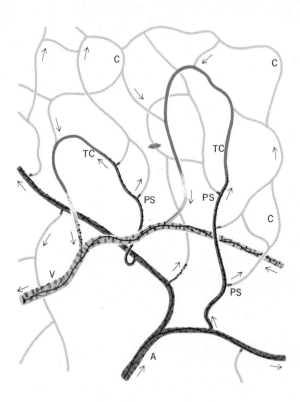

A = arteriole
V = venule
TC = thoroughfare channel

C = capillaries
PS = precapillary
      sphincter

FIGURE 9-34
Diagram of microcirculation. Note the thinning of the smooth
muscle coat in the thoroughfare channels and its complete
absence in the true capillaries. The black lines on the surface
of the vessels are nerve fibers leading to smooth muscle cells.
(Adapted from Zweifach.)

exits is protected by a ring of smooth muscle, the *pre-capillary sphincter,* which continually opens and closes so that flow through any given capillary is usually intermittent. Generally, the more active the tissue, the more precapillary sphincters are open at any moment. The sphincters are best visualized as functioning in concert with arteriolar smooth muscle to regulate not only the total flow of blood through the tissue capillaries but the number of functioning capillaries as well.

Resistance of the capillaries

Since a capillary is very narrow, it offers a considerable resistance to flow, but for two reasons, the resistance is not of critical importance for cardiovascular function: (*1*) Despite the fact that the capillaries are actually narrower than the arterioles, the huge total number of capillaries provides such a great cross-sectional area for flow that the *total* resistance of *all* the capillaries is considerably less than that of the arterioles. (*2*) Because capillaries have no smooth muscle, their radius (and therefore, their resistance) is not subject to active control and simply reflects the volume of blood delivered to them via the arterioles (and the volume leaving via the venules).

Velocity of capillary blood flow

Figure 9-35 illustrates a simple mechanical model of a series of 1-in.-diameter balls being pushed down a single tube which branches into narrower tubes. Although each tributary tube has a smaller cross section than the wide tube, the sum of the tributary cross sections is much greater than the area of the wide tube. Let us assume that in the wide tube each ball moves 3 in./min. If the balls are 1 in. in diameter and they move two abreast, six balls leave the wide tube per minute and enter the narrow tubes. Obviously, then, six balls must be leaving the narrow tubes per minute. At what speed does each ball move in the small tubes? The answer is 1 in./min. This example illustrates the following important generalization: When a continuous stream moves through con-

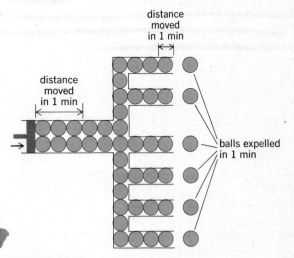

distance
moved
in 1 min

distance
moved
in 1 min

balls expelled
in 1 min

FIGURE 9-35
Relationship between cross-sectional area and velocity of flow.
The total cross-sectional area of the small tubes is three times
greater than that of the large tube. Accordingly, velocity of
flow is one-third as great in the small tubes.

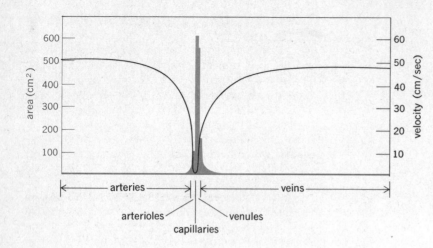

FIGURE 9-36
Relation between cross-sectional area and
velocity of flow in the systemic
circulation. The values are those for a
30-lb dog. Velocity of blood flow
through the capillaries is about
0.07 cm/sec. (*Adapted from Rushmer.*)

secutive sets of tubes, the velocity of flow decreases as the sum of the cross-sectional areas of the tubes increases. This is precisely the case in the cardiovascular system (see Fig. 9-36); the blood velocity is very great in the aorta, progressively slows in the arteries and arterioles, and then markedly slows as it passes through the huge cross-sectional area of the capillaries (600 times the cross-sectional area of the aorta). The speed then progressively increases in the venules and veins because the cross-sectional area decreases. The adaptive significance of this phenomenon is very great; blood flows through the capillaries so slowly (0.07 cm/sec) that there is adequate time for exchange of nutrients and metabolic end products between the blood and tissues.

Diffusion across the capillary wall:
exchanges of nutrients
and metabolic end products

There is no active transport of solute across the capillary wall, materials crossing primarily by simple passive diffusion. As described in the next section, there is some movement of fluid by bulk flow, but it is of negligible importance for the exchange of nutrients and metabolic end products. The factors determining diffusion rates were described in Chap. 2. Because fat-soluble substances penetrate cell membranes easily, they probably actually pass directly through the endothelial capillary cells. In contrast, many ions and molecules are poorly soluble in fat and pass through pores between adjacent endothelial cells. In any case, nearly all nutrients and metabolic end products diffuse across the capillary with great speed.

What is the sequence of events involved in capillary-cell transfers (Fig. 9-37)? Tissue cells do not exchange material *directly* with blood; the interstitial fluid always acts as middleman. Thus, nutrients diffuse across the capillary wall into the interstitial fluid, from which they gain entry to cells. Conversely, metabolic end products move first across cell membranes into interstitial fluid, from which they diffuse into the plasma. Thus, two membrane-transport processes must always be considered, that across the capillary wall and that across the tissue

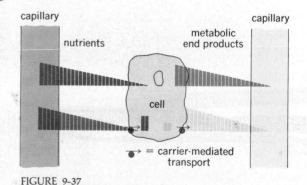

FIGURE 9-37
Movements of nutrients and metabolic end products between
plasma and tissue cells. The substance may enter or leave the
cell by diffusion or carrier-mediated transport, but it moves
across the capillary wall only by diffusion dependent upon
capillary–interstitial-fluid concentration gradients. (Two
capillaries are shown for convenience; actually, of course,
nutrients and end products leave and enter, respectively, the
same capillary.)

cell membrane. The cell-membrane step may be by diffusion or by carrier-mediated transport, but, as described above, the transcapillary movement is always by diffusion. Since to achieve *net* transport of any substance by diffusion, a concentration gradient is required, transcapillary diffusion of nutrients and metabolic end products proceeds primarily in one direction because of diffusion gradients for these substances between the blood and interstitial fluid.

How do these diffusion gradients arise? Let us take two examples, those of glucose and carbon dioxide transcapillary movement in muscle. Glucose is continuously consumed after being transported from interstitial fluid into the muscle cells by carrier-mediated transport mechanisms; this removal from interstitial fluid lowers the interstitial-fluid glucose concentration below that of plasma and creates the gradient for glucose diffusion out of the capillary. Carbon dioxide is continuously produced by muscle cells, thereby creating an increased intracellular carbon dioxide concentration, which causes diffusion of carbon dioxide into the interstitial fluid; in turn, this causes the interstitial carbon dioxide concentration to be greater than that of plasma and produces carbon dioxide diffusion into the capillary. We chose these particular examples to emphasize the fact that net movement of a substance between interstitial fluid and cells is the event which establishes the transcapillary plasma-interstitial diffusion gradients but that it does not matter whether the substance moves across the cell membrane by diffusion or carrier-mediated transport.

When a tissue increases its rate of metabolism, it must obviously obtain more nutrients from the blood and eliminate more metabolic end products. One important mechanism for achieving this is active hyperemia, described above, which increases the blood flow to the tissue. A second important and quite simple mechanism involves alterations of the plasma-interstitial concentration gradient. Let us return to our example of glucose and muscle. When the muscle increases its activity, it also increases its uptake of glucose, thereby lowering the interstitial glucose concentration below normal. This sets up an increased plasma-interstitial glucose concentration gradient which causes the *net* diffusion of glucose out of the capillary to be increased. In other words, this change in concentration gradient has allowed a greater fraction of the total blood glucose to be extracted from the blood as it flows through the capillaries. Thus, there need not be an absolutely strict correlation between a tissue's activity and its blood supply, at least within moderate limits. Similar changes in diffusion gradients permit a

tissue to obtain nutrients and eliminate metabolic end products adequately in spite of modest reductions of blood flow to it.

It should now be clear how the interstitial fluid functions as the true immediate environment for the body's cells and constitutes the body's *internal environment.*

Bulk flow across the capillary wall: distribution of the extracellular fluid

Since the capillary wall is highly permeable to water and to almost all the solutes of the plasma with the exception of the plasma proteins, it behaves like a porous filter through which protein-free plasma moves by bulk flow, known as *ultrafiltration,* under the influence of a hydrostatic pressure gradient. The magnitude of the bulk flow is directly proportional to the hydrostatic pressure difference between the inside and outside of the capillary, i.e., between the capillary blood pressure and the interstitial-fluid pressure. Normally, the former is much larger than the latter, so that a considerable hydrostatic pressure gradient exists to drive the filtration of protein-free plasma out of the capillaries into the interstitial fluid. Why then does all the plasma not filter out into the interstitial space instead of remaining in the capillaries? The explanation was first elucidated by Starling (the same scientist who expounded the law of the heart which bears his name) and depends upon the principles of osmosis.

In Chap. 2 we described how a net movement of water occurs across a semipermeable membrane from a solution of high water concentration to a solution of low water concentration. Recall that the concentration of water depends upon the concentration of solute molecules or ions dissolved in the water. When two solutions *A* and *B*, which are separated by a semipermeable membrane, have identical concentrations of all solutes, the water concentrations are identical and no net water movement occurs. When, however, a quantity of a nonpermeating substance is added to solution *A*, the water concentration of *A* is reduced below that of solution *B* and a net movement of water will occur by osmosis from *B* into *A*. Of great importance is that osmotic flow of water "drags" along with it any dissolved solutes to which the membrane is highly permeable. Thus, a difference in water concentration can result in the movement of both water and permeating solute in a manner virtually indistinguishable from the bulk flow produced by a hydrostatic pressure difference. The difference in water concentration resulting from the presence of the nonpenetrating solute can therefore be expressed in units of pressure (millimeters of mercury).

This analysis can now be applied to capillary fluid movements. The plasma within the capillary and the interstitial fluid outside it contain large quantities of low-molecular-weight solutes (crystalloids), e.g., sodium, chloride, or glucose. Since the capillary lining is highly permeable to all these crystalloids, they all have almost identical concentrations in the two solutions. There are small concentration differences occurring for substances consumed or produced by the cells, but these tend to cancel each other, and, accordingly, no significant water-concentration difference is caused by the presence of the crystalloids. In contrast, the plasma proteins can diffuse across the capillary wall only very slightly and therefore have a very low interstitial-fluid concentration. This difference in protein concentration between plasma and interstitial fluid means that the water concentration of the plasma is lower than that of interstitial fluid, inducing an osmotic flow of water from the interstitial compartment into the capillary. Along with the water are carried all the different types of crystalloids dissolved in the interstitial fluid. Thus, osmotic flow of fluid, like bulk flow, does not alter the concentrations of the low-molecular-weight substances of plasma or interstitial fluid.

In summary, two opposing forces act to move fluid across the capillary: (1) The hydrostatic pressure difference between capillary blood pressure and interstitial-fluid pressure favors the filtration of a protein-free plasma out of the capillary; (2) the water-concentration difference between plasma and interstitial fluid, which results from the protein-concentration differences, favors the osmotic movement of interstitial fluid into the capillary. Accordingly, the movements of fluid depend directly upon four variables: the capillary hydrostatic pressure, interstitial hydrostatic pressure, plasma protein concentration, and interstitial-fluid protein concentration.

We may now consider quantitatively how these variables act to move fluid across the capillary wall (Fig. 9-38). Much of the arterial blood pressure has already been dissipated as the blood flows through the arterioles, so that pressure at the beginning of the capillary is 35 mm Hg. Since the capillary also offers resistance to flow, the pressure continuously decreases to 15 mm Hg at the end of the capillary. The interstitial pressure is so close to zero that it can be ignored. The difference in protein concentration between plasma and interstitial fluid causes a difference in water concentration (plasma water concentration less than interstitial-fluid water concentration), which induces an osmotic flow of fluid into the capillary equivalent to that produced by a hydrostatic pressure difference of 25 mm Hg. It is evident that in the first portion of the capillary the hydrostatic pressure difference is greater than the osmotic forces and a net movement of fluid out of the capillary occurs; in the last portion of the capillary, however, a net force causes fluid movement into the capillary (termed *absorption*). The net result is that the early and late capillary events tend to cancel each other out, and there is little overall net loss or gain of fluid (Fig. 9-38A). In a normal person there is a small net filtration; as we shall see, this is returned to the blood by lymphatics.

The analysis of capillary fluid dynamics in terms of different events occurring at the arterial and venous ends of the capillary is oversimplified. It is likely that any given capillary manifests either net filtration or net absorption along its entire length because the arteriole supplying it is either so dilated or so constricted as to yield a capillary hydrostatic pressure above or below 25 mm Hg along the entire length of the capillary. This does not alter the basic concept that, taken as a unit, a capillary bed manifests net absorption or filtration, depending upon the average levels of hydrostatic pressures within the individual capillaries constituting the bed.

Figure 9-38B and C illustrates, however, the effects on this equilibrium of changing capillary pressure. In B, the arterioles in the organ have been dilated, and the capillary pressure therefore increases since less of the arterial pressure is dissipated in the passage through the arterioles. Outward filtration now predominates, and some of the plasma enters the interstitial fluid. In contrast, marked arteriolar constriction (Fig. 9-38C) produces decreased capillary pressure and net movement of interstitial fluid into the vascular compartment. Figure 9-38D shows how net absorption or filtration can be produced in the absence of capillary pressure changes whenever plasma protein concentration is altered. Thus, in liver disease, protein synthesis decreases, plasma protein concentration is reduced, plasma water concentration is increased, net filtration occurs, and fluid accumulates in the interstitial space.

The transcapillary protein-concentration difference can also be decreased by a quite different event, namely, the leakage of protein across the capillary wall into the interstitium whenever the capillary lining is damaged. This eliminates the protein-concentration difference, and local edema occurs as a result of the unchecked hydrostatic pressure difference still acting across the capillary. The fluid accumulation in a blister is an excellent example.

The major function of this capillary filtration-absorption equilibrium should now be evident: *It determines*

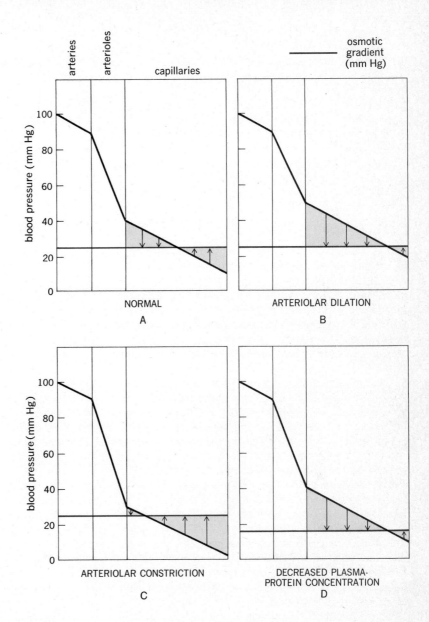

FIGURE 9-38
Relevant filtration-absorption forces acting across the capillary wall in several situations. Arrows down indicate filtration out of the capillary. Arrows up indicate fluid movement from interstitium into capillary. The shaded areas denote relative magnitudes of the fluid movements.

the distribution of the extracellular-fluid volume between the vascular and interstitial compartments. Obviously, the ability of the heart to pump blood depends upon the presence of an adequate volume of blood within the system. Recall that the interstitial-fluid volume is three to four times larger than the plasma volume; therefore, the interstitial fluid serves as a reservoir which can supply additional fluid to the circulatory system or draw off excess. The important role this equilibrium plays in the physiological response to many situations, such as hemorrhage, will be described in a later section.

It should be stressed again that capillary filtration and absorption do not alter *concentrations* of any substance (other than protein) since movement is by bulk flow; i.e., everything in the plasma (except protein) or the interstitial fluid moves together. The reason this process of filtration plays no significant role in the exchange of nutrients and metabolic end products between capillary and tissues is that the total quantity of a substance (such as glucose or oxygen) moving into or out of the capillary during bulk flow is extremely small in comparison with the quantities moving by diffusion. For example, during a single day approximately 20,000 g of glucose crosses the capillary into the interstitial fluid by diffusion but only 20 g enters by bulk flow. Of course, only a small fraction of this glucose is utilized by the cells, the remainder moving back into the blood, again almost entirely by diffusion.

Veins

Most of the pressure imparted to the blood by the heart is dissipated as blood flows through the arterioles and capillaries, so that pressure in the small venules is only approximately 15 mm Hg and only a small pressure remains to drive blood back to the heart. One of the major functions of the veins is to act as low-resistance conduits for blood flow from the tissues back to the heart. This function is performed so efficiently that the total pressure drop from venule to right atrium is only about 10 mm Hg, the right atrial pressure being 0 to 5 mm Hg. The resistance is low because the veins have a large diameter. This completes our description of the pressure changes throughout the vascular tree. The normal pressure profiles for the systemic and pulmonary circulation are given in Fig. 9-39. Note that the pulmonary pressures are considerably smaller than the systemic pressures for reasons shortly to be described. Note also that the resistance offered by the arterioles effectively damps the pulse; by doing so, the arterioles convert the pulsatile arterial flow into a continuous capillary flow.

The veins perform a second extremely important function which has only recently been appreciated: They adjust their total *capacity* to accommodate variations in blood volume. The veins are the last set of tubes through which the blood must flow on its trip to the heart. The force immediately driving this venous return is the venous pressure (more precisely, the pressure gradient between the veins and atria). In turn, the rate of venous return, i.e., inflow to the atria, is one of the most important determinants of atrial pressure. In a previous section on the control of cardiac output, we emphasized that the atrial pressure was the major determinant of ventricular end-diastolic volume and thereby of intrinsic control of stroke volume. Combining these two statements, we now see that venous pressure is a crucial determinant of

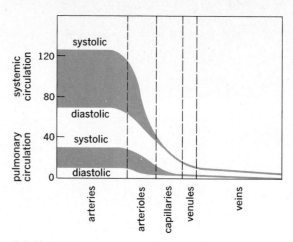

FIGURE 9-39
Summary of pressures in the vascular system.

stroke volume via the intermediation of atrial pressure and ventricular end-diastolic volume.

Determinants of venous pressure

The factors determining pressure in any elastic tube, as we know, are the volume of fluid within it and the distensibility of its wall. Accordingly, total blood volume is one important determinant of venous pressure. The veins differ from the arteries in that their walls are thinner and much more distensible and can accommodate large volumes of blood with little increase of internal pressure. This is illustrated by comparing Fig. 9-40 with Fig. 9-39; approximately 60 percent of the total blood volume is present in the systemic veins at any given moment, but the venous pressure averages less than 10 mm Hg. In contrast, the systemic arteries contain less than 15 percent of the blood at a pressure of approximately 100 mm Hg. This pressure-volume relationship of the veins allows them to act as a reservoir for blood. The walls of the veins contain smooth muscle richly innervated by sympathetic vasoconstrictor nerves, stimulation of which causes venous constriction, thereby increasing the stiffness of the wall, i.e., making it less distensible, and raising the pressure of the blood within the veins. Increased venous pressure drives more blood out of the veins into the right heart. Thus, venous constriction exerts precisely the same effect on venous return as giving a transfusion.

The great importance of this effect can be visualized by the example in Fig. 9-41. A large decrease in

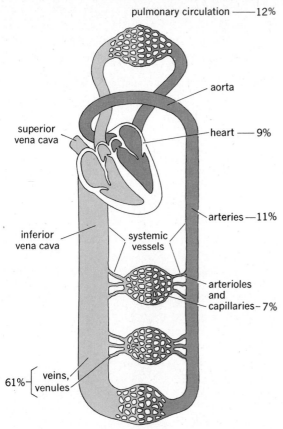

FIGURE 9-40
Distribution of blood in the different portions of the
cardiovascular system. Compare this distribution of blood
volumes with the pressures for the relevant areas shown in
Fig. 9-39. (*Adapted from Guyton.*)

pressed, thereby reducing their diameter and decreasing
venous capacity. As will be described in Chap. 10, during
inspiration, the diaphragm descends, pushes on the abdominal contents, and increases abdominal pressure. The
large veins which pass through the abdomen are partially
compressed by this increased pressure; this facilitates
movement of blood, but only toward the heart because
venous valves in the legs prevent backflow. Simultaneously, the pressure in the chest (thorax) decreases,
and this decrease is transmitted passively to the intrathoracic veins and right atria. The net effect is to increase

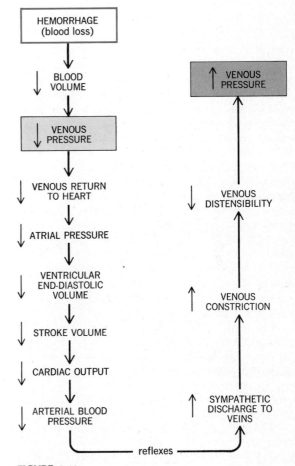

FIGURE 9-41
Role of venoconstriction in maintaining venous pressure
during blood loss. The increased venous smooth muscle
constriction returns the decreased venous pressure toward, but
not to, normal. The reflexes involved in this response are
described later in the chapter.

total blood volume initially reduces the pressures everywhere in the circulatory system, including the veins; venous return to the heart decreases, and cardiac output
decreases. However, reflexes to be described cause increased sympathetic discharge to the venous smooth
muscle, which contracts, thereby returning venous pressure toward normal, restoring venous return and cardiac
output.

   Two other mechanisms can decrease venous capacity, increase venous pressure, and facilitate venous
return; these are the skeletal muscle "pump" and the
effects of respiration upon thoracic and abdominal veins
(respiratory "pump"). During skeletal muscle contraction,
the veins running through the muscle are partially com-

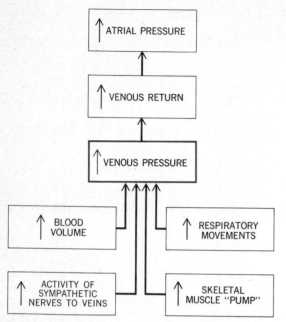

FIGURE 9-42
Major factors determining venous pressure and thereby atrial pressure. The figure as drawn shows how venous and atrial pressures are increased; reversing the arrows in the boxes indicates how these pressures can be reduced.

the pressure gradient between the right atrium and veins outside the thorax; accordingly, venous return to the heart is enhanced.

In summary (Fig. 9-42), the effects of venous muscle tone, the skeletal muscle pump, and the respiratory pump are to facilitate return of blood to the heart. The net result is that atrial pressure and, thereby, cardiac output are determined in large part by these factors.

Effects of venous constriction on resistance to flow

We have seen that decreasing the diameter of the veins increases venous pressure, which increases venous return to the heart. However, this decreased diameter also increases resistance to flow, a phenomenon which would retard venous return if the effect of venous constriction upon resistance were not so slight as to be negligible. The veins have such large diameters that a slight decrease in size (which has great effects on venous capacity) produces little increase in resistance. This is just the opposite of the arterioles, which are so narrow that they

contain little blood at any moment (Fig. 9-40), further decrease having little effect on blood displacement toward the heart but even a slight decrease in diameter producing a marked increase in resistance to flow. Flow back to the heart, therefore, tends to be impaired by arteriolar constriction and enhanced by venous constriction. It should be stressed, however, that abnormally great increases in venous resistance, say from an internal blood clot or a tumor compressing from the outside, may markedly impair blood flow. Under such conditions, blood accumulates behind the lesion, pressures in the small veins and capillaries drained by the occluded vein increase, capillary filtration increases, and the tissue becomes edematous, i.e., swollen with excess interstitial fluid.

The venous valves

Many veins in the body, particularly in the limbs, have valves which close so as to allow flow only toward the heart (Fig. 9-43). Why are these valves necessary if the pressure gradient created by cardiac contraction always moves blood toward the heart? We have seen how two other forces, the muscle pump and inspiratory movements, facilitate flow of venous blood. When these forces squeeze the veins, blood would be forced in both directions if the valves were not there to prevent backward flow. As we shall see, valves also play a critical role in counteracting the effects of upright posture.

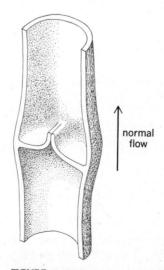

FIGURE 9-43
Normal venous valve. Any tendency toward retrograde flow would immediately push the valve leaflets together.

 Lymphatics

The lymphatics are not part of the circulatory system per se but constitute a one-way route from interstitial fluid to blood. The lymphatic system in man constitutes an extensive network of thin vessels resembling the veins. It arises as a group of blind-end lymph capillaries which are present in almost all organs of the body. These capillaries are apparently permeable to virtually all interstitial-fluid constituents (including protein), which either diffuse or filter into them. The pale yellow lymph then moves through the vessels, which converge to form larger and larger vessels. Ultimately, the largest of these lymphatics drains into veins in the lower neck. Thus, the lymphatics carry fluid from the interstitial fluid into the blood.

Functions of the lymphatic system

Return of excess filtered fluid In a normal person, the fluid filtered out of the capillaries each day slightly exceeds that reabsorbed. This excess is returned to the blood via lymphatics. Partly for this reason, lymphatic malfunction leads to increased interstitial fluid, i.e., *edema*.

Return of protein to the blood Most capillaries in the body have a slight permeability to protein, and accordingly, there is a small steady loss of protein from the blood into the interstitial fluid. The protein returns via the lymphatics. The breakdown of this cycle is without question the most important cause of the marked edema seen in patients with lymphatic malfunction. Because protein (in small amounts) is normally lost from the capillaries, failure of the lymphatics to remove it allows the interstitial protein concentration to increase to that of the plasma. This failure reduces or eliminates the protein-concentration difference and thus the water-concentration difference across the capillary wall and permits the net movement of quantities of fluid out of the capillary into the interstitial space.

Specific transport functions In addition to these nonspecific transport functions, the lymphatics also provide the pathway by which certain specific substances reach the blood. The most important is fat absorbed from the gastrointestinal tract. It is likely that certain high-molecular-weight hormones reach the blood via the lymphatics.

Lymph nodes Besides its transport functions, the lymphatic system plays a critical role in the body's defenses against disease. This function, which is mediated by the lymph nodes located along the larger lymphatic vessels, is described in Chap. 15.

Mechanism of lymph flow

How does the lymph move with no heart to push it? The best explanation at present is that lymph flow depends primarily upon forces external to the vessels, e.g., the pumping action of the muscles through which the lymphatics flow and the effects of respiration on thoracic-cage pressures. Since the lymphatics have valves similar to those in veins, external pressures would permit only unidirectional flow.

SECTION C
INTEGRATION OF CARDIOVASCULAR FUNCTION: REGULATION OF SYSTEMIC ARTERIAL PRESSURE

In Chap. 5 we described the fundamental ingredients of all reflex control systems: (*1*) the internal-environmental variable being regulated, i.e., maintained relatively constant, and the receptors sensitive to it; (*2*) afferent pathways passing information from the receptors to (*3*) a control center, which integrates the different afferent inputs; (*4*) efferent pathways controlling activity of (*5*) effector organs, whose output raises or lowers the level of the regulated variable. The control and integration of cardiovascular function will be described in these terms. *The major variable being regulated is the systemic arterial blood pressure.* The central role of arterial pressure and the adaptive value of keeping it relatively constant should be apparent from the ensuing discussion.

Arterial pressure, cardiac output, and arteriolar resistance

Adequate blood flow through the vital organs (brain and heart) must be completely maintained at all times; the brain, for example, suffers irreversible damage within 3 min of ischemia. In contrast, many areas of the body, e.g., the gastrointestinal tract, the kidney, skeletal muscle, and skin, can withstand moderate reductions of blood flow for longer periods of time or even severe reductions if only for a few minutes. The mean arterial blood pressure is the driving force for blood flow through all the organs. The distribution of flow, i.e., the actual flow through the various organs at any given arterial pressure, depends primarily upon the radii of the arterioles in each vascular bed. A critical relationship has not been emphasized before, although it is implicit in the basic pressure-flow

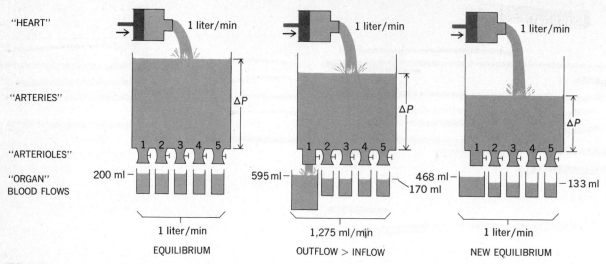

"HEART"   1 liter/min   1 liter/min   1 liter/min

"ARTERIES"   ΔP   ΔP   ΔP

"ARTERIOLES"   1 2 3 4 5   1 2 3 4 5   1 2 3 4 5

"ORGAN"   200 ml —   595 ml —   468 ml —
BLOOD FLOWS   170 ml   — 133 ml

1 liter/min   1,275 ml/min   1 liter/min
EQUILIBRIUM   OUTFLOW > INFLOW   NEW EQUILIBRIUM

FIGURE 9-44
Model to illustrate the dependency of arterial blood pressure upon arteriolar resistance, showing the effects of dilating one arteriolar bed upon arterial pressure and organ blood flow if no compensatory adjustments occur. The middle panel is a transient state before the new equilibrium occurs. In one respect, the illustration of the model is misleading in that the arterial reservoir is shown containing very large quantities of blood. In fact, as we have seen, the volume of blood in the arteries is quite small.

equation; these two factors, arterial pressure and arteriolar resistance, are *not* independent variables; *arteriolar resistance is one of the major determinants of arterial pressure.* This can be illustrated by the simple mechanical model shown in Fig. 9-44. A pump pushes fluid into a cylinder at the rate of 1 liter/min; at steady state, fluid leaves the cylinder via the outflow tubes at 1 liter/min, and the height of the fluid column, which is the driving pressure for outflow, remains stable. Assuming that the radii of the adjustable outflow tubes are all equal so that the flows through them are equal, we disturb the steady state by loosening the cuff on the outflow tube 1, thereby increasing its radius, reducing its resistance, and increasing its flow. The total outflow for the system is now greater than 1 liter/min, more fluid leaves the reservoir than enters via the pump, and the height of the fluid column begins to decrease. In other words, a change in outflow resistance must produce changes in the pressure of the reservoir (unless some compensatory mechanism is brought into play). As the pressure falls, the rate of outflow via all tubes decreases. Ultimately, in our example, a new steady state is reached when the reservoir pressure is low enough to cause only 1 liter/min outflow despite the decreased resistance of tube 1.

This analysis can be applied to the cardiovascular system by equating the pump with the heart, the reservoir with the arteries, and the outflow tubes with various arteriolar beds. An analogy to opening outflow tube 1 is exercise; during exercise, the skeletal muscle arterioles dilate, primarily because of active hyperemia, thereby decreasing resistance. If the cardiac output and the arteriolar diameters of all other vascular beds remain unchanged, the increased runoff through the skeletal muscle arterioles causes a decrease in arterial pressure. This, in turn, decreases flow through all other organs of the body, including the brain and heart. Indeed, even the exercising muscles themselves suffer a lessening of flow (below that seen immediately after they dilated) as arterial pressure falls. Thus, the only way to guarantee the essential flow to the vital organs and the additional flow to the exercising muscle is to prevent the arterial pressure from falling.

This can be accomplished by changing cardiac output or the radii of the other arteriolar vascular beds or both. How these factors contribute can be visualized by returning to Fig. 9-44. When outflow tube 1 is loosened, how can one prevent a drop in the height of the reservoir fluid column, i.e., the driving pressure? Figure 9-45 demonstrates the first major possibility, simultaneously tightening one or more of outflow tubes 2 to 5.

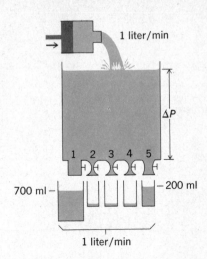

FIGURE 9-45
Compensation for dilation in one bed by constriction in others. When outflow tube 1 is opened, outflow tubes 2 to 4 are simultaneously tightened so that the *total* **outflow resistance remains constant, total rate of runoff remains constant, and reservoir pressure remains constant.**

This partially compensates for the decreased resistance of tube 1, and the total outflow resistance of all tubes can be shifted back toward normal. Therefore, the total outflow remains near 1 liter/min; of course, the distribution of flow is such that flow in tube 1 is increased and all the others are decreased. If, for some reason, tube 5 is declared a vital pathway the flow through which should never be altered, the adjustments can always be made in tubes 1 to 4.

Applied to the body, this process is obviously analogous to control of total vascular resistance. When the skeletal muscle arterioles dilate during exercise, the total resistance of all vascular beds can still be maintained if arterioles constrict in other organs, such as the kidneys and gastrointestinal tract, which can readily suffer moderate flow reductions for at least short periods of time. In contrast, the brain and heart arterioles remain unchanged, thereby assuring constant brain blood supply. (The importance of the absence of significant activation of sympathetic vasoconstrictor fibers to the arterioles of brain and heart should now be apparent.)

This type of resistance juggling, however, can compensate only within limits. Obviously if tube 1 opens very wide, even total closure of the other tubes cannot compensate completely. Moreover, if the closure is prolonged, absence of flow will cause severe tissue damage. There must therefore be a second compensatory mechanism: increasing the inflow by increasing the activity of the pump (Fig. 9-46). When tube 1 widens and total outflow increases, the reservoir column can be completely maintained by simultaneously increasing the inflow

from the pump. Thus, at the new equilibrium, the total outflow and inflow are still equal, the reservoir pressure is unchanged, outflow through tubes 2 to 5 to is unaltered, and the entire increase in outflow occurs through tube 1. Applied to the body, it should be evident that, when the blood vessels dilate, arterial pressure can be maintained constant by stimulating the heart to increase cardiac output. *Thus, the regulation of arterial pressure not only assures blood supply to the vital organs but provides a means for coordinating cardiac output with total tissue requirements.*

In summary, the regulation of arterial blood pressure is accomplished both by control of cardiac output and arteriolar resistance. Figure 9-47 shows both mechanisms in operation simultaneously.

It should now be possible to formalize these qualitative relationships for the entire cardiovascular system, using our basic pressure-flow equation. Flow through any tube is directly proportional to the pressure gradient between the ends of the tube and inversely proportional to the resistance.

$$\text{Flow} = \frac{\Delta P}{R}$$

Rearranging terms algebraically

$$\Delta P = \text{flow} \times R$$

This is simply another way of looking at the same equation, a way which clearly shows the dependence of pressure upon flow and resistance, which we have just described, using our models. Because the vascular tree is a continuous closed series of tubes, this equation holds

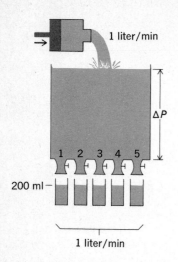

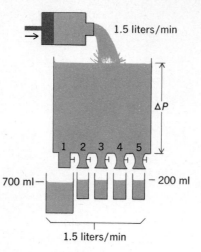

FIGURE 9-46
Compensation for dilation by increasing pump output. When outflow tube 1 is dilated, the total resistance decreases and total rate of runoff increases. Simultaneously, the pump output is increased by precisely the same amount, so that reservoir pressure remains constant.

for the entire system, i.e., from the very first portion of the aorta to the last portion of the vena cava just at the entrance to the heart. Therefore

> Flow = cardiac output
> ΔP = mean aortic pressure − late vena cava pressure
> R = total resistance

where total resistance means the sum of the resistances of all the vessels in the systemic vascular tree. This usually is termed *total peripheral resistance.*

Since the late vena cava pressure is very close to 0 mm Hg, the formula

> ΔP = mean aortic pressure − late vena cava pressure

becomes

> ΔP = mean aortic pressure − 0 or
> ΔP = mean aortic pressure

Moreover, since the mean pressure is essentially the same in the aorta and all large arteries, the pressure term in the equation becomes

> ΔP = mean arterial pressure

The pressure-flow equation for the entire vascular tree now becomes

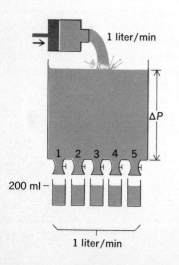

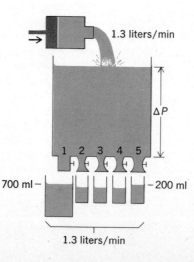

FIGURE 9-47
Compensation for dilation in one vascular bed by a combination of increases in pump output and constriction in other vascular beds, a combination of the compensatory adjustments in Figs. 9-45 and 9-46.

Mean arterial pressure
= cardiac output × total peripheral resistance

Recall that the arteries and veins are so large that they contribute very little to the total peripheral resistance. The major sites of resistance are the arterioles. The capillaries also offer significant resistance, but they have no muscle and their diameter reflects primarily the diameter of the arterioles supplying them. For these reasons, it is convenient to consider changes in arteriolar radius as virtually the only determinant of variations in total peripheral resistance.[1] Thus, the equation formally and quantitatively states the basic relationships described earlier, namely, that *arterial blood pressure can be increased, increasing either by cardiac output or total peripheral resistance.*

This equation is the fundamental equation of cardiovascular physiology. Given any two of the variables, the third can be calculated. For example, we can now explain why pulmonary arterial pressure is much lower than the systemic arterial pressure (Fig. 9-15). The blood flow per minute, i.e., cardiac output through the pulmonary and systemic arteries, is, of course, the same; therefore, the pressures can differ only if the resistances differ. Thus, the pulmonary arterioles must be wider and offer much less resistance to flow than the systemic arterioles. In other words, the total pulmonary vascular resistance is lower than the total systemic peripheral resistance. This permits the pulmonary circulation to function as a low-pressure system.

Figure 9-48 presents the grand scheme of effector mechanisms and efferent pathways which regulates systemic arterial pressure. None of this information is new, all of it having been presented in previous figures. The reader can now appreciate how the function of the heart and various vascular segments are coordinated to achieve this. A change in any single variable shown in the figure will, all others remaining constant, produce a change in mean arterial pressure by altering either cardiac output or total peripheral resistance. Conversely, any such deviation in mean arterial pressure can be eliminated by the reflex alteration of some other variable. It should be evident from the figure that the reflex control of cardiac output and peripheral resistance involves primarily (*1*) sympathetic nerves to heart, arterioles, and veins, (*2*) parasympathetic nerves to the heart, and (*3*) release of epinephrine from the adrenal medulla. In one sense, we have approached arterial pressure regulation backwards, in that the past sections have described the effector sites (heart, arterioles, veins) and motor pathways (autonomic nervous system). Now we must complete the reflexes by describing the monitoring systems (receptors and afferent pathways to the brain) and the control centers in the brain.

Cardiovascular control centers in the brain

The primary cardiovascular control center is in the medulla, the first segment of brain above the spinal cord. The axons of the neurons which this center comprises make synaptic connections with the autonomic neurons and via these connections exert dominant influence over them. The medullary cardiovascular center is absolutely essential for blood-pressure regulation. The relevant medullary neurons are sometimes divided into cardiac and vasomotor centers, which are then further subdivided and classified, but because these areas actually constitute diffuse networks of highly interconnected neurons, we prefer to call the entire area the *medullary cardiovascular center.*

An important aspect of its function is that of reciprocal innervation. The synaptic distribution of the medullary axons and the input to their cell bodies are such that when the parasympathetic nerves to the heart are stimulated, the sympathetic nerves to the heart, as well as to the arterioles and the veins, are usually simultaneously inhibited (Fig. 9-49). Conversely, parasympathetic inhibition and sympathetic stimulation are usually elicited simultaneously.[2] This pattern is important because there is always some continuous discharge of the autonomic nerves. Therefore, the heart can be slowed by two simultaneous events: inhibition of the sympathetic activity to the SA node and enhancement of the parasympathetic activity to the SA node. The converse is also true for accelerating the heart. In contrast, only sympathetic fibers significantly innervate the ventricular muscle itself and the arteriolar and venous smooth muscle. However, the muscle activity can still be decreased below normal by inhibiting the basal sympathetic activity.

Other areas of the brain, particularly in the hypothalamus, have an important influence on blood pressure, but there is good reason to believe that most of them exert their effects via the medullary centers; i.e., nerve impulses from them descend to the medulla and through synaptic connections alter the discharge of the primary medullary neurons. It is through these pathways that

[1] As described earlier, changes in blood viscosity can also contribute to change in flow resistance.

[2] These generalizations are oversimplifications. There is a considerable degree of separateness in the control of sympathetic and parasympathetic discharge, depending upon the precise circumstances eliciting the reflexes.

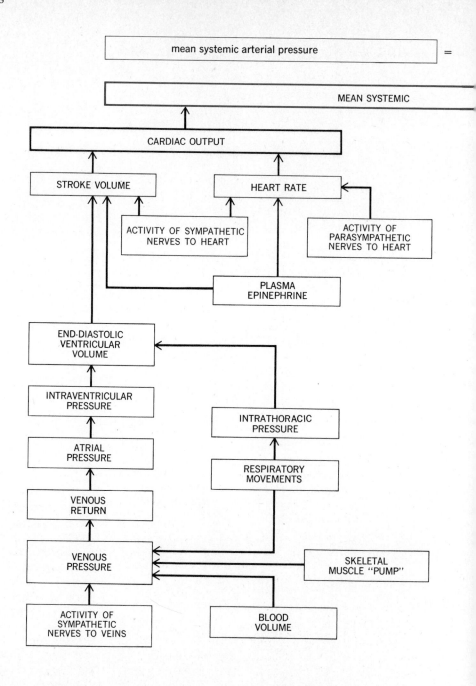

factors such as hunger, pain, anger, body temperature, and many others can alter blood pressure. There is one major exception: The sympathetic vasodilator fibers to skeletal muscle arterioles are apparently not controlled by the medullary centers but are under the direct influ- ence of neuronal pathways originating in the cerebral cortex and hypothalamus. This pathway and the sympathetic vasodilators are activated only during exercise and stress and play no role in any of the many other cardiovascular responses.

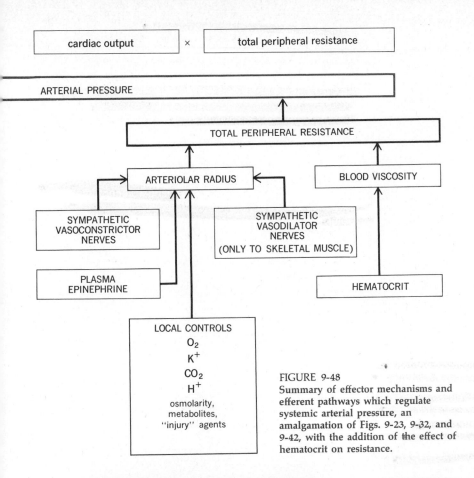

FIGURE 9-48
Summary of effector mechanisms and
efferent pathways which regulate
systemic arterial pressure, an
amalgamation of Figs. 9-23, 9-32, and
9-42, with the addition of the effect of
hematocrit on resistance.

Receptors and afferent pathways

We have now to discuss the last link in arterial pressure regulation, namely, the receptors and afferent pathways bringing information into the medullary centers. The most important of these are the *arterial baroreceptors.*

Arterial baroreceptors

It is only logical that the reflexes which homeostatically regulate arterial pressure originate primarily with receptors within the arteries which are pressure-sensitive. High in the neck each of the major arteries (carotids) supplying the brain divides into two smaller arteries. At this bifurcation, the wall of the artery is thinner than usual and contains a large number of branching, vinelike nerve endings (Fig. 9-50). This small portion of the artery is called the *carotid sinus.* There the nerve endings are apparently highly sensitive to stretch or distortion; since the degree of wall stretching is directly related to the pressure within the artery, the carotid sinus actually serves as a pressure receptor (*baroreceptor*). The nerve endings come together to form afferent neurons which travel to the medulla, where they eventually synapse upon the neurons of the cardiovascular center. An area functionally similar to the carotid sinuses found in the *arch of the aorta* constitutes a second important arterial baroreceptor.

Action potentials recorded in single afferent fibers from the carotid sinus (Figs. 9-50 and 9-51) demonstrate the pattern of response by these receptors. In Fig. 9-50, the arterial pressure within the carotid sinus was artificially controlled. At a steady nonpulsatile pressure of 100 mm Hg, there is a tonic rate of discharge by the nerve. This rate of firing can be decreased or increased by lowering or raising the arterial pressure, respectively. Note that the fiber shows no fatigue or adaptation. Figure 9-51 illustrates the same type of experiment, except that

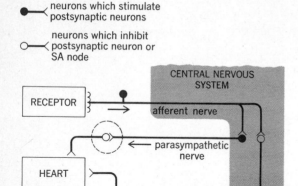

neurons which stimulate postsynaptic neurons

neurons which inhibit postsynaptic neuron or SA node

CENTRAL NERVOUS SYSTEM

RECEPTOR

afferent nerve

parasympathetic nerve

HEART

ARTERIOLES AND VEINS

sympathetic nerve

FIGURE 9-49
Reciprocal innervation in the control of the cardiovascular system. Afferent input, which stimulates the parasympathetic nerves to the heart, simultaneously inhibits the sympathetic nerves to the heart, arterioles, and veins.

pulsatile perfusion is used. The arterial baroreceptors are responsive not only to the mean arterial pressure but to the pulse pressure as well, a responsiveness that adds a further degree of sensitivity to blood-pressure regulation since small changes in certain important factors (such as blood volume) cause changes in pulse pressure before they become serious enough to affect mean pressure.

Our description of the major blood-pressure-regulating reflex is now complete (Fig. 9-51); an increase in arterial pressure increases the rate of discharge of the carotid sinus and aortic arch baroreceptors; these impulses travel up the afferent nerves to the medulla and, via appropriate synaptic connections with the neurons of the medullary cardiovascular centers, induce (1) slowing of the heart because of decreased sympathetic discharge and increased parasympathetic discharge, (2) decreased myocardial contractility because of decreased sympathetic activity, (3) arteriolar dilation because of decreased sympathetic discharge to arteriolar smooth muscle, and (4) venous dilation because of decreased sympathetic discharge to smooth muscle. The net result is a decreased cardiac output (decreased heart rate and stroke volume), decreased peripheral resistance, and return of blood pressure toward normal.

Other baroreceptors

Other portions of the vascular tree contain nerve endings sensitive to stretch, namely, other large arteries, the large

veins, and the cardiac walls themselves. They seem to function like the carotid sinus and aortic arch in that the few from which electric activity has been recorded show increased rates of discharge with increasing pressure. By means of these receptors, the medulla is kept constantly informed about the venous and atrial pressure, and a further degree of sensitivity is gained. Thus, a slight decrease in atrial pressure begins to facilitate the sympathetic nervous system even before the change becomes sufficient to lower cardiac output and arterial pressure far enough to be detected by the arterial baroreceptors. As we shall see in Chap. 11, the atrial baroreceptors are particularly important for the control of body sodium and water.

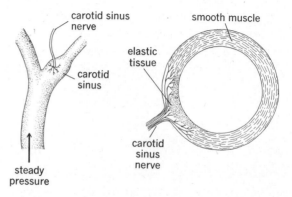

carotid sinus nerve

carotid sinus

steady pressure

smooth muscle

elastic tissue

carotid sinus nerve

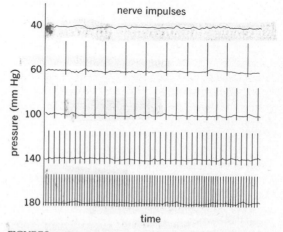

nerve impulses

40

60

100

140

180

pressure (mm Hg)

time

FIGURES 9-50
Action potentials recorded from the carotid sinus nerve during nonpulsatile, i.e., steady, perfusion of an isolated carotid artery. The baroreceptors discharge at greater and greater frequencies as the pressure is increased. (*Adapted from Rushmer.*)

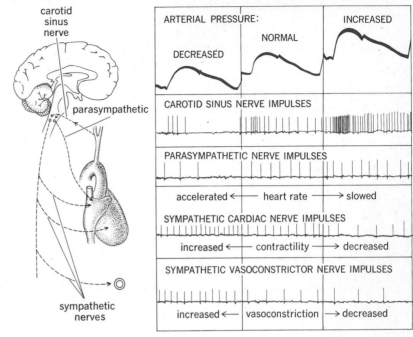

FIGURE 9-51

Carotid sinus reflexes, the major blood-pressure-regulating reflexes. An increase in arterial pressure causes an increase in the rate of discharge of the carotid sinus baroreceptors, which causes a reflex stimulation of the parasympathetic nerves to the heart and an inhibition of the sympathetic nerves to the heart, arterioles, and veins; the net result is a decreased cardiac output and peripheral resistance, both of which cause the arterial blood pressure to decrease. Precisely the opposite events occur in response to a decrease in blood pressure. (*Adapted from Rushmer.*)

Chemoreceptors

The aortic and carotid arteries contain specialized structures sensitive primarily to the concentrations of arterial oxygen but also to those of carbon dioxide and hydrogen ion. Since these receptors are far more important for the control of respiration, they are described in Chap. 10, but they also send information to the medullary cardiovascular centers, the result being that blood pressure tends to be reflexly increased by decreased arterial oxygen. Changes in carbon dioxide and hydrogen-ion concentrations also alter blood pressure reflexly, but the effects are small and the pathways quite complex.

Summary

The medullary cardiovascular control centers are true integrating centers, receiving a wide variety of information from baroreceptors, chemoreceptors, peripheral sensory receptors of all kinds (pain, cold, etc.), and many higher brain centers, particularly the hypothalamus. Therefore, it is not surprising that at every moment arterial pressure reflects the resultant response to all these inputs. Sudden anger increases the pressure; fright may actually cause hypotension (low blood pressure) severe enough to cause fainting, but this complexity should not obscure the important generalization that the primary regulation of

arterial pressure is exerted by the baroreceptors, particularly those in the carotid sinus and aortic arch. Other inputs may alter the pressure somewhat from minute to minute, but the mean arterial pressure in a normal person is maintained by the baroreceptors within quite narrow limits.

SECTION D:
CARDIOVASCULAR PATTERNS IN HEALTH AND DISEASE

In order to demonstrate how the cardiovascular components we have been discussing are integrated, we now examine the responses of the entire systemic circulation to a variety of normal and diseased states. Most of the necessary facts and concepts are already familiar to the reader.

Hemorrhage and hypotension

The decrease in blood volume caused by bleeding produces a drop in blood pressure (*hypotension*) by the sequence of events previously shown in Fig. 9-41. The most

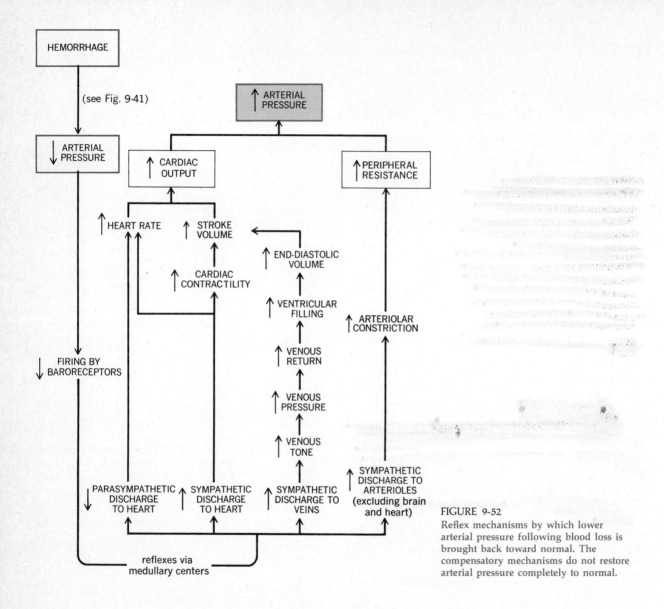

FIGURE 9-52
Reflex mechanisms by which lower arterial pressure following blood loss is brought back toward normal. The compensatory mechanisms do not restore arterial pressure completely to normal.

serious consequences of the lowered blood pressure are the reduced blood flow to the brain and cardiac muscle. Compensatory mechanisms restoring arterial pressure toward normal are summarized in Fig. 9-52; their effects can best be appreciated from the data of Table 9-3. Kidney flow is even lower 5 min after the hemorrhage, despite the improved arterial pressure, but we recall that one of the important compensatory mechanisms is increased arteriolar constriction in many organs; thus, kid-

ney blood flow is reduced in order to maintain arterial blood pressure and thereby brain and heart blood flow.

A second important compensatory mechanism involving capillary fluid exchange results from both the decrease in blood pressure and the increase in arteriolar constriction (Fig. 9-53). Thus, the initial event—blood loss and decreased blood volume—is in large part compensated for by the movement of interstitial fluid into the vascular system. Indeed, as shown in Table 9-4, several

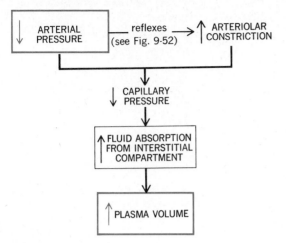

FIGURE 9-53
Mechanisms compensating for blood loss by movement of interstitial fluid into the capillaries. This response is diagramed in Fig. 9-38C.

TABLE 9-3
Cardiovascular effects of hemorrhage

	Pre-hemorrhage	Posthemorrhage Immediate	5 min
Arterial pressure, mm Hg	125/75	80/55	115/75
Left atrial pressure, mm Hg	4	2	2.5
End-diastolic volume, ml	150	75	90
Stroke volume, ml	75	40	53
Heart rate, beats/min	70	70	91
Cardiac output, ml/min	5,250	2,800	4,775
Kidney blood flow, ml/min	1,300	1,000	850
Brain blood flow, ml/min	1,300	1,000	1,275

hours after a moderate hemorrhage, the blood volume may be virtually restored to normal.

The entire compensation is due to expansion of the plasma volume; replacement of the lost erythrocytes requires days. Note that much of the albumin lost in the hemorrhage has already been replaced by synthesis of new protein (by the liver). This phenomenon is of great importance for expansion of plasma volume, as can be seen by considering the capillary filtration-absorption equilibrium (Fig. 9-38). As capillary hydrostatic pressure decreases as a result of the hemorrhage, interstitial fluid enters the plasma; this fluid, however, contains virtually no protein, so that its entrance dilutes the plasma proteins and increases the plasma water concentration. The resulting reduction of the water-concentration difference between the capillaries and the interstitial fluid hinders further fluid reabsorption and would prevent the full compensatory expansion if it were not that the rapid synthesis of new plasma protein minimizes this fall in protein concentration and movement of interstitial fluid into the plasma can continue. It is not known what stimulates the liver to synthesize new protein. We must emphasize that this capillary mechanism has only *redistributed* the extracellular fluid; ultimate replacement of the plasma lost from the body involves the control of fluid ingestion and kidney function, both described in Chap. 11.

The compensatory mechanisms just described are highly efficient; losses of as much as 1 to 1.5 liters of

blood (approximately 20 percent of total blood volume) can be sustained with only slight reduction of mean arterial pressure. When greater losses occur, severe hypotension (*shock*) may be precipitated. If prolonged for several hours, shock becomes irreversible; the person dies even after blood transfusions and other appropriate therapy. Irreversible shock is manifested when the compensatory mechanisms are overridden; thus, a person who has been in mild shock for several hours may suddenly

TABLE 9-4
Fluid shifts after hemorrhage

	Normal	Immediately after hemorrhage	3 hr after hemorrhage
Total blood volume, ml	5,000	4,000 (↓20%)	4,900
Erythrocyte volume, ml	2,300	1,840 (↓20%)	1,840
Plasma volume, ml	2,700	2,160 (↓20%)	3,060
Plasma albumin mass, g	135	108 (↓20%)	125

show a marked worsening of the hypotension. Apparently, the major event is a decrease in cardiac contractility; the mechanism of deterioration leading to irreversible shock is unknown.

Loss from the body of large quantities of extracellular fluid (rather than whole blood) can also cause hypotension. This may occur via the skin, as in severe sweating or burns, via the gastrointestinal tract, as in diarrhea or vomiting, or via unusually large urinary losses. Regardless of the route, the loss decreases circulating blood volume and produces symptoms and compensatory phenomena similar to those seen in hemorrhage.

Hypotension may be caused by events other than blood or fluid loss. Fainting in response to strong emotion is a common form of hypotension. Somehow the higher brain centers involved with emotions act upon the medullary cardiovascular centers to inhibit sympathetic activity and enhance parasympathetic activity (Fig. 9-54) resulting in decreased arterial pressure and brain blood flow. Fortunately, this whole process is usually transient, with no after effects, although a weak heart may suffer damage during the period of reduced blood flow to the cardiac muscle.

Other important causes of hypotension seem to have a common denominator in the liberation within the body of chemicals which relax arteriolar smooth muscle. There the cause of hypotension is clearly excessive arteriolar dilation and reduction of peripheral resistance, an important example being the hypotension which occurs during severe allergic responses.

It may be of interest to point out the physiological reasons for not treating a patient in shock in ways commonly favored by the uninformed, namely, administering alcohol and covering the person with mounds of blankets. Both alcohol and excessive body heat, by actions on the central nervous system, cause profound dilation of skin arterioles, thus lowering peripheral resistance and decreasing arterial blood pressure still further. As shown below, the worst possible thing is to try to get the person to stand up.

The upright posture

The simple act of getting out of bed and standing up is equivalent to a mild hemorrhage, because the changes in the circulatory system in going from a lying, horizontal position to a standing, vertical position result in a decrease in the effective circulating blood volume. The decrease results from the action of gravity upon the long continuous columns of blood in the vessels between the heart and the feet. To understand these changes, one must understand the relationships between weight, gravity, and pressure.

The weight of a substance is the force exerted upon it by gravity; thus weight and force are equivalent terms. Pressure, by definition, is the amount of force acting on a given surface area; pressure = force/area. An object at rest exerts a pressure on the surface beneath as a result of the gravitational force (weight) it exerts on the surface. The greater the weight of the object, the greater is the pressure. Figure 9-55 compares the pressures exerted by 1 cm^3 of blood and 1 cm^3 of mercury. Because of its greater density (mass/volume) 1 cm^3 of mercury weights 13 times as much as the same amount of blood. A column of blood 2 cm high exerts twice the pressure of a column 1 cm high because it weighs twice as much. When a person is standing, the blood in the arteries and veins from the heart to the feet makes up a column approximately 103 cm high, the weight of which exerts pressure within the blood vessels of the legs and feet equivalent to 80 mm Hg (Figs. 9-55 and 9-56).

These facts can now be applied to an analysis of the changes in intravascular pressure when a person goes from the horizontal to the vertical position (Fig. 9-56). All the pressures we have given in previous sections of this chapter were for the horizontal position (Fig. 9-56), in which all the blood vessels are at approximately the same level as the heart, and the weight of the blood produces negligible pressure. In the vertical position (Fig. 9-56C), the intravascular pressure everywhere becomes equal to the usual pressure resulting from cardiac contraction plus an additional pressure equal to the weight of a column of blood from the heart to the point of measurement. In a foot capillary, for example, the pressure increases from 25 to 105 mm Hg (80 + 25).

Earlier we discussed the relationship between pressure, flow, and resistance for the flow of fluids through a tube. Here it must be emphasized that the gravitational hydrostatic pressure produced in the legs upon standing does not enter *directly* into the relationship between blood pressure and the flow of blood through the vessels of the body. A glass U tube 103 cm high is filled with blood (Fig. 9-56B); gravity has the same effect upon these columns of blood as it does on those in the body. Note that the pressure in the bottom of the tube resulting from the weight of the blood is 80 mm Hg but there is no flow of blood because there is no pump to produce a pressure *difference* between the ends of the

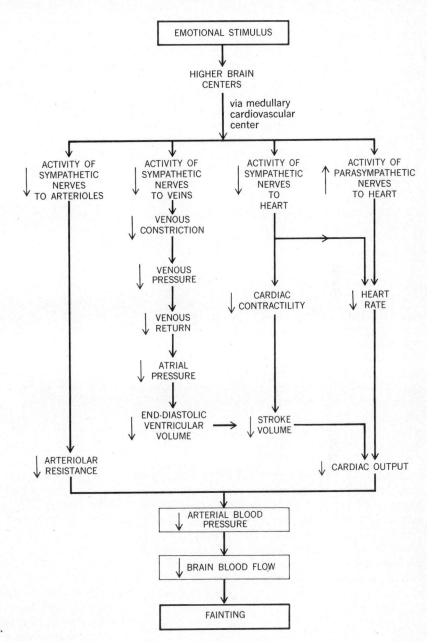

FIGURE 9-54
**Mechanisms inducing fainting in
response to a strong emotional stimulus.**

U tube. In effect, the gravitational pressures in the two arms of the U tube cancel each other out, leaving no net pressure to cause flow. The flow through any curved tube depends upon the difference in pressure between the two ends of the tube and not upon the orientation of the tube

in space. In the body the heart produces the pressure difference between aorta and right atrium which leads to a blood flow, and changing the position of the blood vessels when changing from a lying to a standing position does not directly alter the relationship between pressure

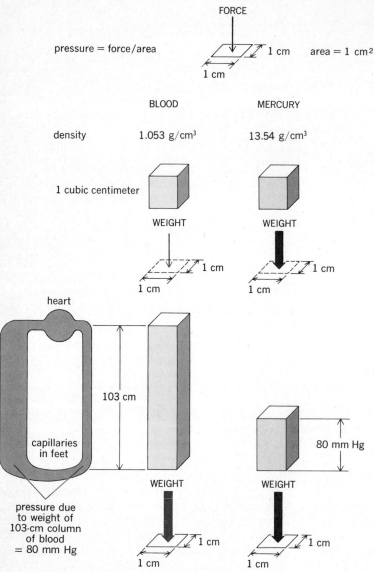

pressure = force/area

area = 1 cm²

pressure = 103 cm of blood = 80 mm Hg

FIGURE 9-55
**Pressure resulting from the weight of a
column of fluid.**

and flow in this system. In other words, gravity does not make it harder for the blood to return from the legs to the heart simply because the blood must travel "uphill." If the blood vessels of the body were composed of impermeable iron pipes, there would be no changes in blood flow when changing from a lying to a standing position.

Yet, the fact is that, upon standing, venous return (blood flow back to the right atrium of the heart) does decrease. This statement appears to contradict what we have just said, but the decrease is due to a decrease in the effective blood volume and not to any direct effects of pressure on flow. Effective blood volume upon standing is reduced by two factors (Fig. 9-56C), both resulting from

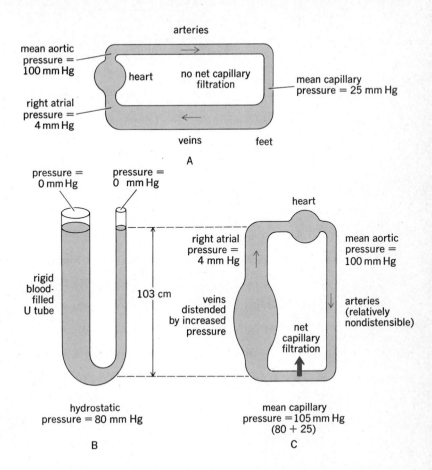

FIGURE 9-56
Effects of standing upon blood pressure, blood flow, and capillary filtration.

the increase in gravitational blood pressure in the legs and feet and the distensibility of blood vessels, namely, increased distension of the veins and increased filtration of fluid from the capillaries into the interstitial space of the legs.

The veins are highly distensible structures, as we have seen. The increased hydrostatic pressure in the veins of the legs which occurs upon standing pushes outward upon the vein walls, causing marked distension with resultant pooling of blood; i.e., much of the blood emerging from the capillaries simply remains in the expanding veins rather than returning to the heart. Simultaneously, the marked increase in capillary pressure caused by the gravitational force produces increased filtration of fluid out of the capillaries into the interstitial space. Most of us have experienced swollen feet after a day's standing. The combined effects of venous pooling and increased capillary filtration are a significant reduc-

tion in the effective circulating blood volume in a manner very similar to a mild hemorrhage. The ensuing decrease in arterial pressure causes reflex compensatory adjustments similar to those shown in Fig. 9-52 for hemorrhage.

Perhaps the most effectve compensation is the contraction of skeletal muscles of the leg, which compresses the veins, thereby diminishing the degree of venous distension and pooling, and causes a marked reduction in capillary hydrostatic pressure, thereby reducing the rate of fluid filtration out of the capillaries (Fig. 9-57). Muscular contraction produces intermittent, complete emptying of veins within the upper leg, so that uninterrupted columns of venous blood from the heart to the feet no longer exist. Thus the skeletal muscle pump reduces both venous pooling and capillary filtration. A common example of the importance of this compensation is when soldiers faint after standing very still, i.e., with minimal contraction of the abdominal and leg muscles,

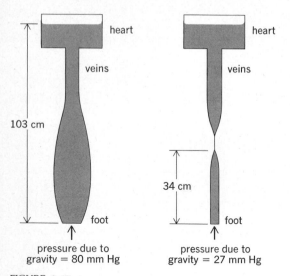

pressure due to gravity = 27 mm Hg

FIGURE 9-57
Role of contraction of the leg skeletal muscles in reducing capillary pressure and filtration in the upright position. The skeletal muscle contraction compresses the veins, causing intermittent complete emptying so that the column of blood is interrupted.

for long periods of time. Here the fainting may be considered a useful compensatory mechanism in that venous and capillary pressure changes induced by gravity are eliminated once the person is prone, the pooled venous blood is mobilized, and the previously filtered fluid is reabsorbed into the capillaries. Thus, the wrong thing to do to anyone in a faint is to hold him upright.

Thus far, we have described only the effects of normal gravitational forces, but modern airplanes and space vehicles have created the further problem of unusually large gravitational forces. During any form of marked change in acceleration—a flyer pulling out of a dive or an astronaut blasting off—the body is subjected to large gravitational forces which cause venous pooling and increased capillary filtration. If the forces are great enough, cardiac output may be so compromised as to produce fainting or blackout. At the other extreme, the phenomenon of weightlessness is encountered in space travel. The major problem encountered so far by astronauts after prolonged periods of weightlessness is a tendency toward hypotension upon returning to earth and standing to leave the spaceship. This is not yet clearly understood but seems to reflect a suppression of the usual compensatory reflexes, as though these reflexes,

used so frequently each day beginning with first getting out of bed in the morning, are temporarily lost during prolonged disuse. This is probably analogous to the dizziness or actual fainting commonly encountered when a long-bedridden patient first tries to arise.

Exercise

In order to maintain muscle activity during exercise, a large increase in blood flow is required to provide the oxygen and nutrients consumed and to carry away the carbon dioxide and heat produced. Thus, cardiac output may increase from a resting value of 5 liters/min to the maximal values of 35 liters/min obtained by trained athletes. The increased skeletal muscle blood flow results from marked dilation of the skeletal muscle arterioles mediated by the sympathetic vasodilator fibers (stimulated by descending pathways from the hypothalamus) and by local factors associated with active hyperemia. In a person just about to begin exercising, the skeletal muscle flow actually increases before the onset of muscular activity and therefore of active hyperemia. This anticipatory response providing a rapid initial supply of blood to the muscle can, in large part, be blocked by cutting the sympathetic nerves or by administering drugs which inhibit the actions of acetylcholine. It must be stressed, however, that once exercise has begun, these sympathetic vasodilator nerves are of little importance and active hyperemia plays the primary role in producing vasodilation.

The cardiovascular response to exercise has already been shown in our series of models (Figs. 9-44 to 9-47). The decrease in peripheral resistance resulting from dilation of skeletal muscle arterioles is partially offset by constriction of arterioles in other organs, particularly the gastrointestinal tract and kidneys. However, the "resistance juggling" is quite incapable of compensating for the huge dilation of the muscle arterioles, and the net result is a marked decrease in total peripheral resistance.

The cardiac output increase during exercise is associated with greater sympathetic activity and less parasympathetic activity to the heart. Thus, heart rate and stroke volume both rise, causing an increased cardiac output. The heart-rate changes are usually much greater than stroke-volume changes. Note (Fig. 9-58) that, in our example, the increased stroke volume occurs without change in end-diastolic ventricular volume; accordingly, the former is ascribable completely to the increased contractility induced by the cardiac sympathetic nerves. The

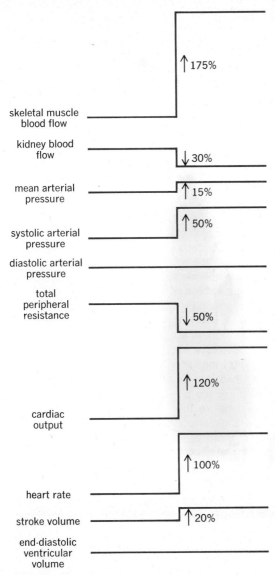

FIGURE 9-58
Summary of cardiovascular changes during mild exercise.

sion that enhanced sympathetic activity to the heart completely accounts for the elevated cardiac output which occurs in exercise, for such is not the case. The fact is that cardiac output could not be increased to high levels unless the venous return to the heart were not simultaneously facilitated to the same degree, for otherwise end-diastolic volume would fall and stroke volume would decrease (because of Starling's law). Therefore, factors promoting venous return during exercise are extremely important. They are (*1*) the marked activity of the skeletal muscle pump, (*2*) the increased "respiratory-pump" activity resulting from increased respiratory movements, (*3*) sympathetically mediated increase in venous tone, and (*4*) the ease with which blood flows from arteries to veins through the dilated skeletal muscle arterioles. These factors may be so powerful that venous return is enhanced enough to cause an increase in end-diastolic ventricular volume; under such conditions, stroke volume (and, thereby, cardiac output) is further enhanced.

Such an effect has been strikingly demonstrated in experiments on greyhounds whose hearts were completely denervated, yet who were able, after recovery, to run a race almost as fast as before. How is this possible? Apparently, the heart deprived of its nerves follows the relationship expressed in Starling's law more closely. As we have seen, various factors facilitate venous return during exercise; normally, increased venous return does not cause an increased end-diastolic pressure because the sympathetically driven heart pumps the blood as fast as it arrives. After cardiac denervation, however, the sympathetically mediated heart-rate and contractility increases do not occur as exercise begins and the increased venous return causes an immediate large increase in end-diastolic pressure. This, by Starling's law, causes a marked increase in stroke volume and cardiac output. Thus, the heart now uses primarily a stroke-volume change rather than a heart-rate increase to raise cardiac output. The net overall result is, however, the same. This problem is far from academic since we can now transplant the heart from a recently deceased person to a person dying of heart disease. Such a heart, of course, lacks nerves, and the entire question of cardiac control in response to exercise, posture, etc., is of the utmost importance.

What happens to arterial blood pressure during exercise? As always, the mean arterial pressure depends only upon the cardiac output and peripheral resistance. During most forms of exercise, the cardiac output tends to increase somewhat more than the peripheral resistance decreases so that mean arterial pressure usually

stability of end-diastolic volume in the face of reduced time for ventricular filling (increased heart rate) is, in part, attributable to the fact, described earlier, that the sympathetic nerves increase the speed of contraction and relaxation as well as inducing a "suctionlike" effect which facilitates filling.

However, it would be incorrect to leave the impres-

increases slightly. However, the pulse pressure may show a marked increase because of greater systolic pressure and a relatively constant diastolic pressure. The former is due primarily to the faster ejection. These changes are all shown schematically in Fig. 9-58.

It is evident from this description that the sympathetic nerves (and inhibition of the parasympathetics) play an important role in the cardiovascular response to exercise. However, a problem arises when we try to understand the mechanisms which control the autonomic nervous system during exercise. The pulsatile (and mean) arterial pressure tends to be elevated, which should cause the arterial baroreceptors to signal the medullary centers to decrease cardiac output and dilate arterioles of the abdominal organs. Obviously, then, the arterial baroreceptors not only cannot be the origin of the cardiovascular changes in exercise but actually oppose these changes. Similarly, other possible inputs, such as oxygen and carbon dioxide, can be eliminated since they show little, if any, change (Chap. 10). We shall meet this same problem again when we describe (Chap. 10) increased respiration during exercise. A major clue is the finding that electric stimulation of certain hypothalamic areas in resting unanesthetized dogs produces all the cardiovascular changes usually observed during exercise. On the basis of these and other experiments, the best present working hypothesis is that a center in the hypothalamus acts, via descending pathways, upon the medullary centers (and directly upon the sympathetic vasodilators to skeletal muscle arterioles) to produce the changes in autonomic function so characteristic of exercise. What is the input to the hypothalamus? We do not know for certain, but it is probably information coming from the same motor areas of the cerebral cortex which are responsible for the skeletal muscle contraction. It is likely that as these fibers descend from cerebral cortex to the spinal-cord motor neurons, they give off branches to the relevant hypothalamic centers. This system would nicely coordinate the skeletal muscle contraction with the blood supply needed to support it.

Hypertension

Hypertension (high blood pressure) is defined as a chronically increased arterial pressure. In general, the dividing line between normal pressure and hypertension is taken to be 140/90 mm Hg, although systolic pressures above 140 are frequently not associated with ill effects. However, the diastolic pressure is really the most important index of hypertension. Hypertension is one of the major causes of illness and death in America today. It is estimated that approximately 6 million people suffer from this disease, the lethal end result of which may be heart failure (see below), brain stroke (occlusion or rupture of a cerebral blood vessel), or kidney damage, all caused by prolonged hypertension and its attendant strain on the various organs.

Theoretically, hypertension could result from an increase in cardiac output or peripheral resistance or both. In fact, at least in well-established hypertension, the major abnormality is increased peripheral resistance due to abnormally reduced arteriolar diameter. What causes arteriolar narrowing? In most cases, we do not know. This has led to the strange label of "essential hypertension," meaning hypertension of unknown cause. In a small fraction of cases, the cause of hypertension is known: (*1*) Certain tumors of the adrenal medulla secrete excessive amounts of epinephrine; (*2*) certain tumors of the adrenal cortex secrete excessive amounts of hormones which lead to hypertension by as yet unknown mechanisms; (*3*) many diseases which damage or decrease the blood supply to the kidneys are associated with hypertension, but despite intensive efforts by many investigators, the actual factor(s) mediating *renal hypertension* remains unknown. In Chap. 11, we shall describe the renin-angiotensin system as the prime controller of adosterone secretion; long before this effect was recognized, it was known that angiotensin is a powerful constrictor of arterioles, and most researchers in the field believed it to be the basis for renal hypertension. Now, the story seems much less certain; regardless of mechanism, however, there is no question that the kidneys frequently play a critical role in hypertension.

Besides the three known causes of hypertension described above and several others, we are left with the vast majority of patients in the essential, or unknown, category. Many hypotheses have been proposed for increased arteriolar constriction but none proved. At present, much evidence seems to point to excessive sodium ingestion or retention within the body as a common denominator of renal, adrenal, and even essential hypertension, but how the sodium is involved in the increased arteriolar constriction remains unknown. In any case, sodium restriction has become a major form of therapy in hypertension. Virtually all other forms of therapy involve drugs which act upon some aspect of autonomic function to produce arteriolar dilation. This does not mean that excessive sympathetic tone was the original cause of the hypertension but only that, whatever the cause, anything which dilates the arterioles reduces the blood pressure.

The perceptive reader might also wonder why the arterial baroreceptors do not, by way of the reflexes they initiate, return the blood pressure to normal. The reason seems to be that, in chronic hypertension, the baroreceptors are "reset" at a higher level; i.e., they regulate blood pressure but at a greater pressure. We have no explanation for this phenomenon.

Congestive heart failure

The heart may become weakened for many reasons; regardless of cause, however, the failing heart induces a similar procession of signs and symptoms grouped under the category of *congestive heart failure*. Patients with early mild heart disease may show at rest no significant abnormalities because of the great safety factor, or reserve, in cardiac function. However, the ability to perform exercise is impaired, as evidenced by shortness of breath and early fatigue. Ultimately, the cardiac reserve becomes inadequate to supply normal amounts of blood even at rest, and the patient becomes bedridden. Finally, the cardiac output may become too low to support life.

The basic defect in heart failure is a decreased contractility of the heart, but the molecular mechanism is unknown. As shown in Fig. 9-59, the failing heart shifts downward to a lower Starling curve. How can this be compensated for? Increased sympathetic stimulation would help to increase contractility, and this does occur. However, an even more striking compensation is increased ventricular end-diastolic volume. The failing heart is generally engorged with blood, as are the veins and capillaries, the major cause being an increase (sometimes massive) in plasma volume. The sequence of events is still not completely understood; decreased cardiac output in some manner (perhaps by reducing the blood flow to the kidneys) leads to failure of normal sodium and water excretion by the kidneys. The retained fluid then causes expansion of the extracellular volume, increasing venous pressure, venous return, and end-diastolic ventricular volume and thus tending to restore stroke volume toward normal.

Another result of elevated venous and capillary pressure is increased filtration out of the capillaries, with resulting edema. This accumulation of tissue fluid may be the chief feature of ventricular failure; the legs and feet are usually most prominently involved (because of the additional effects of gravity), but the same engorgement is occurring in other organs and may cause severe

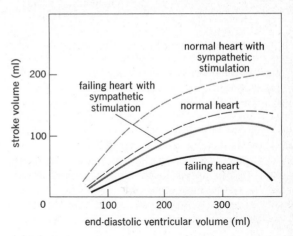

FIGURE 9-59
Relationship between end-diastolic ventricular volume and stroke volume in normal and failing hearts. The normal curves are those shown previously in Figs. 9-20 and 9-21. The failing heart can still eject an adequate stroke volume if the sympathetic activity to it is increased or if the end-diastolic volume increases, i.e., if the ventricle becomes more distended.

malfunction. The most serious result occurs when the left ventricle fails; in this case, the excess fluid accumulates in the lung air sacs (*pulmonary edema*) because of increased pulmonary capillary pressure, and the patient may actually drown in his own fluid. This situation usually worsens at night; during the day, because of the patient's upright posture, fluid accumulates in the legs, but it is slowly absorbed when he lies down at night, the plasma volume expands, and an attack of pulmonary edema is precipitated.

Thus, what began as a useful compensation becomes potentially lethal because the tension-length relationship for muscle holds only up to a point, beyond which further stretching of the muscle may actually cause decreased strength of contraction. Thus, expansion of plasma volume may so increase end-diastolic volume as to decrease contractility (Fig. 9-59) and produce a rapidly progressing downhill course.

The treatment for congestive heart failure is easily understood in these terms: The precipitating cause should be corrected if possible; contractility can be increased by a drug known as digitalis; excess fluid should be eliminated by the use of drugs which increase excretion of sodium and water by the kidneys; the patient should be kept at rest so as to reduce the cardiac output required to fulfill the body's metabolic needs.

"Heart attacks" and arteriosclerosis

We have seen that the myocardium does not extract oxygen and nutrients from the blood within the atria and ventricles but depends upon its own blood supply via the coronary vessels. The coronary arteries exit from the aorta just above the aortic valves and lead to a branching network of small arteries, arterioles, capillaries, venules, and veins similar to those in all other organs. The rate of blood flow depends primarily upon the arterial blood pressure and the resistance offered by the coronary vessels. The degree of arteriolar constriction, or dilation, is almost entirely determined by local metabolic control mechanisms, there being little if any neural control. This is just what one would expect in an organ with varying metabolic requirements which must be met at all times to ensure survival of the entire organism. Insufficient coronary blood flow leads to myocardial damage and, if severe enough, to death of the myocardium (*infarction*), a so-called *heart attack*. This may occur as a result of decreased arterial pressure but is more commonly due to increased vessel resistance following coronary arteriosclerosis.

Arteriosclerosis ("hardening of the arteries") is a disease characterized by a thickening of the arterial wall with connective tissue and deposits of cholesterol. The mechanism by which thickening occurs is not clear, but it is known that smoking, obesity, high-fat diets, nervous tension, and a variety of other factors markedly predispose one to this disease of aging. The suspected relationship between arteriosclerosis and blood concentrations of cholesterol and saturated fatty acids has probably received the most widespread attention, and many studies are presently trying to evaluate the likely hypothesis that high blood concentrations of these lipids increase the rate and the severity of the arteriosclerotic process. Cholesterol is an important physiological substance because it is the precursor of certain hormones and the bile acids (Chap. 12). Since it is found only in animals, ingestion of animal fats (including egg yolks) constitutes the major dietary source. However, the liver (and other body cells) is capable of producing large quantities of cholesterol, particularly from saturated fatty acids, so that even profound reductions of dietary intake frequently do not lower blood cholesterol concentration significantly because the liver responds by producing more. Indeed, it may well be their high content of saturated fatty acids, rather than cholesterol, which causes the ingestion of animal fat to predispose one to arteriosclerosis. Vegetable fat, in contrast, contains primarily unsaturated fatty acids and may actually lower blood cholesterol.

The incidence of coronary arteriosclerosis in the United States is extraordinarily great; it is estimated to cause 500,000 deaths per year. The mechanism by which arteriosclerosis reduces coronary blood flow is quite simple; the fat deposits and fibrous thickening narrow the vessels and increase resistance to flow. This is usually progressive, leading often ultimately to complete occlusion. Acute coronary occlusion may occur because of sudden formation of a clot on the roughened vessel surface or breaking off of a deposit, which then lodges downstream, completely blocking a smaller vessel. If, on the other hand, the arteriosclerotic process causes only gradual occlusion, the heart may remain uninjured because of the development, over time, of new accessory vessels supplying the same area of myocardium. It should be stressed that before complete occlusion many patients experience recurrent transient episodes of inadequate coronary blood flow, usually during exertion or emotional tension. The pain associated with this is termed *angina pectoris.*

The cause of death from coronary occlusion and myocardial infarction may be either severe hypotension resulting from weakened contractility or disordered cardiac rhythm resulting from damage to the cardiac conducting system, but in addition the severe hypotension which may be associated with a heart attack is frequently due to reflex inhibition of the sympathetic nervous system and enhancement of the parasympathetics. The origin of these totally inappropriate and frequently lethal reflexes is not known. Finally, should the patient survive an acute coronary occlusion, the heart may be left permanently weakened, and a slowly progressing heart failure may ensue. On the other hand, many people lead quite active and normal lives for many years after a heart attack.

We do not wish to leave the impression that arteriosclerosis attacks only the coronary vessels, for such is not the case. Most arteries of the body are subject to this same occluding process. For example, cerebral occlusions (*strokes*) are extremely common in the aged and constitute an important cause of sickness and death (200,000 per year). Wherever the arteriosclerosis becomes severe, the resulting symptoms always reflect the decrease in blood flow to the specific area. In recent years, synthetic materials have been developed from which tubes can be made and surgically substituted for a diseased segment of artery.

Organization of the respiratory system

Inventory of steps involved in respiration

Exchange of air between atmosphere and alveoli: ventilation

Exchange and transport of gases in the body

Control of respiration

Hypoxia

Erythrocyte and hemoglobin balance

Respiration

Most cells in the human body obtain the bulk of their energy from chemical reactions involving oxygen. In addition, cells must be able to eliminate the major end product of these oxidations, carbon dioxide. A unicellular organism can exchange oxygen and carbon dioxide directly with the external environment, but this is obviously impossible for most cells of a complex organism like the human body, since only a small fraction of the total cells (skin, gastrointestinal lining, respiratory lining) is in direct contact with the external environment. In order to survive, large animals have had to develop specialized systems for the supply of oxygen and elimination of carbon dioxide. These systems are not the same in all complex animals since evolution often follows several pathways simultaneously. The organs of gas exchange with the external environment in fish are gills; those in man are *lungs*. Specialized blood components have also evolved which permit the transportation of large quantities of oxygen and carbon dioxide between the lungs and cells.

In a man at rest, the body's cells consume approximately 200 ml of oxygen per minute. Under conditions of severe oxygen requirement, e.g., exercise, the rate of

10

oxygen consumption may increase as much as thirtyfold. Equivalent amounts of carbon dioxide are simultaneously eliminated. It is obvious, therefore, that mechanisms must exist which coordinate breathing with metabolic demands. We shall see in Chap. 11 that the control of breathing also plays an important role in the regulation of the acidity of the extracellular fluid.

Before describing the basic processes of oxygen supply, carbon dioxide elimination, and breathing control, we must first define the terms respiration and *respiratory system*. Respiration has two quite different meanings: ❶ the metabolic reaction of oxygen with carbohydrate and other organic molecules and ❷ the exchange of gas between the cells of an organism and the external environment. The various steps of the second process form the subject matter of this chapter; the first process was described in Chap. 3. The term respiratory system refers only to those structures which are involved in the exchange of gases between the blood and external environment; it does not include the transportation of gases in the blood or gas exchange between blood and the tissues. Admittedly, this definition is arbitrary since it includes only half of the processes involved in respiration, but it has become firmly established by long usage. The respiratory system comprises the lungs, the series of passageways leading to the lungs, and the chest structures responsible for movement of air in and out of the lungs.

Organization of the respiratory system

In order for air to reach the lungs, it must first pass through a series of air passages connecting the lungs to the nose and mouth (Fig. 10-1). There are two lungs, the right and left, each divided into several lobes. Together with the heart, great vessels, esophagus, and certain nerves, the lungs completely fill the *chest (thoracic) cavity*. The lungs are not simply hollow balloons but have a highly organized structure consisting of air-containing tubes, blood vessels, and elastic connective tissue. The air passages within the lungs (Fig. 10-2) are actually the continuation of those which connect the lungs to the nose and mouth. Together they are termed the *conducting portion* of the respiratory system and constitute a series of highly branched hollow tubes becoming smaller in diameter and more numerous at each branching, much like arteries and arterioles. The smallest of these tubes end in tiny blind sacs, the *alveoli*, which are the actual sites of gas ex-

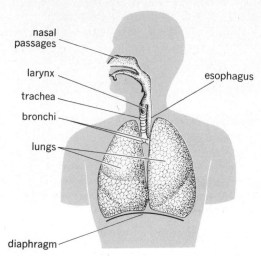

FIGURE 10-1
Organization of the respiratory system.

change within the lungs. All portions of these air passageways and alveoli receive a rich supply of blood by way of blood vessels, which constitute a large portion of the total lung substance (Fig. 10-2). Between the air tubes and blood vessels of the lungs are large quantities of *elastic connective tissue,* which play an important role in breathing. The lungs, however, lack muscle and are therefore passive elastic containers with no inherent ability to increase their volume. Lung expansion is accomplished instead by the action of the *diaphragm* (the muscle which separates the thoracic and abdominal cavities) and the muscles which move the ribs.

Conducting portion of the respiratory system

Air can enter the respiratory passages either by nose or mouth, although the nose is the normal route. It then passes into the *pharynx* (throat), a passage common to the routes followed by air and food. The pharynx branches into two tubes, one (the *esophagus*) through which food passes to the stomach and one through which air passes to the lungs. The first portion of the air passage, called the *larynx*, houses the *vocal cords*. It is protected against the entry of food by closure of the vocal cords across the tracheal opening. The larynx opens into a long tube (the *trachea*), which, in turn, branches into the two *bronchi,* one of which enters each lung. Within the lungs, these major bronchi branch many times into

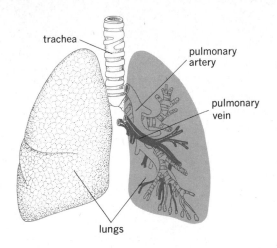

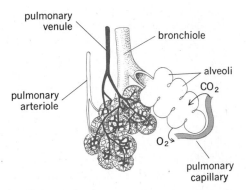

FIGURE 10-2
Relationships between respiratory airways and blood vessels.
(Adapted from McNaught and Callender.)

activity by noxious agents, including substances in cigarette smoke. This, coupled with the stimulation of mucus secretion induced by these same agents, may result in partial or complete airway obstruction by the stationary mucus. (A smoker's early-morning cough is the attempt to clear this obstructive mucus from the airways.) A second protective mechanism is provided by the phagocytic cells, which are present in the respiratory-tract lining in great numbers. These cells, which engulf dust, bacteria, and debris, are also injured by cigarette smoke and other air pollutants.

2 As air flows through the respiratory passages, it is warmed and moistened by contact with the epithelial lining.

3 The vocal cords, two strong bands of elastic tissue, lie stretched across the lumen of the larynx. The movement of air past them causes them to vibrate, initiating the many different sounds which constitute speech.

4 The walls of the respiratory passages contain smooth muscle richly innervated and sensitive to certain circulating hormones, e.g., epinephrine. Contraction or relaxation of this muscle, particularly in the bronchioles, alters resistance to air flow (just as arteriolar diameter is a major determinant of resistance to blood flow).

Site of gas exchange in the lungs: the alveoli

The alveoli are tiny cup-shaped hollow sacs whose open ends are continuous with the lumens of the alveolar ducts (Fig. 10-2). The walls of the alveoli consist of a loose mesh of elastic tissue fibers lined with a thin layer of epithelium. Running within these walls are numerous capillaries, the endothelial lining of which lies snugly against the epithelium lining the alveoli (Fig. 10-4) sepa-

progressively smaller bronchi and, finally, into the smallest tubes of the conducting system, the alveolar ducts from which air passes into the alveoli.

This conducting system of tubes varies in structure and serves several important functions:

1 The epithelial linings of the respiratory tract contain hairlike projections, called *cilia,* which constantly beat toward the pharynx (Fig. 10-3). In addition, the epithelial glands secrete a thick substance (*mucus*), which lines the respiratory passages as far down as the bronchioles. Any particulate matter such as dust contained in the inspired air sticks to the mucus, which is constantly moved by the cilia to the pharynx, and then is swallowed and eliminated in the feces. Besides keeping the lungs clean, this mechanism is important in the body's total defenses against bacterial infection, since many bacteria enter the body on dust particles. A major cause of lung infection is probably paralysis of ciliary

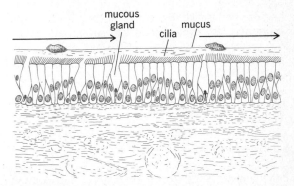

FIGURE 10-3
Epithelial lining of the respiratory tract. The arrows indicate the upward direction in which the cilia move the overriding layer of mucus, to which foreign particles are stuck.

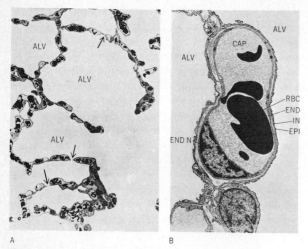

A   B

FIGURE 10-4

A Low-power electron micrograph of dog lung alveoli (Alv).
Note the capillaries in the walls between alveoli (the dark
disklike objects are erythrocytes). The arrows denote pores in
the alveolar walls. **B** Higher magnification of a portion of an
alveolar wall, showing a single capillary (cap) surrounded by
an alveolar epithelial cell. The nucleus (END. N.) and
cytoplasm (END) of a capillary endothelial cell are visible, as
are the nucleus (EPI. N.) and cytoplasm of an alveolar
epithelial cell. Note that the blood in the capillary is separated
from air in the alveoli (Alv.) only by the thin membrane
consisting of endothelium, interstitial fluid (In.), and
epithelium. [*Courtesy of E. R. Weibel. From Physiol. Rev.,* **53**:424
(1973).]

rated from it only by a thin layer of interstitium. Thus the
blood within a capillary is separated from the air within
an alveolus only by an extremely thin barrier (0.2 μm
compared with 7μm, which is the diameter of an average
red blood cell). The total area of alveoli in contact with
capillaries is 70 m² in man (the size of a badminton court).
This immense area combined with the thin barrier permits
the rapid exchange of large quantities of oxygen and
carbon dioxide. In addition to its predominant, extremely
thin cells, the alveolar epithelium also contains phagocytic cells and specialized cells which produce the critical
substance, pulmonary surfactant, to be discussed below.
Finally, there are pores in the alveolar membranes which
permit some flow of air between alveoli. This "collateral
ventilation" can be very important when the duct leading
to an alveolus is occluded by disease, since some air can
still enter this alveolus by way of pores between it and
adjacent alveoli.

Relation of the lungs to the thoracic cage

To understand how we breathe we must know something
about the tissues which make up the chest wall. The
thoracic cage is a closed compartment. It is bounded at
the neck by muscles and connective tissue, and it is
completely separated from the abdomen by a large
dome-shaped sheet of skeletal muscle, the diaphragm.
The outer walls of the thoracic cage are formed by the
breastbone (*sternum*), 12 pairs of *ribs,* and the muscles
which lie between the ribs (the *intercostal muscles*). These
walls also contain large amounts of elastic connective
tissue.

Firmly attached to the entire interior of the thoracic
cage is a delicate sheet of cells, the *pleura,* forming two
completely enclosed sacs within the thoracic cage, one
on each side of the midline. The relationship between the
lungs and pleura can be visualized by imagining what
happens when one punches a fluid-filled balloon (Fig.
10-5): The arm represents the major bronchus leading

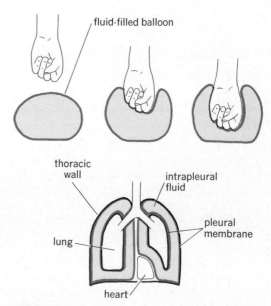

FIGURE 10-5

**Relationship of lungs, pleura, and thoracic cage, analogous to
pushing one's fist into a fluid-filled balloon. Note that there is
no communication between the right and left intrapleural
fluids. The volume of intrapleural fluid is greatly exaggerated;
normally it consists of an extremely thin layer of fluid between
the pleural membrane lining the inner surface of the thoracic
cage and the pleural membrane lining the surface of the lungs.**

to the lung, the fist is the lung, and the balloon is the pleural sac. The outer portion of the fist becomes coated by one surface of the balloon. In addition, the balloon is pushed back upon itself so that its surfaces lie close together. This is precisely the relation between the lung and pleura except that the pleural surface coating the lung is firmly attached to the lung surface. This layer of pleura and the outer layer which lines the interior thoracic wall are so close to each other that they are virtually in contact, being separated only by a very thin layer of *intrapleural fluid*. In a normal person, these two surfaces always maintain this intimate relationship, which is of great importance for breathing.

Inventory of steps involved in respiration

Before discussing in detail the mechanisms by which oxygen and carbon dioxide are exchanged between the body's cells and the external environment, we summarize the steps involved:

1 **Exchange of air between the atmosphere (external environment) and alveoli.** This process includes the movement of air in and out of the lungs and the distribution of air within the lungs. Not only must a large volume of new air be delivered constantly to the alveoli but it must be distributed proportionately to the millions of alveoli within each lung. This entire process is called *ventilation* and occurs by bulk flow.

2 **Exchange of oxygen and carbon dioxide between alveolar air and lung capillaries by diffusion.** The volume and distribution of the pulmonary (lung) blood flow are extremely important for normal functioning of this process.

3 **Transportation of oxygen and carbon dioxide by the blood.** This process forms a second important link between the cardiovascular and respiratory systems (the first being pulmonary blood flow).

4 **Exchange of oxygen and carbon dioxide between the blood and tissues of the body by diffusion as blood flows through tissue capillaries.**

Exchange of air between atmosphere and alveoli: ventilation

Like blood, air moves by bulk flow from a high pressure to a low pressure. We have seen (Chaps. 2 and 9) that bulk flow can be described by the equation

$$F = k(P_1 - P_2)$$

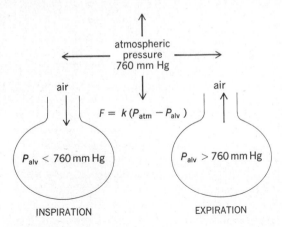

FIGURE 10-6
Relationships required for breathing. When the alveolar pressure P_{alv} is less than atmospheric pressure, air enters the lungs. Flow F is directly proportional to the pressure difference, k being the proportionality constant.

That is, flow is proportional to the pressure difference between two points, k being the proportionality constant. For air flow, the two relevant pressures are the *atmospheric pressure* and the *intraalveolar pressure*, gas flow into or out of the lungs thus being given by

$$F = k(P_{atm} - P_{alv})$$

At sea level the atmospheric pressure is 760 mm Hg and is obviously not subject to control, short of putting a man in a space suit or diving bell. Since atmospheric pressure remains relatively constant, if air is to be moved in and out of the lungs, the air pressure within the lungs, i.e., the intraalveolar pressure, must be made alternately less than and greater than atmospheric pressure (Fig. 10-6).

The concept of intrapleural pressure

If one cuts open the chest of an animal, being careful to cut only the thoracic wall but not the lung, the lung collapses immediately (Fig. 10-7). This is precisely what happens when a person is stabbed in the chest. Normally the highly elastic lungs are stretched within the intact chest, and the force responsible is eliminated when the chest is opened. This force is the subatmospheric pressure within the pleural fluid.

In a newborn child, the lungs and thoracic cage have approximately the same dimensions when unstretched. After birth, the thoracic cage grows more rapidly than the lungs; therefore, the thoracic wall tends to

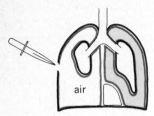

FIGURE 10-7
Lung collapse caused by a stab wound piercing the thoracic cage. Note that the air in the pleural space did *not* come from the lungs since the lung wall is still intact.

move away from the outer lung surface, but separation is prevented by the presence of the intrapleural fluid. An analogy (Fig. 10-8) may help illustrate the mechanism. Imagine two balloons of slightly different size, one inside the other. Since the inner, smaller balloon is open at the top so that there is free communication between its interior and the atmospheric air surrounding the larger balloon, it contains air at atmospheric pressure. The space between the balloons is completely filled with water. The forces acting upon the wall of the inner balloon are the inner air pressure and the water pressure surrounding it. Initially, as shown in the left half of Fig. 10-8, these pressures are approximately equal, and there is no tension in the wall of the inner balloon. When we enlarge the outer balloon by pulling on it in all directions, the inner balloon expands by an almost equal amount and its walls become highly stretched and taut. *This occurs because of a decrease in the fluid pressure of the water surrounding it.* Water is highly *indistensible*; i.e., any attempt to expand or compress a completely water-filled space causes a marked decrease or increase, respectively, of the fluid pressure within the space (it is much more difficult to compress a water-filled balloon than an air-filled balloon of the same size). Thus the pull on the external balloon produces a drop in the fluid pressure surrounding the inner balloon, which now becomes less than the air pressure within the inner balloon, and a transmural pressure gradient pushes out the wall of the inner balloon. As the inner wall moves outward, the internal air pressure falls slightly, but atmospheric air immediately enters through the opening so that atmospheric pressure is maintained. Thus, the external fluid pressure remains lower than the internal air pressure and the inner balloon expands until the force of its elastic recoil becomes great enough to balance this distending pressure difference. The balloon is behaving just like a stretched spring. The crucial event induced by expanding

the outer balloon is a reduction of the fluid pressure between the balloons; this, in turn, causes the inner balloon to expand.

We can apply this analogy to the lungs (air-filled inner balloon), thoracic cage (outer balloon), and intrapleural fluid (the water between the balloons). As the thoracic cage expands during growth, it pulls slightly away from the outer surface of the lungs. This drops the *intrapleural fluid pressure* below that of the intraalveolar air pressure, a pressure difference that forces the lungs to distend. The lungs must expand to virtually the same degree as the thoracic cage, and their elastic walls become greatly stretched. The tendency for the lungs to recoil as a result of this stretch is balanced by the difference between the intraalveolar air pressure and the intrapleural fluid pressure.

Why the lung collapses when the chest wall is opened should now be apparent. The low intrapleural pressure is significantly less than the pressure of the atmospheric air outside the body; i.e., it is subatmospheric. When the chest wall is pierced, atmospheric air rushes into the intrapleural space, the pressure difference across the lung wall is abolished, and the stretched lung collapses. Air in the intrapleural space is known as a *pneumothorax*.

The subatmospheric pressure of the intrapleural fluid, which is generated by the different growth rates of lung and thoracic cage, is maintained throughout

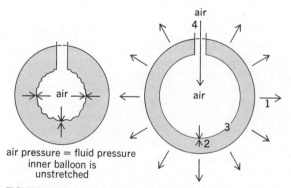

air pressure = fluid pressure
inner balloon is
unstretched

FIGURE 10-8
How fluid between two balloons causes the inner balloon to expand whenever the outer one does. 1 Outer balloon is expanded by an outside force. 2 Expansion of the fluid space causes the fluid pressure to decrease. 3 Fluid pressure is less than internal air pressure; therefore, the wall of the inner balloon is pushed out. 4 As the inner balloon expands, its internal air pressure decreases and air moves in from the atmosphere.

life.[1] Regardless of whether the person is inspiring, expiring, or not breathing at all, the intrapleural pressure is always lower than the air pressure within the lungs, and the lungs are considerably stretched. However, the gradient between intrapleural and intraalveolar pressures does vary during breathing and directly causes the changes in lung size which occur during inspiration and expiration. Since intrapleural pressure is transmitted throughout the intrathoracic fluid surrounding not only the lungs but the heart and other intrathoracic structures as well, it is frequently termed the *intrathoracic pressure*. Recall that (Chap. 9) subatmospheric intrathoracic pressure was mentioned in describing the forces which determine end-diastolic ventricular volume.

Inspiration

The left half of Fig. 10-9 summarizes events which occur during inspiration. Just before the inspiration begins, i.e., at the conclusion of the previous expiration, the respiratory muscles are relaxed and no air is flowing. The intrapleural pressure is subatmospheric (for reasons described above). The intraalveolar pressure, i.e., the air pressure within the alveoli, is exactly atmospheric because the alveoli are in free communication with the atmosphere via the airways. Inspiration is initiated by the contraction of the diaphragm and intercostal muscles (Fig. 10-10). When the diaphragm contracts, its dome moves downward into the abdomen, thus enlarging the volume of the thoracic cage. Simultaneously, the inspiratory intercostal muscles, which insert on the ribs, contract, leading to an upward and outward movement of the ribs and a further increase in thoracic cage size. This expansion of the thoracic cage is merely a speeded-up enlargement of the events described for growth. As the thoracic cage begins to move away from the lung surface, the intrapleural fluid pressure abruptly decreases, i.e., becomes even more subatmospheric. This increases the difference between the intraalveolar and intrapleural pressures, and the lung wall is pushed out. Thus, when the inspiratory muscles increase the thoracic dimensions, the lungs are also forced to enlarge because of the changes in intrapleural pressure. This further stretching of the lung causes an increase in the volumes of all the air-containing passages and alveoli within the lung. As the alveoli enlarge, the air pressure within them drops to less than atmospheric, causing bulk flow of air from the atmosphere through the airways into the alveoli until

---

[1]Only during a forced expiration does intrapleural pressure exceed atmospheric pressure.

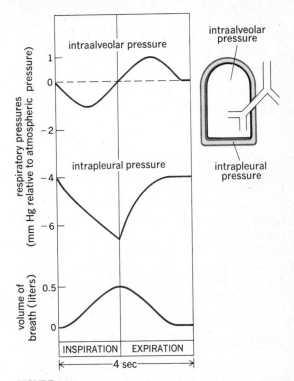

FIGURE 10-9

Summary of intraalveolar and intrapleural pressure changes and air flow during inspiration and expiration of 500 ml of air. Note that, on the pressure scale at the left, normal atmospheric pressure (760 mm Hg) has a scale value of zero.

their pressure again equals atmospheric. Thus, air is literally sucked into the expanding lungs.

Expiration

The expansion of thorax and lungs produced during inspiration by active muscular contraction stretches both lung and thoracic wall elastic tissue. When inspiratory-muscle contraction ceases and these muscles relax, the stretched tissues recoil to their original length since there is no force left to maintain the stretch. An obvious analogy is the snap of a stretched rubber band when it is released. The tissue recoil causes a rapid and complete reversal of the inspiratory process, as shown in the right side of Fig. 10-9. The thorax and lung spring back to their original sizes, alveolar air becomes temporarily compressed so that its pressure exceeds atmospheric, and air flows from the alveoli through the airways out into the atmosphere. Normal expiration is thus completely passive, de-

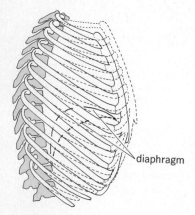

FIGURE 10-10
**Movements of chest wall and diaphragm during breathing.
The contracting intercostal muscles move the ribs upward and
outward during inspiration while the contracting diaphragm
moves downward.** (*Adapted from McNaught and Callender.*)

pending only upon the cessation of inspiratory-muscle
activity and the relaxation of these muscles. Under certain
conditions, however, particularly when resistance to air
flow is abnormally high, expiration can be facilitated by
the contraction of another group of intercostal and abdominal muscles, which actively decrease thoracic dimensions. The abdominal muscles help by increasing
intraabdominal pressure and forcing the diaphragm up
higher into the thorax.

It should be noted that the analysis of Fig. 10-9
treats the lungs as a single alveolus. The fact is that there
are significant regional differences in both intraalveolar
and intrapleural pressures throughout the lungs and thoracic cavity. These differences are due, in part, to the
effects of gravity and to local differences in the elasticity
of the chest structures. They may be of great importance
in determining the pattern of ventilation in disease states.

Quantitative relationship between atmosphere-intraalveolar pressure gradients and air flow: airway resistance

What is the quantitative relationship between the atmos-
phere-intraalveolar pressure gradients and the volume of
air flow? It is expressed by precisely the same equation[1]
given for the circulatory system by

---

[1]Note that, as for blood flow, we have merely transformed the basic
equation, flow = $k(P_1 - P_2)$, by using resistance, which is the
reciprocal of k.

Flow = pressure gradient/resistance

The volume of air which flows in or out of the alveoli is
directly proportional to the pressure gradient between the
alveoli and atmosphere and inversely proportional to the
resistance to flow offered by the airways. Normally the
magnitude of this pressure gradient is increased by taking
deeper breaths, i.e., by increasing the strength of contraction of the inspiratory muscles. This causes, in order,
an increased expansion of the thoracic cage, a greater
drop in intrapleural pressure, increased expansion of the
lungs, lower intraalveolar pressure, and increased flow
of air into the lungs (Fig. 10-11).

What factors determine airway resistance? Resist-

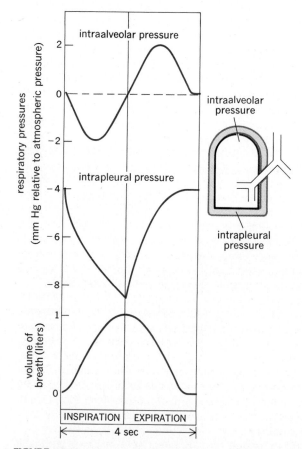

FIGURE 10-11
**Summary of intraalveolar and intrapleural pressure changes
and air flow during inspiration of 1,000 ml of air. Compare
these values with those given in Fig. 10-9.**

ance is (*1*) directly proportional to the magnitude of the interactions between the flowing gas molecules, (*2*) directly proportional to the length of the airway, and (*3*) inversely proportional to the fourth power of the airway radius. These factors, of course, are counterparts of the factors determining resistance in the circulatory system; and in the respiratory tree, just as in the circulatory tree, resistance largely depends upon the radius of the conducting tubes.

The airway diameters are normally so large that they actually offer little total resistance to air flow, and interaction between gas molecules is also usually negligible, as is the contribution of airway length. Therefore, the total resistance remains so small that minute pressure gradients suffice to produce large volumes of air flow. As we have seen (Fig. 10-9), the average pressure gradient during a normal breath at rest is less than 1 mm Hg; yet, approximately 500 ml of air is moved by this tiny gradient.

The diameter of the airways becomes extremely important in certain diseases such as asthma, which is characterized by severe bronchiolar smooth muscle constriction and plugging of the airways by bronchial secretions. Resistance to air flow may become great enough to prevent air flow completely, regardless of the atmosphere-alveolar pressure gradient.

Airway size and resistance may be altered by physical, nervous, or chemical factors. The most important normal physical factor is simply expansion of the lungs; during inspiration, airway resistance decreases because the overall enlargement of the lungs pulls on the airways and widens them. Conversely, during expiration, airway resistance increases. For this reason patients with abnormal airway resistance, as in asthma, have much less difficulty inhaling than exhaling, with the result that air may be trapped in the lungs and the lung volume progressively increased.

Nervous regulation of airway size is mediated by the autonomic nervous system, the sympathetic neurons causing relaxation of the airway smooth muscle (decreased resistance) and the parasympathetics causing smooth muscle contraction (increased resistance). These reflexes are important in causing airway constriction upon inhalation of chemical irritants but their precise contribution to control of airway resistance under normal conditions is unclear.

As would be expected from knowledge of the effects of the sympathetic nerves on airway resistance, circulating epinephrine also causes airway dilation. This is a major reason for administering epinephrine or epi-nephrine-like drugs to patients suffering from airway constriction, as in an asthmatic attack. In contrast, histamine and certain other chemical mediators cause bronchiolar constriction (and increased mucus secretion as well) and may be the cause of the airway constriction observed in allergic attacks. This explains the use of antihistamines to relieve the respiratory symptoms of allergies.

So far we have been discussing *total* airway resistance, but discrete *local* changes in airway resistance are important in promoting efficient gas exchange. The bronchioles contain smooth muscle highly responsive to the carbon dioxide concentration of the medium surrounding it, high carbon dioxide increasing bronchodilation and low carbon dioxide increasing bronchoconstriction. These effects are exerted locally directly upon the smooth muscle and are independent of nerves or hormones. What is the significance of this sensitivity? The lungs are composed of approximately 300 million discrete alveoli, each receiving carbon dioxide from the pulmonary capillary blood. To be most efficient, the right proportion of alveolar air and capillary blood should be available to each alveolus, a pattern local changes in bronchiolar tone help maintain. In Fig. 10-12, for example, the left alveolus is receiving too much air for its blood supply and vice versa on the right. Because of the deficient blood flow, the left alveolus receives little carbon dioxide and the bronchiole supplying it is exposed to a low carbon dioxide concentration and constricts. Conversely, the area around the poorly ventilated right alveolus has a high carbon dioxide concentration, and the bronchiole supplying it dilates. By this completely local mechanism, the ventilation and blood supply are matched.

Control of pulmonary arterioles

The previous chapter described in detail how the systemic arterioles control the distribution of blood to the various organs and tissues. The pulmonary arterioles perform an analogous function in controlling the distribution of blood to different alveolar capillaries, thus providing a second mechanism for matching air flow and blood flow. This control is based, in large part, on the great sensitivity to oxygen of the pulmonary arteriolar smooth muscle; a decreased oxygen concentration causes arteriolar constriction, whereas an increased oxygen concentration causes vasodilation. In addition, the pulmonary vessels are also sensitive to their local hydrogen-ion concentration; an increased hydrogen-ion concentration causes constriction whereas a decreased concentration causes vasodilation. (As will be emphasized subsequently, the hydrogen-ion concentration reflects, in large part, the

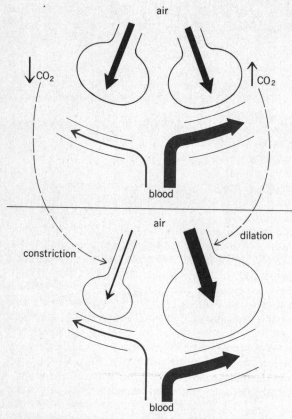

FIGURE 10-12
Adjustment of air flow to blood flow by means of carbon dioxide–induced changes in bronchiolar constriction. Increased carbon dioxide causes bronchiolar dilation; conversely, decreased carbon dioxide results in bronchiolar constriction. (*Top panel*) Initial state; (*bottom panel*) after compensation.

so that the final situation lies somewhere in between. The net result is efficient matching of air flow and blood flow in 300 million alveoli.

Before leaving the subject of the pulmonary circulation, we wish to reemphasize the fact, described in Chap. 9, that the pulmonary circulation is a low-pressure circuit. The normal pulmonary capillary pressure is only 15 mm Hg. This is the major force favoring movement of fluid out of the pulmonary capillaries into the interstitium and alveoli. It is well below the major force favoring absorption, namely, the plasma colloid osmotic pressure of 25 mm Hg. Accordingly, the alveoli normally remain dry, a feature essential for normal gas exchange. Should pulmonary capillary pressure increase greatly, as occurs

partial pressure of carbon dioxide.) These purely local (no nerves or hormones) effects of oxygen and hydrogen ion on pulmonary arterioles are precisely the opposite of those exerted by them on systemic arterioles (Chap. 9); no chemical explanation for this striking difference is known. In Fig. 10-13, with the same uneven distribution pattern as Fig. 10-12, the left alveolar area has high oxygen and low hydrogen ion because it receives much air but little blood; conversely, the right alveolar area has low oxygen and high hydrogen ion. This causes vasoconstriction of the right arteriole and dilation of the left and shunts blood away from the poorly ventilated alveolus and to the richly ventilated one. Of course, the events of Figs. 10-12 and 10-13 occur simultaneously (Table 10-1)

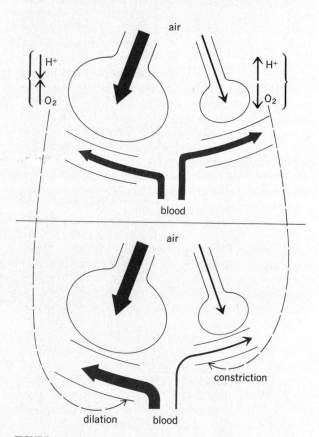

FIGURE 10-13
Adjustment of blood flow to air flow by means of hydrogen ion and oxygen-induced changes in arteriolar constriction. Increased oxygen and decreased hydrogen ion cause arteriolar dilation; conversely, decreased oxygen and increased hydrogen ion result in arteriolar constriction. (*Top panel*) Initial state; (*bottom panel*) after compensation.

when the left ventricle "fails," fluid accumulates in the alveoli (pulmonary edema) with serious consequences for gas exchange across the alveolar wall.

Lung-volume changes during breathing

The volume of air entering or leaving the lungs during a single breath is called the *tidal volume* (Fig. 10-14). For a breath under resting conditions, this is approximately 500 ml. We are all aware that the resting thoracic excursion is small compared with a maximal breathing effort. The volume of air which can be inspired over and above the resting tidal volume is called the *inspiratory reserve* and amounts to 2,500 to 3,500 ml of air. At the end of a normal expiration, the lungs still contain a large volume of air, part of which can be exhaled by active contraction of the expiratory muscles; it is called the *expiratory reserve* and measures approximately 1,000 ml of air. Even after a maximal expiration, some air (approximately 1,000 ml) still remains in the lungs and is termed the *residual volume.*

What, then, is the maximum amount of air which can be moved in and out during a single breath? It is the sum of the normal tidal, inspiratory-reserve, and expiratory-reserve volumes. This total volume is called the *vital capacity.* During heavy work or exercise, a person uses part of both the inspiratory and expiratory reserves (particularly the former) but rarely uses more than 50 percent of his total vital capacity, because deeper breaths than this require exhausting activity of the inspiratory and expiratory muscles. The greater depth of breathing during exercise greatly increases pulmonary ventilation, and still larger increases are produced by speeding the rate of breathing as well.

The *total pulmonary ventilation* per minute (*minute ventilatory volume*) is determined by the tidal volume times the respiratory rate (expressed as breaths per minute). For

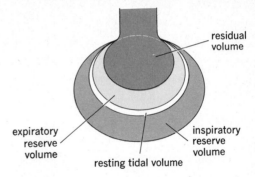

FIGURE 10-14
Lung volumes. (*Adapted from Comroe.*)

example, at rest, a normal person moves approximately 500 ml of air in and out of the lungs with each breath, and he takes 10 breaths each minute. The total minute ventilatory volume is therefore 500 ml $\times$ 10 = 5,000 ml of air per minute. However, as we shall see, not all this air is available for exchange with the blood.

Air distribution within the lungs

Anatomic dead space   The respiratory tract, as we have seen, is composed of conducting airways and the alveoli. Within the lungs, exchanges of gases with the blood occur only in the alveoli and not in the conducting airways, the total volume of which is approximately 150 ml. Picture, then, what occurs during expiration: 500 ml of air is forced out of the alveoli and through the airways. Approximately 350 ml of this air is exhaled at the nose or mouth, but approximately 150 ml still remains in the airways at the end of expiration. During the next inspiration, 500 ml of air flows into the alveoli, but the first 150 ml of entering air is not atmospheric but the 150 ml of alveolar air left behind. Thus, only 350 ml of new atmospheric air enters the alveoli during the inspiration. At the end of inspiration, 150 ml of fresh air also fills the conducting airways, but no gas exchange with the blood can occur there. At the next expiration, this fresh air will be washed out and again replaced by old alveolar air, thus completing the cycle. The end result is that 150 ml of the atmospheric air entering the respiratory system during each inspiration never reaches the alveoli but is merely moved in and out of the airways. Because these airways do not permit gas exchange with the blood, the space within them is termed the *anatomic dead space.*

The volume of fresh atmospheric air entering the alveoli during each inspiration, then, equals the volume

TABLE 10-1
Major factors which control airway and pulmonary vascular resistance

	Airways	Arterioles
Constricted by:	Histamine	$\downarrow O_2$
	Parasympathetic nerves	$\uparrow H^+$
	$\downarrow CO_2$	
Dilated by:	Epinephrine	$\uparrow O_2$
	Sympathetic nerves	$\downarrow H^+$
	$\uparrow CO_2$	

TABLE 10-2
Effect of breathing patterns on alveolar ventilation

Subject	Tidal volume, ml/breath	× Frequency, breaths/min	= Total pulmonary ventilation, ml/min	− Dead space ventilation, ml/min	= Alveolar ventilation, ml/min
A	150	40	6,000	150 × 40 = 6,000	0
B	500	12	6,000	150 × 12 = 1,800	4,200
C	1,000	6	6,000	150 × 6 = 900	5,100

of air in the anatomic dead space subtracted from the total tidal volume. Thus, for a normal breath

$$\text{Tidal volume} = 500 \text{ ml}$$
$$\underline{\text{Anatomic dead space} = 150 \text{ ml}}$$
$$\text{Fresh air entering alveoli} = 350 \text{ ml}$$

This total is called the *alveolar ventilation*. The term is somewhat confusing, because it seems to indicate that only 350 ml air enters and leaves the alveoli with each breath. This is not true—the total is 500 ml of air, but only 350 ml is fresh air.

What is the significance of the anatomic dead space and alveolar ventilation? Total minute ventilatory volume is equal to the tidal volume of each breath multiplied by the number of breaths per minute. But since only that portion of inspired atmospheric air which enters the alveoli, i.e., the alveolar ventilation, is useful for gas exchange with the blood, the magnitude of the alveolar ventilation is of much greater significance than the magnitude of the total pulmonary ventilation, as can be demonstrated readily by the data in Table 10-2.

In this experiment subject A breathes rapidly and shallowly, B normally, and C slowly and deeply. Each subject has exactly the same total ventilatory volume; i.e., each is moving the same amount of air in and out of the lungs each minute. Yet, when we subtract the anatomic-dead-space ventilation from the total ventilatory volume, we find marked differences in alveolar ventilation. Subject A has no alveolar ventilation (and would become unconscious in several minutes), whereas C has a considerably greater alveolar ventilation than B, who is breathing normally. The important deduction to be drawn from this example is that increased depth of breathing is far more effective in elevating alveolar ventilation than an equivalent increase of breathing rate. Conversely, a decrease in depth can lead to a critical reduction of alveolar ventilation. This is because a fraction of *each* tidal volume represents anatomic-dead-space ventilation. If the tidal

volume decreases, this fraction increases until, as in subject A, it may represent the entire tidal volume. On the other hand, any increase in tidal volume goes entirely toward increasing alveolar ventilation. Alveolar ventilation is calculated as follows:

Alveolar ventilation (ml/min)
= frequency (breaths/min)
× [tidal volume (ml) − dead space (ml)]

These concepts have important physiological implications. Most situations, such as exercise, which necessitate an increased oxygen supply (and carbon dioxide elimination) reflexly call forth a relatively greater increase in breathing depth than rate. Indeed, well-trained athletes can perform moderate exercise with very little increase, if any, in respiratory rate. The mechanisms by which rate and depth of respiration are controlled will be described in a later section of this chapter.

Alveolar dead space Some inspired air is not useful for gas exchange with the blood even though it reaches the alveoli because some alveoli, for various reasons, receive too little blood supply for their size. Air which enters these alveoli during inspiration cannot exchange gases efficiently because there is inadequate blood. The inspired air in this *alveolar dead space* must be distinguished from that in the anatomic dead space. It is quite small in normal persons but may reach lethal proportions in several kinds of lung disease. As we have seen, it is minimized by the local mechanisms which match air and blood flows.

The work of breathing

During inspiration, active muscular contraction provides the energy required to expand the thorax and lungs. What determines how much work these muscles must perform in order to provide a given amount of ventilation? First, there is simply the stretchability of the thorax and lungs. To expand these structures they must be stretched. The easier they stretch, the less energy is required for a given

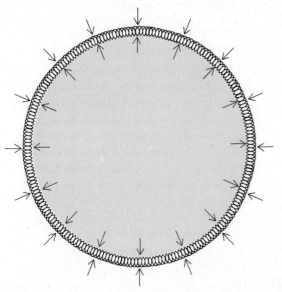

FIGURE 10-15
Forces acting on the surface of a bubble. The springs and dark arrows represent the surface tension resulting from the cohesive forces of water molecules at the air-water interface. This tension is opposed by the air pressure within the bubble (colored arrows).

amount of expansion. Much of the work of breathing goes into stretching the elastic tissue of the lung, but an even larger fraction goes into stretching a different kind of "tissue"—water itself! The air within each alveolus is separated from the alveolar membranes by an extremely thin layer of fluid; in a sense, therefore, the alveoli may be viewed as air-filled bubbles lined with water. At an air-water interface, the attractive forces between water molecules cause them to squeeze in upon the air within the bubble (Fig. 10-15). This force, known as *surface tension*, makes the water lining very like highly stretched rubber which constantly tries to shorten and resists further stretching. Thus inspiration requires considerable energy to expand the lungs because of the difficulty of distending these alveolar bubbles. Indeed, the surface tension of pure water is so great that lung expansion would require exhausting muscular effort and the lungs would tend to collapse. It is extremely important, therefore, that the specialized alveolar cells produce a phospholipoprotein complex, known as *pulmonary surfactant*, which intersperses with the water molecules and markedly reduces their cohesive force, thereby lowering the

surface tension. Surfactant is continuously replenished by the alveolar cells, and normal ventilation of the lungs seems to be the stimulus for its production. A striking example of what occurs when insufficient surfactant is present is provided by the disease known as "respiratory-distress syndrome of the newborn," which frequently afflicts premature infants in whom the surfactant-synthesizing cells are too immature to function adequately. The infant is able to inspire only by the most strenuous efforts which may ultimately cause complete exhaustion, inability to breathe, lung collapse, and death. The recent accidental discovery that the administration of adrenal steroids markedly enhances the maturation process of the surfactant-synthesizing cells may provide an important means of combating this disease.

The second factor determining the degree of muscular work required for a certain amount of ventilation is the magnitude of the airway resistance. When airway resistance is increased by bronchiolar constriction or secretions (as in asthma), the usual pressure gradient does not suffice for adequate air inflow and a deeper breath is required to create a larger pressure gradient.

One might imagine from this discussion (and from observing an athlete exercising hard) that the work of breathing uses up a major portion of the energy spent by the body. Not so; in a normal person, even during heavy exercise, the energy needed for breathing is only about 3 percent of the total expenditure. It is only in disease, when the work of breathing is markedly increased by structural changes in the lung or thorax, by loss of surfactant, or by an increased airway resistance, that breathing itself becomes an exhausting form of exercise.

Exchange and transport of gases in the body

We have completed our discussion of alveolar ventilation, but this is only the first step in the total respiratory process. Oxygen must move across the alveolar membranes into the pulmonary capillaries, be transported by the blood to the tissues, leave the tissue capillaries, and finally cross cell membranes to gain entry into cells. Carbon dioxide must follow a similar path in reverse (Fig. 10-16). At rest, during each minute, body cells consume approximately 200 ml of oxygen and produce approximately the same amount of carbon dioxide. The relative amounts depend primarily upon what nutrients are being used for energy; e.g., when glucose is utilized, one molecule of carbon dioxide is produced for every molecule

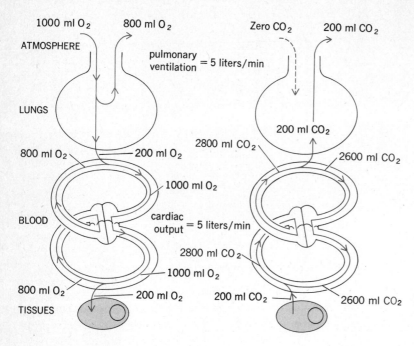

FIGURE 10-16

Summary of exchanges between atmosphere, lungs, blood, and tissues during 1 min. RQ is assumed to be 1.

of oxygen consumed:

$$C_6H_{12}O_6 + 6O_2 \longrightarrow 6CO_2 + 6H_2O + energy$$

The ratio (CO_2 produced)/(O_2 consumed) is known as the *respiratory quotient* (RQ) and accordingly is 1 for glucose. When fat is utilized, only 7 molecules of carbon dioxide are produced for every 10 molecules of oxygen consumed, and RQ = 0.7. For simplicity, Fig. 10-16 assumes that the carbon dioxide and oxygen amounts are equal and the total volumes of air inspired and expired therefore identical.

At rest, the total pulmonary ventilation equals 5 liters of air per minute. Since only 20 percent of atmospheric air is oxygen (most of the remainder is nitrogen), the total oxygen input is 20% × 5 liters = 1 liter of O_2 per minute. Of this inspired oxygen, 200 ml crosses the alveoli into the pulmonary capillaries, and the remaining 800 ml is exhaled. This 200 ml of oxygen is carried by 5 liters of blood, which is the pulmonary blood flow (cardiac output) per minute. Note, however, that blood entering the lungs already contains large quantities of oxygen, to which this 200 ml is added. This blood is then pumped by the left ventricle through the tissue capillaries of the body, and 200 ml of oxygen leaves the blood to be taken up and utilized by cells. Because only a fraction of the

total blood oxygen actually leaves the blood, some oxygen remains in the blood when it returns to the heart and lungs. It is obvious but important that the quantities of oxygen added to the blood in the lungs and removed in the tissues are identical. As shown by Fig. 10-16, the story reads in reverse for carbon dioxide. As we shall see, most of the blood carbon dioxide is actually in the form HCO_3^-, but we have shown it as CO_2 for simplicity.

The pumping of blood by the heart obviously propels oxygen and carbon dioxide between the lungs and tissues, but what forces induce the net movement of these molecules across the alveolar, capillary, and cell membranes? The answer is *diffusion. There is no active membrane transport for oxygen or carbon dioxide; they move solely by passive diffusion.* As described in Chap. 2, diffusion can effect the net transport of a substance only when a concentration gradient exists for it. Understanding the mechanisms involved depends upon familiarity with some basic chemical and physical properties of gases, to which we now turn.

Basic properties of gases

A gas consists of individual molecules constantly moving at great speeds. Since rapidly moving molecules bombard the walls of any vessel containing them, they therefore exert a *pressure* against the walls. The magnitude of the pressure is increased by anything which increases the bombardment. The pressure a gas exerts is proportional to (*1*) the temperature, because heat increases the speed at which molecules move, and (*2*) the concentration of the gas, i.e., the number of molecules per unit volume. In other words, when a certain number of molecules are compressed into a smaller volume, there are more collisions with the walls because the molecules have a shorter distance to travel.

The pressure of a gas is therefore a measure of the concentration and speed of its molecules. Gases move from a region of higher pressure to a region of lower pressure, and diffusion is thus the result of the continuous movement of gas molecules. Some molecules, of course, are moving against the gradient, but many more are moving from the region of higher pressure to that of lower, and we should really speak of *net* diffusion.

Of great importance is the relationship between different gases, i.e., different kinds of molecules, such as oxygen and nitrogen, in the same container. In a mixture of gases, the pressure exerted by each gas is independent of the pressure exerted by the others because gas molecules are normally so far apart that they do not interfere with each other. Since each gas behaves as though the other gas were not present, the total pressure of a mixture of gases is simply the sum of the individual pressures. These individual pressures, termed *partial pressures,* are denoted by a P in front of the symbol for the gas, the partial pressure of oxygen thus being represented by P_{O_2}. Gas pressures are usually expressed in millimeters of mercury, the same units used for the expression of hydrostatic pressure.

Behavior of gases in liquids

Several factors determine the uptake of gases by liquids and the behavior of gases dissolved in liquids. When a free gas comes into contact with a liquid, the number of gas molecules which dissolve in the liquid is directly proportional to the pressure of the gas. This phenomenon is clear from the basic definition of pressure. Suppose, for example, that oxygen is placed in a closed vessel half full of water. Oxygen molecules constantly bombard the surface of the water, some entering the water and dissolving. Since the number of molecules striking the surface is directly proportional to the pressure of the oxygen gas P_{O_2}, the number of molecules entering the water is also directly proportional to P_{O_2}. How many entering mol-

ecules actually stay in the water? Since the dissolved oxygen molecules are also constantly moving, some of them strike the water surface from below and escape into the free oxygen above. The rate of escape from the water and the rate of entry into the water are equal when the rates of bombardment are equal, i.e., when the pressures of the oxygen in the free gas and in the water become identical. Thus, we come back to our earlier statement: The number of gas molecules which will dissolve in a liquid is directly proportional to the pressure of the gas. When the free-gas pressure is higher than the pressure of the gas in a liquid, a number of molecules must dissolve in the liquid for the pressure of the dissolved gas to equal the pressure of the free gas. Conversely, if a liquid containing a dissolved gas at high pressure is exposed to that same free gas the pressure of which is lower, gas molecules leave the liquid and enter the free gas until the free- and dissolved-gas pressures become equal. These are precisely the phenomena occurring between alveolar air and pulmonary capillary blood.

It should also be apparent that dissolved gas molecules diffuse *within* the liquid from a region of higher gas pressure to a region of lower pressure, an effect which underlies the exchange of gases between cells, tissue fluid, and capillary blood throughout the body.

This discussion has been in terms of proportionalities rather than absolute amounts. Ths number of gas molocules which will dissolve in a liquid is *proportional* to the gas pressure, but the *absolute* number also depends upon the *solubility* of the gas in the liquid. Thus, if a liquid is exposed to two different gases at the same pressures, the numbers of molecules of each gas which are dissolved at equilibrium are not identical but reflect the solubilities of the two gases. Nevertheless, doubling the gas pressures doubles the number of gas molocules dissolved.

Pressure gradients of oxygen and carbon dioxide within the body

With these basic gas properties as foundation, we can discuss the diffusion of oxygen and carbon dioxide across alveolar, capillary, and cell membranes. The pressures of these gases in atmospheric air and in various sites of the body are given in Fig. 10-17 for a resting person at sea level. The rest of this section is devoted to an elaboration of this figure.

Atmospheric air consists primarily of nitrogen and oxygen with very small quantities of water vapor, carbon dioxide, and inert gases such as argon. The sum of the partial pressures of all these gases is termed *atmospheric*

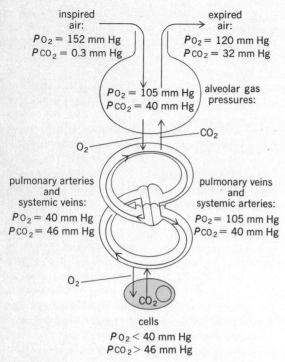

inspired air:
$P_{O_2} = 152$ mm Hg
$P_{CO_2} = 0.3$ mm Hg

expired air:
$P_{O_2} = 120$ mm Hg
$P_{CO_2} = 32$ mm Hg

$P_{O_2} = 105$ mm Hg
$P_{CO_2} = 40$ mm Hg
alveolar gas pressures:

CO_2

O_2

pulmonary arteries and systemic veins:
$P_{O_2} = 40$ mm Hg
$P_{CO_2} = 46$ mm Hg

pulmonary veins and systemic arteries:
$P_{O_2} = 105$ mm Hg
$P_{CO_2} = 40$ mm Hg

O_2

CO_2

cells
$P_{O_2} < 40$ mm Hg
$P_{CO_2} > 46$ mm Hg

FIGURE 10-17
Summary of carbon dioxide and oxygen pressures in the inspired and expired air and various places within the body.

pressure or *barometric pressure*. It varies in different parts of the world as a result of differences in altitude, but at sea level it is 760 mm Hg. Since air is 20 percent oxygen, the P_{O_2} of inspired air is $20\% \times 760 = 152$ mm Hg.

The first question suggested by Fig. 10-17 is why the partial pressures of the constituents of expired air are not identical to those of alveolar air. Recall that approximately 150 ml of the inspired atmospheric air during each breath never gets down into the alveoli but remains in the airways (dead space). This air does not exchange carbon dioxide or oxygen with blood and is expired along with alveolar air during the subsequent expiration. Therefore, the P_{O_2} and P_{CO_2} of the total expired air are higher and lower, respectively, than those of alveolar air.

The next question concerns the alveolar gas pressures themselves. One might logically reason that the alveolar gas pressures must vary considerably during the respiratory cycle, since new atmospheric air enters only during inspiration. In fact, however, the variations in alveolar P_{O_2} and P_{CO_2} during the cycle are so small as to be negligible because, as explained in the section on lung

volumes, a large volume of gas is always left within the lungs after expiration. This remaining alveolar gas contains large quantities of oxygen and carbon dioxide, and when the new air enters, it mixes with the alveolar air already present, lowering its P_{CO_2} and raising its P_{O_2}, but only by a small amount. For this reason, the alveolar-gas partial pressures remain *relatively* constant throughout the respiratory cycle, and we may use the single alveolar pressures shown in Fig. 10-17 in our subsequent analysis of alveolar-capillary exchange, ignoring the minor fluctuations.

Our next question concerns the exchange of gases between alveoli and pulmonary capillary blood. The blood which enters the pulmonary capillaries is, of course, systemic venous blood pumped to the lungs via the pulmonary arteries. Having come from the tissues, it has a high P_{CO_2} (46 mm Hg) and a low P_{O_2} (40 mm Hg). As it flows through the pulmonary capillaries, it is separated from the alveolar air only by an extremely thin layer of tissue. The differences in the partial pressures of oxygen and carbon dioxide on the two sides of this alveolar-capillary membrane result in the net diffusion of oxygen into the blood and of carbon dioxide into the alveoli. As this diffusion occurs, the capillary blood P_{O_2} rises above its original value and the P_{CO_2} falls. The net diffusion of these gases ceases when the alveolar and capillary partial pressures become equal. In a normal person, the rates at which oxygen and carbon dioxide diffuse are so rapid and the velocity of blood flow through the capillaries so slow that complete equilibrium is always reached (at rest, a red blood cell takes 0.75 sec to pass through the pulmonary capillaries). Thus, the blood that leaves the lungs to return to the heart and be pumped into the arteries has the same P_{O_2} and P_{CO_2} as alveolar air. (Actually, the P_{O_2} of arterial blood is slightly less than that of alveolar air, but we shall ignore this discrepancy in our analysis.)

The diffusion of gases between alveoli and capillaries may be impaired in a number of ways. The disease emphysema, which is intimately related to cigarette smoking, is characterized by the breakdown of the alveolar capillary walls with the formation of fewer but larger alveoli. The result is a reduction in the total area available for diffusion. In a different kind of defect, caused by membrane thickening or by pulmonary edema, the area available for diffusion is normal but the molecules must travel a greater distance. Finally, without becoming thicker the alveolar capillary walls may become denser and less permeable, as for example when beryllium is inhaled and deposited on the walls.

As the arterial blood enters capillaries throughout the body, it becomes separated from the interstitial fluid only by the thin, highly permeable capillary membrane. The interstitial fluid, in turn, is separated from intracellular fluid by cell membranes which are also quite permeable to oxygen and carbon dioxide. Metabolic reactions occurring within these cells are constantly consuming oxygen and producing carbon dioxide. Therefore, as shown in Fig. 10-17, intracellular P_{O_2} is lower and P_{CO_2} higher than in blood. As a result, a net diffusion of oxygen occurs from blood to cells, and a net diffusion of carbon dioxide from cells to blood. In this manner, as blood flows through capillaries, its P_{O_2} decreases and its P_{CO_2} increases until, by the end of the capillaries, equilibrium has been reached. This accounts for the venous blood values shown in Fig. 10-17. Venous blood returns to the right ventricle and is pumped to the lungs, where the entire process begins again.

In summary, no active transport mechanisms are required to explain the exchange of gases in the lungs and tissues. The consumption of oxygen in the cells and the supply of new oxygen to the alveoli create P_{O_2} gradients which produce net diffusion of oxygen from alveoli to blood in the lungs and from blood to cells in the rest of the body. Conversely, the production of carbon dioxide by cells and its elimination from the alveoli via expiration create P_{CO_2} gradients which produce net diffusion of carbon dioxide from blood to alveoli in the lungs and from cells to blood in the rest of the body.

Transport of oxygen in the blood: the role of hemoglobin

Table 10-3 summarizes the oxygen content of arterial blood. Each liter of arterial blood contains the same number of oxygen molecules as 200 ml of gaseous oxygen. Oxygen is present in two forms: (1) physically dissolved in the blood water and (2) chemically bound to *hemoglobin* molecules. The amount of oxygen which can be physically dissolved in blood is directly proportional to the P_{O_2} of the blood, but because oxygen is relatively insoluble in water, only 3 ml of oxygen can be dissolved in 1 liter of blood at the normal alveolar and arterial P_{O_2} of 100 mm Hg. In contrast, in 1 liter of blood, 197 ml of oxygen, more than 98 percent of the total, is carried within the red blood cells chemically bound to hemoglobin.

Each milliliter of blood contains approximately 5 billion erythrocytes. Since there is approximately 5,000 ml of blood in the average person, the total number of erythrocytes in the human body is about 25 trillion. The shape

TABLE 10-3
Oxygen content of arterial blood

1 liter arterial blood contains:

$$\begin{array}{ll} & 3 \text{ ml } O_2 \text{ physically dissolved} \\ & \underline{197 \text{ ml } O_2 \text{ chemically bound to hemoglobin}} \\ \text{Total} & 200 \text{ ml } O_2 \end{array}$$

Cardiac output = 5 liters/min
O_2 carried to tissues/min = 5 × 200 = 1,000 ml

of these cells is a biconcave disk, i.e., thicker at the edge than in the middle, like a doughnut with a center depression on each side instead of a hole. Their thickness is approximately 2 μm at the margin of the disk and 1 μm in the center, and their average diameter is slightly greater than 7 μm. This shape and these small dimensions have adaptive value in that oxygen and carbon dioxide can rapidly diffuse throughout the entire cell interior. The outstanding physiological characteristic of erythrocytes is the presence of the iron-containing protein hemoglobin, constituting approximately one-third of the total cell weight. Another erythrocyte substance of great importance is the enzyme *carbonic anhydrase*, which, as we shall see below, facilitates the transportation of carbon dioxide.

The chemical reaction between oxygen and hemoglobin is usually written

$$O_2 + Hb \rightleftharpoons HbO_2 \qquad\qquad \text{10-1}$$

Hemoglobin combined with oxygen (HbO_2) is called *oxyhemoglobin*; not combined (Hb), it is called *reduced hemoglobin* or deoxyhemoglobin. Because the number of sites on the hemoglobin molecule which bind oxygen is finite, there is a maximum quantity of oxygen which can be combined with the hemoglobin molecule. When hemoglobin has been converted completely to HbO_2, it is said to be *fully saturated*. Whem hemoglobin exists as mixed Hb and HbO_2, it is said to be *partially saturated*. The *percentage saturation* of hemoglobin is a measure of the fraction of total hemoglobin which is combined with oxygen, i.e., in the form of HbO_2.

What factors determine the extent to which oxygen will combine with hemoglobin? By far the most important is the P_{O_2} of the blood, i.e., the concentration of physically dissolved oxygen. The blood hydrogen-ion concentration and temperature also play significant roles as do certain chemicals produced by the red blood cells themselves.

Effect of P_{O_2} on hemoglobin saturation   From inspection of

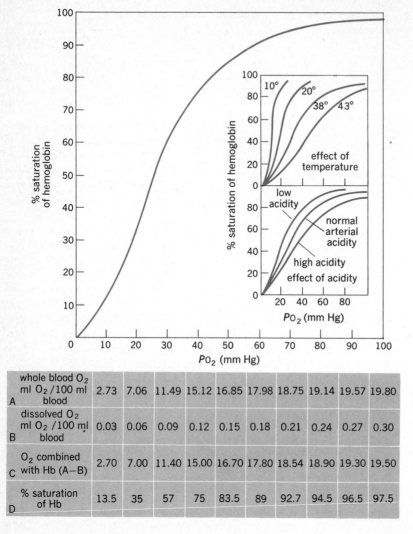

	whole blood O_2 ml O_2/100 ml blood									
A	2.73	7.06	11.49	15.12	16.85	17.98	18.75	19.14	19.57	19.80
B	dissolved O_2 ml O_2/100 ml blood									
	0.03	0.06	0.09	0.12	0.15	0.18	0.21	0.24	0.27	0.30
C	O_2 combined with Hb (A—B)									
	2.70	7.00	11.40	15.00	16.70	17.80	18.54	18.90	19.30	19.50
D	% saturation of Hb									
	13.5	35	57	75	83.5	89	92.7	94.5	96.5	97.5

FIGURE 10-18

Hemoglobin-oxygen dissociation curve. The large curve applies to blood at 38°C and the normal arterial hydrogen-ion concentration (acidity). The inset curves illustrate the effects of altering temperature and acidity on the relationship between Po_2 and hemoglobin saturation with oxygen. (*Adapted from Comroe.*)

Eq. **10-1** and the law of mass action, it is obvious that raising the Po_2 of the blood should increase the combination of oxygen with hemoglobin. The experimentally determined quantitative relationship between these variables is shown in Fig. 10-18. When a sample of blood is placed in a flask and exposed to a large volume of free oxygen, a net diffusion of oxygen occurs from the free gas into the blood until the blood and free-gas partial pressures become equal. If the volume of free gas is extremely large compared with the volume of blood, the final equilibrium Po_2 is very close to that of the original free gas. Using such a procedure, therefore, we can achieve any flask blood Po_2 we wish. By this means, we produce in 10 different samples of blood 10 different oxygen partial pressures ranging from 10 to 100 mm Hg and analyze the effect of Po_2 on hemoglobin saturation by measuring the fraction of hemoglobin combined with oxygen in each flask. Data from such an experiment are plotted in Fig. 10-18, which is called an oxygen-hemoglobin dissociation, i.e., saturation, curve. It is an S-shaped curve with a steep slope between 10 and 60 mm Hg Po_2 and a flat portion between 70 and 100 mm Hg Po_2. In other words, the extent to which hemoglobin combines with oxygen increases very rapidly from 10 to 60 mm Hg so that, at a Po_2 of 60 mm Hg, 90 percent of the total hemoglobin is combined with oxygen.

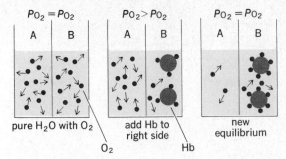

$Po_2 = Po_2$ $Po_2 > Po_2$ $Po_2 = Po_2$

A   B   A   B   A   B

pure H₂O with O₂   add Hb to   new
                    right side   equilibrium

O₂   Hb

FIGURE 10-19
**Effects of adding hemoglobin on the distribution of oxygen
between two compartments separated by a semipermeable
membrane. At the new equilibrium, the Po_2 values are equal to
each other but lower than before the hemoglobin was added;
however, the total oxygen, i.e., both dissolved and chemically
combined with hemoglobin, is much higher on the right side
of the membrane.** (*Adapted from Comroe.*)

From this point on, a further increase in Po_2 produces
only a small increase in oxygen uptake. The adaptive
importance of this plateau at higher Po_2 values is very
great for the following reason. Many situations (severe
exercise, high altitudes, cardiac or pulmonary disease)
are characterized by a moderate reduction of alveolar
and arterial Po_2; even if the Po_2 fell from the normal value
of 100 to 60 mm Hg, the total quantity of oxygen carried
by hemoglobin would decrease by only 10 percent, since
hemoglobin saturation is still close to 90 percent at a Po_2
of 60 mm Hg. The plateau therefore provides an excellent
safety factor in the supply of oxygen to the tissues.

We now retrace our steps and reconsider the
movement of oxygen across the various membranes, this
time including hemoglobin in our analysis. It is essential
to recognize that the oxygen which is chemically bound
to hemoglobin does not contribute to the Po_2 of the blood.
Only gas molecules which are *free in solution,* i.e., physically dissolved, can participate in the bombardment which
creates a gas pressure. Therefore, the diffusion of oxygen
is directly governed only by that moiety which is dissolved, a fact which permitted us to ignore hemoglobin
in discussing transmembrane pressure gradients. However, the presence of hemoglobin plays a critical role in
determining the *total amount* of oxygen which will diffuse,
as illustrated by a simple example (Fig. 10-19). Two solutions separated by a semipermeable membrane contain
equal quantities of oxygen, the gas pressures are equal,
and no net diffusion occurs. Addition of hemoglobin to
compartment B destroys this equilibrium because much

of the oxygen combines with hemoglobin. Despite the fact
that the *total quantity of oxygen* in compartment B is still
the same, the number of molecules *dissolved* has decreased; therefore, the Po_2 of compartment B is less than
that of A, and net diffusion of oxygen occurs from A to
B. At the new equilibrium, the oxygen pressures are once
again equal, but almost all the total oxygen is in compartment B combined with hemoglobin. Thus, hemoglobin
acts as a sink, removing dissolved oxygen, keeping the
Po_2 low, and allowing net diffusion to continue.

Let us now apply this analysis to the lung and tissue
capillaries (Fig. 10-20). The plasma and erythrocytes
entering the lungs have a Po_2 of 40 mm Hg, and the
hemoglobin saturation is 75 percent. Oxygen diffuses
from the alveoli because of its higher Po_2 (100 mm Hg)
into the plasma; this increases plasma Po_2 and induces
diffusion of oxygen into the erythrocytes, elevating erythrocyte Po_2 and causing increased combination of oxygen
and hemoglobin. Thus, the vast preponderance of the
oxygen diffusing into the blood from the alveoli does not
remain dissolved but combines with hemoglobin. In this
manner, the blood Po_2 remains less than that of the
alveolar Po_2 until hemoglobin is virtually completely saturated, and the diffusion gradient favoring oxygen movement into the blood is maintained despite the very large
transfer of oxygen. In the tissue capillaries, the procedure
is reversed: As the blood enters the capillaries, plasma
Po_2 is greater than interstitial fluid Po_2 and net oxygen
diffusion occurs across the capillary membrane; plasma
Po_2 is now lower than erythrocyte Po_2, and oxygen
diffuses out of the erythrocyte into the plasma. The lowering of erythrocyte Po_2 causes the dissociation of HbO_2,
thereby liberating oxygen; simultaneously, the oxygen
which diffused into the interstitial fluid is moving into cells
along the concentration gradient generated by cell utilization of oxygen. The net result is a transfer of large
quantities of oxygen from HbO_2 into cells purely by passive diffusion.

The fact that hemoglobin is still 75 percent saturated at the end of tissue capillaries under resting conditions underlies an important automatic mechanism by
which cells can obtain more oxygen whenever they increase their activity. An exercising muscle consumes
more oxygen, thereby lowering its intracellular Po_2; this
increases the overall blood-to-cell Po_2 gradient and the
diffusion of oxygen out of the blood; in turn, the resulting
reduction of erythrocyte Po_2 causes additional dissociation of hemoglobin and oxygen. An exercising muscle can
thus extract virtually all the oxygen from its blood supply.
This process is so effective because it takes place in the

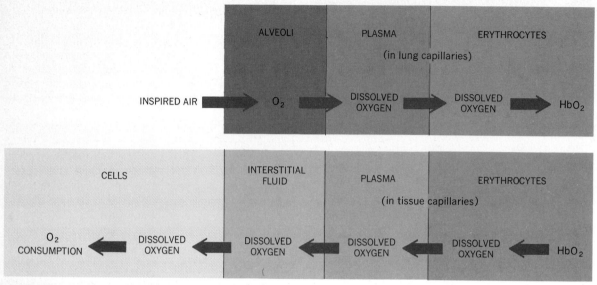

FIGURE 10-20
Oxygen movements in the lungs and tissues. Movement of air into the alveoli is by bulk flow; all movements across membranes are by passive diffusion.

steep portion of the hemoglobin dissociation curve. Of course, an increased blood flow to the muscle also contributes greatly to the increased oxygen supply.

Effect of acidity on hemoglobin saturation The large oxygen-hemoglobin dissociation curve illustrated in Fig. 10-18 is for blood having a specific hydrogen-ion concentration equal to that found in arterial blood. When the same experiments are performed at a different level of acidity, the curve changes significantly. A group of such curves is shown in the inset of Fig. 10-18 for different hydrogen-ion concentrations. It is evident that, regardless of the existing acidity, the percentage saturation of hemoglobin is still determined by the P_{O_2}. However, a change in acidity causes highly significant shifts in the entire curve. An increased hydrogen-ion concentration moves the curve downward and to the right, which means that, at any given P_{O_2}, hemoglobin has less affinity for oxygen when the acidity is high. For reasons to be described later, the hydrogen-ion concentration in the tissue capillaries is greater than in arterial blood. Blood flowing through tissue capillaries becomes exposed to this elevated hydrogen-ion concentration and therefore loses even more oxygen than if the decreased P_{O_2} had been the only factor involved. Conversely, the hydrogen-ion concentration is lower in the lung capillaries than in the

systemic venous blood, so that hemoglobin picks up more oxygen in the lungs than if only the P_{O_2} were involved. Finally, the more active a tissue is, the greater its hydrogen-ion concentration; accordingly, hemoglobin releases even more oxygen during passage through these tissue capillaries, thereby providing the more active cells with additional oxygen. The hydrogen ion exerts this effect on the affinity of hemoglobin for oxygen by combining with the hemoglobin and altering its molecular structure. The importance of this combination for the regulation of extracellular acidity is described in Chap. 11.

Effect of temperature on hemoglobin saturation The effect of temperature on the oxygen-hemoglobin dissociation curve (inset of Fig. 10-18) resembles that of an increase in acidity. The implication is similar: Actively metabolizing tissue, e.g., exercising muscle, has an elevated temperature, which facilitates the release of oxygen from hemoglobin as blood flows through the muscle capillaries.

Effect of DPG on hemoglobin saturation Red cells contain large quantities of the substance 2,3-diphosphoglycerate (DPG) which is present in only trace amounts in other mammalian cells. DPG, which is produced by the red cells during glycolysis, binds reversibly with hemoglobin, causing it to change its conformation and release oxygen.

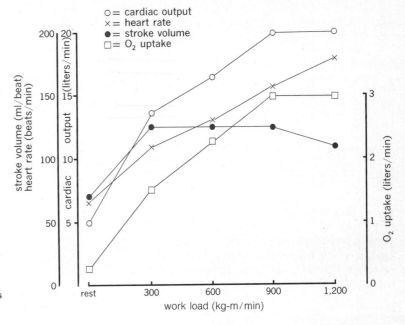

FIGURE 10-21
Changes in heart rate, stroke volume, cardiac output, and oxygen uptake with increasing work loads (measured as kg-m/min). Note that cardiac output plateaus because stroke volume decreases at maximal levels of oxygen uptake.

Therefore, the effect of increased DPG is to shift the curve downward and to the right (just as does an increased temperature or hydrogen-ion concentration). The net result is that whenever DPG is increased there is enhanced unloading of oxygen as blood flows through the tissues. Such an increase is triggered by a variety of conditions associated with decreased oxygen supply to the tissues and helps to maintain oxygen delivery.

In summary, we have seen that oxygen is transported in the blood primarily in combination with hemoglobin. The extent to which hemoglobin binds oxygen is dependent upon the Po_2, hydrogen-ion concentration, temperature, and DPG. The first three of these factors cause the release of large quantities of oxygen from hemoglobin during blood flow through tissue capillaries and virtually complete conversion of reduced hemoglobin to oxyhemoglobin during blood flow through lung capillaries. An active tissue increases its extraction of oxygen from the blood because of its lower Po_2 and higher hydrogen-ion concentration and temperature. DPG also plays a role when oxygen supply is low.

Functions of myoglobin

Myoglobin, an iron-containing protein found in cardiac and skeletal muscle cells, resembles hemoglobin in that it binds oxygen reversibly. Its major function is to act as an intracellular carrier which facilitates the diffusion of oxygen throughout the muscle cell. In addition, it provides a store of oxygen which the cell can call upon during sudden changes in activity.

Maximal oxygen uptake: a synthesis of cardiovascular-respiratory interactions

During exercise the supply of additional oxygen to the exercising muscles and the elimination of carbon dioxide depend upon precise integration of cardiovascular and respiratory functions, and this section reviews and amplifies several of these critical interactions.

When a person is subjected to progressively increasing work loads, there is a linear increase in the amount of oxygen taken up by the muscle cells until a point is reached beyond which a further increase in work load does not result in further oxygen uptake (Fig. 10-21). This is defined as the maximal oxygen uptake. Work loads greater than this can be achieved only for very short periods of time since the muscles are operating by anaerobic glycolysis to a large extent, as manifested by a large increase in lactic acid production. What sets this upper limit for oxygen uptake and therefore work? In a normal person it is usually *not* set by any "saturation" of intracellular oxidative reactions, i.e., by an inability of the muscle cells to use more oxygen. Rather it is set by the maximal ability of the cardiovascular-respiratory systems to provide oxygen.

There are two means by which the muscle cells obtain more oxygen during exercise: (1) increased blood flow to the muscle; (2) extraction of more oxygen from each liter of blood. We have already described the mechanisms responsible for these changes. Local chemical changes cause vasodilation of the arterioles, thereby lowering local vascular resistance and increasing blood flow; the degree to which blood flow can be increased depends ultimately upon the ability to increase cardiac output. Simultaneously, as the contracting muscles use more oxygen, the local tissue P_{O_2} decreases toward zero, thereby causing increased release of oxygen from hemoglobin. Recall that below 60 mm Hg the hemoglobin dissociation curve is very steep so that a large additional amount of oxygen is released for each mm Hg decrease in P_{O_2} below the value seen at rest. Moreover, the more work done by the muscle, the greater is its local temperature and hydrogen-ion concentration, both of which shift the dissociation curve downward and to the right, with the resultant release of even more oxygen. Present evidence suggests that red cell DPG also increases during exercise and contributes to the enhanced release of oxygen. The result of all these factors is that extraction of oxygen by the exercising muscles increases proportionately to work load until virtually all the oxygen is extracted at high work loads.

This analysis reveals that maximal oxygen uptake depends upon the product of maximal cardiac output and arteriovenous oxygen difference:

Maximal O_2 uptake =
maximal cardiac output $\times$ (A-V)O_2 difference

As we shall see, in a normal person at sea level, the increased ventilation triggered by exercise is quite capable of maintaining virtually complete saturation of hemoglobin during even the most severe exercise; accordingly, maximal oxygen uptake is not limited by the respiratory system. Thus, we are left with the fact that cardiac output is the rate-limiting variable in endurance-type exercise, and measurement of maximal oxygen uptake provides a sensitive index to the person's cardiovascular function.

Finally, we ask: What limits cardiac output during exercise? The answer is the interaction between heart rate and stroke volume (Fig. 10-21). Stroke volume increases with work load, although not to the same degree as does heart rate, but then decreases from maximal values at work loads beyond maximal oxygen uptake (Fig. 10-21). The major factors responsible for this decrease are the very rapid heart rate (which decreases diastolic filling time) and failure of the peripheral factors favoring venous return (muscle pump, respiratory pump, venoconstriction, arteriolar dilation; see Chap. 9) to elevate venous pressure high enough to maintain adequate ventricular filling during the very short time available. The net result is a decrease in end-diastolic volume which causes, by Starling's law, a decrease in stroke volume.

A person's maximal oxygen uptake is not fixed at any given value but can be altered (in either direction) by his habitual level of physical activity. For example, prolonged bed rest may decrease it 25 percent whereas intense long-term physical training may increase it a similar amount. It must be emphasized that, to be effective in raising maximal oxygen uptake, the training must be of an endurance type, i.e., involve large muscle groups for an extended period of time. We are still uncertain of the relative combinations of intensity and duration which are most effective but, as an example, jogging 10 to 15 min three times weekly at 5 to 8 mi/hr definitely produces some increase in most people. Training induces an increased stroke volume and decreased heart rate at any given work load. Although still controversial, present evidence suggests that a training program adequate to elevate maximal oxygen uptake may offer some protection against the occurrence of heart attacks.

Thus far, our analysis has been in terms of normal persons. It must be emphasized that disease processes may reduce maximal oxygen uptake by interfering with any step in the transfer of oxygen from the atmosphere to the cells. Thus, maximal oxygen uptake may be decreased in the presence of lung disease which prevents normal alveolar ventilation or alveolar-capillary diffusion and thereby reduces arterial P_{O_2} during exercise; reduced cardiac output secondary to heart disease; anemia or carbon monoxide poisoning; high altitude; failure of venous or arteriolar function. In other words, cardiac output is the limiting variable in normal individuals, but some other factor may become rate-limiting in the presence of disease.

At the other end of the spectrum are the champion marathon runners. They have maximum oxygen uptakes of 5 to 6 liters/min compared with an untrained man's 3 liters/min. Although training can increase the latter's value by approximately 20 percent, it certainly cannot produce the champion's values. Some of this difference may be ascribed to genetic variation, but a certain fraction is likely to be due to the pattern of the individual during early life—another example of the concept of "critical periods" described in Chap. 5.

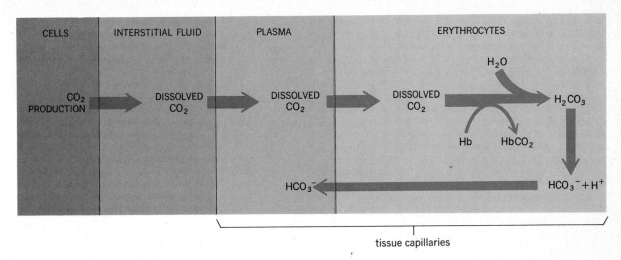

tissue capillaries

FIGURE 10-22
Summary of carbon dioxide movements and reactions as blood flows through tissue capillaries. All movements across membranes are by passive diffusion. Note that most of the CO_2 ultimately is converted to HCO_3^-; this occurs almost entirely in the erythrocytes (because the carbonic anhydrase is located there), but most of the HCO_3^- then diffuses out of the erythrocytes into the plasma.

Transport of carbon dioxide in the blood

As is true for oxygen, the quantity of carbon dioxide which can physically dissolve in blood at physiological carbon dioxide partial pressures is quite small, certainly much smaller than the large volume of carbon dioxide which must be constantly transported from the tissues to the lungs.

Carbon dioxide can undergo the reaction

$$CO_2 + H_2O \rightleftharpoons H_2CO_3$$
Carbonic acid

which goes quite slowly unless it is catalyzed by the enzyme carbonic anhydrase. The quantities of both dissolved carbon dioxide and carbonic acid are directly proportional to the P_{CO_2} of the solution. The actual amount of carbonic acid in blood is small because carbonic acid almost completely ionizes according to the equation

$$H_2CO_3 \rightleftharpoons HCO_3^- + H^+$$

Combining these two equations, we find

$$CO_2 + H_2O \xrightarrow{\text{carbonic anhydrase}}$$
$$H_2CO_3 \rightleftharpoons HCO_3^- + H^+ \quad \text{(10-2)}$$

Thus, the addition of carbon dioxide to a liquid results

ultimately in bicarbonate and hydrogen ions. Carbon dioxide can also react directly with proteins, particularly hemoglobin, to form *carbamino* compounds.

$$CO_2 + Hb \rightleftharpoons HbCO_2 \quad \text{10-3}$$

When arterial blood flows through tissue capillaries, oxyhemoglobin gives up oxygen to the tissues and carbon dioxide diffuses from the tissues into the blood, where the following processes occur (Fig. 10-22):

1 A small fraction (8 percent) of the carbon dioxide remains physically dissolved in the plasma and red blood cells.

2 The largest fraction (67 percent) of the carbon dioxide undergoes the reactions described in Eq. **10-2** and is converted into bicarbonate and hydrogen ions. This occurs primarily in the red blood cells because they contain large quantities of the enzyme carbonic anhydrase but the plasma does not. This explains why tissue capillary hydrogen-ion concentration is higher than that of the arterial blood and increases as metabolic activity increases. The fate of these hydrogen ions will be discussed in Chap. 11. Bicarbonate, in contrast to carbon dioxide, is extremely soluble in blood.

3 The remaining fraction (25 percent) of the carbon dioxide reacts directly with hemoglobin to form $HbCO_2$, as in Eq. **10-3**.

Since these are all reversible reactions, i.e., they can proceed in either direction, depending upon the prevailing

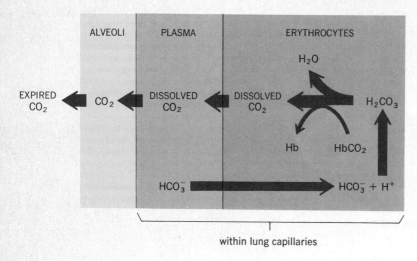

FIGURE 10-23
Summary of carbon dioxide movements and reactions as blood flows through the lung capillaries. All movements across membranes are by passive diffusion. The plasma-erythrocyte phenomena are simply the reverse of those occurring during blood flow through the tissue capillaries, as shown in Fig. 10-22. The breakdown of H_2CO_3 is catalyzed by carbonic anhydrase.

conditions, why do they all proceed primarily to the right, toward generation of HCO_3^- and $HbCO_2$, as blood flows through the tissues? In any chemical reaction, increasing the concentration of any of the reacting substances on the left side of the equation drives the reaction toward the right. The converse, of course, is also true. Once again, the answer is provided by the law of mass action: It is the increase in carbon dioxide concentration which drives these reactions to the right as blood flows through the tissues.

Obviously, a sudden lowering of blood P_{CO_2} has just the opposite effect. HCO_3^- and H^+ combine to give H_2CO_3, which generates carbon dioxide and water. Similarly, $HbCO_2$ generates hemoglobin and free carbon dioxide. This is precisely what happens as venous blood flows through the lung capillaries (Fig. 10-23). Because the blood P_{CO_2} is higher than alveolar, a net diffusion of carbon dioxide from blood into alveoli occurs. This loss of carbon dioxide from the blood lowers the blood P_{CO_2} and drives these chemical reactions to the left, thus generating more dissolved carbon dioxide. Normally, as fast as this carbon dioxide is generated from HCO_3^- and H^+ and from $HbCO_2$, it diffuses into the alveoli. In this manner, all the carbon dioxide delivered into the blood in the tissues now is delivered into the alveoli, from which it is expired and eliminated from the body.

Control of respiration

In dealing with the mechanisms by which the basic respiratory processes are controlled, we shall be concerned primarily with two questions: By what mechanisms are rhythmical breathing movements generated? What factors control the rate and depth of breathing, i.e., the total ventilatory volume?

Neural generation of rhythmic breathing

Like cardiac muscles, the inspiratory muscles normally contract rhythmically; however, the origins of these contractions are quite different. Cardiac muscle has automaticity; i.e., it is capable of self-excitation. The nerves to the heart merely alter this basic inherent rate and are not actually required for cardiac contraction. On the other hand, the diaphragm and intercostal muscles consist of skeletal muscle, which cannot contract unless stimulated by nerves. Thus, breathing depends entirely upon cyclical respiratory muscle excitation by the phrenic nerve (to the diaphragm) and the intercostal nerves (to the intercostal muscles). These nerves originate in the spinal cord at the levels of the neck and thorax. Destruction of them or the spinal cord areas from which they originate (as in poliomyelitis, for example) results in complete paralysis of the respiratory muscles and death, unless some form of artificial respiration can be rapidly instituted.

At the end of expiration, when the chest is at rest, a few impulses are still passing down these nerves. Like other skeletal muscles, therefore, the respiratory muscles have a certain degree of resting tonus. This muscular contraction is too slight to move the chest but plays a role in maintaining normal posture. Inspiration is initiated by an increased rate of firing of these inspiratory motor units. More and more new motor units are recruited, and thoracic expansion increases. In addition, the firing frequency of the individual units increases. By these two measures, the force of inspiration increases as it proceeds. Then almost all these units stop firing, the inspiratory muscles relax, and expiration occurs as the elastic lungs recoil. In addition, when expiration is facilitated by contraction of expiratory muscles, the nerves to these muscles, having been quiescent during inspiration, begin firing during expiration.

By what mechanism are nerve impulses to the respiratory muscles alternately increased and decreased? Control of this neural activity resides primarily in neurons with cell bodies in the lower portion of the brain stem, *the medulla*, which also contains the cardiovascular control centers. If the spinal cord is cut at any point between the medulla and the areas of the spinal cord from which the phrenic and intercostal nerves originate, breathing ceases. This experiment demonstrates that these efferent nerves are controlled by synaptic connections with neurons which descend, within the spinal cord, from the medulla.

By means of tiny electrodes placed in various parts of the medulla to record electric activity, neurons have been found which discharge in perfect synchrony with inspiration and smaller numbers of other neurons which discharge synchronously with expiration, called *inspiratory* and *expiratory neurons*, respectively. Although they are not necessarily clustered together in homogeneous groups, it is convenient to refer to them as the inspiratory and expiratory centers. Electric stimulation of the inspiratory neurons can produce a maximal inspiratory effort which is sustained as long as the stimulus is applied (Fig. 10-24). Conversely, electric stimulation of the expiratory neurons shuts off inspiration abruptly and frequently produces active contraction of the expiratory muscles (Fig. 10-24).

What are the factors which induce firing of the medullary inspiratory neurons? It is quite likely that these neurons have inherent automaticity and rhythmicity, i.e., the capacity for cyclical self-excitation; on the other hand, synaptic input from other neurons plays an essential role in setting their actual rhythm. In other words, this situation resembles that in the heart, where the rate at which the SA node actually fires is normally determined, in large part, by input from the sympathetic and parasympathetic nerves. The inputs to medullary inspiratory neurons which are crucial for maintenance of normal rhythm are (*1*) reciprocal connections with medullary expiratory neurons, (*2*) connections with the *pons*, the area of brain just above the medulla, and (*3*) afferent

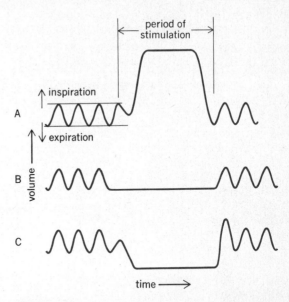

FIGURE 10-24

Respiration during electric stimulation of the medulla of a cat. Stimulation at one point A elicited profound and prolonged inspiration. At B, inspiration was completely inhibited. At C, inspiration was inhibited, and active expiration was induced.

input from *stretch receptors* in the lung. We shall discuss these in order, the letters of Fig. 10-25 serving as guides.

1 The medullary inspiratory and expiratory neurons have connections with each other which are inhibitory in nature (a). Thus, when the inspiratory neurons begin their firing, impulses travel to the expiratory neurons through inhibitory interneurons and restrain the expiratory neurons. Conversely, as the self-limited autonomous discharge of the inspiratory neurons wanes, the expiratory neurons are partially released from inhibition; they begin to fire and, via inhibitory interneurons, help terminate the discharge of the inspiratory neurons. In addition, under certain conditions involving active expiration, descending pathways from the expiratory neurons may stimulate the nerves to the expiratory muscles. Thus, these reciprocal connections help stop inspiration and synchronize it with expiration.

2 Electric stimulation or destruction of areas in the pons, produces profound changes in respiration ranging from marked stimulation to complete suppression. Most of these areas are part of, or closely related to, the reticular system described in Chap. 17. The medullary neurons receive a rich synaptic input from these areas, which exert a tonic effect upon the inspiratory neurons (b). It is also quite likely that they serve as a central relay station (c) for the respiratory inhibition initiated by the lung stretch receptors (d).

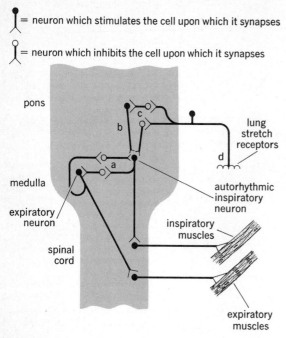

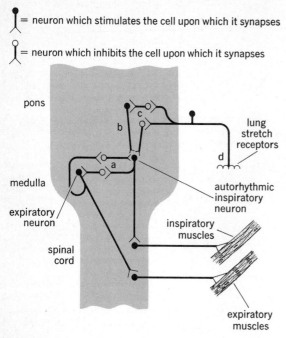 = neuron which stimulates the cell upon which it synapses

 = neuron which inhibits the cell upon which it synapses

pons

medulla

expiratory neuron

spinal cord

b
c
a
d
lung stretch receptors
autorhythmic inspiratory neuron
inspiratory muscles
expiratory muscles

FIGURE 10-25
Diagrammatic summary of neural generation of respiration. The letters denote key points described in the text. The motor nerves to the expiratory muscles are activated only during labored respiration.

3 Stimulation of afferent nerves from the lungs causes a temporary halt in breathing with the lungs remaining in the expiratory position. Conversely, cutting these fibers induces a highly characteristic pattern of slow, very deep breathing. It is evident that there are stretch receptors in the lungs (d), activation of which results in inhibition of the medullary inspiratory neurons, as shown in (c) in Fig. 10-25. As the lungs expand during inspiration, these receptors are stimulated, and impulses travel up the afferent nerves to the brain, where they aid in terminating inspiration. This pathway appears to be a relatively minor one in man.

In summary, the respiratory cycle is controlled primarily by medullary inspiratory neurons which give rise to the descending pathways to the efferent nerves innervating the diaphragm and intercostal muscles. One important factor in initiating inspiration is the spontaneous discharge of these inspiratory neurons. They cease firing as a result of both their own self-limitation and inhibitory impulses from medullary expiratory neurons and pulmonary stretch receptors acting via higher brain centers.

Cessation of inspiratory center activity means a marked decrease in the firing of inspiratory motor units and initiates passive expiration. In addition, active expiratory movements are precisely synchronized with this passive component of expiration as a result of the reciprocal connections between the medullary inspiratory and expiratory centers.

Control of ventilatory volume

In the previous section, we were concerned with the mechanisms that generate rhythmical breathing. It is obvious, however, that the actual respiratory rate is not fixed but can be altered over a wide range. Similarly, the depth and force of breathing movements can also be altered. As we have seen, these two factors, rate and depth, determine the alveolar ventilatory volume. Generally, rate and depth change in the same direction, although there may be important quantitative differences. For simplicity, we shall describe the control of total ventilation without attempting to discuss whether rate or depth makes the greatest contribution to the change.

Depth of respiration depends upon the number of motor units firing and their frequency of discharge, whereas respiratory rate depends upon the length of time elapsing between the bursts of motor unit activity (Fig. 10-26). As described above, the respiratory motor units are directly controlled by descending pathways from the medullary respiratory centers. The efferent pathways for control of ventilation are therefore clear-cut, and the critical question becomes: What is the nature of the afferent input to these centers? In other words, what variables does the control of ventilatory volume regulate?

This may seem a ridiculously complex way of phrasing a question when the answer seems so intuitively obvious. After all, respiration ought to supply oxygen as fast as it is consumed and ought to excrete carbon dioxide as fast as it is produced. *But,* how do the respiratory centers "know" what the body's oxygen requirements are? The logical way to approach this question is to ask what detectable changes would result from imbalance of metabolism and ventilation. Certainly the most obvious candidates are the plasma P_{O_2} and P_{CO_2}. Inadequate ventilation would lower the P_{O_2} because consumption would get ahead of supply and would elevate the P_{CO_2} because production would exceed elimination. Less obvious, perhaps, is the fact that arterial hydrogen-ion concentration also is exquisitely sensitive to changes in ventilation. Recall the equilibrium between H^+, HCO_3^-, and CO_2:

$$CO_2 + H_2O \rightleftharpoons H_2CO_3 \rightleftharpoons HCO_3^- + H^+$$

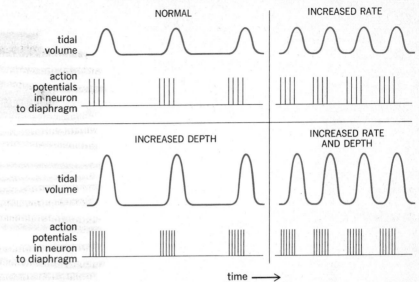

FIGURE 10-26
Recordings of depths of respiration (tidal volume) and frequency of action potentials in motor neurons to the diaphragm. Note that the rate of discharge during the inspiratory "burst" determines the tidal volume, whereas the time between bursts determines the respiratory rate.

As described by the law of mass action any increase in carbon dioxide concentration drives this reaction to the right, thereby liberating additional hydrogen ions. *Any increase or decrease in plasma Pco₂ is accompanied by changes in plasma hydrogen-ion concentration.*

However, these statements of fact in no way prove that any of these plasma concentrations actually are involved in respiratory regulation. The next step is to ask whether there are, indeed, receptors which can detect the levels of these variables in plasma and transmit the information to the respiratory centers. If so, where are these receptors located and what contribution do they make to overall control of ventilation? We shall describe the answers to these questions first for oxygen and then for carbon dioxide and hydrogen ion.

Control of ventilation by oxygen The rationale of the experiments to be described is quite simple: Alteration of inspired P_{O_2} produces changes in plasma P_{O_2}; if plasma P_{O_2} is important in controlling ventilation, we should observe definite changes in ventilation. If a normal person take a single breath of 100 percent oxygen (without being aware of it), after a latent period of about 8 sec, a transient reduction in his ventilation occurs of approximately 10 to 20 percent. Such studies have demonstrated that the normal sea-level plasma P_{O_2} of 100 mm Hg exerts a tonic stimulatory effect adequate to account for approximately 20 percent of total ventilation. Thus, the increase

in plasma P_{O_2} produced by the 100 percent oxygen removed this tonic stimulation and reduced ventilation.

Conversely, does a reduction in plasma P_{O_2} below normal cause a further increase in ventilation? Figure 10-27 illustrates the average response of a group of healthy subjects who breathed low P_{O_2} gas mixtures for 8 min. Note that no significant stimulation of respiration was observed until the oxygen content of the inspired air was reduced to one-half of normal air. Even at this point, some of the subjects failed to increase their ventilation despite an arterial P_{O_2} of 40 mm Hg and hemoglobin saturation of 75 percent. Only when the oxygen lack was

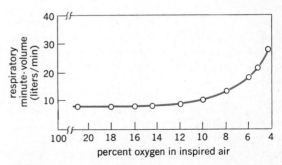

FIGURE 10-27
Effects of altering the oxygen content of inspired air on ventilation in normal man. The points are the average values for all subjects. (*Adapted from Comroe.*)

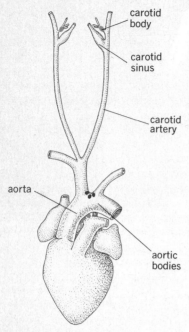

FIGURE 10-28
Location of the carotid and aortic bodies.

severe did all subjects manifest a sustained large increase in ventilation.

More recent studies, however, have demonstrated that our sensitivity to low P_{O_2} is actually greater than was originally concluded from data like those in Fig. 10-27. In the earlier experiments, a second important variable, the plasma P_{CO_2}, was not controlled. As we shall see in the subsequent section, voluntarily increasing one's ventilation causes a reduction of plasma P_{CO_2}, and this chemical change acts as a profound respiratory depressant. What apparently happened to the subjects of Fig. 10-27 was that, although even mild reductions of P_{O_2} actually raised the ventilation at first, this increase lowered their plasma P_{CO_2}, which inhibited ventilation and thereby counteracted the stimulatory effects of the reduced P_{O_2}. When this type of experiment was repeated with the P_{CO_2} held constant, all subjects manifested a considerably greater increase in ventilation when exposed to low P_{O_2} gas mixtures. The new experiments emphasize that man is indeed sensitive to a reduction of his plasma P_{O_2} but the respiratory stimulation may be overridden by other simultaneously occurring changes to which he is even more sensitive, particularly changes in P_{CO_2} and hydrogen-ion concentration. The concept that

respiration is controlled at any instant by multiple factors is of great importance.

The receptors stimulated by low P_{O_2} are located at the bifurcation of the common carotid arteries and in the arch of the aorta, quite close to, but distinct from, the baroreceptors described in Chap. 9 (Fig. 10-28). Known as the *carotid and aortic bodies*, they are composed of epithelial-like cells and neuron terminals in intimate contact with the arterial blood. Afferent nerves, arising from these terminals, ascend and enter the medulla, where they synapse ultimately with the neurons of the medullary centers. A low P_{O_2} increases the rate at which the receptors discharge, resulting in an increased number of action potentials traveling up the afferent nerves and a stimulation of the medullary inspiratory neurons. If an animal is exposed to a low P_{O_2} after the afferent nerves from the carotid and aortic bodies have been cut, the usual rapid increase in alveolar ventilation is not observed, demonstrating that the immediate stimulatory effects of a low P_{O_2} are completely mediated via this pathway.

What is the precise stimulus to these chemoreceptors? It is most likely their own internal P_{O_2}, that is, the concentration of dissolved oxygen within them. Normally, their blood supply is so huge relative to their utilization of oxygen that their internal P_{O_2} is virtually identical to the arterial P_{O_2}. For this reason we can state that, in effect, they monitor arterial P_{O_2}. Thus, any time arterial P_{O_2} is reduced by lung disease, hypoventilation, or high altitude, these chemoreceptors are stimulated and call forth a compensatory increase in ventilation. On the other hand, the oxygen deficiency of anemia (decreased hemoglobin content of the blood) does not stimulate ventilation because, although total oxygen content of the blood is reduced, the arterial P_{O_2} is quite normal. This also explains why there is no respiratory stimulation in carbon monoxide poisoning. Carbon monoxide, a gas which reacts with the same sites on the hemoglobin molecule as oxygen, has such a remarkable affinity for these sites that even small amounts reduce the ability of oxygen to combine with hemoglobin. Since it does not affect the amount of oxygen which can physically dissolve in blood, the P_{O_2} is unaltered, the carotid and aortic bodies are not stimulated[1], and the patient faints and dies of oxygen lack without ever increasing his ventilation.

Control of ventilation by carbon dioxide and hydrogen ion   The preponderant role that CO_2 plays in the control of ventila-

---

[1] Unfortunately, although there is general agreement that carbon monoxide does not stimulate ventilation, recent experiments have suggested that it may, in fact, stimulate the chemoreceptors. An explanation of this discrepancy has not been forthcoming.

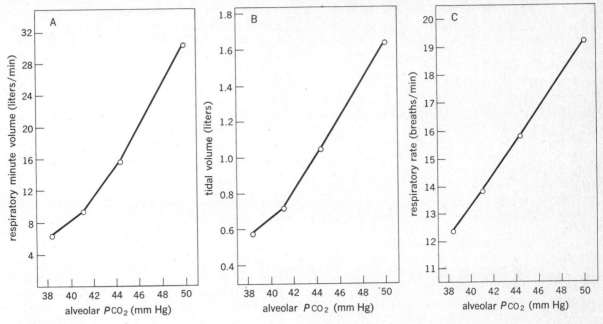

FIGURE 10-29
Effects on respiration of increasing alveolar P_{CO_2} by adding carbon dioxide to inspired air. Note that the increase in respiratory volume is due to an increase in both tidal volume and respiratory rate. This study was performed on normal human subjects. (*Adapted from Lambertsen.*)

tion can be demonstrated by a few relatively simple experiments. (*1*) Figure 10-29 illustrates the effects on respiratory volume of increasing the P_{CO_2} of inspired air. Normally, atmospheric air contains virtually no carbon dioxide. In the experiment illustrated, the subject breathed from bags of air containing variable quantities of carbon dioxide. The presence of carbon dioxide in the inspired air caused an elevation of alveolar P_{CO_2} and thereby an elevation of arterial P_{CO_2} as well. This increased P_{CO_2} markedly stimulated ventilation, an increase of 5 mm Hg in alveolar P_{CO_2} causing a 100 percent increase in ventilation. Obviously, ventilation is acutely dependent upon the P_{CO_2}. (*2*) A subject is asked to breathe as rapidly and deeply as he can. When the period of voluntary hyperventilation is over, the subject is told merely to breathe naturally. All subjects manifest markedly reduced breathing for the next few minutes, and a few stop breathing completely (apnea), often for 1 to 2 min (Fig. 10-30). Ventilation is inhibited because during the period of hyperventilation carbon dioxide was blown off faster than it was produced; accordingly, at the moment when the subject ceased to hyperventilate, his plasma P_{CO_2} was lower than normal, and ventilation was

inhibited until plasma P_{CO_2} returned toward normal as a result of accumulation of metabolically produced carbon dioxide. Note in Fig. 10-30 that the subject began breathing again, at least intermittently, before the P_{CO_2} was completely back to normal; this is due to stimulation of the carotid and aortic bodies by the extremely low P_{O_2}.

The opposite of hyperventilation is holding one's breath. Why can a person hold his breath for only a relatively short time? The lack of ventilation causes an accumulation of carbon dioxide and increased plasma P_{CO_2}; the ability of this increased P_{CO_2} to stimulate the respiratory center is so powerful that it overcomes the voluntary inhibition of the respiratory center (the latter mediated by descending pathways from the cerebral cortex). Unfortunately, underwater swimmers have misguidedly made use of these facts. They voluntarily hyperventilate for several minutes before submerging and are therefore able to hold their breaths for long periods of time. This is a very dangerous procedure, particularly during exercise, when oxygen consumption is high; because of our relative insensitivity to oxygen deficits, a rapidly decreasing P_{O_2} may cause fainting and drowning before ventilation is stimulated.

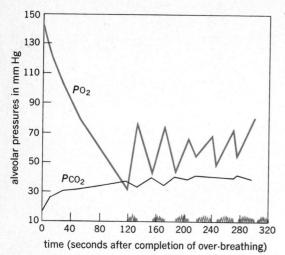

FIGURE 10-30
Variations in alveolar gas pressures after voluntary overbreathing for 2 min. The actual breathing movements are shown by the jagged lines at the bottom of the figure. (*Adapted from Douglas and Haldane.*)

The experiments we have discussed strongly support the hypothesis that P_{CO_2} is the major determinant of respiratory center activity. An increase in P_{CO_2} stimulates ventilation and thereby promotes the excretion of additional carbon dioxide. Conversely, a decrease in P_{CO_2} below normal inhibits ventilation and thereby allows metabolically produced carbon dioxide to accumulate and return the P_{CO_2} to normal. In this manner, the arterial P_{CO_2} is stabilized at the normal value of 40 mm Hg.

The question of the mechanism and afferent pathway by which the P_{CO_2} controls ventilation plunges us into one of the most violent controversies in respiratory physiology, a controversy which now seems close to being resolved, partly as a result of a series of brilliant experiments concerning the hydrogen-ion concentration of cerebrospinal fluid and brain tissue (in goats). The evidence, to date, indicates that the effects of carbon dioxide on ventilation are due not to carbon dioxide itself but to the associated changes in hydrogen-ion concentration. For example, the stimulant effects of breathing mixtures containing large amounts of carbon dioxide are probably due not to the effects of molecular carbon dioxide but to those of the increased hydrogen-ion concentration resulting from the chemical reactions described above. This is why we stressed the relationship between P_{CO_2} and hydrogen-ion concentration.

If we are to ascribe all the respiratory effects of plasma P_{CO_2} alteration to associated changes in hydrogen-ion concentration, we must be able to demonstrate that changes in hydrogen-ion concentration can alter ventilation even when P_{CO_2} remains unchanged (or changes in the opposite direction). This experiment can be readily performed by administering an acid such as hydrochloric acid to an animal; as predicted, the increased hydrogen-ion concentration induces marked stimulation of ventilation (Fig. 10-31). Conversely, reduction of hydrogen-ion concentration by administration of a base inhibits ventilation.

Why, then, the controversy? We must confess that we have presented only one side of the story; numerous other experiments seemed at the time to be inconsistent with this unitary hydrogen-ion concept, but the recent cerebrospinal fluid studies may have given an adequate explanation for these seemingly conflicting results. The critical hydrogen-ion concentration appears to be not that of the arterial blood but rather that of brain extracellular fluid, i.e., cerebrospinal fluid. An increased hydrogen-ion concentration increases the rate of discharge of the inspiratory neurons by acting either directly upon them or upon nearby chemosensitive cells (*central chemoreceptors*) having synaptic input to them; conversely, a decreased hydrogen-ion concentration inhibits their discharge. One is left with the rather startling conclusion that the control

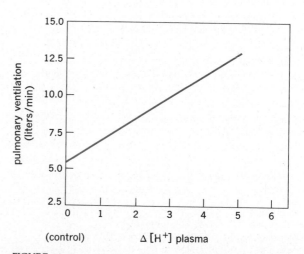

FIGURE 10-31
Change in ventilation in response to an elevation of plasma hydrogen-ion concentration, produced by the administration of hydrochloric acid. (*Adapted from Lambertsen.*)

of breathing (at least during rest) is aimed primarily at the regulation of *brain hydrogen-ion concentration!* However, this actually makes perfectly good sense, in terms of survival of the organism. As emphasized above, the relationship between carbon dioxide and hydrogen ion ensures that the regulation of hydrogen-ion concentration also produces relative constancy of P_{CO_2} as well. To continue the chain, the close relationship between oxygen consumption and carbon dioxide production ensures that ventilation adequate to maintain P_{CO_2} constant by excreting carbon dioxide as fast as it is produced also suffices to supply adequate oxygen (except in certain kinds of lung disease). Moreover, the shape of the hemoglobin dissociation curve minimizes any minor deficiency of oxygen supply, and any major oxygen deficit induces reflex respiratory stimulation via the carotid and aortic bodies. In contrast, brain function is extremely sensitive to changes in hydrogen-ion concentration so that even small increases or decreases in brain hydrogen-ion concentration could induce serious malfunction. We must point out, however, that the longevity of unified field hypotheses is frequently short, and the problem of respiratory control may not be settled quite so neatly as described above.

For the sake of completeness, it should be added that the carotid and aortic bodies, responsible for the low-oxygen reflexes, are also sensitive to changes in plasma hydrogen-ion concentration, but their contribution to the overall ventilatory response to changes in carbon dioxide and hydrogen ion is small, the direct effects on the brain being far more important. Finally, the relationship between ventilation and hydrogen-ion concentration is only one aspect of the regulation of extracellular fluid hydrogen-ion concentration described in Chap. 11.

Throughout this section we have described the stimulatory effects of carbon dioxide on ventilation. It should also be noted that very high levels of carbon dioxide depress the entire central nervous system, including the respiratory centers, and may therefore be lethal. Closed environments, such as submarines and space capsules, must be designed so that carbon dioxide is removed as well as oxygen supplied.

Control of ventilation during exercise

During heavy exercise, the alveolar ventilation may increase ten- to twentyfold to supply the additional oxygen needed and excrete the excess carbon dioxide produced. On the basis of our three variables—P_{O_2}, P_{CO_2}, and hydrogen-ion concentration—it might seem easy to explain the

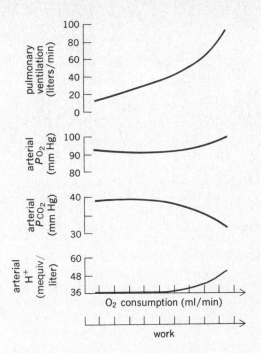

FIGURE 10-32
Relation of ventilation, arterial gas pressures, and hydrogen-ion concentration to the magnitude of muscular exercise. (*Adapted from Comroe.*)

mechanism which induces this increased ventilation. Unhappily, such is not the case.

Decreased P_{O_2} as the stimulus? It would seem logical that, as the exercising muscles consume more oxygen, plasma P_{O_2} would decrease and stimulate respiration. But, in fact, arterial P_{O_2} is *not* significantly reduced during exercise (Fig. 10-32). The alveolar ventilation increases in exact proportion to the oxygen consumption; therefore, P_{O_2} remains constant. Indeed, in exhausting exercise, the alveolar ventilation may actually increase relatively more than oxygen consumption, resulting in an *increased* P_{O_2}.

Increased P_{CO_2} as the stimulus? This is virtually the same story. Despite the marked increase in carbon dioxide production, the precisely equivalent increase in alveolar ventilation excretes the carbon dioxide as rapidly as it is produced, and arterial P_{CO_2} remains constant (Fig. 10-32). Indeed, for the same reasons given above for oxygen, the arterial P_{CO_2} may actually decrease during exhausting exercise.

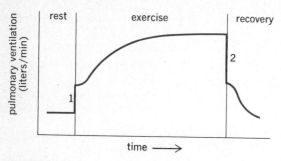

FIGURE 10-33
Ventilation changes in time during exercise. Note the abrupt increase 1 at the onset of exercise and the equally abrupt but larger decrease 2 at the end of exercise.

Increased hydrogen ion as the stimulus?   Since the arterial P_{CO_2} does not change (or decreases) during exercise, there is no accumulation of excess hydrogen ion as a result of carbon dioxide accumulation. Although there is an increase in arterial hydrogen-ion concentration for quite a different reason, namely, generation and release into the blood of lactic acid and other acids during exercise, the changes in hydrogen-ion concentration are not nearly great enough, particularly in only moderate exercise, to account for the increased ventilation.

We are left with the fact that, despite intensive study for more than 100 years by many of the greatest respiratory physiologists, we do not know what input stimulates ventilation during exercise. Our big three—P_{O_2}, P_{CO_2}, and hydrogen ion—appear presently to be inadequate, but many physiologists still believe that they will ultimately be shown to be the critical inputs. They reason that the fact that P_{CO_2} remains constant during moderate exercise is very strong evidence that ventilation is actually controlled by P_{CO_2}. In other words, if P_{CO_2} were not the major controller, how else could it remain unchanged in the face of the marked increase in its production? They reason that there may be a change in sensitivity of the chemoreceptors to CO_2 or hydrogen ion so that changes undetectable by experimental methods might be responsible for the stimulation of ventilation. Of course, this view is based presently on theoretical grounds; it remains to be seen whether experiments can validate it.

Moreover, the problem is even more complicated; as shown in Fig. 10-33, there is an abrupt increase (within seconds) in ventilation at the onset of exercise and an equally abrupt decrease at the end. Clearly, these changes occur too rapidly to be explained by alteration in some chemical constituent of the blood. Several other types of inputs which probably play some role are presented below, but they are also inadequate to explain fully the precise correlation between ventilation and oxygen consumption during exercise.

Control of respiratory-center activity by other factors

Temperature   An increase in body temperature frequently occurs as a result of increased physical activity and contributes to stimulation of alveolar ventilation. This facilitation of the respiratory centers is probably due both to a direct physical effect of increased temperature upon the respiratory-center neurons and to stimulation via pathways from thermoreceptors of the hypothalamus.

Input from the cerebral cortex   In Chap. 9, we described how the cardiovascular responses to exercise were probably mediated by input to the hypothalamus from branches of the neuron pathways descending from the cerebral cortex to the motor nerves. It is likely that these fibers send branches into the medullary respiratory centers as well. However, the magnitude of this contribution to stimulation of ventilation is not known.

Epinephrine   Injection of epinephrine into a person produces stimulation of respiration by unknown mechanisms. Epinephrine secretion is uniformly increased during heavy exercise and probably contributes to the increased ventilation.

Reflexes from joints and muscles   Many receptors in joints and muscles can be stimulated by the physical movements which accompany muscle contraction. It is quite likely that afferent pathways from these receptors play a significant role in stimulating the respiratory centers during exercise. Thus, the mechanical events of exercise help to coordinate alveolar ventilation with the metabolic requirements of the tissues.

Baroreceptor reflexes   We have already discussed the carotid sinus and aortic arch reflexes as the primary controllers of cardiovascular function. These reflexes can also alter alveolar ventilation, but their effect in man is usually minor.

Protective reflexes   A group of responses protect the respiratory tract against irritant materials, most familiar being the *cough* and the *sneeze*. These reflexes originate in receptors which line the respiratory tract. When they are excited, the result is stimulation of the medullary respiratory centers (via afferent nerves) in such a manner as to produce a deep inspiration and a violent expiration. In this manner, particles can be literally exploded out of the respiratory tract. Another example of a protective reflex is the immediate cessation of respiration which is frequently triggered when noxious agents are inhaled.

Pain Painful stimuli anywhere in the body can produce reflex stimulation of the respiratory centers. This is why one spanks a newborn infant to start his respirations. Another common example is the deep inspiration induced by a sudden shock such as entering a cold shower.

Emotion Emotional states are often accompanied by marked respiratory-center stimulation, as evidenced by the rapid breathing rate which characterizes fright and many similar emotions. In addition, the movement of air in or out of the lungs is an absolute requirement for such involuntary expressions of emotion as laughing and crying. In such situations, the respiratory centers must be primarily controlled by descending pathways from higher brain centers.

Voluntary control of breathing Although we have discussed in detail the involuntary nature of most respiratory reflexes, it is quite obvious that we retain considerable voluntary control of respiratory movements. This is accomplished by descending pathways from the cerebral cortex to the medullary respiratory centers. As we have seen, this voluntary control of respiration cannot be maintained when the involuntary stimuli, such as an elevated P_{CO_2} or hydrogen-ion concentration, become intense. Besides the obvious forms of voluntary control, e.g., breath holding, the respiratory centers must also be controlled during the production of complex voluntary actions such as speaking and singing. Obviously, the patterns of respiratory-center control by the cerebral cortex must be highly complex.

Hypoxia

Hypoxia is defined as a deficiency of oxygen at the tissue level. There are many potential causes of hypoxia, but they can be classed in four general categories: (1) *hypoxic hypoxia*, in which the arterial P_{O_2} is reduced; (2) *anemic hypoxia*, in which the arterial P_{O_2} is normal but the total oxygen content of the blood is reduced because of inadequate numbers of red cells, deficient or abnormal hemoglobin, or competition for the hemoglobin molecule by carbon monoxide; (3) *ischemic hypoxia*, in which the basic defect is too little blood flow to the tissues (as in shock, for example); and (4) *histotoxic hypoxia*, in which

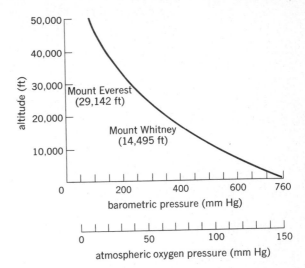

FIGURE 10-34
Relationship between altitude and barometric pressure. The P_{O_2} remains approximately 20 percent of the barometric pressure regardless of altitude.

the quantity of oxygen reaching the tissue is normal but the cell is unable to utilize the oxygen because a toxic agent (cyanide, for example) has interfered with its metabolic machinery.

Clearly, anemic hypoxia is mainly the result of abnormal blood formation or destruction, ischemic hypoxia is caused by failure of the cardiovascular system, and histotoxic hypoxia represents malfunction of the tissue cells, themselves. In contrast to these three categories, hypoxic hypoxia is due either to high-altitude exposure or to malfunction of the respiratory system. The causes are many, and a list of certain of the major ones provides an excellent review of the interplay between the various components of the respiratory system described in this chapter (Table 10-4). We shall describe the response to high altitude as a representative example of the range of reflex adjustments which can be brought into play to ameliorate the reduced arterial P_{O_2} characteristic of hypoxic hypoxia.

Response to high altitude

Barometric pressure progressively decreases as altitude increases (Fig. 10-34). Thus, at the top of Mt. Everest (29,142 ft), the atmospheric pressure is 245 mm Hg. The air is still 20 percent oxygen which means that the P_{O_2} is 49 mm Hg. Obviously, the alveolar and arterial P_{O_2} must

TABLE 10-4
Some causes of hypoxic hypoxia

1 Decreased Po_2 in inspired air (high altitude)†
2 Hypoventilation
   Increased airway resistance (foreign body, asthma)
   Paralysis of respiratory muscles (poliomyelitis)
   Skeletal deformities
   Inhibition of medullary respiratory centers (morphine)
   Decreased "stretchability" of the lungs and thorax
   (deficient surfactant, thickened lung or chest tissues)
   Lung collapse (pneumothorax)
3 Deficient alveolar-capillary diffusion
   Decreased area for diffusion (pneumonia)
   Thickening of alveolar-capillary membranes (beryliosis)
4 Abnormal matching of ventilation and blood flow (emphysema)

† The phrases in parentheses are examples of specific diseases in each category.

decrease as one ascends unless pure oxygen is breathed. The effects of oxygen lack vary with individuals, but most persons who ascend rapidly to altitudes above 10,000 ft experience some degree of mountain sickness, consisting of breathlessness, palpitations, headache, nausea, fatigue, and impairment of a host of mental processes (vision, judgment, etc.). Over the course of several days, these symptoms diminish and ultimately disappear, although maximal physical capacity remains reduced. This process, known as acclimatization, is achieved by the compensatory mechanisms listed below. The highest villages permanently inhabited by man are in the Andes at 18,000 ft. These villagers work quite normally, and apparently the only major precaution they take is that the women come down to lower altitudes during late pregnancy in order to protect the fetus.

In response to the hypoxia, oxygen supply to the tissues is maintained in four ways: (*1*) by increased alveolar ventilation, (*2*) by increased cardiac output, (*3*) by increased erythrocytes and hemoglobin content of blood, and (*4*) by a shift in the hemoglobin dissociation curve downward and to the right. The last phenomenon is probably mediated by an increased DPG secondary to hypoxia-induced stimulation of red cell glycolysis. These four responses have the net effect of supplying normal amounts of oxygen to the tissues with minimal deviation of tissue Po_2 from normal.

The first acute responses to high altitude are increased ventilation and cardiac output, mediated via the carotid and aortic bodies. As we have seen, the arterial Po_2 must generally fall to at least 60 mm Hg in order to elicit increased ventilation although the cardiovascular response is generally more sensitive. The arterial Po_2 does not reach this low level until the altitude is 10,000 ft or greater. However, the increased ventilation produced by low arterial Po_2 induces changes in blood Pco_2 and hydrogen-ion concentration which act as an undesirable brake. Since carbon dioxide is excreted more rapidly than it is produced, plasma (and, therefore, cerebrospinal fluid) Pco_2 and hydrogen-ion concentration decrease below normal and inhibit ventilation. The Pco_2 remains quite low in completely acclimatized persons, but the plasma hydrogen-ion concentration is restored toward normal by means of kidney activity. Of at least equal importance is the fact that, over time, mechanisms are brought into play which return cerebrospinal-fluid hydrogen-ion concentration back to normal despite reduced Pco_2. In this manner, much of the inhibitory effect on respiration of the excess carbon dioxide excretion is eliminated; the net result is an increasing ventilation over the first few days.

As ventilation increases, thereby increasing oxygen supply, the cardiovascular responses simultaneously diminish, and cardiac output may return almost to normal. During this same time, erythrocyte and hemoglobin synthesis have been stimulated by the hormone *erythropoietin* (see below), and the person's total circulating red cell mass increases considerably. Another slowly developing compensation, the mechanism of which is unknown, is the growth of many new capillaries. It has the important effect of decreasing the distance which oxygen must diffuse from blood to cell.

Erythrocyte and hemoglobin balance

Erythrocyte volume and hemoglobin content are not fixed but are subject to physiological control. Erythrocytes are incomplete cells in that they lack both nuclei and the metabolic machinery to synthesize new proteins. Thus, they can neither reproduce themselves nor maintain their normal structure for any length of time. As essential enzymes within them deteriorate and are not replaced, the cell ages and ultimately dies. Fortunately, their oxygen-carrying ability is not significantly diminished during the aging period. The average life of an erythrocyte is approximately 120 days, which means that almost 1 percent of the total erythrocytes in the body are destroyed every day. Destruction of erythrocytes is accomplished by a group of large cells, the *phagocytes,* found in liver, spleen, bone marrow, and lymph nodes, usually lining the blood vessels or lying close to them. These cells, the physiology of which is described in detail in Chap. 15, ingest and destroy the erythrocytes by breaking down their large complex molecules with enzymes. The hemoglobin molecule is broken down and converted to a yellow molecule named *bilirubin,* which is released into the blood. The liver cells pick up this substance from the blood and add it to the bile, but if the liver is damaged or overloaded because of abnormally high rate of erythrocyte destruction, the bilirubin may accumulate in the blood and give a yellow color to the skin. This is called *jaundice.*

Obviously, in a normal person, a quantity of erythrocytes equal to that destroyed must be simultaneously synthesized and released into the circulatory system. The site of erythrocyte production is the soft highly cellular interior of bones called *bone marrow.* In the adult, erythrocyte formation occurs only in the bones of the chest, the base of the skull, and the upper arms and legs. Active bone marrow constitutes approximately 5 percent of the total body weight. The erythrocytes are originally descended from large bone-marrow cells which contain no hemoglobin but do have nuclei and are therefore capable of cell division. After several cell divisions, cells emerge which are identifiable as immature erythrocytes because they contain hemoglobin. As maturation continues, these cells accumulate increased amounts of hemoglobin and their nuclei become progressively smaller until they ultimately disappear completely. The mature erythrocyte leaves the bone marrow via the rich network of marrow capillaries and enters the general circulation, in which it will travel for some 120 days.

This growth process obviously requires a number of different raw materials. The formation of the erythrocyte itself requires the usual nutrients and structural materials: amino acids, lipids, and carbohydrates. In addition, certain growth factors, including vitamin B_{12} and folic acid, are essential for normal erythrocyte formation. Finally, the formation of an erythrocyte requires the materials which go into the making of hemoglobin: iron, amino acids, and the organic molecules which are incorporated into the protein portion of hemoglobin. A lack of any of these growth factors or raw materials results in the failure of normal *erythropoiesis* (erythrocyte formation) and a decreased quantity of effective circulating erythrocytes (*anemia*). The substances which are most commonly lacking are iron and vitamin B_{12}.

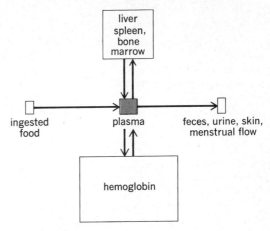

FIGURE 10-35
Summary of iron balance. The sizes of the boxes represent the quantity of iron involved. The magnitude of exchange in a particular direction varies according to conditions.

Iron

Iron is obviously an essential component of the hemoglobin molecule since it is with this element that the oxygen actually combines. The balance of iron and its distribution within the body are shown schematically in Fig. 10-35. About 70 percent of the total body iron is in hemoglobin, and the remainder is stored primarily in the liver, spleen, and bone marrow. As erythrocytes are destroyed, most of the iron released from hemoglobin is returned to these depots. Small amounts of iron, however, are lost each day via the urine, feces, sweat, and cells sloughed from the skin. In addition, women lose a significant quantity of iron via menstrual blood. In order to remain in iron balance, the amount of this metal lost from the body must be replaced by ingestion of iron-containing foods. An upset of this balance results either in iron deficiency and inadequate hemoglobin production or in an excess of iron in the body and serious toxic effects. The control of iron balance is unusual in that it resides primarily in the intestinal epithelium, which absorbs the iron from ingested food. The intestine absorbs only a fraction of the ingested iron, and, what is more important, this fraction is increased or decreased depending upon the state of body iron balance. These fluctuations appear to be mediated by changes in the iron content of the intestinal epithelium itself. When body stores of iron are ample, the epithelial iron store is increased, and its presence somehow prevents the absorption of most of the iron ingested in food. When the body stores of iron drop (as a result of hemorrhage, for example), some of the iron stored in the intestinal epithelium enters the blood and is transported to the bone marrow. The resulting lowering of intestinal epithelial iron content permits increased intestinal absorption of dietary iron and returns total body iron to normal.

Vitamin B$_{12}$

Normal erythrocyte formation requires extremely small quantities (one-millionth of a gram per day) of a cobalt-containing molecule, vitamin B$_{12}$, which, by mechanisms still unknown, permits final maturation of the erythrocyte. This substance is not synthesized within the body, which therefore depends upon its dietary intake. Absorption of vitamin B$_{12}$ from the gastrointestinal tract into the blood requires a substance (still unidentified) normally secreted by the epithelial lining of the stomach. Its absence prevents normal absorption of vitamin B$_{12}$ and causes deficient erythrocyte production and *pernicious anemia.*

Regulation of erythrocyte production

In a normal person, the total volume of circulating erythrocytes remains remarkably constant. Such constancy is required both for delivery of oxygen to the tissues and for maintenance of the blood pressure. From the preceding paragraphs it should be evident that a constant number of circulating erythrocytes can be maintained only by balancing erythrocyte production and destruction (or loss). Wide variations in the latter result from erythrocyte-destroying diseases or hemorrhage, and the balance must therefore be obtained by controlling the rate of erythrocyte production. During periods of severe erythrocyte destruction or hemorrhage, the normal rate of erythrocyte production can be increased more than sixfold.

It is easiest to discuss the control of erythrocyte production by first naming the mechanisms *not* involved. In the previous section, we listed a group of nutrient substances, such as iron and vitamin B$_{12}$, which must be present for normal erythrocyte production. However, none of these substances actually *regulates* the rate of production.

The direct control of erythropoiesis is exerted by a hormone called *erythropoietin.* The best evidence now indicates that the cells producing erythropoietin are located in the kidneys, although other sites in the body may also contribute to its production. Normally, a small quantity of erythropoietin is circulating, which stimulates the bone marrow to produce erythrocytes at a certain basal rate. An increase in circulating erythropoietin further stimulates the bone marrow and increases erythropoiesis, whereas a decrease in circulating erythropoietin below the normal level decreases erythropoiesis.

At the present time, the components of the reflex pathway which controls erythropoietin production remain largely a mystery. The common denominator of changes which increase circulating erythropoietin is precisely what one would logically expect—a decreased oxygen delivery to the tissues. As described earlier, this can result either from decreased erythrocyte volume, decreased blood flow, or decreased oxygen delivery from the lungs into the blood. Thus, hemorrhage, decreased heart pumping, or any of the situations summarized in Table 10-4, all stimulate erythropoietin production. As a result, erythrocyte production is increased, the oxygen-carrying capacity of the blood is increased, and oxygen delivery to the tissues is returned toward normal. Unfortunately, neither the receptors which detect the change in oxygen delivery nor the afferent pathway which relays this information to the erythropoietin-producing cells has been discovered.

Regulation of water and electrolyte balance

water, Na, K, Ca, H

11

One of the major themes of this book is the regulation of the internal environment. A cell's function depends not only upon receiving a continuous supply of organic nutrients and eliminating its metabolic end products but also upon the existence of stable physicochemical conditions in the extracellular fluid bathing it. Among the most important substances contributing to these conditions are water, sodium, potassium, calcium, and hydrogen ion. This chapter is devoted to a discussion of the mechanisms by which the total amounts of these substances in the body and their concentrations in the extracellular fluid are maintained relatively constant.

A substance appears in the body either as a result of ingestion or as a product of metabolism. Conversely, a substance can be excreted from the body or consumed in a metabolic reaction. Therefore, if the quantity of any substance in the body is to be maintained at a constant level over a period of time, the total amounts ingested and produced must equal the total amounts excreted and consumed. This is a general statement of the *balance concept* described in Chap. 5. For water and hydrogen ion, all four possible pathways apply. However, for the mineral

TABLE 11-1
Normal routes of water gain and loss in adults

	Milliliters per day
Intake:	
Drunk	1,200
In food	1,000
Metabolically produced	350
Total	2,550
Output:	
Insensible loss (skin and lungs)	900
Sweat	50
In feces	100
Urine	1,500
Total	2,550

electrolytes balance is simpler since they are neither synthesized nor consumed by cells, and their total body balance thus reflects only ingestion versus excretion.

As an example, let us describe the balance for total body water (Table 11-1). It should be recognized that these are average values, which are subject to considerable variation. The two sources of body water are metabolically produced water, resulting largely from the oxidation of carbohydrates, and ingestion of water in liquids and so-called solid food (a rare steak is approximately 70 percent water). There are four sites from which water is lost to the external environment: skin, lungs, gastrointestinal tract, and kidneys. The loss of water by evaporation from the cells of the skin and the lining of respiratory passageways is a continuous process, often referred to as *insensible loss* because the person is unaware of its occurrence. Additional water can be made available for evaporation from the skin by the production of sweat. The normal gastrointestinal loss of water (in feces) is quite small but can be severe in vomiting or diarrhea.

Under normal conditions, as can be seen from Table 11-1, water loss exactly equals water gain, and no net change of body water occurs. This is obviously no accident but the result of precise regulatory mechanisms. The question then is: Which processes involved in water balance are controlled to make the gains and losses balance? The answer, as we shall see, is voluntary intake (*thirst*) and urinary loss. This does not mean that none of the other processes is controlled but that their control is not primarily oriented toward water balance. Carbohydrate catabolism, the major source of the water of oxidation, is controlled by mechanisms directed toward regula-

tion of energy balance. Sweat production is controlled by mechanisms directed toward temperature regulation. Insensible loss (in man) is truly uncontrolled, and fecal water loss is generally quite small and unchanging.

The mechanism of thirst is certainly of great importance, since body deficits of water, regardless of cause, can be made up only by ingestion of water, but it is also true that our fluid intake is often influenced more by habit and sociological factors than by the need to regulate body water. The control of urinary water loss is the major mechanism by which body water is regulated.

By similar analyses, we find that the body balances of most of the ions determining the properties of the extracellular fluid are regulated primarily by the kidneys. To appreciate the importance of these kidney regulations and the fact that severe kidney malfunction is rapidly fatal, one need only make a partial list of the more important simple inorganic substances which constitute the internal environment and which are regulated by the kidney: water, sodium, potassium, chloride, calcium, magnesium, sulfate, phosphate, and hydrogen ion. It is worth repeating that normal biological processes depend on the constancy of this internal environment, the implication being that the amounts of these substances must be held within very narrow limits, regardless of large variations in intake and abnormal losses resulting from disease (hemorrhage, diarrhea, vomiting, etc.). Indeed, the extraordinary number of substances which the kidney regulates and the precision with which these processes normally occur accounted for the kidney's being the last stronghold of the nineteenth-century vitalists, who simply would not believe that the laws of physics and chemistry could fully explain renal function. By what mechanism does urine flow rapidly increase when a person ingests several glasses of liquid? How is it that the patient on an extremely low salt intake and the heavy salt eater both urinate precisely the amounts of salt required to maintain their sodium balance? What mechanisms decrease the urinary calcium excretion of children deprived of milk?

This regulatory role is obviously quite different from the popular conception of the kidneys as glorified garbage disposal units which rid the body of assorted wastes and "poisons." It is true that several of the complex chemical reactions which occur within cells result ultimately in end products collectively called waste products (primarily because they serve no known biological function in man); e.g., the catabolism of protein produces approximately 30 g of urea per day. Other waste substances produced in relatively large quantities are uric acid (from nucleic acids), creatinine (from muscle creatine), and the end products of hemoglobin breakdown. There are many others, not all of which have been completely identified. Most of these substances are eliminated from the body as rapidly as they are produced, primarily by way of the kidneys. Many of these waste products are harmless, although the accumulation of certain of them within the body during periods of renal malfunction accounts for some of the disordered body functions which eventually kill the patient suffering from severe kidney disease. However, most of the problems which occur in renal disease are due simply to disordered water and electrolyte metabolism.

The kidneys have another excretory function which is presently assuming increased importance, namely, the elimination from the body of foreign chemicals (drugs, pesticides, food additives, etc.). A final kidney function is to act as endocrine glands secreting at least two hormones: erythropoietin (Chap. 10) and angiotensin (to be discussed later in this chapter).

SECTION A
BASIC PRINCIPLES OF RENAL PHYSIOLOGY

Structure of the kidney and urinary system

The kidneys are paired organs which lie in the back abdominal wall, one on each side of the vertebral column (Fig. 11-1). In man, each kidney is composed of approximately 1 million tiny units, all similar in structure and function. One such unit, or *nephron*, is shown in Fig. 11-2. The nephron consists of a *vascular component* (*the glomerulus*) and a *tubular component*. The mechanisms by which the kidneys perform their functions depend on the relationship between these two components.

Throughout its course, the tubule is composed of a single layer of epithelial cells which differ in structure and function from portion to portion. It originates as a blind sac, known as *Bowman's capsule*, which is lined with thin epithelial cells. On one side, Bowman's capsule is intimately associated with the glomerulus; on the other it opens into the first portion of the tubule, which is highly coiled and is known as the *proximal convoluted tubule*. The next portion of the tubule is a sharp hairpinlike loop, called *Henle's loop*. The tubule once more becomes coiled (the *distal convoluted tubule*) and finally runs a straight course as the *collecting duct*. From the glomerulus to the beginning of the collecting duct, each of the million tubules is completely separate from its neighbors, but there

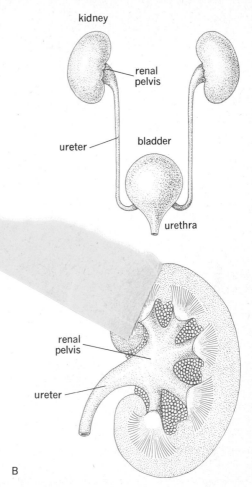

B

FIGURE 11-1
A Urinary system. The urine, formed by the kidney, collects in the renal pelvis and then flows through the ureter into the bladder, from which it is eliminated via the urethra. B Section of a human kidney. Half the kidney has been sliced away. Note that the structure shows regional difference. The outer portion, which has a granular appearance, contains all the glomeruli. The collecting ducts form a large portion of the inner kidney, giving it a striped pyramidlike appearance, and drain into the renal pelvis.

the tiny collecting ducts from separate tubules join to form ducts, which in turn form even larger ducts, which finally empty into a large central cavity, the *renal pelvis*, at the base of each kidney (Fig. 11-1). The renal pelvis is continuous with the *ureter*, which empties into the *urinary bladder*, where urine is temporarily stored and from

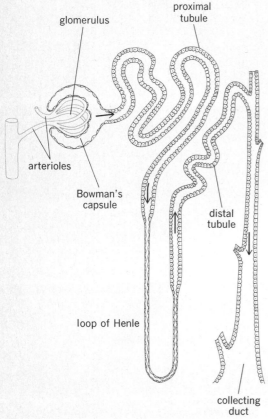

FIGURE 11-2
Basic structure of a nephron. (*Adapted from Pitts.*)

separated from the space within Bowman's capsule only by a thin layer of tissue composed of (*1*) the single-celled capillary lining, (*2*) a layer of basement membrane, and (*3*) the single-celled lining of Bowman's capsule. This thin barrier permits the *filtration* of fluid from the capillaries into Bowman's capsule.

In virtually all other organs, capillaries recombine to form the beginnings of the venous system. The glomerular capillaries instead recombine to form another set of arterioles, the *efferent arterioles*. Thus, blood leaves the glomerulus through an arteriole which soon subdivides into a second set of capillaries (Fig. 11-3). These *peritubular capillaries* are profusely distributed to, and intimately associated with, all the remaining portions of the tubule. They rejoin to form the venous channels, by which blood ultimately leaves the kidney.

which it is intermittently eliminated. The urine is not altered after it leaves the collecting ducts. From the renal pelvis on, the remainder of the urinary system simply serves as plumbing.

To return to the other component of the nephron: What is the origin and nature of the glomerulus? Blood enters the kidney via the renal artery, which then divides into progressively smaller branches. Each small artery gives off, at right angles to itself, a series of arterioles [*afferent arterioles* (Fig. 11-3)], each of which leads to a compact tuft of capillaries. This tuft of capillaries is the glomerulus, which protrudes into the side of Bowman's capsule and is completely covered by the epithelial lining of the capsule. The functional significance of this anatomical arrangement is that blood in the glomerulus is

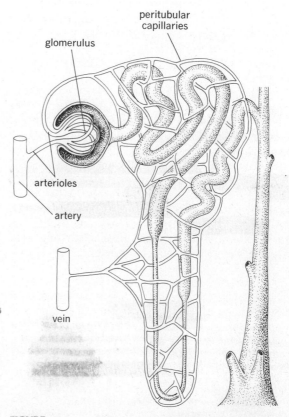

FIGURE 11-3
Relationships between vascular and tubular components of the nephron. (*Adapted from Smith.*)

Basic renal processes

Urine formation begins with the filtration of essentially protein-free plasma through the glomerular capillaries into Bowman's capsule. The final urine which enters the renal pelvis is quite different from the *glomerular filtrate* because, as the filtered fluid flows from Bowman's capsule through the remaining portions of the tubule, its composition is altered. This change occurs by two general processes, *tubular reabsorption* and *tubular secretion*. The tubule is at all points intimately associated with the peritubular capillaries, a relationship that permits transfer of materials between peritubular plasma and the inside of the tubule (*tubular lumen*). When the direction of transfer is from tubular lumen to peritubular capillary plasma, the process is called tubular reabsorption. Movement in the opposite direction, i.e., from peritubular plasma to tubular lumen, is called tubular secretion. (This term must not be confused with *excretion;* to say that a substance has been excreted is to say only that it appears in the final urine.) These relationships are illustrated in Fig. 11-4.

The most common relationships between these basic renal processes—glomerular filtration, tubular reabsorption, and tubular secretion—are shown in Fig. 11-5. Plasma containing substances X, Y, and Z enters the glomerular capillaries. A certain quantity of protein-free plasma containing these substances is filtered into Bowman's capsule, enters the proximal tubule, and begins its flow through the rest of the tubule. The remainder of the plasma, also containing X, Y, and Z, leaves the glomerular capillaries via an efferent arteriole and enters the peritubular capillaries. The cells composing the tubular epithelium can actively transport X (not Y or Z) from the peritubular plasma into the tubular lumen, but not in the opposite direction. By this combination of filtration and tubular secretion all the plasma which originally entered the renal artery is cleared of substance X, which leaves the body via the urine, thus reducing the amount of X remaining in the body. If the tubule were incapable of reabsorption, the Y and Z originally filtered at the glomerulus would also leave the body via the urine, but the tubule can transport Y and Z from the tubular lumen back into the peritubular plasma. The amount of reabsorption of Y is small, so that most of the filtered material does escape from the body, but for Z the reabsorptive mechanism is so powerful that virtually all the filtered material is transported back into the plasma, which flows through the renal vein back into the vena cava. Therefore no Z is lost from the body. Hence the processes of filtra-

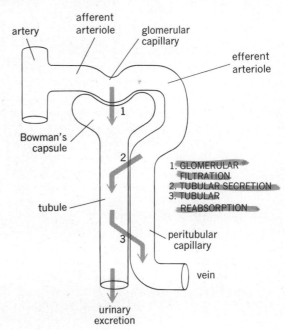

FIGURE 11-4
The three basic components of renal function.

tion and reabsorption have canceled each other out, and the net result is as though Z had never entered the kidney at all.

The kidney works only on plasma (the erythrocytes supply oxygen to the kidney but serve no other function in urine formation). Each substance in plasma is handled in a characteristic manner by the nephron, i.e., by a particular combination of filtration, reabsorption, and secretion. The critical point is that *the rates at which the relevant processes proceed for many of these substances are subject to physiological control.* What is the effect, for example, if the Y filtration rate is increased or its reabsorption rate decreased? Either change means that more Y is lost from the body via the urine. By triggering such changes in filtration or reabsorption whenever the plasma concentration of Y rises above normal, homeostatic mechanisms regulate plasma Y.

Glomerular filtration

As described in Chap. 9, the capillaries of the body are freely permeable to water and solutes of small molecular dimension (crystalloids). They are relatively impermeable

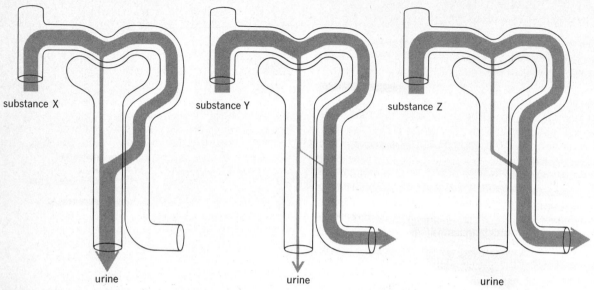

substance X

substance Y

substance Z

urine

urine

urine

FIGURE 11-5
Renal manipulation of three substances X, Y, and Z. X is filtered and secreted but not reabsorbed. Y is filtered, and a fraction is then reabsorbed. Z is filtered and is completely reabsorbed.

to large molecules (colloids), the most important of which are the plasma proteins. The glomerulus thus behaves qualitatively like any other capillary. This may seem surprising since the glomerular barrier is structurally different from most other capillaries in that it is composed of three layers: capillary endothelium, basement membrane, and the epithelium of Bowman's capsule. The presence of this last layer may well account for certain quantitative differences between the glomerulus and other capillaries, but it does not alter the basic process of filtration which occurs from capillary lumen into Bowman's capsule.

Proof of glomerular filtration   The proof that glomerular filtration occurs and is the initial process in urine formation depends upon three criteria: (1) The fluid within Bowman's capsule must be protein-free; (2) this fluid must contain all crystalloids (low-molecular-weight substances) in virtually the same concentrations as the plasma; and (3) the blood hydrostatic pressure in the glomerular capillaries must be large enough to account for the large volume of filtrate normally produced.

The first two criteria were proved by the theoretically simple but technically complicated method of mi-

cropuncture. The nephrons of amphibians are relatively large and can be seen since they are not packed together into an enclosed organ like human kidneys. Tiny pointed tubes (micropipets) were inserted into Bowman's capsule and extremely small quantities of fluid (approximately $\frac{1}{100}$ of an average-sized drop) were withdrawn and analyzed. It was demonstrated that the fluid in Bowman's capsule contains virtually no protein and that all crystalloids measured (glucose, hydrogen ion, chloride, potassium, phosphate, urea, creatinine, and uric acid) were present in the same concentrations as in plasma. Micropuncture has also been used to withdraw fluid from almost all portions of the tubule. Its application to mammals as well as to amphibia has provided most of the fundamental facts of renal physiology.

The third criterion of glomerular filtration can be satisfied by considering the balance of forces involved (Table 11-2). The mean blood pressure in the large arteries of the body is approximately 100 mm Hg, but this is not the hydrostatic pressure of the blood within the glomerular capillaries, since pressure is dissipated as the blood passes through the arterioles connecting the renal artery branches to the glomeruli. The glomerular capillary

TABLE 11-2
Forces involved in glomerular filtration

Forces	Millimeters of mercury
Favoring filtration:	
Glomerular capillary blood pressure	50
Opposing filtration:	
Fluid pressure in Bowman's capsule	10
Osmotic gradient	30
(water concentration difference	
due to protein)	
Net filtration pressure	10

pressure is usually about 50 mm Hg. This is about half of mean arterial pressure and is considerably higher than in other capillaries of the body because the afferent arterioles leading to the glomeruli are wider than most arterioles and therefore offer less resistance to flow. The capillary hydrostatic pressure favoring filtration is not completely unopposed. There is, of course, fluid within Bowman's capsule which results in a capsular hydrostatic pressure of 10 mm Hg resisting further filtration into the capsule. A second opposing force results from the presence of protein in the plasma and its absence in Bowman's capsule. As in other capillaries, this unequal distribution of protein causes the water concentration of the plasma to be less than that of the fluid in Bowman's capsule. Again, as in other capillaries, the water-concentration difference is due completely to the plasma protein since all the crystalloids have virtually identical concentrations in the plasma and Bowman's capsule. The difference in water concentration induces an osmotic flow of fluid (water plus all the crystalloids) from Bowman's capsule into the capillary, a flow that opposes filtration. Its magnitude is equivalent to the bulk flow produced by a hydrostatic pressure difference of 30 mm Hg.

As can be seen from Table 11-2, the net filtration pressure is approximately 10 mm Hg. This, as can be shown on theoretical grounds, is adequate to account for the observed rates of glomerular filtration. This pressure initiates urine formation by forcing an essentially protein-free filtrate of plasma through glomerular pores into Bowman's capsule and thence down the tubule into the renal pelvis. It should be evident from this description that the glomerular membranes serve only as a filtration barrier and play no active, i.e., energy-requiring, role. The energy producing glomerular filtration is the energy transmitted to the blood (as hydrostatic pressure) when the heart contracts.

Before leaving this topic, we must point out the reason for use of the term "essentially protein-free" in describing the glomerular filtrate. In reality there is a very small amount of protein in the filtrate since the glomerular membranes are not perfect sieves for protein. Normally, less than 1 percent of serum albumin and virtually no globulin are filtered; whatever protein is filtered is normally completely reabsorbed so that no protein appears in the final urine. However, in diseased kidneys, the glomerular membranes may become much more leaky to protein so that large quantities are filtered and some of this protein appears in the urine.

Rate of glomerular filtration In man, the average volume of fluid filtered from the plasma into Bowman's capsule is 180 liters/day (approximately 45 gal)! The implications of this remarkable fact are extremely important. When we recall that the average total volume of plasma in man is approximately 3 liters, it follows that the entire plasma volume is filtered by the kidneys some 60 times a day. It is, in part, this ability to process such huge volumes of plasma that enables the kidneys to excrete large quantities of waste products and to regulate the constituents of the internal environment so precisely. The second implication concerns the magnitude of the reabsorptive process. The average person excretes between 1 and 2 liters of urine per day. Since 180 liters of fluid is filtered, approximately 99 percent of the filtered water must have been reabsorbed into the peritubular capillaries, the remaining 1 percent escaping from the body as urinary water.

How is the rate of glomerular filtration measured? To answer this question we use another imaginary substance W (Fig. 11-6). Over a 24-hr period (or any other convenient period of time) the subject's urine is collected. The amount of W in this volume of urine is measured and found to be 720 mg. How did this amount of W get into the urine? From other experiments it has been determined that W is filtered at the glomerulus but is not secreted. Therefore, W can get into the urine only by filtration. Moreover, it also has been previously determined that W is not reabsorbed at all by the tubules. Thus, all the filtered W appears in the final urine. Therefore, in order to excrete 720 mg of W in 24 hr, the subject must have filtered 720 mg of W during the same 24-hr period. The question now becomes: How much glomerular filtrate contains 720 mg of W? Since W is a molecule of small dimensions, it must have the same concentration in the glomerular filtrate as in plasma. Several samples of the person's blood are obtained during the same 24-hr pe-

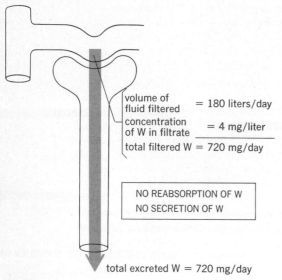

volume of
fluid filtered   = 180 liters/day

concentration
of W in filtrate   = 4 mg/liter

total filtered W = 720 mg/day

NO REABSORPTION OF W
NO SECRETION OF W

total excreted W = 720 mg/day

FIGURE 11-6
Measurement of glomerular filtration. W is filtered but is neither reabsorbed nor secreted.

riod, and his plasma W concentration is found to be 4 mg/liter of plasma. Since 1 liter of plasma contains 4 mg of W, then 720/4, or 180, liters contain 720 mg of W. In other words, 180 liters of plasma having a W concentration of 4 mg/liter must have been filtered during the 24-hr period in order to account for the appearance of 720 mg of W in the final urine. This example is illustrated in Fig. 11-6. The validity of this analysis depends upon the fact that W is freely filterable at the glomerulus and is neither reabsorbed nor secreted by the tubules. A polysaccharide called *inulin* (not insulin) completely fits this description and is used for just such determinations in man and experimental animals.

This type of experiment, called *clearance technic*, has proved invaluable for gaining information about tubular reabsorption and secretion. For example, suppose we were interested in learning whether there is tubular reabsorption of phosphate. Using inulin, we would determine the glomerular filtration rate (GFR). (This must be repeated for every experiment because the GFR is not fixed but varies significantly.) The GFR in this particular experiment is found to be 165 liters/day. The plasma concentration of phosphate is 1 mmole/liter. Since phosphate is completely filterable at the glomerulus, 165 liters/day × 1 mmole/liter = 165 mmoles/day is filtered.

Finally, the amount of phosphate in the urine during this same 24-hr period is found to be 40 mmoles. Therefore, 165 − 40 mmoles = 125 mmoles of phosphate must have been reabsorbed by the tubules per 24 hr. The generalization emerging from this example is that whenever the quantity of a substance excreted in the urine is *less* than the amount filtered during the same period of time, tubular reabsorption must have occurred. Conversely, if the amount excreted in the urine is *greater* than the amount filtered during the same period of time, tubular secretion must have occurred.

To complete this discussion of glomerular filtration we must consider the magnitude of the *total renal blood flow*. We have seen that none of the red blood cells and only a portion of the plasma which enters the glomerular capillaries are filtered into Bowman's capsule, the remainder passing via the efferent arterioles into the peritubular capillaries. Normally, the glomerular filtrate constitutes approximately one-fifth of the total plasma entering the kidney. Thus the total renal plasma flow is equal to 5 × 180 = 900 liters/day or 0.610 liter/min. Since plasma constitutes approximately 55 percent of whole blood, the total renal blood flow, i.e., erythrocytes plus plasma, must be approximately 1.1 liters/min. Thus, the kidneys receive one-fifth to one-fourth of the total cardiac output (5 liters/min) although their combined weight is less than 1 percent of the total body weight! These relationships are illustrated in Fig. 11-7.

Tubular reabsorption

Many filterable plasma components are either completely absent from the urine or present in smaller quantities than were originally filtered at the glomerulus. This fact alone is sufficient to prove that these substances undergo tubular reabsorption. An idea of the magnitude and importance of these reabsorptive mechanisms can be gained from Table 11-3, which summarizes data for a few plasma

TABLE 11-3
Average values for several components handled by filtration and reabsorption

Substance	Amount filtered per day	Amount excreted per day	Percent reabsorbed
Water, liters	180	1.8	99.0
Sodium, g	630	3.2	99.5
Glucose, g	180	0	100
Urea, g	54	30	56

components, all of which are handled by filtration and reabsorption.

These are typical values for a normal person on an average diet. There are at least three important conclusions to be drawn from this table: (1) The quantities of material entering the nephron via the glomerular filtrate are enormous, generally larger than their total body stores; e.g., if reabsorption of water ceased but filtration continued, the total plasma water would be urinated within 30 min. (2) The quantities of waste products, such as urea, which are excreted in the urine are generally sizable fractions of the filtered amounts; thus, in mammals coupling a large glomerular filtration rate with a limited urea reabsorptive capacity permits rapid excretion of the large quantities of this substance produced constantly as a result of protein breakdown. (3) In contrast to urea and other waste products, the amounts of most useful plasma components, e.g., water, electrolytes, and glucose, which are excreted in the urine represent quite smaller fractions of the filtered amounts. For this reason one often hears the generalization that the kidney performs its regulatory function by *completely* reabsorbing all these biologically important materials and thereby preventing their loss from the body. This is a misleading half-truth, the refutation of which serves as an excellent opportunity for reviewing the essential features of renal function and regulatory processes in general.

Let us begin by pointing out the part of the generalization that is true. Certain substances, notably glucose, are not normally excreted in the urine because the amounts filtered are completely reabsorbed by the tubules. But does such a system permit the kidneys to *regulate* the plasma concentration of glucose, i.e., set it at some specific concentration? The following example will point out why the answer is *no*. Suppose the plasma glucose concentration is 100 mg/100 ml of plasma. Since reabsorption of this carbohydrate is complete, no glucose is lost from the body via the urine, and the plasma concentration remains at 100 mg/100 ml. If, instead of 100 mg/100 ml, we set our hypothetical plasma glucose concentration at 60 mg/100 ml, the analysis does not change; no glucose is lost in the urine, and the plasma glucose stays at 60 mg/100 ml. Obviously the kidney is merely maintaining whatever plasma glucose concentration happens to exist and is not involved in the regulatory mechanisms by which the original "setting" of the plasma glucose was accomplished. It is not the kidney but primarily the liver and the endocrine system which set and regulate the plasma glucose concentration. For comparison, consider what happens when a person

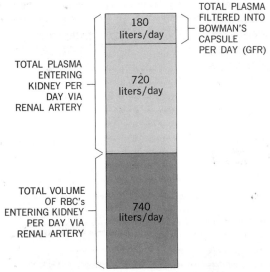

total renal blood flow = 1,640 liters/day

FIGURE 11-7
Magnitude of glomerular filtration rate, total renal plasma flow, and total renal blood flow. Only 20 percent of the plasma entering the kidneys is filtered from the glomerulus into Bowman's capsule. The remaining 80 percent flows through the glomerulus into the efferent arteriole and thence into the peritubular capillaries.

drinks a lot of water: within 1 to 2 hr all the excess has been excreted in the urine, chiefly, as we shall see, as the result of decreased renal tubular reabsorption of water. In this example the kidney is the effector organ of a reflex which maintains plasma water concentration within very narrow limits. The critical point is that for many plasma components, particularly the inorganic ions and water, the kidney does not always completely reabsorb the total amounts filtered. The rates at which these substances are reabsorbed (and therefore the rates at which they are excreted) are constantly subject to physiological control. The ability to *vary* the excretion of water, sodium, potassium, hydrogen ion, calcium, phosphate, and many other substances is really the essence of the kidney's ability to regulate the internal environment.

Types of reabsorption   A bewildering variety of ions and molecules is found in the plasma. With the exception of the proteins (and a few ions tightly bound to protein), these materials are all present in the glomerular filtrate, and most are reabsorbed, to varying extents. It is essential to realize that tubular reabsorption is a qualitatively different process than glomerular filtration. The latter occurs by bulk flow in which water and all dissolved free crystalloids move together. In contrast, tubular reabsorption of various substances is by more or less discrete tubular transport mechanisms and by diffusion, although in many cases a single reabsorptive system transports several different components if they are similar in structure. For example, many of the simple carbohydrates are reabsorbed by the same system. As described in Chap. 2, transport processes can be categorized broadly as *active* or *passive,* and there are many examples of each in the kidney. The process is passive if no cellular energy is directly and specifically involved in the transport of the substance, i.e., if the substance moves downhill by simple or facilitated diffusion as a result of an electric or chemical concentration gradient. Active reabsorption, on the other hand, can produce net movement of the substance uphill against its concentration or electric gradient and therefore requires energy expenditure by the transporting cells.

Transport of any substance across the renal tubule involves a sequence of steps. In Chap. 2 when we described the basic characteristics of transport processes, we were dealing only with transport across a single membrane, i.e., from the outside of the cell to the inside or vice versa. However, to cross the renal tubule, a substance must traverse not just one but a sequence of membranes. For example, to be reabsorbed, a sodium ion must gain entry to the tubular cell by crossing the cell membrane lining the lumen. It must then move through the cell's cytoplasm and cross the opposite cell membrane to enter the interstitial fluid. Finally, it must cross the capillary membranes to enter the plasma. The entire process is known as *transepithelial transport* and occurs not only in the kidney but in the gastrointestinal tract and other epithelial linings of the body.

In transepithelial transport the overall process is called *active* if one or more of the individual steps in the sequence is active. Sodium ions, for example, diffuse across the first tubular cell membrane and the cytoplasm of the cell. They are then actively transported out of the cell into the interstitial fluid; water accompanies the sodium and the entire fluid then gains entry into the capillary by the bulk-flow process typical of all capillaries. Thus, three of the four steps are passive, but the crucial step is mediated by an active carrier process, and the overall process of sodium reabsorption is therefore said to be active. In this chapter we shall ignore the fact that multiple membranes lie between the lumens of the tubule and capillary and treat the tubular epithelium as a single membrane separating tubular fluid and plasma.

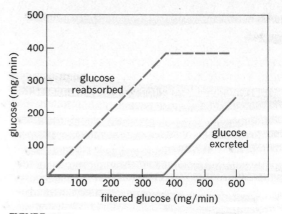

FIGURE 11-8
Saturation of the glucose transport system. Glucose is administered intravenously to a person so that plasma glucose and, thereby, filtered glucose are increased.

Many of the active reabsorptive systems in the renal tubule can transport only limited amounts of material per unit time, primarily because the membrane carrier responsible for the transport becomes saturated. The classic example is the tubular transport process for glucose. As we know, normal persons do not excrete glucose in their urine because tubular reabsorption is complete; but it is possible to produce urinary excretion of glucose in a completely normal person merely by administering large quantities of glucose directly into one of his veins. Recall that the filtered quantity of any freely filterable plasma component such as glucose is equal to its plasma concentration multiplied by the glomerular filtration rate. If we assume that, as we give our subject intravenous glucose, the glomerular filtration remains constant, the filtered load of glucose will be directly proportional to his plasma glucose concentration (Fig. 11-8). We shall find that even after his plasma glucose concentration has doubled, his urine will still be glucose-free, indicating that his *maximal tubular capacity* (T_m) for reabsorbing glucose has not yet been reached. But as the plasma glucose and the filtered load continue to rise, glucose finally appears in the urine. From this point on any further increase in plasma glucose is accompanied by a directly proportionate increase in excreted glucose, because the T_m has now been reached. The tubules are now reabsorbing all the glucose they can, and any amount filtered in excess of this quantity cannot be reabsorbed and appears in the urine. This is precisely what occurs in the patient with diabetes mellitus. Because of a deficiency in pancreatic production of insulin, the patient's plasma glucose may rise to extremely high values. The filtered load of glucose becomes great enough to exceed the T_m, and glucose appears in the urine. There is nothing wrong with his tubular transport mechanism for glucose, which is simply unable to reabsorb the huge filtered load.

Except for our experimental subject receiving intravenous glucose, the plasma glucose in normal persons never becomes high enough to cause urinary excretion of glucose because the reabsorptive capacity for glucose is much greater than necessary for normal filtered loads. However, for certain other substances, e.g., phosphate, the reabsorptive T_m is very close to the normal filtered load. The adaptive value inherent in such a relationship should be readily apparent from the following example. On a normal person the following data are obtained:

Ingested PO_4 = 20 mmoles/day
GFR = 180 liters/day
Plasma PO_4 concentration = 1 mmole/liter
Filtered PO_4 = 1 × 180 = 180 mmole/day
T_m for PO_4 = 160 mmoles/day
Excreted PO_4 = 180 − 160 = 20 mmoles/day

Under these conditions the normal person remains in perfect phosphate balance since he is excreting precisely what he eats, and his plasma phosphate concentration therefore remains constant at 1 mmole/liter. If the next day he eats an unusually large quantity of phosphate and raises his plasma phosphate to 1.1 mmoles/liter, the data are

GFR = 180 liters/day
Filtered PO_4 = 1.1 × 180 = 198 mmoles/day
T_m for PO_4 = 160 mmoles/day
Excreted PO_4 = 198 − 160 = 38 mmoles/day

The very slight increase in plasma phosphate has resulted in a large increase in excreted phosphate and has eliminated the excess phosphate ingested. By this mechanism, depending on neither hormones nor nerves, the kidney can exert control over plasma phosphate concentration. We shall see that, in addition to this simplest of systems, more complex systems involving neural and hormonal components also exist for the regulation of phosphate and other electrolytes, but the basic relationship between filtered load and reabsorptive rate is the underlying principle for many of them.

Just as glucose and phosphate provide excellent examples of actively transported solutes, urea provides an example of passive transport. Since urea is filtered at the glomerulus, its concentration in the very first portion of the tubule is identical to its concentration in peritubular capillary plasma. Then, as the fluid flows along the tubule, water reabsorption occurs, increasing the concentration of any intratubular solute not being reabsorbed at the same rate as the water, with the result that intratubular urea concentration becomes greater than the peritubular plasma concentration. Accordingly, urea is able to diffuse passively down this concentration gradient from tubular lumen to peritubular capillary. Urea reabsorption is thus a passive process and completely dependent upon the reabsorption of water, which establishes the diffusion gradient. In man, urea reabsorption varies between 40 and 60 percent of the filtered urea, the lower figure holding when water reabsorption is low and the higher when it is high.

Passive reabsorption is also of considerable importance for many foreign chemicals. The renal tubular epithelium acts in many respects as a lipid barrier; accordingly, lipid-soluble substances, like urea, can penetrate it fairly readily. Recall from Chap. 1 that one of the major determinants of lipid solubility is the polarity of a molecule—the less polar, the more lipid-soluble. Many drugs and environmental pollutants are nonpolar and, therefore, highly lipid-soluble. This makes their excretion from the body via the urine quite difficult since they are filtered at the glomerulus and then reabsorbed, like urea, as water reabsorption causes their intratubular concentrations to increase. Fortunately, the liver transforms most of these substances to progressively more polar metabolites which, because of their reduced lipid solubility, are poorly reabsorbed by the tubules and can therefore be excreted (polarity does not influence glomerular filtration which is bulk flow, not diffusion).

Tubular secretion

Tubular secretory processes, which transport substances from peritubular capillaries to tubular lumen, i.e., in the direction opposite to tubular reabsorption, constitute a second pathway into the tubule, the first being glomerular filtration. Like tubular reabsorption processes, secretory transport may be either active or passive. Of the large number of different substances transported into the tubules by tubular secretion, only a few are normally found in the body, the most important being hydrogen ion and potassium. We shall see that most of the excreted hydrogen ion and potassium enters the tubules by secretion rather than filtration. Thus, renal regulation of these two important substances is accomplished primarily by mechanisms which control the rates of their tubular secretion. The kidney is also able to secrete a large number of foreign chemicals, thereby permitting their excretion from the body; penicillin is an example.

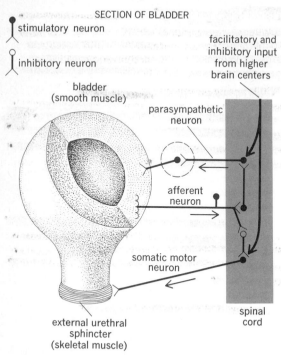

SECTION OF BLADDER

stimulatory neuron

inhibitory neuron

facilitatory and inhibitory input from higher brain centers

bladder (smooth muscle)

parasympathetic neuron

afferent neuron

somatic motor neuron

spinal cord

external urethral sphincter (skeletal muscle)

FIGURE 11-9

Basic structure and nerve supply of the bladder and external urethral sphincter. Stimulation of the mechanoreceptors in the bladder walls causes reflex stimulation of the parasympathetic nerves to the bladder and inhibition of the motor nerves to the external sphincter. The result is bladder contraction and sphincter relaxation. Higher-center input allows voluntary initiation or delay of micturition.

In the remainder of this chapter we shall see how the kidney acts as effector organ in a variety of homeostatic processes and how renal function is coordinated with that of other organs which also serve important regulatory roles as effector organs. Before turning to the individual variables being regulated, we complete our basic story by describing the mechanisms for eliminating urine from the body.

Micturition (urination)

From the kidneys, urine flows to the bladder through the ureters (Fig. 11-1), propelled by peristaltic contractions of the smooth muscle which makes up the ureteral wall and which has inherent rhythmicity, like other smooth muscles. The bladder, of thick layers of smooth muscle, walls squeeze the urine in the bladder, bladder contraction though it is sometimes called a sphincter, it is not a distinct muscle but the last portion of the bladder; however, when the bladder is relaxed, this circular smooth muscle is closed and functions as a sphincter. As the bladder contracts, this sphincter is pulled open simply by changes occurring in bladder shape during contraction. In other words, no special mechanism is required for its relaxation.

Another muscle important in the process of *micturition* (elimination of urine from the bladder) is a circular layer of skeletal muscle which surrounds the urethra farther down from the base of the bladder. When contracted, this *external urethral sphincter* can hold the urethra closed against even strong bladder contractions.

Micturition is basically a local spinal reflex which can be influenced by higher brain centers. The bladder muscle receives a rich supply of parasympathetic nerves, the stimulation of which causes bladder contraction (Fig. 11-9). The external urethral sphincter is innervated by somatic motor nerves just like any other skeletal muscle. The bladder wall contains many stretch receptors whose afferent nerves enter the spinal cord and eventually synapse with these parasympathetic and somatic motor neurons, stimulating the former and inhibiting the latter. Via descending pathways, higher brain centers synaptically facilitate and inhibit these motor pathways.

The following sequence of events leads to bladder emptying in an infant, in whom higher centers have only minor influence. When the bladder contains only small amounts of urine, its internal pressure is low, there is little stimulation of the bladder stretch receptors, the parasympathetics are relatively quiescent, and the somatic nerves are discharging at a moderately rapid rate. As the bladder fills with urine, it becomes distended, and the stretch receptors are gradually stimulated until their output becomes great enough to contract the bladder while simultaneously relaxing the external sphincter. Thus, the entire process is quite analogous to any other spinal reflex.

This process describes micturition adequately in the infant, but it is obvious that adults have the capacity either to delay micturition or to induce it voluntarily. In an adult, the volume of urine in the bladder required to initiate the spinal reflex for bladder contraction is approximately 300 ml. Delay is accomplished via descending pathways from the cerebral cortex which inhibit the bladder parasympathetics and stimulate the motor nerves to the external sphincter, thereby overriding the opposing synaptic input from the bladder stretch receptors. Voluntary initiation of micturition is just the opposite: Descending pathways from the cerebral cortex stimulate the bladder parasympathetics and inhibit the motor nerves to the sphincter, thereby summating with the afferent input from the stretch receptors and initiating micturition. It is by learning to control these pathways that a child achieves the ability to control the timing of micturition.

SECTION B
REGULATION OF SODIUM AND WATER BALANCE

Table 11-1 is a typical balance sheet for water; Table 11-4, for sodium chloride. As with water, the excretion of sodium via the skin and gastrointestinal tract is normally quite small but may increase markedly during severe sweating, vomiting, or diarrhea. Hemorrhage, of course, can result in loss of large quantities of both sodium and water.

Control of the renal excretion of sodium and water constitutes the most important mechanism for the regulation of body sodium and water. The excretory rates of these substances can be varied over an extremely wide range; e.g., a gross consumer of salt may ingest 20 to 25 g of sodium chloride per day whereas a patient on a low-salt diet may ingest only 50 mg. The normal kidney can readily alter its excretion of salt over this range. Similarly, urinary water excretion can be varied physiologically from approximately 400 ml/day to 25 liters/day, depending upon whether one is lost in the desert or participating in a beer-drinking contest.

Basic renal processes for sodium, chloride, and water

None of these substances undergoes tubular secretion; each is freely filterable at the glomerulus and approximately 99 percent is reabsorbed as it passes down the tubules (Table 11-3). Indeed, the vast majority of all renal energy production must be used to accomplish this enormous reabsorptive task. The tubular mechanisms for reabsorption of these substances can be summarized by three generalizations (Fig. 11-10): (*1*) The reabsorption of sodium is an active process, i.e., it is carrier-mediated, requires an energy supply, and can occur against an

TABLE 11-4
Normal routes of sodium chloride intake and loss

	Grams per day
Intake:	
Food	10.5
Output:	
Sweat	0.25
Feces	0.25
Urine	10.0
Total output	10.5

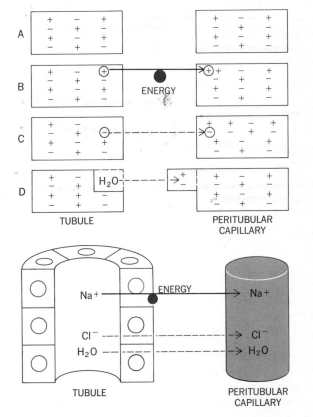

FIGURE 11-10
Diagrammatic representation of sodium, chloride, and water reabsorption. The primary and essential process is the active reabsorption of Na+ (+ in B); the charge separation resulting from this Na+ movement causes the passive reabsorption of Cl- (in C); this reabsorption of solute causes the water concentration of the remaining tubular fluid to rise, thereby inducing the passive reabsorption of water by osmosis (shown in D).

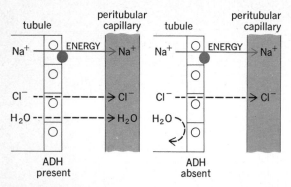

FIGURE 11-11

Effect of ADH on water reabsorption. ADH causes the tubule to be highly permeable to water so that water movement can accompany sodium and chloride reabsorption. Without ADH, the tubule becomes quite impermeable to water, and water reabsorption is impeded. Note that ADH does not alter reabsorption of sodium or chloride ions.

electrochemical gradient; (2) the reabsorption of chloride is by passive diffusion and depends upon the active reabsorption of sodium[1]; (3) the reabsorption of water is also by passive diffusion (osmosis) and also depends upon the active reabsorption of sodium.

Thus, active tubular sodium reabsorption is the primary force, which results in reabsorption of chloride and water as well. The relationship is explained in Fig. 11-10. Since the concentrations of all crystalloids are virtually identical in plasma and in Bowman's capsular fluid, in the very first portion of the proximal tubule no significant transtubular concentration gradients exist for sodium, chloride, or water. As the fluid flows down the tubule, sodium is actively reabsorbed into the peritubular capillaries. Since sodium is a positively charged ion, its movement attracts a negatively charged chloride ion. In other words the sodium ion ''pulls'' the chloride ion with it as a result of its electric charge, thus obviating the need for a separate active-transport system for chloride.[2] (The remainder of this discussion thus implicitly includes chloride when it refers to sodium reabsorption.) Finally, what happens to the water concentration of the tubular fluid as a result of sodium reabsorption? Obviously this re-

[1]Recent evidence suggests that in one part of the tubule—the ascending loop of Henle, generalizations 1 and 2 may be reversed, i.e., chloride may be the actively transported ion with sodium following passively. The end result is the same.

[2]This description is, by necessity, highly simplified. The nature of the electric coupling between sodium and chloride movements is quite complex and still controversial.

moval of solute lowers osmolarity, i.e., raises water concentration, below that of plasma. Thus, a water-concentration gradient is created between tubular lumen and peritubular plasma which constitutes a driving force for water reabsorption via osmosis. If the water permeability of the tubular epithelium is very high, water molecules are reabsorbed passively almost as rapidly as the actively transported sodium ions, so that the tubular fluid is only slightly more dilute than plasma. In this manner almost all the filtered sodium and water could theoretically be reabsorbed and the final urine would still have approximately the same osmolarity as plasma. However, this reabsorption of water can occur only if the tubular epithelium is highly permeable to water. No matter how great the water-concentration gradient, water cannot move if the epithelium is impermeable to it. In reality, the permeability of the last portions of the tubules (the distal tubules and collecting ducts) to water is subject to physiological control. The major determinant of this permeability is a hormone known as *vasopressin* or *antidiuretic hormone* (ADH), the second name being preferable because it describes the effect of the hormone's action, antidiuresis, i.e., against a high urine volume. In the absence of ADH (Fig. 11-11) the water permeability of the distal tubule and collecting duct is very low, sodium reabsorption proceeds normally because ADH *has no effect on sodium reabsorption,* but water is unable to follow and thus remains in the tubule to be excreted as a large volume of urine. On the other hand, in the presence of ADH, the water permeability of these last nephron segments is very great, water reabsorption is able to keep up with sodium reabsorption, and the final urine volume is small.

ADH is an octapeptide (a molecule consisting of eight amino acids) closely related in structure to another hormone, oxytocin, the function of which is discussed in Chap. 14. ADH exerts its action by inducing increased activity of cyclic AMP in the tubular cells; this, in turn, seems to increase the size of pores in tubular membranes. It is interesting that ADH exerts this same effect on other animal tissues, such as frog skin and toad bladder, which actually serve as useful tools for studying its cellular mechanism of action. How ADH production is regulated will be discussed in subsequent sections.

Several crucial aspects of sodium and water reabsorption should be reemphasized:

1 The tubular response to ADH is not all or none, like an action potential, but shows graded increases as the concentration of ADH is elevated, thus permitting fine adjustments of water permeability and excretion.

2 Excretion of large quantities of sodium *always* results in the

excretion of large quantities of water. This follows from the passive nature of water reabsorption, since water can be reabsorbed only if sodium is reabsorbed first. As we shall see, this relationship has considerable importance for the regulation of extracellular volume.

3 In contrast, large quantities of water can be excreted even though the urine is virtually free of sodium. This process we shall find critical for the renal regulation of extracellular osmolarity.

Control of sodium excretion: regulation of extracellular volume

The renal compensation for increased body sodium is excretion of the excess sodium. Conversely, a deficit in body sodium is prevented by reducing urinary sodium to an absolute minimum, thus retaining within the body the amount usually lost via the urine.

Since sodium is freely filterable at the glomerulus and actively reabsorbed but not secreted by the tubules, the amount of sodium excreted in the final urine represents the resultant of two processes, glomerular filtration and tubular reabsorption.

Sodium excretion =
sodium filtered − sodium reabsorbed

It is possible, therefore, to adjust sodium excretion by controlling one or both of these two variables (Fig. 11-12). For example, what happens if the quantity of filtered sodium increases (as a result of a higher GFR) but the rate of reabsorption remains constant? Clearly, sodium excretion increases. The same final result could be achieved by lowering sodium reabsorption while holding the GFR constant. Finally, sodium excretion could be raised greatly by elevating the GFR and simultaneously reducing reabsorption. Conversely, sodium excretion could be decreased below normal levels by lowering the GFR or raising sodium reabsorption or both. Control of GFR and sodium reabsorption is therefore the mechanism by which renal regulation of sodium balance is accomplished.

The reflex pathways by which changes in total body sodium balance lead to changes in GFR and sodium reabsorption include (1) "sodium" receptors (the reasons for the use of quotation marks will soon be apparent) and the afferent pathways leading from them to the central nervous system and endocrine glands; (2) efferent neural and hormonal pathways to the kidneys; (3) renal effector sites: the renal arterioles and tubules.

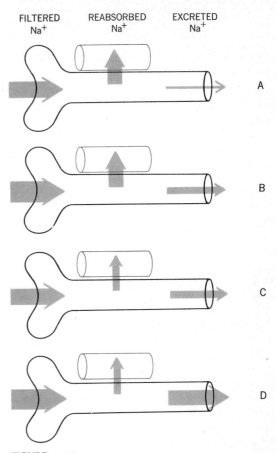

FIGURE 11-12
Sodium excretion is increased by increasing the GFR B, by decreasing reabsorption C, or by a combination of both D. The arrows indicate relative magnitudes of filtration, reabsorption, and excretion.

The first component of the reflexes, the "sodium" receptors, offers a number of theoretical difficulties. Clearly, there is no way that the total body *mass* of sodium can be detected by any imaginable receptor. Therefore, one must look for some other variable which correlates closely with total body sodium and might constitute the critical signal. As shown in the following examples, the ideal candidate is the volume of extracellular fluid.

What happens, for example, when a person ingests a liter of isotonic sodium chloride, i.e., a solution of salt with exactly the same osmolarity as the body fluids? It is absorbed from the gastrointestinal tract; all the salt and water remain in the extracellular fluid (plasma and interstitial fluid) and none enters the cells. Because of the active sodium "pumps" in cell membranes, sodium is effectively barred from the cells and therefore remains in the extracellular fluid. The water, too, remains since only the volume and not the osmolarity of the extracellular compartment has been changed; i.e., no osmotic gradient exists to drive the ingested water into cells.

Another example: A man is given 145 mmoles of sodium chloride to eat but no water. The salt is distributed in the extracellular fluid but is barred from the cells. The addition of this water-free solute to the extracellular fluid causes extracellular osmolarity to rise above intracellular osmolarity; therefore water diffuses out of the cells into the extracellular fluid until the osmolarities are once more equal. The net result is an expansion of extracellular volume (and a decrease of intracellular volume).

These examples lead to the extremely important generalization that the total extracellular fluid volume depends primarily upon the mass of extracellular sodium, which, in turn, correlates directly with total body sodium, since sodium is effectively barred from cells (there are considerable amounts of sodium in bone but this fact does not seriously alter the analysis). *It should now be clear that, normally, reflexes which maintain total body sodium constant simultaneously keep extracellular volume constant.*

Yet, how can there be receptors capable of detecting changes in the total extracellular volume? The answer is almost certainly that there are not any, that total extracellular volume, per se, is also not *directly* monitored. What about its component volumes: plasma volume and interstitial volume? Again it seems unlikely that either of these is *directly* monitored. What seems most likely at present is that closely correlated derivative cardiovascular functions of these volumes, i.e., intravascular and intracardiac pressures, cardiac output, organ blood flow, are the actual variables monitored (Fig. 11-13). Thus a decrease in plasma volume generally tends to lower the hydrostatic pressures within the veins, cardiac chambers, and arteries. These changes are detected by baroreceptors within the blood vessels (for example, the carotid sinus) and the cardiac chambers. Such baroreceptors undoubtedly constitute much of the important input for regulation of sodium excretion. Other derivatives of plasma volume are cardiac output (decreased volume $\longrightarrow$ decreased atrial pressure $\longrightarrow$ decreased output) and, in turn, blood flow through various organs. It is likely that these variables, too, are monitored and reflexly induce changes in sodium excretion. There are probably other variables also dependent on extracellular volume which take part in the reflexes.

In summary, then, regulation of total body sodium

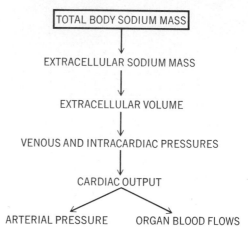

FIGURE 11-13
Dependence of cardiovascular parameters on total body sodium mass. The magnitude of each variable in the flow sheet is determined, in large part, by the magnitude of the one above it. Therefore, receptors which detect changes in pressure, distension, or flow supply the information required for regulation of body sodium.

and extracellular volume depends not upon "sodium" or "volume" receptors per se but rather upon pressure, distension, or flow receptors in various parts of the body. In the normal person, information concerning these variables results in excellent homeostasis of body sodium and extracellular fluid volume, since these parameters are all so dependent upon one another. However, as we shall see, disease states can produce striking discrepancies between them, with a resulting abnormal expansion of extracellular volume and total body sodium.

Control of GFR *Glomerular Filtration Rate*

We have described the three factors which determine the GFR: glomerular capillary blood pressure, Bowman's capsule hydrostatic pressure, and plasma protein concentration. Anything which alters the magnitude of these factors can be expected to change the GFR. Normally the GFR is controlled primarily by the alteration of glomerular capillary pressure. Two factors produce these changes: (1) A fall in arterial blood pressure decreases filtration rate by lowering glomerular capillary pressure. Conversely, an increase in arterial blood pressure has just the opposite effect. (2) A decrease in the diameter of the afferent arterioles, secondary to increased renal sympathetic activity, lowers glomerular capillary pressure

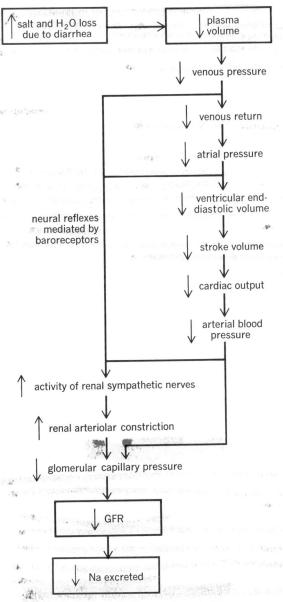

FIGURE 11-14
Pathway by which the GFR is decreased when plasma volume decreases. The baroreceptors which initiate the sympathetic reflex are probably located in large veins and the walls of the heart, as well as in the carotid sinuses and aortic arch.

from the sympathetic nervous system. (They are also very sensitive to the constrictor effects of circulating epinephrine, the hormone from the adrenal medulla.) As described in Chap. 9 these efferent pathways play prominent roles in the reflexes which regulate arterial blood pressure, being activated by a lowered blood pressure and inhibited by an elevated pressure. To take a specific example: What changes in renal hemodynamics occur as a result of severe salt and water loss due to diarrhea (Fig. 11-14)? The decreased plasma volume resulting from salt and water loss prevents adequate venous return, thereby reducing atrial pressure, cardiac output, and arterial blood pressure. These drops in blood pressure are detected by the carotid sinuses and aortic arch, as well as by other baroreceptors in the veins and atria, and the information relayed to the cardiovascular centers, which respond by inhibiting parasympathetic outflow to the heart and by enhancing sympathetic outflow to the heart and arteriolar smooth muscle. The sympathetic stimulation of the renal arterioles (both by the renal nerves and epinephrine from the adrenal medulla) increases constriction of the renal arterioles. This vasoconstriction increases the resistance to blood flow from the renal artery to the glomerular capillaries, lowering the capillary blood pressure and GFR. By this mechanism, both the amount of sodium filtered and the amount excreted are reduced, and further loss from the body is prevented. Conversely, an increased GFR can result from greater plasma volume and contribute to the increased renal sodium loss which returns extracellular volume to normal. This would occur in our subject who drank a liter of isotonic saline. This analysis is in terms of the logical stimulus for GFR control, namely, changes in blood pressure, but the sympathetic outflow via the medullary cardiovascular center can be changed in response to a variety of stimuli (pain, fright, exercise, etc.); accordingly, fluid excretion, at least transiently, may be altered by many situations.

Control of tubular sodium reabsorption

Present evidence indicates that, so far as long-term regulation of sodium excretion is concerned, the control of tubular sodium reabsorption is probably more important than that of GFR. For example, patients with chronic marked reductions of GFR usually maintain normal sodium excretion by decreasing tubular sodium reabsorption. The two major controllers of tubular sodium reabsorption are aldosterone and so-called third factor.

Aldosterone A major clue to the control of sodium reabsorption was the observation that patients whose adre-

and filtration rate because a larger fraction of the arterial pressure is dissipated in overcoming the increased resistance offered by the narrowed arterioles. These arterioles are innervated by a rich supply of vasoconstrictor fibers

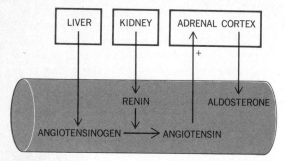

FIGURE 11-15
Summary of the renin-angiotensin-aldosterone system. The plus sign denotes the stimulatory effect of angiotensin on aldosterone secretion.

nal glands are diseased or missing excrete large quantities of sodium in the urine. Indeed, if untreated, they may die because of low blood pressure resulting from depletion of plasma volume. When measurement of the GFR became possible, it was found that this increased sodium excretion often occurs despite lower GFR, thus establishing that decreased tubular reabsorption is the factor responsible for the sodium loss. A large body of evidence clearly indicates that the adrenal cortex produces a hormone, called *aldosterone*, which stimulates sodium reabsorption, specifically by the distal tubules. In the complete absence of this hormone, the patient may excrete 25 g of salt per day, whereas excretion may be virtually zero when aldosterone is present in large quantities. In a normal person, the amounts of aldosterone produced and salt excreted lie somewhere between these extremes. It is interesting that aldosterone also stimulates sodium transport by other epithelia in the body, namely, by sweat and salivary glands and the intestine. The net effect is the same as that exerted on the renal tubules—a movement of sodium out of the luminal fluid into the blood. Thus aldosterone is an "all-purpose" stimulator of sodium retention.

Aldosterone secretion (and thereby tubular sodium reabsorption) is controlled by reflexes involving the kidneys themselves. Specialized cells lining the arterioles within the kidney synthesize and secrete into the blood a protein known as *renin* (not rennin), an enzyme catalyzing the reaction in which a small polypeptide *angiotensin* splits off from a large plasma protein *angiotensinogen* (Fig. 11-15). Angiotensin is a profound stimulator of aldosterone secretion and constitutes the primary input to the adrenal gland controlling production and the release of this hormone.

Angiotensinogen is synthesized by the liver and is always present in the blood; therefore, the rate-limiting factor in angiotensin formation is the concentration of plasma renin which, in turn, depends upon the rate of renin secretion by the kidneys. The critical question now becomes: What controls the rate of renin secretion? The answer to this, the first link in the reflex chain, is presently uncertain, mainly because there seem to be multiple inputs to the renin-secreting cells and it is not yet possible to assign quantitative roles to each of them. It is likely that the renal sympathetic nerves (and circulating epinephrine) constitute the single most important input in normal persons; this makes excellent sense, teleologically, since a reduction in body sodium and extracellular volume triggers off increased sympathetic tone to the kidneys (as shown in Fig. 11-14), thereby setting off the hormonal chain of events which restores sodium balance and extracellular volume to normal (Fig. 11-16).

Third factor   Until recently most renal physiologists believed that the control of sodium excretion could be explained completely in terms of changes in GFR and aldosterone-dependent tubular sodium reabsorption. It is now clear that these two factors do not suffice, since there are circumstances in which GFR is low and aldosterone is high, yet sodium excretion is normal or increased. Therefore, there must exist a *third factor* which importantly influences tubular sodium reabsorption independently of aldosterone. However, the precise identity of this third factor has not yet been determined.

Summary   The control of sodium excretion depends upon the control of two variables of renal function, the GFR and sodium reabsorption. The latter is controlled, at least in part, by the renin-angiotensin-aldosterone hormone system and, in part, by an as yet unidentified factor(s) known as third factor. The reflexes which control both GFR and sodium reabsorption are essentially blood-pressure-regulating reflexes since they are probably most frequently initiated by changes in arterial or venous pressure (or cardiac output). This is only fitting since cardiovascular function depends upon an adequate plasma volume, which, as a component of the extracellular fluid volume, normally reflects the mass of sodium in the body. In normal persons, these regulatory mechanisms are so precise that sodium balance does not vary by more than 2 percent despite marked changes in dietary intake or in losses due to sweating, vomiting, or diarrhea. In several types of diseases, however, sodium balance becomes deranged by the failure of the kidneys to excrete sodium normally. Sodium excretion may fall virtually to zero despite continued sodium ingestion, and the patient may

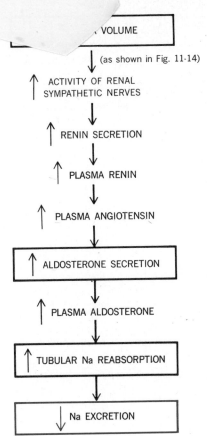

FIGURE 11-16
Pathway by which aldosterone secretion is increased when plasma volume is decreased. (The pathways by which the GFR is reduced are shown in Fig. 11-14.)

retain huge quantities of sodium and water within the body, leading to abnormal expansion of the extracellular fluid and edema. The most important example of this phenomenon is congestive heart failure (Chap. 9). A patient with a failing heart manifests a decreased GFR and increased aldosterone secretion rates, both of which contribute to the virtual absence of sodium from his urine. In addition to aldosterone, third factor also causes decreased sodium excretion by enhancing tubular sodium reabsorption. The net result is expansion of plasma volume, increased capillary pressure, and filtration of fluid into the interstitial space, i.e., edema. Why are these sodium-retaining reflexes all stimulated despite the fact

that the expanded extracellular volume should call forth sodium-losing responses (increased GFR and decreased aldosterone)? We do not know, although it seems fairly certain that the lower cardiac output as a result of cardiac failure is somehow responsible. In addition to treating the basic heart disease, physicians use *diuretics*, or drugs which inhibit tubular sodium reabsorption, thereby leading to greater sodium and water excretion.

ADH secretion and extracellular volume

Although we have spoken of extracellular-volume regulation only in terms of the control of sodium excretion, it is clear that to be effective in altering extracellular volume, the changes in sodium excretion must be accompanied by equivalent changes in water excretion. We have already pointed out that the ability of water to follow when sodium is reabsorbed depends upon ADH. Accordingly, a decreased extracellular volume must reflexly call forth increased ADH production as well as increased aldosterone secretion. What is the nature of this reflex? As described in Chap. 7, ADH is produced by a discrete group of hypothalamic neurons whose axons terminate in the posterior pituitary, from which ADH is released into the blood. These hypothalamic cells receive input from several vascular baroreceptors, particularly a group located in the left atrium (Fig. 11-17). The baroreceptors are *stimulated* by *increased* atrial blood pressure, and the impulses resulting from this stimulation are transmitted via afferent nerves and ascending pathways to the hypothalamus, where they *inhibit* the ADH-producing cells. Conversely, decreased atrial pressure causes less firing by the baroreceptors and stimulation of ADH synthesis and release (Fig. 11-18). The adaptive value of this baroreceptor reflex, one more in our expanding list, should require no comment.

Renal regulation of extracellular osmolarity

We turn now to the renal compensation for pure water losses or gains, e.g., a man drinking 2 liters of water, where no change in total salt content of the body occurs, only total water. The most efficient compensatory mechanism is for the kidneys to excrete the excess water without altering its usual excretion of salt, and this is precisely what they do. ADH secretion is reflexively inhibited, as will be described below, tubular water permeability of the distal tubules and collecting ducts becomes very low, sodium reabsorption proceeds normally but water is unable to follow, and a large volume of extremely dilute

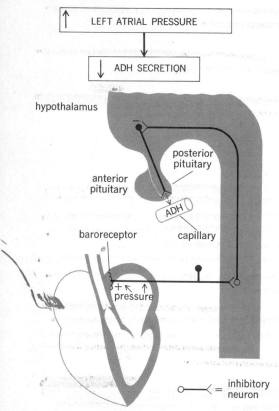

FIGURE 11-17
Pathway by which ADH secretion is decreased when plasma
volume is increased. The greater plasma volume raises
left-atrial pressure, which stimulates the atrial baroreceptors
and inhibits ADH secretion.

urine is excreted. In this manner, the excess pure water
is eliminated.

Thus it is easy to see how the kidneys produce a
final urine having the same osmolarity as that of plasma
or one having a lower osmolarity than plasma (hypo-
osmotic urine), the latter occurring whenever water reabsorption lags behind solute reabsorption, i.e., when
plasma ADH is reduced. Clearly the formation of a hypo-
osmotic urine is a good compensation for an excess of
water in the body.

But how can the kidneys produce a hyperosmotic
urine, i.e., a urine having an osmolarity greater than that
of plasma? For this to occur, does not water reabsorption
have to "get ahead" of solute reabsorption? How can
this happen if water reabsorption is always secondary to

solute (particularly sodium) reabsorption? It would seem
that, if our three generalizations are not to be violated,
the kidneys cannot produce a hyperosmotic urine. Yet
they do. Indeed, the final urine may be as concentrated
as 1,400 milliosmoles/liter compared with a plasma osmolarity of 300 milliosmoles/liter. Moreover, this concentrated urine is produced without violating the three
generalizations.

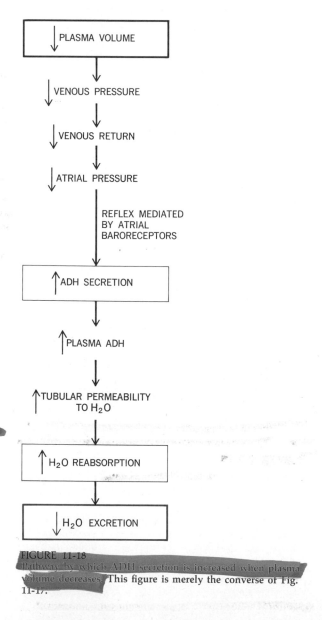

FIGURE 11-18
Pathway by which ADH secretion is increased when plasma
volume decreases. This figure is merely the converse of Fig.
11-17.

Urine concentration:
the countercurrent system

The ability of the kidneys to produce concentrated urine is not merely an academic problem. It is a major determinant of one's ability to survive without water. The human kidney can produce a maximal urinary concentration of 1,400 milliosmoles/liter. The urea, sulfate, phosphate, and other waste products (plus the smaller number of nonwaste ions) which must be excreted each day amount to approximately 600 milliosmoles; therefore, the water required for their excretion constitutes an obligatory water loss and equals

$$\frac{600 \text{ milliosmoles/day}}{1,400 \text{ milliosmoles/liter}} = 0.444 \text{ liter/day}$$

As long as the kidneys are functioning, excretion of this volume of urine will occur, despite the absence of water intake. In a sense, a person lacking access to water may literally urinate himself to death (due to fluid depletion). If the body could produce a urine with an osmolarity of 6,000 milliosmoles/liter then only 100 ml of water need be lost obligatorily each day and survival time would be greatly expanded. A desert rodent, the kangaroo rat, does just that; this animal never drinks water, the water produced in its body by oxidation of foodstuffs being ample for its needs.

The kidneys produce concentrated urine by a complex interaction of events involving the so-called countercurrent multiplier system residing in the loop of Henle. Recall that the loop of Henle, which is interposed between the proximal and distal convoluted tubules, is a hairpin loop extending into the renal medulla. The fluid flows in opposite directions in the two limbs of the loop, thus the name countercurrent. Let us list the critical characteristics of this loop.

1 The ascending limb of the loop of Henle (i.e., the limb leading to the distal tubule) *actively* transports sodium chloride out of the tubular lumen into the surrounding interstitium. (As mentioned earlier, in the ascending loop of Henle it is possible that chloride is the actively transported ion with sodium following passively; for simplicity, we shall refer to the process as "sodium chloride transport.") It is *relatively* impermeable to sodium chloride and water, so that *passive* fluxes into or out of it are small.

2 The descending limb of the loop of Henle (i.e., the limb into which drains fluid from the proximal tubule) does *not actively* transport sodium chloride; it is the only tubular segment that does not. Moreover, it is relatively permeable to both ions and water, so that *passive* fluxes into or out of it are large.

Given these c... Henle filled with a sta... the proximal tubule. At... where would be 300 millios... the proximal tubule is isos... ...quivalent amounts of sodium chloride an... ...er having been reabsorbed by the proximal tubule.

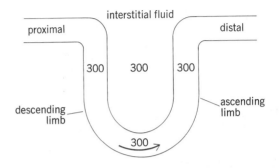

Now let the active pump in the ascending limb transport sodium chloride into the interstitium until a limiting gradient (say 200 milliosmoles/liter) is established between ascending-limb fluid and interstitium.

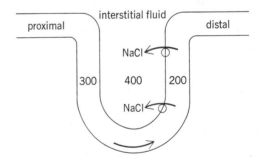

A limiting gradient is reached because the ascending limb is *not completely impermeable* to sodium and chloride; accordingly, passive backflux of ions into the lumen counterbalances active outflux and a steady-state limiting gradient is established.

Given the relatively high permeability of the descending limb to sodium, chloride, and water, what net fluxes now occur between interstitium and descending limb? There is a net diffusion of sodium and chloride into the descending limb and a net diffusion of water out until the osmolarities are equal. The interstitial osmolarity is maintained at 400 milliosmoles/liter during this equilibration because of active sodium transport out of the ascending limb.

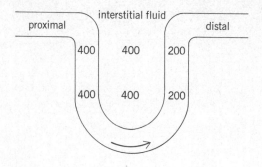

Note that the osmolarities of the descending limb and interstitium are equal and both are higher than that of the ascending limb. So far we have held the fluid stationary in the loop, but of course it is actually continuously flowing. Let us look at what occurs under conditions of flow (Fig. 11-19), simplifying the analysis by assuming that flow through the loop, on the one hand, and ion and water movements, on the other, occur in discontinuous out-of-phase steps. During the stationary phase as described above, sodium chloride is actively transported out of the ascending limb to establish a gradient of 200 milliosmoles, and sodium chloride passively diffuses into and water diffuses out of the descending limb until descending limb and interstitium have the same osmolarity. During the "flow" phase, fluid leaves the loop via the distal tubule and new fluid enters the loop from the proximal tubule.

Note that the fluid is progressively concentrated as it flows down the descending limb and then is progressively diluted as it flows up the ascending limb. While only a 200-milliosmole/liter gradient is maintained across the ascending limb at any given *horizontal level* in the medulla, there exists a much larger osmotic gradient from the top of the medulla to the bottom (312 milliosmoles/liter versus 700 milliosmoles/liter). In other words, the 200-milliosmole/liter gradient established by active sodium chloride transport has been *multiplied* because of the *countercurrent* flow within the loop (i.e., flow in opposing directions through the two limbs of the loop). It should be emphasized that the active sodium chloride transport mechanism within the ascending limb is the essential component of the entire system; without it, the countercurrent flow would have no effect whatever on concentrations.

The highest concentration achieved at the tip of the loop depends upon many factors, particularly the length of the loop (the kangaroo rat has extremely long loops) and the strength of the sodium chloride pump. In human

beings, the value reached is 1,400 milliosmoles/liter, which, you will recall, is also the maximal concentration of the excreted urine. But what has this system really accomplished? Certainly, it concentrates the loop fluid to 1,400 milliosmoles/liter, but then it immediately redilutes the fluid so that the fluid entering the distal tubule is actually more dilute than the plasma. Where is the *final urine* concentrated and how?

The site of final concentration is in the collecting ducts. Recall that the collecting ducts course through the renal medulla parallel to the loops of Henle, and are bathed by the interstitial fluid of the medulla. In the presence of maximal levels of ADH, fluid leaves the distal tubules isosmotic to plasma (that is, 300 milliosmoles/liter) because it has reequilibrated with peritubular plasma. As this fluid then flows through the collecting ducts it equilibrates with the ever-increasing osmolarity of the interstitial fluid. Thus, the real function of the loop countercurrent multiplier system is to concentrate the *medullary interstitium*. Under the influence of ADH, the collecting ducts are highly permeable to water which diffuses out of the collecting ducts into the interstitium as a result of the osmotic gradient (Fig. 11-20). The net result is that the fluid at the end of the collecting duct has equilibrated with the interstitial fluid at the tip of the medulla. By this means, the final highly concentrated urine contains relatively less of the filtered water than solute, which is precisely the same as adding pure water to the extracellular fluid, and thereby compensating for a pure water deficit.

In contrast, in the presence of low plasma–ADH concentration, the collecting ducts, like the distal tubules, become relatively impermeable to water and the interstitial osmotic gradient is ineffective in inducing water movement out of the collecting ducts; therefore a large volume of dilute urine is excreted, thereby compensating for a pure water excess.

Osmoreceptor control of ADH secretion

To reiterate, pure water deficits or gains are compensated for by partially dissociating water excretion from that of salt through changes in ADH secretion. What receptor input controls ADH under such conditions? The answer is changes in extracellular osmolarity. The adaptive rationale should be obvious, since osmolarity is the variable most affected by pure water gains or deficits. What are the pathways by which osmolarity controls the hypothalamic ADH-producing cells? If osmolarity is the parameter being regulated, it follows that receptors must exist which are sensitive to extracellular osmolarity. These osmo-

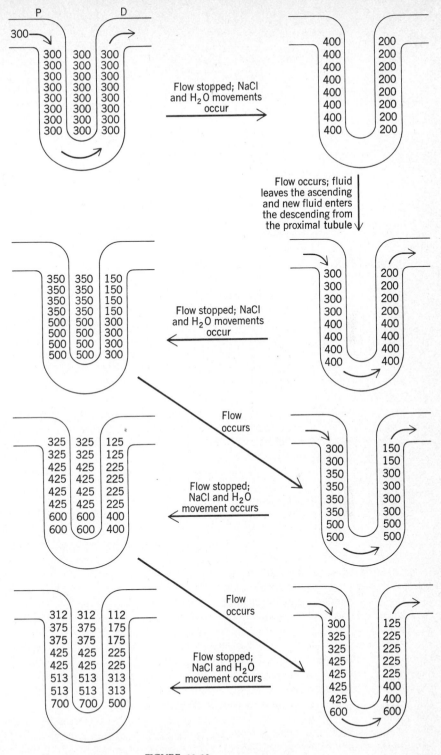

FIGURE 11-19
Countercurrent multiplier system in the loop of Henle. (See text for explanation.)

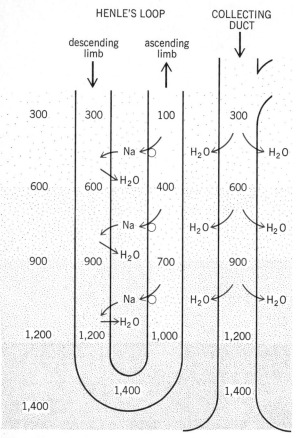

HENLE'S LOOP
descending limb   ascending limb

COLLECTING DUCT

FIGURE 11-20
Operation of countercurrent multiplication of concentration in the formation of hypertonic urine. (*Adapted from R. F. Pitts.*)

ity cause maximal inhibition of ADH secretion; conversely, the opposite changes produce maximal stimulation. But what happens in the following situation? A person suffering from severe diarrhea loses 3 liters of salt and water during the same time that he drinks 2 liters of pure water. His total extracellular volume is decreased, but his osmolarity is also decreased. As a result, the ADH-producing cells receive opposing input from the baroreceptors and osmoreceptors. Which predominates depends completely upon the strength of the two inputs.

To add to the complexity, these cells receive synaptic input from many other brain areas, so that ADH

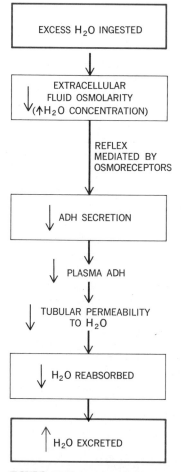

FIGURE 11-21
Pathway by which ADH secretion is lowered and water excretion raised when excess water is ingested.

receptors are located in the hypothalamus, but the mechanism by which they detect changes in osmolarity are unknown. The hypothalamic cells which secrete ADH receive neural input from these osmoreceptors. Via these connections an increase in osmolarity stimulates them and increases their rate of ADH secretion, and, conversely, decreased osmolarity inhibits ADH secretion (Fig. 11-21).

We have now described two different afferent pathways controlling the ADH-secreting hypothalamic cells, one from baroreceptors and one from osmoreceptors. These hypothalamic cells are therefore true integrating centers whose rate of activity is determined by the total synaptic input. Thus, a simultaneous increase in extracellular volume and decrease in extracellular osmolar-

secretion (and therefore urine flow) can be altered by pain, fear, and a variety of other factors. However, these effects are usually short-lived and should not obscure the generalization that ADH secretion is determined primarily by the states of extracellular volume and osmolarity. Alcohol is a powerful inhibitor of ADH release and probably accounts for much of the large urine flow accompanying the ingestion of alcohol.

The disease *diabetes insipidus,* which is distinct from diabetes mellitus, i.e., sugar diabetes, illustrates what happens when the ADH system is disrupted. This disease is characterized by the constant excretion of a large volume of highly dilute urine (as much as 25 liters/day). In most cases, the flow can be restored to normal by the administration of ADH. These patients apparently have lost the ability to produce ADH, usually as a result of damage to the hypothalamus. Thus, renal tubular permeability to water is low and unchanging regardless of extracellular osmolarity or volume. The very thought of having to urinate (and therefore to drink) 25 liters of water per day underscores the importance of ADH in the control of renal function and body water balance.

Figure 11-22 shows all these factors which control renal sodium and water excretion in operation in response to severe sweating, as in exercise; the renal retention of fluid helps to compensate for the water and salt lost in the sweat.

Thirst and salt appetite

Now we must turn to the other component of the balance, control of intake. It should be evident that large deficits of salt and water can be only partly compensated by renal conservation and that ingestion is the ultimate compensatory mechanism. The subjective feeling of thirst, which drives one to obtain and ingest water, is stimulated both by a lower extracellular volume and a higher plasma osmolarity, the adaptive significance of both being self-evident. Note that these are precisely the same changes which stimulate ADH production. Indeed, the centers which mediate thirst are also located in the hypothalamus very close to those areas which produce ADH. Damage to these centers abolishes water intake completely. Conversely electric stimulation of them may induce profound and prolonged drinking (these water-intake centers are very close to, but distinct from, the food-intake centers to be described in a later chapter). Because of the similarities between the stimuli for ADH secretion and thirst, it is tempting to speculate that the receptors (osmorecep-

tors and atrial baroreceptors) which initiate the ADH-controlling reflexes are identical to those for thirst. Much evidence indicates that this is, indeed, the case.

A recent finding of considerable interest is that angiotensin stimulates thirst by a direct effect on the brain. Thus, the renin-angiotensin system is not only an important regulator of sodium balance but of water balance as well and constitutes one of the pathways by which thirst is stimulated when extracellular volume is decreased.

There are also other pathways controlling thirst. For example, dryness of the mouth and throat causes profound thirst, which is relieved by merely moistening them. It is fascinating that animals such as the camel (and man, to a lesser extent) which have been markedly dehydrated will rapidly drink just enough water to replace their previous losses and then stop; what is amazing is that when they stop, the water has not yet had time to be absorbed from the gastrointestinal tract into the blood. Some kind of "metering" of the water intake by the gastrointestinal tract has occurred, but its nature remains a mystery.

This last phenomenon may be only one example of the larger area concerning "learned," "feedforward," or "anticipatory" control of thirst. As described above, there exist powerful osmoreceptor and "volume receptor" reflexes controlling water intake; yet a person may normally regulate his water intake with only minimal use of these reflexes. Much of our drinking is done in association with eating and is therefore called *prandial drinking.* The quantity of fluid drunk with each meal is, in large part, a learned response determined by past experience. Thus, as described in Chap. 5, such "anticipatory" responses help to *prevent* dehydration; in contrast to these "feedforward" phenomena, the osmoreceptor and volume-receptor reflexes go into operation only *after* a deficit has already occurred.

The analog of thirst for sodium, salt appetite, is also an extremely important component of sodium homeostasis in most mammals. It is clear that salt appetite is innate and consists of two components: "hedonistic" appetite and "regulatory" appetite; i.e., animals "like" salt and eat it whenever they can, regardless of whether they are salt-deficient, and, in addition, their drive to obtain salt is markedly increased in the presence of deficiency. The significance of these animal studies for man is unclear. Salt craving seems to occur in human beings who are severely salt-depleted but the contribution of regulatory salt appetite to everyday sodium homeostasis in normal persons is probably slight. On the other hand, man seems to have a strong hedonistic appetite for salt as manifested

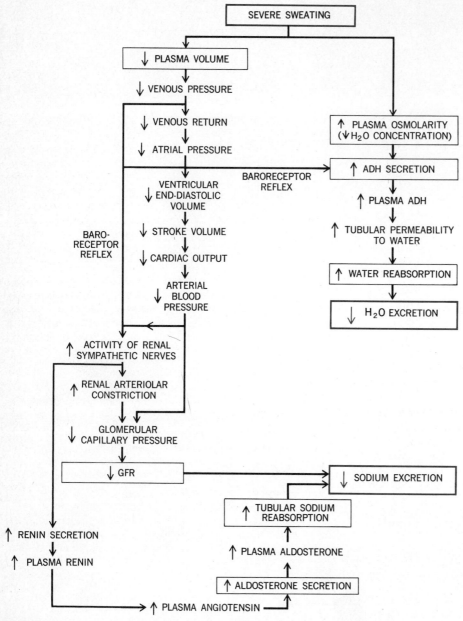

FIGURE 11-22

Pathways by which sodium and water excretion are decreased
in response to severe sweating. This figure is an amalgamation
of Figs. 11-14, 11-16, and 11-18 and the converse of Fig. 11-21.
"Third factor" is not shown in the figure.

by almost universally large intakes of sodium whenever it is cheap and readily available. Thus, the average American intake of salt is 10 to 15 g/day despite the fact that human beings can survive quite normally on less than 0.5 g/day. Present evidence strongly suggests that a large salt intake may be an important contributor to the pathogenesis of hypertension.

SECTION C:
REGULATION OF POTASSIUM, CALCIUM, AND HYDROGEN-ION CONCENTRATIONS

Potassium regulation

The potassium concentration of extracellular fluid is a closely regulated quantity. The importance of maintaining this concentration in the internal environment stems primarily from the role of potassium in the excitability of nerve and muscle. Recall that the resting-membrane potentials of these tissues are directly related to the ratio of intracellular to extracellular potassium concentration. Raising the external potassium concentration lowers the resting-membrane potential, thus increasing cell excitability. Conversely, lowering the external potassium hyperpolarizes cell membranes and reduces their excitability.

Since most of the body's potassium is found within cells, primarily as a result of active ion-transport systems located in cell membranes (Chap. 2), even a slight alteration of the rates of ion transport across cell membranes can produce a large change in the amount of extracellular potassium. Unfortunately, very little is known about the physiological control of these transport mechanisms, and obviously our understanding of the regulation of extracellular potassium concentration will remain incomplete until further data are obtained on this critical subject.

The normal person remains in total-body potassium balance (as is true for sodium balance) by daily excreting an amount of potassium equal to the amount ingested minus the small amounts eliminated in the feces and sweat. Normally potassium losses via sweat and the gastrointestinal tract are small (although large quantities can be lost by the latter during vomiting or diarrhea). Again the control of renal function is the major mechanism by which body potassium is regulated.

Renal regulation of potassium

Potassium is completely filterable at the glomerulus. The amounts of potassium excreted in the urine are generally a small fraction (10 to 15 percent) of the filtered quantity, thus establishing the existence of tubular potassium reabsorption. However, it has also been demonstrated that under certain conditions the excreted quantity may actually exceed that filtered, thus establishing the existence of tubular potassium secretion. The subject is therefore complicated by the fact that potassium can be both reabsorbed and secreted by the tubule. Present evidence suggests, however, that normally almost all the filtered potassium is reabsorbed (by active transport) regardless of changes in body potassium balance. In other words, the reabsorption of potassium does not seem to be controlled so as to achieve potassium homeostasis. The important result of this phenomenon is that *changes* in potassium *excretion* are due to *changes* in potassium *secretion* (Fig. 11-23). Thus, during potassium depletion when the homeostatic response is to reduce potassium excretion to a minimal level, there is no significant potassium secretion, and only the small amount of potassium escaping reabsorption is excreted. In all other situations, to this same small amount of unreabsorbed potassium is added a variable amount of secreted potassium. Thus, in describing the homeostatic control of potassium excretion

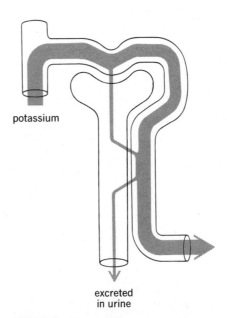

FIGURE 11-23
Basic renal processing of potassium. Since virtually all the filtered potassium is reabsorbed, potassium excreted in the urine results from tubular secretion.

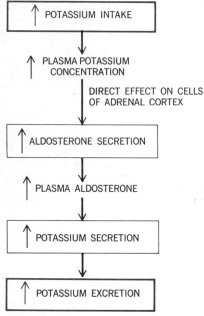

FIGURE 11-24
Pathway by which an increased potassium intake induces greater potassium excretion mediated by aldosterone.

we may ignore changes in GFR or reabsorption and focus only on the factors which alter the rate of tubular potassium secretion.

One of the most important of these factors is the potassium concentration of the renal tubular cells themselves. When a high-potassium diet is ingested, potassium concentration in most of the body's cells increases, including the renal tubular cells. This higher concentration facilitates potassium secretion into the lumen and raises potassium excretion. Conversely, a low-potassium diet or a negative potassium balance, e.g., from diarrhea, lowers renal tubular cell potassium concentration; this reduces potassium entry into the lumen and decreases potassium excretion, thereby helping to reestablish potassium balance.

A second important factor controlling potassium secretion is the hormone aldosterone, which besides assisting tubular sodium reabsorption enhances tubular potassium secretion. The reflex by which changes in extracellular volume control aldosterone production is completely different from the reflex initiated by an excess or deficit of potassium. The former constitutes a complex pathway, involving renin and angiotensin; the latter, how-

ever, seems to be much simpler (Fig. 11-24): The aldosterone-secreting cells of the adrenal cortex are apparently themselves sensitive to the potassium concentration of the extracellular fluid bathing them. For example, an increased intake of potassium leads to an increased extracellular potassium concentration, which in turn directly stimulates aldosterone production by the adrenal cortex. This extra aldosterone circulates to the kidney, where it increases tubular potassium secretion and thereby eliminates the excess potassium from the body. Conversely, a lowered extracellular potassium concentration would decrease aldosterone production and thereby inhibit tubular potassium secretion. Less potassium than usual would be excreted in the urine, thus helping to restore the normal extracellular potassium concentration. Again, complete compensation depends upon the ingestion of additional potassium.

The control and renal tubular effects of aldosterone are summarized in Fig. 11-25. The reader should realize that, regardless of whether aldosterone production is altered by changes in extracellular volume or potassium concentration, the renal effects on both sodium and potassium occur. It should be evident that a conflict will arise if increases in extracellular potassium and extracellular volume occur simultaneously, since these two changes drive aldosterone production in opposite directions. The adaptive value, if any, of such a potential conflict is at present unknown. We mention the problem primarily as a reminder that different homeostatic mechanisms frequently conflict with each other and that the ability of any living organism to survive ultimately depends upon whether it can resolve such conflicts by integrating the opposing demands.

Calcium regulation

Extracellular calcium concentration is also normally held relatively constant, the requirement for precise regulation stemming primarily from the profound effects of calcium on neuromuscular excitability. A low calcium concentration increases the excitability of nerve and muscle cell membranes so that patients with diseases in which low calcium occurs suffer from *hypocalcemic tetany*, characterized by skeletal muscle spasms which can be severe enough to cause death by asphyxia. Calcium is also important in blood clotting, but low calcium is never a cause of abnormal clotting clinically because the levels required for this function are considerably below those which produce fatal tetany. Hypercalcemia is also dan-

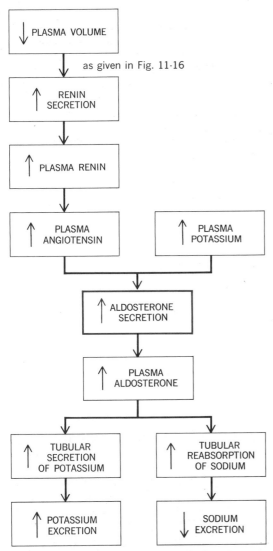

FIGURE 11-25
Summary of the control of aldosterone and its functions.

gerous in that it causes cardiac arrhythmias as well as depressed neuromuscular excitability.

Effector sites for calcium homeostasis

At least three effector sites are involved in the regulation of extracellular calcium concentration: bone, the kidney, and the gastrointestinal tract.

Bone Approximately 99 percent of total body calcium is contained in bone, which consists primarily of a framework of organic molecules upon which calcium phosphate crystals are deposited. Contrary to popular opinion, bone is not an absolutely fixed, unchanging tissue but is constantly being remolded and, what is more important, is available for either the withdrawal or deposit of calcium from extracellular fluid.

Gastrointestinal tract We have previously commented on the indiscriminate nature of most gastrointestinal absorptive processes. This is certainly not true for calcium absorption, which is subject to quite precise hormonal control.

Kidney The kidney handles calcium by filtration and reabsorption. In addition, as we shall see, the renal handling of phosphate also plays an important role in the regulation of extracellular calcium.

Parathormone

All three of the effector sites described above are subject to control by a protein hormone, called *parathormone* (also called parathyroid hormone), produced by the parathyroid glands. Parathormone production is controlled directly by the calcium concentration of the extracellular fluid bathing the cells of these glands. Lower calcium concentration stimulates parathormone production and release, and a higher concentration does just the opposite. It should be emphasized that extracellular calcium concentration acts directly upon the parathyroids (just as was true of the relation between extracellular potassium and aldosterone production) without any intermediary hormones or nerves.

Parathormone exerts at least four distinct effects on the sites described earlier (Fig. 11-26):

1 It increases the movement of calcium (and phosphate) from bone into extracellular fluid by stimulating cells (called *osteoclasts*) which break down bone structure, thus liberating calcium phosphate crystals. In this manner the immense store of calcium contained in bone is made available for the regulation of extracellular calcium concentration.

2 It increases gastrointestinal absorption of calcium by stimulating the active-transport system which moves the ion from gut lumen to blood. This is an important mechanism for elevating plasma calcium concentration since, under normal conditions, considerable amounts of ingested calcium are not absorbed from the intestine but are eliminated via the feces.

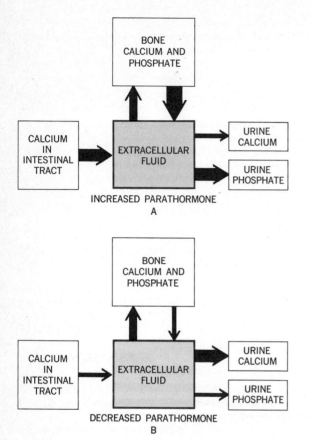

FIGURE 11-26
Effects of parathormone on the gastrointestinal tract, kidneys, and bone, the arrows signifying relative magnitudes. Note that when parathormone is decreased there is net movement of calcium and phosphate into bone, urine calcium is raised, and gastrointestinal absorption of calcium is reduced.

3. It increases renal tubular calcium reabsorption, thus decreasing urinary calcium excretion.
4. It reduces the renal tubular reabsorption of phosphate, thus raising its urinary excretion and lowering extracellular phosphate concentration.

The adaptive value of the first three should be obvious: They all result in a higher extracellular calcium concentration, thus compensating for the lower concentration which originally stimulated parathormone production. Conversely, an increase in extracellular calcium concentration inhibits normal parathormone production, thereby producing increased urinary and fecal calcium loss and net movement of calcium from extracellular fluid into bone (Fig. 11-26).

The adaptive value of the fourth effect requires further explanation. Because of the solubility characteristics of undissociated calcium phosphate, the extracellular concentrations of ionic calcium and phosphate bear the following relationship to each other: The product of their concentrations, i.e., calcium times phosphate, is approximately a constant. In other words, if the extracellular concentration of phosphate increases, it forces the deposition of some extracellular calcium in bone, lowering the calcium concentration and keeping the calcium phosphate product a constant. The converse is also true. Imagine now what happens when parathormone causes the breakdown of bone. Both calcium and phosphate are released into the extracellular fluid. If the phosphate concentration is allowed to increase, further movement of calcium from bone is retarded; but in addition to this effect on bone, we have seen that parathormone also decreases tubular reabsorption of phosphate, thus permitting the excess phosphate to be eliminated in the urine. Indeed, extracellular phosphate may actually be reduced by this mechanism, which would allow even more calcium to be mobilized from bone.

Parathormone has other functions in the body, notably its role in milk production, but the four effects discussed above constitute the major mechanisms by which it integrates various organs and tissues in the regulation of extracellular calcium concentration.

Vitamin D

Vitamin D plays an important role in calcium metabolism, as attested by the fact that its deficiency results in poorly calcified bones. Vitamin D really should be called a hormone since it can be produced by the skin in the presence of sunlight, but since people's clothing prevents this reaction and they are dependent on dietary intake for its supply, it is classed as a vitamin. The major action of vitamin D is to stimulate active calcium absorption by the intestine. It is of greater importance in this regard than parathormone; indeed the action of parathormone is, itself, dependent upon the presence of adequate quantities of vitamin D. Thus, the major event in vitamin D deficiency is decreased gut calcium absorption, resulting in decreased plasma calcium. In children, the newly formed bone protein matrix fails to be calcified normally because of the low plasma calcium, leading to the disease *rickets*.

The activity of vitamin D in the body is controlled ultimately by plasma calcium. The vitamin D which is ingested or produced by the skin is relatively inactive and requires biochemical alteration first by the liver and then by the kidneys before it is fully able to stimulate gut calcium absorption. This activation is stimulated by para-thormone; accordingly, a decreased plasma calcium stimulates the secretion of parathormone which, in turn, enhances the activity of vitamin D which, by its actions, helps to restore plasma calcium to normal. Thus, the feedback loops for parathormone and vitamin D activity are closely intertwined.

Calcitonin

A hormone known as calcitonin (also known as thyro-calcitonin) has recently been discovered which has sig-nificant effects on plasma calcium. Calcitonin is secreted by cells within the thyroid gland which surround but are completely distinct from the thyroxine-secreting follicles. Calcitonin lowers plasma calcium primarily by inhibiting bone resorption. Its secretion is controlled directly by the calcium concentration of the plasma supplying the thyroid gland; increased calcium causes increased calcitonin secretion. Thus, this system constitutes a second feed-back control over plasma calcium concentration, one that is opposed to the parathormone system. However, its overall contribution to calcium homeostasis is minor compared with that of parathormone.

Hydrogen-ion regulation

Most metabolic reactions are exquisitely sensitive to the hydrogen-ion concentration[1] of the fluid in which they occur. This sensitivity is due primarily to the marked influence on enzyme function exerted by the hydrogen ion. Accordingly, the hydrogen-ion concentration of the extracellular fluid is one of the most critical and closely regulated chemical quantities in the entire body.

Basic definitions

Dissociation and ionization were described in Chap. 1. The hydrogen ion is an atom of hydrogen which has lost its only electron. When dissolved in water, many com-pounds dissociate reversibly to produce negatively charged ions (anions) and hydrogen ions, e.g.,

$$\text{Lactic acid} \rightleftharpoons H^+ + \text{lactate}^-$$
$$H_2CO_3 \rightleftharpoons H^+ + HCO_3^-$$

Carbonic acid   Bicarbonate

[1]Hydrogen-ion concentration is frequently expressed in terms of pH, which is defined as the negative logarithm to the base 10 of the hydrogen-ion concentration: $pH = -\log H^+$. This can be confusing for several reasons, not the least of which is that pH decreases as H^+ increases. We have chosen not to use pH in this text.

The double arrows mean that the reaction can proceed in either direction, depending upon conditions. Any com-pound capable of liberating a hydrogen ion in this manner is called an *acid*. Conversely, any substance which can accept a hydrogen ion is termed a *base*. Thus, in the reactions above, lactate and bicarbonate are bases since they can bind hydrogen ions. The hydrogen-ion concen-tration often is referred to in terms of *acidity*: the higher the hydrogen-ion concentration, the greater the acidity (it must be understood that the hydrogen-ion concen-tration of a solution is a measure only of the hydrogen ions which are *free* in solution).

Strong and weak acids

Strong acids dissociate completely when they dissolve in water. For example, hydrochloric acid added to water gives

$$\underset{\substack{\text{Hydrochloric} \\ \text{acid}}}{HCl} \longrightarrow H^+ + Cl^-$$

Virtually no HCl molecules exist in the solution, only free hydrogen ions and chloride ions. On the other hand, a large number of *weak acids* do not dissociate completely when dissolved in water. For example, when dissolved, only a fraction of lactic acid molecules dissociate to form lactate and hydrogen ions, the other molecules remaining intact. This characteristic of weak acids forms the basis of an important chemical and physiological phenomenon, *buffering*.

Buffer action and buffers

Figure 11-27 pictures a solution made by dissolving lactic acid and sodium lactate in water. The sodium lactate dissociates completely into sodium ions and lactate ions, but only a very small fraction of the lactic acid molecules dissociate to generate hydrogen ion. Accordingly, the solution has a relatively low concentration of hydrogen ion and relatively high concentrations of undissociated lactic acid molecules, sodium ions, and lactate ions. Lactic acid, hydrogen ion, and lactate are in equilibrium with each other:

$$\text{Lactic acid} \rightleftharpoons \text{lactate}^- + H^+$$

By the mass-action law, an increase in the concentration of any substances on one side of the arrows forces the reaction in the opposite direction. Conversely, a decrease in the concentration of any substances on one side of the arrows forces the reaction toward that side, i.e., in the direction which generates more of that substance.

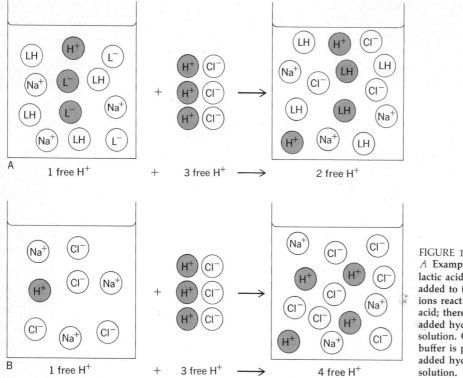

FIGURE 11-27

A **Example of a buffer system. LH =
lactic acid, L⁻ = lactate. When HCl is
added to the beaker, two of the hydrogen
ions react with lactate to give lactic
acid; therefore, only one of the three
added hydrogen ions remains free in
solution. Contrast this to** *B,* **in which no
buffer is present and all three of the
added hydrogen ions remain free in
solution.**

What happens if we add hydrochloric acid to this solution? Hydrochloric acid, a strong acid, completely dissociates and liberates hydrogen ions. This excess of hydrogen ions drives the lactic acid reaction to the left, and many of the hydrogen ions liberated from the dissociation of hydrochloric acid thus combine with lactate to give undissociated lactic acid. As a result, many of the hydrogen ions generated by the dissociation of hydrochloric acid do not remain free in solution but become bound in lactic acid molecules. The final hydrogen-ion concentration is therefore smaller than if the hydrochloric acid had been added to pure water (Fig. 11-27).

Conversely, if we add a chemical which *removes* hydrogen ions instead of adding them, the lactic acid reaction is driven to the right, lactic acid molecules dissociate to generate more hydrogen ions, and the original fall in hydrogen-ion concentration is minimized. This process preventing large changes in hydrogen-ion concentration when hydrogen ion is added or removed from a solution is termed *buffering,* and the chemicals (in this case lactate and lactic acid) are called *buffers* or *buffer systems.*[1]

Generation of hydrogen ions in the body

There are three major sources of hydrogen ion in the body:
1 Phosphorus and sulfur are present in large quantities in many proteins and other biologically important molecules. The catabolism of these molecules releases phosphoric and sulfuric acids into the extracellular fluid. These acids, to a large extent, dissociate into hydrogen ions and anions (phosphate and sulfate). For example,

$$H_2PO_4^- \longrightarrow H^+ + HPO_4^{2-}$$
$$H_2SO_4 \longrightarrow 2 H^+ + SO_4^{2-}$$

[1]Of necessity, this description of acids and buffering is oversimplified and incomplete; e.g., no mention has been made of the ionization of water itself or of the importance of the hydroxyl ion. The interested reader can find excellent descriptions of acid-base chemistry in any textbook of biochemistry.

2 Many organic acids, e.g., fatty acids and lactic acid, are produced as end products of metabolic reactions and also liberate hydrogen ion by dissociation.

3 We have already described (Chap. 10) the major source of hydrogen ion, namely, liberation of hydrogen ion from metabolically produced carbon dioxide via the reactions

$$CO_2 + H_2O \longrightarrow H_2CO_3 \longrightarrow H^+ + \underset{\text{Bicarbonate}}{HCO_3^-}$$

As described in Chap. 10, the lungs normally eliminate carbon dioxide from the body as rapidly as it is produced in the tissues. As blood flows through the lung capillaries and carbon dioxide diffuses into the alveoli, the chemical reactions which originally generated HCO_3^- and H^+ from carbon dioxide and water in the venous blood are reversed:

$$CO_2 + H_2O \longleftarrow H_2CO_3 \longleftarrow H^+ + HCO_3^-$$

As a result, all the hydrogen ion generated from carbonic acid is completely reincorporated into water molecules. Therefore, there is *normally* no net gain or loss of hydrogen ion in the body from this source, but what happens when lung disease prevents adequate elimination of carbon dioxide? The retention of carbon dioxide within the body means an elevated extracellular P_{CO_2}. Some of the hydrogen ion generated in the venous blood from the reaction of this carbon dioxide with water would also be retained in the body and would raise the extracellular hydrogen-ion concentration. This hydrogen ion must be eliminated from the body (via the kidneys) in order for hydrogen-ion balance to be maintained.

Buffering of hydrogen ions within the body

Between the generation of hydrogen ions in the body and their excretion (to be described) what happens? The hydrogen-ion concentration of the extracellular fluid is extremely small, approximately 0.00004 mmole/liter. What would happen to this concentration if just 2 mmoles of hydrogen ion remained free in solution following its dissociation from an acid? Since the total extracellular fluid is approximately 12 liters, the hydrogen-ion concentration would increase by 2/12, or approximately 0.167 mmole/liter. Since the original concentration was only 0.00004, this would represent more than a 4,000-fold increase. Obviously, such a rise cannot really occur, for we have already observed that the extracellular hydrogen ion is kept remarkably constant. Therefore, of the 2 mmoles of hydrogen ion liberated in our problem, only an extremely small portion can have remained free in

solution. The vast majority has been bound (buffered) by other ions. The kidneys ultimately eliminate excess hydrogen ions produced in the body, but it is buffering which minimizes hydrogen-ion concentration changes until excretion occurs.

The most important body buffers are bicarbonate—CO_2, large anions such as plasma protein and intracellular phosphate complexes, and hemoglobin. Recall that only free hydrogen ions contribute to the acidity of a solution. These buffers all act by binding hydrogen ions according to the general reaction

$$Buffer^- + H^+ \rightleftharpoons H\text{-buffer}$$

It is evident that H-buffer is a weak acid in that it can exist as the undissociated molecule or can dissociate to $buffer^- + H^+$. When hydrogen-ion concentration increases, the reaction is forced to the right and more hydrogen ion is bound. Conversely, when hydrogen-ion concentration decreases, the reaction is forced to the left and hydrogen ion is released. In this manner, the body buffers stabilize hydrogen-ion concentration against changes in either direction.

Bicarbonate–CO_2 In this section and Chap. 10 we have already described the relationships between HCO_3^-, H^+, and CO_2. Let us once more write the pertinent equations in their true forms as reversible equations:

$$H^+ + HCO_3^- \rightleftharpoons H_2CO_3 \rightleftharpoons H_2O + CO_2$$

The basic mechanism by which this system acts as a buffer should be evident: An increased extracellular hydrogen-ion concentration drives the reaction to the right, H^+ and HCO_3^- combine, hydrogen ion is thereby removed from solution, and the hydrogen-ion concentration turns toward normal. Conversely, a decreased extracellular hydrogen-ion concentration drives the reaction to the left, CO_2 and H_2O combine to generate hydrogen ion, and this additional hydrogen ion turns hydrogen-ion concentration toward normal. One reason for the importance of this buffer system is that the extracellular bicarbonate concentration is normally quite high and is closely regulated by the kidney. A second and even more important reason stems from the relationship between extracellular hydrogen-ion concentration and carbon dioxide elimination from the body.

When additional hydrogen ion is added to the extracellular fluid, i.e., when the H^+ combines with HCO_3^-, the extent to which this reaction can restore hydrogen-ion concentration to normal depends upon precisely how much additional H^+ actually combines with HCO_3^- and

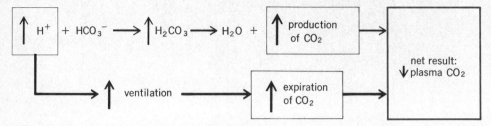

FIGURE 11-28
Effects of excess hydrogen ions on plasma carbon dioxide. The direct effect, by mass action, is to increase the production of carbon dioxide, but the indirect effect is to lower carbon dioxide by reflexly stimulating breathing. Since the latter effect predominates, the net effect is a reduction of plasma carbon dioxide.

is thereby removed from solution. Complete compensation (which never actually occurs) can be obtained only if the HCO_3^-—CO_2 reaction proceeds to the right until all the additional H^+ has combined with HCO_3^-. However, this reaction obviously generates CO_2, which seriously hinders the further buffering ability of HCO_3^- because, as can be seen from the equation, any increase in the concentration of CO_2, by the mass-action law, tends to drive the reaction back to the left, thus preventing further net combination of $H^+ + HCO_3^-$. In reality, however, this expected increase in extracellular CO_2 does not occur. Indeed, during periods of increased body hydrogen-ion production from organic, phosphoric, or sulfuric acids, the extracellular CO_2 is actually *decreased*! What causes this? We have already studied the mechanism, in Chap. 10: A greater extracellular hydrogen-ion concentration stimulates the respiratory centers to increase alveolar ventilation and thereby causes greater elimination of carbon dioxide from the body (Fig. 11-28). Thus, although an increased combination of H^+ and HCO_3^- generates more carbon dioxide, respiratory stimulation produced by the higher extracellular hydrogen-ion concentration results in the elimination of carbon dioxide even faster than it is generated. As a result, the extracellular carbon dioxide decreases, and, by the mass-action law, the further combination of H^+ and HCO_3^- is actually facilitated. It is this control of carbon dioxide elimination exerted by the extracellular hydrogen-ion concentration that allows the HCO_3^-—CO_2 system to function so efficiently as a buffer.

A lower extracellular hydrogen-ion concentration resulting from either decreased hydrogen-ion production or increased hydrogen-ion loss from the body is compensated for by just the opposite buffer reactions: (1) The lower hydrogen-ion concentration drives the HCO_3^-—CO_2 reaction to the left, carbon dioxide and water combine to generate $H^+ + HCO_3^-$, and this additional hydrogen ion turns hydrogen-ion concentration toward normal; (2) the lower hydrogen-ion concentration decreases alveolar ventilation and carbon dioxide elimination by inhibiting the medullary respiratory centers. The elevated extracellular carbon dioxide resulting from this process also serves to drive the HCO_3^-—CO_2 reaction to the left, thus allowing further generation of hydrogen ion.

When lung disease is itself the cause of the increased hydrogen-ion concentration, the efficiency of this buffer system is obviously greatly impaired. Actually a large fraction of any hydrogen-ion excess or deficit is always buffered by the other buffers listed.

Hemoglobin as a buffer   In Chap. 10 we pointed out that most metabolically produced carbon dioxide is carried from the tissues to the lungs in the form of HCO_3^-. These bicarbonate ions are generated by the hydration of CO_2 to form H_2CO_3 and the dissociation of H_2CO_3 to HCO_3^- and H^+. As blood flows through the lungs, the reaction is reversed, and H^+ and HCO_3^- recombine. However, these hydrogen ions must be buffered while they are in transit from the tissues to the lungs, a function performed primarily by hemoglobin. Its suitability for this role depends upon a remarkable characteristic of the hemoglobin molecule: Reduced hemoglobin has a much greater affinity for hydrogen ion than oxyhemoglobin does. As blood flows through the tissues, a fraction of oxyhemoglobin loses its oxygen and is transformed into reduced hemoglobin. Simultaneously, a large quantity of carbon dioxide enters the blood and undergoes (primarily in the red blood cells) the reactions which ultimately generate HCO_3^- and H^+. Because reduced hemoglobin has a strong affinity for hydrogen ion, most of these hydrogen ions become bound to hemoglobin (Fig. 11-29). In this manner only a few hydrogen ions remain free, and the acidity of venous blood is only slightly greater than that of arterial blood. As the venous blood passes through the lungs, all these reactions are reversed. Hemoglobin becomes saturated with oxygen, and its ability to bind hydrogen ions decreases. The hydrogen ions are released, whereupon they react with HCO_3^- to give CO_2, which diffuses into the alveoli and is expired.

In Chap. 10 we described how the hydrogen-ion concentration of the blood is an important determinant of the ability of hemoglobin to bind oxygen. Now we have shown how the presence of oxygen is an important determinant of hemoglobin's ability to bind hydrogen ion. The adaptive value of these phenomena is enormous. Their combination in one molecule marks hemoglobin as a remarkable evolutionary development.

It should be emphasized that none of these buffer systems eliminates hydrogen ion from the body, instead causing the hydrogen ion to be bound in some molecule (H_2O, hemoglobin, etc.). Binding thus removes the hydrogen ion from solution, preventing it from contributing to the free hydrogen-ion concentration; the actual elimination of hydrogen ion from the body is normally performed only by the kidneys.

Renal regulation of extracellular hydrogen-ion concentration

In a normal person the quantity of phosphoric, sulfuric, and organic acids formed depends primarily upon the type and quantity of food ingested. A high protein diet, for example, results in increased protein breakdown and release of large quantities of sulfuric acid. The average American diet results in the liberation of 40 to 80 mmoles of hydrogen ion each day. If hydrogen-ion balance is to be maintained, the same quantity must be eliminated from the body each day. This loss occurs via the kidneys. In addition, the kidneys must be capable of *altering* their hydrogen-ion excretion in response to changes in body hydrogen-ion production, regardless of whether the source is carbon dioxide (in lung disease) or phosphoric, sulfuric, or organic acids. The kidneys must also be able to compensate for any gastrointestinal loss or gain of hydrogen ion resulting from disease.

Virtually all the hydrogen ion excreted in the urine enters the tubules via tubular secretion, the mechanism and its control being quite complex. Suffice it to say that the presence of excess acid in the body induces an increased urinary hydrogen-ion excretion. Conversely, the kidney responds to a decreased amount of acid in

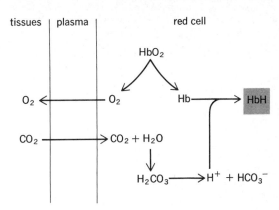

FIGURE 11-29
Buffering of hydrogen ions by hemoglobin as blood flows through tissue capillaries.

the body by lowering urinary hydrogen-ion excretion. The controlling effect of the acid appears to be primarily exerted directly upon the tubular cells, with no nerve or hormone intermediates. Obviously, such a system is effective in stabilizing hydrogen-ion concentration at the normal value.

The ability of the kidneys to excrete H^+ depends both upon tubular hydrogen-ion secretion and upon buffers in the urine. When there are no buffers, the free hydrogen-ion concentration of the tubular fluid rises so high that hydrogen ion diffuses back into the peritubular capillary plasma as fast as it is actively secreted into the lumen, and no net transfer of hydrogen ion from plasma to lumen can result. The major urinary buffers are HPO_4^{2-} and ammonia NH_3:

$$HPO_4^{2-} + H^+ \longrightarrow H_2PO_4^-$$
$$NH_3 + H^+ \longrightarrow NH_4^+$$

The HPO_4^{2-} in the tubular fluid has been filtered and not reabsorbed. In contrast, the ammonia of the tubular fluid is formed by the tubular cells themselves by the deamination of certain amino acids transported into the renal tubular cells from the peritubular capillary plasma. From the cells the ammonia diffuses into the lumen. Of great importance is that the amount which *remains* in the lumen depends upon the rate of hydrogen-ion accumulation there, since the tubular cell membrane is quite permeable to ammonia but not to NH_4^+. Accordingly, the more hydrogen ion there is in the lumen, the greater the conversion of NH_3 to NH_4^+, increased secretion of hydrogen ion automatically inducing an increased excretion of NH_3

in the form of NH_4^+. Another important feature of this system is that, by unknown mechanisms, the rate of ammonia production by the renal tubular cells increases whenever extracellular hydrogen-ion concentration remains elevated for more than 1 to 2 days; this extra ammonia provides the additional buffers required for increased hydrogen-ion secretion.

Kidney disease

The term kidney disease is no more specific than car trouble, since many diseases affect the kidneys. Bacteria cause kidney infections, most of which are collectively called *pyelonephritis.* A common type of kidney disease, *glomerulonephritis,* results from an allergy incident to throat infection by a specific group of bacteria. Congenital defects, stones, tumors, and toxic chemicals are possible sources of kidney damage. Obstruction of the urethra or a ureter may cause injury due to a buildup of pressure and may predispose the kidneys to bacterial infection.

Disease can attack the kidney at any age. Experts estimate that there are at present more than 3 million undetected cases of kidney infection in the United States and that 25,000 to 75,000 Americans die of kidney disease each year.

Early symptoms of kidney disease depend greatly upon the type of disease involved and the specific part of the kidney affected. Although many diseases are self-limited and produce no permanent damage, others progress if untreated. The end stage of progressive diseases, regardless of the nature of the damaging agent (bacteria, toxic chemical, etc.), is a shrunken, nonfunctioning kidney. Similarly, the symptoms and signs of profound renal malfunction are independent of the damaging agent and are collectively known as *uremia,* literally "urine in the blood."

The severity of uremia depends upon how well the impaired kidneys are able to preserve the constancy of the internal environment. Assuming that the patient continues to ingest a normal diet containing the usual quantities of nutrients and electrolytes, what problems arise? The key fact to keep in mind is that the kidney destruction has markedly reduced the glomerular filtration rate. Accordingly, the many substances which gain entry to the tubule primarily by filtration are filtered in diminished

amounts. Of the substances described in this chapter, this category includes sodium chloride, water, calcium, and a number of waste products. In addition, the excretion of potassium, hydrogen ion, and certain other substances is impaired because the diseased kidneys have a diminished capacity for tubular secretion. The buildup of these substances in the blood causes the symptoms and signs of uremia.

The artificial kidney

The artificial kidney is an apparatus that eliminates the excess ions and wastes which accumulate in the blood when the kidneys fail. Blood is pumped from one of the patient's arteries through tubing which is bathed by a large volume of fluid. The tubing then conducts the blood back into the patient by way of a vein. The tubing is generally made of a cellophane, which is highly permeable to most solutes but relatively impermeable to protein—characteristics quite similar to those of capillaries. The bath fluid, which is constantly replaced, is a salt solution similar in ionic concentrations to normal plasma. The basic principle is simply that of dialysis, or diffusion. Because the cellophane is permeable to most solutes, as blood flows through the tubing, solute concentrations tend to equilibrate in the blood and bath fluid. Thus, if the plasma potassium concentration of the patient is above normal, potassium diffuses out of the blood into the bath fluid. Similarly, waste products and excesses of other substances diffuse across the cellophane tubing and thus are eliminated from the body.

Until recently patients were placed on the artificial kidney only when there was reason to believe that their kidney damage was only temporary and that recovery would occur if the patient could be kept alive during temporary renal failure. However, in recent years, technical improvements have permitted many patients to utilize the artificial kidney several times a week for unlimited periods, and thus patients with permanent kidney failure can be kept alive and functioning, many of them performing the dialysis in their homes.

The other major hope for patients with permanent renal failure is kidney transplantation. Although great strides have been made, the major problem remains the frequent rejection of the transplanted kidney by the recipient's body (see Chap. 15).

Digestion and absorption of food

12

The function of the gastrointestinal system (Fig. 12-1), which includes the gastrointestinal tract (mouth, esophagus, stomach, small and large intestines), salivary glands, and portions of the liver and pancreas, is to transfer food and water from the external environment to the internal environment, where they can be distributed to the cells of the body by the circulatory system. Most food is taken into the mouth as large pieces of matter, consisting of high-molecular-weight substances such as proteins and polysaccharides which are unable to cross cell membranes. Before these substances can be absorbed, they must be broken down into smaller molecules, such as amino acids and monosaccharides. This breaking-down process, *digestion*, is accomplished by the action of acid and enzymes secreted into the gastrointestinal tract. The small molecules resulting from digestion cross the cell membranes of the intestinal cells and enter the blood and lymph, a process known as *absorption*. While these processes are taking place, contractions of the smooth muscle lining the walls of the gastrointestinal tract move the luminal contents through the tract. Contrary to popular belief, the gastrointestinal system is not a major excretory organ for eliminating wastes from the body. It is true that small amounts of some end products, such as the breakdown products of hemoglobin, are normally eliminated in the feces, but the elimination of most metabolic end products (wastes) from the internal environment

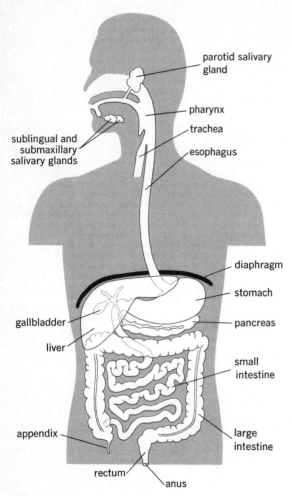

FIGURE 12-1
Anatomy of the gastrointestinal system.

parotid salivary gland

pharynx

trachea

esophagus

sublingual and submaxillary salivary glands

diaphragm

stomach

pancreas

gallbladder

liver

small intestine

appendix

large intestine

rectum

anus

is performed not by the gastrointestinal tract but by the lungs and kidneys. Feces consist primarily of bacteria and ingested material which failed to be digested and absorbed during its passage along the gastrointestinal tract, i.e., material that was technically never in the internal environment of the body. In this chapter we examine four aspects of gastrointestinal function: (1) secretion, (2) digestion, (3) absorption, and (4) motility (Fig. 12-2). In Chap. 13 the mechanisms which control the distribution and utilization of absorbed food products as well as hunger and thirst will be considered.

Structure of gastrointestinal tract

The gastrointestinal tract consists of a tube of variable diameter, 15 ft in length, running through the body from mouth to anus. The lumen of this tube, like the hole in a doughnut, is continuous with the external environment, which means that its contents are technically outside of the body. This fact is relevant to an understanding of some of the tract's properties. For example, the lower portion of the intestinal tract is inhabited by millions of living bacteria, most of which in this location are harmless and even beneficial, but if the same bacteria enter the blood, as may happen as a result of a ruptured appendix, they are extremely harmful and even lethal.

The gastrointestinal tract has the same general structure throughout most of its length as the segment of intestine illustrated in Fig. 12-3. From the esophagus to the anus, it is surrounded by two layers of smooth muscle, an outermost layer with fibers oriented into sheets running longitudinally along the tube and an inner layer with a circular orientation. A smaller third layer of smooth muscle, the *muscularis mucosae,* is located between the mucosa and submucosa and consists of both longitudinal and circular fibers. The contractions of these smooth muscle layers exert pressures on the contents of the lumen, causing material to flow. The layer of cells between the muscularis mucosae and the lumen of the tube is known as the *mucosa.* This layer contains most of the exocrine gland cells secreting material into the lumen and the epithelial cells involved in the absorption of material from the lumen into the blood vessels and lymphatics passing through the mucosal and submucosal layers. The submucosal layer of cells contains large amounts of connective tissue and some exocrine gland cells as well as blood vessels and lymphatics. The luminal surface of the tube is generally not flat and smooth but highly convoluted, with many ridges and valleys which greatly increase the total surface area available for absorption. The degree of folding varies in different areas of the gastrointestinal tract, being most extensive in the small intestine.

Regulation of the gastrointestinal system

Gland secretion and smooth muscle contraction in the gastrointestinal system are regulated by nerves and hor-

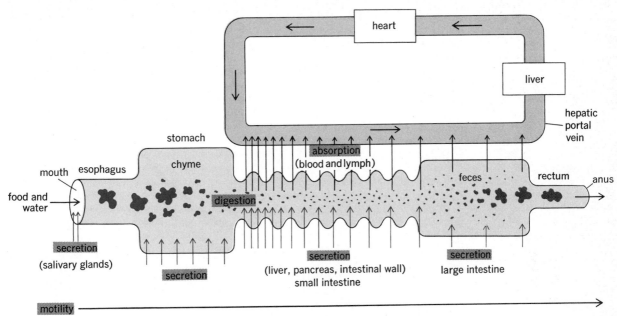

FIGURE 12-2
Summary of gastrointestinal activity involving motility,
secretion, digestion, and absorption.

mones. The glands include both endocrine and exocrine
types. Some of the exocrine glands are outside the walls
of the gastrointestinal tract, notably the salivary glands,
the liver, and the pancreas, whereas others are in the
mucosal and submucosal layers of the tract wall. All these
exocrine glands secrete material through ducts which
empty into the lumen of the gastrointestinal tract (Fig.
12-4). Endocrine glands in the mucosa secrete hormones
into the blood which flows through the capillaries in the
walls of the gastrointestinal tract; the hormones are
carried by the venous blood back to the heart and return
to the gastrointestinal tract by way of the arterial system.
Thus these hormones can affect smooth muscle activity
or gland secretion in an area of the gastrointestinal tract
far removed from the site where they are produced. The
three major gastrointestinal hormones are *gastrin* (secreted by the stomach), *secretin,* and *cholecystokinin* (the
latter two secreted by the first segment of the small intestine).

The contractile and secretory activity of the gastrointestinal tract is regulated in large part by its own local
nervous system. The gastrointestinal tract has in its walls
two major nerve plexuses, the *myenteric plexus,* between
the longitudinal and circular layers of smooth muscle, and
the *submucous nerve plexus,* in the submucosa (Fig. 12-3).
These two plexuses are found throughout the length of
the gastrointestinal tract from esophagus to anus. They
are composed of neurons forming synaptic junctions with
other neurons in the plexus or ending in the regions
around smooth muscles and glands. Many axons leave
the myenteric plexus and synapse with neurons in the
submucous plexus and vice versa, so that neural activity
in one plexus influences the activity in the other. The
axons in both plexuses branch profusely, and electric
stimulation at one point in the plexus is found to lead
to electric activity that is conducted both up and down
the gastrointestinal tract. Thus, activity initiated in the
plexus in the upper part of the small intestine may affect
smooth muscle and gland activity in the stomach as well
as in the lower part of the intestinal tract.

Nerve fibers from both the sympathetic and parasympathetic branches of the autonomic nervous system
enter the intestinal tract and synapse with neurons in the
internal nerve plexuses. Thus, these autonomic fibers

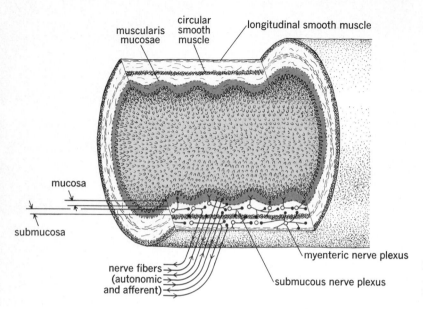

FIGURE 12-3
Anatomy of a segment of the small intestine.

(which we shall refer to as *external fibers* to distinguish them from the internal plexuses) exert many of their effects upon the glands and muscles of the gastrointestinal tract through the internal plexuses. Frequently, however, the external fibers (particularly the sympathetic fibers) bypass the plexuses and end in the immediate vicinity of gland cells and vascular smooth muscle. The major autonomic nerve supplying the gastrointestinal tract is the *vagus nerve*, which sends branches to the stomach, small intestine, and upper portion of the large intestine. This nerve is composed of efferent parasympathetic fibers and many afferent fibers from receptors and the nerve plexuses in the gastrointestinal wall.

The afferent pathways of the gastrointestinal tract are also complicated by the presence of the internal plexuses. The walls of the tract contain numerous sensory receptors, many of which are the dendritic endings of the internal plexus neurons, so that action potentials arising in them are conducted directly to the internal plexuses. In some cases the situation is similar to that of other afferent neurons in the body in that the receptors are the endings of neurons whose cell bodies are outside the gastrointestinal tract in ganglia. In these cases, the afferent information bypasses the internal plexuses and is conducted directly to the central nervous system via the vagus nerve.

The general pattern of gastrointestinal control systems can now be summarized (Fig. 12-5). The smooth muscle and exocrine glands are directly influenced by three efferent pathways: (*1*) external autonomic nerves, (*2*) internal plexus nerve fibers, (*3*) hormones secreted by the gastrointestinal tract itself. The endocrine gland cells which secrete these hormones are controlled not only by these same three types of efferent input but respond directly to changes in the composition of the gastrointestinal contents as well.

These neural and hormonal inputs constitute the efferent pathways of reflexes initiated by receptors in the gastrointestinal tract itself or in other parts of the body. The gastrointestinal receptors respond to changes in the chemical composition of the luminal contents and to the degree of wall distension. As noted above, the information from these receptors is conducted to the internal plexuses as well as to the central nervous system. In the former case, all the components of the reflex arc are located completely within the walls of the gastrointestinal tract; these so-called *short* reflexes confer a considerable degree of self-regulation upon the tract. The point should not be overemphasized, however, since most gastrointestinal reflexes initiated by receptors in the tract have some degree of central nervous system control. Of course, whenever the reflexes are initiated by receptors outside the tract, e.g., by the sight of food, the central nervous system must be involved in the response. Similarly, complex behavioral influences, e.g., emotion, operate through the central nervous system.

ENDOCRINE GLAND

blood flow

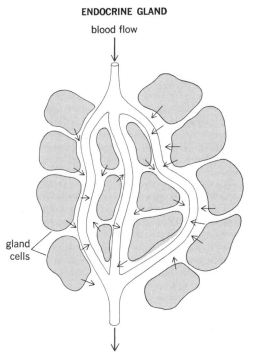

gland
cells

EXOCRINE GLAND

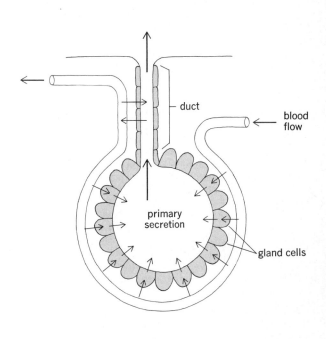

duct

blood
flow

primary
secretion

gland cells

FIGURE 12-4

General structure of an endocrine and an exocrine gland. Note that as the solution originally secreted by the exocrine gland cells (the primary secretion) moves along the duct its composition may be altered by active transport or diffusion across the duct walls.

The mouth, pharynx, and esophagus

Chewing

The primary function of the teeth is to bite off and grind down chunks of food into pieces small enough to be swallowed. The incisors of an adult man can exert forces of 25 to 50 lb, and the molars the 200 lb required to crack a walnut or support a trapeze artist by his teeth. Prolonged chewing of food in the mouth, so characteristic of man, does not appear to be essential to the digestive process (many animals, such as the dog and cat, swallow their food almost immediately). Although chewing prolongs the subjective pleasure of taste, it does not appreciably alter the rate at which the food is digested and absorbed. On the other hand, attempting to swallow a particle of food too large to enter the esophagus may lead to choking, as the particle lodges over the trachea, blocking the entry of air into the lungs. A surprising number of preventable deaths occur each year from choking, the symptoms of which are often confused with the sudden onset of a heart attack so that no attempt is made to remove the obstruction from the airway. The rhythmic act of chewing is a combination of voluntary and reflex activation of the skeletal muscles of the mouth and jaw. In animals whose cerebral cortex has been destroyed, the reflex rhythmic movements of chewing still occur when food is placed in the mouth.

Secretion of saliva

Saliva is secreted by three pairs of exocrine glands, the parotid, the submaxillary, and the sublingual (Fig. 12-1). Water accounts for 99 percent of the secreted fluid, the remaining 1 percent consisting of various salts and a few proteins. The major proteins of saliva are the *mucins,* which contain small amounts of carbohydrate attached to the amino acid side chains of the protein. When mixed with water, the mucins form a highly viscous solution

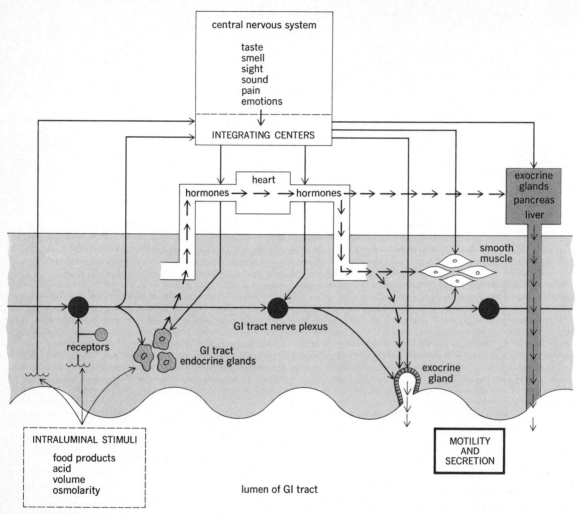

FIGURE 12-5
Summary of pathways regulating gastrointestinal activity.

known as *mucus*. Mucins are also secreted by gland cells located throughout the gastrointestinal tract. In the mouth the watery solution of mucus moistens and lubricates the food particles, allowing them to slide easily along the esophagus into the stomach. (Imagine what it would be like trying to swallow a dry cracker.)

Another protein secreted by the salivary glands is the enzyme *amylase*, which catalyzes the breakdown of polysaccharides into disaccharides. Although salivary amylase starts the digestive process in the mouth, food does not remain there long enough for much digestion

to occur; however, amylase continues its digestive activity in the stomach until inhibited by the hydrochloric acid there.

A third function of saliva is to dissolve some of the molecules in the food particle; only in this dissolved state can they react with the chemoreceptors in the mouth, giving rise to the sensation of taste (Chap. 16).

The secretion of saliva is controlled by the autonomic nerves to the glands. The salivary glands are innervated by branches from both the sympathetic and parasympathetic nervous systems; however, unlike their

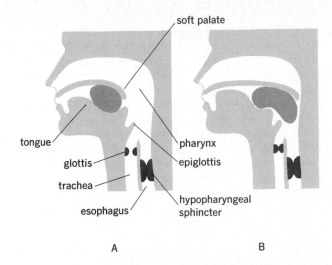
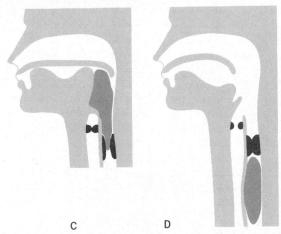

soft palate

tongue

glottis

trachea

esophagus

pharynx

epiglottis

hypopharyngeal
sphincter

A   B   C   D

FIGURE 12-6
Movement of a bolus of food through the pharynx and upper
esophagus during swallowing.

antagonistic activity in most organs, both systems stimulate secretion although the parasympathetic branch causes by far the greatest increase in fluid volume. During sleep very little saliva is secreted. In the awake state a basal rate of about 0.5 ml/min keeps the mouth moist. Food in the mouth increases the rate of salivary secretion. This response is mediated by nerve fibers from chemoreceptors and pressure receptors in the walls of the mouth and tongue. Chewing tasteless wax or marbles increases the rate of salivary secretion even though they have no taste or nutritional value since they activate the pressure receptors. The receptors send fibers to the brain medulla, which contains the integrating center controlling salivary secretion. The most potent stimuli for salivary secretion are acid solutions, e.g., fruit juices and lemons, which may lead to a maximal secretion of 4 ml of saliva per minute. Salivation initiated by the sight, sound, or smell of food is very slight in man in contrast to the marked increase produced by these stimuli in dogs.

During the course of a day, between 1 and 2 liters of saliva is secreted, most of which is swallowed. The proteins in saliva are broken down into amino acids by the digestive enzymes in the stomach and intestinal tract, and the amino acids, salts, and water absorbed into the circulation. This is a typical pattern for most of the secretions of the gastrointestinal tract. Although large amounts of fluid may be secreted into the tract during the course of a day, most of it, together with its salt and protein

content, is digested and reabsorbed. If these secretions are not reabsorbed, large quantities of fluid and salt can be lost from the body through the gastrointestinal tract.

Swallowing

Swallowing is a complex reflex initiated when the tongue forces a bolus of food into the rear of the mouth, where pressure receptors in the walls of the pharynx are stimulated. They send afferent impulses to the swallowing center in the medulla, which coordinates the swallowing process via efferent impulses to the 25 different skeletal muscles in the pharynx, larynx, and early esophagus and the smooth muscles of the lower esophagus. Once swallowing has been initiated, it cannot be stopped voluntarily even though it involves skeletal muscles. This reflex is a stereotyped all-or-none response, the entire coordination of which resides in the swallowing center in the medulla. The swallowing reflex is an example of a triggered reflex in which multiple responses occur in a regular temporal sequence that is predetermined by the synaptic connections in the coordinating center.

As the bolus of food moves into the pharynx, the soft palate is elevated and lodges against the back wall of the pharynx, sealing off the nasal cavity and preventing food from entering this area (Fig. 12-6). The swallowing center inhibits respiration, raises the larynx, and closes the glottis (the opening between the vocal cords), keeping food from getting into the trachea. As the tongue

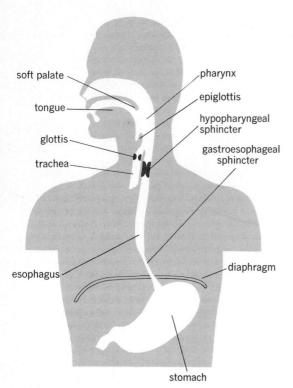

FIGURE 12-7
Anatomy of the mouth, pharynx, and esophagus.

Labels (left to right, top to bottom): soft palate, tongue, glottis, trachea, esophagus, pharynx, epiglottis, hypopharyngeal sphincter, gastroesophageal sphincter, diaphragm, stomach

forces the food further back into the pharynx, the bolus tilts the epiglottis backward to cover the closed glottis. It is the closure of the glottis, however, and not the folding of the epiglottis which is primarily responsible for preventing food from entering the trachea.

The upper third of the esophagus in man is surrounded by skeletal muscle; the lower two-thirds consist of smooth muscle. At rest, the opening into the esophagus is closed, this region of the esophagus forming the *hypopharyngeal sphincter* (Fig. 12-7). The skeletal muscles in this region are so arranged that when they are relaxed, the sphincter is closed by passive elastic tensions in the walls. The swallowing center initiates contraction of these muscles, opening the hypopharyngeal sphincter and allowing the bolus to pass into the esophagus. Immediately after the bolus has passed, the sphincter muscles relax, the sphincter closes, the glottis opens, and breathing resumes. The pharyngeal phase of swallowing lasts about 1 sec.

Once within the esophagus, the bolus is moved

along it by a wave of contraction that passes along the walls of the esophagus and moves toward the stomach. Such waves of contractile activity in the muscle layers surrounding a tube are known as *peristaltic waves* and are found in various parts of the body, including other portions of the gastrointestinal tract. A peristaltic wave takes a total of about 9 sec to travel the length of the esophagus. The progression of the wave is controlled·by autonomic nerves coordinated by the swallowing center in the medulla, unlike the peristaltic waves in other portions of the gastrointestinal tract, which are coordinated primarily through the internal nerve plexuses. If a large or sticky bolus of food, e.g., peanut butter, becomes stuck so that it is not carried along the esophagus by the initial wave, the distension of the esophagus triggers a second more forceful peristaltic wave, which begins without movement of the mouth or pharynx. Distension of the esophagus also reflexly stimulates increased salivary secretion. The combination of increased salivary fluid and repeated peristaltic waves helps dislodge the trapped bolus. Swallowing can occur while a person is standing on his head since it is not primarily gravity but the peristaltic wave which moves the contents of the esophagus toward the stomach.

The last 4 cm of the esophagus before it enters the stomach is known as the *gastroesophageal sphincter*. Anatomically, this region appears no different from adjacent regions, but its functional activity is distinct. When swallowing is not taking place, this sphincter remains tonically contracted, forming a barrier between the contents of the stomach and the esophagus. Most of the esophagus lies in the thoracic cavity and is subject to a subatmospheric intrathoracic pressure of about −5 to −10 mm Hg. The stomach, however, lying below the diaphragm, has an internal pressure slightly above atmospheric, +5 to +10 mm Hg, due to its compression by the contents of the abdominal cavity. Thus, without the gastroesophageal sphincter, the pressure gradient from the stomach to the esophagus would tend to force the contents of the stomach into the esophagus. As the peristaltic wave begins in the esophagus, the gastroesophageal sphincter relaxes, allowing the bolus upon its arrival to enter the stomach. After the bolus has passed into the stomach, the sphincter contracts, resealing the junction.

The ability of the gastroesophageal sphincter to maintain a barrier between the stomach and esophagus is helped by the fact that its last portion lies below the diaphragm and is subject to the same pressures in the abdominal cavity as the stomach. Thus, if the pressure of the abdominal cavity is raised, e.g., during cycles of respiration or by contraction of the abdominal muscles, the pressures of both the stomach contents and this terminal segment of the esophagus are raised together and there is no change in the pressure gradient between the two. During pregnancy the growth of the fetus increases the pressure on the abdominal contents and displaces the terminal segment of the esophagus through the diaphragm into the thoracic cavity. The gastroesophageal barrier is therefore no longer assisted by changes in abdominal pressure, and during the last 5 months of pregnancy there is a tendency for the increased pressures in the abdominal cavity to force some of the contents of the stomach up into the esophagus. The hydrochloric acid from the stomach contents irritates the walls of the esophagus, causing contractile spasms of the smooth muscle which is experienced as pain and known generally as *heartburn* because the sensation appears to be located in the region over the heart. Heartburn often subsides in the last weeks of pregnancy as the uterus descends prior to delivery, decreasing the pressure on the abdominal organs. A newborn child has no intra-abdominal segment of the esophagus and thus has a tendency to regurgitate.

In the abnormal condition called *achalasia,* the gastroesophageal sphincter fails to relax, and a whole meal may become lodged in the esophagus, entering the stomach very slowly. Distension of the esophagus results in pain in the chest which is often confused with pain originating from the heart. Achalasia appears to be the result of damage to, or absence of, the myenteric nerve plexus in the region of the gastroesophageal sphincter.

Besides solids and semisolids, both liquids and air (as much as 500 ml) are swallowed during the normal course of a meal. Most of the air travels no further than the esophagus, where it is eventually expelled by belching. Some of the air, however, reaches the stomach and may even be passed on into the intestine (100 ml/day).

The stomach

The stomach is a chamber located between the end of the esophagus and the beginning of the small intestine (Fig. 12-8). The stomach secretes a very strong acid, hydrochloric acid, and several enzymes which along with salivary amylase begin the process of digestion. The degree to which food is digested in the stomach is limited, however, to breaking down large lumps of food into a solution of individual molecules and fragments of large molecules, most of which are still too large to be ab-

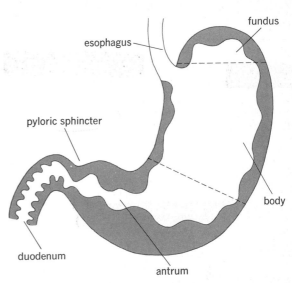

FIGURE 12-8
Anatomy of the stomach.

sorbed. This mixture of fluids, partially digested food particles, and enzymes has the consistency of a thick soup and is known as *chyme.* The most important function of the stomach is to regulate the rate at which chyme enters the small intestine, where most of the process of digestion and absorption takes place. In the absence of the stomach, a normal-sized meal moves so rapidly through the small intestine that only a fraction of the food has time to be digested and absorbed.

Gastric motility

The empty stomach has a volume of about 50 ml, and its lumen is little larger than that of the small intestine. The surface of the interior of the stomach is highly folded into ridges. Upon being filled with food, the stomach relaxes and the folds get smaller; thus the wall tension and intraluminal pressure change only slightly.

The walls of the stomach are surrounded by the two layers of smooth muscle found throughout the gastrointestinal tract, the external longitudinal layer and the internal circular layer. These muscle layers are relatively thin around the fundus and body but are thicker and more powerful in the lower portion of the stomach, the *antrum* (see Fig. 12-8). Emptying the stomach contents depends primarily upon the contractile activity of these smooth muscle layers.

The membrane potential of the longitudinal smooth

muscle cells undergoes rhythmic oscillations of about 10 to 15 mV, repeating at a rate of about three per minute. Known as the *basic electrical rhythm*, its magnitude and rate of propagation control the mechanical activity of the stomach wall. The oscillation in membrane potential is initiated by pacemaker cells in the region near the entry of the esophagus into the stomach. These potential changes are then conducted through gap junctions along the longitudinal layer of smooth muscle. When membrane depolarization is greatest during these cycles, action-potential spikes often occur and are associated with muscle contraction. The propagation of the slow-wave electric activity along the longitudinal smooth muscle produces a peristaltic wave of contraction which spreads across the stomach from the esophagus to the small intestine. The circular smooth muscle cells do not show a basic electrical rhythm, but since action potentials appear in these cells coincident with action potentials in the longitudinal layer, there must be some form of electrical connection between the layers, possibly through the internal nerve plexuses.

The basic electrical rhythm is continuously active in the empty stomach, but only occasionally is the depolarization great enough to give rise to action-potential spikes and a resulting wave of peristaltic contraction. Moreover, when these contractions do occur, they cause only a slight ripple in the wall. As fasting is prolonged the magnitude of the stomach contractions becomes greater and greater. These changes appear to be mediated by parasympathetic input from higher brain centers which increase the sensitivity of the smooth muscle cells to the basic electrical rhythm. Attempts have been made to correlate these strong stomach contractions with the sensation of hunger pangs, but the correlation is generally poor.

During the first half hour after a meal, peristaltic activity in the stomach is very weak; thereafter, the waves increase in intensity and slightly in frequency. The waves start at the esophagus and proceed as a weak ripple over the body of the stomach at about 1 cm/sec. As the wave approaches the large mass of muscle in the walls of the antrum, it speeds up, reaching velocities of 3 to 4 cm/sec. The higher conduction velocity in the antrum means that a large part of the smooth muscle in this region undergoes an intense contraction almost simultaneously. This strong antral contraction is responsible for emptying material from the stomach into the *duodenum* (the first segment of the small intestine). Because of the size of the stomach and the frequency of contractions, two to three waves may be proceeding over the surface of the stomach simultaneously (Fig. 12-9).

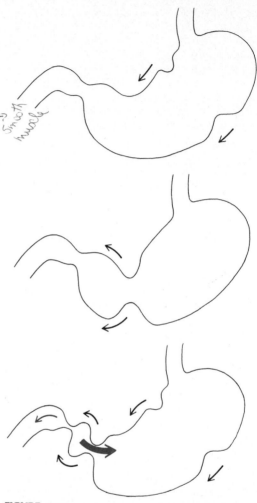

FIGURE 12-9
Peristaltic waves passing over the stomach empty a small amount of material into the duodenum. Most of the material is forced back into the antrum.

The *pyloric sphincter* is a ring of smooth muscle and connective tissue between the terminal antrum and the duodenum. Since at rest the pressure in the lumen of the duodenum is equal to or slightly greater than the pressure in the stomach, there is normally no pressure gradient to cause material to move from the stomach into the duodenum. The pyloric sphincter exerts only a slight pressure at the junction and is normally open most of the time. When a strong peristaltic wave arrives at the antrum, the pressure of the antral contents is increased,

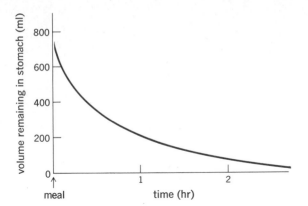

FIGURE 12-10
Typical rate of emptying the stomach contents into the duodenum.

forcing some of the material into the duodenum. However, the antral contraction also closes the pyloric sphincter, with the result that the pressure in the antrum forces most of the antral contents back into the body of the stomach and only small amounts pass into the duodenum. The strong contractions of the antral region are the primary force acting to mix the contents of the stomach; little mixing occurs in the body of the stomach, which is subjected only to weak peristaltic waves.

Control of gastric emptying   The amount of material forced into the duodenum from the stomach depends upon the strength with which the antral muscles contract. One of the factors which influences this contraction is the amount of food in the stomach. The stomach empties at a rate proportional to the volume of material in it at any given time (Fig. 12-10). This effect may be mediated by the internal nerve plexus or may be a direct effect of stretching the smooth muscle, which partially depolarizes the membrane, thus requiring less depolarization by the basic electrical rhythm to reach threshold and fire an action potential that will lead to contraction. In addition, stretch receptors present in the walls of the stomach influence smooth muscle activity through both short and long reflex pathways.

The most important factor controlling gastric emptying is not gastric volume, however, but the chemical composition and amount of chyme in the duodenum. When the duodenum contains fat, acid, or hypertonic solutions, or when it is distended, gastric motility is reflexly inhibited. These reflexes are initiated by chemoreceptors, osmoreceptors, and pressure receptors lo-

cated in the walls of the duodenum (Fig. 12-11). The efferent pathways are external nerves, the internal nerve plexuses, and a group of hormones secreted by cells in the wall of the duodenum. Two of these hormones, *secretin* and *cholecystokinin* (abbreviated CCK), have been isolated and found to be small proteins. As will be described in subsequent sections, secretin and CCK affect the contractile activity of smooth muscles and glandular secretion, not just in the stomach but in many other locations of the gastrointestinal system. Their effects on the gastrointestinal system are not identical; some effectors respond much more strongly to one or the other of the two hormones. These two hormones are secreted by separate cells distributed throughout the duodenum. The stimulus for the release of secretin and CCK is the chemical composition of the chyme entering the intestine from the stomach. The only effective chemical stimulus for the release of secretin appears to be acid. In contrast, the products of protein and triglyceride digestion—certain essential amino acids and long-chain fatty acids—are the most potent stimuli for CCK release. Carbohydrates are ineffective in releasing either hormone. Finally, many of the stimuli which release the duodenal hormones also stimulate nerve cells in the internal nerve plexus which lead to reflex neural inhibition of gastric motility. This completely neural pathway is known as the *enterogastric reflex*. Let us look at each of these hormonal and neural reflexes more closely.

Fat in the duodenum is the most potent stimulus for the inhibition of gastric motility. Part of this inhibitory effect is mediated by CCK, but other, as yet unidentified, duodenal hormones appear to be released, since the relatively weak inhibitory activity of cholecystokinin cannot account for the magnitude of the inhibition that occurs in the presence of fat. When gastric emptying is slowed down, less fat enters the intestine per unit time, giving more time for digestion and absorption. The high fat content of eggs and milk may inhibit gastric emptying to the point where some of the meal may be found in the stomach after 6 hr. In contrast, a meal of meat and potatoes (protein and carbohydrate) may empty in 3 hr.

Hydrochloric acid, secreted by the stomach, empties into the duodenum, where, as we shall see, it is neutralized by sodium bicarbonate secreted by the pancreas. Unneutralized acid in the duodenum inhibits the emptying of more acid by way of the enterogastric reflex as well as by the release of secretin.

As molecules of protein and starch are digested in the duodenum to form large numbers of molecules of amino acids and glucose, the osmolarity of the solution

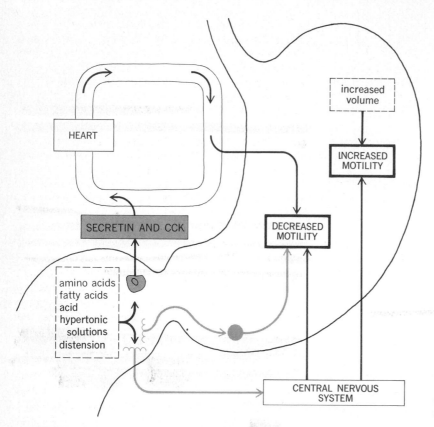

FIGURE 12-11
Summary of pathways controlling gastric motility.

rises. If the rate of absorption does not keep pace with the rate at which osmotically active molecules are formed by digestion, the osmolarity of the chyme becomes greater than blood and large volumes of water may enter the intestine by osmosis, causing significant lowering of blood volume. It is important, therefore, that reflex inhibition of gastric emptying by hypertonic solutions in the duodenum decrease the rate at which osmotically active molecules are formed in the intestine by reducing the amount of material available for digestion. This response is mediated primarily by the enterogastric reflex.

In addition to the control of gastric emptying by the gastric volume and duodenal contents, motility is influenced by external nerve fibers to the stomach from higher centers in the nervous system. Emotions such as sadness, depression, and fear tend to decrease motility; aggression or anger tend to increase it. These relationships are not always predictable, however, and different people show different responses to apparently similar emotional states. Intense pain from any part of the body

tends to inhibit gastric motility. These effects mediated through external nerves from higher centers in the nervous system usually affect not only the stomach but the entire gastrointestinal tract. Inhibition of motility is mediated by decreased parasympathetic activity and increased sympathetic activity.

Vomiting   Vomiting is the forceful expulsion of the contents of the stomach and upper intestinal tract through the mouth. It is a complex reflex coordinated by the vomiting center in the medulla and is usually preceded by increased salivation, sweating, faster heart rate, and feelings of nausea—all characteristic of a general discharge of the autonomic nervous system. Vomiting begins with a deep inspiration, closure of the glottis, and elevation of the soft palate. The abdominal and thoracic muscles contract, raising the intraabdominal pressure, which is transmitted to the contents of the stomach; the gastroesophageal sphincter relaxes, and the high abdominal pressure forces the contents of the stomach into the esophagus. When the pressure is great enough, the contents of the esophagus are forced through the hypopharyngeal sphincter into the mouth. Vomiting is also accompanied by strong contractions of the upper portion of the small intestine which tend to force some of the intestinal contents back into the stomach.

Input to the vomiting center in the medulla comes from a number of receptors throughout the body. The primary stimuli are tactile stimulation of the back of the throat, e.g., sticking a finger in the back of the throat; excessive distension of the stomach or duodenum; high pressures within the skull; rotating movements of the head producing dizziness; and intense pain as well as certain chemical agents which act on chemoreceptors in the brain and upper parts of the gastrointestinal tract. Excessive vomiting can lead to large losses of secreted fluids and salts from the body which would normally be reabsorbed. This can result in severe dehydration, upset the salt balance of the body, and produce circulatory problems due to a decrease in plasma volume.

Gastric secretions

HCl Secretion   The normal human stomach secretes about 2 liters of hydrochloric acid solution a day. Hydrochloric acid is a strong acid which completely dissociates in water into hydrogen and chloride ions. The concentration of hydrogen ions in the lumen of the stomach may reach 150 mM (isotonic HCl) compared with the concentration of hydrogen ion in the blood, 0.00004 mM. The secretion of acid is not absolutely essential to gastrointestinal function, since digestion and absorption occur in the small intestine in the absence of a stomach; however, the acid performs several functions which assist digestion. A high acidity denatures proteins and breaks intermolecular bonds, thereby breaking up connective tissue and cells, releasing ionized molecules into solution. Hydrochloric acid thus continues the process begun by chewing, namely, reducing large particles of food to smaller particles and individual molecules. However, the acid has relatively little ability to break down proteins and polysaccharides into amino acids and glucose. A second function is to kill most of the bacteria that enter along with food. This process is not 100 percent effective, and some live bacteria enter the intestinal tract where they may continue to multiply, especially in the large intestine. Many of these bacteria are harmless; others may release disease-producing toxins which reach the blood even after the bacteria themselves have been killed (Chap. 15). A third function of hydrochloric acid is to activate some of the enzymes secreted by the stomach.

Hydrochloric acid is secreted by the *parietal* ("in a wall") cells lying in gastric pits in the body of the stomach (Fig. 12-12). These cells have many large mitochondria, and their cell membranes are indented at the luminal surface to form minute intracellular channels (*canaliculi*) which penetrate deep into the cell. The canaliculi greatly enlarge the surface area of the cell available for secretion, and their intracellular distribution brings them into close association with the mitochondria which produce the ATP necessary for operating the active transport systems involved in acid secretion.

The molecular mechanism of acid secretion is unknown. Hydrogen ions must be moved against a very large concentration gradient, the concentration of hydrogen ions in the lumen of the stomach being as much as 3 million times greater than in the blood. Chloride ions are moved actively but against a much smaller concentration gradient, since their concentration in the stomach lumen is only about 1.5 times that of the blood. The two ions appear to be actively transported by separate pumps in the luminal membrane of the parietal cell. The enzyme carbonic anhydrase, similar to the enzyme in red blood cells, is present in the gastric mucosa and is believed to be involved in the secretion of acid, probably by way of the general pathway illustrated in Fig. 12-13. Whenever an acid is formed, an equivalent amount of base is also formed. Here the base is the bicarbonate ion, which enters the blood. The venous blood leaving the stomach is therefore more alkaline than the arterial blood entering it because of its higher bicarbonate concentration and lower hydrogen-ion concentration. Whenever there is considerable loss of the acid contents of the stomach through vomiting, the blood may become progressively more alkaline, requiring adjustments in the acid-base balance of the body by the kidneys and lungs. Under normal conditions acid secretion produces little change in the acid-base balance of the body because an amount of bicarbonate equal to that released into the blood by the stomach is secreted by the pancreas into the duodenum.

Enzyme secretion and digestion   Although several different enzymes have been detected in gastric secretions, the only ones with any significant effect upon the normal digestive process are the proteolytic enzymes known as *pepsins.* At least seven separate pepsins have been isolated from human gastric mucosa. These enzymes are similar in their activity and will be referred to simply as pepsin. Pepsin is secreted by the chief cells, which contain numerous zymogen granules (Fig. 12-12). Acid is

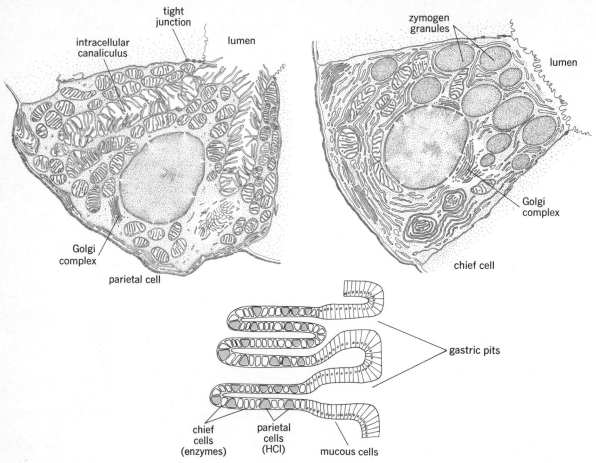

FIGURE 12-12
Structure of parietal and chief cells and their location in the gastric glands. (*Adapted from Ito.*)

secreted only by the glands in the body of the stomach, but pepsin is secreted by both the gastric glands in the body and the pyloric glands in the antrum.

Pepsin is secreted in an inactive form known as *pepsinogen*, which is converted to pepsin by splitting off a small fragment of the molecule. The hydrochloric acid in the stomach initiates this process, and once pepsin is formed, it can act upon other molecules of pepsinogen to form more pepsin. Thus a small amount of pepsin acts autocatalytically to activate more pepsin.

Pepsin catalyzes the splitting of bonds between particular types of amino acids in protein chains. The products of pepsin digestion are primarily small peptide fragments composed of several amino acids. As we shall see, these fragments are one of the major chemical stimuli controlling hydrochloric acid production and pancreatic secretion. Pepsin is active only in an environment having the high hydrogen-ion concentration provided by the hydrochloric acid in the stomach. In the duodenum, where the stomach acid is neutralized by bicarbonate ions from the pancreas, pepsin is inactive.

The ability of acid and pepsin to break down particles of food depends upon their coming in contact with the food in the stomach. There is very little mixing of solid food in the stomach until it reaches the antrum, where strong peristaltic waves compress and mix the contents.

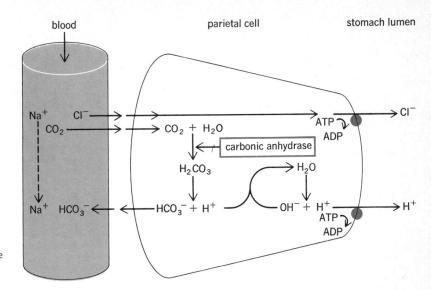

FIGURE 12-13
Secretion of HCl by parietal cells and the release of bicarbonate ions into the blood.

Food in the body of the stomach remains as a semisolid mass which is attacked by acid and pepsin only at its surface, the interior of the mass being free of acid. On the other hand, although salivary amylase is inactivated by hydrochloric acid, it may continue to act upon starch for long periods of time in the stomach because it is inside the bolus, where acid has not yet reached. Bacteria in the same location may also escape the sterilizing action of hydrochloric acid.

Control of gastric secretion Acid and enzyme secretions in the stomach generally vary in a parallel manner. Most of the factors described below which stimulate the secretion of acid also stimulate enzyme secretion.

When no food is in the stomach, acid secretion occurs at a basal rate of about 0.5 ml/min. After a meal the rate of acid secretion increases considerably, reaching maximal values of about 3 ml/min. Figure 12-14 illustrates the characteristic rise in acid secretion following a meal and its decline over a period of hours. For reasons to be discussed shortly, the concentration of acid in the stomach decreases immediately following a meal and then rises as the rate of acid secretion declines.

Cephalic phase Enhanced acid secretion begins even before food reaches the stomach. This is known as the *cephalic phase* of gastric secretion. Sight, smell, and taste all contribute to the stimulation, mediated by parasympathetic nerve fibers traveling to the stomach in the vagus nerve. Central nervous system factors other than

those normally associated with the process of eating (such as emotions) may also stimulate secretion by this pathway.

Gastric phase Food placed directly in the stomach through a tube elicits an increase in acid and enzyme secretion, the amount depending upon the chemical composition of the food and its volume. The chemical stimulation of gastric secretion is mediated by a hormone, *gastrin,* released by cells located in the walls of the terminal segment of the stomach.

Gastrin is the third major gastrointestinal hormone and, like secretin and CCK, has multiple sites of action throughout the gastrointestinal system. Because the last five amino acids in CCK are identical to those in gastrin, they are both able to bind to the same receptor sites throughout the gastrointestinal tract; therefore, CCK shows weak gastrinlike activity and gastrin has weak CCK-like activity. However, because the affinity of the hormones for a given receptor site is determined by the remaining amino acids in their respective structures, the magnitude of the response produced at any given site differs between the two. For example, as we shall see, CCK is a more potent stimulator of pancreatic secretion than is gastrin.

In addition to gastrin's ability to stimulate hydrochloric acid secretion in the stomach, it also stimulates the contraction of the smooth muscle in the gastroesophageal sphincter. Thus, at the same time that gastrin

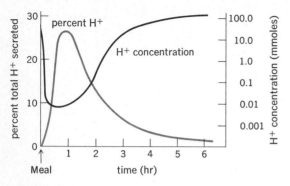

FIGURE 12-14
Changes in hydrogen-ion secretion and concentration in the stomach following a meal.

is stimulating acid secretion, it helps to prevent the entry of acid into the esophagus. Patients who have had the lower portion of their stomach removed (e.g., because of cancer) often have an increased incidence of heartburn (irritation of the esophagus by acid) because of the loss of gastrin and its ability to keep the esophageal sphincter closed.

The major chemical stimulus for the release of gastrin is the presence of protein in the food entering the stomach. Intact protein, however, is much less effective than the peptide fragments produced by the action of pepsin upon these proteins. In addition, alcohol and caffeine (present in coffee, tea, and cola drinks) are potent stimuli for gastrin release and thus acid secretion. A cocktail before dinner and a bowl of soup, i.e., protein broth, before the main course both stimulate acid secretion and prepare the stomach for the main meal. On the other hand, the intake of large quantities of alcohol and coffee in the absence of food produces the secretion of very concentrated acid, which can irritate the linings of the esophagus, stomach and duodenum. In addition to these chemical stimuli, gastrin is also released in response to stimulation of the parasympathetic fibers in the vagus nerves to the stomach.

It should be emphasized that gastrin is not the unique final common pathway for stimulating acid secretion. Distension of the parietal region of the stomach or stimulation of that portion of the vagus supplying the parietal region enhances acid secretion independently of gastrin release. A summary of the pathways involved in controlling the increased secretion by the gastric phase of acid secretion is given in Fig. 12-15.

Is there any inhibitory influence on acid secretion during the gastric phase? Obviously a lessening of any of the stimuli causing acid secretion will reduce the secretion. All the stimuli mentioned above lessen as the stomach empties its contents into the duodenum. In addition, acid itself is a powerful inhibitor of acid secretion. A high concentration of hydrogen ions directly inhibits the release of gastrin by the antrum and thus inhibits gastric secretion by reducing the amount of gastrin available to stimulate acid secretion. In Fig. 12-14, the increase in acid secretion following a meal is shown along with the changes in the hydrogen-ion concentration in the stomach. Before food enters the stomach, the hydrogen-ion concentration is high, inhibiting gastrin release. When food enters the stomach, the hydrogen-ion concentration drops, removing the inhibition of gastrin release and thereby acid secretion. Why does the hydrogen-ion concentration drop and remain low for some time in spite of increased acid secretion? Food, particularly protein, buffers the hydrogen ions, which combine with the ionized carboxyl groups and un-ionized amino groups on the protein molecules, thus removing them from solution and lowering the concentration of free hydrogen ions. Acid secretion continues as long as gastrin is not inhibited by a rise in hydrogen-ion concentration. As more and more acid is secreted, the buffering capacity of the proteins in food is exceeded and the hydrogen-ion concentration begins to rise, inhibiting gastrin release and turning off acid secretion. Thus, for two reasons, the total amount of acid secreted during a meal is directly proportional to the amount of protein in the meal: (1) The greater the amount of protein, the greater the buffering capacity, and thus the more acid can be secreted without appreciably raising the hydrogen-ion concentration and inhibiting gastrin release; (2) the greater the protein content of the meal, the more peptide fragments are formed, which directly stimulate the release of gastrin.

Intestinal phase   Like gastric motility, gastric secretion is inhibited by reflexes initiated in the duodenum. Thus distension, hypertonic solutions, fatty acids, amino acids, and acid in the duodenum all inhibit gastric secretion. The pathways (Fig. 12-16) are the same as those inhibiting motility: the internal and external nerves and the duodenal hormones secretin and CCK.[1] Thus, during the intestinal phase of gastric activity, as chyme passes into

---

[1]CCK inhibits gastric secretion, in spite of its chemical similarity to gastrin, because when it is bound to the receptor sites for acid secretion in the stomach it prevents the more potent hormone, gastrin, from combining with the same receptor sites.

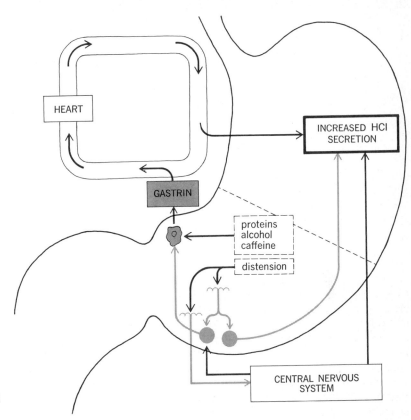

FIGURE 12-15
Pathways regulating increased acid secretion in the stomach.

the duodenum it initiates reflexes which decrease acid secretion and gastric emptying, allowing time for digestion and absorption in the intestine.

Ulcers Considering the high concentration of acid and proteolytic enzymes secreted by the stomach, it is natural to wonder why the stomach does not digest itself. Several factors protect the walls of the stomach and duodenum. The surface of the mucosa is lined with cells which secrete a slightly alkaline mucus, which in man forms a layer of 1.0 to 1.5 mm thick over the stomach surface. The protein content of mucus and its alkalinity tend to neutralize hydrogen ions in the immediate area of the epithelial cell layer, thus forming a chemical barrier between the highly acid contents of the lumen and the cell surface. In addition, the cell membranes lining the stomach have a very low permeability to hydrogen ions, preventing their entry into the underlying mucosa. Moreover, the lateral surfaces of the epithelial cells near the lumen are joined together by tight junctions (Fig. 12-12), so that

there is no extracellular passage between the cells by which material could diffuse from the lumen into the mucosa. Finally, the epithelial mucous cells lining the walls of the stomach are continually being replaced over a period of 1 to 3 days through cell division. The mucous layer, the cell membranes, and cell replacement all contribute to maintaining a barrier between the contents of the lumen and the underlying tissues.

Yet, in some people these protective mechanisms are inadequate, and erosions (*ulcers*) of the gastric or duodenal mucosa occur. If severe enough, the ulcer may damage the underlying blood vessels and cause bleeding into the lumen. On occasion, the ulcer may penetrate the entire wall, with the leakage of luminal contents into the abdominal cavity. About 10 percent of the population of the United States are found at autopsy to have ulcers, which are about 10 times more frequent in the walls of the duodenum than in the stomach itself.

What causes the breakdown of this barrier and the formation of ulcers is unknown. In one unusual disease,

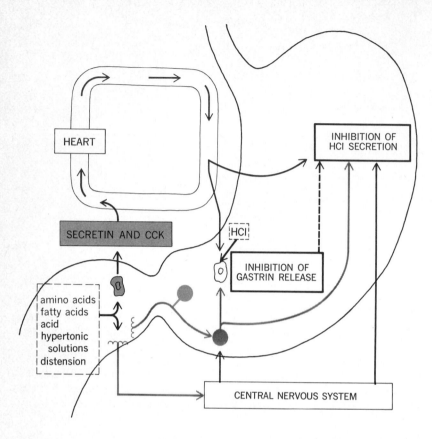

FIGURE 12-16
Pathways inhibiting HCl secretion. All except the local effect of HCl on gastrin release are involved in the intestinal phase of gastric secretion. Inhibition of gastrin release decreases HCl secretion.

huge quantities of acid are continually secreted by the stomach, and the patient has multiple ulcers in the esophagus and duodenum. The high acid secretion appears to be the result of a tumor in the pancreas which produces a gastrinlike hormone. However, many ulcer patients are found to have normal or below-normal rates of acid secretion. A genetic factor appears to contribute to ulcer formation, since ulcers are often found in many members of the same family. Emotional stress and worry have often been implicated as contributing factors. Emotions are known to influence gastric motility and secretion by way of the autonomic nerves, and cutting the vagus nerves contributes to the relief of some, but not all, ulcer patients.

The pain associated with ulcers probably is the result of irritation of exposed nerve fibers and smooth muscle cells in the region of the ulcer wound as well as increased tension in the walls resulting from the contractile spasms of the smooth muscles irritated by acid. The pain from a duodenal ulcer often subsides following a meal, which buffers the acid secreted by the stomach and inhibits the emptying of acid from the stomach into the duodenum. The pain is most intense during the early hours of the morning, when unbuffered acid is entering the duodenum.

Absorption by the stomach   Very little food is absorbed from the stomach into the blood. There are no special transport systems for salts, amino acids, and sugars in the stomach walls like those found in the intestine, and most of the digestion products in the stomach are large, highly charged, and ionized so that they cannot diffuse across cell membranes. The lipids in the food reaching the stomach are not soluble in water and tend to separate into large droplets which do not mix with the acid contents of the stomach. Since little of this lipid comes into contact with the membranes lining the stomach, little is absorbed.

Several classes of molecules, however, can be absorbed directly by the stomach. A prime example is alcohol, CH_3CH_2OH. Alcohol is water-soluble, but since

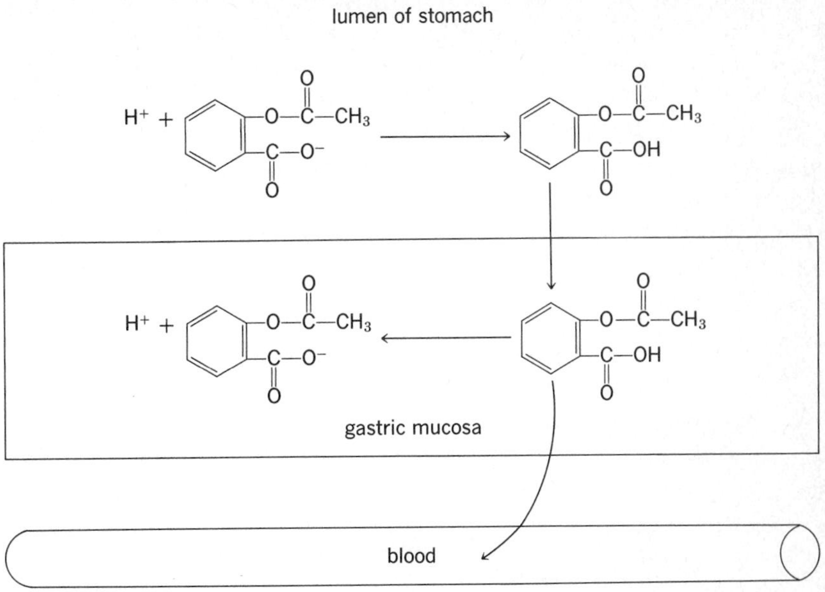

lumen of stomach

$$H^+ +$$

gastric mucosa

$$H^+ +$$

blood

FIGURE 12-17
Absorption of the weak acid, aspirin, by the stomach.

it is not ionized, it also has some degree of lipid solubility, allowing it to diffuse across lipid membranes and reach the bloodstream through the walls of the stomach. Although alcohol is absorbed from the stomach, it is absorbed more rapidly from the intestinal tract, which has a greater membrane surface area available for absorption. Accordingly, drinking a glass of milk before a cocktail party or eating cheese dip and high-fat hors d'oeuvres inhibits the rate of gastric emptying through the reflexes from the duodenum and slows down the rate of alcohol absorption but does not stop it.

Weak acids, most notably acetylsalicylic acid (aspirin), are also absorbed directly across the walls of the stomach. A weak acid is fully ionized in solutions of low hydrogen-ion concentration and in its ionized form is unable to diffuse through cell membranes. In the highly acid environment of the stomach, however, a weak acid is almost totally converted to its un-ionized form, which is lipid-soluble and can cross the cell membrane. Figure 12-17 illustrates the absorption of aspirin. Once in the low-acid environment of the cell, the weak acid again ionizes, liberating a hydrogen ion, which tends to make the cell interior more acid. Therefore, if enough weak acid enters in a short time, the intracellular acidity may rise sufficiently to damage the cell. Although most people show few ill effects from the normal dosage of aspirin, small amounts of blood can be found in the stomachs of most individuals after taking aspirin, and in some individuals it can produce severe hemorrhaging. Since alcohol increases acid secretion, the combination of aspirin and alcohol increases the damage to the gastric mucosa.

Pancreatic secretions

The pancreas, just below the stomach, is a mixed gland containing both endocrine and exocrine portions. The endocrine cells secrete the hormones insulin and glucagon into the blood (they will be discussed in Chap. 13). The exocrine portion of the pancreas secretes two solutions involved in the digestive process, one containing a high concentration of sodium bicarbonate and the other containing a large number of digestive enzymes. These solutions are secreted into ducts that converge into a single pancreatic duct, which joins the bile duct from the liver just before entering the duodenum (Fig. 12-18). The enzyme secretions of the pancreas are released from the acinar cells at the base of the exocrine glands. These cells have a high density of granular endoplasmic reticulum and many zymogen granules, indicating a high capacity for synthesizing and secreting proteins. The bicarbonate solution appears to be secreted by the cells lining the early portions of the ducts leading from the acinar cells.

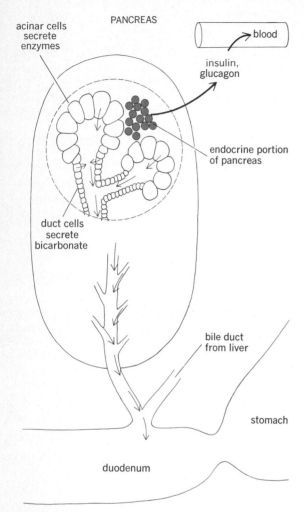

PANCREAS

acinar cells secrete enzymes

blood

insulin, glucagon

endocrine portion of pancreas

duct cells secrete bicarbonate

bile duct from liver

stomach

duodenum

FIGURE 12-18
Structure of the pancreas. The gland areas are greatly enlarged relative to the entire pancreas and ducts.

Bicarbonate secretion

The concentration of sodium bicarbonate in the pancreatic secretions rises with the rate of secretion and may approach values of 150 mM at maximal rates of secretion, 6 to 7 ml/min. Since the concentration of bicarbonate in the blood is about 27 mM, the secretion of bicarbonate by the pancreas is an active process. During the course of a day, 1.5 to 2.0 liters of solution is secreted into the duodenum, most of which is reabsorbed.

The mechanism of bicarbonate secretion may be similar to the process of hydrochloric acid secretion by the stomach, the crucial difference being a reverse orientation of the transport systems, such that the bicarbonate ions are released into the lumen rather than the blood. The pancreas, like the stomach, contains a high concentration of carbonic anhydrase, and inhibition of this enzyme reduces bicarbonate secretion, as it does hydrogen-ion secretion in the stomach.

Secretion of an alkaline solution of bicarbonate ions necessitates the formation of an equivalent amount of acid, and the blood leaving the pancreas is therefore more acid than the blood entering it. Accordingly, loss of large quantities of bicarbonate ions from the intestinal tract during periods of prolonged diarrhea leads to acidification of the blood, just as loss of acid from the stomach by vomiting leads to alkalinization of the blood. Normally there is no change in the acidity of the blood since the increase in blood bicarbonate by the stomach is balanced by the increase in blood acidity in the pancreas, and the acid secreted by the stomach is neutralized by the bicarbonate secreted by the pancreas (Fig. 12-19).

When a person takes a bicarbonate solution to relieve an upset stomach, the bicarbonate ions neutralize the acid in the stomach just as bicarbonate ions secreted by the pancreas neutralize acid in the duodenum. However, neutralization of the acid in the stomach actually leads to an *increase* in acid secretion by the stomach since it removes the inhibition of gastrin release. Provided adequate bicarbonate is ingested, the increased acid secreted is still neutralized by the bicarbonate. Since, as we shall see, acid in the duodenum is a major stimulant for the pancreatic secretion of bicarbonate, there is little pancreatic secretion and the bicarbonate formed during acid secretion in the stomach accumulates in the blood. In this case the acid-base balance of the body is restored by the elimination of the excess bicarbonate ions through the kidneys.

Enzyme secretion

Most of the enzymes which digest triglycerides, polysaccharides, and proteins to fatty acids, sugars, and amino acids are secreted by the pancreas. A partial list of these enzymes and their activity is given in Table 12-1. Many are secreted in an inactive form, similar to the secretion of pepsinogen by the stomach, and are then activated by ions and other enzymes in the duodenum. The secretion of enzymes in an inactive form is one mechanism for preventing these potent enzymes from digesting the cells in which they are formed. Over short periods of time, the relative proportions of the enzymes

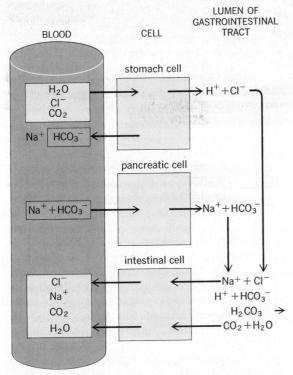

BLOOD   CELL   LUMEN OF
GASTROINTESTINAL
TRACT

FIGURE 12-19
Overall pathway of HCl production in the stomach,
neutralization by sodium bicarbonate secreted by the pancreas,
and reabsorption in the small intestine of the products formed
by the neutralization. Note that there is no net gain or loss of
any substance.

secreted by the pancreas do not vary with the composition of the diet. However, there is some evidence to suggest that if the diet contains a high proportion of protein for a period of weeks or months, the relative proportions of the proteolytic enzymes in the pancreatic secretion are increased. The mechanisms responsible for these changes are unknown.

Control of pancreatic secretion

The exocrine secretions of the pancreas are controlled by the autonomic nerves to the pancreas (primarily the vagus nerve) and the three gastrointestinal hormones: secretin, CCK, and gastrin. As was true for gastric secretion, there is a cephalic phase to pancreatic secretion during which the sight and smell of food or its presence in the mouth produces increased pancreatic secretion mediated by the vagus nerve. This is primarily an enzy-

matic secretion containing little bicarbonate. The gastric phase of pancreatic secretion is mediated by the release of gastrin which stimulates primarily bicarbonate secretion. The major portion of pancreatic secretion occurs during the intestinal phase and is mediated by the release of secretin and CCK in response to chyme in the intestine.

Recall that the stimulus for secretin release is the presence of acid in the duodenum. Secretin[1] elicits a marked increase in the amount of bicarbonate and the volume of fluid secreted by the pancreas but only a slight stimulation of enzymatic secretion. Thus, the action of secretin is primarily on the duct cells of the pancreas which secrete bicarbonate. The bicarbonate ions neutral-

[1]Secretin has the honor of being the first hormone ever to be discovered. In 1902 Bayliss and Starling, in England, noted that when food was placed in an isolated segment of the duodenum of a dog pancreatic secretion was increased and that extracts of the duodenum also increased secretion, indicating the release of a chemical substance from one tissue (the duodenum) which affected the activity of another tissue (the pancreas). Within a few years the term hormone was introduced into the biological literature to designate such chemical mediators.

TABLE 12-1
Pancreatic enzymes

Enzyme	Substrate	Action
Trypsin, chymotrypsin	Proteins	Breaks amino acid bonds in the interior of proteins, forming peptide fragments.
Carboxypeptidase	Proteins	Splits off terminal amino acid from end of protein containing a free carboxyl group.
Lipase	Lipids (triglycerides)	Splits off fatty acids from positions 1 and 3 of triglycerides, forming free fatty acids and 2-monoglycerides.
Amylase	Polysaccharides	Similar to salivary amylase: splits polysaccharides into a mixture of glucose and maltose.
Ribonuclease, deoxyribonuclease	RNA, DNA	Splits nucleic acids into free mononucleotides.

ize the acid entering the intestine from the stomach; it is therefore appropriate that the most potent stimulus for bicarbonate secretion is secretin, which is released by acid in the duodenum, thus providing a negative-feedback control system to maintain the neutrality of the intestinal contents.

The second duodenal hormone, CCK,[1] produces a marked increase in pancreatic enzyme secretion but little increase in bicarbonate secretion. Recall that the stimulus for CCK release is the presence of the organic components in chyme—amino acids and fatty acids—rather than acid, and it is thus appropriate that CCK stimulates primarily enzyme secretion leading to the digestion of fat and protein.

In spite of the multiple number of hormones, nerves, and varieties of interactions between the stomach, duodenum, and pancreas, the overall adaptive significance of these interactions should be reemphasized. As chyme empties from the stomach into the duodenum, its content of acid, amino acids, and fatty acids stimulates the release of secretin and CCK from the wall of the duodenum; these hormones stimulate the secretion of bicarbonate and enzymes from the pancreas. The pancreatic enzymes act upon the large molecules in the chyme to produce amino acids, sugars, and fatty acids which can be absorbed and also maintain the chemical stimuli responsible for hormone release until such time as the nutrients are absorbed. The bicarbonate from the pancreas neutralizes the acid from the stomach, preventing ulcerative damage to the walls of the intestine, and also provides an environment in which the enzymes from the pancreas, which are inactive in a highly acid environment, can be active. Gastrin, which strongly stimulates acid secretion, also stimulates the pancreatic secretion of bicarbonate, assuring neutralization of the acid upon its arrival in the duodenum. The release of secretin and CCK also inhibits gastric secretion and motility, allowing time for the intestinal contents to be neutralized, digested, and absorbed. Figure 12-20 summarizes these pathways controlling pancreatic secretion.

Secretion of bile

Bile, which is secreted by the liver, is essential for the digestion of fat in the small intestine. The digestion of

[1] Prior to its isolation, the duodenal hormone stimulating the secretion of enzymes by the pancreas was called pancreozymin. However, upon purification it was found to be identical to the hormone cholecystokinin.

fat presents special problems because fat is insoluble in water. As fat is released from the breakdown of food particles in the stomach, the molecules of triglyceride aggregate together, forming large globules of fat that are immiscible with the chyme in the stomach. The function of bile is to break down these large globules into a suspension of very fine droplets about 1 μm in diameter, a process known as *emulsification*. This emulsion of fat can then be digested and absorbed. We shall discuss the details of this process further in the section on the absorption of fat later in the chapter.

Bile consists of a salt solution containing four primary ingredients: (*1*) bile salts, (*2*) cholesterol, (*3*) lecithin (a phospholipid), and (*4*) bile pigments. The first three of these are involved in the emulsification of fat in the small intestine. The fourth, bile pigments, is one of the few substances that are normally excreted from the body through the intestinal tract. The major bile pigment is bilirubin, which is a breakdown product of hemoglobin (Chap. 10). The cells of the liver extract bilirubin from the circulation and secrete it into the bile by an active process. Bile pigments are yellow and give bile its golden color. The color of these pigments is modified in the intestinal tract by the digestive enzymes, and the mixture of these pigments gives feces their brown color. In the absence of bile secretion, feces are grayish white. Some of the pigments are reabsorbed during their passage through the intestinal tract and are eventually excreted in the urine, giving urine its yellow color.

Bile is secreted by the liver cells into a number of small ducts which converge to form the common bile duct which finally empties into the duodenum (Fig. 12-21), often at the same site at which the pancreatic duct enters the duodenum. In man, a small sac branches off from the bile duct on the underside of the liver, forming the *gallbladder*. The gallbladder stores, concentrates, and releases bile into the intestinal tract. The gallbladder can be surgically removed without impairing the function of bile in the intestinal tract; in fact many animals which secrete bile do not even have a gallbladder, e.g., the rat and horse.

Bile is continually secreted by the liver, although the rate of secretion may vary, as described in the next section. Surrounding the bile duct at the point where it enters the duodenum is a ring of smooth muscle known as the *sphincter of Oddi*. When this sphincter is closed, the bile secreted by the liver is shunted into the gallbladder. The cells lining the gallbladder actively transport sodium from the bile back into the plasma. As solute is pumped out of the bile, water follows by osmosis. The

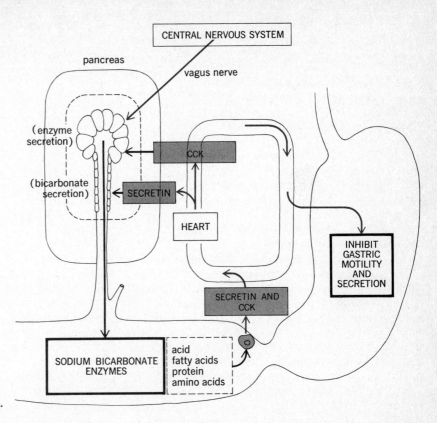

FIGURE 12-20
Pathways regulating pancreatic secretion.

net result is a five- to tenfold concentration of the organic constituents of bile.

Cholesterol, which is normally insoluble in water, is found in very high concentration in the bile. This is made possible by the presence of the bile salts and lecithin which solubilize cholesterol just as they solubilize fat in the intestine. If insufficient bile salts or lecithin is present in the bile, or if there is excessive cholesterol, the cholesterol precipitates out of solution, forming a *gallstone*. In man the concentration of bile salts, lecithin, and cholesterol in the bile is normally very near the point at which cholesterol precipitates out of solution. Slight changes in the concentrations of any of these may precipate cholesterol, especially during the concentrating process in the gallbladder. If the gallstone is small, it may pass through the bile duct into the intestine with no complications. A larger stone may become lodged in the neck of the gallbladder, causing contractile spasms of the smooth muscle and pain. A more serious complication arises when the gallstone lodges in the bile duct, thereby preventing bile from entering the intestine and resulting

in a failure to digest and absorb fat. In addition, a buildup of pressure in the blocked duct prevents further secretion of bile, and the bile pigments accumulate in the blood and tissues, giving them the deep yellow color known as jaundice. It is also possible for a gallstone to lodge at the junction of the bile and pancreatic duct in the duodenum, blocking the entry of both bile and pancreatic juices. Under these conditions, little digestion or absorption of any substance occurs in the intestinal tract.

From the standpoint of gastrointestinal function, the bile salts are the most important components of bile since they are involved in the digestion and absorption of fats. The total amount of bile salt in the body is about 3.6 g; yet during the digestion of a single fatty meal, as much as 4 to 8 g of bile salts may be emptied into the duodenum. This is possible because most of the bile salts entering the intestinal tract are reabsorbed in the lower part of the small intestine and returned to the liver, where they are again secreted into the bile. Recall that the venous system coming from the intestinal tract does not return directly to the heart but passes first through the

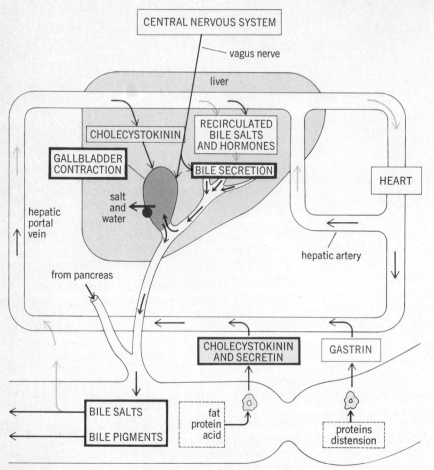

FIGURE 12-21
Pathways regulating bile secretion and release from the gallbladder.

hepatic portal vein to the liver; there it flows through a second capillary network from which it leaves the liver and returns to the heart. Thus, material absorbed into the blood from the intestinal tract passes through the liver before returning to the general circulation. This circulatory pathway from the intestines to the liver provides the route generally known as the *enterohepatic circulation* (Fig. 12-21), by which such substances as the bile salts can be recycled through the intestinal tract.

Control of bile secretion

Bile is secreted at the rate of 250 to 1,000 ml/day. Two components of bile secretion are under separate controls: the secretion of bile salts and the secretion of the isotonic fluid containing sodium, chloride, and bicarbonate in which the bile salts and other organic constituents are dissolved. The rate of bile-salt secretion by the liver is primarily determined by the concentration of the bile salts in the plasma; this concentration rises during a meal as a result of bile-salt reabsorption from the intestinal tract. Between meals, when there is little bile salt in the intestinal tract to be absorbed, the rate of secretion of bile salt by the liver is low.

In contrast, the secretion of fluid (with little increase in the rate of bile-salt secretion) is increased by secretin, cholecystokinin, and gastrin (as well as by the vagus nerves), secretin being the most active in this regard. Thus all the factors discussed previously which cause the

TABLE 12-2
Activities of the gastrointestinal hormones

	Secretin	Cholecystokinin	Gastrin
Secreted by:	Duodenum	Duodenum	Antrum of stomach
Primary stimulus for hormone release	Acid in duodenum	Amino acids and fatty acids in duodenum	Peptides in stomach
			Parasympathetic nerves to stomach
Effect on:			
Gastric motility	Inhibits	Inhibits	Stimulates
Gastric HCl secretion	Inhibits	Inhibits	STIMULATES †
Pancreatic secretion			
Bicarbonate	STIMULATES	Stimulates	Stimulates
Enzymes	Stimulates	STIMULATES	Stimulates
Bile secretion of bicarbonate	STIMULATES	Stimulates	Stimulates
Gallbladder contraction	Stimulates	STIMULATES	Stimulates

† STIMULATES denotes that this hormone is quantitatively more important than the other two.

release of these gastrointestinal hormones affect bile secretion.

Shortly after a meal, especially if it contains fat, the sphincter of Oddi relaxes and the gallbladder contracts, discharging concentrated bile into the duodenum. This response is mediated by both the vagus nerves and the three hormones, cholecystokinin being the most active. (It is from this ability to cause contraction of the gallbladder that cholecystokinin first received its name: *chole,* bile; *kystis,* bladder; *kinin,* to move.) The net result of these neural and hormonal influences is to assure the secretion and discharge of bile into the intestinal tract when food is present, the amount of bile discharge depending on the composition and amount of food entering the duodenum. Figure 12-21 summarizes the factors controlling the release of bile.

We have now discussed the major activities of the three gastrointestinal hormones, secretin, cholecystokinin, and gastrin. A summary of their activities is provided in Table 12-2.

The small intestine

The small intestine consists of about 9 ft of 1½-in. -diameter tubing coiled in the abdomen and leads from the stomach to the large intestine. It is divided by anato- mists into three segments: an initial short 8-in. segment, the *duodenum,* and two much longer segments, the *jejunum* (3 ft) and the terminal segment, the *ileum.* It is in the small intestine that most digestion and absorption occur.

Motility

When the motion of the intestinal tract is observed by x-ray fluoroscopic examination, the contents of the lumen are seen to move back and forth with little apparent net movement toward the large intestine. The net flow of material toward the large intestine is normally quite slow; chyme begins to enter the large intestine at about the time that chyme from the next meal is entering the duodenum from the stomach. In contrast to the waves of peristaltic contraction that sweep over the surface of the stomach, the primary motion of the small intestine is an oscillating contraction and relaxation of the smooth muscle in the intestinal wall, known as *segmentation.* Rings of smooth muscle contract and relax at intervals along the intestine, dividing the chyme into segments (Fig. 12-22). The chyme in the lumen of a contracting segment is forced both up and down the intestine. This rhythmical contraction and relaxation of the intestinal segments produce a continuous division and subdivision of the intestinal contents which thoroughly mixes the chyme in the lumen and brings it into contact with the intestinal wall where absorption can occur.

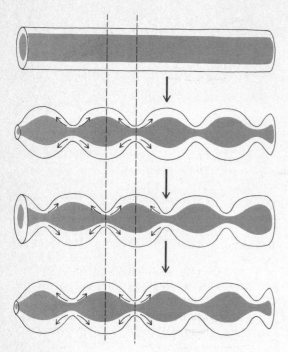

FIGURE 12-22
Segmentation movements of the small intestine. The small arrows indicate the movements of the luminal contents.

These movements of the small intestine are initiated by electric activity generated by pacemaker cells located in the longitudinal smooth muscle. Just as in the stomach, these pacemaker cells produce a basic electrical rhythm that is conducted through the gap junctions between smooth muscle cells. The frequency of segmentation follows the frequency of the basic electrical rhythm. The primary factor responsible for moving the chyme along the intestine is a gradient in the frequency of segmentation along the length of the intestine. Segmentation in the duodenum occurs at a frequency of about 12 contractions per minute whereas in the terminal portion of the ileum segmentation occurs at a rate of only 9 contractions per minute. Since the frequency of contraction is greater in the upper portion of the small intestine than in the lower, on the average, more chyme is forced downward than is forced upward. The gradient in segmentation frequency results from a sequence of pacemaker regions along the intestine. Each successive pacemaker has a slightly lower frequency than the one above.

The degree of contractile activity in the intestine can be altered by a number of factors which initiate reflexes in the internal nerve plexus or through external nerves and hormones. These reflexes produce changes in the intensity of the smooth muscle contraction and the rate of propagation of the basic electrical rhythm but do not change the natural frequency of the pacemaker regions. Local distension of the intestine produces a characteristic response. The distended portion contracts and the region just ahead (toward the large intestine) relaxes. The contracted segment then progresses several centimeters down the intestine, producing a short peristaltic wave. The coordination of this response appears to be mediated by the internal nerve plexus. These short peristaltic waves always proceed in the direction of the large intestine (a property known as the law of the intestine); this accounts for some of the net propulsion of chyme along the intestine. However, these peristaltic waves die out after traveling a very short distance; peristaltic waves which travel the entire length of the small intestine do not occur in man except under abnormal conditions.

Parasympathetic stimulation of the intestine increases contractile activity, and sympathetic stimulation decreases it. A person's emotional state can affect the contractile activity of the intestine and thus the rate of propulsion of chyme and the time available for digestion and absorption. Fear tends to decrease motility whereas hostility increases it, although these responses vary greatly in different individuals.

The contractile activity of an empty intestine is weak. Following a meal, distension increases the intensity of contractions but the actual rate of propulsion is decreased because the contracting segments narrow the lumen of the intestine, producing a greater resistance to flow. These changes in contractile activity are mediated by local reflexes, such as that produced by distending the lumen with chyme, and by external nerves to the intestine. In man, contractile activity in the ileum also increases during periods of gastric emptying. This is known as the *gastroileal reflex* (conversely, distension of the ileum produces decreased gastric motility, the *ileo-gastric reflex*). Large distensions of the intestine, injury to the intestinal wall, and various bacterial infections in the intestine lead to a complete cessation of motor activity, the *intestino-intestinal reflex*. All these reflexes appear to be mediated primarily by external nerves.

Structure of the intestinal mucosa

The mucosa of the small intestine is highly folded (Fig. 12-3), and the surface of these folds is further convoluted by fingerlike projections known as *villi*. The surface of each villus is covered with numerous epithelial cells, and the surface area of each cell is increased by small projections known as *microvilli* (Fig. 12-23). The combination of folded mucosa, villi, and microvilli increases the total surface area of the small intestine available for absorption about 600-fold over that of a flat-surfaced tube of the same length and diameter. The total surface area of the human small intestine has been estimated to be about 2,000 ft², or equivalent to the area of a tennis court. The cells of the intestinal epithelium are joined together by tight junctions near their luminal border; thus the only route available for entry from lumen to the blood and lymph is by crossing a cell membrane.

The structure of a villus is illustrated in Figure 12-24. The center of the villus is occupied by a capillary network which branches from an arteriole and drains into a venule. The blood flow to the small intestine of a normal man at rest averages about 1 liter/min, or one-fifth the resting cardiac output. Intestinal blood flow increases during periods of digestive activity as a result of local autoregulation produced by the increased metabolic activity of the intestinal cells and by reflexes triggered by mechanical distension of the lumen by chyme. A single blind-ended lymph vessel, a *lacteal*, occupies the center of the villus. Nerve fibers and smooth muscle cells are also present. The epithelial cells at the tip of the villus have a much greater number of microvilli per cell than the cells at the base of the villus. These microvilli contain several enzymes important for digestion.

The villi of the intestine move back and forth independently of each other, presumably as a result of the contraction of the smooth muscle in them. The motion of the villi is increased following a meal, and associated with their increased motion is a greater flow of lymph through the lacteals. The folds of the intestinal mucosa are not permanent structures but can change their pattern with contraction of the underlying smooth muscle layer, the muscularis mucosae. Irritation of the mucosal wall by lumps of food or stimulation of the sympathetic nerves to the intestine causes contraction of the muscularis mucosae and greater mucosal folding.

The cells lining the intestinal tract are continually being replaced as a result of the high mitotic activity of the cells at the base of the villi. The new cells migrate from the base of the villi to the top, replacing older cells, which disintegrate and are discharged into the lumen of the intestine. The entire epithelium of the intestine is replaced every 36 hr. The continuous discharge of cells into the lumen amounts to about 250 g of cells every day.

The surface of the intestinal tract, along with the blood-forming regions of the bone marrow, because of their high rate of cell division, is very sensitive to damage by radiation during the period when DNA is being replicated. These areas are the first to be severely damaged by excessive x-ray radiation or atomic radiation.

Secretions

Glands located in the mucosa of the small intestine secrete about 2,000 ml of mucus and salt solutions into the lumen each day. These secretions were once believed to contain digestive enzymes, but it now appears that the enzymes found in the secretions of isolated intestinal segments are derived from the disintegration of the epithelial cells constantly being discharged from the tips of the villi. The true glandular secretions appear to be nearly enzyme-free.

Intestinal secretion rises following a meal, and mechanical stimulation of the intestinal wall enhances secretion. Secretion released in response to acid in the duodenum appears to be the primary stimulus for intestinal secretions. These secretions contain bicarbonate which may help to protect the surface of the intestine from the damaging effects of acid.

Digestion and absorption

Almost all digestion and absorption of food and water occur in the small intestine. About 50 percent of the small intestine can be removed without interfering with the digestive and absorptive processes. Normally, most of the contents of the small intestine have been absorbed by the time the chyme has reached the middle of the jejunum.

During the course of a day, the average adult consumes about 2 lb of solid food and 2.5 lb of water (Table 12-3), but this is only a fraction of the total material entering the gastrointestinal tract each day. To the approximately 2,000 ml of ingested food and drink are added about 7,000 ml of fluid from the salivary glands, stomach, pancreas, liver, and intestinal tract (Fig. 12-25). Of this total volume of 9,000 ml, only about 500 ml passes into the large intestine each day, 94 percent being absorbed across the walls of the small intestine.

Carbohydrate The daily intake of carbohydrate varies considerably, ranging from 250 to 800 g/day in a typical American diet. Most of the carbohydrate is in the form of the vegetable polysaccharide, starch. Small amounts of the disaccharides sucrose (glucose-fructose, table sugar) and lactose (glucose-galactose, milk sugar) may

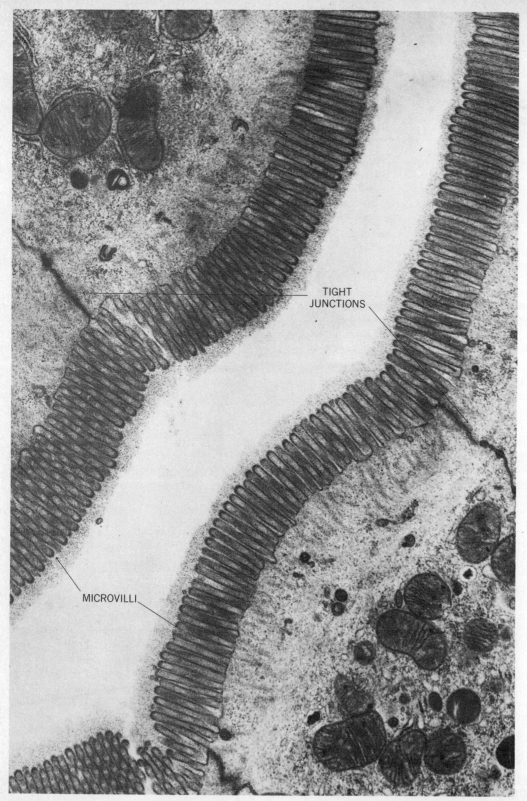

TIGHT
JUNCTIONS

MICROVILLI

FIGURE 12-23

Microvilli on the surface of intestinal epithelial cells. [*From
D. W. Fawcett, J. Histochem. Cytochem,* **13**:75–91 (1965). *Courtesy of
Dr. Susumu Ito.*]

TABLE 12-3
Typical food and water intake per day

	Average per day	Percent solids
Carbohydrate, g	500	62.5
Protein, g	200	25
Fat, g	80	10
Salt, g	20	2.5
Water, ml	1,200	

be present. In infants, lactose from milk makes up the majority of the carbohydrate in the diet. Only very small amounts of monosaccharides are normally present. The polysaccharide cellulose, which is present in plant cells, cannot be broken down into glucose units by the enzymes secreted by the gastrointestinal tract. Cellulose passes through the small intestine in undigested form and enters the large intestine, where it is partially digested by the bacteria which inhabit this region.

About 50 percent of the starch is digested by salivary amylase during passage through the stomach, and the remainder is digested in the duodenum by pancreatic amylase. Blockage of pancreatic secretions does not appreciably affect the digestion and absorption of carbohydrates because of the presence of salivary amylase. The primary product formed by these enzymes is the disaccharide maltose (glucose-glucose). Enzymes that split disaccharides into monosaccharides are located

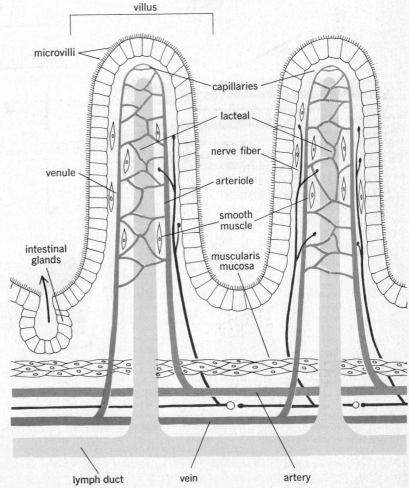

FIGURE 12-24
Structure of intestinal villi.

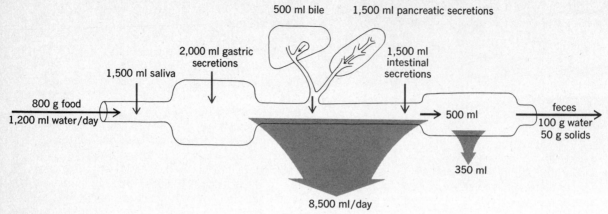

FIGURE 12-25
Average amounts of food and fluid ingested, secreted, absorbed, and excreted from the gastrointestinal tract daily.

both in the microvilli of the cells lining the intestinal tract and in the luminal contents as a result of the continuous disintegration of intestinal epithelial cells. The monosaccharides glucose and galactose liberated from the breakdown of polysaccharides and disaccharides are actively transported across the intestinal epithelium into the blood. The presence of sodium in the lumen is required for monosaccharide transport. Apparently sodium and sugar combine with some carrier molecule in the epithelial cell membrane. The six-carbon sugar fructose, however, is not actively absorbed by the intestinal epithelium but probably crosses the membrane by a facilitated-diffusion process. The rapid digestion of polysaccharides and the transport of the resulting monosaccharides into the blood lead to the absorption of most of the dietary carbohydrate by the end of the jejunum.

Some children cannot hydrolyze or absorb lactose (milk sugar). Most of this sugar appears in the feces or is metabolized by bacteria in the large intestine. The production of acids by the fermenting bacteria stimulate motility resulting in diarrhea and dehydration. Because milk constitutes the major diet of infants, they may also become undernourished as a result of their inability to absorb lactose. Lactose intolerance has been shown to be a genetically determined error of metabolism in which the enzyme lactase is not synthesized by the intestinal epithelium.

Protein An intake of about 50 g of protein each day is required by an adult to supply essential amino acids and replace amino acid nitrogen loss in the urine. A typical American diet contains about 200 g of protein. In addition to the protein in the diet, 10 to 30 g of protein is secreted into the gastrointestinal tract by the various glands, and about 25 g of protein is derived from the disintegration of epithelial cells. The feces contain 10 to 20 g of protein, most of which comes from bacteria and disintegrated cells rather than from the diet.

Proteins are broken down to peptide fragments by pepsin and the proteolytic enzymes secreted by the pancreas. These peptide fragments are further digested to free amino acids by carboxypeptidase from the pancreas and aminopeptidase in the intestinal villi, which split off amino acids from the carboxyl and amino ends of the peptide chains, respectively. The free amino acids are actively transported across the walls of the intestine from lumen to blood.

Several different carrier systems are available for transporting different classes of amino acids. Some of these require sodium, just as in the case of carbohydrate transport. Several genetic diseases have been found in which there is a defective absorption of certain amino acids. These appear to result from the absence of specific carriers or their malfunction, and the same individual may show defective amino acid transport of the kidney tubules as well. In the absence of pancreatic secretion, protein digestion and absorption are decreased about 50 percent.

Fat The amount of fat in the diet varies from about 25 to 160 g/day. Most of it is in the form of triglycerides

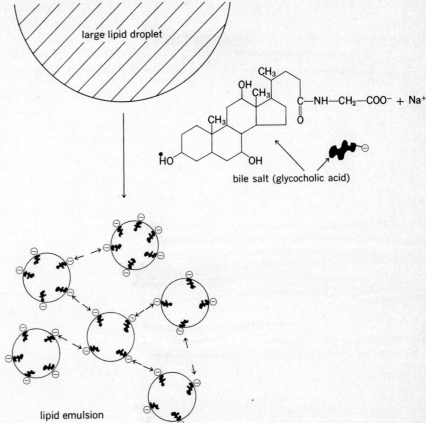

FIGURE 12-26   lipid emulsion
Emulsification of fat by bile salts.

(neutral fat), the remainder being primarily phospholipids and cholesterol. There are very few free fatty acids in the average diet. Almost all the fat entering the digestive tract is absorbed, and the 3 to 4 g of fat in the feces is derived primarily from the bacteria in the large intestine.

Most fat digestion occurs in the small intestine under the influence of pancreatic lipase. In order to digest fat, the water-soluble lipase molecule must come into contact with a molecule of triglyceride, but neutral fats are not soluble in water and tend to separate from the water phase into droplets of lipid. It is in this form that they enter the small intestine from the stomach. If the triglycerides remained aggregated in droplets, the only available area for lipase action would be the surface of the droplet, which contains only a small fraction of all the molecules in the droplet. Digestion of lipid in this state is very slow.

Bile salts along with the lecithin and cholesterol

secreted by the liver speed up lipid digestion by emulsifying fat and increasing the surface area of the lipid droplets accessible to pancreatic lipase. Bile salts consist of a large steroid portion which is lipid-soluble and a small carbon chain containing an ionized carboxyl group (Fig. 12-26). The steroid portion dissolves in the large lipid droplets, leaving the ionized carboxyl groups on the surface, where they interact with polar water molecules. This produces a negative charge on the surface of the lipid droplet. Mechanical agitation in the intestine breaks up the large droplets into smaller ones, and the negatively charged groups on the droplet surface produce an electric repulsion between droplets, preventing them from recoalescing into large drops (Fig. 12-26).

Pancreatic lipase acts on the surface of the small droplets, forming free fatty acids and mixtures of mono- and diglycerides with only small amounts of free glycerol. The free fatty acids and monoglycerides, having polar and nonpolar segments like the bile salts, assist in the emulsification of the lipid droplets.

As digestion proceeds, the bile salts perform a second important function in promoting aggregation of the liberated free fatty acids and monoglycerides into water-soluble particles known as *micelles*. The structure of a micelle is similar to that of the lipid emulsion, only smaller. The nonpolar portions of the fatty acids, monoglycerides, and bile salts are oriented to the center of the micelle, with the polar portions at the surface. A single micelle is only about 30 to 100 Å in diameter and consists of only a few thousand molecules. Whereas a lipid emulsion appears cloudy because of the relatively large size of the emulsion droplets, a solution of micelles is perfectly clear.

Although free fatty acids and monoglycerides are essentially insoluble in water, a few free molecules can exist in solution, and it is in this form that they are absorbed across the cell membrane by simple diffusion because of their high degree of lipid solubility in the lipid cell membrane. The small amounts of free fatty acids and monoglycerides that are present in solution are in equilibrium with the micelles. As the free components diffuse into the cells, some of the micelles liberate their fatty acids and monoglycerides into solution, thus maintaining a saturated solution with respect to the free fatty acids and monoglycerides. Since their concentration in free solution is very, very low and absorption takes place by diffusion down a concentration gradient, the rate of fat absorption is limited by diffusion. The ability of the micelles to maintain this solution in a saturated state allows fat to be absorbed relatively rapidly. In the absence of micelles, fat absorption is very slow.

If 90 percent of the pancreas is destroyed, the intestine loses little of its ability to digest and absorb fat; but if all the pancreas is destroyed so that there is no lipase, only about one-third of the dietary lipid is absorbed and two-thirds appears in the feces. The little digestion that does occur is probably the result of enzymes from the disintegrating epithelial cells. If, on the other hand, the bile duct is blocked so that bile salts do not enter the intestine to promote emulsification and micelle formation, about 50 percent of ingested fat appears in the feces.

Although fatty acids are the primary form of fat entering the epithelial cells, very little free fatty acid is released into the circulation. During their passage through the epithelial cells, fatty acids are resynthesized into triglycerides, and it is the triglycerides which are released into the circulation. Electron micrographs of the intestinal epithelium during fat absorption show the accumulation of fat droplets in the endoplasmic reticulum of these cells. The isolated membranes of the endoplasmic reticulum contain the enzymes involved in triglyceride synthesis. The droplets of lipid become larger as they proceed through the cell. Release of the lipid droplet is believed to be similar to the release of protein secretory granules: The membrane of the endoplasmic reticulum surrounding the lipid droplet fuses with the cell membrane, freeing the droplet into the extracellular space.

The triglycerides enter the lacteals (lymph vessels) rather than the capillaries, primarily in the form of small lipid droplets, 0.1 to 3.5 μm in diameter, known as *chylomicrons*. These small droplets contain about 90 percent triglyceride and small amounts of phospholipid, cholesterol, free fatty acids, and protein. The presence of chylomicrons in the blood after a fatty meal gives the plasma a milky appearance.

Entrance of the chylomicrons into the lacteals rather than into the capillaries appears to depend upon the relative permeabilities of the two vessels to the lipid droplets released from the intestinal epithelium. As in other capillaries of the body, a basement membrane composed of polysaccharides covers the outer surface of the capillary but not the lacteal and may be the barrier keeping the chylomicrons out of the capillaries. Figure 12-27 summarizes the pathway taken by fat in moving from the lumen into the lymphatic system.

Vitamins Most of the vitamins readily diffuse across the walls of the intestine into the blood. The fat-soluble vitamins, vitamins A, D, E, and K, are dissolved in micelles and their absorption is markedly decreased in the absence of bile. Vitamin B_{12}, a charged molecule with a high molecular weight, is unable to diffuse across cell membranes. A special system is required to move vitamin B_{12} across the wall of the ileum, where most of its absorption occurs. Upon entering the stomach, the vitamin combines with a special protein secreted by the stomach, known as the *intrinsic factor*. In the absence of the intrinsic factor, very small amounts of B_{12} are absorbed. Patients with pernicious anemia lack the intrinsic factor and thus do not absorb adequate amounts of vitamin B_{12}, which is required for the formation of red blood cells (Chap. 10). During the absorption process, vitamin B_{12} is released from the intrinsic factor and enters the blood. What role the intrinsic factor plays in the entry of the vitamin across the intestinal wall is not known. It has been suggested that vitamin B_{12} may cross the intestinal wall by pinocyto-

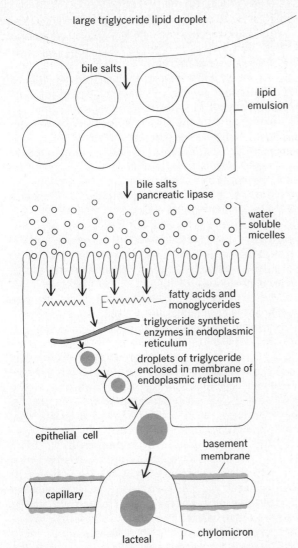

large triglyceride lipid droplet

bile salts

lipid emulsion

bile salts pancreatic lipase

water soluble micelles

fatty acids and monoglycerides

triglyceride synthetic enzymes in endoplasmic reticulum

droplets of triglyceride enclosed in membrane of endoplasmic reticulum

epithelial cell

basement membrane

capillary

lacteal

chylomicron

FIGURE 12-27
Summary of fat absorption across the walls of the small intestine.

sis. Vitamin B_{12} absorption is the only case known where a large molecule combines with a still larger molecule, the intrinsic factor, in order to cross a cell membrane. Vitamin B_{12} is probably the largest essential nutrient absorbed from the intestinal tract without first being digested into smaller molecules. The loss of intrinsic factor and the eventual development of pernicious anemia are consequences of the surgical removal of the stomach;

in this sense the stomach is essential for the life of the individual, but not because of its digestive role. Normally, a large amount of B_{12} is stored in the liver; and the symptoms of pernicious anemia often do not develop for several years after the removal of the stomach.

Water and salt   The normal volume of 9,000 ml of fluid entering the intestinal tract each day is absorbed at the rate of about 200 to 400 ml/hr. Chyme entering the duodenum from the stomach is nearly isotonic. In the small intestine, sodium ions derived from the diet and from the gastrointestinal secretions are actively transported across the intestinal wall. Besides sodium, monosaccharides and amino acids are also actively transported from lumen to blood, often in combination with sodium, as we have seen. As a result of the active absorption of these solutes, the total concentration of solute in the lumen tends to drop below that of plasma and thus becomes relatively hypotonic. However, the intestinal wall has a very high permeability to water and thus cannot maintain a water-concentration gradient between the lumen and blood. Water rapidly diffuses from the hypotonic solution in the lumen into the isotonic solution of the plasma. The movement of water is so rapid that it virtually accompanies the movement of solute, and thus the contents of the lumen remain essentially isotonic as they decrease in volume because of the net water and solute movement into the blood. This phenomenon is clearly similar to that for sodium and water reabsorption by the proximal renal tubules.

Water diffuses rapidly in both directions across the intestinal wall. Therefore, if a hypertonic solution is present in the lumen, there is a net movement of water from blood to lumen, expanding the luminal contents until they become isotonic. Although the chyme entering the duodenum from the stomach is approximately isotonic, the duodenal contents may become hypertonic as a result of the rapid digestion of polysaccharides and proteins, forming a large number of osmotically active molecules from a few large molecules. If osmotically active molecules are formed by digestion faster than they are absorbed, the osmolarity of the solution increases and water flows from the blood into the lumen. In some cases, the movement of water into the intestine may be large enough to cause a significant decrease in blood volume, leading to cardiovascular complications. For example, when a small child consumes a lot of candy and soda pop, the high carbohydrate content has little inhibitory effect on gastric emptying, and the contents of the stomach soon enter the duodenum. Water rapidly diffuses from blood to lumen, diluting the high sugar concentration and increasing the volume of the lumen; distension of the intestinal wall and the decrease in blood volume lead to nausea, vomiting, pallor, sweating, and possibly fainting as a number of autonomic reflexes are triggered in the cardiovascular and gastrointestinal systems. The distension of the intestine triggers a reflexive increase in blood flow to the intestine which further aggravates the circulatory problems resulting from a decrease in blood volume.

A similar reaction, known as the *dumping syndrome,* is common in patients who have had large portions of their stomachs surgically removed because of disease. In these patients food may enter the duodenum directly from the esophagus without being retained by the stomach. The rapid increase in the osmolarity of the duodenum's contents, as a result of digestion and the accompanying movement of water into the small intestine, produces the responses described above. Such patients must be fed small quantities of food many times a day.

Another example of fluid accumulation in the intestine occurs when the small intestine is obstructed. The rate of salt and water movement out of the intestine into the blood above the site of the obstruction is markedly decreased, with little changes in the rate at which salt and water enter the lumen from the blood. This may result from marked distension of the intestinal wall, which blocks the blood flow to the affected region, damaging the cells and altering membrane permeability. The result is a large net movement of fluid into the intestine and further distension of the intestinal walls. Sufficient quantities of fluid may be withdrawn from the blood under these conditions to reduce blood pressure and cause severe complications, even death.

In addition to the active absorption of sodium, other inorganic ions such as Cl^-, K^+, Mg^{2+} and Ca^{2+} are absorbed into the blood. Some are absorbed by active transport; others diffuse into the blood down their electrochemical gradients. The absorption of iron was described in Chap. 10.

The large intestine

The colon (large intestine), a tube about 2.5 in. in diameter, forms the last 4 ft of the gastrointestinal tract. The *cecum* forms a blind-ended pouch below the junction of the small and large intestines. The *appendix* (Fig. 12-1), a small fingerlike projection from the end of the cecum, has no known function. The colon is not coiled but consists of three relatively straight segments, the ascending,

transverse, and descending portions. The terminal portion of the descending colon is S-shaped, forming the sigmoid colon, which empties into a short section of tubing, the *rectum.* Although the large intestine has a greater diameter than the small intestine and is about half as long, its epithelial surface area is only about $\frac{1}{30}$ that of the small intestine because the mucosa of the large intestine lacks villi and is not convoluted. The large intestine secretes no digestive enzymes and is responsible for the absorption of only about 4 percent of the total intestinal contents per day. Its primary function is to store and concentrate fecal material prior to defecation.

Chyme enters the colon through the ileocecal sphincter separating the ileum from the colon. This sphincter is normally closed, but after a meal when the gastroileal reflex increases the contractile activity of the ileum, the sphincter relaxes each time the terminal portion of the ileum contracts, allowing chyme to enter the large intestine. Distension of the colon, on the other hand, produces a reflexive contraction of the sphincter, preventing further material from entering.

Secretion and absorption

About 500 ml of chyme from the small intestine enters the colon each day. Most of this material is derived from the secretions of the small intestine, since most of the ingested food has been absorbed before reaching the large intestine. The secretions of the colon are very scanty and consist mostly of mucus.

The primary absorptive process in the large intestine is the active transport of sodium from the lumen to blood with the accompanying osmotic reabsorption of water. If fecal material remains in the large intestine for a long time, almost all the water is reabsorbed, leaving behind dry fecal pellets. The cells lining the large intestine are unable to actively transport either glucose or amino acids. There is a small net leakage of potassium into the colon, and severe depletion of total body potassium can occur as a result of repeated enemas and diarrhea.

The large intestine also absorbs some of the products synthesized by the bacteria in it. For example, small amounts of vitamins are synthesized by intestinal bacteria and absorbed into the body. Although this source of vitamins generally provides only a small part of the normal vitamin requirement per day, it may make a significant contribution when dietary intake of vitamins is low. The intestinal bacteria digest cellulose and utilize the glucose released for their own growth and reproduction.

Other bacterial products contribute to the production of intestinal gas (*flatus*). This gas is a mixture of nitrogen and carbon dioxide with small amounts of the inflammable gases hydrogen, methane, and hydrogen sulfide. Bacterial fermentation produces gas in the colon at the rate of about 400 to 700 ml/day. In the cow, where bacterial fermentation makes a major contribution to the digestive process, as much as 300 to 600 liters of flatus may be produced each day.

Motility

The longitudinal smooth muscle in the human colon is incomplete, and the walls of the large intestine are folded into sacs called *haustra* by the contraction of the circular smooth muscle. Contractions of the circular smooth muscle produce a segmentation motion which is not propulsive. This movement is considerably slower than in the small intestine, and a contraction may occur only once every 30 min. Because of this slow movement, material entering the colon from the small intestine remains for 18 to 24 hr. Bacteria have time to grow and accumulate in the large intestine because of its slow movements; in the small intestine they do not have sufficient time to accumulate before being swept into the large intestine. During sleep and most of the day there is generally little or no movement in the large intestine, but three to four times a day, generally after meals, a marked increase in motility occurs. This usually coincides with the gastroileal reflex, described earlier, and probably has similar reflex mechanisms. This increased motility may lead to the phenomenon known as *mass movement,* in which large segments of the ascending and transverse colon contract simultaneously, propelling fecal material one-third to three-fourths of the length of the colon in a few seconds.

Defecation   About 150 g of feces is eliminated from the body each day. This fecal material consists of about 100 g of water and 50 g of solid matter. The solid matter is made up mostly of bacteria, undigested cellulose, cell debris from the turnover of the intestinal epithelium, bile pigments, and small amounts of salt. The feces contain more potassium than the fluid entering the colon from the small intestine because some potassium is secreted by the cells lining the colon.

The sudden distension of the walls of the rectum produced by the mass movement of fecal material is the normal stimulus for defecation. It initiates the defecation reflex, which is mediated primarily by the internal nerve plexuses but can be reinforced by external nerves to the terminal end of the large intestine. The reflex response consists of a contraction of the rectum, relaxation of the internal and external anal sphincters, and increased peri-

staltic activity in the sigmoid colon. This activity is sufficient to propel the feces through the anus. Defecation is normally assisted by a deep inspiration followed by closure of the glottis and contraction of the abdominal and chest muscles, causing a marked increase in intra-abdominal pressure, which is transmitted to the contents of the large intestine and assists in the elimination of feces. This maneuver also causes a rise in intrathoracic pressure which leads to a sudden rise in blood pressure followed by a fall as the venous return to the heart is decreased (Chap. 9). In elderly people the cardiovascular stress resulting from the strain of defecation may precipitate a stroke or heart attack.

The *internal anal sphincter* is composed of smooth muscle, but the *external anal sphincter* is skeletal muscle under voluntary control. Higher brain centers in the central nervous system may, via descending pathways, override the afferent input from the rectum, thereby keeping the external sphincter closed and allowing a person to delay defecation.

The conscious urge to defecate accompanies the initial distension of the rectum. If defecation does not occur, the tension in the walls of the rectum slackens as the muscle relaxes, and the urge to defecate subsides until the next mass movement propels more feces into the rectum, increasing its volume and again initiating the defecation reflex.

Constipation and diarrhea   *Constipation* is the condition in which defecation is delayed for a variety of reasons. It may be due to consciously ignoring or preventing defecation or to decreased colonic motility, which most commonly is secondary to aging, emotion, or a low-bulk diet. Bulk refers to the content of cellulose or other undigested materials in the diet, the volume of which is not decreased by absorption. The longer fecal material remains in the large intestine, the more water is reabsorbed, and the harder and drier the feces become, making defecation more difficult and sometimes painful. During this period additional material from the small intestine continues to enter the colon, progressively increasing the volume of its contents.

Many people have a mistaken belief that unless there is a bowel movement every day retention of fecal material and bacteria in the large intestine will somehow poison the body because of toxic products produced by the bacteria. Attempts to isolate such toxic agents from intestinal bacteria have been totally unsuccessful. In unusual cases where defecation has been prevented for a year or more by blockage of the rectum no ill effects from accumulated feces were noted except for the discomfort of carrying around the extra weight of 50 to 100 lb of feces retained in the large intestine. The symptoms of nausea, headache, loss of appetite, and general feeling of discomfort sometimes accompanying constipation appear to come from the distension of the rectum and large intestine. Experimentally inflating a balloon in the rectum of a normal individual produces similar sensations. Thus, there is no physiological necessity for having bowel movements regulated by a clock; whatever maintains a person in a comfortable state is physiologically adequate, whether this means a bowel movement after every meal, or once a day, or once a week.

Cathartics, or laxatives, are sometimes necessary to relieve constipation. Several types are in common use. Cellulose in vegetable matter is a natural cathartic because of its ability to increase intestinal motility by providing bulk which stretches the smooth muscle of the intestinal wall, increasing its sensitivity to the basic electrical rhythm, and thus increasing its contractile activity. Castor oil acts by irritating the smooth muscle of the intestinal tract, increasing its motility. Some cathartics, such as mineral oil, act by lubricating hard, dry fecal material, thus easing defecation. Such agents as milk of magnesia are not absorbed or absorbed only slowly by the intestinal wall; the presence of nonabsorbable solute causes water to be retained in the intestinal tract and along with the increased motility resulting from the increased volume helps to flush out the large intestine.

Diarrhea, the opposite of constipation, is characterized by frequent defecation, usually of highly fluid fecal matter. A primary cause is greater intestinal motility with less time for absorption and thus the delivery of a large volume of fluid to the large intestine overloading its capacity to absorb salt and water. Certain foods, such as prunes, stimulate intestinal motility and tend to produce diarrhea. Disease-producing bacteria often irritate the intestinal wall, increase motility of the intestinal tract, and lead to diarrhea. Prolonged diarrhea can result in a serious loss of fluid and salt, especially potassium, from the body as well as upsetting the acid-base balance of the body due to loss of bicarbonate.

SECTION A
CONTROL AND INTEGRATION OF CARBOHYDRATE, PROTEIN, AND FAT METABOLISM

In Chap. 3, we described the basic chemistry of living cells and their need for a continuous supply of nutrients. Although a certain fraction of these organic molecules is used in the synthesis of structural cell components, enzymes, coenzymes, hormones, antibodies, and other molecules serving specialized functions, most of the molecules in the food we eat are used by cells to provide the chemical energy required to maintain cell structure and function.

Essential for an understanding of organic metabolism is the remarkable ability of most cells, particularly those of the liver, to convert one type of molecule into another. These interconversions permit the human body to utilize the wide range of molecules found in different foods, but there are limits, and certain molecules must be present in the diet in adequate amounts. Enough protein must be ingested to provide the nitrogen needed for synthesis of protein and other nitrogenous substances

Regulation of organic metabolism and energy balance

13

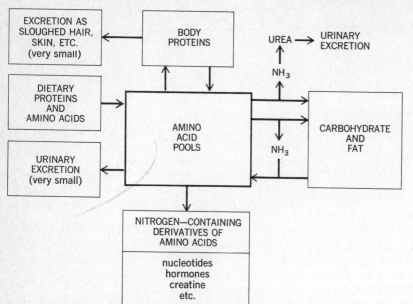

FIGURE 13-1
Amino acid pools and major pathways of protein metabolism.

within the body and it must contain an adequate quantity of specific amino acids, called *essential* because they cannot be formed within the body by conversion from another molecule type. The other essential organic nutrients are a small group of fatty acids and the vitamins. The last group was discovered when diets adequate both in total calories and essential amino acids and fatty acids were still found to be incapable of maintaining health. As described in Chap. 3, the vitamins are not utilized for energy or for synthesis of structural components but serve as cofactors or coenzymes in chemical reactions.

The concept of a dynamic catabolic-anabolic steady state is also a critical component of organic metabolism. With few exceptions, e.g., DNA, virtually all organic molecules are being continuously broken down and rebuilt, usually at a rapid rate. The turnover rate of body protein is approximately 100 g/day; i.e., this quantity is broken down into amino acids and resynthesized each day. Few of the atoms present in a person's skeletal muscle a month ago are still there today.

With these basic concepts of *molecular interconvertibility* and *dynamic steady state* as foundation, we can discuss organic metabolism in terms of total body interactions. Figure 13-1 summarizes the major pathways of protein metabolism. The *amino acid pools,* which constitute the body's total free amino acids, are derived primarily from ingested protein (which is degraded to amino acids

during digestion) and from the continuous breakdown of body protein. These pools are the source of amino acids for resynthesis of body protein and a host of specialized amino acid derivatives, such as nucleotides, epinephrine, etc. A very small quantity of amino acid and protein is lost from the body via the urine, skin, hair, and fingernails. The interactions between amino acids and the other nutrient types, carbohydrate and fat, are extremely important: Amino acids may be converted into carbohydrate (or fat) by removal of ammonia (Fig. 13-2); one type of amino acid may participate in the formation of another by passing its nitrogen group to a carbohydrate (Fig. 13-3). Both these processes were described in greater detail in Chap. 3 and are mentioned here to emphasize the interconvertibility of protein, carbohydrate, and fat. The ammonia, NH_3, formed during the first process is converted by the liver into urea, which is then excreted by the kidneys as the major end product of protein metabolism. Not all the events relating to amino acid metabolism occur in all cells; urea is formed in one organ (the liver) but excreted by another (the kidneys), but the concept of a pool is valid because all cells are interrelated by the vascular system and blood.

If any of the essential amino acids is missing from the diet, negative nitrogen balance always results. Apparently, the proteins for which that amino acid is essential cannot be synthesized and the other amino acids

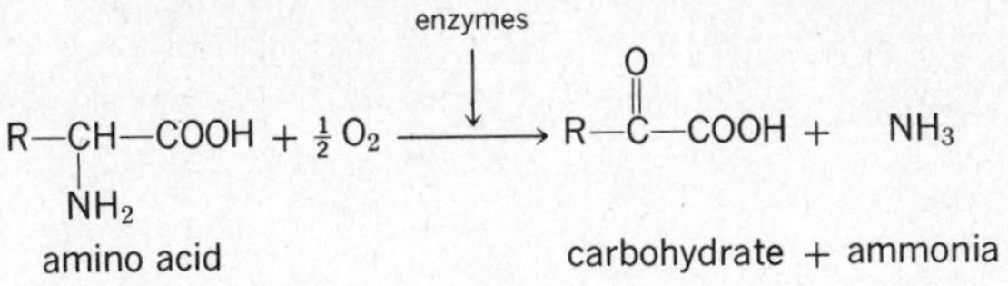

$$R-\underset{\underset{NH_2}{|}}{CH}-COOH + \tfrac{1}{2}O_2 \xrightarrow{\text{enzymes}} R-\underset{\underset{}{\overset{O}{\|}}}{C}-COOH + \quad NH_3$$

amino acid   carbohydrate + ammonia

FIGURE 13-2
Transformation of an amino acid into a carbohydrate (a keto acid).

which would have been incorporated into the proteins are deaminated, their nitrogen being excreted as urea. It should be obvious, therefore, why a dietary requirement for protein cannot be specified without regard to the amino acid composition of that protein. Protein is graded in terms of how closely its ratio of essential amino acids approximates the ideal, which is their relative proportions in body protein. The highest-quality proteins are those found in animal products whereas the quality of most plant proteins is lower. Nevertheless, it is quite possible to obtain adequate quantities of all essential amino acids from plant protein alone although the total quantity of protein ingested must be larger.

Figure 13-4 summarizes the metabolic pathways for carbohydrate and fat, which are considered together because of their high potential rate of interconversion, particularly carbohydrate to fat (see below and Chap. 3). The similarities between Figs. 13-1 and 13-4 are obvious, but there are several critical differences: (1) The major fate of both carbohydrate and fat is catabolism to yield energy, whereas amino acids can supply energy only after they are converted to carbohydrate or fat; and (2) excess carbohydrate and fat can be stored as such, whereas excess amino acids are not stored as protein but are converted to carbohydrate and fat.

In discussing the mechanisms which regulate the magnitude and direction of these molecular interconversions, we shall see that the *liver, adipose tissue* (the storage tissue for fat), and *muscle* are the dominant effectors and

that the major controlling input to them is a group of hormones and the sympathetic nerves to adipose tissue and the liver. At this point, the reader should review the biochemical pathways described in Chap. 3, particularly those dealing with glucose and the interconversions of carbohydrate, protein, and fat.

Events of the absorptive and postabsorptive states

When food is readily available human beings can get along by eating small amounts of food all day long if they wish; however, this is clearly not true for other animals, for early man, or for most persons today, and animals have been forced to evolve mechanisms for survival during alternating periods of plenty and fasting. We speak of two functional states: the *absorptive state,* during which ingested nutrients are entering the blood from the gastrointestinal tract, and the *postabsorptive* (or *fasting) state,* during which the gastrointestinal tract is empty and energy must be supplied by the body's endogenous stores. Since an average meal requires approximately 4 hr for complete absorption, our usual three-meal-a-day pattern places us in the postabsorptive state during the late morning and afternoon and almost the entire night. The average person can easily withstand a fast of many weeks (so long as water is provided), and extremely obese patients have been fasted for many months, being given only water and vitamins.

The absorptive state can be summarized as follows: During absorption of a normal meal, glucose provides the major energy source; only a small fraction of the absorbed amino acids and fat is ultimately utilized for energy; another fraction of amino acids and fat is used to resynthesize the continuously degraded body proteins and structural fat, respectively; most of the amino acids and fat as well as the large quantity of carbohydrate not oxidized for energy are transformed into adipose-tissue fat.

FIGURE 13-3
Formation of an amino acid through interaction of a carbohydrate and a different type of amino acid, the original amino acid being converted into a carbohydrate, and vice versa. The crucial event is the transfer of the nitrogen group from one molecule to the other.

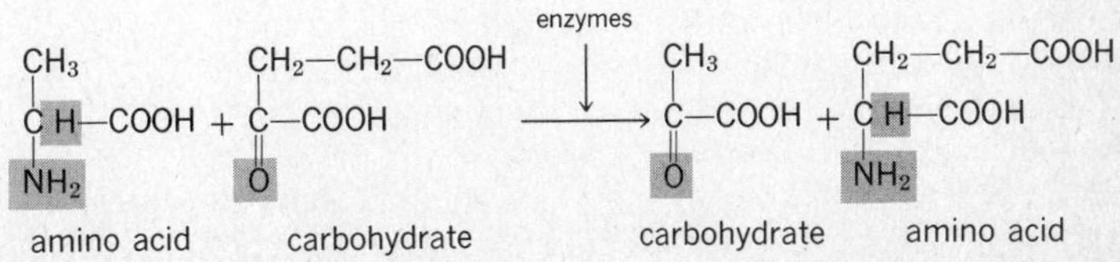

$$\underset{\text{amino acid}}{\underset{\underset{NH_2}{|}}{\overset{CH_3}{|}}{CH}-COOH} + \underset{\text{carbohydrate}}{\underset{\underset{O}{\|}}{\overset{CH_2-CH_2-COOH}{}}{C}-COOH} \xrightarrow{\text{enzymes}} \underset{\text{carbohydrate}}{\underset{\underset{O}{\|}}{\overset{CH_3}{|}}{C}-COOH} + \underset{\text{amino acid}}{\underset{\underset{NH_2}{|}}{\overset{CH_2-CH_2-COOH}{}}{CH}-COOH}$$

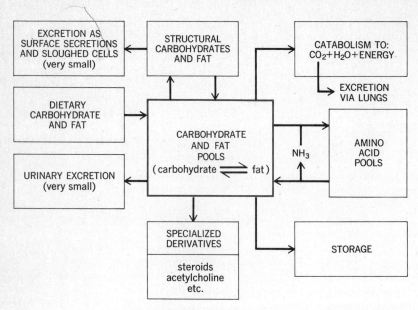

FIGURE 13-4
Major pathways of carbohydrate and fat metabolism. "Carbohydrate and fat pools" include the simple unspecialized carbohydrates and fats dissolved in the body fluids. Note that structural carbohydrate and fat are constantly being broken down and resynthesized.

In the postabsorptive state, carbohydrate is synthesized in the body, but its utilization for energy is greatly reduced; the oxidation of endogenous fat provides most of the body's energy supply; fat and protein synthesis are curtailed and net breakdown occurs.

Figures 13-5 and 13-6 summarize the major pathways to be described. Although they may appear formidable at first glance, they should give little difficulty after we have described the component parts, and they should be referred to constantly during the following discussion. Mastery of this material is essential for an understanding of the hormonal mechanisms which control and integrate metabolism.

Absorptive state

We shall assume an average meal to contain approximately 65 percent carbohydrate, 25 percent protein, and 10 percent fat. Recall from Chap. 12 that these nutrients enter the blood and lymph from the gastrointestinal tract primarily as monosaccharides, amino acids, and triglycerides, respectively. The first two groups enter the blood, which leaves the gastrointestinal tract to go directly to the liver, allowing this remarkable biochemical factory to alter the composition of the blood before it is pumped to the rest of the body. In contrast, the fat droplets are absorbed into the lymph and not into the blood.

Glucose A large fraction of absorbed carbohydrate is galactose and fructose, but since the liver converts most of these carbohydrates immediately into glucose (and because fructose enters essentially the same metabolic pathways as does glucose), we shall simply refer to these sugars as glucose. As shown in Fig. 13-5, much of the absorbed carbohydrate enters the liver cells, but little of it is oxidized for energy, instead being built into the polysaccharide glycogen or transformed into fat. The importance of glucose as a precursor of fat cannot be overemphasized; note that glucose provides both the glycerol and the fatty acid moieties of triglycerides. Some of this fat synthesized in the liver may be stored there, but most is transported into the blood, from which it enters adipose-tissue cells. Much of the absorbed glucose which did not enter liver cells but remained in the blood enters adipose-tissue cells, where it is transformed into fat; another fraction is stored as glycogen in skeletal muscle and certain other tissues, and a very large fraction enters the various cells of the body and is oxidized to carbon dioxide and water, thereby providing the cells' energy requirements. Glucose is the body's major energy source during the absorptive state.

Triglycerides Almost all ingested fat is absorbed into the lymph as fat droplets (*chylomicrons*) containing primarily triglycerides, which enter adipose-tissue cells, where they are stored. Thus, there are three prominent sources of adipose-tissue triglyceride (TG): (*1*) ingested TG, (*2*) TG synthesized in adipose tissue from glucose, (*3*) TG synthesized in the liver and transported via the blood to the

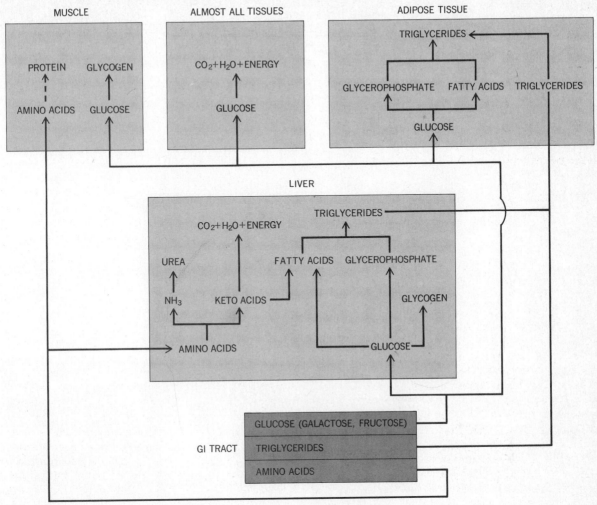

FIGURE 13-5
Major metabolic pathways of the absorptive phase.

adipose tissue. For simplicity, we have not shown in Fig. 13-5 that a fraction of fat is also oxidized during the absorptive state by various organs to provide energy. The actual amount utilized depends upon the content of the meal and the person's nutritional status.

Amino acids Many of the absorbed amino acids enter liver cells and are entirely converted into carbohydrate (keto acids) by removal of the NH_3 portion of the molecule. The ammonia is converted by the liver into urea, which diffuses into the blood and is excreted by the

kidneys. The keto acids can then enter the Krebs tricarboxylic acid cycle and be oxidized to provide energy for the liver cells; indeed the liver is unusual in that by this interconversion amino acids provide much of its energy during the absorptive state. Finally, the keto acids can also be converted to fatty acids, thereby participating in fat synthesis by the liver. The ingested amino acids not taken up by the liver cells enter other cells of the body (Fig. 13-5). Although virtually all cells require a constant supply of amino acids for protein synthesis, we have simplified the diagram by showing only muscle be-

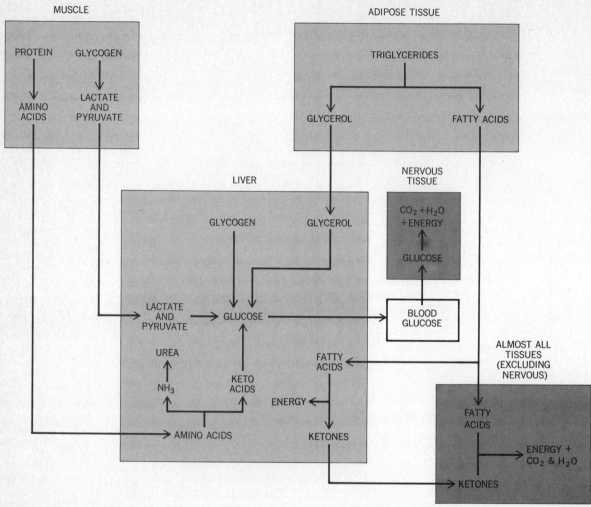

FIGURE 13-6
Major metabolic pathways of the postabsorptive (fasting) phase. The central focus is regulation of the blood glucose concentration.

cause it constitutes the great preponderance of body mass and therefore contains the most important store, quantitatively, of body protein. Other organs of course participate, but to much lesser degree, in the amino acid exchanges occurring during the absorptive and post-absorptive states. After entering the cells, most of the amino acids are synthesized into protein. This process is represented by the dotted line in Fig. 13-5 to call attention to an important fact: Excess amino acids are *not* stored as protein, in the sense that glucose and fat are stored as fat and to a lesser degree as glycogen.

Eating large amounts of protein does not significantly increase body protein; the excess amino acids are merely converted into carbohydrate or fat. On the other hand, a minimal supply of ingested amino acids is essential to maintain normal protein stores by preventing net protein breakdown in muscle and other tissues. During the usual alternating absorptive and postabsorptive states, the fluctuations in total body protein are relatively small.[1]

[1] This discussion applies, of course, only to the adult; the growing child manifests a continuous increase of body protein.

Summary   During the absorptive period, energy is provided primarily by glucose, protein stores are maintained, and excess calories (regardless of source) are stored mostly as fat. Glycogen constitutes a quantitatively less important storage form for carbohydrate. The use of fat to store excess calories is an excellent adaptation for mobile animals, since 1 g of TG contains more than twice as many calories as 1 g of protein or glycogen, and the weight of fat storage depots is minimal because there is very little water in adipose tissue.

Postabsorptive state

The essential problem during this period is that no glucose is being absorbed from the intestinal tract, yet the plasma glucose concentration must be maintained because the nervous system is an obligatory glucose utilizer; i.e., it is unable to oxidize any other nutrient for energy.[1] Lack of adequate glucose supply to the brain causes damage, coma, and death within minutes. Perhaps the most convenient way of viewing the events of the postabsorptive state is in terms of how the blood glucose concentration is maintained. These events fall into two categories: sources of glucose; and glucose sparing and fat utilization.

Sources of blood glucose   The sources of blood glucose during fasting (Fig. 13-6) are as follows:

1 Glycogen stores in the liver are broken down to liberate glucose but are adequate only for a short time. After the absorptive period is completed, the normal liver contains less than 100 g of glycogen; at 4 kcal/g, this provides 400 kcal, enough to fulfill the body's total caloric need for only 4 hr.

2 Glycogen in muscle (and to a lesser extent other tissues) provides approximately the same amount of glucose as the liver. A complication arises because muscle lacks the necessary enzyme to form free glucose from glycogen.[2] But glycolysis breaks the glycogen down into pyruvate and lactate, which are then liberated into the blood, circulate to the liver, and are synthesized into glucose. Thus, muscle glycogen contributes to the blood glucose indirectly via the liver.

3 As shown in Fig. 13-6, the catabolism of triglycerides yields glycerol and fatty acids. The former can be converted into glucose by the liver, but the latter cannot. Thus, a potential source of glucose is adipose-tissue TG breakdown, in which glycerol is liberated into the blood, circulates to the liver, and is converted into glucose.

4 The major source of blood glucose during prolonged periods of fasting comes from protein. Large quantities of protein in muscle and to a lesser extent other tissues are not absolutely essential for cell function; i.e., a sizable fraction of cell protein can be catabolized, as during prolonged fasting, without serious cellular malfunction. There are, of course, limits to this process, and continued protein loss ultimately means functional disintegration, sickness, and death. Before this point is reached, protein breakdown can supply large quantities of amino acids which are converted into glucose by the liver.

To summarize, for survival of the brain, plasma glucose concentration must be maintained. Glycogen stores, particularly in the liver, form the first line of defense, are mobilized quickly, and can supply the body's needs for several hours, but they are inadequate for longer periods. Under such conditions, protein and, to a lesser extent, fat supply amino acids and glycerol, respectively, for production of glucose by the liver. Hepatic synthesis of glucose from pyruvate, lactate, glycerol, and amino acids is known as *gluconeogenesis*, i.e., new formation of glucose. During a 24-hr fast, it amounts to approximately 180 g of glucose. The liver was once believed to be the only human organ capable of glucose synthesis, but it is now known that the kidneys are capable of similar activity, particularly in a prolonged fast (several weeks) at the end of which they may be contributing as much glucose as the liver.

Glucose sparing and fat utilization   A simple calculation reveals that even the 180 g of glucose per day produced by the liver during fasting cannot possibly supply all the body's energy needs: 180 g/day × 4 kcal/g = 720 kcal/day, whereas normal total energy expenditure equals 1,500 to 3,000 kcal/day. The following essential adjustment must therefore take place during the transition from absorptive to postabsorptive state: The nervous system continues to utilize glucose normally, but virtually all other organs and tissues markedly reduce their oxidation of glucose and depend primarily on fat as their energy source, thus sparing the glucose produced by the liver to serve the obligatory needs of the nervous system. The essential step is the catabolism of adipose-tissue TG to liberate fatty acids into the blood. These fatty acids are picked up by virtually all tissues (excluding the nervous system), enter the Krebs cycle, and are oxidized to carbon dioxide and water, thereby providing energy. The liver, too, utilizes fatty acids for its energy source, thereby sparing amino acids (its usual energy source) for glucose synthesis. However, the liver's handling of fatty acids during fasting is unique; it oxidizes them to acetyl CoA, which instead of being oxidized further via the Krebs cycle is processed into a group of compounds called

---

[1] There is an important exception to be described subsequently.
[2] Muscle glycogen is broken down in several steps to glucose 6-phosphate rather than free glucose, which is then catabolized via glycolysis and the Krebs cycle for energy.

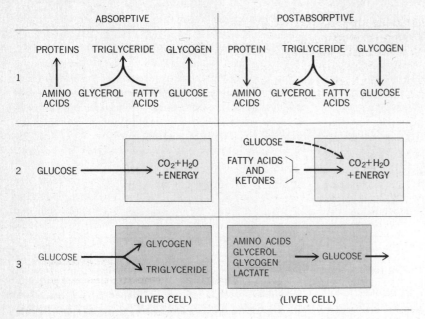

FIGURE 13-7
Summary of critical shifts in transition from absorptive to postabsorptive states.

ketone bodies. One of these substances is acetone, some of which is expired and accounts for the distinctive breath odor of persons undergoing prolonged fasting or suffering from severe untreated diabetes mellitus. The significance of this process is that the ketone bodies are released into the blood and provide an important energy source for the many tissues capable of oxidizing them via the Krebs cycle.

The net result of adipose-tissue breakdown during fasting (as much as 160 g/day) is provision of energy for the body and sparing of glucose for the brain. The combined effects of gluconeogenesis and the switch over to fat utilization are so efficient that, after several days of complete fasting, the plasma glucose concentration is reduced only by a few percent. After one month, it is decreased only 25 percent.

Recent studies have revealed an important change in brain metabolism with prolonged starvation. Apparently, after 4 to 5 days of fasting, the generalization that brain is an obligatory glucose utilizer is no longer valid, for the brain begins to utilize large quantities of ketone bodies, as well as glucose, for its energy source. This is made possible by activation of several brain enzymes required for utilization of the ketone bodies. The survival value of this phenomenon is very great; if the brain significantly reduces its glucose requirement (by utilizing ke-

tones instead of glucose), much less protein need be broken down to supply the amino acids for gluconeogenesis. Accordingly, the protein stores will last longer, and the ability to withstand a long fast without serious tissue disruption is enhanced.

Thus far, our discussion has been purely descriptive; we now turn to the factors which so precisely control and integrate these metabolic pathways and transformations. Without question, the most important single factor is insulin. As before, the reader should constantly refer to Figs. 13-5 and 13-6. We shall focus primarily on the following questions (Fig. 13-7) raised by the previous discussion: (*1*) What controls the shift from the net anabolism of protein, glycogen, and TG to net catabolism? (*2*) What induces primarily glucose utilization during absorption and fat utilization during fasting; i.e., how do cells "know" they should start oxidizing fatty acids and ketones instead of glucose? (*3*) What drives net hepatic glucose uptake during absorption but net glucose synthesis (gluconeogenesis) and release during fasting?

Insulin

Insulin is a protein hormone secreted by the islets of Langerhans, clusters of endocrine cells in the pancreas. It acts directly or indirectly on most tissues of the body,

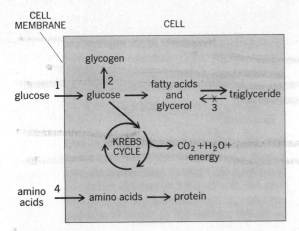

FIGURE 13-8
Major effects of insulin upon organic metabolism. The numbers denote distinct direct effects, the others being the indirect effects of increased glucose and amino acid entry. The X on arrow 3 denotes inhibition of TG breakdown.

with the notable exception of brain. The effects of insulin are so important and widespread that an injection of this hormone into a fasting person duplicates the absorptive state pattern of Fig. 13-5 (except, of course, for the absence of gastrointestinal absorption); conversely, patients suffering from insulin deficiency (diabetes mellitus) manifest the postabsorptive pattern of Fig. 13-6. From these statements alone, it might appear (correctly) that secretion of insulin is stimulated by eating and inhibited by fasting.

Effects of insulin

Glucose uptake Glucose enters most cells by the carrier-mediated mechanism which we described as facilitated diffusion in Chap. 2. The most important single effect exerted by insulin is to stimulate this facilitated diffusion of glucose into certain cells, particularly muscle and adipose tissue (Fig. 13-8). The reason should be apparent; greater glucose entry into cells increases the availability of glucose for all the reactions in which glucose participates. Thus, glucose oxidation, fat synthesis, and glycogen synthesis are all stimulated, in short, the major events of the absorptive state. It is important to note that insulin does *not* alter glucose uptake by the brain, nor does it influence the active transport of glucose across the renal tubule and gastrointestinal epithelium.

This single effect once seemed so allpowerful that many experts believed it to be the only one insulin ex-

erted, but it has now been demonstrated that insulin has other direct effects which reinforce its role as the dominant hormone of the absorptive period.

Stimulation of glycogen synthesis As shown above, increased glucose uptake per se stimulates glycogen synthesis. In addition, insulin also increases the activity of the enzyme which catalyzes the rate-limiting step in glycogen synthesis. Thus, insulin ensures glucose transformation into glycogen by a double-barreled effect.

Inhibition of TG breakdown Increased entry of glucose into adipose tissue facilitates fatty acid and glycerophosphate synthesis, which, by mass action, drives TG synthesis. In addition, insulin inhibits the enzyme which catalyzes TG breakdown. Insulin thus increases TG stores by a double effect: driving TG synthesis by facilitating glucose entry and at the same time inhibiting TG breakdown via the enzyme.

Stimulation of protein synthesis Net protein synthesis is also increased by insulin, which stimulates the active membrane transport of amino acids, particularly into muscle cells. Thus, in a manner analogous to that described for glucose, greater amino acid entry shifts the intracellular protein equilibrium toward net synthesis. Insulin also has important effects on the ribosomal protein-synthesizing machinery.

Effects on other liver enzymes Insulin causes changes in the activities or concentrations of almost all the critical liver enzymes involved in the utilization or synthesis of glucose. As might be predicted the former are all stimulated whereas the latter are inhibited. The precise mechanisms by which insulin induces these changes are still poorly understood.

The other side of the coin We have thus far dealt with the positive effects of insulin. Clearly insulin deficit will have just the opposite results. High rates of glucose entry and oxidation (except in brain) and the net anabolism of glycogen, protein, and TG all depend upon the presence of high blood concentrations of insulin; when the blood concentration decreases, the metabolic pattern is shifted toward decreased glucose entry and oxidation and a net catabolism of glycogen, protein, and TG. In other words, these metabolic pathways are in a dynamic state, capable of proceeding, in terms of net effect, in either direction. For this reason, energy metabolism can be shifted from the absorptive to the postabsorptive pattern merely by lowering the rate of insulin secretion: Glucose entry and oxidation decrease; glycogen breakdown increases; net protein catabolism liberates amino acids into the blood;

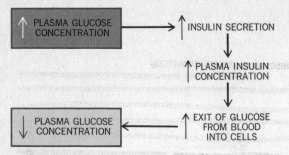

FIGURE 13-9
Negative-feedback nature of plasma glucose control over insulin secretion.

net TG catabolism liberates glycerol and fatty acids into the blood, and the resulting higher fatty acid concentration in blood facilitates cellular uptake of fatty acids, which in turn, stimulates fatty acid oxidation; and gluconeogenesis is stimulated not only by the increased availability of precursors (amino acids and glycerol) but by enzyme changes in the liver itself. The glucose can be utilized by the brain since its glucose uptake is not insulin-dependent. Despite its great importance, insulin is not the only hormone controlling these patterns. The role of the other hormones will be discussed below.

Control of insulin secretion

Insulin secretion is directly controlled by the glucose concentration of the blood flowing through the pancreas, a simple system requiring no participation of nerves or other hormones. An increase in blood glucose concentration stimulates insulin secretion; conversely, a reduction inhibits secretion. The feedback nature of this system is shown in Fig. 13-9. A rise in plasma glucose stimulates insulin secretion; insulin induces rapid entry of glucose into cells; this transfer of glucose out of the blood reduces the blood concentration of glucose, thereby removing the stimulus for insulin secretion, which returns to its previous level.

Figure 13-10 illustrates typical changes in plasma glucose and insulin concentrations following a normal carbohydrate-rich meal. Note the close association between the rising blood concentration (resulting from gastrointestinal absorption) and the plasma insulin increase induced by the glucose rise. The low postabsorptive values for plasma glucose and insulin concentrations are not the lowest attainable, and prolonged fasting induces even further reductions of both variables until insulin is barely detectable in the blood.

Although it was once believed that plasma glucose constitutes the sole control over insulin secretion, such is not the case, for insulin secretion is sensitive to numerous other types of input. One of the most important is the plasma concentration of certain amino acids, an elevated amino acid concentration causing enhanced insulin secretion. This is easily understandable since amino acid concentrations increase after eating, particularly after a high-protein meal. The increased insulin stimulates cell uptake of these amino acids. There is also some direct neural and hormonal control over insulin secretion; e.g., the gastrointestinal hormones described in Chap. 12 may stimulate insulin secretion.

Diabetes mellitus

The name diabetes, meaning syphon or running through, was used by the Greeks over 2,000 years ago to describe the striking urinary volume excreted by certain people. Mellitus, meaning sweet, distinguishes this urine from the large quantities of insipid urine produced by persons suffering from ADH deficiency (Chap. 11). This sweetness of the urine was first recorded in the seventeenth century, but in England the illness had long been called the pissing evil. Because of the marked weight loss despite huge

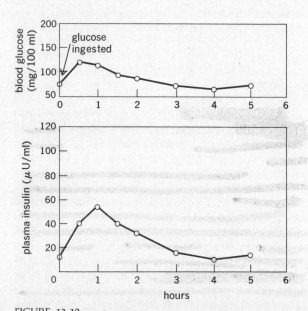

FIGURE 13-10
Blood concentrations of glucose and insulin following ingestion of 100 g of glucose. Study performed on normal human subjects. (*Adapted from Daughaday and Kipnis.*)

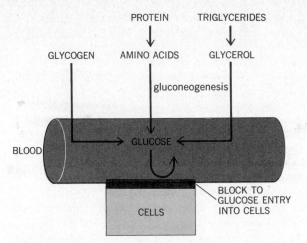

FIGURE 13-11
Factors which elevate blood glucose concentration in insulin deficiency.

food intake, the body's substance was believed to be dissolving and pouring out through the urinary tract, a view not far from the truth. In 1889, experimental diabetes was produced in dogs by surgical removal of the pancreas, and 32 years later, in 1921, Banting and Best discovered insulin.

A tendency toward diabetes can be inherited. We say tendency because diabetes often is not an all-or-none disease but may develop slowly, and overt signs may all but disappear with appropriate measures, e.g., weight reduction. The cause of diabetes is relative insulin deficiency. We have described how a lowered insulin concentration induces virtually all the metabolic changes characteristic of the fasting state (Fig. 13-6). The picture presented by an untreated diabetic is a gross caricature of this state (Fig. 13-11). The catabolism of triglyceride with resultant elevation of plasma fatty acids and ketones is an appropriate response because these substances must provide energy for the body's cells, which are prevented from taking up adequate glucose by the insulin deficiency. In contrast, glycogen and protein catabolism and the marked gluconeogenesis so important to maintain plasma glucose during fasting are completely inappropriate in the diabetic since his plasma glucose is already high because it cannot enter into cells. These reactions serve only to raise the plasma glucose still higher, with disastrous consequences. This paradox is the essence of the diabetic situation: cell starvation in the presence of a markedly elevated plasma glucose. Only the brain

is spared glucose deprivation since its uptake of glucose is not insulin-dependent. The obvious consequences of these catabolic processes is progressive loss of weight despite the increased food intake induced by constant hunger. The ingested carbohydrate and protein simply are converted to glucose, which further increases the plasma glucose to no avail.

The elevated plasma glucose of diabetes is per se relatively innocuous, but it induces changes in renal function of serious consequence. In Chap. 11, we pointed out that a normal person does not excrete glucose because all glucose filtered at the glomerulus is reabsorbed by the tubules. However, the elevated plasma glucose of diabetes may so increase the filtered load of glucose that the maximum tubular reabsorptive capacity is exceeded, and large amounts of glucose may be excreted. For the same reasons, large amounts of ketones may also appear in the urine. These urinary losses, of course, only aggravate the situation by further depleting the body of nutrients. Far worse, however, is the effect of these solutes on sodium and water excretion. In Chap. 11, we saw how tubular water reabsorption is a passive process induced by active solute reabsorption. In diabetes, the osmotic force exerted by unreabsorbed glucose and ketones holds water in the tubule, thereby preventing its reabsorption. For several reasons (the mechanisms are beyond the scope of this book) sodium reabsorption is also retarded. The net result is marked excretion of sodium and water, which leads, by the sequence of events shown in Fig. 13-12, to hypotension, brain damage, and death.

Another serious abnormality in diabetes is a markedly increased hydrogen-ion concentration, due primarily to the accumulation of ketone bodies, which as moderately strong acids generate large amounts of hydrogen ion by dissociation. The kidneys respond to this increase by excreting more hydrogen ion and are generally able to maintain balance fairly well, at least until the volume depletion described above interferes with renal function. What effect does the increased hydrogen-ion concentration have on respiration? A marked increase in ventilation occurs in response to stimulation of the medullary respiratory centers by hydrogen ion, and the resulting overexcretion of carbon dioxide further helps to keep the hydrogen-ion concentration below lethal limits.

Associated with diabetes are arteriosclerosis, small-vessel and nerve disease, susceptibility to infection, and a variety of other abnormalities. It was once argued that these secondary problems resulted from insulin deficiency, but this view is now seriously questioned, since it has been shown that they can be delayed or lessened

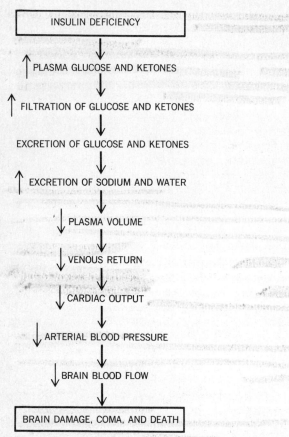

INSULIN DEFICIENCY

↑ PLASMA GLUCOSE AND KETONES

↑ FILTRATION OF GLUCOSE AND KETONES

EXCRETION OF GLUCOSE AND KETONES

↑ EXCRETION OF SODIUM AND WATER

↓ PLASMA VOLUME

↓ VENOUS RETURN

↓ CARDIAC OUTPUT

↓ ARTERIAL BLOOD PRESSURE

↓ BRAIN BLOOD FLOW

BRAIN DAMAGE, COMA, AND DEATH

FIGURE 13-12
Effects of severe untreated insulin deficiency on renal function.

but not abolished by treatment with insulin. Many experts now believe that defects other than insulin deficiency contribute to the total diabetic picture.

Presently being investigated is the observation that many diabetics actually manifest normal or elevated plasma insulin concentrations! Thus, in these cases, the diabetes is due not to an *absolute* insulin deficiency but to a *relative* deficiency. In other words, there must be some additional factor(s) present which antagonizes certain of the actions of insulin. Several hormones exert effects which oppose those of insulin, but in only a small number of patients can the disease be ascribed to excessive amounts of these hormones. An important clue may be the close relationship between obesity and development of diabetes in later life. Coupled with the fact that insulin exerts profound effects on adipose tissue, this association has stimulated much theorizing and research into the possible sequence of events. In any case, there is no question that simple weight reduction is frequently sufficient to eliminate the chemical manifestations of diabetes.

The treatment of diabetes is aimed at maintaining plasma glucose at a relatively normal value. The avoidance of concentrated sugars (candy, for example) helps, but the major therapy is administration of insulin, which must be given by injection since as a protein it is broken down by gastrointestinal enzymes. The dose must be determined carefully since an overdose abnormally lowers the plasma glucose concentration and causes brain damage, coma, and even death. Another type of therapy has proved useful; noninsulin drugs which can be taken by mouth act upon the islet cells to stimulate insulin secretion. Thus, therapy is actually accomplished with the patient's own insulin. Unfortunately, these drugs are effective in only a fraction of diabetics, the others having such severe islet-cell malfunction that drug stimulation of endogenous insulin secretion is not possible. Other types of oral medication are presently being introduced.

Epinephrine, glucagon, and growth hormone

Metabolic effects

We have devoted so much space to the physiology of insulin because of its central role in regulating organic metabolism. Now we must turn to the three other hormones which play primary roles in controlling the metabolic adjustments required for feasting or fasting: epinephrine, the major hormone secreted by the adrenal medulla; glucagon, secreted by the α cells of the pancreatic islets; and growth hormone (GH), from the anterior pituitary. Most of the major effects of these hormones on organic metabolism are opposed to those of insulin and are listed in Table 13-1. The intensity of any given effect varies between hormones, and there still is controversy concerning the relative importance of some of these effects, but, for present purposes, we need not go into these details. Note that the overall results of all three hormones' effects are just the opposite of those of insulin: elevation of plasma glucose and fatty acid concentrations. The latter facilitates cellular utilization of fatty acids, glucose sparing, and maintenance of plasma glucose.

TABLE 13-1
Major effects of epinephrine, glucagon, and growth hormone on carbohydrate and lipid metabolism

	Increased glycogen breakdown (glycogenolysis)	Increased liver gluconeogenesis	Increased breakdown of adipose-tissue triglyceride (fat mobilization)	Decreased glucose uptake by muscle and other tissues ("insulin antagonism")†
Result:	↑Plasma glucose	↑Plasma glucose	↑Plasma fatty acids and glycerol	↑Plasma glucose
Epinephrine‡	Yes	Yes	Yes	No
Glucagon	Yes	Yes	Yes	No
Growth hormone	No	Yes	Yes	Yes

†This term has been used traditionally to denote the fact that GH interferes with insulin's stimulatory effect on the glucose transport system, but the mechanism of this effect is controversial.

‡Activation of the sympathetic nerves to liver and adipose tissue produces effects virtually identical to those of epinephrine shown in the table.

Control of secretion

From a knowledge of these effects, one would logically suppose that the secretion of these hormones (and the activity of the sympathetic nerves to adipose tissue and hepatic cells) should be increased during the post-absorptive period and prolonged fasting, and such is the case. The stimulus is the same for all three, a decreased or decreasing plasma glucose concentration (although the receptors or pathways, or both, differ). The adaptive value of such reflexes is obvious; a decreasing plasma glucose stimulates increased release of the hormones which, by their effects on metabolism, serve to restore normal blood glucose levels and at the same time supply fatty acids for cell utilization. Conversely, an increased or increasing plasma glucose inhibits their secretion, thereby helping to return plasma glucose toward normal.

What are the receptors and pathways for these reflexes? That for glucagon is the simplest: The glucagon-secreting cells in the pancreas respond to changes in the glucose concentration of the blood perfusing the pancreas, no other nerves or hormones being involved. Thus the α and β cells of the pancreas constitute a push-pull system for regulating plasma glucose. As described in Chap. 6, epinephrine release is controlled entirely by the preganglionic sympathetic nerves to the adrenal medulla; the receptors initiating increased activity in these neurons (and the sympathetic pathways to the liver and adipose tissue) in response to changes in glucose are glucose receptors in the brain, probably in the hypothalamus (Fig. 13-13). As described in Chap. 7, growth hormone secretion is directly controlled by a

hypothalamic releasing factor. It is likely that the same brain glucose receptors described above for epinephrine communicate neurally with the hypothalamic neurons secreting GH-releasing factor so that a decreased glucose stimulates the release of GH-releasing factor, which then stimulates GH secretion by the anterior pituitary (Fig. 13-14); conversely, increased glucose inhibits the secretion of GH by decreasing the secretion of GH-releasing factor.

Thus far, the story is quite uncomplicated; the body produces two sets of hormones (insulin versus epinephrine, glucagon, and growth hormone) whose actions and controlling inputs are just the opposite of each other. However, we must now point out a complicating feature: A second major control of glucagon and growth hormone secretion is the plasma amino acid concentration (acting on the α cells and brain amino acid receptors, respectively), and in this regard the effect is identical rather than opposite to that for insulin; glucagon and GH secretion, like that of insulin, is strongly stimulated by a rise in plasma amino acid concentration such as occurs following a protein-rich meal. Thus, during absorption of a carbohydrate-rich meal containing little protein, there occurs an increase in insulin secretion alone, caused by the rise in plasma glucose, but during absorption of a low-carbohydrate–high-protein meal, all three hormones increase, under the influence of the increased plasma amino acid concentration. The usual meal is somewhere between these extremes and is accompanied by a rise in insulin and relatively little change in glucagon and GH since the simultaneous increases in blood glucose and amino acids counteract each other so far as glucagon

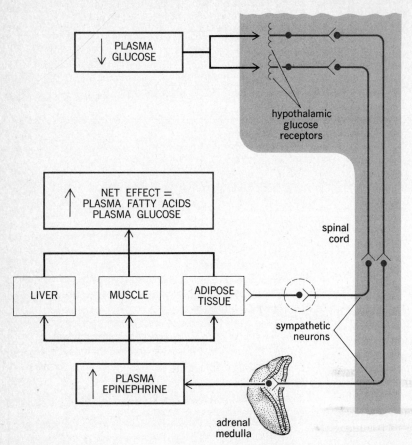

FIGURE 13-13
Control of epinephrine secretion and sympathetic nerves to adipose tissue by plasma glucose concentration. A decrease in plasma glucose stimulates the hypothalamic glucose receptors and, via the reflex chain shown, restores plasma glucose to normal while at the same time increasing plasma fatty acids.

and GH secretion is concerned. Of course, regardless of the type of meal ingested, the postabsorptive period is always accompanied by a rise in glucagon and GH secretion.

What is the adaptive value of the amino acid–glucagon and GH relationship? Imagine what might occur were glucagon and GH secretion not part of the response to a high-protein meal: Insulin secretion would be increased by the amino acids but, since little carbohydrate was ingested and therefore available for absorption, the increase in plasma insulin could cause a marked and sudden drop in plasma glucose. In reality, the rise in glucagon and GH secretion caused by the amino acids permits the hyperglycemic effects of these hormones to counteract the hypoglycemic actions of insulin, and the net result is a stable plasma glucose. Thus, a high-protein meal virtually free of carbohydrate can be absorbed with little change in plasma glucose despite a marked increase

in insulin secretion; moreover, the glucose derived from amino acids (under the influence of glucagon and GH) can be converted into fat for caloric storage (under the influence of insulin).

There is another, perhaps even more important adaptive value to having GH respond to amino acid level as well as glucose. In addition to the effects shown in Table 13-1, all of which oppose those of insulin, GH has another major effect (not shown in the table) identical to that of insulin, namely, stimulation of amino acid uptake and protein synthesis by most cells of the body. This constitutes one of the major growth-promoting effects of GH and is synergistic with that of insulin. Therefore, during absorption of a protein meal, insulin levels are high and GH levels are either high or unchanged (depending upon the carbohydrate content of the meal); accordingly, insulin and GH together strongly stimulate amino acid uptake and synthesis, i.e., growth (in a child) or mainte-

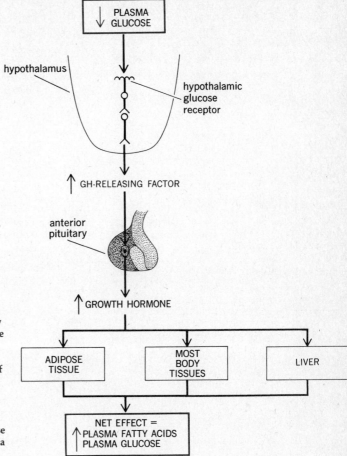

FIGURE 13-14
Control of growth hormone secretion by plasma glucose concentration. A decrease in plasma glucose stimulates the hypothalamic glucose receptors, which in turn stimulate the increased release of GH-releasing factor into the hypothalamus–anterior pituitary portal capillaries. The resulting stimulation of growth hormone secretion has the net effect of restoring normal plasma glucose while at the same time increasing plasma fatty acid concentration.

nance of protein stores in an adult. In contrast, during the postabsorptive state, net protein synthesis is impossible because of the decrease in plasma insulin occurring at this time, but the elevated GH may at least act as a partial brake on protein catabolism so that amounts of protein greater than that required for gluconeogenesis are not broken down.

Finally, it should be noted that the secretion of growth hormone, epinephrine, and glucagon is stimulated by a variety of nonspecific "stresses," both physical and emotional. This constitutes a mechanism for the mobilization of energy stores for coping with a fight-or-flight situation. The GH-induced interference with glucose uptake might, at first thought, seem a maladaptive response since it might seem to reduce the uptake of glucose in exercising ("fighting" or "flighting") muscle; however, for reasons still poorly understood, glucose uptake by exercising muscle is very rapid and independent of circulating hormones. Moreover, mobilized fatty acids are also available as a major energy source.

Summary

The effects of epinephrine (and the sympathetic nerves to liver and adipose tissue), glucagon, and growth hormone are mainly opposed, in various ways, to the effects of insulin. To a great extent, insulin may be viewed as the "hormone of plenty" and the others as "hormones of fasting" (although, as described, this oversimplistic view must be qualified in several important ways). Insulin is increased during the absorptive period and decreased during fasting. In contrast, the other three hormones are increased during the postabsorptive period (and show

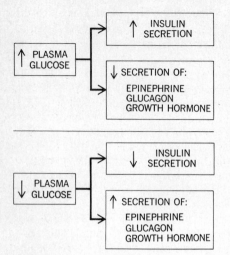

FIGURE 13-15
Summary of hormonal changes induced by changes in plasma glucose concentration.

varied responses to eating, depending upon the content of the meal). The influence of plasma glucose concentration is paramount in this regard, producing opposite effects on insulin secretion and on the secretion of the other three (Fig. 13-15).

In addition to these four hormones, there are others which have important effects on organic metabolism and the flow of nutrients. However, the secretion of these other hormones—cortisol, thyroxine, and the sex steroids—bears little or no relationship to the absorptive and postabsorptive states. This is not to say that they are secreted at constant rates but only that the rates are not determined by the state of glucose metabolism or by other indicators of the absorptive-postabsorptive state. Their physiology is described in detail in other sections of this book.

Control of growth

A human being begins as a single cell and by repeated cell divisions ultimately reaches a total in the trillions. In the process, cells differentiate and become grouped into specialized tissues. Obviously, growth and development are extraordinarily complex processes, and our discussion will attempt to present only a few scattered but important generalizations concerning their control.

A simple gain in body weight does not necessarily mean true growth since it may represent retention of either excess water or adipose tissue. In contrast, true growth usually involves lengthening of the long bones and increased cell division, but the real criteria are increased synthesis and accumulation of protein. Man manifests two periods of rapid growth (Fig. 13-16), one during the first 2 years of life, which is actually a continuation of rapid fetal growth, and the second during adolescence. Note, however, that total body growth may be a poor indicator of the rate of growth of specific organs (Fig. 13-16). We know relatively little about the control of many of these individual-organ growth patterns.

Another important implication of differential growth rates is that the so-called critical periods of development vary from organ to organ. Thus, a period of severe malnutrition during infancy when the brain is growing extremely rapidly may produce stunting of brain development which is irreversible, whereas reproductive organs would be little affected.

External factors influencing growth

An individual's growth capacity is genetically determined, but there is no guarantee that the maximum capacity will be attained. Adequacy of food supply and freedom from disease are the primary external factors determining growth. Lack of sufficient amounts of any of the essential amino acids, essential fatty acids, vitamins, or minerals interferes with growth, and total protein and total calories must be adequate. No matter how much protein is ingested, growth cannot be normal if caloric intake is too low, since the protein is simply oxidized for energy.

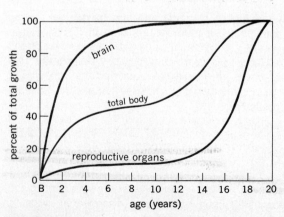

FIGURE 13-16
Rate of growth.

Many studies have demonstrated that the growth-inhibiting effects of malnutrition are most profound when they occur very early in life. Indeed, malnutrition during infancy may cause irreversible stunting of total body growth (and brain development, as mentioned above). The individual seems "locked in" to a younger developmental age. A related question is whether maternal malnutrition may cause retardation of growth in the fetus. One reason for the great importance of this question is that low birth weight is strongly associated with increased infant mortality; accordingly, it is possible that prenatal malnutrition might cause increased numbers of prenatal and early postnatal deaths. The possibility of irreversible stunting of brain development due to prenatal malnutrition is another important consideration. Present data for human beings are inadequate to answer these questions.

Conversely, the profound stimulatory effects of good nutrition are well illustrated by the present teenage Japanese, who already seem to tower above their parents. On the other hand, it is important to realize that one cannot stimulate growth beyond the genetically determined maximum by eating more than adequate vitamins, protein, or total calories; this produces obesity, not growth.

Sickness can stunt growth, most likely because protein catabolism is enhanced by cortisol and other factors. If the illness is temporary, upon recovery the child manifests a remarkable growth spurt which rapidly brings him up to his normal growth curve. The mechanisms which control this important phenomenon are unknown, but the process illustrates the strength and precision of a genetically determined sequence.

Hormonal influences on growth

The hormones most important to human growth are growth hormone, thyroxine, insulin, androgens, and estrogen, which exert widespread effects. In addition, ACTH, TSH, prolactin, and FSH and LH selectively influence the growth and development of their target organs, the adrenal cortex, thyroid gland, breasts, and gonads, respectively.

Growth hormone   Removal of the pituitary in young animals arrests growth. Conversely, administration of large quantities of growth hormone to young animals causes excessive growth. When excess growth hormone is given to adult animals after the actively growing cartilaginous areas of the long bones have disappeared, it cannot lengthen the bones further, but it does produce the disfiguring bone thickening and overgrowth of other organs known as *acromegaly*. These experiments have spontaneously occurring human counterparts, as shown in Fig. 13-17. Thus growth hormone is essential for normal growth and in abnormally large amounts can cause excessive growth.

The effects of growth hormone on organic metabolism have already been described; its growth-promoting effects are due primarily to its ability to stimulate protein synthesis. This it does by increasing membrane transport of amino acids into cells and also by stimulating the synthesis of RNA, two events essential for protein synthesis. Growth hormone also causes large increases of mitotic activity and cell division, the other major components of growth.

The effects on bone are dramatic. Bone is a living tissue consisting of a *protein matrix,* upon which *calcium salts* (particularly phosphates) are deposited. The cells responsible for laying down this matrix are *osteoblasts.* Growth of a long bone depends upon actively proliferating layers of cartilage (the *epiphyseal plates*) at the ends of the bone. The osteoblasts at the edge of the epiphyseal plates convert the cartilaginous tissue into bone while new cartilage is simultaneously formed in the plates (Fig. 13-18). Growth hormone promotes this lengthening by stimulating protein synthesis in both the cartilaginous center and bony edge of the epiphyseal plates as well as by increasing the rate of osteoblast mitosis. It has now become clear that these effects on bone are mediated not by growth hormone, per se, but by a substance *somatomedin,* which growth hormone causes to be released by the liver.

A recent finding of considerable interest is that African Pygmies have normal blood concentrations of growth hormone but show virtually no metabolic response to this hormone. Thus, their small stature is due to a genetically determined nonresponsiveness to growth hormone.

Thyroxine   Infants and children with deficient thyroid function manifest retarded growth, which can be restored to normal by administration of physiological quantities of thyroxine. Administration of excess thyroxine, however, does not cause excessive growth (as was true of growth hormone) but marked catabolism of protein and other nutirents, as willl be explained in the section on energy balance. The essential point is that normal amounts of thyroxine are necessary for normal growth, the most likely explanation apparently being that thyroxine somehow promotes the effects of growth hormone on protein synthesis; certainly the absence of thyroxine significantly reduces the ability of growth hormone to stimulate amino

FIGURE 13-17

Progression of acromegaly. (*Top left*) Normal, age nine years; (*top right*) age sixteen years, with possible early coarsening of features; (*bottom left*) age thirty-three years, well-established acromegaly; (*bottom right*) age fifty-two years, end stage, acromegaly with gross disfigurement. [*From William H. Daughaday, The Adenohypophysis, in Robert H. Williams (ed.), "Textbook of Endocrinology," 4th ed., p. 74, fig. 2-28, W. B. Saunders Company, Philadelphia, 1968.*]

acid uptake and RNA synthesis. Thyroxine also plays a crucial role in the closely related area of organ development, particularly that of the central nervous system. Hypothyroid infants (*cretins*) are mentally retarded, a defect that can be completely repaired by adequate treatment with thyroid hormone although if the infant is untreated for long, the developmental failure is largely irreversible. The defect is probably due to failure of nerve myelination which occurs as a result of thyroid deficiency.

Insulin It should not be surprising that adequate amounts of insulin are necessary for normal growth since insulin is, in all respects, an anabolic hormone. Its effects on amino acid uptake and protein are particularly important in favoring growth.

Androgens and estrogen In Chap. 14 we shall describe in detail the various functions of the sex hormones in directing the growth and development of the sexual organs and the obvious physical characteristics which distinguish male from female. Here we are concerned only with the effects of these hormones on general body growth.

Sex hormone secretion begins in earnest at about the age of eight to ten and progressively increases to reach a plateau within 5 to 10 years. The testicular hormone, testosterone, is the major male sex hormone, but other androgens similar to it are also secreted in significant amounts by the adrenal cortex of both sexes. Females manifest a sizable increase of adrenal androgen secretion during adolescence. However, the adrenal androgens are not nearly so potent as testosterone. During adolescence the large increases in secretion of estrogen, the dominant female sex hormone, are virtually limited to the female. Thus the relative quantities of androgen and estrogen are very different between the sexes.

Androgens strongly stimulate protein synthesis in many organs of the body, not just the reproductive organs, and the adolescent growth spurt in both sexes is due, at least in part, to these anabolic effects. Similarly, the increased muscle mass of men compared with women may reflect their greater amount of more potent androgen. Androgens stimulate bone growth but also ultimately stop bone growth by inducing complete conversion of the epiphyseal plates. This accounts for the pattern seen in adolescence, i.e., rapid lengthening of the bones culminating in complete cessation of growth for life, and explains several clinical situations: (*1*) Unusually small children treated before puberty with large amounts of testosterone may grow several inches very rapidly but then stop completely and (*2*) eunuchs may be very tall because bone growth, although slower, continues much longer due to persistence of the epiphyseal plates.

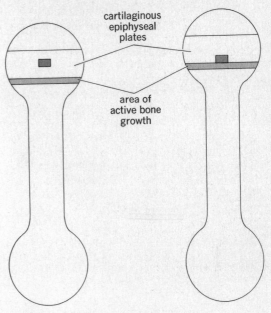

■ = "MARKER" PIECE OF CARTILAGE

cartilaginous epiphyseal plates

area of active bone growth

FIGURE 13-18
Diagram of long-bone lengthening. The edge of the cartilaginous plate is converted into bone, but the cells of the plate keep producing more cartilage to take its place. Thus the marker, once in the center of the plate, is now at the edge and about to be converted into bone; yet the marker itself has not moved.

Estrogen profoundly stimulates growth of the female sexual organs and sexual characteristics during adolescence (Chap. 14), but unlike the androgens, it has relatively little direct anabolic effect on nonsexual organs and tissues. Present evidence suggests that the major function of estrogen in the adolescent growth spurt and ultimate closure of the epiphyses is to stimulate the adrenal secretion of androgens which then directly mediate these responses.

Compensatory hypertrophy

We have dealt thus far only with growth during childhood. During adult life, maintenance of the status quo is achieved by the mechanisms described earlier in this chapter. In addition, a specific type of organ growth, known as *compensatory hypertrophy,* can occur in many human organs and is actually a type of regeneration. For example, within 24 hr of the surgical removal of one kidney, the cells of the other begin to manifest increased mitotic activity, ultimately growing until the total mass approaches the initial mass of the two kidneys combined.

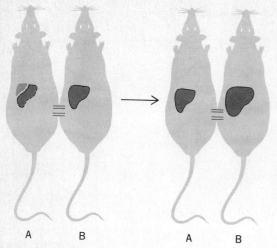

FIGURE 13-19
Evidence for hormone participation in compensatory hypertrophy. Removal of a large portion of rat A's liver induces compensatory hypertrophy in rat A and liver enlargement in parabiotic rat B.

What causes this compensatory growth? It certainly does not depend upon the nerves to the organ since it still occurs after their removal or destruction. Nor has it been possible to attribute it to any known hormone. Several types of experiments indicate, however, that some un-identified blood-borne agent is responsible. For example, if one removes 75 percent of the liver from one of two parabiotic rats, i.e., surgically created Siamese twins, the livers of both rats increase in size (Fig. 13-19).

SECTION B
REGULATION OF TOTAL
BODY ENERGY BALANCE

Basic concepts of energy expenditure and caloric balance

The breakdown of organic molecules liberates the energy locked in their intramolecular bonds (Chap. 3). This is the source of energy utilized by cells in their performance of the various forms of biological work (muscle contraction, active transport, synthesis of molecules, etc.). As described in Chap. 3, the first law of thermodynamics states that energy can neither be created nor destroyed but can be converted from one form to another. Thus, internal energy liberated (ΔE) during breakdown of an

organic molecule can either appear as heat (H) or be used for performing work (W).

$$\Delta E = H + W$$

In all animal cells, most of the energy appears immedi-ately as heat, and only a small fraction is used for work. (As described in Chap. 3, the energy used for work must first be incorporated into molecules of ATP, the subse-quent breakdown of which serves as immediate energy source for the work.) It is essential to realize that the body is not a heat engine since it is totally incapable of con-verting heat into work. The heat is, of course, valuable for maintaining body temperature.

It is customary to divide biological work into two general categories: (*1*) *external work*, i.e., movement of external objects by contracting skeletal muscles, and (*2*) *internal work*, which comprises all other forms of biological work, including skeletal muscle activity not moving exter-nal objects. As we have seen, most of the energy liber-ated from the catabolism of nutrients appears immediately as heat, only a small fraction being used for performance of external or internal work. What may not be obvious is that all internal work is ultimately transformed into heat except during periods of growth (Fig. 13-20). Several examples will illustrate this essential point:

1 Internal work is performed during cardiac contraction, but this energy appears ultimately as heat generated by the resistance (friction) to flow offered by the blood vessels.

2 Internal work is performed during secretion of HCl by the stomach and $NaHCO_3$ by the pancreas, but this work appears as heat when the H^+ and HCO_3^- react in the small intestine.

3 The internal work performed during synthesis of a plasma protein is recovered as heat during the inevitable

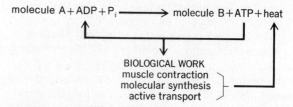

FIGURE 13-20
General pattern of energy liberation in a biological system. Most of the energy released when nutrients, such as glucose, are broken down appears immediately as heat. A smaller fraction goes to form ATP, which can be subsequently broken down and the released energy coupled to biological work. Ultimately, the energy which performs this work is also completely converted into heat.

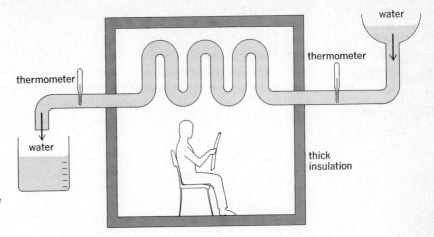

FIGURE 13-21
Direct method for measuring metabolic rate. The water flowing through the calorimeter carries away the heat produced by the person's body. The amount of heat is calculated from the total volume of water and the difference between inflow and outflow temperatures.

catabolism of the protein, since, with few exceptions, all bodily constituents are constantly being built up and broken down. However, during periods of net synthesis of protein, fat, etc., energy is stored in the bonds of these molecules and does not appear as heat.

Thus, the total energy liberated when organic nutrients are catabolized by cells may be transformed into body heat, appear as external work, or be stored in the body in the form of organic molecules, the latter occurring only during periods of growth (or net fat deposition in obesity). The total energy expenditure of the body is therefore given by the equation

Total energy expenditure =
heat produced + external work + energy storage

The units for energy are kilocalories (Chap. 3), and total energy expenditure per unit time is called the *metabolic rate.*[1]

In man the metabolic rate can be measured directly or indirectly. In either case the measurement is much simpler if the person is fasting and at rest; total energy expenditure then becomes equal to heat production since energy storage and external work are eliminated. The direct method is simple to understand but difficult to perform; the subject is placed in a *calorimeter,* an instrument large enough to accommodate people, and his heat production is measured by the temperature changes in the calorimeter (Fig. 13-21). This is an excellent method in that it measures heat production directly, but calorimeters are found in only a few research laboratories; ac-

cordingly, a simple, indirect method has been developed for widespread use.

Using the indirect procedure, one simply measures the subject's oxygen uptake per unit time (by measuring total ventilation and P_{O_2} of inspired and expired air). From this value one calculates heat production based on the fundamental principle (Fig. 13-22) that the energy liberated by the catabolism of foods in the body must be the same as when they are catabolized outside the body. We know precisely how much heat is liberated when 1 liter of oxygen is consumed in the oxidation of fat, protein, or carbohydrate outside the body; this same quantity of heat must be produced when 1 liter of oxygen is consumed in the body. Fortunately, we do not need to know precisely which type of nutrient is being oxidized internally, because the quantities of heat produced per liter of oxygen consumed are reasonably similar for the oxidation of fat, carbohydrate, and protein, and average 4.8 kcal/liter of oxygen. When more exact calculations are required, it is possible to estimate the relative quantity of each nutrient. Figure 13-22 presents values obtained by the indirect method for a normal fasted resting adult male. This man's rate of heat production is approximately equal to that of a single 100-W bulb.

Determinants of metabolic rate

Since many factors cause the metabolic rate to vary (Table 13-2), when one wishes to compare metabolic rates of different people, it is essential to control as many of the variables as possible. The test used clinically and experimentally to find the *basal metabolic rate* (BMR) tries to accomplish this by standardizing conditions: The sub-

[1]In the field of nutrition, 1 Calorie implies, by convention, 1 *large calorie,* which is actually 1 kilocalorie.

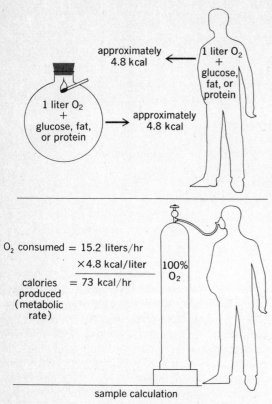

$$O_2 \text{ consumed} = 15.2 \text{ liters/hr}$$
$$\times 4.8 \text{ kcal/liter}$$
calories
produced $= 73 \text{ kcal/hr}$
(metabolic
rate)

sample calculation

FIGURE 13-22
Indirect method for measuring metabolic rate. The calculation depends upon the basic principle that when 1 liter of oxygen is utilized in the oxidation of organic nutrients approximately 4.8 kcal is liberated.

ject is at mental and physical rest, in a room at comfortable temperatures, and has not eaten for at least 12 hr. These conditions are arbitrarily designated basal, the metabolic rate during sleep being actually less than the BMR. The measured BMR is then compared with previously determined normal values for a person of the same weight, height, age, and sex.

BMR is often appropriately termed the metabolic *cost of living.* Under these conditions, most of the energy is expended, as might be imagined, by the heart, liver, kidneys, and brain. Its magnitude is related not only to physical size but to age and sex as well. The growing child's resting metabolic rate, relative to his size, is considerably higher than the adult's because he expends a great deal of energy in net synthesis of new tissue. On the other end of the age scale, the metabolic cost of living

gradually decreases with advancing age, for unknown reasons. The female's resting metabolic rate is generally less than that of the male (even taking into account size differences) but increases markedly, for obvious reasons, during pregnancy and lactation. The greater demands upon the body by infection or other disease generally increase total energy expenditure; moreover the presence of fever directly stimulates metabolic reaction rates and increases metabolic rate.

The ingestion of food also increases the metabolic rate, as shown by measuring the oxygen consumption or heat production of a resting man before and after eating; the metabolic rate is 10 to 20 percent higher after eating. This effect of food on metabolic rate is known as the *specific dynamic action* (SDA). Protein gives the greatest effect, carbohydrate and fat much less. The cause of SDA is not what one might expect, namely, the energy expended in the digestion and absorption of ingested food. These processes account for only a small fraction of the increased metabolic rate: Intravenous administration of amino acids produces almost the same SDA effect as oral ingestion of the same material. Most of the increased heat production appears to be secondary to the processing of exogenous nutrients by the liver, since it does not occur in an animal whose liver has been removed. On an average diet, SDA contributes to the total metabolic rate approximately one-tenth as much as the basal metabolism does. In contrast to eating, prolonged fasting causes a decrease in metabolic rate. This is due, in part, simply to reduction of body mass, but even when expressed on a per weight basis, metabolic rate is reduced.

TABLE 13-2
Factors affecting the metabolic rate

Age
Sex
Height, weight, and surface area
Growth
Pregnancy, menstruation, lactation
Infection or other disease
Body temperature
Recent ingestion of food (SDA)
Prolonged fasting
Muscular activity
Emotional state
Sleep
Environmental temperature
Circulating levels of various hormones, especially epinephrine and thyroxine

The mechanism is unclear, but the adaptive value of this change may be considerable since it decreases the amount of nutrient stores which must be catabolized each day.

All these influences on metabolic rate are small compared with the effects of *muscular activity* (Table 13-3). Even minimal increases in muscle tone significantly increase metabolic rate, and severe exercise may raise heat production more than fifteenfold. Changes in muscle activity also explain part of the effects on metabolic rate of sleep (decreased muscle tone), reduced environmental temperature (increased muscle tone and shivering), and emotional state (unconscious changes in muscle tone).

Metabolic rate is strongly influenced by the hormones *epinephrine* and *thyroxine.* The intravenous injection of epinephrine may promptly increase heat production by more than 30 percent. As we have seen, epinephrine has powerful effects on organic metabolism, and its calorigenic, i.e., heat-producing, effect is probably related to its stimulation of glycogen and triglyceride catabolism since ATP splitting and energy liberation occur in both the breakdown and the subsequent resynthesis of these molecules. Regardless of the mechanism, whenever epinephrine secretion is stimulated, the metabolic rate rises. This probably accounts for part of the greater heat production associated with emotional stress, although increased muscle tone is also contributory.

Thyroxine also increases the oxygen consumption and heat production of most body tissues, a notable exception being the brain. In contrast to epinephrine, this calorigenic effect does not begin for 6 to 12 hr but lasts for many days, even after a single injection. So powerful is this effect that long-term excessive thyroxine, as in patients with hyperthyroidism, induces a host of effects secondary to the hypermetabolism which well illustrate the interdependence of bodily functions. The increased metabolic demands markedly increase hunger and food intake; the greater intake frequently remains inadequate to meet the metabolic needs, and net catabolism of endogenous protein and fat stores leads to loss of body weight; excessive loss of skeletal muscle protein results in muscle weakness; catabolism of bone protein weakens the bones and liberates large quantities of calcium into the extracellular fluid, resulting in increased plasma and urinary calcium; the hypermetabolism increases the requirement for vitamins, and vitamin deficiency diseases may occur; respiration is increased to supply the required additional oxygen; cardiac output is also increased and, if prolonged, the enhanced cardiac demands may cause heart failure; the greater heat production activates heat-

TABLE 13-3
Energy expenditure during different types of activity for a 70-kg man

Form of activity	kcal/hr
Awake, lying still	77
Sitting at rest	100
Typewriting rapidly	140
Dressing or undressing	150
Walking level at 2.6 mi/hr	200
Sexual intercourse	280
Bicycling on level, 5.5 mi/hr	304
Walking 3 percent grade at 2.6 mi/hr	357
Sawing wood or shoveling snow	480
Jogging (5.3 mi/hr)	570
Rowing 20 strokes/min	828
Maximal activity (untrained)	1,440

dissipating mechanisms, and the patient suffers from marked intolerance to warm environments. These are only a few of the many results induced by thyroxine's calorigenic effect. The important effects of thyroxine relating to growth and development, described earlier, appear to be quite distinct from the calorigenic effect. The mechanisms by which thyroxine exerts this profound effect on oxygen consumption and heat production are presently unclear.

One adaptive value of this ability of epinephrine and thyroxine to drive metabolism lies in the role these hormones play in the regulation of body temperature (described in detail in a subsequent section). Finally, we must at least mention the intimate relationship between thyroxine and the sympathetic nervous system. For example, in cases of hyperthyroidism, the administration of epinephrine (or norepinephrine) causes much more profound effects on the cardiovascular system and metabolism than in a normal person; conversely, after administration of drugs which block the sympathetic nervous system, even large doses of thyroxine produce only minor effects. The precise nature of this relationship is not known, but it provides an excellent example of the importance of hormone-hormone interactions.

Determinants of total body caloric balance

Using the basic concepts of energy expenditure and metabolic rate as a foundation, we can consider total body caloric-fuel balance in much the same way as any other balance, i.e., in terms of input and output. The laws of thermodynamics dictate that, in the steady state, the total caloric expenditure of the body equals total body caloric-fuel input. We have already identified the ultimate forms of energy expenditure: internal heat production, external work, and net molecular synthesis (energy storage). The source of input, of course, is the energy contained in ingested food. Therefore the caloric-balance equation is

$$\text{Food energy intake} = \begin{array}{c} \text{internal} \\ \text{heat} \\ \text{produced} \end{array} + \begin{array}{c} \text{external} \\ \text{work} \end{array} + \begin{array}{c} \text{energy} \\ \text{storage} \end{array}$$

Our equation includes no term for loss of fuel from the body via urinary excretion of nutrients. In a normal person, almost all the carbohydrate, amino acids, and lipids filtered at the glomerulus are reabsorbed by the tubules, so that the kidneys play no significant role in the regulation of caloric balance. In certain diseases, however, the most important being diabetes, urinary losses of organic molecules may be quite large and would have to be included in the equation. In all normal persons very small losses occur via the urine, feces, and as sloughed hair and skin, but we can ignore them as being negligible.

As predicted by this caloric-balance equation, three states are possible:

Food intake = internal heat production
+ external work
(body weight constant)

Food intake > internal heat production
+ external work
(body weight increases)

Food intake < internal heat production
+ external work
(body weight decreases, i.e., negative energy storage)

In most adults, body weight remains remarkably constant over long periods of time, implying that precise physiological regulatory mechanisms operate to control (1) food intake or (2) internal heat production plus external work or (3) both. Actually, all these variables are subject to control in man, but the amount of food intake is the dominant factor. Control mechanisms for heat production are aimed primarily at regulating body temperature, rather than total caloric balance. For example, when someone is cold, his body produces additional heat by shivering even if he is starving; conversely, a fat person is not automatically impelled by his hypothalamus to run around the block—quite the reverse in most cases. It is essential to understand that as shown by the caloric-balance equation, an individual's degree of activity, i.e., heat production plus external work, *is* one of the essential determinants of total body energy balance, but its automatic physiological control is not aimed primarily at achieving such a balance. Moreover, a man's total activity generally reflects the kind of work he does, his inclination toward sports, etc. The important generalization is that food intake is the major factor being automatically controlled so as to maintain caloric balance and constant body weight. To alter the two examples cited above: When exposure to cold or running around the block causes increased energy expenditure, the individual automatically increases his food intake by an amount sufficient to match the additional energy expended (this example, however, will be qualified later in the section on obesity).

Despite the fact that the control of food intake is the major mechanism by which caloric balance is *normally* regulated, it should be emphasized that balance can also be achieved in the presence of a markedly deficient intake by means of several physiological adaptations. These responses to partial chronic malnutrition offer an excellent summary of many of the factors which determine metabolic rate. Perhaps the most striking experiment designed to study this problem was the reduction of caloric intake in normal volunteers from 3492 kcal/day to 1570 kcal/day for 24 weeks (these men were soldiers performing considerable physical activity). Initially, weight loss was very rapid and ultimately averaged 24 percent of the men's original body weights, but the important fact is that the weights did stabilize, i.e., caloric balance was reestablished despite the continued caloric intake of 1570 kcal/day. Clearly, metabolic rate must have decreased an identical amount. This was due to a reduction in both BMR (40 percent) and physical activity as the men became apathetic and reluctant to engage in any activity. The decrease in BMR was due both to a decreased total body mass and to a decrease in metabolism of the liver and other organs out of proportion to their changes in weight (the mechanism of the latter change is unknown but seems not to be due to changes in circulating epinephrine or thyroxine). Thus, people can adjust to marked caloric restriction but only at a price. Moreover, in children some of the changes caused by malnutrition are irreversible.

Control of food intake

Hypothalamic integration centers

The structures primarily concerned in the control of food intake are several clusters of nerve cells (nuclei) in the hypothalamus. The general pattern has been elucidated

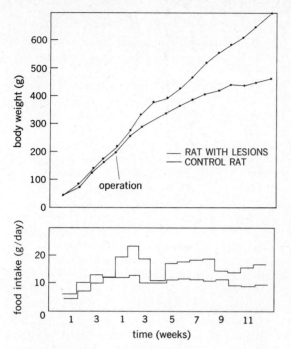

FIGURE 13-23
Effect of midline hypothalamic destruction upon food intake and weight gain in growing rats. (*Adapted from Brobeck et al.*)

by a series of experiments in which various areas of the hypothalamus were either stimulated or destroyed. Destruction of small areas in the midline *ventromedial* portion of the hypothalamus induces profound overeating and obesity, as shown in Fig. 13-23. In contrast, small lesions in the outer or *lateral hypothalamus* inhibited eating and food-intake behavior of any kind completely (Fig. 13-24); these animals starved to death, never touching the food in their cages. Finally, destruction of the outer hypothalamic areas in animals which previously had undergone midline hypothalamic destruction caused these obese rats to stop eating. Experiments in which the same hypothalamic areas were stimulated electrically in unanesthetized animals produced virtually the opposite results; lateral hypothalamic stimulation induced not only eating, but licking, chewing, salivating, and active searching for food; ventromedial stimulation caused cessation of eating even in a previously fasted animal. These types of experiment have led to the present concept of hypothalamic food intake control diagramed in Fig. 13-25. The lateral hypothalamus contains a *feeding center* which stimulates the efferent output controlling both the final motor acts

of eating and such associated behavior as food seeking; the ventromedial hypothalamic neurons can, via synaptic input, inhibit the activity of this feeding center; thus, eating proceeds unless the midline centers are stimulated so as to inhibit the outer centers; the stimulatory inputs to the midline centers are appropriately known as *satiety signals,* and the centers themselves are termed the *satiety centers.* Figure 13-25 illustrates in these terms the behavior of the experimental rats described above.

Satiety signals

It should be evident that the hypothalamic centers involved in control of food intake serve only as integrating centers processing afferent input and controlling efferent output. The afferent input must provide the critical information about the body's need for food. Although obviously food intake is often stimulated by the sight or smell of food, our present thinking is that food intake is basically a tonic process which continues unless turned off by an input signaling satiety. In a sense, we start eating not because we become hungry but because we stop being satiated.

Early observations that hunger was generally associated with contractions of the empty stomach and that gastric distension could lessen hunger led to the hypothesis that hunger and satiety were signaled by afferent pathways from the stomach and other areas of the gastrointestinal tract. However, experiments have since proved that complete denervation of the upper gastrointestinal tract does not interfere with normal mainte-

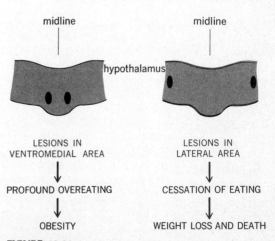

FIGURE 13-24
Effects on food intake of damaging different portions of the hypothalamus.

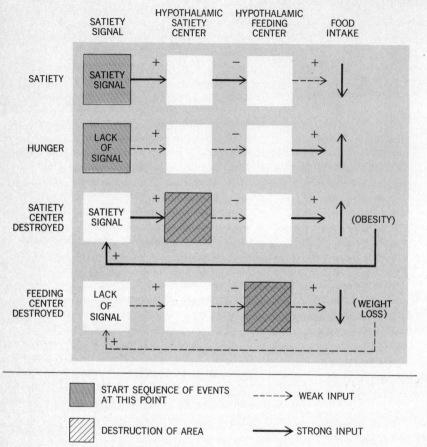

FIGURE 13-25
Basic patterns of food intake during
satiety, hunger, and hypothalamic
lesions. Food intake is reduced when
satiety signals stimulate the
hypothalamic satiety centers to inhibit
the hypothalamic feeding centers.
Obesity results when the satiety center is
destroyed, so that satiety signals are no
longer effective in inhibiting the feeding
center. Food intake is stimulated
normally by the absence of satiety
signals (hunger). (*Adapted from Tepperman.*)

nance of energy balance, and thus gastrointestinal signals cannot constitute the major long-term regulators of
food intake although they may play a modifying role. Nor
does it seem possible, on purely theoretical grounds, that
a bulk-detecting system could maintain energy balance,
since the caloric content of food may bear no relationship
to its bulk.

   This leads us to the perplexing problem of what
characteristics a receptor and environmental signal must
have in order to detect total body energy content, which,
after all, is the variable actually being regulated. Wide
experimentation has led to three quite different but by
no means mutually exclusive hypotheses, termed the
glucostatic, lipostatic, and thermostatic theories. Most
experts concede that none of them can explain all the
observations and that probably all their postulated pathways contribute to the overall control of food intake.

Glucostatic theory   There is no doubt that some type of
glucose receptor exists in the brain. Present evidence
suggests that it is in the ventromedial hypothalamus and
is perhaps the same group of neurons which constitute
the satiety center. It is also possible that these hypothalamic glucose receptors are the same ones which
initiate the reflexes leading to release of epinephrine and
growth hormone when plasma glucose is reduced. For
reasons already discussed, the plasma glucose concentration and, more specifically, the rate of cellular glucose
utilization increase during or after eating and decrease
during fasting. Detection of greater glucose utilization
could signal satiety and inhibit eating (Fig. 13-26). Conversely, fasting would decrease glucose utilization, remove the input signaling satiety, and thereby promote
eating. Note that this description is in terms of glucose
utilization, not plasma glucose itself. Apparently, the

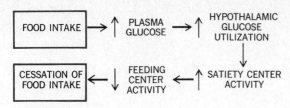

FIGURE 13-26
Glucostatic theory of food-intake control. Greater glucose utilization is the satiety signal.

receptors detect changes in their own rate of glucose uptake and utilization which is, in turn, dependent not only upon the plasma glucose concentration but also upon plasma insulin levels (this tiny area of the hypothalamus, unlike most brain areas, is sensitive to insulin); thus, patients with diabetes mellitus have increased appetite despite their increased plasma glucose. A detailed critique of this theory is beyond the scope of this book; suffice it to say that despite its attractiveness and the almost certain participation of brain glucose receptors in energy balance, several types of experiments indicate that it cannot be the sole controller involved.

Lipostatic theory Basically this theory postulates that an ideal indicator of total body energy content would be a substance released from fat stores (adipose tissue) in direct proportion to their total mass. Thus, a positive energy balance would increase the amount of adipose tissue, which in turn would signal satiety. Such a mechanism would constitute an excellent long-term regulator, but it must be confessed that the concept is still highly theoretical with little experimental evidence either to support or disprove it.

The idea that some blood-borne humoral agent plays a role is certainly suggested by experiments on parabiotic rats. The capillary connections between such animals mean that hormones or parahormones secreted by one animal cross over and affect the other, whereas the total amount of nutrients crossing is negligible. When the hypothalamic satiety center is damaged in only one rat, he overeats and grows obese but his twin progressively eats less and becomes thin. The satiety-signaling hormone suggested by this experiment remains unidentified.

Thermostatic theory The specific dynamic action of food, i.e., the increase in metabolic rate induced by eating, tends to raise body temperature, and it seems likely that temperature elevation may constitute a satiety signal.

Such a mechanism would also be consistent with the fact that people eat more in colder climates than in warm ones.

Psychological components of food-intake control

Thus far we have described the control of food intake in a manner identical to that for any other homeostatic control system, such as those regulating sodium or potassium. Thus, the automatic "involuntary" nature of the system was emphasized despite the obvious fact that there are psychological correlates of hunger and satiety. The problem may be stated simply: Are the psychological correlates merely epiphenomena which accompany the regulation but do not cause it, or are they essential components of the regulatory process? This same problem recurs in every case in which behavioral voluntary actions (obtaining and eating food, for example) are the ultimate responses in a homeostatic system. Thus, thirst, temperature regulation, salt appetite, and specific appetites for other nutrients all involve voluntary consummatory acts.

We describe here some of the interesting food-intake data that have revealed how overly simplistic the homeostatic system described above really is. Let us return to the animals with hypothalamic lesions and look more closely. First, we find that the rats with ventromedial lesions do not maintain their gross overeating (hyperphagia) indefinitely; rather, their food intake gradually returns toward normal so that their body weight ultimately stabilizes, albeit at a much higher value than normal. Thus, they regain the ability to regulate their total caloric content but at a higher value. Perhaps even more revealing are the data relative to their "motivation" for food and their responses to different foods. The paradoxical facts are that these hyperphagic animals are relatively unwilling to work for their food (run a maze, press a bar, etc.) and are extremely finicky eaters; for example, addition of a small amount of bitter-tasting quinine to their food may cause them to stop eating entirely although such a manipulation has no effect on control animals. Conversely, addition of a small amount of sweetener to their food causes a weight-stabilized animal to launch into another prolonged period of overeating.

The story on rats with lateral hypothalamic lesions is also more complex than originally described. If these animals, which would die from starvation if left alone, are force-fed by stomach tube for several months, they ultimately regain their ability to eat (and drink) spontaneously and regulate their body weights, albeit at a rate somewhat below normal. During the recovery period (and persisting

TABLE 13-4
Summary of factors which influence hypothalamic centers controlling food intake

Plasma glucose
Total body adipose tissue
Body temperature
State of gastrointestinal distension
Psychological, social, and economic factors

after recovery to a considerable degree) these rats will eat only moistened and highly palatable foods; it is as though they are driven to the food by taste, smell, or other reinforcing inputs rather than by the homeostatic inputs described above.

Clearly, these data demand a more complex view of the function of the ventromedial and lateral hypothalamic nuclei and of the homeostatic reflexes controlling food intake than that described above. Although total caloric balance unquestionably reflects, in large part, the reflex input from some combination of glucostats, lipostats, and thermostats, it also is strongly influenced by the reinforcement (both positive and negative) of such things as smell, taste, texture, psychological associations, etc. Thus, the behavioral concepts of reinforcement, drive, and motivation to be described in a later chapter must be incorporated into any comprehensive theory of food-intake control. Obviously, these psychological factors having little to do with energy balance are of very great importance in obese persons. It should be emphasized, however, that most people whose obesity is ascribed to psychological factors do not continuously gain weight; their automatic homeostatic control mechanisms are operative but maintain total body energy content at supranormal levels. (See Table 13-4.)

Specific-nutrient appetites

An important question distinct from the regulation of total caloric balance is whether there exist homeostatic mechanisms for control of intake of specific nutrients. In other words, does a person crave and seek out foods containing a specific essential nutrient (thiamine, niacin, etc.) when his body is deficient in that substance? There unquestionably exist such regulatory systems for water (*thirst*) and sodium (*salt appetite*), as described in Chap. 11, but present evidence indicates that such may not be the case for essential organic substances. Rather, behavior which appears to be the result of specific appetite actually is the result of nonspecific reinforcing effects of sickness and cure. Let us take thiamine as an example:

As animals become ill because of continuously eating thiamine-deficient food, they avidly try any new food (because of the aversive consequences associated with their old food). If the new food contains thiamine, they begin to recover within hours and the positive-reinforcing effects of this improvement in health maintain their preference for the new food. Thus, despite its nonspecific nature, such a system serves nicely in animals to direct the consumption of adequate amounts of the essential nutrients when they are available.

In human beings, of course, the question is far more difficult to study and little reliable information is available. It is probably safe to conclude, however, that human food selection is mainly dictated by social and cultural factors. There is no question that many food faddists become ill or die as a result of poor eating habits, despite the easy availability of other food sources containing all essential nutrients.

Obesity

Obesity has been called the most common disease in America. The term disease is perfectly justified since obesity predisposes to illness and premature death from a multitude of causes. The seriousness of being overweight is underlined by statistics, which show a mortality rate more than 50 percent greater than normal in overweight persons in the same age groups. Our view of the etiology of obesity has undergone radical changes both in the past and recently. Most obesity was once ascribed to "glandular conditions." Later, the dictum became "You're fat because you eat too much." This is an unassailable fact but completely evades the issue of how much is "too much." The energy-balance equation clearly shows that too much simply means more than is needed to supply the energy needs of the body. Since beyond this point, additional food is stored as fat, the amount one eats must be viewed in relationship to one's activity pattern. For example, a study of obese high school girls revealed that they ate, on the average, *less* than a control group of normal weight girls. The obese girls had much less physical activity than the control group and were eating "too much" only in relationship to their physical activity.

This example stresses the fact, so easily forgotten, that there are two sides to the energy-balance equation. Moreover, recent studies have revealed a startling physiological relationship, namely, that low levels of physical activity may cause increased eating. The caloric intakes and body weights of large numbers of workers in the same factory in India were studied after grouping the men

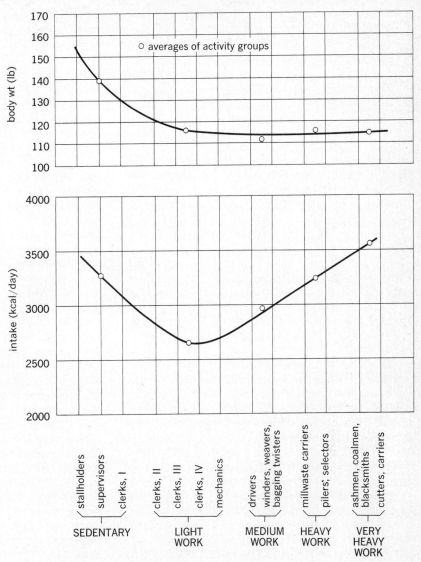

FIGURE 13-27
Body weight and caloric intake as functions of physical activity in workers at an Indian factory. (*Adapted from Mayer et al., Am. J. Clin. Nutrition.**)

according to the physical exertion required by their jobs. Levels of activity below a certain arbitrary minimum were classified as sedentary. As shown in Fig. 13-27, men performing work loads above the sedentary range displayed the expected pattern; caloric intake was directly proportional to work level, and body weights for all groups

of men were similar. The unexpected finding was that for men performing small work loads in the sedentary range, caloric intake varied inversely with work load, i.e., the less physical activity the men performed, the more they ate. Accordingly, these men were considerably fatter, on the average, than the other men. Since this kind of study is

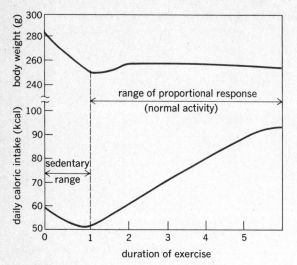

FIGURE 13-28

Voluntary food intake and body weight as functions of duration of controlled exercise (hours per day) in normal adult rats. (*Adapted from Mayer et al., Am. J. Physiol.*)

difficult to interpret in terms of cause and effect, a series of experiments was performed on rats forced to remain sedentary or to exercise. Figure 13-28 shows data similar to those for the factory workers. It is therefore apparent that very low levels of activity do not induce similar reductions of food intake but actually stimulate eating. The implications of these findings for energy balance in a society where so many of us fall into the sedentary category are obvious, and the elucidation of the factors responsible will be of considerable importance.

Having stressed the importance of physical activity (or rather its lack) in the etiology of obesity, we must admit that most cases of obesity can be explained only in part on this basis. *Ultimately, all obesity represents failure of normal food-intake control mechanisms.* The inappropriate effects of low activity levels on food intake merely represent one example of this failure. In rare cases obese persons have hypothalamic lesions similar to those produced experimentally in the rat. With greater understanding of the effects of various hormones on adipose tissue, the cry of "glandular disorder" may be heard again; cortisol and insulin, in particular, may play a role in cases characterized by multiple metabolic disorders. These latter types of obesity, which can be understood in light of our knowledge of the normal control mechanisms, constitute a tiny group, and we are left with the conclusion that psychological factors, habit, and social customs are the predominant causes of hypothalamic mal-

adjustment. Most experts feel that, little by little, the discovery of specific causes will reduce this category.

As an exercise in energy balance, let us calculate how rapidly a person can expect to lose weight on a reducing diet. Suppose a woman whose metabolic rate per 24 hr is 2000 kcal goes on a 1000 kcal/day diet. How much of her own body fat will be required to supply this additional 1000 kcal/day? Fat contains 9 kcal/g; therefore,

$$1000 \text{ kcal/day} \div 9 \text{ kcal/g} = 111 \text{ g/day or } 777 \text{ g/week}$$

Approximately another 77 g of water is lost from the adipose tissue along with this fat (adipose tissue is 10 percent water) so that the grand total for one week's loss equals 854 g or 1.8 lb. Thus, during her diet she can reasonably expect to lose approximately this amount of weight each week. Actually, the amount of weight lost during the first week might be considerably greater since, for reasons poorly understood, a large amount of water may be lost early in the diet, particularly when the diet contains little carbohydrate. This early loss, which is really of no value so far as elimination of excess fat is concerned, often underlies the wild claims made for fad diets (indeed, to enhance this effect, drugs which cause the kidneys to excrete even more water are sometimes included in the diet). Clearly, weight loss is a slow process requiring patience and a meaningful reshaping of eating patterns.

Another aspect of obesity, the potential importance of which is just being recognized, is the temporal pattern of food intake. When rats were given several large meals each day instead of being allowed to nibble frequently, as is their habit, their body weights were similar to those of the nibblers. However, the meal eaters had relatively more body fat and less protein than the nibblers, indicating that their bodies' total caloric content was increased. Moreover, they manifested a diabetes-like state with hyperglycemia and increased arteriosclerosis. Similar studies have been reported for monkeys as well. The significance of such data for man is unknown, but we mention these studies, despite the fact that the data are very few, to stress the many virtually untouched areas for future research.

Another example of such an area concerns the possibility that influences early in life may induce physiological changes which predispose to adult obesity. Adipose-tissue growth is due to both an increased size and number of cells. The number of cells stabilizes sometime during early adulthood and does not change thereafter. Present evidence suggests that overfeeding early in life, particularly during infancy, causes the generation

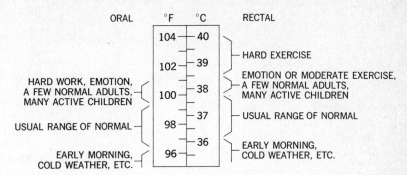

FIGURE 13-29
Ranges of body temperatures in normal persons. (*Adapted from Dubois.*)

of an abnormally large number of adipose-tissue cells which persists throughout later life; the presence of these extra cells may well act as a stimulus for increased food intake although the mechanism is unknown. In any case, the view that "a fat baby is a healthy baby" may well require reevaluation in light of the potential effects of early obesity on later obesity with its attendant consequences.

Finally, it should be recognized that obesity is only one form of "overnutrition" endemic to westernized societies. There is no question that deficits of essential nutrients produce disease, but the question of whether excessive amounts of these same nutrients might be harmful has received far less attention.

Regulation of body temperature

Animals capable of maintaining their body temperatures within very narrow limits are termed *homeothermic*. The adaptive significance of this ability stems primarily from the marked effects of temperature upon the rate of chemical reactions in general and enzyme activity in particular. Homeothermic animals are spared the slowdown of all bodily functions which occurs when the body temperature falls. However, the advantages obtained by a relatively high body temperature impose a great need for precise regulatory mechanisms since even moderate elevations of temperatures begin to cause nerve malfunction, protein denaturation, and death. Most people suffer convulsions at a body temperature of 106 to 107°F, and 110°F is the absolute limit for life. In contrast, most body tissues can withstand marked cooling (to less than 45°F), which has found an important place in surgery when the heart must be stopped, since the dormant cold tissues require little nourishment.

Figure 13-29 illustrates several important generali-

zations about normal body temperature in man: (*1*) Oral temperature averages about 1.0°F less than rectal; thus, all parts of the body do not have the same temperature. (*2*) Internal temperature is not absolutely constant but varies several degrees in perfectly normal persons in response to activity pattern and external temperature, and in addition, there is a characteristic diurnal fluctuation, so that temperature is lowest during sleep and slightly higher during the awake state even if the person remains relaxed in bed. An interesting variation in women is a higher temperature during the last half of the menstrual cycle caused by progesterone (Chap. 14).

If temperature is viewed as a measure of heat "concentration," temperature regulation can be studied by our usual balance methods. In this case, the total heat content of the body is determined by net difference between heat produced and heat lost from the body. Maintaining a constant body temperature implies that, overall, heat production must equal heat loss. Both these variables are subject to precise physiological control.

Temperature regulation offers a classic example of a biological control system; its generalized components are shown in Fig. 13-30. The balance between heat production and heat loss is continuously being disturbed, either by changes in metabolic rate (exercise being the most powerful influence) or by changes in the external environment which alter heat loss. The resulting small changes in body temperature reflexly alter the output of the effector organs, which drive heat production or heat loss and restore normal body temperature.

Heat production

The basic concepts of heat production have already been described. Recall that heat is produced by virtually all chemical reactions occurring in the body and that the cost-of-living metabolism by all organs sets the basal level of heat production, which can be increased as a result

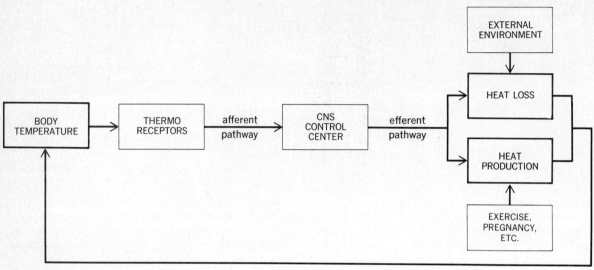

FIGURE 13-30
Summary of temperature regulation. Heat loss from the body depends directly upon the external environment and upon changes controlled by temperature-regulating reflexes.

of skeletal muscular contraction or the action of several hormones.

Changes in muscle activity   The first muscle changes in response to cold are a gradual and general increase in skeletal *muscle tone.* This soon leads to *shivering,* the characteristic muscle response to cold, which consists of oscillating rhythmic muscle tremors occurring at the rate of about 10 to 20 per second. So effective are these contractions that body heat production may be increased severalfold within seconds to minutes. Because no external work is performed, all the energy liberated by the metabolic machinery appears as internal heat. As always, the contractions are directly controlled by the efferent motor neurons to the muscles. During shivering these nerves are controlled by descending pathways under the primary control of the hypothalamus. It is important to note that this "shivering pathway" can be suppressed, at least in part, by input from the cerebral cortex since a cold man ceases to shiver when he starts to perform voluntary activity. Besides increased muscle tone and shivering, which are completely reflex in nature, man also uses voluntary heat-production mechanisms such as foot stamping, hand clapping, etc.

Thus far, our discussion has focused primarily on the muscular response to cold; the opposite reactions occur in response to heat. Muscle tone is reflexly decreased and voluntary movement is also diminished ("It's

too hot to move"). However, these attempts to reduce heat production are relatively limited in capacity both because muscle tone is already quite low normally and because of the *direct* effect of a body temperature increase on metabolic rate. As shown in Fig. 13-31, heat production rose as environmental temperature was cooled below normal, this response being due to increased muscle tone and shivering. As room temperature

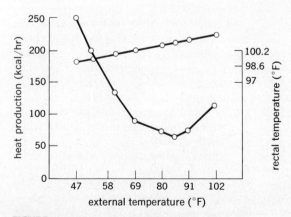

FIGURE 13-31
Effects of altering external temperature upon metabolic rate (heat production) and body (rectal) temperature. The subject was lightly clothed and seated quietly.

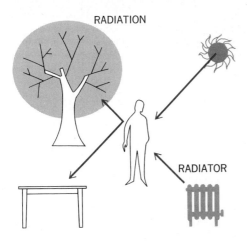

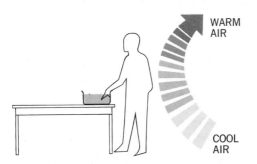

FIGURE 13-32
Mechanisms of heat transfer. By radiation, heat is transferred by electromagnetic waves (solid arrows); in conduction and convection, heat moves by direct transfer of thermal energy from molecule to molecule (dashed arrow).

was heated above normal, there was a small drop in metabolic rate (decreased muscle tone), but the highest temperatures again elicited a higher metabolic rate, certainly an inappropriate response as far as temperature regulation is concerned. The explanation is that *body* temperature rose slightly at this point and directly increased the rate of all metabolic reactions; man's ability to regulate temperature by *decreasing* metabolic rate is seen to be quite limited.

Nonshivering ("chemical") thermogenesis In most experimental animals chronic cold exposure induces an increase in metabolic rate, which is not due to increased muscle activity. Indeed, as this so-called nonshivering thermogenesis increases over time, it is associated with a decrease in the degree of shivering. The cause of

nonshivering thermogenesis has been the subject of considerable controversy; present evidence suggests that it is due mainly to an increased secretion of epinephrine and activity of the sympathetic nerves to adipose tissue. (Thyroxine may also be involved but this is much less clear.) Equally controversial is the question whether nonshivering thermogenesis is a significant phenomenon in man. Regardless of the outcomes of these debates, it seems clear that human hormonal changes and nonshivering thermogenesis are of secondary importance and that changes in muscle activity constitute the major control of heat production for temperature regulation, at least in the early response to temperature changes.

Heat-loss mechanisms

The surface of the body exchanges heat with the external environment by radiation, conduction and convection (Fig. 13-32), and water evaporation.

Radiation, conduction, convection The surface of the body constantly emits heat, in the form of electromagnetic waves. Simultaneously, all other dense objects are radiating heat. The rate of emissions is determined by the temperature of the radiating surface. Thus, if the surface of the body is warmer than the *average* of the various surfaces in the environment, net heat is lost, the rate being directly dependent upon the temperature difference. The sun, of course, is a powerful radiator, and direct exposure to it greatly decreases heat loss by radiation or may reverse it.

Conduction is the exchange of heat not by radiant energy but simply by transfer of thermal energy from atom to atom or molecule to molecule. Heat, like any other quantity, moves down a concentration gradient, and thus the body surface loses or gains heat by conduction only through direct contact with cooler or warmer substances, including, of course, the air.

Convection is the process whereby air (or water) next to the body is heated, moves away, and is replaced by cool air (or water), which in turn follows the same pattern. It is always occurring because warm air is less dense and therefore rises, but it can be greatly facilitated by external forces such as wind or fans. Thus, convection aids conductive heat exchange by continuously maintaining a supply of cool air. In the absence of convection, negligible heat would be lost to the air, and conduction would be important only in such unusual circumstances as immersion in cold water or sitting nude on a cold chair. (Because of the great importance of air movement in aiding heat loss, attempts have been made to quantitate the cooling effect of combinations of air speed and tem-

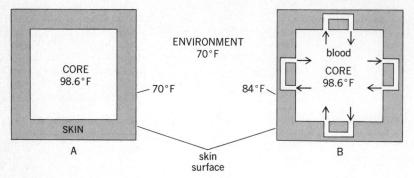

FIGURE 13-33
Relationship of skin's insulating capacity to its blood flow. A Skin as a perfect insulator, i.e., with zero blood flow, the temperature of the skin outer surface equaling that of the external environment. When the skin blood vessels dilate B, the increased flow carries heat to the body surface, i.e., reduces the insulating capacity of the skin, and the surface temperature becomes intermediate between that of the core and the external environment.

perature; the most useful tool has been the wind-chill index.) Henceforth we shall also imply conduction when we use the term convection.

It should now be clear that heat loss by radiation and convection is largely determined by the temperature gradient between the body surface and the external environment. It is convenient to view the body as a central core surrounded by a shell consisting of skin and subcutaneous tissue (for convenience, we shall refer to the complex shell of tissues simply as skin) whose insulating capacity can be varied. It is the temperature of the central core which is being regulated at approximately 99°F; in contrast, as we shall see, the temperature of the outer surface of the skin changes markedly. If the skin were a perfect insulator, no heat would ever be lost from the core; the outer surface of the skin would equal the environmental temperature (except during direct exposure to the sun), and net convection or radiation would be zero. The skin, of course, is not a perfect insulator, so that the temperature of its outer surface generally lies somewhere between that of the external environment and the core. Of profound importance for temperature regulation of the core is that the skin's effectiveness as an insulator is subject to physiological control by changing the blood flow to the skin. The more blood reaching the skin from the core, the more closely the skin's temperature approaches that of the core. In effect, the blood vessels diminish the insulating capacity of the skin by carrying heat to the surface (Fig. 13-33). These vessels are controlled primarily by vasoconstrictor sympathetic nerves. Vasoconstriction may be so powerful that the skin of the finger, for example, may undergo a 99 percent reduction in blood flow during exposure to cold.

The pattern of vasomotor response to cold and heat is diagrammed in Fig. 13-34. Exposure to cold increases the gradient between core and environment; skin vasoconstriction increases skin insulation, reduces skin tem-

perature, and lowers heat loss. Exposure to heat decreases (or may even reverse) the gradient between core and environment; in order to permit the required heat loss, skin vasodilation occurs, the gradient between skin and environment increases, and heat loss increases. Although we have spoken of skin temperature as if it were uniform throughout the body, certain areas participate much more than others in the vasomotor responses; accordingly, skin temperatures vary with location.

What are the limits of this type of process? The lower limit is obviously the point at which maximal skin vasoconstriction has occurred; any further drop in environmental temperature increases the gradient and causes excessive heat loss. At this point, the body must increase its heat production to maintain temperature. The upper limit is set by the point of maximal vasodilation, the environmental temperature, and the core temperature itself. As shown in Fig. 13-34, at high environmental temperatures, even maximal vasodilation cannot establish a core-environment gradient large enough to eliminate heat as fast as it is produced. Another heat-loss mechanism, therefore, is brought strongly into play, sweating. Thus, the skin vasomotor contribution to temperature is highly effective in the midrange of environmental temperature (70 to 85°F), but the major burden is borne by increased heat production at lower temperatures and by increased heat loss via sweating at higher temperatures.

Two other important mechanisms for altering heat loss by radiation and convection remain: changes in surface area and clothing. Curling up into a ball, hunching the shoulders, and similar maneuvers in response to cold reduce the surface area exposed to the environment, thereby decreasing radiation and convection. In man, clothing is also an important component of temperature regulation, substituting for the insulating effects of feathers in birds and fur in other mammals. The principle is similar in that the outer surface of the clothes now

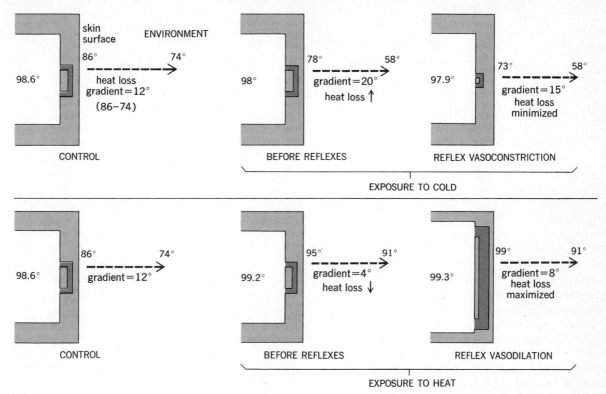

FIGURE 13-34
Effects of reflex vasoconstriction or vasodilation in maintaining the heat-loss gradient for radiation and convection. Had the reflexes not occurred, core temperatures would have changed much more. However, the compensations are incomplete, and other mechanisms (changes in heat production and sweating) are required to stabilize body temperature completely during exposure to marked changes in environmental temperature.

forms the true "exterior" of the body surface. The skin loses heat directly to the air space trapped by the clothes; the clothes in turn pick up heat from the inner air layer and transfer it to the external environment. The insulating ability of clothing is determined primarily by its type and thickness as well as by the thickness of the trapped air layer. We have spoken thus far only of the ability of clothing to reduce heat loss; the converse is also desirable when the environmental temperature is greater than body temperature, since radiation and conduction then produce heat gain. Man therefore insulates himself against temperatures which are greater than body temperature by wearing clothes. The clothing, however, must be loose so as to allow adequate movement of air to permit evaporation, the only source of heat loss under such conditions. White clothing is cooler since it reflects radiant energy, which dark colors absorb. Contrary to popular belief, loose-fitting light-colored clothes are far more cooling than going nude during direct exposure to the sun.

Evaporation Evaporation of water from the skin and lining membranes of the respiratory tract is the second major process for loss of body heat. Thermal energy must be supplied in order to transform water from the liquid to the gaseous state. Thus, whenever water vaporizes from the body's surface, the heat required to drive the process is absorbed from the surface, thereby cooling it. Even in the absence of sweating, there is still loss of water by diffusion through the skin, which is not completely waterproof. A like amount is lost during expiration from the respiratory lining. This insensible water loss amounts to approximately 600 ml/day in man and accounts for a significant fraction of total heat loss. In contrast to this passive water diffusion, *sweating* requires the active secretion of fluid by sweat glands and its extrusion into ducts, which carry it to the skin surface. The sweat is pumped to the surface by periodic contraction of cells resembling smooth muscle in the ducts. Production and delivery of sweat to the surface are stimulated by the

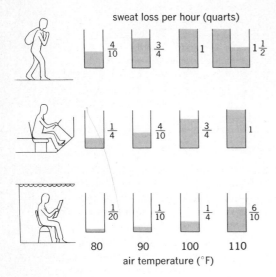

sweat loss per hour (quarts)

air temperature (°F)

FIGURE 13-35
Average sweating rates at various air temperatures and activities. (*Redrawn from Adolph.*)

sympathetic nerves. Sweat is a dilute solution containing primarily sodium chloride. The loss of this salt and water during severe sweating can cause diminution of plasma volume adequate to provoke hypotension, weakness, and fainting. It has been estimated that there are over 2.5 million sweat glands spread over the adult human body, and production rates of over 4 liters/hr have been reported. This is 9 lb of water, the evaporation of which would eliminate almost 1200 kcal from the body! It is essential to recognize that sweat must evaporate in order to exert its cooling effect. The most important factor determining evaporation is the water vapor concentration of the air, i.e., the *humidity*. The discomfort suffered on humid days is due to the failure of evaporation; the sweat glands continue to secrete, but the sweat simply remains

on the skin or drips off. Most other mammals differ from man in lacking sweat glands. They increase their evaporative losses primarily by panting, thereby increasing pulmonary air flow and increasing water losses from the lining of the respiratory tract, and they deposit water for evaporation on their fur or skin by licking.

Heat loss by evaporation of sweat gradually dominates as environmental temperature rises since radiation and convection decrease linearly as the body–environment temperature gradient diminishes. At environmental temperatures above that of the body, heat is actually gained by radiation and conduction, and evaporation is the sole mechanism for heat loss. Man's ability to survive such temperatures is determined by the humidity and by his maximal sweating rate (Fig. 13-35). For example, when the air is completely dry, man can survive a temperature of 266°F for 20 min or longer, whereas very moist air at 115°F is not bearable for even a few minutes.

Changes in sweating determine man's chronic adaptation to high temperatures. A person newly arrived in a hot environment has poor ability to do work initially, his body temperature rises, and severe weakness and illness may occur. After several days, there is great improvement in work tolerance with little increase in body temperature, and the person is said to have *acclimatized* to the heat. Body temperature is kept low because there is an earlier onset of sweating and because of increased rates and lower sodium content of the sweat. The mechanisms remain unknown although heightened aldosterone secretion plays an important role in reducing the sodium.

Summary of effector mechanisms in temperature regulation

Table 13-5 summarizes the mechanisms regulating temperature, none of which is an all-or-none response but calls for a graded progressive increase or decrease in activity. As we have seen, heat production via skeletal

TABLE 13-5
Summary of effector mechanisms in temperature regulation

Stimulated by cold		Stimulated by heat	
Decrease heat loss	Vasoconstriction of skin vessels; reduction of surface area (curling up, etc.)	Increase heat loss	Vasodilation of skin vessels; sweating
Increase heat production	Shivering and increased voluntary activity; (?) increased secretion of thyroxine and epinephrine; increased appetite	Decrease heat production	Decreased muscle tone and voluntary activity; (?) decreased secretion of thyroxine and epinephrine; decreased appetite

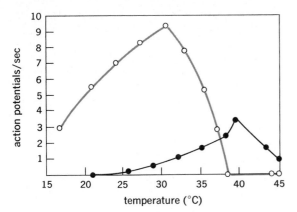

FIGURE 13-36
Discharge rates of a typical skin cold receptor (open circles) and warm receptor (closed circles) in response to changes in temperature. (*Adapted from Dodt and Zotterman.*)

muscle activity becomes extremely important at the cold end of the spectrum, whereas increased heat loss via sweating is critical at the hot end.

Hypothalamic centers involved in temperature regulation

Removal of the cerebral cortex does not hinder an animal's ability to regulate temperature, but destruction of the hypothalamus severely disturbs temperature regulation. Neurons in this latter area, via descending pathways, control the output of somatic motor nerves to skeletal muscle (muscle tone and shivering) and of sympathetic nerves to skin arterioles (vasoconstriction and dilation), sweat glands, and the adrenal medulla. In animals in which thyroxine is an important component of the response to cold, these centers also control the output of hypothalamic TSH-releasing factor (TRF).

Afferent input to the hypothalamic centers

The final component of the human thermostat is the afferent input. Obviously, these temperature-regulating reflexes require receptors capable of detecting changes in the body temperature. There are two groups of receptors, one in the skin (*peripheral thermoreceptors*) and the other in deeper body structures (*central thermoreceptors*).

Peripheral thermoreceptors In the skin (and certain mucous membranes) are nerve endings usually categorized as *cold* and *warm receptors*. In one sense, these are misleading terms since cold is not a separate entity but a lesser

degree of warmth. Really there are two populations of temperature-sensitive skin receptors, one stimulated by a lower and the other by a higher range of temperatures (Fig. 13-36). Information from these receptors is transmitted via the afferent nerves and ascending pathways to the hypothalamus, which responds with appropriate efferent output; in this manner, the firing of cold receptors stimulates heat-producing and heat-conserving mechanisms, whereas enhanced firing of warmth receptors accomplishes just the opposite.

Central thermoreceptors It should be clear that the skin thermoreceptors alone would be highly inefficient regulators of body temperature for the simple reason that it is the core temperature, not the skin temperature, which is actually being regulated. On theoretical grounds alone it was apparent that core, i.e., central, receptors had to exist somewhere in the body, and numerous experiments have localized them, to a large extent, to the hypothalamus itself (important thermoreceptors also exist in other internal locations). In unanesthetized dogs, local warming of hypothalamic neurons through previously implanted thermodes causes them to fire rapidly (Fig. 13-37) and

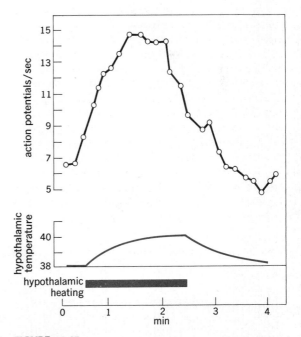

FIGURE 13-37
Effect of local heating of a discrete area of hypothalamus on the discharge rate of a single thermosensitive hypothalamic neuron. (*Adapted from Nakayama et al.*)

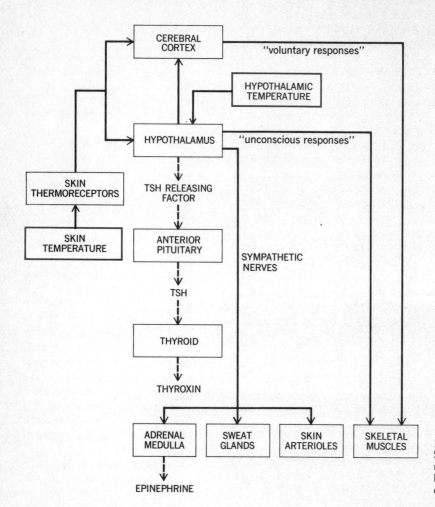

FIGURE 13-38
Summary of temperature-regulating mechanisms. The dashed lines are hormonal pathways, which are probably of minor importance in man.

reproduces the entire picture of the dog's usual response to a warm environment: he becomes sleepy, stretches out to increase his surface area, pants heavily, salivates, and licks his fur. Conversely, local cooling induces vasoconstriction, intensive shivering, fluffing out of the fur, and curling up. These hypothalamic thermoreceptors have synaptic connections with hypothalamic integrating centers, which also receive input from other central thermoreceptors as well as from the skin thermoreceptors. The precise relative contributions of the various thermoreceptors remain the subject of considerable debate. Temperature-regulating reflexes are summarized in Fig. 13-38.

Fever

The elevation of body temperature so commonly induced by infection is due not to a breakdown of tempera-

ture-regulating mechanisms but to a resetting of the hypothalamic thermostat. Thus, a person with a fever regulates his temperature in response to heat or cold but always at a higher set point. The onset of fever is frequently gradual but it is most striking when it occurs rapidly in the form of a chill. It is as though the thermostat were suddenly raised; the person suddenly feels cold, and marked vasoconstriction and shivering occur. This association of heat conservation and increased heat production serves to drive body temperature up rapidly. All this time, the person may be putting on more blankets because he feels cold. Later, the fever breaks as the thermostat is reset to normal; the person feels hot, throws off the covers, and manifests profound vasodilation and sweating.

What is the basis for the resetting? As is described in the chapter on resistance to infection, certain endogenously produced chemicals known as *pyrogens* are released in the presence of infection or inflammation. These pyrogens act directly upon the thermoreceptors in the hypothalamus, altering their rate of firing and their input to the integrating centers. Recent evidence suggests that this effect of pyrogens is mediated via local release of prostaglandins which then directly alter thermoreceptor function. Consistent with this hypothesis is the fact that aspirin, which reduces fever by restoring thermoreceptor activity toward normal, inhibits the synthesis of prostaglandins. Aspirin does not lower the body temperature of a normal nonfebrile person, presumably because prostaglandin is not liberated in significant amounts, except when stimulated by pyrogen.

In addition to infection and inflammation, in which fever is induced by pyrogens, as described above, there are other situations in which fever is produced by quite different mechanisms. Excessive blood levels of epinephrine or thyroxine resulting from diseases of the adrenal medulla or thyroid gland elevate the body temperature by direct actions on heat-producing metabolic reactions rather than by altering the hypothalamic thermoreceptor setting. Certain lesions of the brain do not reset the hypothalamus but rather completely destroy its normal regulatory capacity; under such conditions lethal hyperthermia may occur very rapidly. *Heat stroke* is also characterized by a similar breakdown in function of the regulatory centers. It is frequently a positive-feedback state in which, because of inadequate balancing of heat loss and production, body temperature becomes so high that the hypothalamic regulatory centers are put out of commission and body temperature therefore rises even higher. Thus a patient suffering from heat stroke manifests a dry skin (absence of sweating) despite a markedly elevated body temperature.

In this regard, it is instructive to calculate how rapidly body temperature can rise when heat-loss mechanisms are completely shut down. The amount of heat required to raise the body temperature 1°C is 0.83 kcal/kg. Therefore, in a 70-kg man, the body temperature is increased 1°C by the retention of 0.83 kcal × 70 kg = 58 kcal. Thus, a lethal 6°C rise in body temperature occurs when 6°C × 58 kcal/1°C rise = 348 kcal has been retained. Normal metabolism produces this much heat in approximately 5 hr (actually less since metabolic rate is greatly increased as body temperature rises).

Heat stroke, the attainment of a body temperature at which vital bodily functions are endangered, should be distinguished from *heat exhaustion*. The former is due to a breakdown in heat-regulating mechanisms, whereas the latter is not the result of failure of heat regulation but rather of the inability to meet the price of heat regulation. Heat exhaustion is a state of collapse due to hypotension brought on by depletion of plasma volume (secondary to sweating) and by extreme dilation of skin blood vessels, i.e., by decreases in both cardiac output and peripheral resistance. Thus, heat exhaustion occurs as a direct consequence of the activity of heat-loss mechanisms; because these mechanisms have been so active, the body temperature is only modestly elevated. In a sense, heat exhaustion is a safety valve which, by forcing cessation of work when heat-loss mechanisms are overtaxed, prevents the larger rise in body temperature which would precipitate the far more serious condition of heat stroke.

Reproduction

14

For many years, the fields of sexual physiology and behavior were the least advanced of all branches of human physiology, one of the primary reasons being the differences which exist between species. Most mammals, including man, demonstrate similar physiological mechanisms of cardiac, renal, respiratory, etc., function, but when it comes to reproductive physiology, this is frequently not the case.

Before beginning detailed descriptions of male and female reproductive physiology, it may be worthwhile to summarize some of the important terminology and patterns of classification. The primary reproductive organs are known as the *gonads*, the *testes* in male and the *ovaries* in female. In both sexes, the gonads serve dual functions: (1) production of the reproductive cells, *sperm* or *ova*, and (2) secretion of the *sex hormones*. The system of ducts (and the glands lining them) through which the sperm or ova are transported is collectively known as the *accessory reproductive organs*, and in the female the breasts are also usually included in this category. Finally, the *secondary sexual characteristics* comprise the many external differences (hair, body contours, etc.) between male and female which are not directly involved in reproduction. The gonads and accessory reproductive organs are present at birth but remain relatively small and nonfunctional until the onset of *puberty*, at about ten to fourteen years of age.

Control of ovarian function

Control of follicle and ovum development / Control of ovulation / Control of the corpus luteum / Summary

Uterine changes in the menstrual cycle

Nonuterine effects of estrogen and progesterone

Female sexual response

Pregnancy

Ovum transport / Sperm transport / Entry of the sperm into the ovum / Early development, implantation, and placentation / Hormonal changes during pregnancy / Parturition (delivery of the infant) / Lactation / Birth control

SECTION C   THE CHRONOLOGY OF SEX DEVELOPMENT
Sex determination

Sex differentiation

Differentiation of the gonads / Differentiation of internal and external genitalia / Sexual differentiation of the central nervous system

Puberty

Menopause

The secondary sexual characteristics are virtually absent until puberty. The term puberty actually means the attainment of sexual maturity in the sense that conception becomes possible; as commonly used, it usually refers to the entire period of sexual development culminating in the attainment of sexual maturity. The term *adolescence* has a much broader meaning and includes the total period of transition from childhood to adulthood in all respects, not just sexual.

SECTION A
MALE REPRODUCTIVE PHYSIOLOGY

The essential male reproductive functions are the manufacture of sperm (*spermatogenesis*) and the deposition of the sperm in the female. The organs which produce sperm, the *testes*, also serve an endocrine function, the manufacture of the primary male sex hormone, *testosterone*. Each testis is composed of different types of tissues subserving the spermatogenic and endocrine activities, namely, the *seminiferous tubules* and *interstitial cells*, respectively. Of great importance is the relationship between these two basic functions; the process of spermatogenesis requires testosterone, but testosterone production does *not* depend upon spermatogenesis. In other words, testosterone

deficiency produces sterility by interrupting spermatogenesis, but interference with the function of the seminiferous tubules does not alter normal testosterone production by the interstitial cells. The great importance of this relationship is that a simple, effective method of sterilizing the male is vasectomy, surgical ligation and removal of a segment of the large duct which carries sperm from the testes. This procedure prevents the delivery of sperm but does not appear to alter secretion of testosterone, which is the primary determinant of male sexual drive and maleness in general.

Spermatogenesis

From Fig. 14-1, a diagrammatic cross section of a human testis, it is evident that the testis is composed primarily of the many highly coiled seminiferous tubules, the combined length of which is approximately 750 ft. In Fig. 14-2, a microscopic section of an adult human testis, it can be seen that the tubules have a complex lining of cells, the vast majority of which are in various stages of division; these are the cells which give rise to the sperm cells.

The outer layer of dividing cells in the tubules are the undifferentiated germ cells termed *spermatogonia*, which, by dividing mitotically, provide a continuous

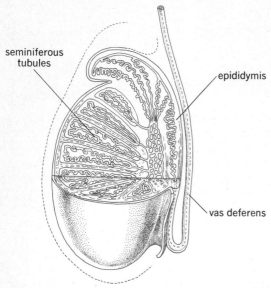

FIGURE 14-1
Diagrammatic cross section of a human testis. The highly coiled seminiferous tubules are the sites of sperm production.

source of new cells. Some spermatogonia move away from the membrane lining the tubule and increase markedly in size. Each large cell, now termed a *primary spermatocyte,* divides to form two *secondary spermatocytes;* each of which in turn divides into two *spermatids,* the latter ultimately being transformed into mature *spermatozoa* (Fig. 14-3). The division of the primary spermatocytes by *meiosis* differs from the ordinary mitotic division. During mitosis (Chap. 4) each daughter cell receives the full number of chromosomes; in meiosis, however, each daughter cell receives only half of the chromosomes present in the parent cell. Therefore each human secondary spermatocyte contains only 23 chromosomes rather than the 46 present in other cells of the body. The division of the secondary spermatocyte into spermatids is, in principle, an ordinary mitotic division except that only 23 chromosomes are involved. Since development of the female germ cell follows a similar pattern, the eventual combination of sperm and ovum results in the reestablishment of the normal complement of 46 chromosomes.

The final phase of spermatogenesis, the transformation of the spermatids into mature spermatozoa, involves no further cell divisions. The head of a mature spermatozoan (Fig. 14-4) consists almost entirely of the nucleus, a dense mass of DNA bearing the sperm's ge-

netic information. The tip of the nucleus is covered by a cap known as the *acrosome,* consisting of a protein-filled vesicle derived from the Golgi apparatus and believed to contain several lytic enzymes which enable the sperm to enter the ovum. The tail comprises a group of actomyosinlike contractile filaments, contraction of which produces a whiplike movement of the tail capable of propelling the sperm at a velocity of 1 to 4 mm/sec. Finally, the mitochondria of the spermatid have become rearranged to form the midpiece of the tail and probably provide its energy.

Throughout the entire process of maturation, the developing germ cells remain intimately associated with another type of cell, the *sertoli cell.* Each sertoli cell extends from the basement membrane of the seminiferous tubule all the way to the lumen. Apparently, they serve as the route by which nutrients reach the developing germ cells as the latter move away from the basement membrane; indeed, large portions of the spermatids are actually embedded in the cytoplasm of the sertoli cells. Study of these cells has recently taken on new impetus with the demonstration that they may well produce steroids which act locally to stimulate spermatogenesis.

The mechanisms which guide the remarkable cellular transformation from spermatid to mature sperm remain uncertain. In any small segment of seminiferous tubules, the entire process of spermatogenesis proceeds in a regular sequence. For example, at any given time, virtually all the primary spermatocytes in one portion of the tubule are undergoing division, whereas in an adjacent segment, the secondary spermatocytes may be dividing. The entire process in a single area takes approximately 72 days. In mammals which breed seasonally, spermatogenesis is periodic, activity being followed by degeneration of the spermatogonia and shrinking of the seminiferous tubules. In contrast, nonseasonal breeders, such as man, manifest continuous activity, the cycles following each other without a break. Perhaps the most amazing characteristic of spermatogenesis is its sheer magnitude: the normal human male may manufacture several hundred million sperm per day (the ram produces billions).

Delivery of sperm

The anatomical organization of the male genital duct system is shown in Fig. 14-5. From the seminiferous tubules, the sperm pass through a network of interconnected highly coiled ducts which join to form the *epididymis,* which in turn, leads to the large, thick-walled *vas*

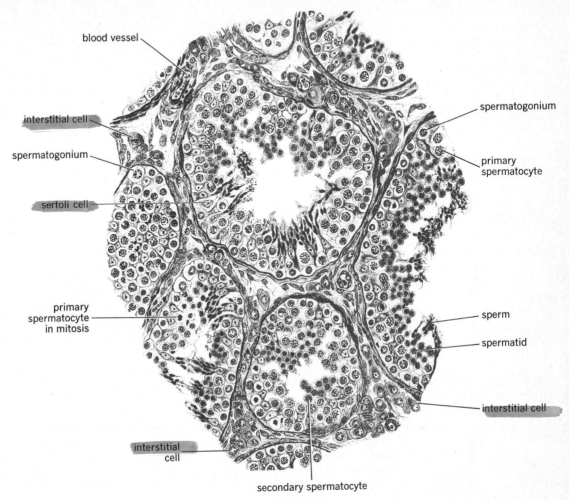

FIGURE 14-2

Section of human testis. The seminiferous tubules show various stages of spermatogenesis. (*Adapted from Bloom and Fawcett.*)

deferens. Movement through these ducts is accomplished by means of peristaltic action of contractile cells in the duct walls; the sperm themselves are nonmotile at this time. Besides serving as a route for sperm exit, this system performs several important functions: (*1*) The epididymis and first portion of the vas deferens store sperm prior to ejaculation; (*2*) during passage through the epididymis or storage there, a final maturation confers the capacities for both motility and fertility upon the sperm; this appears to be mediated by fluid secreted by the epithelium of the epididymis, but little is known about this process or the factors which influence it; (*3*) at ejaculation, the sperm are expelled from the epididymis and vas deferens by strong contractions of the smooth muscle lining the duct walls.

Several large glands, the *seminal vesicles*, drain into the vas deferens just before it passes through the body of the *prostate* gland to open by a small slit into the urethra. The prostate and seminal vesicles, as well as other smaller glands lining the gential ducts, secrete small

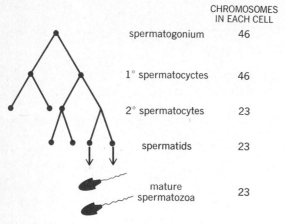

	CHROMOSOMES IN EACH CELL
spermatogonium	46
1° spermatocyctes	46
2° spermatocytes	23
spermatids	23
mature spermatozoa	23

FIGURE 14-3
Summary of spermatogenesis. Each spermatogonium yields eight mature sperm, each containing 23 chromosomes.

quantities of fluid continuously and much larger quantities during sexual intercourse (*coitus*). The secretions constitute the bulk of the ejaculated fluid, the *semen*, which contains a large number of different chemical substances, the functions of which are presently being worked out.

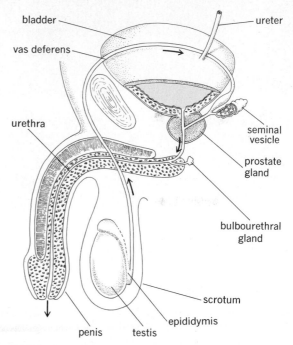

FIGURE 14-5
Anatomical organization of the male reproductive tract.

FIGURE 14-4
Human mature sperm.

One function of the seminal fluid is that of sheer dilution of the sperm (in human beings, sperm constitutes only a few percent of the total ejaculated semen); without such dilution, motility is impaired. In addition, specific chemicals also contribute to the motility of the sperm; for example, the seminal vesicles secrete large quantities of the carbohydrate fructose utilized by the sperm contractile apparatus for energy. We shall describe later the possible contributions of the prostaglandins, a group of fatty acids found in very large concentrations in seminal fluid.

Erection

The primary components of the male sexual act are *erection* of the penis, which permits entry into the female vagina, and *ejaculation* of the sperm-containing semen into the vagina. Erection is a vascular phenomenon which can be understood from the structure of the penis (Fig. 14-6). This organ consists almost entirely of three cylindrical cords of *erectile tissue*, which are actually vascular spaces. Normally the arterioles supplying these chambers are constricted so that they contain little blood and the penis is flaccid; during sexual excitation, the arterioles dilate,

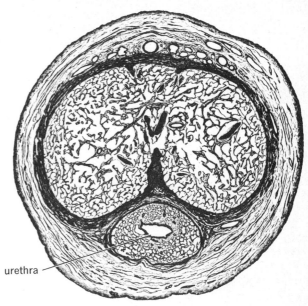

FIGURE 14-6
Cross section of a human penis. The large central area and that surrounding the urethra are vascular spaces which become filled with blood to cause erection. (*Adapted from Bloom and Fawcett.*)

urethra

the chambers become engorged with blood, and the penis becomes rigid. Moreover, as the erectile tissues expand, the veins emptying them are compressed, thus minimizing outflow and contributing to the engorgement. This entire process occurs rapidly, complete erection sometimes taking only 5 to 10 sec. The vascular dilation is accomplished by stimulation of the parasympathetic nerves and inhibition of the sympathetic nerves to the arterioles of the penis (Fig. 14-7). This appears to be one of the few cases of direct parasympathetic control over high-resistance blood vessels. In addition to these vascular effects, the parasympathetic nerves stimulate urethral glands to secrete a mucuslike material which aids in lubrication. What receptors and afferent pathway initiate these reflexes? The primary input comes from highly sensitive mechanoreceptors located in the tip of the penis. The afferent fibers carrying the impulses synapse in the lower spinal cord and trigger the efferent outflow. It must be stressed, however, that higher brain centers, via descending pathways, may exert profound facilitative or inhibitory effects upon the efferent neurons. Thus, thoughts or emotions can cause erection in the complete absence of mechanical stimulation of the penis; conversely, failure of erection (*impotence*) may frequently be due to psychological disturbances. The ability of alcohol to interfere with erection is probably due to its effects on higher brain centers.

Ejaculation

This process is basically a spinal reflex, the afferent pathways apparently being identical to those described for erection. When the level of stimulation reaches a critical peak, a patterned automatic sequence of efferent discharge is elicited to both the smooth muscle of the genital ducts and the skeletal muscle at the base of the penis. The precise contribution of various pathways is complex, but the overall response can be divided into two phases: (*1*) The genital ducts contract, as a result of sympathetic stimulation to them, emptying their contents into the urethra (*emission*), and (*2*) the semen is then expelled from the penis by a series of rapid muscle contractions. During ejaculation the sphincter at the base of the bladder is closed so that sperm cannot enter the bladder nor can urine be expelled. The rhythmical contractions of the penis during ejaculation are associated with intense pleasure, the entire event being termed the *orgasm*. A simultaneous marked skeletal muscle contraction throughout the body is followed by the rapid onset of muscular and psychological relaxation. At orgasm there is also a marked increase in heart rate and blood pressure. Once ejaculation has occurred, there is a so-called latent period during which a second erection is not possible; this period is quite variable but may be hours in perfectly normal men.

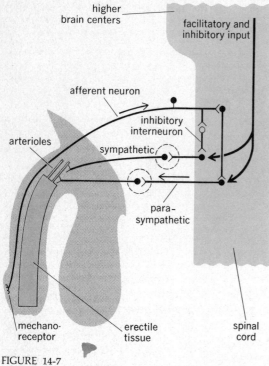

FIGURE 14-7
Reflex pathway for erection. The reflex is initiated by mechanoreceptors in the penis. Input from higher centers can facilitate or inhibit this reflex.

The average volume of fluid ejaculated is 3 ml, containing approximately 300 million sperm. However, the range of normal values is extremely large, and older ideas of the minimal concentration of sperm required for fertility are now being reevaluated. Although quantity is important, it is obvious that the quality of the sperm is another critical determinant of fertility.

Hormonal control of male reproductive functions

Virtually all aspects of male reproductive function are either directly controlled or indirectly influenced by testosterone or the anterior pituitary gonadotropins, *follicle-stimulating hormone* (FSH) and *luteinizing hormone* (LH). These pituitary hormones were named for their effects in the female, but their molecular structures are precisely the same in both sexes. (LH, in the male, is frequently called interstitial-cell-stimulating hormone, ICSH.) FSH

and LH exert their effects only upon the testes, whereas testosterone manifests a broad spectrum of actions not only on the testes but on the accessory reproductive organs, the secondary sexual characteristics, sexual behavior, and organic metabolism in general.

Effects of testosterone

Spermatogenesis Adequate amounts of the steroid hormone testosterone are essential for spermatogenesis, and sterility is an invariable result of testosterone deficiency. The cells which secrete testosterone are known as the *interstitial cells;* as shown in Fig. 14-2, they lie scattered between the seminiferous tubules. It seems likely that the stimulatory effects of testosterone on spermatogenesis are exerted locally by the hormone diffusing from the interstitial cells into the seminiferous tubules. Testosterone is not the only hormone required for spermatogenesis; the pituitary gonadotropins are also required, and the relationship with testosterone will be described subsequently.

Accessory reproductive organs The morphology and function of the entire male duct system, lining glands, and penis all depend upon testosterone. Following removal of the testes (*castration*) in the adult, all the accessory reproductive organs decrease in size (Fig. 14-8), the glands markedly reduce their rates of secretion, and the smooth-muscle activity of the ducts is inhibited. Erection and ejaculation are usually deficient. These defects disappear upon the administration of testosterone.

Behavior Most of our information comes from experiments on animals other than human beings, but even from our fragmentary information about man, there is little doubt

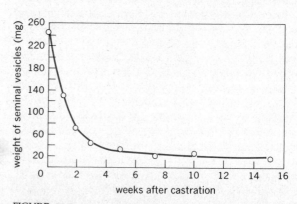

FIGURE 14-8
Atrophy of the seminal vesicles of the adult mouse after castration. (*Adapted from Deanesley and Parkes.*)

that the development and maintenance of normal sexual drive and behavior in man are testosterone-dependent and may be seriously impaired by castration. However, it is a mistake to assume that deviant male sexual behavior must therefore be due to testosterone deficiency or excess. For example, most (but not all) male homosexuals have normal rates of testosterone secretion; although administration of exogenous testosterone may sometimes increase sexual activity in these men, it remains homosexual. To date, no clear-cut correlation has been established between homosexuality or hypersexuality and hormonal status in either men or women.

A question which has recently become the subject of enormous controversy is whether testosterone influences other human behavior in addition to sex, i.e., are there any inherent male-female differences or are the observed differences in behavior all socially conditioned. There is little doubt that biological differences based on sex do exist in other mammals; perhaps the best-studied behavior is that of aggression which is clearly greater in males and is testosterone-dependent. Obviously, it will be difficult to answer such questions in human beings but attempts are now being made to study them in a controlled scientific manner.

Secondary sex characteristics   In the animal world secondary sex characteristics range from the exotic courtship dances of salamanders to the mane of the lion to the stag's antlers. In many species they are important for normal sexual functions; antlers used in fighting for the female and the deep attracting voice of the tree toad are two examples from literally thousands.

In man, virtually all the obvious masculine secondary characteristics are testosterone-dependent. For example, a male body castrated before puberty does not develop a beard or axillary or pubic hair. A strange and unexplained finding is that baldness, although genetically determined in part, does not occur in castrated men. Other testosterone-dependent secondary sexual characteristics are the deepening of the voice (resulting from growth of the larynx), skin texture, thick secretion of the skin oil glands (predisposing to acne), and the masculine pattern of muscle and fat distribution. This leads us into an area of testosterone effects usually described as *general metabolic effects* but very difficult to separate from the secondary sex characteristics. It is obvious that the bodies of men and women (even excepting the breasts and external genitals) have very different appearances; a woman's curves are due in large part to the feminine distribution of fat, particularly in the region of the hips and lower abdomen but in the limbs as well. A castrated male gradually develops this pattern; conversely, a woman treated with testosterone loses it. A second very obvious difference is that of skeletal muscle mass; testosterone exerts a profound effect on skeletal muscle to increase its size. The overall relationship of testosterone to general body growth was described in Chap. 13.

Mechanism of action of testosterone   The fact that testosterone exerts such extraordinarily widespread effects should not obscure what appears to be a common denominator, namely, that its major action is to promote growth. Testosterone, like other steroid hormones, acts upon the cell nucleus to promote transcription or translation, or both, of genetic information, but the precise mechanisms still are unclear. Moreover, we are still left with the problem of explaining the specificity of action, i.e., the fact that only certain types of cells respond to testosterone and then in a distinctive qualitative and quantitative manner. Finally, how it affects behavior is completely unknown.

Testosterone is not a uniquely male hormone. Testosterone or testosteronelike substances, i.e., *androgens,* can be found in the blood of normal women. In the female the site of production is primarily the adrenal gland, which also contributes some androgens in the male. The quantity and potency of the androgens secreted by the adrenal are normally small. These androgens play several important roles in the female, specifically stimulation of general body growth (Chap. 13) and maintenance of sexual drive. In several disease states, the female adrenal may secrete abnormally large quantities of androgen, which produce a remarkable virilism; the female fat distribution disappears, a beard appears along with the male body-hair distribution, the voice lowers in pitch, the skeletal muscle mass enlarges, the clitoris (homologue of the male penis) enlarges, and the breasts diminish in size. These changes illustrate the sex-hormone dependency of secondary characteristics. Sterility also results.

Anterior pituitary and hypothalamic control of testicular function

FSH and LH are essential for normal spermatogenesis and testosterone secretion. Following removal of the anterior pituitary, the testes decrease greatly in weight, and spermatogenesis and testosterone secretion almost cease. FSH directly stimulates spermatogenesis, whereas LH stimulates testosterone secretion, but because testosterone is also required for spermatogenesis, it is evident that LH is indirectly involved in this process as well.

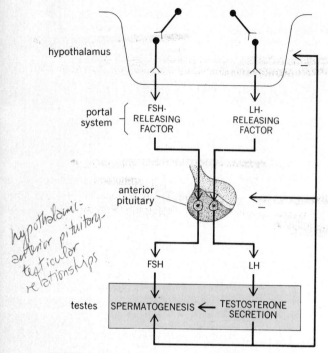

hypothalamic–anterior pituitary–testicular relationships

FIGURE 14-9
Summary of hormonal control of testicular function. The negative signs indicate that testosterone inhibits LH secretion in both the hypothalamus and the anterior pituitary. Testosterone reaches the seminiferous tubules to stimulate spermatogenesis both by local diffusion and by release into the blood and recirculation to the testes. See text for a discussion of FSH control.

No discussion of the anterior pituitary is complete without inclusion of the hypothalamus. As described in Chap. 7, all anterior pituitary hormones are controlled by releasing factors secreted by discrete areas of the hypothalamus and reaching the pituitary via the hypothalamo-pituitary portal blood vessels. This input is essential for sexual function since destruction of the relevant hypothalamic areas stops spermatogenesis and markedly reduces testosterone secretion. Figure 14-9 summarizes the hypothalamic–anterior pituitary–testicular relationships.

What controls the secretion of the hypothalamic releasing factors for FSH and LH? It is known that testosterone exerts a negative-feedback inhibition of LH secretion via both the anterior pituitary and the hypothalamus. Testosterone has less effect on the secretion of FSH, and it is presently unclear just how the testes exert their negative-feedback inhibition of FSH secretion. The solution to the latter question is not merely of academic interest since an agent which inhibits FSH only and not LH would constitute an ideal male contraceptive; spermatogenesis, but not testosterone secretion, would be blocked. As emphasized in Chap. 7, such negative feedbacks are only modifiers and not the primary controllers of secretion. Recall that the hypothalamic cells which produce the releasing factors are nerve cells; we are left with the generalization that these nerve cells secrete their releasing factors either as a result of some inherent automaticity or secondary to stimuli reaching them via synaptic connections with other neurons. It is important to realize that, regardless of the triggering event, the secretion of releasing factors for LH and FSH in the male probably proceeds normally at a rather fixed, continuous rate during adult life; accordingly, FSH and LH release, spermatogenesis, and testosterone secretion also occur at relatively unchanging rates. This is unusual for hormonal systems and, more important, is completely different from the large cyclical swings of activity so characteristic of the female reproductive hormones. A word of caution may be in order here, however, for recent work in nonprimate mammals has indicated that testosterone levels can be made to vary in response to various sexual and social stimuli; the relevance of these studies for man is unknown.

SECTION B
FEMALE REPRODUCTIVE PHYSIOLOGY

Unlike the continuous sperm production of the male, the maturation and release of the female germ cell, *the ovum,* are cyclical and intermittent. This pattern is true not only for ovum development but for the structure and function of virtually the entire female reproductive system (Fig. 14-10). In human beings and other primates these cycles are called *menstrual cycles.*

Ovarian function

The human ovary, like the testis, serves a dual purpose: (1) production of ova and (2) secretion of the female sex hormones, *estrogen* and *progesterone.*

Ovum and follicle growth

In Fig. 14-11, a cross section of an ovary taken from a rhesus monkey, note the many discrete cell clusters

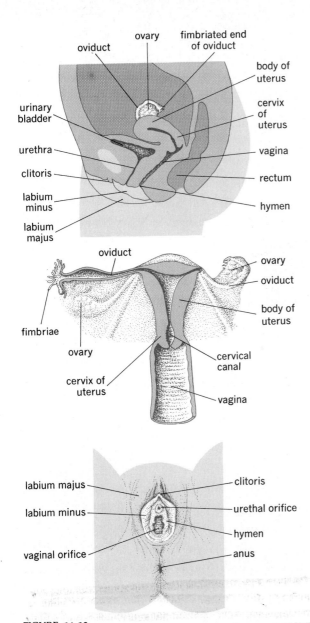

FIGURE 14-10
Anatomical organization of the female reproductive organs.
The clitoris is a small stalk of tissue similar in structure to the
penis. The hymen is a fold of tissue at the lower end of the
vagina; it does not close off the vaginal opening completely.
The fingerlike fimbria projecting at the ends of the oviducts
trap the ovum after its release from the ovary.

known as *primordial follicles.* Each follicle is composed of
one ovum surrounded by a single layer of flattened cells.
At birth, normal human ovaries contain an estimated
400,000 such follicles, and no new ones appear after
birth. Thus, in marked contrast to the male, the newborn
female already has all the germ cells she will ever have.
Only a few, perhaps 400, are destined to reach full maturity during her active sexual life. All the others degenerate
starting from birth on, so that few, if any, remain by the
time she reaches menopause. One result of this is that
the ova which are released (ovulated) near the end of
a woman's sex life are 30 to 35 years older than those
ovulated just after puberty; it has been suggested that
certain congenital defects much commoner among children of older women are the result of aging changes in
the ovum.

In Fig. 14-11 there are still a large number of primordial follicles, mainly on the periphery of the ovary,
but there are also more mature follicles in different stages
of development. The development of the follicle is characterized by an increase in size of the ovum and a proliferation of the surrounding follicle cells. The ovum becomes separated from the follicle cells by a thick
membrane, the *zona pellucida,* which is probably formed
by the follicle cells. As the follicle grows, new cell layers
are formed, not only from mitosis of the original follicle
cells but from the growth of specialized ovarian connective-tissue cells. When the follicle reaches a certain diameter, a fluid-filled space, the *antrum,* begins to form between the follicle cells as a result of fluid they secrete.
We do not know what mechanisms stimulate development
of certain primary follicles, leaving most unstimulated to
degenerate without ever showing a growth phase.

By the time the antrum begins to form, the ovum
has reached full size. From this point on, the follicle grows
in part because of continued follicular cell proliferation
but largely because of the expanding antrum. Ultimately,
the ovum, surrounded by the zona pellucida and a thin
layer of follicle cells, occupies a thin ridge projecting into
the antrum. The antrum becomes so large that the thin
wall of the completely mature follicle actually balloons out
on the surface of the ovary. *Ovulation* occurs when the
wall at the site of ballooning ruptures and the ovum,
surrounded by its tightly adhering zona pellucida and
follicle cells, is carried out of the ovary by the antral fluid
(Fig. 14-12). Many women experience varying degrees
of abdominal pain at approximately the midpoint of their
menstrual cycles, which has generally been presumed to
represent abdominal irritation induced by the entry of
follicular contents at ovulation. However, recent evidence

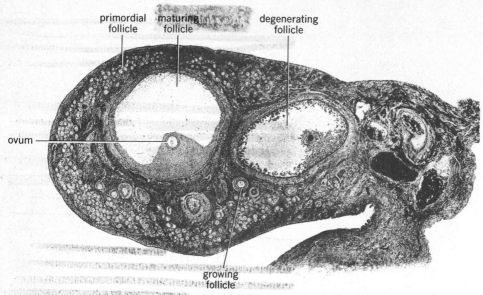

primordial follicle   maturing follicle   degenerating follicle

ovum

growing follicle

FIGURE 14-11
Cross section of an ovary from an adult monkey. (*Adapted from Bloom and Fawcett.*)

on the precise timing of ovulation has indicated that this time-honored concept may be wrong, and the cause of this discomfort remains unclear.

In the human adult ovary, there are always several antrum-containing follicles of varying sizes, but during each cycle, normally only one follicle reaches the complete maturity just described, the process requiring approximately 2 weeks. All the other partially matured antral follicles undergo degeneration at some stage in their growth, the mechanism being unknown. On occasion (1 to 2 percent of all cycles), two or more follicles reach maturity, and more than one ovum may be ovulated. This is the commonest cause of multiple births; in such cases the siblings are nonidentical, or fraternal.

Ovum division   An essential aspect of ovum development is its pattern of cell division. The ova present at birth are the result of numerous mitotic divisions which occurred during intrauterine life. After birth, no further division occurs until just before ovulation, at which time the mature ovum about to be ovulated divides. This division is meiotic and analogous to the division of the primary spermatocyte because each daughter cell receives only 23 chromosomes instead of the usual 46. However, in this division one cell retains virtually all the cytoplasm,

the other being only small and rudimentary. In this manner, the already full-size ovum loses half of its chromosomes but almost none of its nutrient-rich cytoplasm. A second cell division in the oviduct (fallopian tube) after ovulation (indeed, after penetration by a sperm) follows the usual mitotic pattern, and the daughter cells each retain 23 chromosomes. Once again, one daughter cell retains nearly all the cytoplasm. The net result is that each primitive ovum is capable of producing only one mature fertilizable ovum (Fig. 14-13); in contrast, each primary spermatocyte produces four viable spermatozoa.

Formation of corpus luteum

After rupture of the follicle and discharge of the antral fluid and the ovum, a transformation occurs within the follicle, which collapses, and the antrum fills with partially clotted fluid. The follicular cells enlarge greatly and become filled with lipid; the entire glandlike structure is known as the *corpus luteum*. If the discharged ovum is not fertilized, i.e., if pregnancy does not occur, the corpus luteum reaches its maximum development within approximately 10 days and then rapidly degenerates. If pregnancy does occur, the corpus luteum grows and persists until near the end of pregnancy.

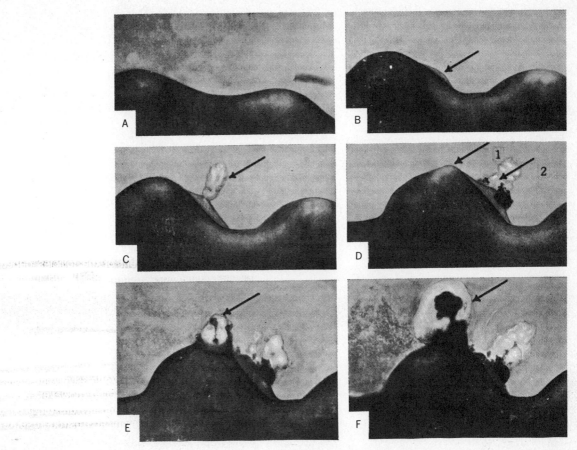

FIGURE 14-12
Enlargements of single frames of a time-lapse motion picture recording of ovulation in the rabbit. A About 1½ hr before ovulation; B ½ hr before rupture; C follicle has just ruptured; D arrow 1 points to the apex of large follicle and arrow 2 points at first ruptured follicle; E arrow points at follicle near the breaking point; F extrusion of follicular fluid with darker blood. Time lapse between E and F is 8 sec. [*From R. T. Hill, E. Allen, and T. C. Cramer, Anat. Rec.,* **63:**245 (1935).]

Ovarian hormones

The female sex hormones secreted by the ovary are the steroids, estrogen[1] and progesterone. Estrogens are secreted to some extent by various ovarian cell types but primarily by the follicle cells (*not* the ovum) and corpus luteum. Progesterone may be secreted in very minute amounts by the follicle cells, but its major source is the corpus luteum. The detailed physiology of these hormones will be described subsequently.

[1] As mentioned in Chap. 7, there are actually multiple estrogenic hormones, but we shall refer to them all as estrogen.

Cyclical nature of ovarian function

The length of a menstrual cycle varies considerably from woman to woman, averaging about 28 days. Day 1 is the first day of menstrual bleeding, and in a typical 28-day cycle, ovulation occurs around day 14. In terms of ovarian function, therefore, the menstrual cycle may be divided into two approximately equal phases: (*1*) the *follicular phase,* during which a single follicle and ovum develop to full maturity and (*2*) the *luteal phase,* during which the corpus luteum is the active ovarian structure. It must be

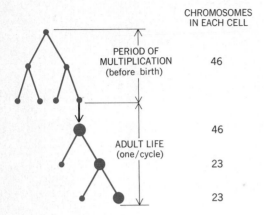

CHROMOSOMES
IN EACH CELL

PERIOD OF
MULTIPLICATION   46
(before birth)

ADULT LIFE   46
(one/cycle)
23

23

FIGURE 14-13
Summary of ovum development. Compare with the male
pattern of Fig. 14-3. Each primitive ovum produces only one
mature ovum containing 23 chromosomes.

stressed that the day of ovulation varies from woman to
woman and frequently in the same woman from month
to month.

Control of
ovarian function

The basic pattern controlling ovum development, ovulation, and formation of the corpus luteum is analogous to
the controls described for testicular function in that the
anterior pituitary gonadotropins, FSH and LH, and the
gonadal sex hormone, estrogen, play the primary roles.
However, the overall schema is more complex in the female since it includes a second important gonadal hormone (progesterone) and a hormonal cycling, quite
different from the stable continuous rates of male hormone secretion.

For purposes of orientation, let us look first at the
changes in the blood concentrations of all four participating hormones during a normal menstrual cycle (Fig.
14-14). Note that FSH is slightly higher in the early part
of the follicular phase of the menstrual cycle and then
steadily decreases throughout the remainder of the period except for a small transient midcycle peak. LH is
quite constant during most of the follicular phase but then
shows a very large midcycle surge (approximately 12 to
24 hr before ovulation) followed by a progressive slow
decline during the luteal phase. The estrogen pattern is

more complex. After remaining fairly low and stable for
the first week (as the follicle develops), it rises to reach
a peak just before LH starts off on its surge. This peak
is followed by a dip, a second peak (due to secretion
by the corpus luteum) and, finally, a rapid decline during
the last days of the cycle. The progesterone pattern is
simplest of all; virtually no progesterone is secreted by
the ovaries during the follicular phase, but very soon after
ovulation, the developing corpus luteum begins to secrete
progesterone, and from this point the progesterone pattern is similar to that for estrogen. It is hoped that, after
the ensuing discussion, the reader will understand how
these changes are all interrelated to yield a self-cycling
pattern.

Control of follicle and
ovum development

Growth and development of the follicles depend upon
follicle-stimulating hormone (FSH) and luteinizing hormone (LH). Accordingly, the ovary from an animal whose
pituitary has been removed shows numerous early follicles but no late antral ones.

A second requirement for follicle development is
estrogen, which may act, in large part, locally within the
ovary. Estrogen is secreted largely by the follicle cells,
so that its secretion rate progressively increases as the
follicle enlarges. This secretion, however, also requires
FSH and LH. Thus, FSH and LH are required both for
normal follicle development and estrogen secretion, as
summarized in Fig. 14-15.

Control of ovulation

If one administers small quantities of FSH and LH each
day to a woman whose pituitary has been removed (because of disease), she manifests normal follicle development, ovum maturation, and estrogen secretion, but she
does not ovulate. If, on the other hand, after approximately 14 days of this therapy, she is given one or two
larger injections of LH, ovulation occurs. This is precisely
what happens in the normal woman; the ovum matures
for 2 weeks under the influence of FSH, LH, and estrogen,
and ovulation is triggered by a rapid brief outpouring from
the pituitary of larger quantities of LH (Fig. 14-16). The
specific mechanism by which LH then causes ovulation
is unclear; best present evidence indicates that the hormone induces increased synthesis of enzymes which
catalyze the chemical dissolution of the thin follicular and
ovarian membranes at the bulge of the mature follicle.

Thus, the midcycle surge of LH emerges as perhaps the single most decisive event of the entire men-

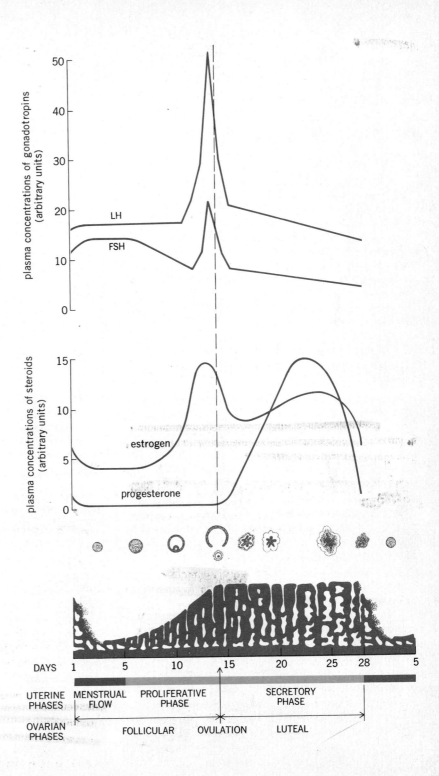

FIGURE 14-14
Summary of plasma hormone concentrations, ovarian events, and uterine change during the menstrual cycle.

strual cycle; indeed, it is the presence of this surge which most strikingly distinguishes the female pattern of pituitary secretion from the relatively unchanging pattern of the male. What causes the LH surge? The explanation is relatively simple in animals which ovulate only after copulation; receptors in the female genital tract are excited, and afferent neurons transmit these impulses ultimately to the hypothalamus, where neurons which secrete LH-releasing factor are stimulated. This is not the case for human beings and most other mammals since ovulation occurs regularly each cycle unrelated to intercourse.[1] Because this surge of LH is the crucial factor in ovulation, a large number of studies have been performed, but major interpretative difficulties have existed, both because there is considerable species difference in this regard and because the relevant hormones could not be measured accurately in the blood until quite recently. The question is far from settled, and we shall present what seems to be the most likely mechanism in women. Our reason for going into this problem in some detail is that it is of enormous practical importance, not just to the relatively few women who fail to ovulate but to the entire world's population, since more precise knowledge of the mechanism triggering the normal LH surge might provide the basis for birth-control methods more physiological than those now available.

It seems very likely that there exist two regulatory brain centers whose neurons synapse ultimately with the cells secreting LH-releasing factor. One of these centers—the "tonic" center—is located in the hypothalamus and exerts a continuous stimulatory effect on the secretion of LH-RF (and FSH-RF), the magnitude of which is dependent on the degree of negative-feedback inhibition exerted by estrogen on the center. It is this center which controls secretion of the gonadotropins during all the cycle except the period of the midcycle LH surge. The location of the other center—the "cyclic" center—is still uncertain but is probably the hypothalamus or the area just beyond it; its neurons are stimulated (rather than inhibited, as is the tonic center) by high levels of estrogen.[2] Moreover, this cyclic center is strongly inhibited by progesterone.

These facts nicely explain the LH surge and ovula-

[1] Scattered bits of evidence suggest, however, that intercourse may indeed influence the time of ovulation in human beings, particularly if there are relatively long intervals between intercourse. Thus, the rate of conception observed for wives of soldiers home on leave is much greater than that predicted if ovulation were not related to intercourse. This seems to be true for rape victims, also.

[2] This stimulatory effect of estrogen on LH release is often called a "positive-feedback" effect to distinguish it from the "negative feedback" usually exerted.

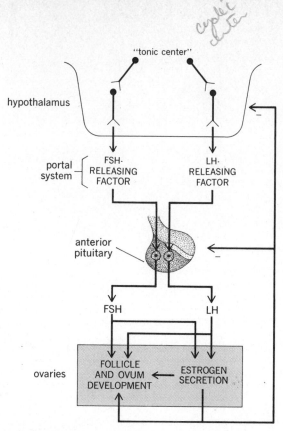

FIGURE 14-15
Summary of hormonal control of follicle and ovum development and estrogen secretion during the follicular phase of the menstrual cycle. Compare with the analogous pattern for the male (Fig. 14-9). The negative signs indicate that estrogen inhibits both the hypothalamus and the anterior pituitary. Estrogen reaches the developing ovum and follicle both by local diffusion and by release into the blood and recirculation to the ovaries.

tion. As estrogen secretion rises rapidly during the last half of the follicular phase, its blood concentration eventually becomes high enough to stimulate the cyclic center, which, by way of its synaptic connections with the neurons secreting LH-RF, causes an outpouring of LH, which, in turn, induces ovulation. Why is there only one surge per cycle, i.e., why does not the elevated estrogen existing throughout most of the luteal phase keep inducing LH surges? The most likely answer is that, as a result of ovulation and succeeding corpus luteum formation, both induced by the LH surge, progesterone secretion begins and progesterone inhibits the cyclic center.

This discussion has so far dealt only with the effects of the sex hormones on the brain. There is, in addition, much evidence to support the view that estrogen exerts

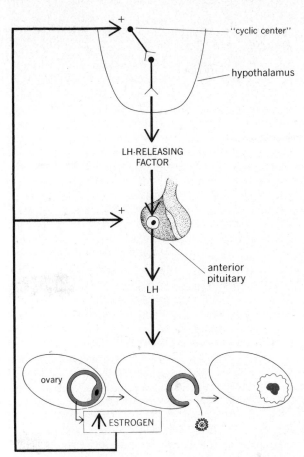

FIGURE 14-16
Ovulation and corpus luteum formation are induced by the markedly increased LH, itself induced by the stimulatory effects of high levels of estrogen.

much of its inhibitory and stimulatory effects on LH secretion via a direct effect on the pituitary cells themselves. Thus, low levels of estrogen may inhibit not only the tonic brain center but the pituitary LH-secreting cells as well. Conversely, high levels of estrogen probably stimulate these pituitary cells directly (perhaps by enhancing their sensitivity to LH-RF) as well as indirectly through the cyclic center. The end result is, of course, the same.[1]

[1] This question takes on added significance in light of evidence that there may be only a single releasing factor for both FSH and LH. If such turns out to be the case, then alteration by estrogen of the relative sensitivities of FSH- and LH-secreting cells to the single releasing factor would be critical. Answers to these questions should be forthcoming shortly.

Control of the corpus luteum

In certain species, e.g., the rat, it is clear that formation and maintenance of the corpus luteum depend upon a third pituitary gonadotropin, prolactin. This does not seem to be the case in women. Most experts now feel that the induction of corpus luteum formation in women depends only upon ovulation, the formation itself occurring autonomously, i.e., depending on no external stimuli. Once formed, maintenance of the corpus luteum requires some stimulatory support from LH (rather than prolactin), but the amount of LH needed is quite small.

What causes the corpus luteum to degenerate if no pregnancy results? Here again the situation is very unclear. The blood concentration of LH shows no sudden decrease during the late luteal phase so that it is difficult to blame regression on any sudden withdrawal of LH. In some mammalian species, it seems that regression is actively induced by some substance (perhaps prostaglandin) produced by the nonpregnant uterus, but such seems not to be the case in women. There are other hypotheses, including the idea that the corpus luteum has a "built-in" life span of approximately 10 to 14 days and that it "self-destructs" unless prevented from doing so by the onset of pregnancy. How pregnancy does this will be described subsequently.

During its short life, the corpus luteum secretes large quantities of estrogen and progesterone. These hormones, particularly estrogen, exert a powerful negative-feedback inhibition (via the hypothalamus and anterior pituitary) of tonic FSH and LH secretion. Accordingly, during the luteal phase of the cycle, pituitary gonadotropin secretion is reduced, which explains the diminished rate of follicular maturation during this second half of the cycle. With degeneration of the corpus luteum, blood estrogen and progesterone concentrations decrease, FSH and LH increase, and a new follicle is stimulated to mature.

Summary

The events described thus far in this section are summarized in Fig. 14-14, showing the ovarian and hormonal changes during a normal nonpregnant menstrual cycle:

1 Under the influence of FSH, LH, and estrogen, a single follicle and ovum reach maturity at about 2 weeks.
2 During the second of these two weeks, under the influence of LH and FSH, estrogen secretion by the follicle cells and other ovarian cell types progressively increases.
3 For several days near midperiod production of LH (and FSH, to a lesser extent) increases sharply as a result

of the stimulatory effects of high levels of estrogen on the cyclic center and the pituitary.

4 The high concentration of LH induces rupture of the follicular ovarian membranes, and ovulation occurs (it is not known what role if any the increased FSH plays).

5 Estrogen secretion decreases for several days after ovulation, perhaps in part because of disruption of the follicle.

6 The ruptured follicle is rapidly transformed into the corpus luteum, which secretes large quantities of both estrogen and progesterone.

7 The high blood concentration of estrogen inhibits the release of LH and FSH, thereby lowering their blood concentrations and preventing the development of a new follicle or ovum during the last two weeks of the period; in addition, progesterone prevents any additional LH surge by inhibiting the cyclic center.

8 Failure of ovum fertilization leads, by unknown mechanisms, to the degeneration of the corpus luteum during the last days of the cycle.

9 The disintegrating corpus luteum is unable to maintain its secretion of estrogen and progesterone, and their blood concentrations drop rapidly.

10 The marked decrease of estrogen and progesterone removes the inhibition of FSH secretion.

11 The blood concentration of FSH begins to rise, follicle and ovum development are stimulated, and the cycle begins anew.

Uterine changes in the menstrual cycle

Profound changes in uterine morphology occur during the menstrual cycle and are completely attributable to the effects of estrogen and progesterone. Estrogen stimulates growth of the uterine smooth muscle (*myometrium*) and the glandular epithelium (*endometrium*) lining its inner surface. Progesterone acts upon this estrogen-primed endometrium to convert it to an actively secreting tissue: The glands become coiled and filled with secreted glycogen; the blood vessels become spiral and more numerous; various enzymes accumulate in the glands and connective tissue of the lining. All these changes are ideally suited to provide a hospitable environment for implantation of a fertilized ovum.

Estrogen and progesterone also have important effects on the mucus secreted by the *cervix*, the first

portion of the uterus. Under the influence of estrogen alone, this mucus is abundant, clear, and nonviscous; all these characteristics are most pronounced at the time of ovulation and facilitate penetration by sperm. In contrast, progesterone causes the cervical mucus to become thick and sticky, in essence a "plug" which may constitute an important blockade against the entry of bacteria from the vagina—a further protection for the fetus should conception occur.

The endometrial changes throughout the normal menstrual cycle should now be readily understandable (Fig. 14-14).

1 The fall in blood progesterone and estrogen, which results from regression of the corpus luteum, deprives the highly developed endometrial lining of its hormonal support; the immediate result is profound constriction of the uterine blood vessels, which leads to diminished supply of oxygen and nutrients. Disintegration starts, the entire lining begins to slough, and the *menstrual flow* begins, marking the first day of the cycle.

2 After the initial period of vascular constriction, the endometrial arterioles dilate, resulting in hemorrhage through the weakened capillary walls; the menstrual flow consists of this blood mixed with endometrial debris. (Average blood loss per period equals 50 to 150 ml of blood.)

3 The *menstrual phase* continues for 3 to 5 days, during which time blood estrogen levels are low.

4 The menstrual flow ceases as the endometrium repairs itself and then grows under the influence of the rising blood estrogen concentration; this period, the *proliferative phase*, lasts for the 10 days between cessation of menstruation and ovulation.

5 Following ovulation and formation of the corpus luteum, progesterone, acting in concert with estrogen, induces the secretory type of endometrium described above.

6 This period, the *secretory phase*, is terminated by disintegration of the corpus luteum, and the cycle is completed.

It is evident that the phases of the menstrual cycle can be named either in terms of the ovarian or uterine events. Thus, the ovarian follicular phase includes the uterine menstrual and proliferative phases; the ovarian luteal phase is the same as the uterine secretory phase. The essential point is that the uterine changes simply reflect the effects of varying blood concentrations of estrogen and progesterone throughout the cycle. In turn, the secretory pattern of these hormones reflects the complex hypothalamic–anterior pituitary–ovarian interactions described previously.

TABLE 14-1
Effects of female sex steroids

Effects of estrogens

1 Growth of ovaries and follicles
2 Growth and maintenance of the smooth muscle and epithelial linings of the entire reproductive tract
　Also: *a* Oviducts: increased motility and ciliary activity
　　　　b Uterus:　increased motility
　　　　　　　　　secretion of abundant, clear cervical mucus
　　　　c Vagina:　increased "cornification" (layering of epithelial cells)
3 Growth of external genitalia
4 Growth of breasts (particularly ducts)
5 Development of female body configuration: narrow shoulders, broad hips, converging thighs, diverging arms
6 Stimulation of fluid sebaceous gland secretions ("anti-acne")
7 Pattern of pubic hair (actual growth of pubic and axillary hair is androgen-stimulated)
8 Stimulation of protein anabolism and closure of the epiphyses (? due to stimulation of adrenal androgens)
9 Sex drive and behavior (? role of androgens)
10 Reduction of blood cholesterol
11 Vascular effects (deficiency → "hot flashes")
12 Feedback effects on hypothalamus and anterior pituitary

Effects of progesterone

1 Stimulation of secretion by endometrium; also induces thick, sticky cervical secretions
2 Stimulation of growth of myometrium (in pregnancy)
3 Decrease in motility of oviducts and uterus
4 Decrease in vaginal "cornification"
5 Stimulation of breast growth (particularly glandular tissue)
6 Inhibition of effects of prolactin on the breasts
7 Elevation of body temperature
8 Feedback effects on hypothalamus and anterior pituitary

Nonuterine effects of estrogen and progesterone

The uterine effects of the sex hormones described above represent only one set of a wide variety exerted by estrogen and progesterone (Table 14-1); all these effects are discussed in this chapter and are listed here for reference. The effects of estrogen in the female are analogous to those of testosterone in the male in that estrogen exerts dominant control over all the accessory sex organs and secondary sex characteristics. Estrogenic stimulation maintains the entire female genital tract—uterus, oviducts, vagina—the glands lining the tract, the external genitalia, and the breasts. It is responsible for the female body-hair distribution and the general female body configuration: narrow shoulders, broad hips, and the characteristic female "curves," the result of fat deposition in the hips, abdomen, and other places. Estrogen has much less general anabolic effect on nonreproductive tissues than testosterone but probably contributes to the general body growth spurt at puberty. Finally, as described above, estrogen is required for follicle and ovum maturation; its increased secretion at puberty, in concert with that of the pituitary gonadotropins, permits ovulation and the onset of menstrual cycles. For the rest of the woman's reproductive life, estrogen continues to support the ovaries, accessory organs, and secondary characteristics. Because its blood concentration varies so markedly throughout the cycle, associated changes in all these dependent functions occur, the uterine manifestations being the most striking.

As was true for testosterone, estrogen acts on the cell nucleus, and its biochemical mechanism of action appears to be at the level of the genes themselves. It should be reemphasized that estrogen is not a uniquely female hormone. Small and usually insignificant quantities of estrogen are secreted by the male adrenal and the testicular interstitial cells (Fig. 14-17), the latter probably being responsible for the breast enlargement so commonly observed in pubescent boys; apparently, the rapidly developing interstitial cells release significant quantities of estrogen along with the much larger amounts of testosterone.

Progesterone is present in significant amounts only during the luteal phase of the menstrual cycle, and its effects are less widespread than those of estrogen, the endometrial changes being the most prominent. Progesterone also exerts important effects on the breasts, the oviducts, and the uterine smooth muscle, the significance of which will be described later. One interesting property of progesterone is its ability to elevate body temperature, apparently by a direct effect on the hypothalamus. The phenomenon serves as a useful, although crude, indicator of the time of ovulation since the increased progesterone secretion associated with formation of the corpus luteum causes the body temperature to rise slightly and to remain elevated throughout the luteal phase. Progesterone also causes a transformation of the cells lining the vagina, and the microscopic examination of some of these cells provides another indicator that ovulation has or has not occurred.

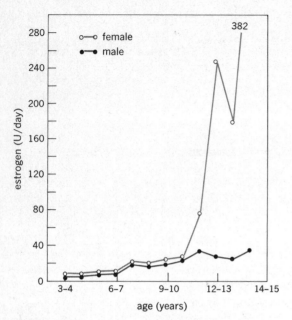

FIGURE 14-17
Excretion of estrogen in the urine of children, an indicator of the blood concentration of estrogen. (*Adapted from Nathanson et al.*)

Female sexual response

Until very recently, the taboos regarding sexual response, particularly in the female, have been so severe that objective scientific investigation has been retarded, but the question is now being studied and knowledge should advance rapidly. Present evidence indicates that the female response to coitus is very similar to that of the male in that it is characterized by marked vasocongestion and muscular contraction in many areas of the body. For example, mounting sexual excitement is associated with engorgement of the breasts and erection of the nipples, resulting from contraction of muscle filaments in them. During coitus, the vaginal epithelium also becomes highly congested and secretes a mucuslike lubricant.

The major sites of female sexual sensation are the vagina and the area of the external genitalia just above the vaginal opening, particularly the *clitoris* (Fig. 14-10). This small shaft is a homologue of the penis and is composed primarily of erectile tissue and endowed with a rich supply of sensory nerve endings. During sexual excitation, the clitoris becomes erect much as the penis does, and massaging it during coitus constitutes a primary

source of sexual tension. Recent direct observations of clitoral function during human intercourse illustrate the unfounded basis of most of the information to be found in the ubiquitous drugstore sex manuals, which almost invariably devote whole chapters to directing the male how to maintain penile contact with the clitoris during coitus. In fact during most of coitus, the position of the erect clitoris completely precludes penile contact, and massage is automatically maintained indirectly via penile movement of the labia, which do touch the clitoris.

The final stage of female sexual excitement may be the process of orgasm, as in the male; if no orgasm occurs, there is a slow resolution of the physical changes and sexual excitement. The female has no counterpart to male ejaculation, but with this exception, the physical correlates of orgasm are very similar in the sexes: there is a sudden increase in skeletal muscle activity involving almost all parts of the body, followed by rapid relaxation; the heart rate and blood pressure increase; the female counterpart of male genital contraction is transient rhythmical contraction of the vagina and uterus. Psychologically, the woman feels a sudden instant of release, followed immediately by an intense awareness of clitoral and pelvic sensation radiating upward, often reported as a feeling of opening up. Finally, a feeling of warmth and relaxation ensues. As in the male, the mechanisms which underlie these physical and psychological components of sexual response remain virtually unknown. An important question is whether the female orgasm plays an important role in assuring fertilization. It was formerly believed that the orgasmic uterine contractions might provide a suction pulling sperm out of the vagina; it seems evident now, however, that uterine contractions are expulsive (as in labor) instead. (The problem of sperm transport within the female genital tract is discussed in a later section.)

A final question related to the female sexual response is sex drive. Incongruous as it may seem, sexual desire in adult women is more dependent upon androgens than estrogen. Thus libido is usually not altered by removal of the ovaries (or its physiological analog, menopause). In contrast, sexual desire is greatly reduced by adrenalectomy, since these glands are the major source of androgens in women. Finally, women receiving large doses of testosterone (for the treatment of breast cancer) generally report a large increase in sexual desire.

This completes our survey of normal reproductive physiology in the nonpregnant female. In weaving one's way through this maze, it is all too easy to forget the prime function subserved by this entire system, namely, reproduction. Accordingly, we must now return to the mature ovum we left free in the abdominal cavity, find it a mate, and carry it through pregnancy, delivery, and breast feeding.

Pregnancy

Following ejaculation into the vagina, the sperm live approximately 48 hr; after ovulation, the ovum remains fertile for 10 to 15 hr. The net result is that for pregnancy to occur coitus must be performed no more than 48 hr before or 15 hr after ovulation (these are only average figures, and there is probably considerable variation in the survival time of both sperm and ovum). However, even these short time limits are probably too generous since, although fertile, the older ova manifest a variety of malfunctions after fertilization, frequently resulting in their rapid death. There seems little doubt that reflex ovulators like the rabbit are far more efficient than human beings are in ensuring a viable pregnancy.

Ovum transport

At ovulation, the ovum is extruded onto the surface of the ovary, and its first mission is to gain entry into the *oviduct*. The end of the oviduct has long, fingerlike projections lined with ciliated epithelium (Fig. 14-10). At ovulation, the smooth muscle of these projections causes them to sweep over the ovary while the cilia beat in waves toward the interior of the duct; these motions immediately suck in the ovum as it emerges from the ovary and start it on its trip toward the uterus. The functional adaptability of this system has been strikingly demonstrated by the fact that women with a single ovary and a single oviduct on the opposite side, e.g., right ovary and a left oviduct, can become pregnant.

Once in the oviduct, the ovum moves rapidly for several minutes, propelled by cilia and by contractions of the duct's smooth muscle coating; the contractions soon diminish, and ovum movement becomes so slow that it takes several days to reach the uterus. Thus, fertilization must occur in the oviduct because of the short life span of the unfertilized ovum. In this regard, the inhibitory effect of progesterone on the oviduct smooth muscle is probably of considerable importance in that ovum movement through the oviduct would otherwise be too rapid. Estrogen, in contrast, enhances oviduct motility so that, during the luteal phase, the actual degree of motility represents a subtle interplay between the opposing effects of the sex steroids.

Sperm transport

Transport of the sperm to the site of fertilization within the oviduct is so rapid that the first sperm arrive within 30 min of ejaculation. This is far too rapid to be accounted for by the sperm's own motility; indeed, the movement produced by the sperm's tail is probably essential only for the final stages of approach and penetration of the ovum. The act of coitus itself provides some impetus for transport out of the vagina into the uterus because of the fluid pressure of the ejaculate and the pumping action of the penis during orgasm. After coitus, the primary transport mechanism is the contractions of the uterine and oviduct musculature. The factors controlling these muscular contractions remain obscure but may involve certain fatty acid derivatives (*prostaglandins*) in the semen which cause smooth muscle to contract. If the reader is puzzled by how the oviduct can move the ovum in one direction and simultaneously aid sperm movement in the other, so are physiologists. The mortality rate of sperm during the trip is huge; of the several hundred million deposited in the vagina, only a few thousand reach the oviduct. This is one of the major reasons that there must be so many sperm in the ejaculate to permit pregnancy; it may result in the selecting out of the most "fit" sperm.

In addition to aiding transport of sperm, the female reproductive tract exerts a second critical effect on them, namely, the conferring upon them of the capacity for fertilizing the egg. Although, as we have mentioned, sperm gain some degree of maturity during their stay in the epididymis, they are still not able to penetrate the zona pellucida surrounding the ovum until they have resided in the female tract for some period of time. The mechanism by which this process, known as *capacitation*, occurs is still very poorly understood.

Entry of the sperm into the ovum

The sperm makes initial contact with the ovum presumably by random motion, there being no good evidence for the existence of "attracting" chemicals. Having made contact, the sperm rapidly moves between adhering follicle cells and through the zona pellucida by releasing from its acrosomal cap, and perhaps other sites as well, enzymes which break down cell connections and intermolecular bonds.

Once through the zona pellucida, the sperm makes contact with the ovum cell membrane and by obscure mechanisms involving fusion of the membranes slowly passes through into the cytoplasm, frequently losing its tail in the process.[1] The nuclei of the sperm and ovum unite; the cell now contains 46 chromosomes, and *fertilization* is complete. However, viability depends upon stopping the entry of additional sperm, which inevitably prevents development of the fertilized egg. Although the mechanism is not known, it is clear that entry of a sperm causes a marked and rapid chemical transformation of the zona pellucida, making it impenetrable to other sperm.

The fertilized egg is now ready to begin its development as it continues its passage down the oviduct to the uterus. If fertilization had not occurred, the ovum would slowly disintegrate and usually be phagocytized by the lining of the uterus. Rarely a fertilized ovum remains in the oviduct, where implantation may take place. Such tubal pregnancies cannot succeed because of lack of space for the fetus to grow, and surgery may be necessary.

Early development, implantation, and placentation

During the leisurely 3- to 4-day passage through the oviduct, the fertilized ovum undergoes a number of cell divisions[2] and after reaching the uterus, it floats free in the intrauterine fluid (from which it receives nutrients) for several more days, all the while undergoing cell division. This entire time span corresponds to days 14 to 21 of the typical menstrual cycle; thus while the ovum is undergoing fertilization and early development, the uterine lining is simultaneously being prepared by estrogen and progesterone to receive it. On approximately the twenty-first day of the cycle, i.e., 7 days after ovulation, *implantation* occurs. The fertilized ovum has by now developed into a ball of cells surrounding a recently formed central fluid-filled cavity and is known as a *blastocyst*.

A section of the blastocyst is shown in Fig. 14-18; note the disappearance of the zona pellucida, an event necessary for implantation. The inner cell mass is destined to develop into the fetus itself, whereas the outer lining of cells is already differentiating into the specialized cells, *trophoblasts*, which will form the nutrient membranes for the fetus, as described below. Once the zona pellucida has disintegrated, the trophoblastic layer rapidly enlarges and makes contact with the uterine wall. The

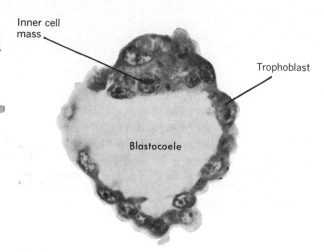

FIGURE 14-18
Blastocyst recovered from the uterus. The zona pellucida is completely absent. [*From A. T. Hertig et al., Carnegie Contrib. Embryol.,* **35:**299 (1954).]

trophoblast cells of blastocysts recovered from the uterus have been found to be quite sticky, particularly in the region overlying the inner cell mass; it is this portion which adheres to the endometrium upon contact and initiates implantation (Fig. 14-19). This initial contact somehow induces rapid development of the trophoblasts, which liberate lytic enzymes capable of breaking down the endometrial tissue. By this means, the *embryo* literally eats its way into the endometrium and is soon completely embedded within it (Fig. 14-20). The destructive powers of the trophoblastic layer serve a second important function: The breakdown of the nutrient-rich endometrial cells provides the metabolic fuel and raw materials for the developing embryo, and during this period, the embryo is completely dependent upon direct absorption of materials from this cell debris. This system, however, is adequate to provide for the embryo only during the first few weeks when it is very small. The system taking over after this is the fetal[3] circulation and *placenta*.

The placenta is a combination of interlocking fetal and maternal tissues which serves as the organ of exchange between mother and fetus. The expanding trophoblastic layer breaks down endometrial capillaries, allowing maternal blood to ooze into the spaces surrounding it; clotting is prevented by the presence of some anti-

---

[1] Upon penetration by the sperm, the ovum completes its last cell division, and the one daughter cell with practically no cytoplasm is extruded.

[2] Identical twins result when a single fertilized ovum at a very early stage of development becomes completely divided into two independently growing cell masses.

[3] After the end of the second month, the embryo is known as the *fetus*.

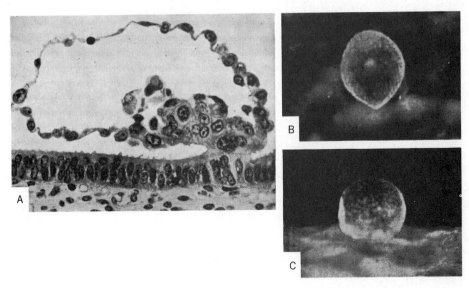

FIGURE 14-19
Photomicrographs showing initial implantation of 9-day
monkey blastocysts. A Cross section of the embryo and uterine
lining. B An entire blastocyst attached to the uterus, viewed
from above. C The same embryo viewed from the side. [*From
C. H. Heuser and G. L. Streeter, Carnegie Contrib. Embryol.,* **29**:15
(1941).]

coagulant substance produced by the trophoblasts. Soon
the trophoblastic layer completely surrounds and projects
into these oozing areas (Fig. 14-20). By this time, the
developing embryo has begun to send out into these
trophoblastic projections blood vessels which are all
branches of larger vessels, the *umbilical arteries and veins,*
communicating with the main intraembryonic arteries and
veins via the *umbilical cord* (Fig. 14-21). Five weeks after
implantation, this system has become well established, the
fetal heart has begun to pump blood, and the entire
mechanism for nutrition of the fetus is in operation. Waste
products move from the fetal blood across the placental
membranes into the maternal blood; nutrients move in the
opposite direction. Many substances, such as oxygen and
carbon dioxide, move by simple passive diffusion,
whereas other substances are carried by active-transport
mechanisms in the placental membranes.

At first, as described above, the trophoblastic projections simply lie in endometrial spaces filled with blood,
lymph, and some tissue debris. This basic pattern (Fig.
14-21) is retained throughout pregnancy, but many
structural alterations have the net effect of making the
system more efficient; e.g., the trophoblastic layer thins,
the distance between maternal and fetal blood thereby
being reduced. It must be emphasized that there is ex-

change of materials between the two blood streams but
no actual mingling of the fetal and maternal blood; the
maternal blood enters the placenta via the uterine artery,
percolates through the spongelike endometrium, and then
exits via the uterine veins; similarly, the fetal blood never
leaves the fetal vessels. In several ways, the system is
analogous to the artificial kidney described in Chap. 11
with the endometrial vascular spaces serving as the bath
through which the fetal vessels course like the dialysis
tubing. A major difference is that active transport is an
important component of placental function. Moreover, the
placenta must serve not only as the embryo's kidney but
as its gastrointestinal tract and lungs as well. Finally, as
will be discussed below, the placenta (probably the
trophoblasts) secretes several hormones of crucial importance for maintenance of pregnancy.

The fetus, floating in its completely fluid-filled cavity
and attached by the umbilical cord to the placenta (Fig.
14-22), develops into a viable infant during the next 9
months (babies born prematurely, often as early as 7
months, frequently survive). (Description of intrauterine
development is beyond the scope of this book but can
be found in any standard textbook of human embryology.)

A point of great importance is that the fetus, during
these 9 months, is subject to considerable influence by

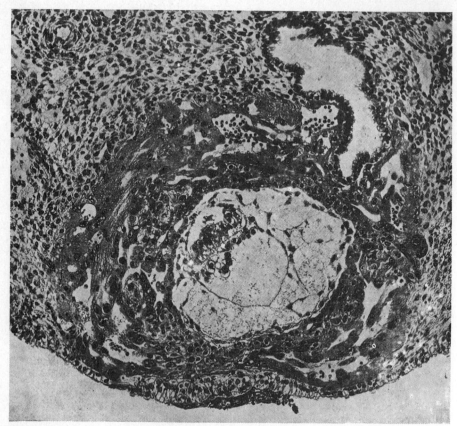

FIGURE 14-20

Eleven-day human embryo, completely embedded in the
uterine lining. [*From A. T. Hertig and J. Rock, Carnegie Contrib.*
Embryol., **29:**127 (1941).]

a host of factors (noise, chemicals, etc.) affecting the mother. Via the placenta, drugs taken by the mother can reach the fetus and influence its growth and development. The thalidamide disaster was our major reminder of this fact in recent years, as is the growing number of babies who suffer heroin withdrawal symptoms after birth as a result of their mothers' drug use during pregnancy. Two other examples: Lead and DDT cross the placenta very easily; we do not know the potential effects, if any, on the fetus of many agents present in the environment.

Hormonal changes during pregnancy

Throughout pregnancy, the specialized uterine structures and functions depend upon high concentrations of circu-lating estrogen and progesterone (Fig. 14-23). During approximately the first 3 months of pregnancy, almost all these steroid hormones are supplied by the extremely active corpus luteum formed after ovulation. Recall that if pregnancy had not occurred, this glandlike structure would have degenerated within 2 weeks after ovulation; in contrast, continued corpus luteum growth and steroid secretion is associated with a developing fetus. Persist-ence of the corpus luteum is essential since continued secretion of estrogen and progresterone is required to sustain the uterine lining and prevent menstruation (which does not occur during pregnancy). We have mentioned our lack of knowledge of the factors causing corpus luteum degeneration during a nonpregnant cycle; its persistence during pregnancy is due, at least in part, to

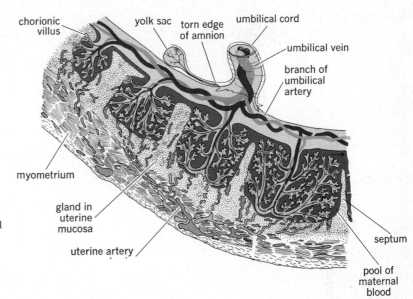

FIGURE 14-21
Schematic diagram: interrelations of fetal and maternal tissues in formation of placenta. The placenta becomes progressively more developed from left to right. (*From B. M. Patten, "Human Embryology," 3d ed., © 1968 by McGraw-Hill, Inc., New York.*)

a hormone from the placenta called *chorionic gonadotropin* (CG). Almost immediately after beginning their endometrial erosion, the trophoblastic cells start to secrete CG into the maternal blood. This protein hormone has properties very similar to those of LH although it is chemically different. What leads to its secretion is unknown, but CG strongly stimulates steroid secretion by the corpus luteum.

Recall that tonic secretion of both LH and FSH is powerfully inhibited by estrogen (it is also likely that CG itself inhibits the release of LH and FSH); therefore, the blood concentrations of these pituitary gonadotropins remain extremely low throughout pregnancy; by this means, further follicle development, ovulation, and menstrual cycles are eliminated for the duration of the pregnancy. In contrast, placental secretion of CG is inhibited to a much lesser degree by these steroids.

The LH-like activity of CG previously formed the basis of most pregnancy tests; blood or urine (CG is excreted in the urine) of a woman late in the first month of pregnancy already contains enough CG to produce detectable reactions when injected into test animals, e.g., (1) ovulation with formation of corpora lutea in the mouse ovary, (2) ovulation in the rabbit, (3) release of sperm by the male frog. These tests have proved highly accurate (98 percent when performed properly) but are now being replaced by a quick and highly sensitive chemical determination of CG.

The secretion of CG increases rapidly during early pregnancy, reaching a peak at 60 to 80 days after the end of the last menstrual period; it then falls just as rapidly, so that by the end of the third month, it has reached a low but definitely detectable level which remains relatively constant for the duration of the pregnancy. Associated with this falloff of CG secretion, the placenta itself begins to secrete large quantities of estrogen and progesterone. The very marked increases in blood steroids during the last 6 months of pregnancy are due almost entirely to the placental secretion. The corpus luteum remains, but its contribution is dwarfed by that of the placenta; indeed, removal of the ovaries during the last 6 months has no effect at all upon the pregnancy, whereas removal during the first 33 months causes immediate loss of the fetus (*abortion*).

An important and clinically useful aspect of placental steroid secretion is that the placenta, unlike the other steroid-producing organs of the body, is not capable of performing alone the entire sequence of reactions leading to the finished steroids. Nor are the fetal steroid-producing organs, but the steps that the one lacks are present in the other and vice versa. Therefore, the placenta and fetus complement each other by the transport of intermediates between them, with a supply of basic compounds coming from the maternal circulation and a supply of finished steroids entering it. Therefore, it is possible to monitor the well-being of the fetus by measur-

FIGURE 14-22
Seven-week embryo. (*Photograph by Chester Reather of Carnegie embryo 8537A, from B. M. Patten, "Human Embryology," 3d ed., © 1968 by McGraw-Hill, Inc., New York.*)

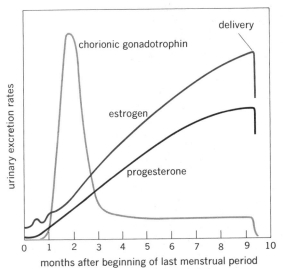

urinary excretion rates

chorionic gonadotrophin

delivery

estrogen

progesterone

months after beginning of last menstrual period

FIGURE 14-23
Urinary excretion of estrogen, progesterone, and chorionic gonadotropin during pregnancy. Urinary excretion rates are an indication of blood concentrations of these hormones.

ing in the maternal blood the concentrations of those steroids mainly arising from the fetus; an example is the estrogen, estriol.

Finally, this remarkable organ not only produces steroids and CG but several other hormones as well. Indeed, just as CG closely mimics the activity of LH, it may well be that the placenta produces a complete set of analogs of the anterior pituitary hormones. The best-documented one at present is a hormone which has effects very similar to those of growth hormone (and prolactin as well). This hormone may play an important role in maintaining, in the mother, a positive protein balance, mobilizing fats for energy, and stabilizing plasma glucose at relatively high levels to meet the needs of the fetus.

Of the numerous other physiological changes, hormonal and nonhormonal, in the mother during pregnancy, many, such as increased metabolic rate and appetite, are obvious results of the metabolic demands placed upon her by the growing fetus. Of great importance is salt and water metabolism because of its relationship to *toxemia of pregnancy*. During a normal pregnancy, body sodium and water increase considerably, the extracellular volume alone rising by approximately 1 liter. At present, it appears that the major factor is greater secretion of renin, aldosterone, and ADH, but the mechanisms responsible

remain obscure. Some women retain abnormally great amounts of fluid and manifest protein in the urine and hypertension, which if severe enough may cause convulsions. These are the symptoms of the disease known as toxemia of pregnancy. Since it can usually be well controlled by salt restriction, in the United States it is seen primarily in the lower socioeconomic segments of the population which do not obtain adequate medical care during pregnancy. Despite the obvious association with salt retention, all attempts to determine the factors responsible for the disease have failed: e.g., a causal role for aldosterone seems unlikely since full-blown toxemia can occur in adrenalectomized women.

Parturition
(Delivery of the infant)

A normal human pregnancy lasts approximately 40 weeks, although many babies are born 1 to 2 weeks earlier or later. Delivery of the infant, followed by the placenta, is produced by strong rhythmical contractions of the uterus. Actually, beginning at approximately 30 weeks, weak and infrequent uterine contractions occur, gradually increasing in strength and frequency. During the last month, the entire uterine contents shift downward so that the baby is brought into contact with the outlet of the uterus, the *cervix*. In over 90 percent of births, the baby's head is downward and acts as the wedge to dilate the cervical canal. By the onset of labor, the uterine contractions have become coordinated and quite strong (although usually painless at first) and occur at approximately 10- to 15-min intervals. Usually during this period or before, the membrane surrounding the fetus ruptures, and the intrauterine fluid escapes out the vagina. As the contractions, which begin in the upper portion and sweep down the uterus, increase in intensity and frequency, the cervical canal is forced open to a maximum diameter of approximately 10 cm. Until this point, the contractions have not moved the fetus out of the uterus but have served only to dilate the cervix. Now the contractions move the fetus through the cervix and vagina. At this time the mother, by bearing down to increase abdominal pressure, can help the uterine contractions to deliver the baby. The umbilical vessels and placenta are still functioning, so that the baby is not yet on its own, but within minutes of delivery both the infant's and mother's placental vessels completely contract, the entire placenta becomes separated from the underlying uterine wall, and a wave of uterine contractions delivers the placenta (the afterbirth) as well. Ordinarily, the entire process from beginning to end proceeds automatically and requires no real medical intervention, but in a small percentage of cases, the position of the baby or some maternal defect can interfere with normal delivery. The position is important for several reasons: (1) If the baby is not oriented head first, another portion of his body is in contact with the cervix and is generally a far less effective wedge; (2) because of its large diameter compared with the rest of the body, if the head went through the canal last, it might be obstructed by the cervical canal, leading to obvious problems when the baby attempts to breathe; and (3) if the umbilical cord becomes caught between the birth canal and the baby, mechanical compression of the umbilical vessels can result. Despite these potential difficulties, however, most babies who are not oriented head first are born normally and with little difficulty.

What mechanisms control the events of parturition? Let us consider a set of fairly well-established facts:

1 The uterus is composed of smooth muscle capable of autonomous contractions and having inherent rhythmicity, both of which are facilitated by stretching the muscle.

2 The efferent neurons to the uterus are of little importance in parturition since anesthetizing them in no way interferes with delivery.

3 Progesterone exerts a powerful inhibitory effect upon uterine contractility. Shortly before delivery, the secretion of progesterone (and estrogen) sometimes drops, perhaps due to "aging" changes in the placenta.

4 Oxytocin, one of the hormones released from the posterior pituitary, is an extremely potent uterine-muscle stimulant. Oxytocin is reflexly released as a result of afferent input into the hypothalamus from receptors in the uterus, particularly the cervix.

5 The pregnant uterus contains several prostaglandins, one of which is a profound stimulator of uterine smooth muscle; an increase in the release of this substance has been demonstrated during labor.

These facts can now be put together in a unified pattern, as shown in Fig. 14-24. The precise contributions of each of these factors are unclear; moreover, we cannot answer the crucial question: Which factor (if any) actually initiates the process? Once started, the uterine contractions exert a positive-feedback effect upon themselves via reflex stimulation of oxytocin and local facilitation of their inherent contractility; but what *starts* the contractions? The decrease in progesterone cannot be essential since it simply does not occur in most women. Nor is uterine distension or the presence of a fetus a requirement, as attested by the remarkable fact that typical "labor" begins at the expected time in animals from which the fetus

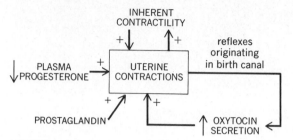

FIGURE 14-24
Factors stimulating uterine contractions during parturition. Note the positive-feedback nature of several of the inputs. What initiates parturition is not known.

has been removed weeks previously. A primary role for prostaglandin has recently been touted but the evidence is far from conclusive. Regardless of their relative contributions to normal parturition, both prostaglandins and oxytocin are useful clinically in artificially inducing labor.

Lactation

Perhaps no other process so clearly demonstrates the intricate interplay of various hormonal control mechanisms as milk production. The endocrine control has been established by numerous investigations and observations, none more striking than that in 1910 of Siamese twins: one twin became pregnant, and both women lactated after delivery.

The *breasts* are formed of epithelium-lined ducts which converge at the *nipples*. These ducts branch all through the breast tissue and terminate in saclike structures typical of exocrine glands, called *alveoli*. The alveoli, which secrete the milk, look like bunches of grapes with stems terminating in the ducts. The alveoli and the ducts immediately adjacent to them are surrounded by specialized contractile cells called *myoepithelial cells*. Before puberty, the breasts are small with little internal structure. With the onset of puberty, estrogen and progesterone act in concert upon the ductile tissue and alveoli to produce the basic architecture of the adult breast. However, in addition to these steroids, normal breast development at puberty requires *prolactin* and *growth hormone*, both secreted by the anterior pituitary.

During each menstrual cycle, breast morphology fluctuates in association with the changing blood concentrations of estrogen and progesterone, but these changes are small compared with the marked breast enlargement which occurs during pregnancy as a result of the stimulatory effects of estrogen and progesterone on both ducts and alveoli. Indeed, there also frequently occurs development *in utero* of the fetal breasts and production of so-called witches' milk; this infantile development is, of course, short-lived and quickly disappears when delivery removes the child from the effects of sex hormone stimulation.

The single most important hormone promoting milk production is prolactin from the anterior pituitary. Yet, despite the fact that prolactin is elevated and the breasts markedly enlarged as pregnancy progresses, there is no secretion of milk. What occurs at delivery to permit the onset of lactation? Experiments on other mammals have shown that removing the fetus during pregnancy without interfering with the placenta does not induce lactation; in contrast, removal of the placenta any time late in pregnancy results in the onset of lactation. Most evidence indicates that the inhibitory effect of the placenta is due to its secretion of estrogen and progesterone, which in large concentration appear to inhibit milk production by a direct action on the breasts. Thus, delivery removes the sources of sex steroids and, thereby, the inhibition.

The stimulus to increased prolactin secretion during pregnancy is unclear (stimulation by estrogen is one factor), but the major factor maintaining the secretion during lactation is quite clear: reflex input to the hypothalamus from receptors in the nipples which are stimulated by suckling (Fig. 14-25). Thus, milk production ceases soon after the mother stops nursing her infant but continues uninterrupted for years if nursing is continued.

One final reflex process is essential for nursing. Milk is secreted into the lumen of the alveoli (Fig. 14-26), but because of their structure the infant cannot suck the milk out. It must first be moved into the ducts, from which it can be sucked. This process is called *milk let-down* and is accomplished by contraction of the myoepithelial cells surrounding the alveoli; the contraction is directly under the control of oxytocin, which is reflexly released by suckling (Fig. 14-27), just like prolactin.[1] Many women experience uterine contractions during nursing because of the uterine effects of oxytocin. Higher brain centers also exert important influence over oxytocin release: many nursing mothers actually leak milk when the baby cries. In view of the central role of the nervous system in lactation reflexes, it is no wonder that psychological factors can interfere with a woman's ability to nurse.

The end result of all these processes, the milk, contains four major constituents: water, protein, fat, and

[1] The functions of prolactin and oxytocin in males are unknown.

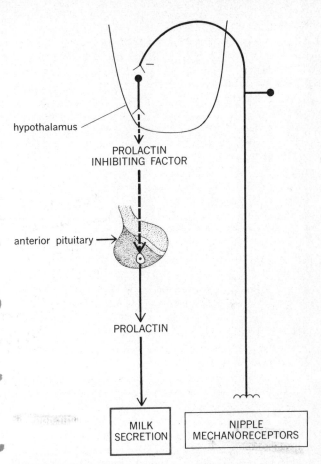

FIGURE 14-25
Nipple suckling reflex. The neural pathway is schematic; actually multiple interneurons are involved.

the carbohydrate lactose. The mammary alveolar cells must be capable of extracting the raw materials—amino acids, fatty acids, glycerol, glucose, etc.—from the blood and build them into the higher-molecular-weight substances. These synthetic processes require the participation of prolactin, insulin, growth hormone, cortisol, and probably other hormones—an amazing coordination.

Another important neuroendocrine reflex triggered by suckling is inhibition of FSH and LH release by the pituitary, with subsequent block of ovulation. This inhibition apparently is relatively short-lived in many women, and approximately 50 percent begin to ovulate despite continued nursing. Pregnancy is common in women lulled into false security by the mistaken belief that failure to ovulate is always associated with nursing.

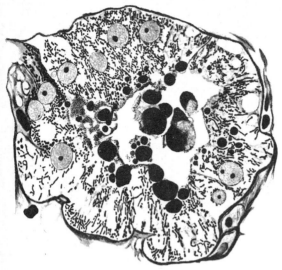

FIGURE 14-26
Alveolus of a lactating mammary gland of a rabbit. The cells contain fat droplets (stained black), which, together with the adjacent cytoplasm, are extruded into the lumen. (*Adapted from Bloom and Fawcett.*)

Birth control

Only recently has it become apparent that birth control is necessary not for personal reasons but for the general survival of mankind. If present birth rates do not change, the world's population is expanding so fast that ultimately mass deaths will result from starvation and pollution. The world's population in 6000 B.C. is estimated at 8 million. At the time of Christ's birth, the number had increased to 250 million, or approximately five doublings in 6,000 years. The numbers were still fairly low, 500 million, by 1600. The primary reason for this relatively slow growth was the enormous death rate; most of the children died within the first few years, and the average life expectancy even in 1800 was approximately 40 years. Figure 14-28 illustrates population growth since 1650; the population doubles more and more frequently, and the next doubling from now will take less than 50 years. The earth's population will increase from 3.2 to 6.4 billion with no end in sight. There is not the slightest chance that science can force the earth to accommodate such numbers for very long. Actually, the doubling would soon cease and the population stabilize but only because the death rate will soar to equal the birth rate! The only way to avert this

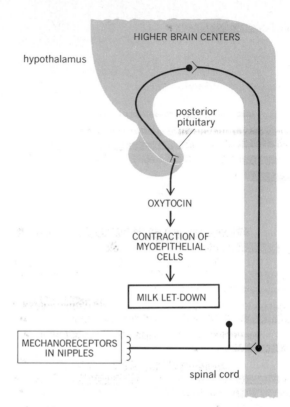

FIGURE 14-27
Suckling-reflex control of oxytocin secretion and milk let-down.

catastrophe is to reduce the birth rate. Thus, birth control is no longer a purely personal decision but a problem of global dimensions.

Table 14-2 summarizes possible means of preventing fertility. We present it because it should serve to summarize a majority of the information presented in this chapter (there are no new facts in the table). For each possibility listed there are many possible preventive aspects; for example, implantation is extremely complex and requires a long sequence of events. Thus, for each possibility, scientific investigation revolves around the question: What are the normal physiological events that make the process possible and how can we intervene to prevent them?

Until recently, techniques of birth control (*contraception*) were primarily those which prevent sperm from

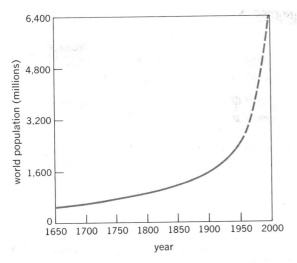

FIGURE 14-28
Growth of world population since 1650.

TABLE 14-2

Possible means of preventing fertility in men†

I Interference with sperm survival
  a Prevention of maturation process in epididymis
  b Prevention of function of accessory glands
      1 Prevention of androgen action on accessory glands
      2 Prevention of formation of accessory-gland secretion
      3 Prevention of accessory-gland secretion from entering urethra
  c Creation of hostile environment to sperm in vas deferens or urethra

II Interference with testicular function at testicular level
  a Prevention of androgen action on seminiferous tubules
  b Prevention of action of FSH on seminiferous tubules
  c Prevention of sperm division

III Interference with pituitary function
  a Prevention of FSH secretion at a pituitary level
  b Prevention of action of FSH-releasing factor on pituitary
  c Prevention of secretion of FSH-releasing factor
  d Prevention of unknown seminiferous-tubule inhibition of FSH secretion
  e Abolishment of extrahypothalamic central nervous factors required for normal reproductive function

Possible means of preventing fertility in women‡

I Interference with ovarian function at ovarian level
  a Prevention of initiation of follicle growth (no response to early FSH rise)
  b Prevention of response to ovulatory LH surge
  c Prevention of maturation of follicle or maturation of ova (e.g., inhibition of meiotic division)
  d Prevention of corpus luteum formation
  e Prevention of ovarian estrogen secretion
  f Prevention of ovarian progesterone secretion
  g Prevention of the maintenance of the corpus luteum during early pregnancy

II Interference with pituitary function
  a Prevention of FSH or LH secretion, or both, at a pituitary level
  b Prevention of action of releasing factors on pituitary
  c Prevention of secretion of releasing factors
  d Alteration of estrogen or progesterone action on tonic center (too much or too little)
  e Prevention of estrogen action on cyclic center
  f Abolishment of extrahypothalamic central nervous factors required for normal reproductive hypothalamic function

III Interference with sperm action
  a Prevention of sperm entrance to vagina
  b Prevention of sperm entrance to uterus
  c Prevention of sperm entrance to oviducts
  d Creation of hostile environment to sperm in vagina, uterus, or oviducts
  e Prevention of sperm capacitation
  f Prevention of sperm penetration of ovum by action on sperm

IV Interference with ovum action
  a Prevention of ova release from ovary
  b Prevention of ova entrance to oviducts
  c Creation of hostile environment for ova in oviducts
  d Prevention of sperm penetration of ovum by action on ova

V Prevention of survival of fertilized ova
  a Prevention of fertilized ova from undergoing mitosis
  b Alteration of migration along fallopian tube (too fast or too slow)
  c Prevention of implantation of blastocyst in uterine wall
  d Destruction or expulsion of embryo after implantation in uterine wall
  e Prevention of CG secretion by placenta
  f Prevention of sex steroid secretion by placenta

  † Not listed are surgical interventions removing the source of sperm, such as castration, or preventing egress of sperm, such as vasectomy.

  ‡ Not listed is surgical removal of any of the components (e.g., uterus, fallopian tubes, ovaries).

reaching the ovum; vaginal diaphragms, sperm-killing jellies, and male condoms. Each of these methods has aesthetic drawbacks, and, more important, they are frequently ineffective (Table 14-3). Another widely used method is the so-called rhythm method, in which couples merely abstain from coitus near the time of ovulation. Unfortunately, it is difficult to time ovulation precisely even with laboratory techniques; e.g., the progesterone-associated rise in body temperature or change in cervical mucus and vaginal epithelium, all of which are indicators of ovulation, occur only *after* ovulation. This problem, combined with the marked variability of the time of ovulation in many women, explains why this technique is only partially effective.

Since 1950, an intensive search has been made for a simple, effective contraceptive method, the first fruit of these studies being "the pill," the *oral contraceptive*. Its development was based on the knowledge that combinations of estrogen and progesterone inhibit pituitary gonadotropin release, thereby preventing ovulation. The most commonly used agents, at least at first, were combinations of an estrogen- and a progesterone-like substance (progestogens). Each month, the pill is taken for 20 days, then discontinued for 5 days; this steroid withdrawal produces menstruation, and the net result is a menstrual cycle without ovulation. The monthly withdrawal is required to avoid "breakthrough" bleeding which would occur if the steroids were administered continuously. Another type of regimen is the so-called sequential method in which estrogen is administered alone for 15 days followed by estrogen plus progestogen for 5 days, followed by withdrawal of both steroids. As with the combination pills, this regimen interferes with the orderly secretion of gonadotropins and prevents ovulation; no LH

surge occurs, at least in part because of the constant estrogen levels, i.e., the absence of a rising estrogen level capable of exerting positive feedback. Finally, another type of contraceptive pill is progestogen alone, administered either continuously or intermittently.

Study of the mechanism of action of this last so-called minipill has revealed that it does not usually prevent ovulation. Rather it interferes with other steps required for fertility. Recall that progesterone causes a thickening of the cervical mucus; this is done by progestogen pills and probably prevents sperm entry into the uterus. Moreover, it has now become clear that all the oral contraceptives have multiple antifertility effects. In other words, the hormonal milieu required for normal pregnancy is such that these exogenous steroids interfere with many of the steps between coitus and implantation of the blastocyst. Taken correctly, they are almost 100 percent effective (the minipill is somewhat less effective). Serious side effects, such as intravascular clotting, have been reported but only in a small number of women; however, only time can show whether undesirable effects will ultimately appear as a result of chronic alteration of normal hormonal balance.

Another type of contraceptive which is highly effective (although not 100 percent) and which illustrates the dependency of pregnancy upon the right conditions is the *intrauterine device* (Table 14-3). Placing one of these small objects in the uterus prevents pregnancy, apparently by somehow interfering with the endometrial preparation for acceptance of the blastocyst.

The search goes on for an effective method which will reduce even further the possibility of unwanted side effects and will still be easy to use. Almost every possible process shown in Table 14-2 is worth following up. Two recent developments are postcoital medications: prostaglandins and the estrogenlike diethylstilbestrol (DES). Both cause increased contractions of the female genital tract and may, therefore, cause expulsion of the fertilized ovum or failure to implant. Enthusiasm for DES has been tempered by the possibility that it may have cancer-producing properties.

Fertility prevention is of great importance, but the other side of the coin is the problem of unwanted infertility. Approximately 10 percent of the married couples in the United States are infertile. There are many reasons—some known, some unknown—for infertility; indeed, Table 14-2 also serves as a list of possible causes of infertility. Careful investigation of infertile couples frequently permits diagnosis and therapy of the basic problem.

TABLE 14-3
Effectiveness of contraceptive methods

Method	Pregnancies per 100 women per year
None	115
Douche	31
Rhythm	24
Jelly alone	20
Withdrawal	18
Condom	14
Diaphragm	12
Intrauterine device	5
Oral contraceptive	1
correctly used	0

SECTION C
THE CHRONOLOGY OF SEX DEVELOPMENT

Sex determination

Sex is determined by the genetic inheritance of the individual, specifically by two chromosomes called the *sex chromosomes*. With appropriate tissue-culture techniques all the chromosomes in human cells can be made visible; such studies have demonstrated the presence of 46 chromosomes, 22 pairs of somatic (nonsex) chromosomes and 1 pair of sex chromosomes. The larger of the sex chromosomes is called the X chromosome and the smaller the Y chromosome. Genetic males possess one X and one Y, whereas females have two X chromosomes

(Fig. 14-29). Thus the genetic difference between male and female is simply the difference in one chromosome. The reason for the approximately equal sex distribution of the population should be readily apparent (Fig. 14-30): the female can contribute only an X chromosome, whereas the male, during meiosis, produces sperm, half of which are X and half of which are Y. When the sperm and ovum join, 50 percent should have XX and 50 percent XY.

Interestingly, however, sex ratios at birth are not 1:1. Rather, there tends to be a slight preponderance of male births (in England, the ratio of male to female births is 1.06). Even more surprising, the ratio at the time of conception seems to be much higher. From various types of evidence, it has been estimated that there may

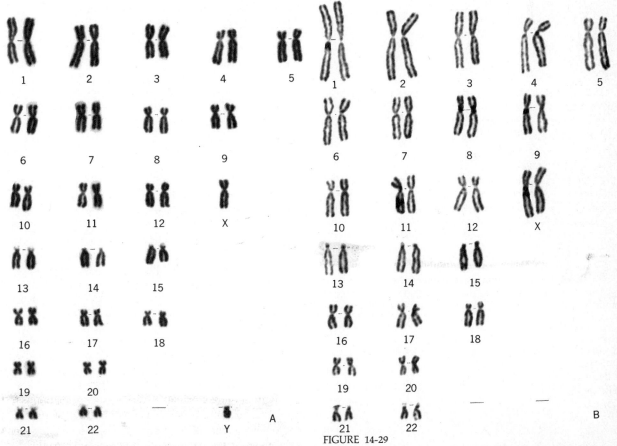

FIGURE 14-29

Normal chromosomes in male (A) and female (B) cells. (*Courtesy of J. Lejeune, Chaire de Génétique Fondamentale, Paris.*)

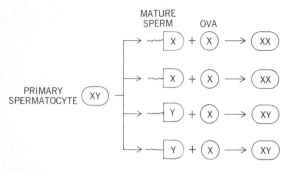

FIGURE 14-30
Basis of genetic sex determination.

be 30 percent more male conceptions than female. There are several implications of these facts. First, there must be a considerably larger *in utero* death rate for males. Second, the "male," i.e., XY, sperm must have some advantage over the "female" sperm in reaching and fertilizing the egg. It has been suggested, for example, that since the Y chromosome is lighter than the X, the "male" sperm might be able to travel more rapidly. There are numerous other theories, but we are far from an answer. Moreover, it should be pointed out that conception and birth ratios show considerable variation in different parts of the world and, indeed, in rural and urban areas of the same country.

The methods of making human chromosomes visible are quite difficult. However, an easy method for distinguishing between the sex chromosomes was found quite by accident; the cells of female tissue (scrapings from the cheek mucosa are convenient) contain a readily detected nuclear mass believed to derive from the XX chromosomal combination. This has been called the *sex chromatin* and is not usually found in male cells. The method has proved valuable when genetic sex was in doubt. Its use and that of the more exacting tissue-culture visualization have revealed a group of genetic sex abnormalities characterized by such bizarre chromosomal combinations as XXX, XXY, X, and many others. Just how these combinations arise remains obscure, but the end result is usually tragic, the failure of normal anatomical and functional sexual development. For example, patients with only one sex chromosome, an X, show no gonadal development; apparently both X chromosomes are required for normal ovarian growth during embryological development.

Sex differentiation

It is not surprising that persons with abnormal genetic endowment manifest abnormal sexual development, but careful study has also revealed patients with normal chromosomal combinations but abnormal sexual appearance and function. For example, a genetic male (XY) may have testes and female internal genitalia (vagina and uterus); such people are termed male pseudohermaphrodites. This kind of puzzle leads us into the realm of sex differentiation, i.e., the process by which the fetus develops the male or female characteristics directed by its genetic makeup. The genes *directly* determine only whether the individual will have testes or ovaries; virtually all the rest of sexual differentiation depends upon this genetically determined gonad.

Differentiation of the gonads

The male and female gonads derive embryologically from the same site in the body. Until the sixth week of life, there is no differentiation of this site. During the seventh week, in the genetic male, the testes begin to develop; in the genetic female, this does not take place, and several weeks later ovaries begin to develop instead. The embryonic gonad, testis or ovary, then regulates the remainder of the individual's sexual development.

Differentiation of internal and external genitalia

The very early fetus is sexually bipotential so far as its internal duct system and external genitalia are concerned. During differentiation, either the male or female duct system will fully develop, the other becoming vestigial. Similarly, externally, either a penis or clitoris will develop, and the tissue near it will either fuse, in the male, to become a scrotum or remain separate, in the female, as the labia.

Shortly after the gonadal development, the genital ducts and glands begin to differentiate and later still the external genitalia. In many species it is possible to remove or damage the area from which the gonads derive without interrupting the pregnancy. When this is done, regardless of genetic sex, the gonadless embryo develops *female* internal genitalia (uterus, vagina, etc.). The conclusion is that normally the presence of functioning testes represses the development of the female duct system and induces development of the male organs; in contrast, a female gonad need not be present for the female organs to

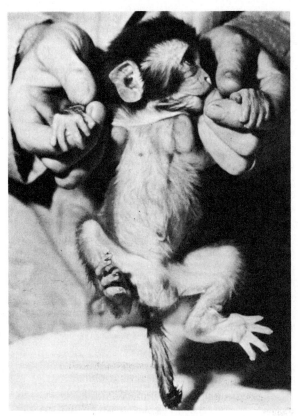

FIGURE 14-31
Masculinization of the external genitalia of a genetic female by treatment with testosterone prior to birth. Photograph taken shortly after birth. (*From R. W. Goy, Reproductive Behavior in Mammals, in C. W. Lloyd (ed.), "Human Reproduction and Sexual Behavior," Lea and Febiger, Philadelphia, 1964.*)

· develop. The testes exert these effects by secreting an unknown substance (probably not testosterone), which acts upon the developing genital tissue. This concept of inducers applies in virtually all areas of embryological development. A similar analysis holds for the later development of the external genitalia except that ·here the inducer does appear to be testosterone secreted by the interstitial cells as a result of stimulation by chorionic gonadotropin from the placenta. The converse of this type of experiment can also be done; Fig. 14-31 shows a genetic female monkey whose mother had been given large doses of testosterone during the critical period of external genital development; she has a scrotum and ·

markedly enlarged clitoris. To return to our previously described XY patient with the female duct system, it seems reasonable to suspect a failure of his gonadal function during the period of duct differentiation. Note that, depending upon the timing of gonadal failure, one could develop external and internal genitalia of opposite sexes.

Sexual differentiation of the central nervous system

As we have seen, the male and female hypothalamus differ in that there is cyclical secretion of LH-releasing factor in the female but rather fixed continuous release in the male. In primates (as opposed to rats) it appears that this difference does not reflect any qualitative difference in hypothalamic "imprinting" between male and female; rather it is due simply to the fact, described previously, that large amounts of the dominant female sex hormone, estrogen, can stimulate LH release whereas testosterone can only inhibit it. Thus, it has been demonstrated recently that the administration of estrogen to male castrated monkeys elicited LH surges indistinguishable from those shown by females.

The situation may be quite different for sexual behavior in that qualitative differences may be imprinted during development; genetic female monkeys given testosterone during late *in utero* life manifest not only masculinized external genitalia but pronounced masculine sex behavior as adults. Related to this is the question of whether exposure to androgens during *in utero* existence is necessary for development of other behavior patterns in addition to sex. Again, for other primates, the answer seems a clear-cut *yes;* for example, virilized female monkey offspring manifest a high degree of male-type play behavior during growth. The evidence for human beings is very scanty; perhaps the best-studied group has been 10 young women who were accidentally virilized during late *in utero* development because their mothers were given synthetic hormones (to prevent abortion) not known to be androgenic at the time. At birth they had male external genitalia (enlarged clitoris and fused empty scrotum) but normal internal genitalia; this was corrected surgically very early in life, and their behavior for the next 5 to 15 years was studied carefully and compared with a control group matched to them in every possible way. Of the 10 girls, 9 were tomboys; they preferred boys' toys to dolls, outdoor activities (climbing trees, baseball, football), and playing with boys and wearing boys' clothes.

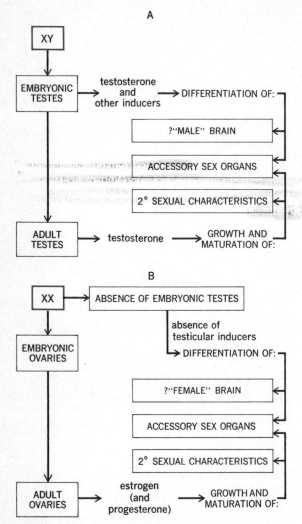

FIGURE 14-32
Summary of sex-organ differentiation in (A) male and (B) female.

Their high levels of physical energy expenditure and self-assertiveness were quite analogous to the rough-and-tumble play of prenatally masculinized female monkeys. They were much more interested in future careers and less interested in children or marriage. However, although their interest in dating and boyfriends was somewhat delayed, there were no indications of lesbianism.

On the basis of this study and several others like

it, the present tentative conclusion is that the most powerful factor determining gender identity and behavior are experiential and social but that these influences play upon a range of potentials which is, in some way, influenced by prenatal exposure to (in the male) or lack of exposure to (in the normal female) testosterone.

In summary (Fig. 14-32), it is evident that the male gonadal secretions play a crucial role in determining normal sexual differentiation *in utero* or just after birth. Shortly after delivery, when the gonad is deprived of the stimulant action of chorionic gonadotropin, testosterone secretion becomes very low until puberty, when it once again increases to stimulate the organs it had previously helped to differentiate. In contrast, estrogen probably takes little active part in *in utero* development, female differentiation requiring only the absence of testicular secretion. Estrogen is, of course, the stimulating agent for the female sex organs during puberty and adult life. Finally, the X and Y chromosomes appear to dictate directly only whether testes or ovaries develop.

Puberty

Puberty is the period, usually occurring sometime between the ages of ten and fourteen, during which the reproductive organs mature and reproduction becomes possible. In the male, the seminiferous tubules begin to produce sperm; the genital ducts, glands, and penis enlarge and become functional; the secondary sex characteristics develop; and sexual drive is initiated. All these phenomena are effects of testosterone, and puberty in the male is the direct result of the onset of testosterone secretion by the testes (Fig. 14-33). Although significant testosterone secretion occurs during late fetal life, within the first days of birth the interstitial cells disappear and testosterone secretion becomes very low until puberty.

The critical question is: What stimulates testosterone secretion at puberty? Or conversely: What inhibits it before puberty? Experiments with testis or pituitary transplantation in other mammals and studies of hormone injections in man have suggested the hypothalamus as the critical site of control. Before puberty, the hypothalamus fails to secrete significant quantities of releasing factors to stimulate secretion of FSH and LH (Fig. 14-34); deprived of these gonadotropic hormones, the testes fail to produce sperm or large amounts of testosterone. Puberty is initiated by an unknown alteration of brain function which permits secretion of the hypothalamic releasing factors for LH and FSH; the mechanism of this change

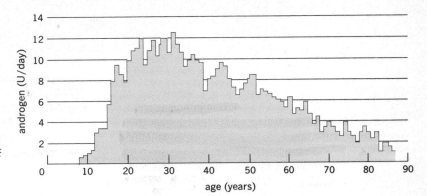

FIGURE 14-33
Excretion of androgen in the urine of normal boys and men, an indicator of the blood concentration of androgen. *(Adapted from Pedersen-Bjergaard and Tonnesen, Acta Med. Scand.)*

remains unknown. It is interesting that children with brain tumors or other lesions of the hypothalamus or pineal may undergo precocious puberty, i.e., sexual maturation at an unusually early age, sometimes within the first 5 years of life, but the interpretation of these "experiments of nature" remains controversial. In particular, the possible role of that mysterious gland, the pineal, in control of reproduction is presently the subject of much investigation.

The picture for the female is completely analogous to that for the male. Throughout childhood, estrogen is secreted at very low levels (Fig. 14-17). Accordingly, the female accessory sex organs remain small and nonfunctional; there are minimal secondary sex characteristics,

and ovum maturation does not occur. As for the male, prepuberal dormancy is probably due mainly to deficient secretion of the hypothalamic releasing factors controlling FSH and LH secretion, and the onset of puberty is occasioned by an alteration in brain function which raises secretion of these releasing factors. Production of larger amounts of releasing factors at puberty raises secretion of pituitary gonadotropins and estrogen. The increased estrogen induces the striking changes associated with puberty and, through its stimulatory effect on the hypothalamic cyclic center, permits menstrual cycling to begin. Precocious puberty also occurs in females; the youngest mother on record gave birth to a full-term healthy infant by cesarean section at 5 years, 8 months.

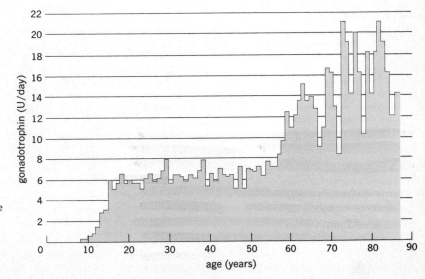

FIGURE 14-34
Excretion of anterior pituitary gonadotropins (FSH and LH) in the urine of normal boys and men, an indicator of the blood concentration of gonadotropins. *(Adapted from Pedersen-Bjergaard and Tonnesen, Acta Med. Scand.)*

It is fascinating to learn that light, which has a powerful effect on the sex life of other animals, also exerts some effect in human beings: Blind girls undergo puberty significantly earlier than normal girls, and airline hostesses, whose normal day-night patterns are disrupted by long jet flights, also have frequent menstrual irregularities.

It should be recognized that the maturational events of puberty usually proceed in an orderly sequence but that the ages at which they occur may vary among individuals. In boys, the first sign of puberty is acceleration of growth of testes and scrotum; pubic hair appears a trifle later, and axillary and facial hair still later. Acceleration of penis growth begins on the average at 13 (range 11 to 14.5 years) and is complete by 15 (13.5 to 17). But note that, because of the overlap in ranges, some boys may be completely mature whereas others at the same age (say 13.5) may be completely prepubescent. Obviously, this can lead to profound social and psychological problems. In girls, appearance of "breast buds" is usually the first event (average age = 11) although pubic hair may, on occasion, appear first. Menarche, the first menstrual period, is a later event (average = 13) and occurs almost invariably after the peak of the total body growth spurt has passed. The early menstrual cycles are usually not accompanied by ovulation so that conception is generally not possible for 12 to 18 months after menarche. One of the most striking facts concerning menarche is the remarkable decrease over the past 150 years in the age at which it occurs in all industrialized countries. For example, the age of menarche in Norway has decreased from near 17.5 in 1830 to 13 at present. Improved nutrition may have played an important causal role in this phenomenon but other factors, as yet unknown, almost certainly contribute. Again, one can imagine the social and psychological impact of such a change on young people.

Menopause

Ovarian function declines gradually from a peak usually reached before the age of thirty, but significant problems, if they occur at all, do not usually arise until the forties. Figures 14-35 and 14-36 demonstrate that the cause of the decline is decreasing ability of the aging ovaries to respond to pituitary gonadotropins. Estrogen secretion drops despite the fact that the gonadotropins, partially released from the negative-feedback inhibition by estrogen, are secreted in greater amounts. Ovulation and the menstrual periods become irregular and ultimately cease

completely. Some secretion of estrogen generally continues beyond these events but gradually diminishes until it is inadequate to maintain the estrogen-dependent tissues: the breasts and genital organs gradually atrophy, the decrease in protein anabolism causes thinning of the skin and bones; however, sexual drive is frequently not diminished and may even be increased. Severe emotional disturbances are not uncommon during menopause and are generally ascribed not to a direct effect of estrogen deficiency but to the disturbing nature of the entire

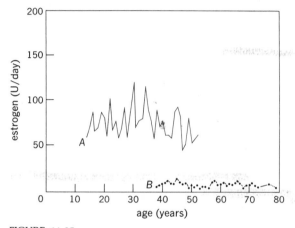

FIGURE 14-35

Excretion of estrogen in the urine of women from puberty to senescence, an indicator of blood concentration of estrogen: A **before menopause and** B **menopause.** (*Adapted from Pedersen-Bjergaard and Tonnesen.*)

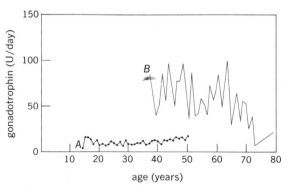

FIGURE 14-36

Excretion of gonadotropins in the urine of nonpregnant women, an indicator of blood concentration of gonadotropins: A **before menopause and** B **after menopause.** (*Adapted from Pedersen-Bjergaard and Tonnesen.*)

period—the awareness that reproductive potential is ended, the hot flashes, etc. The hot flashes, so typical of menopause, result from dilation of the skin arterioles, causing a feeling of warmth and marked sweating; why estrogen deficiency causes this is unknown. Many of these symptoms of the menopause can be reduced by the administration of small amounts of estrogen.

Another important aspect of menopause is the relationship between estrogen and plasma cholesterol. Estrogen significantly lowers the plasma cholesterol, and this or some related effect on lipid metabolism may explain why women have much less arteriosclerosis than men until after the menopause, when the incidence becomes similar in both sexes.

Male changes with aging are much less drastic. Once testosterone and pituitary gonadotropin secretions are initiated at puberty, they continue throughout adult life (Figs. 14-33 and 14-34). A steady decrease in testosterone secretion in later decades apparently reflects slow deterioration of testicular function. The mirror-image rise in gonadotropin secretion is due to diminishing negative-feedback inhibition from the decreasing plasma testosterone concentration. Despite the significant decrease, testosterone secretion remains high enough in most men to maintain sexual vigor throughout life, and fertility has been documented in men in their eighties. Thus, there is usually no complete cessation of reproductive function analogous to menopause.

Defense mechanisms of the body

15

SECTION A
HEMOSTASIS: THE PREVENTION OF BLOOD LOSS

All animals with a vascular system must be able to minimize blood loss consequent to vessel damage. In man,
blood coagulation is only one of several important mechanisms for hemostasis. The hemostatic mechanism which
predominates varies, depending upon the kind and number of vessels damaged and the location of the injury.

The basic prerequisites for bleeding are (1) loss of
vessel continuity or marked increase in permeability so
that cells can leak out; (2) a pressure inside the vessel
greater than outside, since hemorrhage occurs by bulk
flow. Accordingly, bleeding ceases if at least one of two
requirements is met: The pressure difference favoring
blood loss is eliminated, or the damaged portion of vessel
is blocked. All hemostatic processes accomplish one of
these two requirements. We shall discuss the probable

sequence of events in damage to small vessels—arterioles, capillaries, venules—because they are the most common source of bleeding in everyday life and because the hemostatic mechanisms are most effective in dealing with such injuries. In contrast, the bleeding from a severed artery of medium or large size is not usually controllable by the body and requires radical aids such as application of pressure and ligatures. Venous bleeding is less dangerous because of the vein's low hydrostatic pressure; indeed, the drop in hydrostatic pressure induced by simple elevation of the bleeding part may stop the hemorrhage. In addition, if the venous bleeding is into

the tissues, the accumulation of blood (*hematoma*) may increase interstitial pressure enough to eliminate the pressure gradient required for continued blood loss.

The hemostatic events in small vessels (Table 15-1) do not occur in neat orderly sequence but overlap in time and are closely interrelated functionally. The *platelets* play a critical role in the last four hemostatic components of our synopsis. In man, the circulating platelets (Fig. 15-1) are colorless corpuscles much smaller than erythrocytes and containing numerous granules. It is difficult to tell how many individual platelets are present in blood since as soon as blood is removed for study, they form clumps

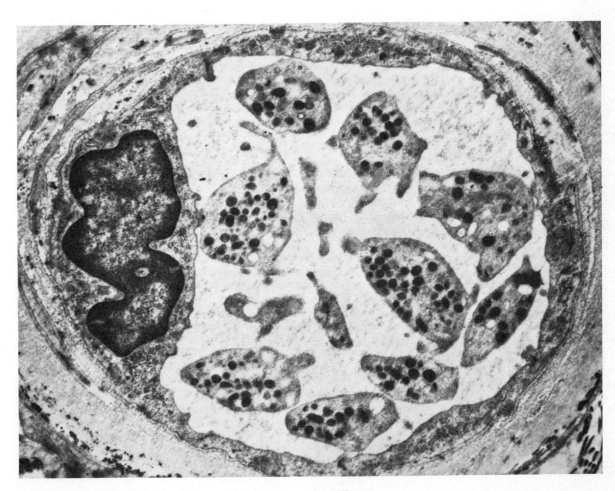

FIGURE 15-1
Electron micrograph of a capillary containing several blood platelets in its lumen. Note the numerous granules and lack of nuclei. (*From W. Bloom and D. W. Fawcett, "A Textbook of Histology," 9th ed., W. B. Saunders Company, Philadelphia, 1968.*)

TABLE 15-1
Synopsis of hemostatic events in small blood vessels

1 Contraction of smooth muscle in wall of damaged vessel
2 Sticking together of injured endothelium
3 Clumping of platelets to form a plug
4 Facilitation of the initial vasoconstriction
5 Blood coagulation, i.e., formation of a fibrin clot
6 Retraction of the clot

and stick to most surfaces; however, a figure generally accepted is approximately 250 million per milliliter of blood (compare this with other blood cell values given in Table 15-2). The platelets are not complete cells since they lack nuclei. They originate from certain large cells (*megakaryocytes*) found in the bone marrow and apparently are portions of the cytoplasm of these cells which are pinched off and enter the circulation. The factors which control the rate of platelet formation are unknown.

Hemostatic events prior to clot formation

Contraction of the injured vessel

When a blood vessel is severed or injured, its immediate response is to constrict. This spasm is particularly long-lasting in larger veins and arteries and may be intense enough to close the severed end completely. Its adaptive value is obvious, but what is the mechanism? We know this initial spasm to be independent of platelets or the coagulation process in general, and local humoral substances have not been implicated. It seems most likely that the rupture of the vessel wall in some manner directly stimulates the smooth muscle or nerves supplying it.

TABLE 15-2
Numbers and distribution of erythrocytes and white cells in normal human blood

Total erythrocytes = 5,000,000,000 cells per milliliter of blood
Total white cells = 7,000,000 cells per milliliter of blood
Percent of total white cells:
  Polymorphonuclear granulocytes:
    Neutrophils   50–70
    Eosinophils   1–4
    Basophils   0.1
  Mononuclear cells
    Monocytes   2–8
    Lymphocytes   20–40

Sticking of the endothelial surfaces

This event occurs as a direct result of the initial spasm, which presses the opposed endothelial surfaces of the vessel together. This contact induces a stickiness capable of keeping them "glued" together against high pressures even after the active vasoconstriction has begun to wane. The injury itself also alters the surface properties of the vessel and facilitates the adhesiveness of the damaged membranes. This entire process is probably of great importance only in the very smallest vessels of the microcirculation. It also occurs independently of platelets.

Formation of a platelet plug

The involvement of platelets in hemostatic events requires their adhesion to a surface. Although platelets have a propensity for adhering to many foreign or rough surfaces, they do not adhere to the normal endothelial cells lining the blood vessels. However, injury to a vessel disrupts the endothelium and exposes the underlying connective tissue with its collagen[1] molecules. Platelets adhere strongly to collagen, and this attachment somehow triggers the release from the platelets' granules of potent chemical agents, including *adenosine diphosphate* (*ADP*). This ADP then causes the surface of the adhered platelets to become extremely sticky so that new platelets adhere to the old ones and an aggregate or plug of platelets is rapidly built up by this self-perpetuating (positive-feedback) process. In general, this platelet plug may occlude small vessels so as to slow or even stop bleeding from them, but it is relatively unstable by itself and cannot withstand for long the high pressures of the larger vessels. However, regardless of its role as a blood stauncher, platelet adhesion is one of the crucial events in hemostasis because, as we shall see below, the release of the platelet chemicals triggered by adhesion also induces vasoconstriction and blood coagulation.

Humoral facilitation of vasoconstriction

As we have seen, the initial vascular spasm following vessel injury is a direct response of the smooth muscle itself or its innervation. The maintenance of vasoconstriction in small vessels, however, is due to the local release from the aggregated platelets of *serotonin* and *epinephrine*, both of which are powerful vasoconstrictors. This secondary, prolonged vasoconstriction does not occur unless there have been platelet adhesion and granule dis-

---

[1]Collagen molecules are fibrous proteins which constitute a major component of the interstitial matrix.

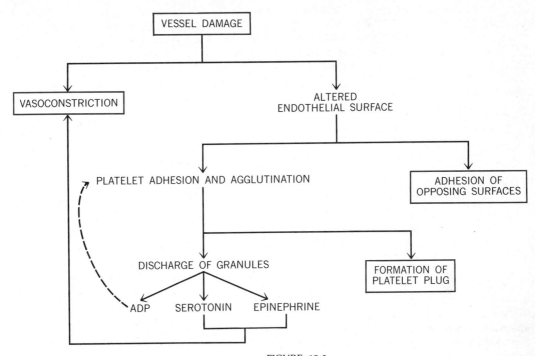

FIGURE 15-2
Summary of hemostatic mechanisms not dependent upon blood coagulation. The dashed line indicates the positive-feedback effect of ADP on platelet adhesion and agglutination.

charge. The first four events in hemostasis are summarized in Fig. 15-2.

Blood coagulation: clot formation

Despite the participation of the four mechanisms just described, blood coagulation is the dominant hemostatic defense in man, as attested by the fact that, with few exceptions, abnormal bleeding is associated with some clotting defect. Pure vascular defects, interfering with the preclot hemostatic mechanisms, do occur but are only rarely the cause of abnormal bleeding.

The event transforming blood into a solid gel is the conversion of the plasma protein *fibrinogen* to *fibrin*. Fibrinogen is a soluble, large, rod-shaped protein (molecular weight approximately 340,000) produced by the liver and always present in the plasma of normal persons. Its conversion to fibrin is catalyzed by the enzyme *thrombin*

$$\text{Fibrinogen} \xrightarrow{\text{thrombin}} \text{fibrin}$$

In this reaction, several small negatively charged polypeptides are split from fibrinogen, conferring upon the remaining large molecule a high degree of attraction for molecules like it. They join each other end to end and side by side to form fibrin (Fig. 15-3). This polymerization causes the fluid portion of the blood to gel, rather like a gelatin dessert. In addition, all the cellular elements of the blood become entangled in the meshwork and contribute to its strength. It must be emphasized that the clot is basically due to fibrin and can occur in the absence of blood cells (except the platelets).

Since fibrinogen is always present in the blood, the enzyme thrombin must normally be absent and its formation must be triggered by vessel damage. The generation of thrombin follows the same general principle as fibrin, in that an inactive precursor, *prothrombin,* is produced by the liver and is normally present in the blood, being enzymatically converted to thrombin during clot formation:

$$\text{Prothrombin} \xrightarrow{?} \text{thrombin}$$
$$\text{Fibrinogen} \xrightarrow{} \text{fibrin}$$

A

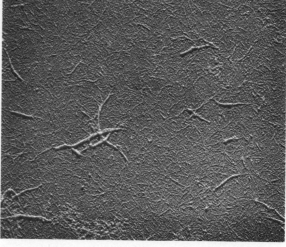

B

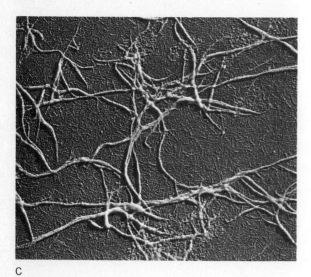

C

FIGURE 15-3

Electron micrographs of fibrin formation from fibrinogen. A A droplet of fibrinogen solution shows the particulate nonfibrillar character of fibrinogen. B Thirty seconds after addition of thrombin, the strands are combining to form thicker fibers. C Small fibers begin to associate to form a meshwork. [*From K. R. Porter and C. van Zandt Hawn, J. Exp. Med.,* **90**:235 (1949).]

We have now only pushed the essential question one step further back: What catalyzes the conversion of the pro-thrombin to thrombin? (This is the question mark in the

equation.) The answer is that this reaction, too, is cata-lyzed by another plasma factor which itself is activated by another plasma factor, which. . . . Thus, we are deal-ing with a *cascade* of plasma protein factors, each normally inactive until activated by the previous one in the se-quence (Fig. 15-4). Ultimately, the final factor in the se-quence is activated and in turn catalyzes the activation of prothrombin, i.e., its conversion to thrombin. We must point out that the designations A and B in our model are arbitrarily chosen to demonstrate the general principle. The first factor is called *Hageman factor* and, as will be described in subsequent sections, it has several other important functions in addition to initiating clotting.

What is the adaptive value of such a chain? It certainly is *not* analogous to the electron transport chain, which yields a little energy at several steps, since no clot formation occurs until the final step. We have no answer to the question. There is at least one disadvantage in that many potential defects, either hereditary or disease-induced, can block the entire system, thereby interfering with clot formation.

In addition to these protein plasma factors, calcium is required as cofactor for several steps. However, cal-cium deficiency is never a cause of clotting defects in man since only very small concentrations are required.

Figure 15-4 reveals that we have still not answered the basic question of what *initiates* clotting. What activates the first protein (Hageman factor or factor A in the figure) in the catalytic sequence? The answer is *contact of this*

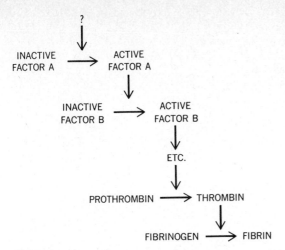

FIGURE 15-4
Cascade theory of blood clotting. Each substance left of an arrow is normally present in plasma but requires activation by the action of the previous substance in the sequence. The name of factor A is Hageman factor.

protein with a damaged vessel surface, most likely with the collagen fibers underlying the damaged endothelium (as was true for platelet aggregation). At last we have reached the connecting link between vessel injury and initiation of clotting.

However, even contact activation of this first factor cannot produce clotting in the absence of platelets. A phospholipid substance exposed on the surface of platelets during their adhesion and agglutination is required as cofactor for several of the steps in the catalytic sequence. Thus, the critical event initiating clot formation is contact of the blood with a damaged surface for two reasons: (1) It activates the first factor in the activation sequence, and (2) it causes platelet adhesion and exposure of a phospholipid cofactor. One final detail is that thrombin markedly enhances the adhesion and agglutination of platelets; thus, once thrombin formation has begun, the overall reaction progresses explosively owing to the positive-feedback effect of thrombin. Our growing figure can now be completed (Fig. 15-5).

This entire process occurs only locally at the site of vessel damage. Each active component is formed, functions, and is rapidly inactivated without spilling over into the rest of the circulation. Otherwise, because of the chain-reaction nature of the response, the appearance in the overall circulation of platelet phospholipid and any single activated factor would induce massive widespread clotting throughout the body.

The reason should also be clear why blood coagulates when it is taken from the body and put in a glass tube. This has nothing whatever to do with exposure to air, as popularly supposed, but happens because the glass surface induces precisely the same effects as a

FIGURE 15-5
Summary of blood-clotting mechanism. The dashed line indicates the positive-feedback effect of thrombin on platelet adhesion (recall that ADP exerts a similar positive feedback).

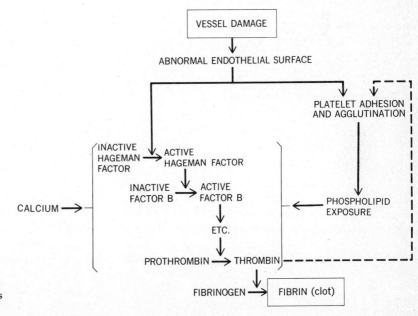

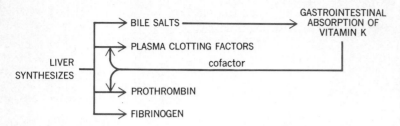

FIGURE 15-6
Role of the liver and vitamin K in the synthesis of plasma clotting factors, prothrombin, and fibrinogen.

damaged vessel surface. A silicone coating markedly delays clotting by reducing the chemical activity of the glass surface.

The liver plays several important indirect roles in the overall functioning of the clotting mechanism (Fig. 15-6). First, it is the site of production for many of the plasma factors, prothrombin, and fibrinogen, although the reflex mechanisms controlling their rates of synthesis are unknown. Second, the bile salts produced by the liver are required for normal gastrointestinal absorption of the fat-soluble *vitamin K,* which is an essential cofactor in the hepatic synthesis of prothrombin and the plasma factors. For these two reasons, patients with liver disease or defective gastrointestinal fat absorption frequently have serious bleeding problems.

Finally, a word must be said about the contribution of a tissue (rather than blood) factor to clotting. If one extracts almost any of the body's tissues and injects the extract into unclotted normal blood in a siliconized tube, clotting occurs within seconds. The explanation is that the tissues contain a substance known as *tissue thromboplastin* which can substitute for both platelet phospholipid and several of the plasma factors, and thus an abnormal surface is no longer required to initiate clotting. This is known as the *extrinsic clotting pathway* to distinguish it from the *intrinsic pathway* described above. Its quantitative contribution to normal intravascular clotting is unclear but it may play an essential role in the response to many bacterial infections by initiating interstitial fibrin clots which may block further spread of the bacteria.

This completes our discussion of the events leading to clot formation. The initiating event is the presence of a damaged vessel surface, which induces platelet adhesion and activation of the first plasma factor in the catalytic sequence leading to generation of thrombin. Calcium and a platelet phospholipid are required for the overall reaction. As the final step, thrombin enzymatically splits off several small polypeptides from fibrinogen, leading to

clot formation by generating polymerizing fibrin strands. A defect or lack of any of these components may induce inadequate clotting and prolonged bleeding.

Clot retraction

When blood is carefully collected and placed in a glass test tube, clotting usually occurs within 5 to 8 min, the entire volume of blood appearing as a coagulated gel. However, during the next 30 min, a striking transformation occurs; the clot literally retracts, squeezing out the fluid which constituted a large fraction of the gel. The end result is a small hard clot at the bottom of the tube with a large volume of serum floating on top (plasma without fibrinogen is called serum). The fibrin meshwork with its entangled cells has thus become denser and stronger. This same process in the body is known as *clot retraction.* Besides increasing the strength of the clot, it has the advantage of pulling the vessel walls adhering to the clot closer together.

Clot retraction is due to the platelets. As fibrin strands form around them during clotting, the agglutinated mass of platelets sends out adhering pseudopods along them. The pseudopods then contract, pulling the fibrin fibrils together and squeezing out the serum. It has been shown that the platelets contain actomyosinlike contractile protein and that they split ATP during clot retraction. The various roles of platelets in hemostasis are summarized in Fig. 15-7.

The anticlotting system

It has long been observed that clots frequently disappear after lengthy standing. Blood also fails to clot in a variety of special circumstances, including endometrial invasion by the fetal placental cells. This is due to a proteolytic

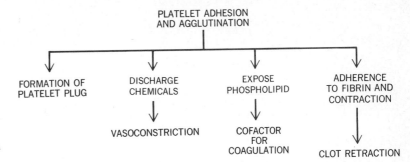

FIGURE 15-7
Summary of platelet functions in hemostasis.

enzyme called *plasmin,* which is able to decompose fibrin, thereby dissolving a clot. The physiology of plasmin bears some striking similarities to that of the coagulation factors in that it circulates in blood in an inactive form which is enzymatically converted to active plasmin by the action of activated Hageman factor as well as by other substances found in tissue fluid (Fig. 15-8). It may seem paradoxical that the same substance, Hageman factor, triggers off simultaneously the clotting and anticlotting systems. Yet, in fact, this makes sense since the generated plasmin becomes trapped within the newly formed clot and very slowly dissolves it, thereby contributing to tissue repair at a time when the danger of hemorrhage is past.

The anticlotting system no doubt has other functions as well. It may be that small amounts of fibrin are constantly being laid down throughout the vascular tree and that plasmin acts on this fibrin to prevent clotting. The lung tissue, for example, contains a substance which activates plasmin; this probably explains the lung's ability to dissolve the fibrin clumps which its capillaries filter from

the blood. Moreover, the uterine wall is extremely rich in a similar activator, and thus normal menstrual blood generally does not clot.

A second naturally occurring anticoagulant is *heparin.* This substance, found in various cells of the body, especially mast cells, acts by interfering with the ability of thrombin to split fibrinogen. Despite its presence in the body and the fact that it is the most powerful anticoagulant known, there is really no good evidence to prove that it plays a physiological role in clot prevention. On the other hand, heparin is widely used as an anticoagulant in medicine.

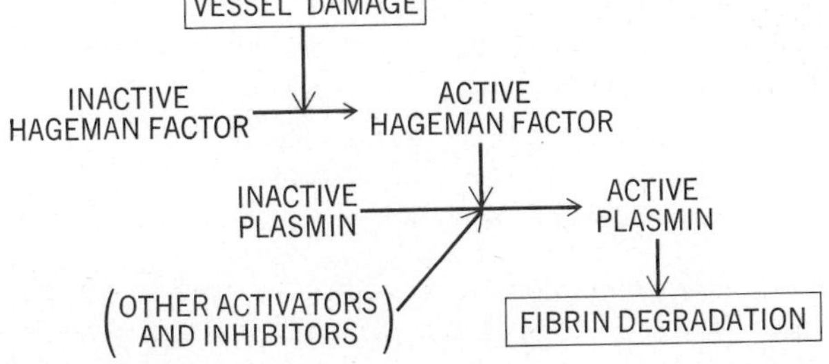

Excessive clotting: intravascular thrombosis

Formation of a clot in a bleeding vessel is obviously a homeostatic physiological response, but the formation of clots within intact vessels is pathological. It may occur in the veins, the microcirculation, or arteries. Coronary arterial occlusion secondary to thrombosis (thrombus means clot) is one of the major killers in the United States today. The sequence of events leading to thrombosis is the subject of intensive study, and numerous theories have been proposed. A brief synopsis of some of them is in order because of the great importance of the subject and its illustration of the basic physiological processes.

One of the dominant theories today postulates that the clotting mechanism in persons prone to thrombosis is hyperactive, as manifested by the reduced time it takes for withdrawn blood to clot in a test tube. Perhaps one of the plasma factors is present in excessive amounts, or a normally occurring anticoagulant is deficient. This theory emphasizes that the blood itself is the cause of excessive clotting. However, there seems little question

FIGURE 15-8
Summary of the plasmin anticlotting system.

that hypercoagulability is not always essential for thrombosis, since hemophiliacs have been known to suffer from coronary thrombosis.

A second category of theories puts the blame on the blood vessels. Since initiation of blood clotting is primarily dependent upon the state of the blood vessel lining, even minor transient alterations in the endothelial surface could trigger the autocatalytic sequence leading to clot formation. These vessel-oriented theories can explain many of the situations associated with vascular thrombosis (Fig. 15-9): (1) Stasis, i.e., decreased movement, of blood in veins, which occurs during quiet standing, valve malfunction, or cardiac insufficiency, may induce damage in the vein wall as a result of oxygen lack. (2) Inflammation of veins and other vessels caused by bacteria, allergic reactions, or toxic substances may cause vessel damage. (3) Deposition of lipids and connective tissue in arterial walls (arteriosclerosis) causes marked thickening and irregularity of the arterial lining, although there is certainly no agreement as to which comes first, the deposits or the clot.

Thus, the three major conditions predisposing to clot formation are consistent with the damaged-lining theory. Association does not prove causality, however, and it should be noted that this concept and the hypercoagulability theory are not mutually exclusive. Both probably are valid, depending upon the circumstances.

Regardless of the initiating event, there is no question that a clot, no matter how small, provides a suitable surface upon which more clot can form. Thus the thrombus grows and may eventually occlude the entire vessel, thereby leading to damage of the tissue supplied or drained by the vessel. A second important factor in vessel closure during clot growth is the release of vasoconstrictors from freshly adhered platelets. Finally, the chances are greatly increased of clot fragments breaking off and being carried to the lungs (if from a vein) or other organs (if from an artery). These *emboli* not only plug the microcirculation but in the lung may induce totally inappropriate cardiovascular reflexes culminating in hypotension, disturbance of the cardiac rhythm, and death.

The prevention of *new* clot growth with its associated consequences is the major reason for giving patients anticoagulant drugs. These drugs do not dissolve a clot once it has been formed. Anticoagulants now in use include heparin and a group of drugs which interfere with the synthesis of the plasma clotting factors, including prothrombin, by blocking the action of vitamin K. Indeed, one of the important observations leading to the discovery of vitamin K was that animals on a diet of spoiled sweet clover hay, which contains these anticoagulants, manifested serious bleeding tendencies. Of course, a patient receiving any anticoagulant is prone to bleeding.

SECTION B
IMMUNOLOGY: THE BODY'S DEFENSES AGAINST FOREIGN MATERIALS

Immunity constitutes all the physiological mechanisms which allow the body to recognize materials as foreign to itself and to neutralize or eliminate them; in essence, these mechanisms maintain uniqueness of "self." Classically, immunity referred to the resistance of the body to microbes: viruses, bacteria, and other unicellular and multicellular organisms. It is now recognized, however, that the immune system has more diverse functions than this. It is involved both in the elimination of "worn-out" or damaged body cells (such as old erythrocytes) and in the destruction of abnormal or mutant cell types which arise within the body. This last function, known as *immune surveillance,* apparently constitutes one of the body's major defenses against cancer.

It has also become evident that immune responses are not always beneficial but may result in serious damage to the body. Such noxious effects may contribute to the development of a variety of diseases collectively known as allergies (or hypersensitivity). In addition, the immune system seems to be involved in the process of aging. Finally, it constitutes the major obstacle to successful transplantation of organs. Because of these broad relationships, few other research areas of biology have grown so rapidly in the past 5 to 10 years or have pro-

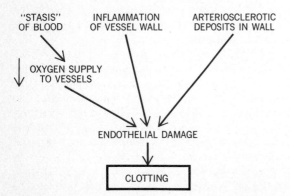

FIGURE 15-9
Damaged-vessel theory of abnormal intravascular clotting.

TABLE 15-3
Classification of immune mechanisms

1 Nonspecific defense mechanisms
2 Specific immune mechanisms
  a Humoral (antibodies)
  b Cell-mediated ("sensitized" lymphocytes)

duced such a wealth of new and exciting, albeit sometimes bewildering, information.

Immune responses may be classified (Table 15-3) into two categories: (*1*) *nonspecific defense mechanisms* and (*2*) *specific immune mechanisms.* Specific immune responses depend upon prior exposure to the specific foreign substance, recognition of it upon subsequent exposure, and reaction to it. In contrast, the nonspecific or innate defense mechanisms do not depend upon previous exposure to the particular foreign substance; rather they nonselectively protect against foreign substances without having to recognize their specific identities. They are particularly important during the initial exposure to a foreign organism before the specific immune responses have been activated.

Specific immunity may be divided into two primary categories: *humoral* and *cell-mediated.* Both are the function of the *lymphoid cells,* and a critical difference between them, as we shall see, is that the former is mediated by circulating *antibodies* and the latter by *sensitized lymphocytes.*

Immune responses can be viewed in the same way as other homeostatic processes in the body, i.e., as stimulus-response sequences of events. In such an analysis, the groups of cells which mediate the final responses are effector cells, but before we begin our detailed description of immunology by introducing the various effector cells involved, let us first at least mention their major opponents: microbes (or microorganisms)—bacteria, viruses, and fungi.

Bacteria are unicellular organisms belonging to the plant kingdom. As is typical of plant cells, they have not only a cell membrane but also an outer coating, the cell wall. Most bacteria are self-contained complete cells in that they have all the machinery required to sustain life and to reproduce themselves. In contrast, the *viruses,* as described in Chap. 4, are essentially nucleic acid cores surrounded by a protein coat. They lack both the enzyme machinery for energy production and the ribosomes essential for protein synthesis. Thus, they cannot survive by themselves but must "live" inside other cells whose metabolic apparatus they make use of; i.e., they are obligatory parasites. Other types of microorganisms and multicellular parasites are potentially harmful to man, but we shall devote most of our attention to the body's defense mechanisms against the bacteria and viruses.

How do microorganisms cause damage and endanger health? There are many answers to this question, depending upon the specific bacterium or virus involved. Some cause cellular destruction directly by locally releasing enzymes which break down cell membranes and organelles. Others give off toxins which may act throughout the body to disrupt the functions of the neuromuscular system and other organs and tissues. Moreover, the presence of microorganisms may constitute a continuous drain on the body's energy supplies. The viruses, in particular, often capture the metabolic and reproductive machinery of the cell which they inhabit—indeed, some forms of cancer in man may be caused by viruses. It must be admitted, however, that the damage-producing mechanisms in many infectious diseases are unknown.

Effector cells of the immune system

The effector cells of the immune system are distributed throughout the organs and tissues of the body, but many are housed within the so-called lymphoid tissues: *thymus, lymph nodes, spleen,* and areas in the lining of the gastrointestinal tract (Fig. 15-10). The major cell types are *granulocytes, lymphocytes, monocytes, macrophages,* and *plasma cells.* The first three types circulate in the blood and constitute the white blood cells or *leukocytes.*

If one takes a drop of blood, adds appropriate dyes, and examines it under a microscope, the various cell types illustrated in Fig. 15-11 and listed in Table 15-2 can be seen. Over 99 percent of all the cells are erythrocytes, the functions of which were described in Chap. 10. The remaining cells, the *white blood cells,* are classified according to their structure and affinity for various dyes. The name *polymorphonuclear granulocytes* refers to the three types of cells with lobulated nuclei and abundant cytoplasmic granules. The granules of one group show no dye preference, and the cells are therefore called *neutrophils.* The granules of the second group take up the red dye eosin, thus giving the cells their name *eosinophils.* Cells of the third group have an affinity for a basic dye and are called *basophils.* All three types of granulocytes are produced in the bone marrow and released into the circulation. It should be understood that, unlike the erythrocytes, the major functions of leukocytes are exerted not within the blood vessels but in the interstitial fluid; i.e., leukocytes

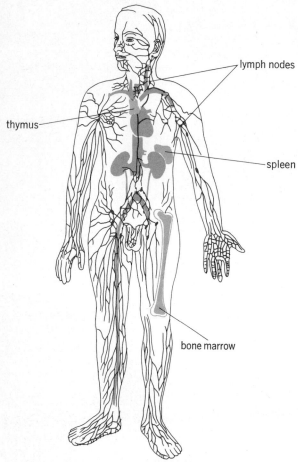

FIGURE 15-10
Location of various lymphoid organs; lymph nodes, thymus,
and spleen. Not shown are the lymphoid patches in the
various epithelial linings of the body. The bone marrow is not
a lymphoid organ but is the site of production of the cells
which come to reside in the lymphoid organs.

phil, in contrast, is not a phagocytic cell; rather it contains
powerful chemicals, such as histamine, which it releases
locally, a response which, as we shall see, may importantly contribute to tissue damage and allergy. The basophil is virtually identical to the group of cells known as
mast cells which are found in connective tissue throughout
the body and do not circulate; i.e., the basophil is, in
essence, a circulating mast cell.

A second type of leukocyte quite different in appearance from the granulocytes is the *monocyte*. These
cells, which are also produced by the bone marrow, are
somewhat larger than the granulocytes, with a single oval
or horseshoe-shaped nucleus and relatively few cytoplasmic granules. Upon entering an area which has been
invaded or wounded, monocytes are transformed into
macrophages. Thus, the function of circulating monocytes
is to provide to the tissues a source of new macrophages,
the function of which is described below.

The final class of leukocyte is the *lymphocyte* (in a
normal person the white blood cells have approximately
the distribution shown in Table 15-2). Their outstanding
structural features are a relatively large nucleus and
scanty surrounding cytoplasm. The circulating pool of
lymphocytes continually travels from the blood across
capillary membranes to the lymph, thence to lymph
nodes, and finally, back to the blood. However, circulating
lymphocytes constitute only a very small fraction of the
total body lymphocytes, most of which are found at any
instant in the lymphoid tissues. The life history of lymphocytes and the interactions between them and the various
lymphoid organs listed above are quite complex and will
be described in a subsequent section. Suffice it for now
to point out that the lymphocytes (and a daughter cell
line, the *plasma cells*) are responsible for specific immunity.
They are the cells which either secrete antibodies (humoral immunity) or participate in cell-mediated immune
responses.

The last type of effector cell is the macrophage,
the major function of which, like that of the neutrophil,
is phagocytosis. Macrophages are found scattered
throughout the tissues of the body. In liver, spleen, lymph
nodes, and bone marrow they form part of the lining of
vascular and lymphatic channels. The structure of these
cells varies from place to place, but their common distinctive features are numerous cytoplasmic granules and
the ability to ingest almost any kind of foreign particle.
Indeed, the method of identifying them is to inject a dye
and observe which tissue cells display marked uptake of
it. Tissue macrophages are capable of mitotic activity, but
division is not the only mechanism for increasing their

utilize the circulatory system only as the route for reaching a damaged or invaded area. Once there, they leave
the blood vessels to enter the tissue and perform their
functions.

The primary function of the neutrophil is *phagocytosis,*
the ingestion and digestion of particulate material. It has
a life span of only a few days and is incapable of division,
so that its supply must be continuously replenished by
the bone marrow. The eosinophil is also involved in phagocytosis but its precise function is unknown. The baso-

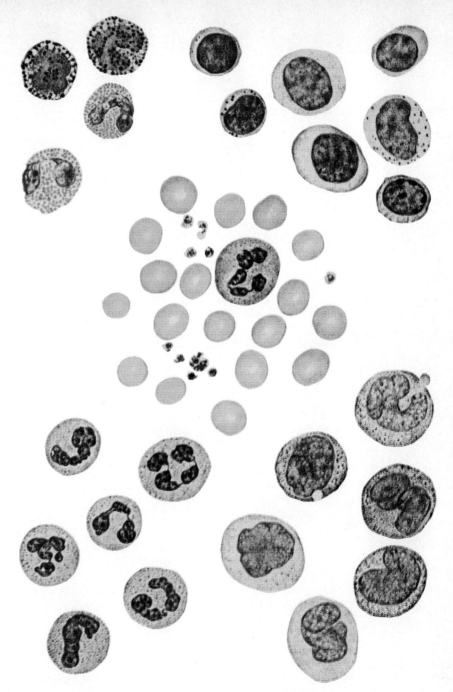

FIGURE 15-11
Normal human blood cells. In the center are erythrocytes,
platelets, and polymorphonuclear neutrophils. At left above are
two basophils; and, just below them, two eosinophils. At right
above are three large and four small lymphocytes. At left
below are neutrophils, and at right below are six monocytes.
(From Roy O. Greep and Leon Weiss, "Histology," 3d ed. © 1973 by
McGraw-Hill, Inc., New York.)

number; in response to invasion by microorganisms or other foreign material, monocytes can be transformed into macrophages indistinguishable from the usual tissue macrophages.

Nonspecific defense mechanisms

External anatomic and chemical "barriers"

The body's first lines of defense against infection are the barriers offered by surfaces exposed to the external environment. Very few microorganisms can penetrate the intact skin, and the sweat, sebaceous, and lacrymal glands secrete chemical substances which are highly toxic to certain forms of bacteria. The mucous membranes also contain antimicrobial chemicals, but more important, mucus is sticky. When particles adhere to it, they can be swept away by ciliary action, as in the upper respiratory tract, or engulfed by phagocytic cells. Other specialized mechanisms related to the surface barriers are the hairs at the entrance to the nose, the cough reflexes, and the acid secretion of the stomach. Finally, a major "barrier" to infection is the normal microbial flora of the skin and other linings exposed to the external environment; these microbes suppress the growth of other potentially more virulent microorganisms.

Nonspecific inflammatory response

Despite the effectiveness of the external barriers, small numbers of microorganisms penetrate them every day. Think of all the small breaks produced in the skin or mucous membranes by tooth brushing, shaving, tiny scratches, etc. In addition, we now recognize that many viruses are somehow able to penetrate seemingly intact healthy skin or mucous membranes. Also, under certain conditions and against certain microbes, the external barriers may simply be ineffective.

Once the invader has gained entry, it triggers off *inflammation,* the basic response to injury. The local manifestations of the inflammatory response are a complex sequence of highly interrelated events, the overall functions of which are to bring plasma proteins and phagocytes into the damaged area so that they can destroy the foreign invaders and set the stage for tissue repair. The nature of the sequence of events which constitute the inflammatory reaction varies, depending upon the injurious agent (bacteria, cold, heat, trauma, etc.), the site of injury, and the state of the body. But since the similarities are in many respects more striking than the differences, inflammation can be viewed as a relatively stereotyped response to tissue damage, the precise manifestations of which differ according to the specific injurious agent and other important variables. It should be emphasized that, in this section, we describe inflammation in its most basic form, i.e., the nonspecific innate response to foreign material. As we shall see, inflammation remains the basic scenario for the acting out of specific immune responses as well, the difference being that the entire process is amplified and made more efficient by the participation of antibodies and sensitized lymphocytes.

The sequence of events in a local infection is briefly as follows:

1 Initial entry of microbes
2 Vasodilation of all the vessels of the microcirculation leading to increased blood flow
3 A marked increase in vascular permeability to protein
4 Filtration of fluid into the tissue with resultant swelling
5 Exit of neutrophils (and later, monocytes) from the vessels into the tissues
6 Phagocytosis and destruction of the microbes
7 Tissue repair

The familiar gross manifestations of this process are redness, swelling, heat, and pain, the latter being the result both of distension and the direct effect of released substances on afferent nerve endings.

Vasodilation and increased permeability to protein   Immediately upon tissue damage, increased blood flow and vascular permeability to proteins occur. The latter is easily demonstrated experimentally by the leakage of colloidal dyes through the walls of capillaries and small venules. The chemical mediators of these responses dilate most of the vessels of the microcirculation and somehow alter the intercellular material so as to make it quite leaky to large molecules. The causes of the tissue swelling should be evident from knowledge of the factors which determine capillary fluid exchange (Chap. 9). Recall that the major factor favoring movement of interstitial fluid into the capillary is the water-concentration gradient resulting from the presence of protein in the plasma but not in the interstitium. During inflammation the protein which leaks out of the vessels as a result of increased permeability builds up locally in the interstitium, thereby eliminating the protein difference between plasma and interstitium. A second factor favoring filtration is the arteriolar dilation associated with inflammation, which increases capillary hydrostatic pressure by reducing arteriolar resistance.

The adaptive value of all these vascular changes is twofold: (*1*) The increased blood flow ensures an adequate supply of phagocytic leukocytes and plasma proteins crucial for immune responses (to be described below) to the inflamed area, and (*2*) the increased capillary permeability to protein ensures that the relevant plasma proteins—all normally restrained by the capillary membranes—can gain entry to the inflamed area.

The major direct chemical mediators of these vascular changes in nonspecific inflammatory responses are *histamine* and a group of polypeptides known as *kinins*. Histamine is present in many tissues of the body but is particularly concentrated in mast cells, circulating basophils, and platelets. Release of histamine is induced by a wide variety of factors. In an inflammatory response in which specific immune responses are absent, two of the major factors are simple mechanical disruption of histamine-containing cells and chemicals secreted by neutrophils attracted to the site (Fig. 15-12). The released histamine, in addition to its vasodilating and permeability-increasing effects, has profound effects on nonvascular smooth muscle, perhaps the most significant being its constriction of the respiratory airways.

The kinins are small polypeptides whose vascular effects are similar to those of histamine. The kinins are generated in plasma in the following ways (Fig. 15-13): There is present in plasma an enzyme, known as *kallikrein,* which exists normally in an inactive form but which, when activated, catalyzes the splitting off of the kinins from another normally occurring plasma protein known appropriately as *kininogen*. What is it that activates this enzyme?

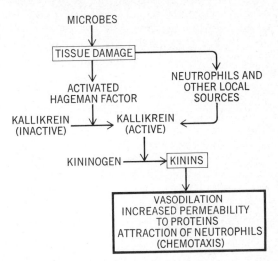

FIGURE 15-13
Pathways for kinin generation in nonspecific inflammatory reactions.

Again we find that multiple factors are capable of doing so but that, in the absence of specific immune responses, the most important is a chemical substance we already met in the section of this chapter on hemostasis: activated Hageman factor. Here is one of many connections between the hemostatic and immune systems. Recall that Hageman factor is, itself, inactive until activated locally by contact with altered vascular surfaces; once activated, it catalyzes the first steps in the cascade sequences leading to both clotting and plasmin formation. Now we see that, in addition, it leads to the activation of the key enzyme of the kinin-generating system, thereby contributing to the first phases of the inflammatory process.

In addition to being generated from its inactive form in blood, active kallikrein is also found in many different tissues of the body as well as in neutrophils. Once inflammation has begun, the kallikrein present in these sources contributes to the generation of kinin. In essence, this is a positive feedback since plasma kininogen is usually protected from the action of tissue kallikrein because the former is too large a molecule to pass through capillaries; with enhanced vascular permeability, it penetrates into the tissues, thereby permitting the generation of more kinin and more inflammation. It is a positive feedback in a second respect, also; as we shall see in the next section, the kinins actually attract neutrophils to the area. It should also be noted that, by their effects on afferent neuron terminals, the kinins account for much of the pain associated with inflammation.

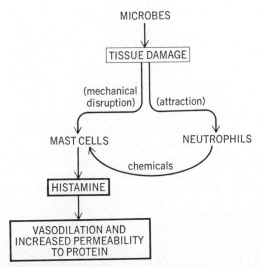

FIGURE 15-12
Pathways for histamine release in nonspecific inflammatory reactions.

Neutrophil exudation Within 30 to 60 min after the onset of inflammation, a remarkable interaction occurs between the vascular endothelium and circulating neutrophils. First, the blood-borne neutrophils begin to stick to the inner surface of the endothelium. The process is quite specific since erythrocytes show no tendency to stick and other leukocytes do so only later, if at all. How the endothelium is made sticky by injury remains a mystery.

Following their surface attachment, the neutrophils begin to manifest considerable amoebalike activity. Soon a narrow amoeboid projection is inserted into the space between two endothelial cells (Fig. 15-14), and the entire

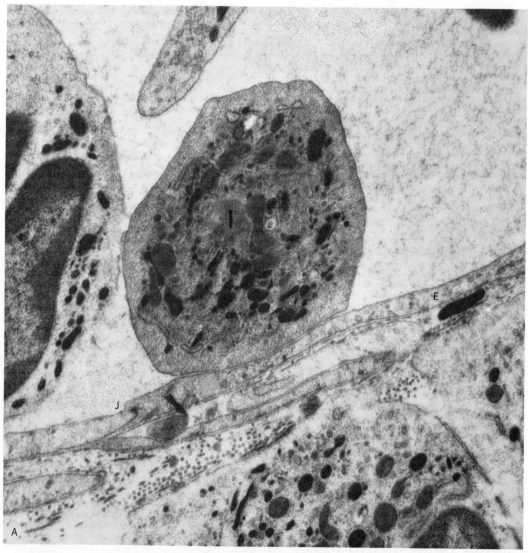

FIGURE 15-14
White blood cell emigration. A A polymorphonuclear granulocyte (at center) in the lumen of a capillary has just adhered to the endothelium (E); note the intact intercellular junction (J). B This cell is now protruding a pseudopod through the intercellular junction and out into the interstitial space. The entire cell will move through in this manner. [*From V. T. Marchesi, Q. J. Exp. Physiol.*, **46**:115 (1961).]

neutrophil then squeezes through into the interstitium. The alterations of vessel structure induced by the released agents described above may facilitate this process by loosening the intercellular connections, or the neutrophil may simply pry the connection apart by the force of its amoeboid movement.

By this process (neutrophil exudation), huge numbers of neutrophils move into the inflamed areas of tissue. This event is dependent upon *chemotaxis,* the attraction of neutrophils to an inflamed area by certain chemicals. Once again, we find a multitude of chemotactic chemicals from a variety of sources, including the bacteria, and

neutrophils themselves. The kinins are also potent chemotactic agents. Again, we see the positive-feedback nature of the inflammatory process (Fig. 15-15).

Leukocyte exudation is usually not limited only to neutrophils. Monocytes follow, but much later (although their mode of entry is unclear), and once in the tissue are transformed into macrophages. Meanwhile the macrophages normally present in the tissue have begun to multiply by mitosis and to become motile. Thus, all the phagocytic types are present in the inflamed area. Usually the neutrophils predominate early in the infection but tend to die off more rapidly than the others, thereby yielding

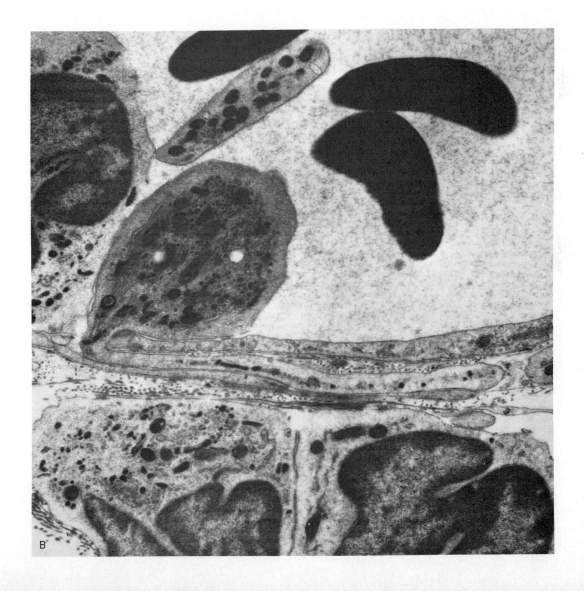

B

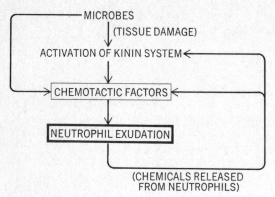

FIGURE 15-15
Pathways for release of chemotactic factors in nonspecific inflammatory reactions.

a predominantly mononuclear picture later. In certain types of allergic or parasitic inflammatory responses, the eosinophils are in striking preponderance. Indeed, one of the tests for allergically induced "runny nose" is to study a small amount of nasal discharge for the presence of eosinophils.

Phagocytosis Phagocytosis is the primary function of the inflammatory response, and the increased blood flow, vascular permeability, and leukocyte exudation serve only to ensure the presence of adequate numbers of phagocytes and to provide the milieu required for the performance of their function. The process of phagocytosis is demonstrated in Fig. 15-16. The phagocyte engulfs the organism by membrane invagination and pouch formation. Once inside, the microbe remains in its own pouch, a layer of phagocyte cell membrane separating it from the phagocyte cytoplasm.

The next step observed in vitro is known as *degranulation* (Fig. 15-17). The membrane surrounding the pouch makes contact with the phagocyte granules, which are *lysosomes* filled with a variety of hydrolytic enzymes; the two membranes fuse, and the still intact granules enter the pouch; the lysosomal membranes separating the contents of the pouch and lysosome break down, and the powerful enzymes are discharged into the pouch. This process of degranulation, as observed microscopically, occurs explosively, and all these events subsequent to engulfment frequently require less than 10 min.

The substances released from the lysosomes kill the microorganism and catabolize it into low-molecular-weight products which can then be safely released from the phagocyte or actually utilized by the cell in its own

metabolic processes. Nondegradable foreign particles (such as wood, tattoo dyes, or metal) may be retained indefinitely within macrophages. This entire process need not kill the phagocyte, which may repeat its function over and over before dying.

The neutrophils may also release entire lysosomal granules into the extracellular fluid; the enzymes released from these granules attack extracellular debris at the injury site, making it easier for the macrophages to phagocytize it at the battle's end, thus paving the way for repair of the damaged area. Figure 15-18 summarizes the basic events of the local inflammatory response to infection.

Thus far we have presented the role of the phagocytes, but what are the microorganisms doing all this time to protect themselves? Most of them do very little. Certain kinds, however, release substances which diffuse into the phagocyte cytoplasm and disrupt the membranes of the lysosomes, thereby allowing these potent chemicals to destroy the phagocyte itself.

Tissue repair The final stage of the inflammatory process is tissue repair. Depending upon the tissue involved, regeneration of organ-specific cells may or may not occur (for example, regeneration occurs in skin and liver but not in muscle or the central nervous system; the latter two tissues are incapable of cell division in adults). In addition, fibroblasts in the area divide rapidly and begin to secrete large quantities of collagen which endows the area with great tensile strength.

The end result may be complete repair (with or without a scar), *abscess* formation, or *granuloma* formation. An abscess is basically a bag of pus (microbes, leukocytes, and liquefied debris) walled off by fibroblasts and collagen. This occurs when tissue breakdown is very severe and when the microbes cannot be eliminated but only contained. When this stage has been reached, the abscess must be drained for it will not resolve spontaneously.

A granuloma occurs when the inflammation has been caused by certain microbes (such as the bacteria causing tuberculosis) which are engulfed by phagocytes but survive within them. It also occurs when the inflammatory agent is a nonmicrobial substance which cannot be digested by the phagocytes. The granuloma consists of numerous layers of phagocytic-type cells, the central ones of which contain the offending material. The whole thing is, itself, usually surrounded by a fibrous capsule. Thus, a person may harbor live tuberculosis-producing bacteria within his body for many years and show no ill effects as long as the microbes are contained within the granuloma and not allowed to escape.

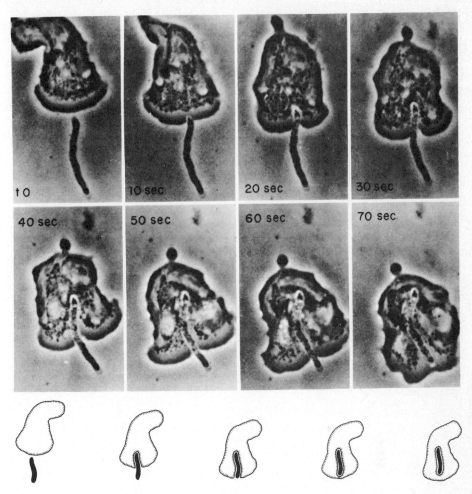

FIGURE 15-16
Prints from a motion picture of a human neutrophil engulfing a bacterium. The bottom shows the process in diagrammatic form. (*From R. J. Dubos and J. G. Hirsch, "Bacterial and* *Mycotic Infections of Man," 4th ed., J. B. Lippincott Company, Philadelphia, 1965.*)

Systemic manifestations of inflammation   We have thus far described the local inflammatory response. What are the systemic, overall body responses? Probably the single most common and striking systemic sign of injury is fever. The substance primarily responsible for the resetting of the hypothalamic thermostat, as described in Chap. 13, is a protein released by the neutrophils (and perhaps other cells) participating in the inflammatory response. When an animal pretreated with a drug which eliminates most of its neutrophils is infected with bacteria, its body temperature hardly rises. The mechanism which stimulates the synthesis and release of this pyrogen by neutro-

phils is not known. Although this unitary theory of fever can explain most experimental findings, there is also a strong likelihood that certain proteins released by the bacteria themselves act directly upon the brain to cause fever. Extract of these bacteria injected directly into the cerebrospinal fluid promptly induces fever. However, the present belief is that the primary effect of such bacterial toxins is to damage neutrophils, thereby releasing neutrophil pyrogen into the blood or lymph. One would expect fever, being such a consistent concomitant of infection, to play some important protective role; yet, to date, there is little evidence that an elevated body temperature

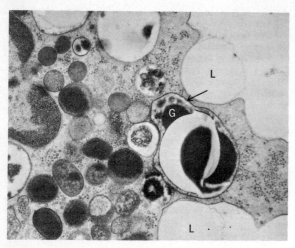

FIGURE 15-17
Electron micrograph of a rabbit granulocyte following incubation with latex particles (L). The membrane of the granule has fused with the membrane surrounding the latex particle (*arrow*) and the granular contents (G) are being discharged into the vacuole. [*From D. Zucker-Franklin, Seminars Hematol.*, **5**:109 (1968).]

enhances the body's resistance. In contrast, fever may be quite harmful, particularly its effects on the functioning of the central nervous system, and convulsions are not infrequent in highly febrile young children. Moreover, elevated body temperature, by increasing the rates of all chemical reactions, places a greater burden on the circulatory system and the tissues involved generally in metabolism.

Another systemic manifestation of many bacterial diseases is *leukocytosis,* a marked increase in the synthesis and release of neutrophils by the bone marrow. In contrast, viral infections frequently are associated with decreased neutrophils. The factors controlling these phenomena are unknown.

With few specific exceptions, we do not know what is responsible for the generalized malaise, aching, and weakness so frequently associated with infection. Our ignorance extends even to the most obvious symptoms, such as loss of appetite. It is hoped that this important and fertile general area will attract more researchers in the future.

Interferon

Interferon is a nonspecific defense mechanism against viral infection. It is a protein which inhibits viral growth and replication and can be produced by several different cell types of the body in response to a viral infection of the particular cell. Its production (or lack of production) can be understood from the basic principles of protein synthesis (Chap. 4). When there is no virus within the host cell, the potential for interferon synthesis exists but the actual synthesis is repressed. Entry of a virus into the cell in some manner eliminates the repressor, thereby inducing interferon synthesis by the usual DNA-RNA-ribosomal mechanisms. The inducer (or derepressor) may well be the viral nucleic acid. Once synthesized, interferon may leave the cell, enter the circulation, and be picked up by another cell, despite the fact that it is a protein. Some interferon is always present in plasma (and in epithelial secretions) but a large increase occurs during viral infection.

To reiterate, interferon is not specific; all viruses induce the same kind of interferon synthesis, and interferon in turn can inhibit the multiplication of many different viruses. Several other properties of interferon differ from those of antibodies (to be described below) and confer great adaptive value. (*1*) The system is a rapidly reacting one, interferon synthesis beginning within hours of the onset of a viral infection. This is a time at which antibody formation is just barely getting underway, particularly in a first infection. (*2*) Antibodies cannot enter intact human cells, so that once having gained entry to its host cell, a virus is safe from antibody attack; in contrast, interferon functions inside the infected cell to prevent further growth, multiplication, and spread of the virus from cell to cell. It is not inconceivable that, in the near future, physicians will be administering interferon to patients suffering from viral illnesses in order to enhance their resistance.

Specific immune responses

There are two parts to the specific immune system, one mediated by antibodies secreted by a given population of lymphoid cells and the other mediated directly by a second distinct population of lymphocytes. The two cell populations have been labeled *B cells* and *T cells,* respectively, for reasons related to their life histories.

All lymphocytes stem originally from precursors in the bone marrow. After their release from the bone marrow (Fig. 15-19), cells destined to be T cells travel to the thymus (thus the name T cell) which in some manner confers upon them the ability to differentiate and mature into cells competent to carry out the activities associated

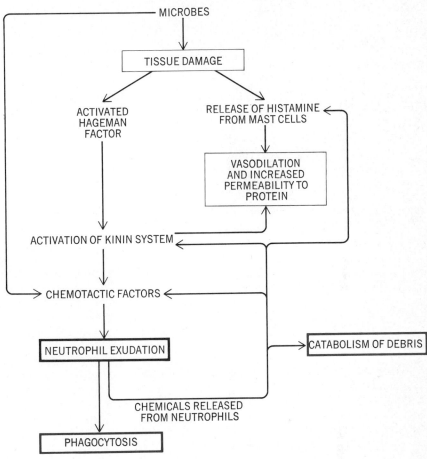

FIGURE 15-18
Summary of the nonspecific local inflammatory response to infection. This figure is an amalgam of Figs. 15-12, 15-13, and 15-15.

with cell-mediated immunity. The T cells then leave the thymus and take up residence in lymph nodes and other lymphoid tissues, but the thymus continues to influence them by means of a hormone which its epithelial cells secrete. Although most of this activity occurs prior to puberty when the thymus is largest and most active, most of the T cells are quite long-lived, so that removal of the thymus during adult life has little effect on a person's resistance.

In contrast, the B cells do not enter the thymus after their release from the bone marrow. In the chicken, they enter a gastrointestinal organ known as the bursa of Fabricius (thus the name B cell) which confers upon them the ability to differentiate into antibody-producing cells. From the bursa they then migrate to lymph nodes and other lymphoid tissues (Fig. 15-19). Human beings have no bursa, and it is not yet certain which human organ serves its function (for this reason, the name B cell has been retained). In any case it should be clear that whether a lymphocyte, newly synthesized and released from the bone marrow, is destined to function as a T cell (cell-mediated immunity) or as a B cell (humoral immunity) depends upon whether it passes through the thymus or the human counterpart of the bursa.

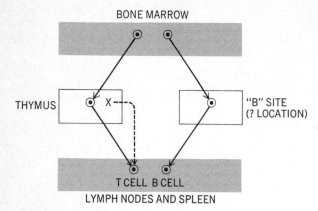

BONE MARROW

THYMUS

X

"B" SITE
(? LOCATION)

T CELL B CELL
LYMPH NODES AND SPLEEN

X = HORMONE

FIGURE 15-19
Life histories of B and T cells.

The scanning electron microscope has now made possible a structural differentiation between the two kinds of lymphocytes (Fig. 15-20). The B cells have many fingerlike projections on their surface whereas the T cells have only a few. In blood samples from healthy people, about 20 percent of the lymphocytes are B cells, the remainder T cells.

In general, the B cells and T cells serve different functions. The B cells, through their secreted antibodies, confer specific immune resistance against most bacteria. The T cells are the major carriers of specific immunity against fungi, viruses, parasites, and the few bacteria which, to survive, must live inside cells. The T cells also mediate the destruction of cancer cells and the rejection of solid-tissue transplants. This division of labor between the two cell populations is vividly demonstrated by the following clinical observations: Patients with deficient B cells produce few antibodies and are extremely sensitive to infection by most bacteria. However, they are fairly resistant to infection by nonbacterial invaders and intracellular bacteria. Moreover, they are able to reject solid-tissue grafts fairly normally. In contrast, another group of patients who have normal B cells but who lack T cells can secrete antibodies normally and show just the opposite pattern; their ability to tolerate grafts is particularly striking. Moreover, these patients show a much higher incidence of cancer than either the normal population or those patients with B-cell and antibody deficiency.

In the discussion to follow, we emphasize the separation of function of humoral immunity (B cells) and cell-

mediated immunity (T cells). However, it is rapidly becoming apparent that a complete separation is not warranted. In ways still poorly understood, B cells and T cells influence the activity of each other in both synergistic and inhibitory ways. These interactions will certainly prove to be of great importance, but we do not discuss them to any great extent because of the present paucity of definite information.

Antigens and antibodies

An *antibody* is a specialized protein capable of combining chemically with the specific *antigen* which stimulated its production. Conversely, antigens are substances induc-

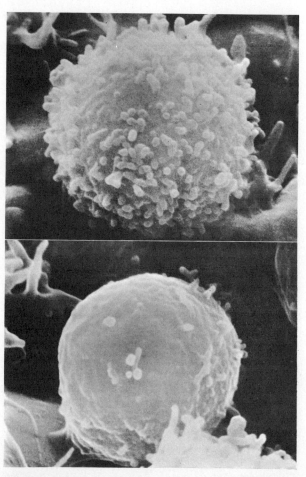

FIGURE 15-20
Electron micrographs of B cells (top) **and T cells** (bottom).
(Courtesy of Robert A. Good, Memorial Sloan-Kettering Cancer Center.)

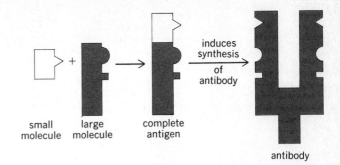

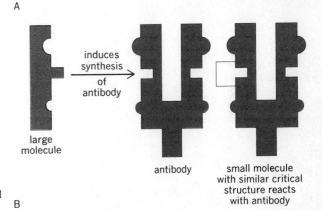

A

B

FIGURE 15-21
Mechanisms by which a small molecule can form an antigen-antibody complex. Only in case *A* does the small molecule play a role in the induction of the antibody.

ing the synthesis of antibodies, with which they can then specifically combine. The word specifically is essential in the definition since an antigenic substance reacts only with the type of antibodies elicited by its own kind or an extremely closely related kind of molecule. Specificity is thus related to the chemical structure of the antigen and its antibody. The same lock-and-key analogy used in discussing the enzyme-substrate combination probably applies.

Antibodies are all composed of polypeptide chains and are identical except for a relatively small number of amino acids occupying the first positions in the chains. These differences constitute the antibody's specificity. The three-dimensional structure conferred upon the tip of the chains by these amino acids permits the antibody to recognize the complementary structure in the antigen and combine with it.

Antibodies all belong to the family of proteins known as gamma globulins and are also known as immunoglobulins. They may be further subdivided into five classes according to differences in chemical structure and biological function. These are designated by the letters G, A, M, D, and E after the symbol Ig (for immunoglobulin). IgG and IgM antibodies provide the bulk of specific immunity against infectious microbes. The other class we shall be much concerned with is IgE, for these antibodies mediate certain allergic responses. IgA antibodies are produced by lymphoid tissue lining the GI, respiratory, and genitourinary tracts and exert their major activities in the secretions of these tracts. The function of the IgD class is presently uncertain.

Since an essential determinant of antigenic capacity is size (most antigens have molecular weights greater than 10,000), many chemicals injected into the body do not induce antibody formation. However, many smaller molecules act as antigens after first attaching themselves to one of the host's proteins, thus forming a complex large enough to induce antibody formation (Fig. 15-21A). Still other low-molecular-weight substances incapable of inducing antibody synthesis because of their small size can combine with antibodies induced by another antigen; in such cases the structural unit of the large true antigen which was critical in the induction process must have been similar or identical to that of the small molecule (Fig.

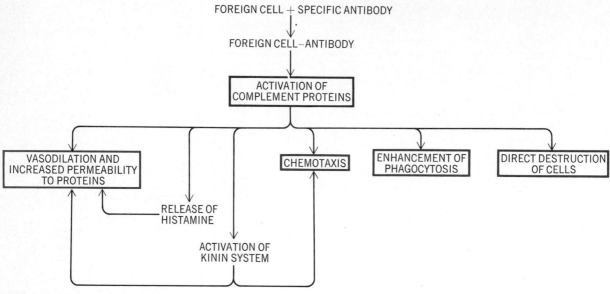

FIGURE 15–22
Functions of complement. Note how certain of these actions amplify the nonspecific inflammatory response illustrated in Fig. 15-18.

15-21B). These last two phenomena explain why many small molecules can cause allergic attacks.

The antigens we shall be most concerned with occur on the surface of microbes or are microbial products, such as bacterial toxins. However, components of almost any foreign cell or molecule not normally present in the body, e.g., penicillin, may act as an antigen. Indeed, even normal body components can induce antibody formation under unusual circumstances. Once the antibodies have been formed (the mechanism will be discussed in a subsequent section), they are released into the blood, reach the site where the antigen is located, and combine with it.

Functions of antibodies

Activation of complement system   In a previous section we described how a local inflammatory response is induced nonspecifically by any tissue damage. Now we shall see that the presence of antigen-antibody complexes triggers off events which profoundly amplify the inflammatory response. In other words, the major function of humoral immune mechanisms is to enhance and make more efficient the inflammatory response and cell elimination al-

ready initiated in a nonspecific way by the invaders. Thus, vasodilation, increased vascular permeability to protein, neutrophil exudation, and phagocytosis are all markedly enhanced; in addition, as we shall see, microbes may be destroyed without prior phagocytosis.

How does the presence of antigen-antibody complexes enhance inflammation? There are several mechanisms, but the single most important involves the *complement* system. It is yet another example (the clotting, plasmin, and kinin systems are others) of a "system" consisting of a group of plasma proteins which normally circulate in the blood in an inactive state; upon activation of the first protein of the group, there occurs a sequential cascade in which active molecules are generated from the inactive precursors. The activators of the initial step in the complement sequence are antigen-antibody complexes, although how this occurs is presently unclear. The active complement components generated in the sequence then act as the mediators (Fig. 15-22) for enhancing the various aspects of the inflammatory response (there are eleven proteins in the complement system mediating different effects, but we shall refer to the group collectively as complement).

Certain of the activated complement proteins enhance vasodilation and increased permeability by direct effects on the vasculature as well as by stimulating release of histamine from mast cells and platelets and by activating plasma kallikrein. They also facilitate neutrophil exudation by acting as powerful chemotactic agents. Other complement components enhance phagocytosis by coating the microbe and somehow permitting the phagocyte to gain a better "grip" on it. This is a particularly important function because many virulent bacteria have a thick polysaccharide capsule which strongly resists engulfment by phagocytes.[1] Thus, complement acts as an *amplification system* for inflammation. Once activated by the presence of antigen-antibody complexes, it becomes the single most important mediator of the local inflammatory response.

Moreover, certain complement proteins mediate still another effect not usually a part of the basic nonspecific inflammatory response: direct killing of microbes without prior phagocytosis. The complement apparently acts as an enzyme which combines with the microbes' surface and catalyzes the breakdown of the walls' lipid structure, thereby making the microbe leaky and killing it (this is quite distinct from the other action of complement on cell walls which results in enhancement of phagocytosis). These complement components combine only with cells to which antibody has already become attached; thus the antibody is said to "fix" complement to the cell surface.

It must be emphasized that the specificity in all these responses resides in the antigens and antibodies, not in complement. Complement is activated by almost any antigen-antibody complex (at least when the antibody is of the IgG or IgM class). In other words, there is only one set of complement molecules, and, once activated, they do essentially the same thing, regardless of the specific identity of the invader. In contrast, the formation of antibodies to antigens on the invader and their subsequent combination are highly specific. The function of the antibodies is to identify (i.e., "mark") the invading cells as foreign (by combining with antibody-specific antigens on the cells' surface) and to activate the complement system, which then mediates the actual attack. The "identification" function of the antibodies serves to "guide" those complement components which facilitate phagocytosis or kill the microbes outright; i.e., the antibodies must ensure that these complement components

---

[1] Activation of complement is not the only mechanism by which antigen-antibody complexes enhance phagocytosis. Merely the presence of antibody alone has some enhancing effect.

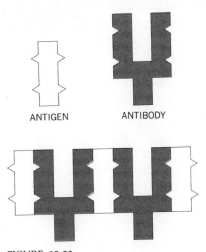

FIGURE 15-23

Interlocking complex of antigens and antibodies. This type of chain, or clump, can be formed because antibodies have more than one potential site for combination with antigen.

combine only with the invading cells and not randomly with the body's own cells (otherwise the latter might be phagocytized or destroyed). Somehow, the presence of the antibody combined with the antigen on the surface of the invading cell permits complement also to combine with surface sites and exert its effects.

Until quite recently, it was believed that antigen-antibody complexes were the only activators of the complement system. Present evidence suggests, however, that such may not be the case and has opened up the important possibility that the complement system may participate in the inflammatory response and cell killing in a variety of normal and pathological states even in the absence of antigen-antibody complexes. Evaluation of the physiological significance of nonspecific activators of complement must await further investigation. One of the most interesting possibilities is that the previously discussed Hageman factor may activate the complement system as well as the clotting, plasmin, and kinin systems.

To summarize, complement performs several functions important for immunity: (*1*) enhancement of the entire local inflammatory response culminating in phagocytosis; (*2*) direct killing of microorganisms without prior phagocytosis.

Direct neutralization of bacterial toxins and viruses Bacterial toxins and certain viral components act as antigens to induce specific antibody production. The antibodies then

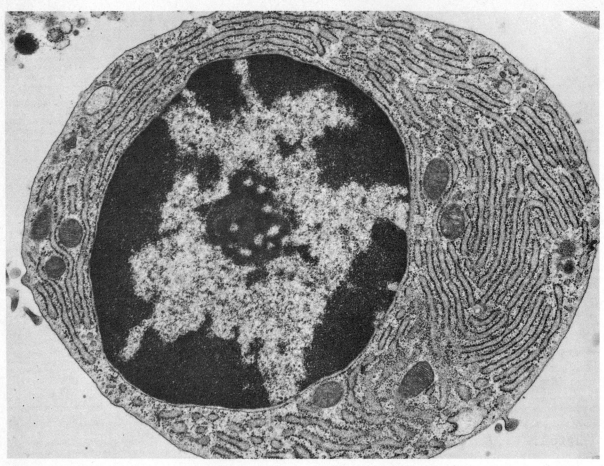

FIGURE 15-24
Electron micrograph of a guinea pig plasma cell. Note the extensive endoplasmic reticulum. *(From W. Bloom and D. W. Fawcett, "A Textbook of Histology," 9th ed., W. B. Saunders Company, Philadelphia, 1968.)*

combine chemically with the toxins and viruses to neutralize them. Neutralization referred to a virus means that the combined antibody somehow prevents attachment of the virus to host cell membranes, thereby preventing virus entry into the cell. Since viruses are obligatory intracellular parasites, i.e., can live only in the host's cells, the antibody kills the virus indirectly by cutting off its access to possible host cells. Similarly, antibodies neutralize bacterial toxins by combining chemically with them, thus preventing the interaction of the toxin with susceptible cell-membrane sites. Antibodies generally have more than one potential site for combination with antigen so

that an aggregate, or chain, of antigen-antibody complexes is formed (Fig. 15-23) which is then phagocytized.

Antibody production

When a foreign antigen reaches lymph nodes, the spleen, or the lymphoid patches in the body's epithelial linings, it triggers off antibody synthesis. The antigen may reach the spleen via the blood, but much more commonly it is carried from its site of entry via the lymphatics to lymph nodes. There it stimulates a tiny fraction of the B lymphocytes to undergo rapid cell division, most of the progeny of which then differentiate rapidly into plasma cells, which

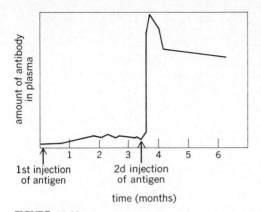

FIGURE 15-25
Rate of antibody production following the initial contact with an antigen and a subsequent contact with the same antigen.

are the active antibody producers. The most striking aspect of this transformation is a marked expansion of the cytoplasm, which consists almost entirely of the granular type of endoplasmic reticulum (Fig. 15-24) found in other cells which manufacture protein for export; after synthesis, the antibodies are released into the blood or lymph. Those B-cell progeny which do not fully differentiate into plasma cells constitute the "memory" of the occurrence, ready to respond more rapidly and forcefully should the antigen ever reappear at a future time.

Note that we stated that only a tiny fraction of the total B cells respond to any given antigen. It can also be demonstrated that different antigens stimulate entirely different populations (*clones*) of B cells. This is because the cells of any one lymphocyte clone (and the plasma cells it gives rise to) are capable of secreting only one kind of antibody. This limited synthetic capacity was probably determined by random mutations (during embryonic life) of the genes coding for the variable amino acids in the terminal portions of the antibody chains. According to this clonal theory, different antigens do not direct a single cell to produce different antibodies; rather each specific antigen triggers activity in the clone of cells already predetermined to secrete only antibody specific to that antigen. The antigen selects this particular clone and no other because the B cell displays on its surface the antibody molecules which it is capable of producing. These surface antibodies act as receptor sites with which the antigen can combine, thereby triggering off the entire process of division, differentiation, and antibody secretion just described. The staggering but statistically possible

implication is that there must exist millions of different clones, one for each of the possible antigens an individual *might* encounter during his life.

In concluding this section on antibody synthesis, it should be pointed out that interaction between antigen and B-cell receptor site must be considerably more complex than that just described above, for it seems clear that macrophages also play a crucial role. During antigen activation of the B cells, there occurs a clustering of macrophages around the relevant B-cell clone. It is likely that the macrophages process the antigen in some way so as to allow it to activate the receptor sites, but the actual events remain unknown. Some cooperation from T cells may also be required for the synthesis of certain antibodies.

Active and passive immunity

We have been discussing antibody formation without regard to the course of events in time. The response of the antibody-producing machinery to invasion by a foreign antigen varies enormously, depending upon whether it has previously been exposed to that antigen. Antibody response to the first contact with a microbial antigen occurs slowly over several days, with some circulating antibody remaining for long periods of time, but a subsequent infection elicits an immediate and marked outpouring of specific antibodies (Fig. 15-25). It is evident that this type of "memory" confers a greatly enhanced resistance toward subsequent infection with a particular microorganism. This resistance, built up as a result of actual contact with microorganisms and their toxins or other antigenic components, is known as *active immunity*. Until modern times, the only way to develop active immunity was actually to suffer an infection, but now a variety of other medical technics are used, i.e., the injection of vaccines or microbial derivatives. The actual material injected may be small quantities of living or weakened microbes, e.g., polio vaccine, small quantities of toxins, or harmless antigenic materials derived from the microorganism or its toxin. The general principle is always the same: Exposure of the body to the agent results in the induction of the antibody-synthesizing machinery required for rapid, effective response to possible future infection by that particular organism. However, it must be mentioned that not all microorganisms induce active immunity. For many microorganisms the memory component of the antibody response does not occur, and antibody formation follows the same course in time regardless of how often the body has been infected with the particular microorganism.

A second kind of immunity, known as *passive immunity,* is simply the direct transfer of actively formed antibodies from one person (or animal) to another, the recipient thereby receiving preformed antibodies. This exchange normally occurs between fetus and mother across the placenta and is an important source of protection for the infant during the first months of life, when his own antibody-synthesizing capacity is relatively poor. The same principle is used clinically when specific antitoxin or pooled gamma globulin is given a person exposed to or actually suffering from certain infections, such as measles, hepatitis, or tetanus. The protection afforded by this passive transfer of antibodies is relatively short-lived, usually lasting only a few weeks. The procedure is not without danger since the injected antibodies (often of nonhuman origin) may themselves serve as antigens, eliciting active antibody production by the recipient and possibly severe allergic responses.

Cell-mediated immunity

The T lymphocytes are responsible for cell-mediated immunity. Upon initial exposure to an appropriate antigen, a clone of T cells becomes "sensitized" to that particular antigen. The mechanism by which this occurs is unclear but it confers upon the lymphocyte the capacity to release locally a powerful battery of chemicals when the lymphocyte encounters that antigen again and combines with it (the antigen combines with receptor sites on the surface of the sensitized T cell). We must emphasize an important geographic difference between T-cell and B-cell function. Antibodies are secreted by B-cell progeny located in lymph nodes and other lymphoid organs far removed from the invasion site and reach the site via the blood; in contrast, the T cells travel to the invasion site where, upon combination with antigen, they release their chemicals.

As is true for the B system, some of the sensitized lymphocytes (this term specifically denotes T cells) do not actually participate in the immune response but serve as a "memory bank" which greatly speeds up and enhances the immune response if the person is ever exposed to the specific antigen again. Thus, active immunity exists for cell-mediated immune responses just as for antibody responses. How can one induce passive immunity in this system? This can be done by administering sensitized lymphocytes taken from a previously infected person (or animal).

The chemicals released when sensitized lymphocytes combine with specific antigen kill foreign cells directly and, in addition, act as an amplification system for the facilitation of the inflammatory response and phagocytosis. Thus, cell-mediated immunity is analogous to humoral immunity in that it serves, in large part, to enhance and make more efficient the nonspecific defense mechanisms already elicited by the foreign material. The major difference is that humoral immunity utilizes a circulating group of plasma proteins (the complement system) as its major amplification system whereas the T cells literally produce and secrete their own chemical amplification system.

As might be predicted, some of these chemicals are chemotactic factors. These serve to attract some neutrophils but many more monocytes to the area. The monocytes are converted to macrophages and begin their job of phagocytosis. But the lymphocytes go one step further: They not only secrete chemotactic factors to attract the macrophages-to-be; they secrete another substance which keeps the macrophages in the area and stimulates them to greater phagocytic activity (indeed, such revved-up macrophages are known as "angry" macrophages). In addition to facilitating the killing of target cells in this manner, the lymphocytes secrete so-called cytotoxins which are able to kill target cells directly, i.e., without phagocytosis. Here is another analogy to the complement system, with its ability to destroy cells directly as well as to facilitate phagocytosis.

This list of substances secreted by sensitized lymphocytes upon combination with specific antigen is by no means complete. There appear to be many other chemicals whose functions are presently being worked out. For example, it is known that interferon is released in large quantity from these cells.

We may now summarize the interplay between nonspecific and specific immune mechanisms in resisting infection. When a microbe is encountered for the first time, nonspecific defense mechanisms resist its entry and, if entry is gained, attempt to eliminate it by phagocytosis. Simultaneously, the antigens on the foreign matter induce the final development of specific cell clones capable of antibody production or cell-mediated immune responses or both. If the nonspecific defenses are rapidly successful, these specific immune responses may never play an important role. If only partly successful, the infection may persist long enough for significant amounts of antibody or sensitized T cells, or both, to reach the scene; antibody activates its chemical amplification system—complement—which both enhances phagocytosis and directly destroys the foreign cells. Similar functions are served by the chemicals released from sensitized T cells. In either case, all subsequent encounters with that microbe will be associated with the same sequence of events, with the crucial difference that the specific immune responses are brought into play much sooner and with greater force; i.e., the person would enjoy active immunity against that microbe.

Factors which alter the body's resistance to infection

Let us examine two seemingly opposed statements: (*1*) Tuberculosis is *caused* by the tubercle bacillus; (*2*) tuberculosis is *caused* by malnutrition. The first statement seems the more accurate in the sense that the disease, tuberculosis, will not occur in the absence of infection by tubercle bacilli. Yet we also known that many people harbor these bacteria, but never develop tuberculosis. There must exist, therefore, other factors which, by upsetting the balance between host and microbe, permit invasion, multiplication, and production of symptoms. In a sense, then, malnutrition "causes" tuberculosis by doing just this. In fact, there need be no quibbling over semantics once one realizes that the presence of the microbe is the necessary but frequently not sufficient cause of the disease. Therefore, it becomes very important to define those influences which determine the body's capacity to resist infection. We offer here only a few examples.

The hormone cortisol is commonly given in large doses specifically to inhibit inflammation (in patients with arthritis, for example), which it does in a variety of ways. This administration of cortisol can, on occasion, lower resistance enough for an infection to occur. Moreover, as we shall see, the body normally secretes large amounts of cortisol when a person is under psychic or physical stress; it is likely that this could cause lowered resistance.

A person's general nutritional status is also extremely important, but there is no indication that *excess* vitamins, etc., can confer increased resistance. A preexisting disease (infectious or noninfectious) can predispose the body to infection. Diabetics, for example, suffer from a propensity to numerous infections, at least partially explainable on the basis of defective leukocyte function. Many other illnesses are similarly associated with decreased resistance although not necessarily on the same basis. Moreover, any injury to a tissue lowers its resistance, perhaps by altering the chemical environment or interfering with blood supply.

In numerous examples one of the basic resistance mechanisms itself is deficient. A striking case is that of congenital deficiency of plasma gamma globulin, i.e., failure to synthesize antibodies; these patients cannot survive without frequent intravenous injections of gamma globulin. A similar lethal complement deficiency may also exist. A decrease in the production of leukocytes is also an important cause of lowered resistance, as, for example, in patients receiving tissue or organ transplants who must be given drugs specifically to inhibit rejection. The total quantity of leukocytes circulating is not necessarily critical; patients with leukemia, for example, may have tremendous numbers of blood neutrophils or monocytes, but these cells are almost all immature or otherwise incapable of normal function; such patients are extremely prone to infection. Reduced functional activity of lymphocytes is also seen in the elderly and may account for their decreased resistance.

Finally, we must mention the most important of the external agents we employ in altering resistance to infection, the *antibiotics,* such as penicillin. Use of these antibacterial agents is made possible because they are harmful to microbes but *relatively* innocuous to the body's cells. This characteristic distinguishes them from the *disinfectants,* which are highly effective antibacterial agents but equally toxic to the body. The "relatively" is quite important since all antibiotics are toxic to a lesser or greater degree and must not be used indiscriminantly. A second reason for judicious use is the problem of drug resistance. Most large bacterial populations contain mutants which are not sensitive to the drug and which are thus selected out by the drug. These few are capable of multiplying into large populations resistant to the effects of that particular antibiotic. Perhaps even more important, resistance can be transferred from one microbe directly to another previously nonresistant microbe by means of chemical agents ("resistance factors") passed between them. A third reason for the judicious use of antibiotics is that these agents may actually contribute to a new infection by altering the normal flora so that overgrowth of an antibiotic-resistant species occurs.

Antibiotics exert a wide variety of effects, but the common denominator is interference with the synthesis of one or more of the bacteria's essential macromolecules. For example, the actinomycins can prevent DNA and RNA synthesis, thus blocking bacterial division and protein synthesis. As might be expected, such antibiotics are also highly effective against viruses, but they cannot be used in man because their toxicity is not selective. This is true of all viral antibiotics thus far developed, and we still lack effective antiviral chemical therapy.

Immune surveillance: defense against cancer

A major function of cell-mediated immunity is to recognize and destroy cancer cells. This is made possible by the fact that virtually all cancer cells have some surface antigens different from those of other body cells and can, therefore, be recognized as "foreign." It is likely that cancer arises as a result of genetic alteration (by viruses, chemicals, radiation, etc.) in previously normal body cells. One manifestation of the genetic change is the appearance of the new surface antigens. Circulating T cells encounter and become sensitized to these foreign cells, combine with the antigens on their surface, and release the effector chemicals which destroy the cells by the mechanisms described above. It is presently believed that such transformations occur very frequently, i.e., that we may "get cancer once a day" (one expert's estimate), but that the cells are destroyed as fast as they arise. According to this view, only when the cell-mediated system is ineffective in either recognizing or destroying the cells do they multiply and produce clinical cancer.

In this last regard, there may occur an important interaction between the humoral and cell-mediated systems which actually protects the cancer cell. We have just pointed out that the tumor-specific antigens stimulate the development of sensitized lymphocytes against themselves. Simultaneously, they may also stimulate the production by B cells of circulating antibodies (of the IgG class). These antibodies combine with the antigen sites on the cell surface but, for some reason, are unable to activate complement. Therefore, the antigen-antibody complex causes no damage to the cancer cell. But because the antibodies are combined with the surface antigens, the sensitized lymphocytes cannot get at the antigens; accordingly, the T cell is not stimulated to release its chemicals. Such protecting antibodies are called "blocking" antibodies, and the process is known as *immune enhancement*. Clearly, the relative amounts of blocking antibody and sensitized T cells elicited by the antigenic stimulus are a major determinant of whether the emergent cancer cell is destroyed or not. Of great potential importance is the finding that dietary protein deficiency markedly impairs the production of IgG antibodies and, therefore, favors destruction of tumor cells. Does our large protein intake favor tumor growth by enhancing IgG synthesis? Can growth of certain tumors be medically controlled by restriction of certain amino acids? Such unanswered questions and many others have obvious practical significance, and a more complete understanding of immune processes may permit the development of effective anticancer weapons.

Rejection of tissue transplants

The cell-mediated immune system is also mainly responsible for the recognition and destruction, i.e., *rejection,* of tissue transplants. On the surfaces of all nucleated cells of an individual's body are protein molecules which are antigenic, known as *histocompatibility antigens*. The genes which code for these proteins are, of course, inherited from one's parents, so that the offspring's group of antigens are, in part, similar to his parents but not identical. Clearly, the more closely related two people are the more similar these antigens will be, but no two people (other than identical twins) have identical groups. Thus, the surface antigens constitute the basis for immune individuality.

When tissue is transplanted from one individual to another, those surface antigens which differ from the recipient's are recognized as foreign and are destroyed by circulating T cells which become sensitized to them. As was true for the response to cancer cells, the foreign cells may also stimulate the secretion of circulating "blocking" antibodies; if this occurs, the chances for graft survival are enhanced. Note that, in immunology, whether a phenomenon is desirable or undesirable depends on the point of view; tumor enhancement is undesirable whereas graft enhancement is desirable.

Some of the most valuable tools aimed at reducing graft rejection are radiation and drugs which kill actively dividing lymphocytes and, thereby, decrease the T-cell population. Unfortunately, this also results in depletion of B cells as well so that antibody production is diminished and the patient becomes highly susceptible to infection. A more discriminating method presently being tried is to prepare and inject into the recipient antibodies against his T cells; by this means, his T cells would be destroyed but not the B cells. Many other tools are presently being developed, all of which depend upon an understanding of the basic physiology of specific immune responses. As might be predicted from the previous section, graft recipients receiving medication to reduce their T-cell activity manifest an increased tendency to develop certain types of cancer.

Related to the general problem of graft rejection is one of the major unsolved questions of immunology: How does the body avoid producing antibodies or sensitized lymphocytes to its own cells; i.e., how does it distinguish self-antigens from nonself-antigens? In general, it appears that any antigens present during embryonic and very early neonatal life are recognized as self and no antibodies or sensitized lymphocytes are formed against them later in life, following maturation of specific immune mechanisms. This can be shown by fooling the embryo in the following manner: Foreign mouse cells are injected into an embryo mouse during intrauterine life; months later, when the mature mouse is given a graft from the same foreign species, the graft is not rejected.

Of course, the generalization stated above is empirical fact, not explanation. Present evidence warrants the generalization that the thymus, as might be predicted, is very much involved in this "imprinting" of self-recognition but just how is simply not known. The question is clearly of more than academic interest since, if we understood the mechanism by which tolerance for one's own tissues is established, then we might be able to confer tolerance for transplants.

Transfusion reactions and blood types

Transfusion reactions are a special example of tissue rejection; moreover, they illustrate the fact that antibodies rather than sensitized T cells can sometimes be the major factor in leading to the destruction of nonmicrobial cells. It was very early recognized that the transfusion of blood into a person was, more often than not, rapidly followed by clumping and hemolysis of erythrocytes with the appearance of hemoglobin in the plasma. If severe, this was associated with jaundice, fever, and a variety of tissue damage because of liberation of erythrocyte contents. With the identification of erythrocyte surface antigens, it was ultimately demonstrated that the red cell damage was caused by antigen-antibody reaction.

Among the large numbers of erythrocyte membrane antigens, we still recognize those designated A, B, and O as most important. These antigens are inherited, A and B being dominant. Thus, an individual with the genes for either A and O or B and O will develop only the A or B antigen. Accordingly, the possible blood types are A, B, O, and AB. If the typical pattern of antibody induction were followed, one would expect that a type A person would develop antibodies against type B cells only if the B cells were introduced into his body. However, what is atypical of this system is that even without initial exposure the type A person always has a high plasma concentration of anti-B antibody. The sequence of events during early life which lead to the presence of the so-called natural antibodies in all type A persons is unknown. Similarly, type B persons have high levels of anti-A antibodies; type AB persons obviously have neither anti-A nor anti-B antibody; type O persons have both; anti-O antibodies are usually not present in anyone.

With this information as background, what will happen if a type A person is given type B blood? There are two incompatibilities: (1) The recipient's anti-B antibody causes the transfused cells to be attacked and (2) the anti-A antibody in the transfused plasma causes the recipient's cells to be attacked. The latter is generally of little consequence, however, because the transfused antibodies become so diluted in the recipient's plasma that they are ineffective. It is the destruction of the transfused cells which produces the transfusion reaction. The range of possibilities is shown in Table 15-4. It should be evident why type O people are frequently called universal donors whereas type AB people are universal recipients. These terms, however, are misleading and dangerous since there are a host of other incompatible erythrocyte antigens and plasma antibodies besides those of the ABO type. Therefore, except in dire emergency, the blood of donor and recipient must be matched carefully.

Another antigen of great medical importance is the so-called *Rh factor* (because it was first studied in rhesus monkeys) now known to be a group of erythrocyte membrane antigens. The Rh system follows the classic immunity pattern in that no one develops anti-Rh antibodies unless exposed to Rh-type cells (usually termed Rh-positive cells) from another person. Although this can be a problem in an Rh-negative person, i.e., one whose cells have no Rh antigen, subjected to multiple transfusions with Rh-positive blood, its major importance is in the mother-fetus relationship (Fig. 15-26). When an Rh-negative mother carries an Rh-positive fetus, apparently some of the fetal erythrocytes may cross the placental barriers into the maternal circulation, inducing her to synthesize anti-Rh antibodies. These, in turn, cross the placenta into the fetus against whose erythrocytes they cause an attack to be launched. The resulting hemolysis may be severe enough to kill the fetus or produce serious anemia. Moreover, irreversible brain damage may result in these infants. Fortunately, only about 5 percent of Rh-negative mothers actually produce anti-Rh antibodies while carrying an Rh-positive child. Moreover, the first baby is almost always safe, later pregnancies becoming

TABLE 15-4
Summary of ABO blood-type incompatibilities

Recipient	Donor	Incompatible ?
A or AB	A	No
B or O	A	Yes
B or AB	B	No
A or O	B	Yes
A, B, or AB	AB	No
O	AB	Yes
A, B, AB, or O	O	No

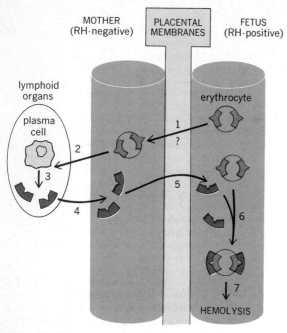

MOTHER
(RH-negative)

PLACENTAL
MEMBRANES

FETUS
(RH-positive)

lymphoid
organs

erythrocyte

plasma
cell

1
?

2

3

4

5

6

7

HEMOLYSIS

FIGURE 15-26

Sequence of events leading to hemolysis of fetal erythrocytes in Rh incompatibility between mother and fetus. The question mark indicates the unknown mechanism of erythrocyte movement across the placental membranes (1). The Rh antigen induces antibody formation by the mother's plasma cells (2 and 3), and these antibodies enter the mother's blood (4) and cross the placental membranes to enter the fetal blood (5). They then react (6) with the antigens on the erythrocyte membrane with resulting damage and hemolysis of the cell (7). The last step also involves complement, which is not shown.

more dangerous because of the memory component of immune mechanisms. In addition to the Rh problem, other maternal-fetal erythrocyte incompatibilities can produce a similar picture.

Allergy and tissue damage

Immune responses obviously evolved to protect the body against invasion by foreign matter. Unfortunately, they frequently cause malfunction or damage to the body itself. The term *allergy* or *hypersensitivity* refers to an acquired reactivity to an antigen which can result in bodily damage upon subsequent exposure to that particular antigen. Allergic responses may be due to activation of either the humoral or cell-mediated system. There are a variety of types of allergic responses, and we shall de-

scribe only two categories: atopic allergy and autoimmune disease.

Atopic allergy

This is the type of allergy popularly associated with the term allergy. A certain portion of the population is susceptible to sensitization by environmental antigens such as pollen, dusts, foods, etc. Initial exposure to the antigen leads to some antibody synthesis but, more important, to the memory storage which characterizes active immunity. Upon reexposure, the antigen elicits a more powerful antibody response. So far, none of this is unusual. What is it then that leads to body damage? The fact is that these particular antigens stimulate the production of the IgE class of antibodies which, upon their release from plasma cells, circulate to various parts of the body and attach themselves to mast cells (and basophils). When the antigen then combines with the IgE attached to the mast cell, the complex triggers, by unknown mechanisms (complement is not involved), a release of the mast cell's histamine and other vasoactive chemicals. These chemicals then initiate a local inflammatory response. Thus, the symptoms of atopic allergy are due to the various effects of these chemicals and the body site in which the antigen–IgE–mast cell combination occurs. For example, when a previously sensitized person inhales ragweed pollen, the antigen combines with IgE–mast cells in the respiratory passages. The histamine released causes increased mucus secretion, increased blood flow, leakage of protein, and constriction of the smooth muscle lining the airways. Thus, there follow the symptoms of congestion, running nose, sneezing, and difficulty in breathing which characterize hayfever.

In this manner, the symptoms of atopic allergy may be localized to the site of entry of the antigen. However, sometimes systemic symptoms may result if very large amounts of the vasoactive chemicals released enter the circulation and cause severe hypotension and bronchiolar constriction. This sequence of events, which can actually cause death, can be elicited in some sensitized people by the antigen in a single bee sting.

A major puzzle to biologists is the inappropriate nature of most atopic allergic responses, which are usually far more damaging to the body than the antigen triggering it. In other words, we clearly see the maladaptive nature of antigen-IgE reactions, but we do not know why such a system should have evolved, i.e., what normal physiological function is subsumed by IgE antibodies. Similarly, it is not known why only a certain portion of the population is susceptible to atopic allergies.

As may well be imagined, antihistamines offer some

relief in atopic allergies; it is usually incomplete, however, since other vasoactive agents are also released from the mast cells. In severe cases, the anti-inflammatory powers of large doses of cortisol are employed. The therapy known as *desensitization* is also of great interest because it offers another example of so-called blocking antibodies. After the specific antigen to which a person is sensitive has been identified, it is injected frequently into him in minute but increasing quantities. This procedure induces the synthesis of IgG antibodies against the antigen, so that when the antigen is subsequently encountered in the normal way, it will be complexed by these IgG antibodies, thereby preventing it from combining with the IgE. Because the IgG antibodies are not fixed to mast cells, no allergic response results.

Autoimmune disease

We must qualify a generalization made previously by pointing out that the body does, all too often, produce antibodies or sensitized cells against its own tissues, the result being cell damage or destruction. A growing number of diseases in man are being recognized as *autoimmune* (a better word would be *autoallergic*) in origin.

There are multiple causes for the body's failure to recognize its own cells: (*1*) Normal antigens may be altered by combination with haptens (drugs or environmental pollutants, for example) from the external environment; (*2*) the cell may be infected by a virus whose DNA codes for a new protein (antigen); (*3*) genetic mutations may yield new antigens; (*4*) the body may encounter microbes whose antigens are so close in structure to certain self-antigens that the antibodies or sensitized lymphocytes produced against these antigens cross-react with the self-antigens; (*5*) components of certain tissues might never be exposed during embryonic life to whatever organs (? the thymus) must recognize and memorize the self-antigens; if they appear in the blood later in life, as the abnormal result of tissue disruption following injury or infection, they are treated as foreign. This list of possibilities is by no means complete, but whatever the cause, a breakdown in self-recognition results in turning the body's immune mechanisms against its own tissues.

The above description centers on the production of antibodies or sensitized lymphocytes against the body's own cells. However, autoimmune damage may also be brought about in several other quite different ways. An overzealous response (too much generation of complement or release of chemicals from platelets, neutrophils, or sensitized lymphocytes) may cause damage not only to invading foreign cells but to neighboring nor-

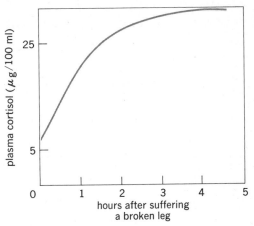

FIGURE 15-27
Effects of trauma on plasma cortisol concentration.

mal cells or membranes as well. For example, were a circulating antigen-antibody complex to be trapped within capillary membranes, the generation of complement or release of chemicals into the area might cause damage to the adjacent membranes. As might be predicted, the kidney glomeruli, with their large filtering surface, are prime targets for such autoimmune destruction.

SECTION C
RESISTANCE TO STRESS

Much of this book has been concerned with the body's response to stress in its broadest meaning of an environmental change which must be adapted to if health and life are to be maintained. Thus, any change in external temperature, water intake, etc., sets into motion mechanisms designed to prevent a significant change in some physiological variable. In this section, however, we describe the basic stereotyped response to stress in the more limited sense of noxious or potentially noxious stimuli. These comprise an immense number of situations, including physical trauma, prolonged heavy exercise, infection, shock, decreased oxygen supply, prolonged exposure to cold, pain, fright, and other emotional stresses. It is obvious that the overall response to cold exposure is very different from that to infection, but in one respect the response to all these situations is the same: Invariably secretion of cortisol is increased (Fig. 15-27); indeed, the term stress has come to mean any event which elicits increased cortisol secretion. Also, sympathetic nervous activity is usually increased.

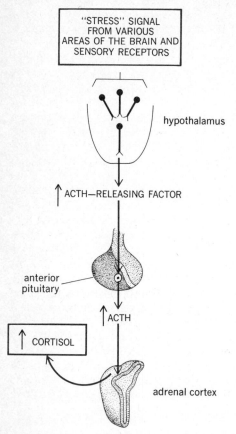

FIGURE 15-28
Pathway by which stressful stimuli elicit increased cortisol secretion.

Historically, activation of the sympathetic nervous system was the first overall response to stress to be recognized and was labeled the fight-or-flight response. Only later did further work clearly establish the contribution of the adrenal cortical response. The increased cortisol secretion is mediated entirely by the hypothalamus-anterior pituitary system (Fig. 15-28) and does not occur in animals lacking a pituitary or given a lesion in the hypothalamic area which secretes ACTH-releasing factor. Thus, afferent input to the hypothalamus induces secretion of ACTH-releasing factor, which is carried by the hypothalamopituitary portal vessels to the anterior pituitary and stimulates ACTH release. The ACTH, in turn, circulates to the adrenal and stimulates cortisol release. As described in Chap. 7, the hypothalamus receives input

from virtually all areas of the brain and receptors of the body, and the pathway involved in any given situation depends upon the nature of the stress; e.g., ascending pathways from the arterial baroreceptors carry the input during hypotension, whereas pathways from other brain centers mediate the response to emotional stress. The destination is always the same, namely, synaptic connection with the hypothalamic neurons which secrete ACTH-releasing factor.

These same pathways also converge on the hypothalamic areas which control sympathetic nervous activity (including release of epinephrine from the adrenal medulla). However, it should be evident that certain types of stress can stimulate the sympathetic nervous system without the participation of the hypothalamus; e.g., the increased activity induced by hypotension requires the integrating activity only of the medullary cardiovascular centers (Chap. 9). Such reflexes are of such crucial and universal importance that the lower brain centers have maintained this control even in higher animals.

Functions of cortisol in stress

Many of cortisol's most important effects are on organic metabolism. Cortisol (1) stimulates protein catabolism, (2) stimulates liver uptake of amino acids and their conversion to glucose (gluconeogenesis), (3) is permissive for stimulation of gluconeogenesis by other hormones (glucagon, growth hormone, etc.), and (4) inhibits glucose uptake and oxidation by many body cells ("insulin antagonism") but not by the brain. Indeed, so striking are these effects that cortisol is often called a glucocorticoid to distinguish it from the other major adrenal steroid, aldosterone, called a mineralcorticoid because its major effects are on sodium and potassium metabolism.

These effects are ideally suited to meet a stressful situation. First, an animal faced with a potential threat is usually forced to forgo eating, and these metabolic changes are essential for survival during fasting—indeed, an adrenalectomized animal rapidly dies of hypoglycemia and brain dysfunction during fasting. Second, the amino acids liberated by catabolism of body protein stores provide not only energy, via gluconeogenesis, but also constitute a potential source of amino acids for tissue repair should injury occur.

A few of the many medically important implications of these cortisol-induced metabolic effects associated with stress are as follows: (1) Any patient ill or subjected to surgery catabolizes considerable quantities of body protein; (2) a diabetic who suffers an infection requires

TABLE 15-5
Psychosocial situations shown to be associated with increased plasma concentration or urinary excretion of adrenal cortical steroids

Experimental animals

1 Any "first experience" characterized by novelty, uncertainty, or unpredictability.
2 Conditioned emotional responses; anticipation of something previously experienced as unpleasant.
3 Involvement in situations in which the animal must master a difficult task in order to avoid or forestall aversive stimuli. The animal must really be "trying."
4 Situations in which long-standing rules are suddenly changed so that previous behavior is no longer effective in achieving a goal.
5 Socially subordinate animals. (Dominant animals have decreased cortisol.)
6 Crowding (increased social interactions).
7 Fighting or merely observing other animals fighting.

Human beings

I Normal persons
  a Acute situations
    1 Aircraft flight
    2 Awaiting surgical operation
    3 Final exams (college students)
    4 Novel situations
    5 Competitive athletics
    6 Anticipation of exposure to cold
    7 Decreased during weekends
    8 Many job experiences
  b Chronic life situations
    1 Predictable personality-behavior profile: aggressive, ambitious, time-urgency
    2 Discrepancy between levels of aspiration and achievement
  c Experimental techniques
    1 "Stress" or "shame" interview
    2 Many motion pictures
II Psychiatric patients
  a Acute anxiety
  b Depression, but only when patient is aware of and involved in a struggle with it
  c Decreased in the manic state

much more insulin than usual; (3) a child subjected to severe stress of any kind manifests retarded growth. The explanations for these phenomena should be evident.

Cortisol has important effects other than those on organic metabolism. One of the most important is that of enhancing vascular reactivity. A patient lacking cortisol faced with even a moderate stress may go into shock and die if untreated. This is due primarily to a marked decrease in total peripheral resistance. For unknown reasons, stress induces widespread arteriolar dilation, despite massive sympathetic nervous system discharge, unless large amounts of cortisol are present. A large part of its counteracting effect is ascribable to the fact that moderate amounts of cortisol permit norepinephrine to induce vasoconstriction, but this can be only part of the story since considerably larger amounts of cortisol are required to prevent stress-induced hypotension completely. In other words, the normal cardiovascular response to stress requires *increased* cortisol secretion, not just permissive quantities.

Thus far we have presented the adaptive value of the stress-induced cortisol increase mainly in terms of its role in preparing the body physically for fight or flight, and there is no doubt that cortisol does function importantly in this way. However, in recent years, it has become apparent that cortisol may have other important functions. Table 15-5 is a partial list of the large variety of psychosocial situations demonstrated to be associated with increased cortisol secretion. Common denominators of many of them are novelty and challenge. Of great interest, therefore, are recent experiments which suggest that cortisol enhances learning in experimental animals, probably through direct actions on the brain. Thus, it may well be that the rise in cortisol secretion induced by psychosocial stress helps one to cope with the stress by facilitating the learning of appropriate responses.

Cortisol's pharmacological effects and disease

There are several situations in which adrenal corticosteroid levels in human beings become abnormally elevated. Patients with excessively hyperactive adrenals (there are several causes of this disease) represent one such situation, but the common occurrence is that of steroid administration for medical purposes. When cortisol is present in very high concentration, the previously described effects on organic metabolism are all magnified, but in addition there may appear one or more new effects, collectively known as the *pharmacological effects* of cortisol. The most obvious is a profound reduction in the inflammatory response to injury or infection (indeed, reducing the inflammatory response in allergy, arthritis, or other diseases is the major reason for administering the cortisol to patients). Large amounts of cortisol inhibit almost every step of inflammation (vasodilation, increased vascular permeability, phagocytosis) and may decrease antibody production as well. As might be expected, this decreases the ability of the person to resist infections. In addition, large amounts of cortisol may accelerate development of hypertension, atherosclerosis, and gastric ulcers, and may interfere with normal menstrual cycles.

As emphasized above, these pharmacological effects are known to be elicited when cortisol levels are extremely elevated. Yet an unsettled question of great importance is whether long-standing lesser elevations of cortisol may do the same thing, albeit more slowly and less perceptibly. Put in a different way, do the psychosocial stresses, noise, etc., of everyday life contribute to disease production via increased cortisol?

Functions of the sympathetic nervous system in stress

A list of the major effects of increased general sympathetic activity almost constitutes a guide on how to meet emergencies. Since all these actions have been discussed in other sections of the book, they are listed here with little or no comment:

1 Increased hepatic and muscle glycogenolysis (provides a quick source of glucose)
2 Increased breakdown of adipose tissue triglyceride (provides a supply of glycerol for gluconeogenesis and of fatty acids for oxidation)
3 Increased central nervous system arousal and alertness
4 Increased skeletal muscle contractility and decreased fatigue
5 Increased cardiac output secondary to increased cardiac contractility and heart rate as well as increased venous return (venous constriction)
6 Shunting of blood from viscera to skeletal muscles by means of vasoconstriction in the former beds and vasodilation in the latter
7 Increased ventilation
8 Increased coagulability of blood

The adaptive value of these responses in a fight-

or flight situation is obvious. But what purpose do they serve in the psychosocial stresses so common to modern life when neither fight nor flight is appropriate? As for cortisol, a question yet to be answered is whether certain of these effects, if prolonged, might not enhance the development of certain diseases, particularly atherosclerosis and hypertension. For example, one can easily imagine the increased blood fat concentration and cardiac work contributing to the former disease. Considerable work remains to be done to evaluate such possibilities.

Other hormones released during stress

Other hormones which are definitely released during many kinds of stress are aldosterone, antidiuretic hormone, and growth hormone. The increases in ADH and aldosterone ensure the retention of sodium and water within the body, an important adaptation in the face of potential losses by hemorrhage or sweating. Growth hormone reinforces the insulin antagonism effects of cortisol and the fat-mobilizing effects of epinephrine. Moreover, it probably stimulates the uptake of amino acids by an injured tissue and thereby facilitates tissue repair if needed; but since it cannot counteract the generalized protein catabolic effects of the increased cortisol, gluconeogenesis is not hampered.

Finally, recent evidence suggests that this list of hormones whose secretion rates are altered by stress is by no means complete. It is likely that the secretion of almost every known hormone may be influenced by stress. For example, prolactin, thyroxine, and glucagon are often increased whereas the pituitary gonadotropins (LH and FSH), insulin, and the sex steroids (testosterone or estrogen) are decreased. The adaptive significance of many of these changes is unclear but their possible contribution to stress-induced disease processes may be very important.

Processing sensory information

16

Man's awareness of the world is determined by the physiological mechanisms involved in the processing of afferent information, including such steps as the conversion of stimulus energy into coded neural activity indicating the quality, intensity, location, and duration of the stimulus. The action potentials are coded in different temporal patterns along different nerve fibers. This code represents the information from the external world even though, as is frequently the case with symbols, it differs vastly from the information it represents. The coded afferent information may or may not have a conscious correlate; i.e., it may or may not be incorporated into a conscious awareness of the physical world. Afferent information which does have a conscious correlate is called *sensory information* and, for the purposes of this book, that conscious experience of objects and events of the external world which we acquire from the neural processing of afferent information is called *perception*.

Intuitively, it might seem that sensory systems operate like electrical equipment, but this is true only up to a point. As an example, let us compare telephone transmission with our auditory sensory system. The telephone changes sound waves into electric impulses, which are then transmitted along wires to the receiver; thus far the analogy holds. (Of course, the mechanisms by which electric currents and action potentials are transmitted are quite different, but this does not affect our argument.) The telephone then changes the coded electric impulses *back into sound waves*. Here is the crucial difference, for our brain does not physically translate the code into sound; rather the coded information itself or some correlate of it is what we perceive as sound. At present there is absolutely no understanding how coded action potentials or composites of them can be associated with conscious sensations.

Basic characteristics of sensory coding

It is worthwhile restating the fact that all the information transmitted by the nervous system over distances greater than a few millimeters is signaled in the form of action potentials traveling over specific neural pathways. Several different kinds of information must be relayed by this code: stimulus quality, intensity, and localization.

Stimulus quality

As described in Chap. 6, receptors possess differential sensitivities; i.e., each receptor type responds more readily to one form of energy than to others. Therefore, the type of receptor activated by a stimulus constitutes the first step in the coding of different types (*modalities*) of stimuli. If, however, the differential sensitivity of receptors is to play an important role in the separation of the various sensory modalities, the afferent nerve fibers and at least some of the ascending spinal cord and brain pathways activated by the receptors must retain the same degree of specificity, carrying information that pertains to only one sensory modality. As expected, therefore, there are specific pathways ("labeled lines," as it were) for the different modalities. In tracing these pathways we shall begin at the receptor.

A single afferent neuron plus all the receptor units it innervates make up a *sensory unit*. In a few cases the afferent neuron innervates a single receptor, but generally the peripheral end of an afferent neuron divides into many fine branches, each terminating at a receptor (Fig. 16-1). When the sensory unit contains more than one receptor, all the receptors are differentially sensitive to the same stimulus energy form. The *receptive field* of a neuron is that area which, if stimulated, leads to activity in the neuron (Fig. 16-1).

The central processes of the afferent neurons terminate in the central nervous system, often diverging to terminate on several (or many) interneurons (Fig. 16-2A). The central processes of the afferent neurons also overlap, so that the processes of many afferent neurons converge upon a single interneuron (Fig. 16-2B). If the interneuron is to retain the specificity of the receptors, the sensory units converging upon it must be of the same modality.

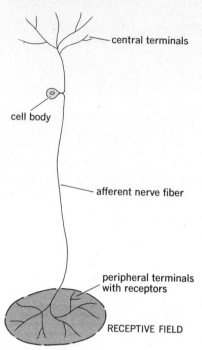

FIGURE 16-1
Sensory unit and receptive field.

The parallel chains of interneurons, which are grouped together to form the ascending pathways of the central nervous system, are of two kinds, *specific* and *nonspecific*. Each chain of neurons in the specific pathways consists of three to five synaptically connected neuronal links. Several sensory units may converge upon a given chain of neurons, but all these sensory units respond to the same stimulus energy, and specificity is maintained. The specific pathways (except for the olfactory pathways) pass to the thalamus of the brain and synapse there with neurons which go to the cerebral cortex.

In contrast to the specific pathway, chains of neurons in the nonspecific pathways are activated by sensory units of several different modalities and therefore convey only general information about the level of excitability; i.e., they indicate that *something* is happening, usually without specifying just what (or where). The nonspecific pathways feed into areas of the brain which are not highly discriminative but are important in determining states of consciousness such as sleep and wakefulness. More will be said of them in Chap. 18.

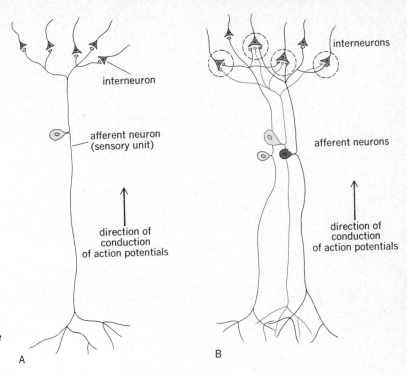

FIGURE 16-2
A **Divergence of afferent-neuron (sensory unit) terminals.** B **Convergence of afferent neurons onto single interneurons.**

Pathways carrying specific information about the different sensory modalities go to different areas of the cortex (Fig. 16-3). The fibers subserving the somatic sensory modalities (touch, temperature, etc.) synapse at cortical levels in a strip of cortex which lies in the parietal lobe just behind the junction between the parietal and frontal lobes (*somatosensory cortex*). The specific pathways which originate in receptors of the taste buds, after synapsing in brainstem and thalamus, probably pass to cortical areas adjacent to the face region of the somatosensory strip. However, the specific pathways from the ears, eyes, and nose do not pass to the somatosensory cortex but go to other primary receiving areas (Fig. 16-3). The pathways subserving olfaction are different from all the others in that they do not pass through thalamus, have no representation in cerebral cortex, and pass instead into parts of the limbic system (Fig. 18-5).

Thus stimulus quality is indicated by the specific sensitivity of individual receptors and the pathways conveying the information to the primary sensory areas of the brain. It is not known how the activation of neurons in different parts of the cortex results in the different sensations.

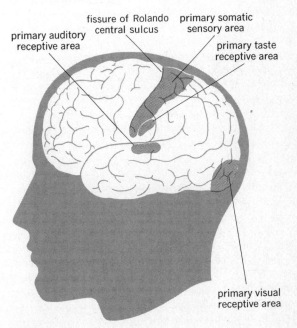

FIGURE 16-3
Primary sensory areas of the cerebral cortex.

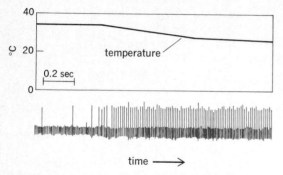

FIGURE 16-4
A single cold receptor signals a drop in temperature from 34 to 26°C with an increase in firing rate of action potentials in its afferent-nerve fiber. (*Adapted from Hensel and Bowman.*)

Stimulus intensity

The second kind of information contained in the code is stimulus intensity or quantity. As described in Chap. 6, one important mechanism for signaling intensity is the number of sensory units activated; the greater the intensity of the stimulus, the greater is the number of sensory units activated. A second mechanism is the frequency at which the sensory unit fires. Action-potential frequency correlates with stimulus intensity for afferent pathways leading to the sensory experiences of touch, temperature, limb position (joint extension or flexion), taste, sound (loudness), and light (brightness). Figure 16-4 shows the afferent impulses in a single cold fiber as the cutaneous receptor is gradually cooled from 34 to 26°C (normal body temperature is close to 37°C).

The action-potential frequency in temperature-sensitive sensory units is linearly related to the perceived intensity, but in other sensory systems the relation between stimulus intensity and perceived intensity is more complicated. Examples will be given in the discussions of vision and hearing.

Stimulus localization

A third factor to be relayed in the code is the *location of the stimulus*. The specific pathways channel the afferent information in a relatively unmixed manner and indicate stimulus location as well as stimulus quality. Since only sensory units from a restricted area converge upon any one interneuron, the specific afferent pathway which begins with that particular interneuron transmits exclusively information about that restricted area. Before discussing the terminations of the specific afferent pathways, let us examine more closely how information is actually fed into them. The branching peripheral terminals of the afferent neuron spread over areas of variable size (2 to 200 mm² in skin). The thresholds of the receptors of a single sensory unit vary within the area covered by the peripheral terminals (the receptive field of the afferent neuron), usually being lowest at the geometric center. Thus, a stimulus of a given intensity causes more activity in an afferent neuron if it occurs at the center of the receptive field (point *A*, Fig. 16-5) than at the periphery (point *B*). Since the peripheral terminations of afferent neurons also overlap to a great extent (Fig. 16-6), the placement of a stimulus determines not only the rate at which a single afferent nerve fiber fires but also the balance of activity within a group of sensory units. In the example in Fig. 16-6, neurons *A* and *C*, stimulated near the edge of their receptive fields, where the thresholds are higher, fire at a lower frequency than neuron *B*, stimulated at the center of its receptive field. Because of this gradient of sensitivity across the receptive field, the information content of the pattern of activity in a population of afferent neurons is great. As we have seen above, stimulus strength is related to the firing frequency of the afferent neuron, but a high frequency of impulses in the single afferent fiber of Fig. 16-5 could mean either that a stimulus of moderate intensity was applied at the center of the receptive field (point *A*) or that a strong stimulus was applied at the periphery (point *B*). Neither the intensity nor the localization of the stimulus can be detected precisely. But in a group of sensory units (Fig. 16-6), a high frequency of action potentials in neuron *B* arriving simultaneously with a lower frequency of action potentials in neurons *A* and *C* permits accurate localization of the stimulus. Once the location of the stimulus within the receptive field of neuron *B* is known, the firing frequency of neuron *B* can be taken as a meaningful measure of stimulus intensity.

The precision with which a stimulus can be localized and differentiated from an adjacent stimulus depends on the size of the receptive field covered by a single afferent neuron and the amount of overlap of nearby receptive fields. For example, the ability to discriminate between two adjacent mechanical stimuli to the skin is greatest on the thumb, fingers, lips, nose, and cheeks, where the sensory units are small and overlap considerably. The localization of visceral sensations is less precise than that of somatic stimuli because there are fewer

noreceptors within this entire area are activated (Fig. 16-7A). This potential information is discarded by mechanisms of lateral inhibition, and the pencil tip is accurately localized (Fig. 16-7B and C). Lateral inhibition occurs in

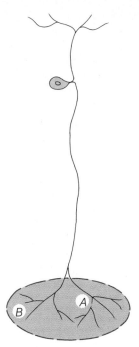

FIGURE 16-5
Two stimulus points, *A* and *B*, in the receptive field of a single afferent neuron.

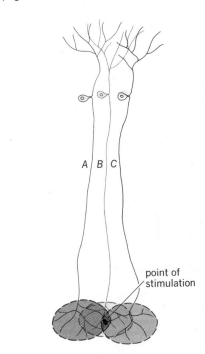

point of stimulation

afferent fibers and each has a larger receptive field. (*Somatic* refers to the framework or outer walls of the body as opposed to the viscera.)

We have given examples stressing the role played by receptive-field size and distribution in accurate stimulus localization; now we demonstrate the importance of functional interaction between the neurons of the afferent pathways.

One important type of such interaction, in which the afferent neurons and interneurons themselves inhibit other parallel afferent components, is *lateral inhibition*. The localized area of sensation is surrounded by an area of inhibition and decreased sensitivity; this serves to refine and clarify the information about stimulus localization on its way to higher levels of the central nervous system and to intensify contrasts. Lateral inhibition can be demonstrated in the following way. While pressing the tip of a pencil against the finger with one's eyes closed, one can localize the pencil point quite precisely, even though the region around the pencil tip is also indented and mecha-

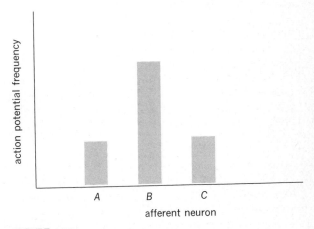

FIGURE 16-6
A stimulus point falls within the overlapping receptive fields of three afferent neurons.

the pathways of virtually all sensory modalities and is of great importance for the detection and emphasis of contrast, serving to lessen the weaker responses and collect the stronger ones into a common pathway.

As will be described later, stimulus localization is quite complicated in the visual, auditory, and olfactory systems since the perception's origin must be projected to an external source, but in the somatic sensory systems, localization of a stimulus simply means identification of the point at which the stimulus was applied. Stimulus localization is indicated by activity in selected specific pathways by the mechanisms just described. These pathways pass through the brainstem and thalamus to somatosensory cortex. The specific pathways cross from their side of entry to the opposite side of the central nervous system in the spinal cord or brainstem; thus the sensory pathways from receptors on the left side of the body go to the somatosensory strip of the right cerebral hemisphere and vice versa. In somatosensory cortex, the terminations of the individual components of the specific somatic pathways are grouped according to the location of the receptors. The pathways which originate in the foot end nearest the longitudinal dividing line between the two cerebral hemispheres. Passing laterally over the surface of the brain, one finds the terminations of the pathways from leg, trunk, arm, hand, face, tongue, throat, and viscera (Fig. 16-8). The parts with the greatest sensitivity (fingers, thumb, and lips) are represented by the largest

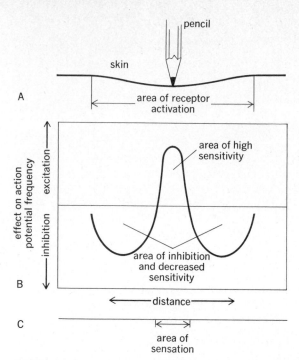

FIGURE 16-7
A A pencil tip pressed against the skin depresses surrounding tissues. Receptors are activated under the pencil tip and in the adjacent tissue. B Because of lateral inhibition, the central area of excitation is surrounded by an area of inhibition. C The sensation is localized to a more restricted region than that in which mechanoreceptors were actually activated.

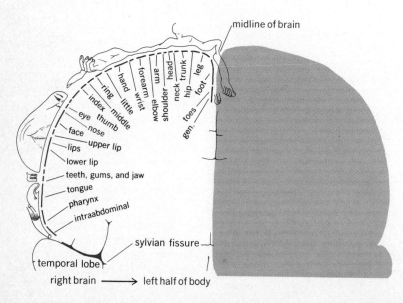

FIGURE 16-8
Location of pathway terminations for different parts of the body in the somatosensory cortex. This pattern is duplicated on the opposite cerebral hemisphere. The left half of the body is represented on the right hemisphere of the brain, and the right half of the body is represented on the left cerebral hemisphere.

areas of somatosensory cortex. The sensory strip of one cerebral hemisphere is duplicated in the opposite hemisphere.

We now turn to the receptor mechanisms and specialized patterns of coding of the specific sensory systems.

Specific sensory systems

Somatic sensation

Somatic receptors respond to mechanical stimulation of the skin or hairs and underlying tissues, rotation or bending of joints, temperature changes, and possibly some chemical changes. Their activation gives rise to the sensations of touch, pressure, heat, cold, the awareness of the position and movement of the parts of the body, and pain. Recalling that by *receptor* we mean the afferent-nerve fiber ending plus the specialized nonneural cells associated with it, we can say that probably each of these sensations is associated with a specific type of receptor; i.e., there are distinct receptors for heat, cold, touch, pressure, joint position, and pain.

Touch-Pressure One of the best examples of how receptor specificity is determined by the characteristics of the surrounding tissue is the pacinian corpuscle, discussed in Chap. 6. The nerve terminal of the pacinian corpuscle is surrounded by alternating layers of cells and extracellular fluid in such a way that a mechanical stimulus reaches the central nerve terminal only after displacing the layers of the surrounding capsule (Fig. 16-9). Rapidly applied pressures are transmitted to the nerve terminal without delay and give rise to a generator potential, presumably by the mechanisms described in Chap. 6. The energy of slowly administered or sustained forces is, in part, absorbed by the elastic tissue components of the capsule and is therefore partially dissipated before it reaches the nerve terminal at the core. Thus, the receptor fires only at the fast onset—and perhaps again at the release—of the mechanical stimulus but not under sustained pressure. What the pacinian corpuscle signals is not pressure, but *changes* in pressure with time. It can effectively discriminate stimuli vibrating up to frequencies of 300 cycles per second, or hertz (Hz). Other mechanoreceptors which adapt more slowly than the pacinian corpuscles provide information about both the *rate* of stimulus application and the stimulus *intensity*.

Joint position The activation of receptors in the joints, associated ligaments, and tendons and their pathways through the nervous system give rise to the conscious

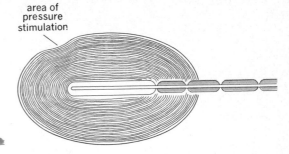

area of pressure stimulation

FIGURE 16-9
Pacinian corpuscle.

awareness of the position and movement of the joints. Input from these receptors is integrated with visual and other information to provide awareness of the position of the body in space. The different receptors in the joints and ligaments are activated by mechanical stimuli such as stretching, twisting, or compressing or by painful stimuli caused by damage. Their combined sensitivities signal movement and final position of the joint. Some receptors fire rapidly at the initiation of the movement, the action-potential frequency indicating the speed of the movement, and then slow their rate of firing down to a frequency dependent upon the final joint position (Fig. 16-10). Figure 16-11 shows the response of a single joint receptor during flexion and extension of the joint. Other receptors respond oppositely, firing faster during extension and slower during flexion.

Temperature Several hypothetical receptor mechanisms have been proposed for thermal receptors. Those responding to high-temperature stimuli might work in the following way: Increased temperature and the associated thermal agitation of molecular bonds cause configurational changes in protein molecules in the nerve ending; these alter the membrane permeability, causing a generator potential, which leads to the formation of an action potential. Although this mechanism has not been established, it is the most plausible of those currently proposed. The so-called cold receptors are activated by lower temperatures; their mechanism is unknown.

Pain A stimulus which causes or is on the verge of causing tissue damage often elicits a sensation of pain and a reflex escape or withdrawal response as well as a gamut of physiological changes which resemble the effects of activation of the sympathetic nervous system

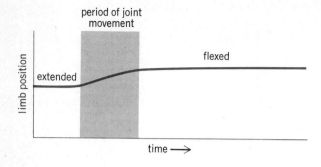

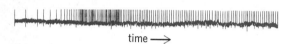

FIGURE 16-10
Response of a stretch receptor in the knee of a cat as the limb is flexed. (*Redrawn from Boyd and Roberts.*)

in fear, rage, or fight or flight. These physiological changes usually include faster heart rate, higher blood pressure, greater secretion of epinephrine into the bloodstream, increased blood sugar, less gastric secretion and motility, decreased blood flow to the viscera and skin, dilated pupils, and sweating. Moreover, the experience of pain includes an emotional component of fear, anxiety, and sense of unpleasantness as well as information about the stimulus's location, intensity, and duration. And probably more than any other type of sensation, the experience of pain can be altered by past experiences, suggestion, emotions (particularly anxiety), and the simultaneous activation of other sensory modalities.

This complex nature of pain can be accounted for by saying that the stimuli which give rise to pain result in a sensory experience *plus* a reaction to it, the reaction including the emotional response (anxiety, fear) and behavioral response (withdrawal or other defensive behavior). Both the sensation and the reaction to the sensation must be present for tissue-damaging stimuli to cause suffering. The sensation of pain can be dissociated from the emotional and behavioral reactive component by drugs, e.g., morphine, or by selective brain operations which interrupt pathways connecting the frontal lobe of the cerebrum with other parts of the brain. When the reactive component is no longer associated with the sensation, pain is felt, but it is not necessarily disagreeable; the patient does not mind as much. Thus, satis-

factory pain relief can be obtained even though the perception of painful stimuli is not reduced.

The receptors whose stimulation gives rise to pain are high-threshold receptors at the ends of certain small unmyelinated or lightly myelinated afferent neurons. These receptors fire specifically in response to tissue-damaging pressure, intense heat, or irritating chemicals, their firing frequency increasing as the severity of the stimulus rises.

The central connections of these afferent "pain fibers" are not well understood, but it is postulated that the afferent information is transmitted via two classes of ascending pathways simultaneously. The specific pathways, which go to the thalamus and cerebral cortex, are involved in the perception of pain, i.e., they convey information about where, when, and how strongly the stimulus

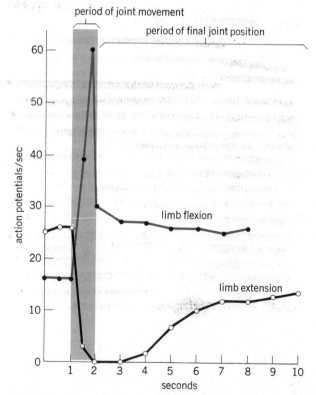

FIGURE 16-11
Two different responses of a single joint stretch receptor to movements in opposite directions, the upper curve during flexion and the lower curve during extension. (*Redrawn from Boyd and Roberts.*)

was applied; the nonspecific pathways, which go to the brainstem reticular formation and a part of the thalamus, different from that supplied by the specific pathways, arouse the aversive, reactive response to the stimulus. These same neurons which are activated by the nonspecific pathway are interconnected with the hypothalamus and other areas of the brain which play major roles in integrating autonomic and endocrine stress responses and the behavioral patterns of aggression and defense.

Descending pathways capable of altering the transmission of information in the afferent neurons, spinal pathways, or brain centers are known to exist in most sensory systems, but they are particularly important in pain. They are thought to be one means by which emotions, past experiences, state of attention, etc., can alter sensitivity to pain. When the descending pathways reduce the activity in the pain pathways, the unpleasant emotions and response behavior, as well as the specific pain perception, are diminished.

Vision

Light The receptors of the eye are sensitive to only that tiny portion of the vast spectrum of electromagnetic radiation which we call light (Fig. 16-12). Electromagnetic

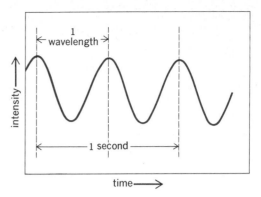

FIGURE 16-13
Properties of a wave. The frequency of this wave is 2 Hz.

radiation has both particlelike and wavelike properties. The radiant energy is propagated in the form of small discrete "packets" called *photons*. Radiant energy is described, however, in terms of wavelengths and frequencies. The *wavelength* is the distance between two successive wave peaks (Fig. 16-13) and varies from several miles long at the top of the spectrum to minute fractions of a millimeter at the bottom end. Those wavelengths capable of stimulating the receptors of the eye are between 400 and 700 nm (a nanometer is one-billionth of a meter). Light of different wavelengths is associated with different color sensations; for example, light having a wavelength of about 540 nm gives rise to the sensation of green and light having a wavelength of about 565 nm gives rise to the sensation of red. The *frequency* is the number of wave peaks (or cycles) that passes a point in a given period of time and is expressed in cycles per second or hertz (Hz) (Fig. 16-13). As they follow their wave path, the photons of visible light oscillate 4×10^{14} to 7×10^{14} times each second. The energy of a photon is proportional to the frequency of its oscillation.

The wavelength times the frequency equals the velocity, and the velocity of light in free space is one of the fundamental constants of nature. Thus, as wavelength decreases, the frequency increases.

A light wave can be represented most simply by a ray or line drawn in the direction in which the wave is traveling. At a boundary between two substances, such as the cornea of the eye (Fig. 16-14) and the air outside it, the rays are bent so that they travel in a new direction. The degree of bending depends upon the frequency of the light and the angle at which it enters the second medium.

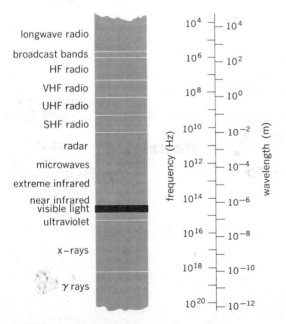

FIGURE 16-12
Electromagnetic spectrum.

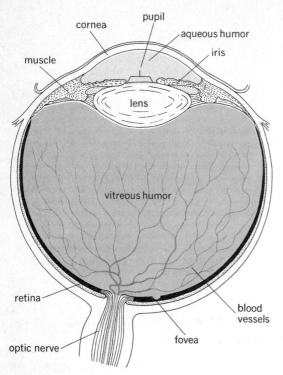

cornea
pupil
aqueous humor
iris
muscle
lens
vitreous humor
retina
blood vessels
fovea
optic nerve

FIGURE 16-14
Human eye. The blood vessels depicted run along the back of the eye between the retina and the vitreous humor.

The optics of vision Light waves are propagated in all directions from every point of a light source. These divergent light waves must pass through an optical system which focuses them back into a point before an accurate image of the light source is achieved. In the eye itself, the image of the object being viewed must be focused upon the *retina*, a thin layer of neural tissue lining the back of the eyeball (Fig. 16-14), where the light-sensitive receptor cells of the eye are located. The *lens* and *cornea* of the eye (Fig. 16-14) are the optical systems which focus the image of the object upon the retina. The cornea plays a larger role than the lens in focusing the image upon the retina because light rays are bent more in passing from air into the cornea than in passing into and out of the lens.

The surface of the cornea is curved so that light rays coming from a single point source hit the cornea at different angles and are bent different amounts, but all in such a way that they are directed to a point after emerging from the lens (Fig. 16-15A). Notice what hap-

pens to the image when the object being viewed has more than one dimension (Fig. 16-15B); the image on the retina is upside down relative to the original light source. It is also reversed left to right.

The shape of the cornea and lens and the length of the eyeball determine the point where light rays reconverge. Although the cornea performs the greater part quantitatively of focusing the visual image on the retina, all adjustments for distance are made by changing the shape of the lens. Such changes are called *accommodation*. The shape of the lens is controlled by a muscle which flattens the lens when distant objects are to be focused upon the retina and allows it to assume a more spherical shape to provide additional bending of the light rays when near objects are viewed (Fig. 16-16).

Cells are added to the lens throughout life but only to the outer surface. This means that cells at the center of the lens are both the oldest and the farthest away from the nutrient fluid which bathes the outside of the lens (if capillaries ran through the lens, they would interfere with its transparency). These central cells age and die first, and with death they become stiff, so that accommodation of the lens for near and far vision becomes more difficult. This is one reason why many people who never needed glasses before start wearing them in middle age.

Cells of the lens can also become opaque so that detailed vision is impaired; this is known as *cataract*. The defective lens can usually be removed surgically from persons suffering from cataract, and with the addition of compensating eyeglasses, effective vision can be restored.

Defects in vision occur if the eyeball is too long in relation to the lens size, for then the images of near objects fall on the retina but the images of far objects are focused in front of the retina. This is a *nearsighted*, or *myopic*, eye, which is unable to see distant objects clearly. If the eye is too short for the lens, distant objects are focused on the retina while near objects are focused behind it (Fig. 16-17); this eye is *farsighted*, or *hyperopic*, and near vision is poor. Defects in vision also occur where the lens or cornea does not have a smoothly spherical surface. The improperly shaped eyeball or irregularities in the cornea (astigmatism) or lens can usually be compensated for by eyeglasses (Fig. 16-17).

The amount of light entering the eye is controlled by a ringlike pigmented muscle known as the *iris*, the color being of no importance as long as the tissue is sufficiently opaque to prevent the passage of light. The hole in the center of the iris through which light enters the eye is the *pupil*. The iris muscle reflexly contracts in

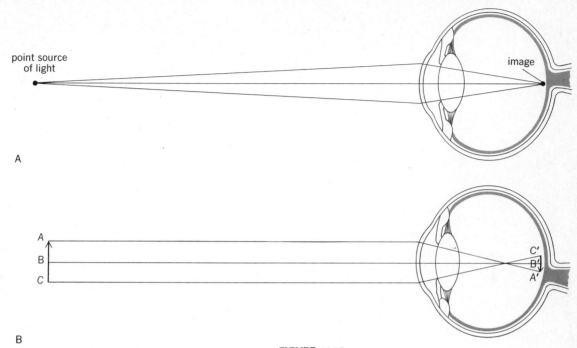

A

B

FIGURE 16-15
Refraction (bending) of light by the lens system of the eye.
The light source is A a point and B an object consisting of
many point sources.

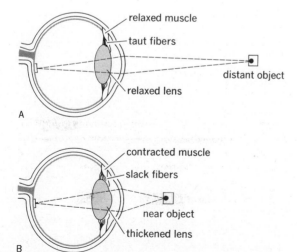

A

B

FIGURE 16-16
Accommodation for distant and near vision by the pliable
lens. A The lens is stretched for distant vision so that it adds
the minimum amount of focusing power. B The lens thickens
for near vision to provide greater focusing power.

bright light, decreasing the diameter of the pupil; this not
only reduces the amount of light entering the eye but also
directs the light to the central and most optically accurate
part of the lens. Conversely, the iris relaxes in dim light,
when maximal sensitivity is needed.

Receptor cells   The receptor cells in the retina are called
either *rods* or *cones* because of their microscopic appearance (Fig. 16-18). Both cell types contain light-sensitive
molecules called *photopigments*, whose prime function is to
absorb light. Light energy causes the photopigments to
change their molecular configuration which, in turn, alters
the properties of the receptor-cell membranes in which
they are situated. Unlike other receptor cells that have
been studied, in response to stimulation the membrane
decreases its permeability to sodium ions, which *hyperpolarizes* the receptor-cell membrane. This seems to release other neurons in the visual pathway from inhibition.

   There are four kinds of photopigments: one, *rhodopsin*, which is very sensitive to low levels of illumination,
and three, *erythrolabe, chlorolabe,* and *cyanolabe,* which are

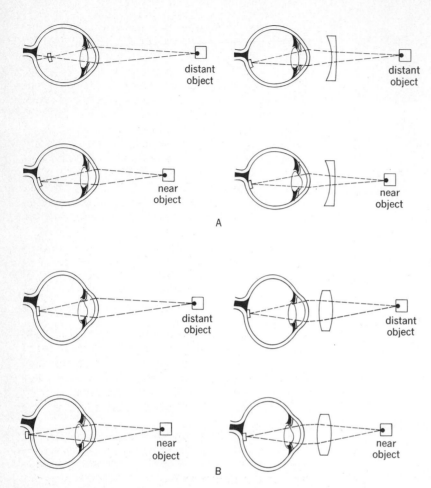

FIGURE 16-17

A In the nearsighted eye, light rays from a distant source are focused in front of the retina. A concave lens placed before the eye bends the light rays out sufficiently to move the focused image back onto the retina. When near objects are viewed through concave lenses, the eye accommodates to focus the image on the retina. B The farsighted eye must accommodate to focus the image of distant objects upon the retina. (The normal eye views distant objects with a flat, stretched lens.) The accommodating power of the lens of the eye is sufficient for distant objects, and these objects are seen clearly. The lens cannot accommodate enough to keep images of near objects focused on the retina, and they are blurred. A convex lens converges light rays before they enter the eye and allows the eye's lens to work in a normal manner.

sensitive to light wavelengths of the three primary colors, red, green, and blue, respectively. All four photopigments are made up of a protein (*opsin*) bound to a *chromophore* molecule. The chromophore is always the same slight variant of vitamin A, but the opsin differs in each of the four cell types and confers the specific light sensitivities upon the photopigment, i.e., determines whether it responds to all light or selectively to red, blue, or green. The photic energy (light) acts upon the chromophore, which then splits away from the opsin, changing the molecular configuration. After this breakdown of the photopigment in the presence of light, the chromophore molecule is rearranged and rejoined to opsin to restore the photopigment. Thus, the only action of light in vision is to change the chromophore; everything else in the sequence leading to vision—whether chemical, physio-logical, or psychological—is a "dark" consequence of this one light reaction.

Because the rod receptor cells contain rhodopsin, they are very sensitive, being able to detect very small amounts of light and acting as the photoreceptors during conditions of poor illumination and for night vision. Their responses do not indicate color, showing only shades of gray; they do indicate brightness. Their *acuity*, i.e., their ability to distinguish one point in space from another nearby point, is very poor. Rods are most numerous in the peripheral retina, i.e., that part closest to the lens, and are absent from the very center of the retina (the *fovea*) (Fig. 16-14). There are three types of cones, each containing one of the three photopigments for color vision. Cones operate only at high levels of illumination and are the photoreceptors for day vision. Cone visual acuity

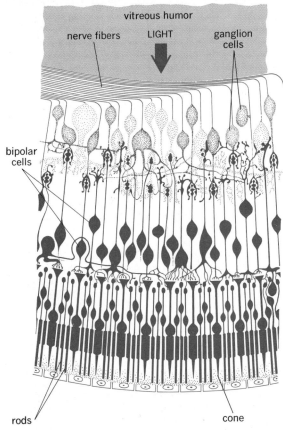

vitreous humor

nerve fibers   LIGHT   ganglion cells

bipolar cells

rods   cone

FIGURE 16-18

Human retina. Light entering the eye must pass through the fibers and cells of the retina before reaching the sensitive tips of the rods and cones. (*Adapted from Gregory.*)

cone) synapses upon a second neuron (a *bipolar* cell, Fig. 16-19) which in turn synapses upon a *ganglion* cell. The axons of the ganglion cells form a bundle called the optic nerve (Fig. 16-14) which passes directly into the brain. Generally, cone receptor cells have relatively direct lines to the brain; i.e., each bipolar cell receives synaptic input from relatively few cones, and each ganglion cell receives synaptic input from relatively few bipolar cells. This relative lack of convergence provides precise information about the area of the retina that was stimulated, but it offers little opportunity for the summation of subthreshold events to fire the ganglion cell. Conversely, many rod cells converge on bipolar and ganglion cells, and, although acuity is poor, opportunities for spatial and temporal summation are good. Therefore, a relatively low-intensity light stimulus that would cause only a subthreshold response in a cone ganglion cell can cause an action potential in a rod ganglion cell. Thus, the difference in acuity and light sensitivity between rod and cone vision is due, at least in part, to the anatomical wiring patterns of the retina.

These differences explain why objects in a darkened theater are indistinct and appear only in shades of gray; with such low illumination the cones do not reach threshold and fail to fire, so that all vision is supplied by the more sensitive but less accurate rod vision. The loss of visual acuity in dim light is due in part to the shift from cone to rod receptors.

The sensitivity of the eye improves after being in the dark for some time, due to *dark adaptation*. The modern theory of dark adaptation still has many unsolved problems but, in general, states that the excitability of the rod visual pathways depends on the number of intact rhodopsin molecules in the rods. In bright light so many rod rhodopsin molecules are broken down that the rods are ineffective, and vision is chiefly due to cone activation. When one moves from bright light to a darkened room, there are relatively few intact rhodopsin molecules, but as the rhodopsin slowly regenerates in the dark, visual sensitivity improves.

is very high, and because cones are concentrated in the center of the retina, it is that part which we use for finely detailed vision.

Each receptor cell in the retina (whether rod or

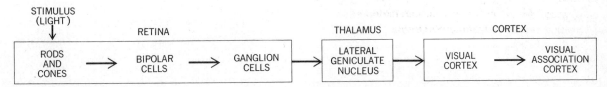

FIGURE 16-19

Diagrammatic representation of the cells in the visual pathway.

Visual system coding   The codes for afferent information leading to vision have been worked out in relatively great detail. Therefore, we have chosen to describe visual coding in considerable detail in order to illustrate general coding mechanisms, to show the degree of information processing occurring at each level of the afferent pathways, and to demonstrate the complexity of the coding mechanisms. This discussion is based on research done chiefly on frogs, cats, and monkeys, but it almost certainly applies to man as well. In most of these experiments simple visual shapes such as white bars against a black background were projected onto a screen in front of the anesthetized animal while the activity of single cells in the visual system was recorded. Different parts of the retina could be stimulated by varying the position of the bar on the screen. We shall constantly refer to receptive fields of neurons within the visual pathway and to the responses of these neurons to light. It is essential to recognize that only the rods and cones respond directly to light; all other components of the pathway are influenced only by the synaptic input to them. Thus when we speak of the receptive field of a neuron in the visual pathway, we really mean that area of the retina which when stimulated can influence the activity of that neuron. Similarly the neuron's "response to light" is really its response to neural activity within the visual pathway initiated by light falling upon the rods and cones. We shall follow the information processing elucidated by these experiments through the stages of the visual pathway, starting at the level of the ganglion cells (Fig. 16-19).

Retina   It is found that even at this early stage in the visual system an amazing amount of data processing has occurred. The retinal ganglion cells discharge spontaneously; i.e., they fire in the absence of any light stimulus. This spontaneous activity gives the cell an important second signal with which to work; it can either increase or decrease its rate of firing. Each receptor-cell–bipolar-cell–ganglion-cell chain is synaptically connected to other similar chains by cells which conduct laterally through the retina. These interconnections, which occur at both bipolar- and ganglion-cell levels, provide the pathways by which many receptor cells converge upon a single ganglion cell. The greater the degree of convergence, the larger is the area of the retina that can influence the ganglion cell.

The receptive fields of the ganglion cells are circular; i.e., any light falling within a specific circular area of the retina influences the activity of a given ganglion cell. The response of the ganglion cell varies markedly, depending on the region of the receptive field stimulated.

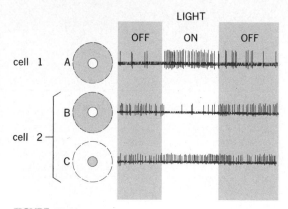

FIGURE 16–20
Recordings of the activity of a single ganglion cell.
A **Response of an** *on*-**center ganglion cell (cell 1) which increased its activity when light stimulated the center of its receptive field. B Activity of an** *off*-**cell (cell 2) is suppressed when the center of its receptive field is stimulated.** C **The same** *off*-**center cell increases its activity when the light is restricted to the periphery.** (*Adapted from Hubel and Wiesel, J. Physiol.,* **154.**)

Some ganglion cells speed up their rate of firing when a spot of light is directed at the center of their receptive field and slow down their firing when the periphery is stimulated. Such a cell is said to have an *on* center. The activity of a ganglion cell of this type is shown in Fig. 16-20A. Other ganglion cells, such as cell 2 (Fig. 16-20), have just the opposite response, decreasing activity when the center of the receptive field is stimulated (Fig. 16-20B) and increasing activity when the light stimulates the periphery (Fig. 16-20C). Notice the spontaneous activity of the *off*-center cell and the abrupt inhibition of its activity when the light is turned on. Often when the intermediate region of the receptive field is stimulated, the cell responds both when the light is turned on and again when it is turned off; i.e., it has an *on-off* response. Therefore, one ganglion cell can give an *on, off,* or *on-off* response, depending upon which region of its receptive field is stimulated.

Moreover, the basic pattern of ganglion-cell activity can be greatly modified. An *on* response increases, i.e., the frequency of firing action potentials increases, if the intensity of the light spot is greater, if the diameter of the spot is larger, or if the spot is moved. The *on* response decreases if the diameter of the spot becomes so much larger that it encroaches upon adjacent *off* regions or if a second spot is shown simultaneously on a nearby *off* region.

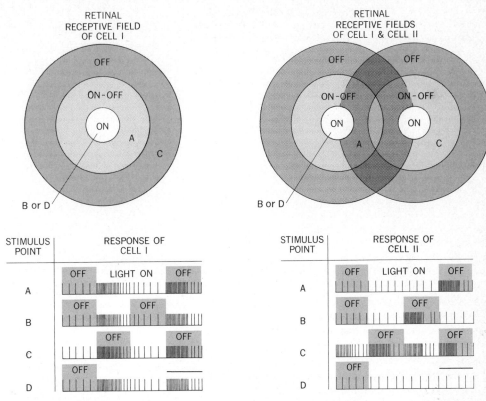

FIGURE 16-21
Letters A, B, C, and D on the receptive fields of cell I and cell II represent points of light stimulus. A particular *on*-center ganglion cell (cell I) responds to the four different stimulus arrangements with only one pattern; therefore it conveys no specific information. The simultaneous activity in two neurons discriminates between the four different events. In this case cell II alone provides specific information, but in other stimulus conditions cell II could fail to provide specific information, and activity of an additional neuron would be required. Horizontal bars (D) indicate increased stimulus.

This great flexibility in the response pattern of a single ganglion cell indicates that a complex series of events has taken place before the information even leaves the retina. Yet one ganglion cell alone cannot convey specific information. A single ganglion cell (Fig. 16-21, cell I) can respond with an *on-off* pattern, i.e., two bursts of activity separated by a short pause, if it is stimulated in the appropriate region of its receptive field (Fig. 16-21, cell IA). The same activity pattern occurs when the light spot is aimed at the *on* center of the receptive field and turned on and off twice (Fig. 16-21, cell IB), when the light spot is aimed at the *off* periphery of the receptive field and turned on and off twice (Fig. 16-21, cell IC), and when the spot is aimed at the *on* center, turned on, and then suddenly increased in intensity (Fig. 16-21, cell ID). The pattern of activity transmitted to the brain is the same in all four cases, but many ganglion cells simultaneously or almost simultaneously responding to the stimulus pattern on their individual receptive fields can convey a great deal of highly specific information. Notice (Fig. 16-21) how easily the four stimulus conditions are differentiated when the activities of cell I and cell II are evaluated simultaneously.[1]

The axons of the ganglion cells form the optic nerve, which passes to the brain. The optic nerves from the two eyes meet near the center of the head where

[1] Although it looks as though the responses of cell II alone could differentiate satisfactorily between the four stimulus conditions, this is not so. For example, the firing pattern transmitted to the brain by cell II would have been similar to that indicated in part A (Fig. 16-21, cell IIA) if the light stimulus had been directed to the center of the receptive field and turned on a bit later than it actually was.

some of the fibers cross over to the opposite side of the brain. This partial crossover provides both cerebral hemispheres with input from both eyes. After entering the brain, the visual pathways pass to the *lateral geniculate nucleus* in the thalamus.

Lateral geniculate nucleus   The receptive field of a single lateral geniculate cell, i.e., that area of the retina which when stimulated can influence the activity of the lateral geniculate neuron, resembles that of a retinal ganglion cell in being concentric with either an *on*-center–*off*-periphery pattern or vice versa. Movement of the stimulus across the receptive field of a cell always produces a stronger response than a stationary stimulus does, but the increase does not depend on the direction of movement.

Visual cortex   The partially processed visual information is transmitted along the axons of the lateral geniculate neurons to *primary visual cortex,* where the processing continues. Although the receptive fields of retinal ganglion cells and lateral geniculate neurons are usually concentric, with *on* or *off* centers, the receptive fields of cells in the visual cortex vary widely in organization. The cortical cells are classified as either simple or complex according to the stimuli to which they respond. The *simple* cells have receptive fields which are divided into *off* and *on* regions, but the divisions are no longer concentric (Fig. 16-22), all having a side-by-side arrangement of excitatory and inhibitory areas with straight boundaries rather than circular ones. Diffuse light over the entire receptive field generally gives little or no response because the effects of the simultaneously stimulated *on* and *off* areas cancel out. The most effective stimulus is one which covers the *on* area but does not encroach upon the *off* area, e.g., long, narrow slits of light; dark, rectangular bars against a light background (lines); or straight line

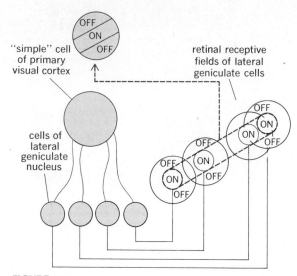

FIGURE 16-23
A possible way in which many *on*-center retinal receptive fields could give rise to a slit-shaped receptive field in cortex. (*Adapted from Hubel and Wiesel, J. Physiol.,* **160.**)

borders between areas of different brightness. The orientation of the optimum stimulus varies from cell to cell. How is this profound difference in receptive-field organization achieved? At present, there is no direct evidence to show how the visual cortex transforms incoming information, but some suggestions have been made. For example (Fig. 16-23), a simple cortical cell with a receptive field similar to that shown in Fig. 16-22A might receive excitatory input from many lateral geniculate cells with receptive-field *on* centers arranged along a straight line on the retina.

A second group of cells in primary visual cortex has more complex receptive fields. As for simple cells, some of the complex cells respond optimally to a line in a particular orientation across the receptive field but, unlike simple cells, the complex cells respond only when the stimulus is in motion across the visual field (Fig. 16-24). The *complex cells* have no separation of their receptive fields into excitatory and inhibitory parts. Most cells at cortical levels can be influenced from either eye with the most effective stimulus form, orientation, and rate of movement similar for both eyes. The cortical response increases when the two eyes are stimulated simultaneously. The complex cells, in turn, activate others which are hypercomplex.

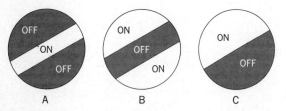

FIGURE 16-22
Retinal receptive fields of simple cortical cells are no longer arranged concentrically but are organized to provide information about lines and borders. (*Adapted from Hubel and Wiesel, J. Physiol.,* **160.**)

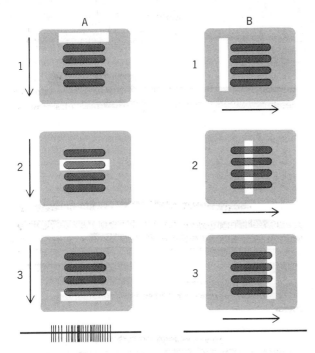

FIGURE 16-24
Complex cortical cells increase their rate of firing action potentials only when a bar of light moves across the visual field in a specific direction (e.g., vertically, as in A) and not in any other direction (e.g., horizontally, as in B). Recorded activity of the cells is shown in traces at bottom.

The multiple interconnections in the visual pathways are there to provide for active data processing rather than the simple transmission of action potentials. By means of these intricate cellular hookups, cells of the visual pathways respond only to selected features of the visual world. They are organized to handle information about line, contrast, movement, and color, but they are not very good intensity detectors. Note, also, that the visual system does not form a picture in the brain but through the simultaneous activation of many neurons forms a specifically coded electrical statement.

Color vision   Light is the source of all colors. Pigments, such as those mixed by a painter, serve only to reflect, absorb, or transmit different wavelengths of light, yet the nature of the pigments determines how light of different wavelengths will react. For example, an object appears red because all wavelengths other than those of 565 nm are absorbed by the material; light of 565 nm is reflected to excite the red-catching photopigment of

the retina. Light perceived as white is a mixture of all wavelengths, and black is the absence of all light. Sensation of any color can be obtained by the appropriate mixture of three lights, red, blue, and green (Fig. 16-25). Light and pigments are properties of the physical world, but color exists only as a sensation in the mind of the beholder. The problem for the scientist is to discover how the perception of a brilliantly colored world results from packets of photic energy of varying wavelengths.

Color vision begins with the activation of the photopigments in the cone receptor cells. Normal human retinas, as we have seen, have cones which contain either red-, green-, or blue-sensitive photopigments, responding optimally to light of 565-, 540-, and 435-nm wavelengths, respectively. Although each type of cone is ex-

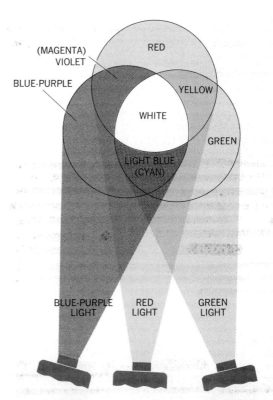

FIGURE 16-25
All known colors can be produced by different combinations of the three primary wavelengths of light, those giving rise to the color sensations of red, blue, and green. [See Plate I of color insert. (*Color plate from R. L. Gregory, "Eye and Brain: The Psychology of Seeing," p. 120, fig. 8-2, © 1966 by McGraw-Hill, Inc., New York.*)]

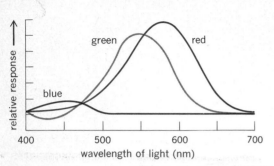

FIGURE 16-26
Response of the three photopigments to light of different wavelengths. [See Plate II of color insert. (*Color plate from R. L. Gregory, "Eye and Brain: The Psychology of Seeing," p. 121, fig. 8-3, © 1966 by McGraw-Hill, Inc., New York.*)]

cited most effectively by light of one particular wavelength, it responds to other wavelengths as well; thus, for any given wavelength, the three cone types are excited to different degrees. For example (Fig. 16-26), in response to a light of 540-nm wavelength, the green cones fire maximally, the red cones at about two-thirds of their maximum rate, and the blue cones not at all. Our sensation of color depends upon the ratios of these three cone outputs.

The fact that there are three different kinds of cone cells explains the various types of color blindness. Most people (over 90 percent of the male population and over 99 percent of the female population) have normal color vision; i.e., their color vision is determined by the differential activity of the three types of cones. Most color-blind (or better, color-defective) people appear to lack one of the three photopigments, and their color vision is therefore formed by the differential activity of the remaining two types of cones. For example, people with green-defective vision see as if they have only red- and blue-sensitive cones.

The cells processing color vision follow the pathways described earlier for the line-contrast processors (Fig. 16-19). The cones synapse upon bipolar cells, and the bipolar cells synapse upon ganglion cells, single ganglion cells dealing simultaneously with information from two sets of cones. One type of ganglion cell receives input from red and green cones. The inputs are additive so that the ganglion cell responds more briskly when it is receiving input from activated red and green cones than it does to input from red or green cones alone. The response of these cells is also greatly influenced by light

intensity or brightness. In fact, when the intensity is great enough, these cells respond to wavelengths throughout the entire spectrum. They code brightness rather than specific colors. A second type of ganglion cell codes specific colors and is called the *opponent color cell*. These cells receive excitatory input from one of the three cone types and inhibitory input from another. For example, the cell in Fig. 16-27 fired with a strong *on* response when stimulated by a blue light, but when a red light replaced the blue, the firing was greatly suppressed. The cell gave a weak *off* response when stimulated with a white light because the light contained both excitatory blue and inhibitory red wavelengths.

The retinal ganglion cells project to the lateral geniculate cells of the thalamus where a large proportion of the cells are of the opponent color type. Some of the cells have a center-surround arrangement to their receptive fields. For example, a red light in the center of their receptive field increases their activity and a green light at the periphery of the receptive field decreases cell activity. Figure 16-28 gives examples of some of the different arrangements of the opponent-type cells' receptive fields. There are also nonopponent cells in the lateral geniculate whose receptive fields are activated by light of many wavelengths in a broad spectrum of color.

This opponent aspect of color vision means that perception of one member of the pair is associated with decreased sensitivity to the other. For example, the neu-

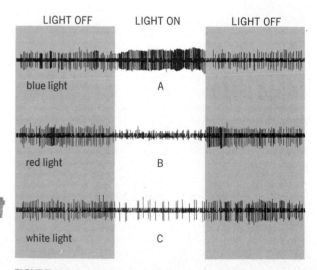

FIGURE 16-27
Response of a single ganglion cell to A **blue light,** B **red light, and** C **white light.** (*Adapted from Hubel and Wiesel, J. Physiol., 154.*)

ronal activity which results in the perception of green inhibits the pathways to the perception of red. This mechanism of the opponent cell explains the induced colors apparent in afterimages. Stare at the black dot in the center of Plate III for 20 sec and then look immediately at a white surface. The afterimage is that of a red square bordered in blue because withdrawal of the green and yellow images allows the red and blue members of the pair to be perceived for a moment or two as the inhibition is removed. As mentioned earlier, the relatively common red color-blindness in man has been attributed to a photopigment that is abnormally formed or lacking in one of the cone cells, but it may be due to a paucity of red-green opponent cells.

Thus, a push-pull type of behavior, in which paired neurons respond in opposite directions, is a common feature of the visual coding system. One cell responds when the light goes on, the other when it is turned off; one cell is sensitive to illumination at the center of the receptive field, the other to light at the periphery; one is sensitive to red, the other to green; one is excitatory, the other inhibitory. Pairs of neurons having opposing behaviors emphasize *contrasts* in the visual stimuli; much of our visual perception depends on such contrasts.

Eye-movement control

The cones are highly concentrated in a specialized area of the retina known as the *fovea,* and images focused there are seen with the greatest acuity. In order to keep the visual image focused on the fovea, the eye muscles perform four main types of movements.

1 **Search for visual targets** This type of movement is a small rapid jerk called a *saccade.* In addition to search of the visual field, saccades move the visual image over the receptors, preventing adaptation. In fact, if such movements of the eye are stopped, all color and most detail fade away in a matter of seconds. Saccades are among the fastest movements in the body. This pattern of periods of steady fixation interrupted by quick changes in fixation occurs, for example, while examining an object or reading. Although these movements are generally modified by visual information, they also occur during certain periods of sleep when the eyes are closed. Perhaps then they are associated with "watching" the visual imagery of dreams.

2 **Tracking of visual objects** These smooth movements cause the eyes to follow an object if it moves throughout the visual field. They require continual feedback of visual information about the moving object.

3 **Compensation for movements of the head** If a stationary visual object is focused on the fovea and the head is moved to

FIGURE 16-28
Opponent-type cells' visual fields can be arranged in many different ways.

the left, the eyes must be moved an equal distance to the right if the object's image is to remain focused on the fovea; if the head moves up, the eyes must move down.

4 **Convergence** This type of movement is used to track a visual object in depth through the visual field, turning the eyes inward as the object comes closer and outward as it moves farther away.

The four movement types seem to be controlled by separate neurological systems, yet they cooperate in almost all eye movements. Except for those eye movements which compensate for movements of the head, the control systems depend upon information from the retina. The compensating movements obtain their information about the movement of the head from the semicircular canals of the vestibular system, which will be described shortly.

Eye movements thus serve several important functions, yet they compound the problem of stimulus localization. For example, when the eyes and head remain stationary, the image of a moving object moves across the receptors and gives rise to signals from the retinas (Fig. 16-29A). Recall that most neurons in the visual pathways fire at different rates in response to moving and stationary stimuli. When the eyes follow a moving object, however, the image remains relatively stationary on the retina, and the retinas cannot signal movement (Fig. 16-29B). Since we still see the movement of the object, the rotation of the eyes in the head evidently can give

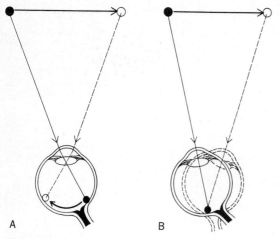

FIGURE 16-29

Motion of an object is perceived A when the eye remains stationary and the image of the object sweeps across the retina and B when the eye moves and the image remains stationary upon the retina.

rise to perception of movement and even fairly accurate estimates of velocity. In addition, when there is movement, a decision must be made about what is moving and what is stationary with respect to some reference frame. An obvious example occurs whenever we change position by walking or driving. We usually know that the change is due to the movement of our bodies and not to that of the external world, but it involves a decision. If the information concerning movement is known by vision alone, we generally assume that the larger objects in the visual field are stationary.

Hearing

Sound energy is transmitted through air as a disturbance of air molecules. When there are no air molecules, as in a vacuum, there can be no sound. The disturbance of air molecules that makes up a sound wave consists of regions of compression, in which the air molecules are close together and the pressure is high, alternating with areas of rarefaction, where the molecules are farther apart and the pressure is lower. Anything capable of creating such disturbances can serve as a sound source. A tuning fork at rest emits no sound (Fig. 16-30), but if it is struck sharply, it gives rise to a pure tone. As the arms of the tuning fork move, they push air molecules ahead of them, creating a zone of compression, and pull apart the molecules behind them, leaving a zone of rarefaction (Fig. 16-30 B). As they move in the opposite direction, they again create pressure waves of compression

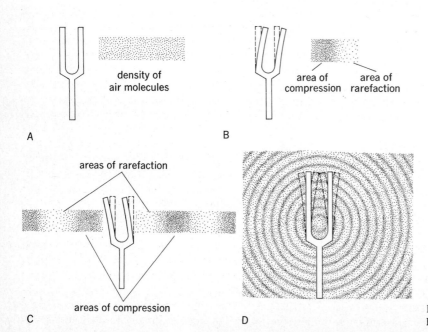

density of
air molecules

area of   area of
compression   rarefaction

areas of rarefaction

areas of compression

FIGURE 16-30

Formation of sound waves.

of rarefaction (Fig. 16-30C). The molecules in an area of compression, pushed together by the vibrating prong of the tuning fork, bump into the molecules ahead of them, push them together, and create a new region of compression. Individual molecules travel only short distances, but the disturbance passed from one molecule to another can travel many miles; and it is in these disturbances (sound waves) that sound energy is transmitted. The sound dies out only when so much of the original sound energy has been dissipated that one sound wave can no longer disturb the air molecules around it. The tone emitted by the tuning fork is said to be *pure* because the waves of rarefaction and compression are regularly spaced. The waves of speech and many other common sounds are not regularly spaced but are complex waves made up of many frequencies of vibration.

The sounds heard most keenly by human ears are those from sources vibrating at frequencies between 1,000 and 4,000 Hz, but the entire range of frequencies audible to man extends from 20 to 20,000 Hz. The *frequency* of vibration of the sound source is related to the pitch we hear; the faster the vibration, the higher the pitch. We can also detect loudness and tonal quality, or timbre, of a sound. The difference between the packing (or pressure) of air molecules in a zone of compression and a zone of rarefaction, i.e., the *amplitude* of the sound wave, is related to the loudness of the sound that we hear. The number of sound frequencies in addition to the fundamental tone, i.e., the degree of *purity* of the sound wave, is related to the quality or timbre of the sound. We can distinguish some 400,000 different sounds. We can distinguish the note A played on a piano from the same note played on a violin, and we can identify voices heard over the telephone. We can also selectively *not* hear sounds, tuning out the babel of a party to concentrate on a single voice. How is this accomplished by an apparatus small enough to fit into a teacup?

The first step in hearing is usually the entrance of pressure waves into the *ear canal* (Fig. 16-31). The waves reverberate from the side and end of the ear canal so that it is filled with the continuous vibrations of pressure waves. The *tympanic membrane (eardrum)* is stretched across the end of the ear canal. The air molecules, under slightly higher pressure during a wave of compression, push against the membrane, causing it to bow inward. The distance the membrane moves, although always very small, is a function of the force and velocity with which the air molecules hit it and is therefore related to the loudness of the sound. During the following wave of rarefaction, the membrane returns to its original position.

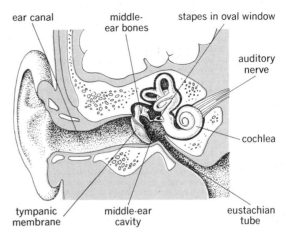

FIGURE 16-31
Anatomy of the human ear.

The exquisitely sensitive tympanic membrane responds to all the varying pressures of the sound waves, vibrating slowly in response to low-frequency sounds and rapidly in response to high tones. It is sensitive to pressures to which the most delicate touch receptors of the skin are totally insensitive.

The tympanic membrane separates the ear canal from the *middle-ear cavity* (Fig. 16-31). The pressures in these two air-filled chambers are normally equal, but a difference can be produced with sudden changes in altitude, as in an elevator or airplane. This difference distorts the tympanic membrane and causes pain. The outer ear canal is, of course, normally at atmospheric pressure. The middle ear is exposed to atmospheric pressure only through the *eustachian tube,* which connects the middle ear to the pharynx and nose or mouth. The slitlike ending of the eustachian tube in the pharynx is normally closed; but during yawning, swallowing, or sneezing, when muscle movements of the pharynx open the entire passage, the pressure in the middle ear equilibrates with atmospheric pressure.

The second step in hearing is the transmission of sound energy from the tympanic membrane, through the cavity of the middle ear, and then to the receptor cells in the *inner ear,* which are surrounded by fluid. The major function of the middle ear (Fig. 16-31) is to transfer movements of the air in the outer ear to the fluid-filled chambers of the inner ear. The tympanic membrane is coupled by a chain of three small, *middle-ear bones* to a membrane-covered opening (the *oval window*), which sep-

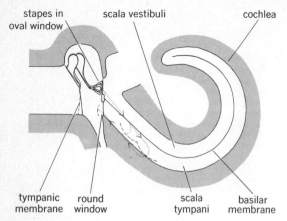

stapes in oval window
scala vestibuli
cochlea
tympanic membrane
round window
scala tympani
basilar membrane

FIGURE 16-32

Auditory portions of the middle and inner ear. The position of the membranes and middle-ear bones is shown at rest (solid lines) and following the inner displacement of the tympanic membrane by a sound wave (dashed line). Arrows show the different paths of two sound pressure waves of different frequency.

arates the middle and inner ear. The total force on the tympanic membrane is transferred to the much smaller oval window. The *total* force on the oval window is the same as that on the tympanic membrane, but because the oval window is so much smaller, the *force per unit area* (i.e., pressure) is increased 15 to 20 times. Additional advantage is gained through the lever action of the three middle-ear bones. Thus, the tiny amounts of energy involved are transferred to the inner ear with relatively small loss. The amount of energy transmitted to the inner ear can be modified by the contraction of two small muscles in the middle ear which alter the tension of the tympanic membrane and the position of the third middle-ear bone (*stapes*) in the oval window. These muscles protect the delicate receptor apparatus from intense sound stimuli and possibly aid intent listening over certain frequency ranges.

Thus far, the entire system has been concerned with the transmission of the sound energy into the inner ear, where the receptors are located. The inner ear, or *cochlea*, is a coiled passage in the temporal bone. It is almost completely divided lengthwise by the *basilar membrane* (Figs. 16-32 and 16-33). As the pressure wave pushes in on the tympanic membrane, the chain of bones rocks the footplate of the stapes against the oval window, causing it to bow into the cochlea. As this occurs, a wave of pressure is produced in the inner-ear compartment (*scala vestibuli*) on the other side of the oval window (Figs.

16-32 and 16-33). The wall of the scala vestibuli is largely bone, but there are two paths by which the pressure waves can be dissipated. One path is to the end of the scala vestibuli, where the waves pass around the end of the basilar membrane into a second compartment, the *scala tympani*, and back to another membrane-covered window, which they bow out into the middle-ear cavity. However, most of the pressure waves are transmitted to the basilar membrane, which is deflected into the scala tympani.

The pattern by which the basilar membrane is deflected is important because this membrane contains the sensitive receptor cells which transform sound energy, i.e., the pressure wave, into action potentials. At the end of the cochlea closest to the middle-ear cavity, the basilar membrane is narrow and relatively stiff, but it becomes wider and more elastic as it extends throughout the length of the cochlear spiral. The stiff end nearest the middle-ear cavity vibrates immediately in response to the pressure changes transmitted to the scala vestibuli, but the responses of the more distant parts are slower. Thus, with each change in pressure in the inner ear, a wave of vibrations is made to travel down the basilar membrane (Fig. 16-34).

The region of maximal displacement of the basilar membrane varies with the frequency of vibration of the sound source. The properties of the membrane nearest the oval window and middle ear are such that this region resonates best with high-frequency tones and undergoes the greatest amplitude of vibration when high-pitched tones are heard. The traveling wave soon dies out once it is past this region. Lower tones also cause the basilar membrane to vibrate near the middle-ear cavity, but the vibration wave travels out along the membrane for greater distances. The more distant regions of the basilar membrane vibrate maximally in response to low tones. Thus the frequencies of the incoming sound waves are in effect sorted out along the length of the basilar membrane (Fig. 16-35).

Where the displacement of the basilar membrane is a maximum the stimulation of the receptors (*hair cells,* Fig. 16-33) which ride upon the membrane is the greatest. The fine hairs on the top of the receptor cells are in contact with the overhanging *tectorial membrane*, which projects inward from the side. As the basilar membrane is displaced by pressure waves in the scala vestibuli, the hair cells move in relation to the tectorial membrane, and, consequently, the hairs are displaced. In this interaction, the incoming sound energy is transformed from the vibrating molecules of pressure waves to electric events in the

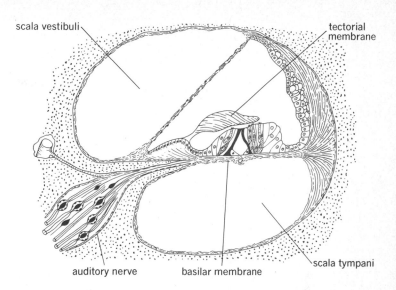

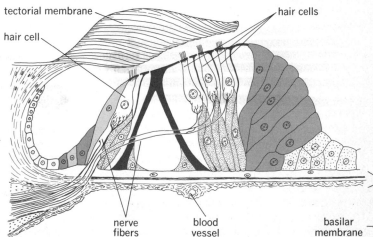

FIGURE 16-33
Cross section of the membranes and compartments of the inner ear with a detailed view of the hair cells and other structures upon the basilar membrane. (*Adapted from Rasmussen.*)

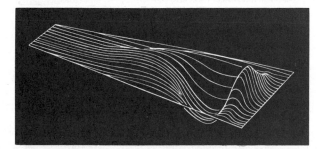

FIGURE 16-34
Wave motion of the basilar membrane in response to pressure changes in the inner ear. (*From M. Alpern, M. Lawrence, and D. Wolsk, "Sensory Processes," Brooks/Cole Publishing Company, Belmont, Calif., 1968.*)

hair cells, for in some way (the precise mechanism is not known) movements of the hairs cause a depolarization of the hair cells which is similar to the generator potential of the receptors. The hair cells are easily damaged by exposure to high-intensity noises such as the typical live amplified rock music concerts and engines of jet planes and revved-up motorcycles. The damaged sensory hairs form giant, abnormal hair structures or are lost altogether and, in cases of long exposure to loud sounds, areas of the organ of Corti itself completely degenerate (Fig. 16-36).

In normal hearing, the generator potentials formed by the activated hair cells lead ultimately to the production of action potentials in the peripheral endings of the

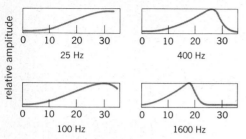

FIGURE 16-35
The point along the basilar membrane where the traveling wave peaks is different with different sound frequencies. The region of maximal displacement of the basilar membrane occurs near the end of the membrane for low-pitched (low-frequency) tones and near the oval window and middle ear for high-pitched tones. (*Adapted from von Békésy and Rosenblith.*)

afferent nerves, the resulting action potentials being transmitted into the central nervous system. The greater the energy of the sound wave (loudness), the greater is the movement of the basilar membrane, the greater the amplitude of the generator potential, and the greater the frequency of action potentials in the afferent nerve.

This completes our analysis of the auditory receptor mechanisms. We should like to be able to follow the coding of information from the organ of Corti to auditory cortex, as we did for vision. However, the precise mechanisms are considerably less clear than is the present case for vision and we therefore will not do so, except to describe sound localization.

Times of occurrence of sounds are important clues used to detect the location of a sound source. In hearing (and in vision and smell) there is the problem in stimulus localization of projecting the stimulus to an external source. Although the receptors stimulated by sound are located in the cochlea of the inner ears, the source of the sound is perceived to be the phonograph speaker across the room. Comparison of the times of onset and intensities of sounds at each of the two ears are the two most important clues in finding a sound source. In fact, the head is turned when localizing sounds to emphasize the difference in this information. The difference in time of onset helps particularly to localize low-frequency sounds (low pitch). A sound originating on the left stimu-
~tes the left ear slightly before it stimulates the right (Fig.
~7), periods of condensation and rarefaction of each
~wave occurring slightly earlier on the left. The

sound is also of slightly greater intensity on the left, and the difference in sound intensity becomes an increasingly important clue as the sound frequency increases and pitch rises. The sound is localized to the side where it is louder.

A simple experiment can demonstrate the role of inequality of sound intensity upon sound localization. A quietly hummed tone is localized at the middle of the head. If, while the humming continues, one ear is lightly plugged, the sound intensity in that ear increases because of the greater sound reverberations in the blocked ear canal. As the tone increases in loudness, it becomes localized in the closed ear. Even this slight imbalance in intensity or loudness is sufficient to cause large changes in localization of the stimulus. The first step in this comparison occurs in the brainstem at the level of entry of the auditory nerves before there has been much chance for synaptic alteration and delay to modify the information, but in man the cerebral cortex is necessary for actual sound localization.

Vestibular system

The vestibular system contains mechanoreceptors specialized to detect changes in both the motion and position of the head. The receptors are part of the *vestibular apparatus* which is housed in the bony channels of the inner ear, one on each side of the head. The vestibular apparatus is a membranous sac within a bony tunnel in the temporal bone of the skull. It forms three *semicircular canals* and a slight bulge for the *utricle* and *saccule* (Fig. 16-38A), all of which are filled with a specialized fluid, the *endolymph*.

The three semicircular canals on each side of the skull are arranged at right angles to each other (Fig. 16-38B). The actual receptors of the semicircular canals are hair cells which sit at the ends of the afferent neurons. The sensory hairs are closely ensheathed by a gelatinous mass which blocks the channel of the semicircular canal at that point.

The receptor system in the semicircular canals works in the following way. Whenever the head is moved, the bony-tunnel wall, its enclosed membranous semicircular canal, and the attached bodies of the hair cells, of course, turn with it. The endolymph fluid filling the membranous semicircular canal, however, is neither attached to the skull nor automatically pulled with it; instead, because of inertia, the fluid tends to retain its original position, i.e., to be "left behind." As the bodies of the hair cells move with the skull, the hairs are pulled against the relatively stationary column of endolymph and

A

B

FIGURE 16-36

Injury to the inner ear by intense noise. A Normal organ of Corti (guinea pig) showing the three rows of outer hair cells and single row of inner hair cells. B Injured organ of Corti after 24-hr exposure to noise levels typical of very loud rock music (2,000-Hz-octave band at 120 dB). Several outer hair cells are missing, and the cilia of others no longer form the orderly W pattern of the normal ear. Note also the increased number and size of small villi on the cell surfaces. (*Scanning electron micrograph by Robert E. Preston. Courtesy Joseph E. Hawkins, Kresge Hearing Research Institute.*)

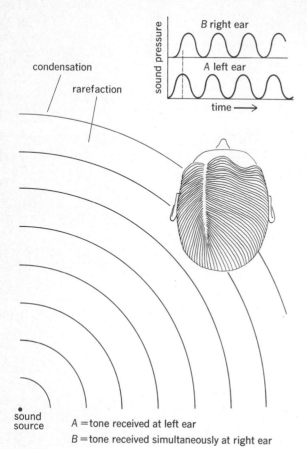

A = tone received at left ear

B = tone received simultaneously at right ear

FIGURE 16-37

Difference in input between the two ears helps to localize sound sources.

are bent (Fig. 16-39). The speed and magnitude of the movement of the head determine the degree to which the hairs are bent and the hair cells stimulated. As the inertia is overcome, the hairs slowly return to their resting position; for this reason, the hair cells are stimulated only during *changes* in rate of motion, i.e., during acceleration of the head. During motion at a constant speed, stimulation of the hair cells ceases.

The transducing mechanisms of these receptor cells, by which bending of the hairs gives rise to action potentials in the afferent nerve, are not known. Although the junction between the hair cell and afferent-nerve fiber has the anatomic features of a chemically mediated synapse, the mechanism of synaptic transmission and the nature of the transmitter substance are unknown.

Even when the head is motionless, the afferent nerve fibers are activated at a relatively low resting frequency (Fig. 16-40A). The explanation for this resting activity is unknown—perhaps it is due to the spontaneous leak of chemical transmitter from the receptor cell—but it allows the receptor cells to signal information by either increasing or decreasing the frequency of action potentials in the afferent-nerve fiber.

Thus, in some way, the shearing force bending the hairs on the receptor cells is related to the frequency of action potentials in the afferent nerve; when the hairs are bent one way, the rate of firing speeds up; when the hairs are bent in the opposite direction, the firing frequency slows down (Fig. 16-40B and C). Consider what is taking place in the corresponding semicircular canal on the opposite side of the head (Fig. 16-41). If the head is turned to the left, the hairs of the left semicircular canal move in that direction causing a higher rate of firing action potentials in the left vestibular nerve. But in the right semicircular canal, the movement of the hairs is in the opposite direction, and the rate of firing in the right vestibular nerve drops below its resting level. There is an imbalance of input to the central nervous system, and it is the imbalance which is significant.

Whereas the semicircular canals signal the rate of change of motion of the head, the utricle and saccule contain the receptors of the vestibular system which provide information about the position of the head relative to the direction of the forces of gravity. The receptor cells here, too, are mechanoreceptors sensitive to the movement of projecting cilia, or hairs. The hair cells of the utricle and saccule are collected into groups from which the hairs protrude into a gelatinous substance. In the utricle and saccule tiny calcium carbonate stones, or *otoliths,* are embedded in the gelatinous covering of the hair cells, making the gelatinous substance heavier than the surrounding endolymph. When the head is tipped, the gelatinous-otolith material changes its position, pulled by gravitational forces to the lowest point in the utricle or saccule. The shearing forces of the gelatinous-otolith substance against the hair cells bend the hairs and stimulate the receptor cells.

The information from the vestibular apparatus is used for two purposes. The first is to control the muscles which move the eyes so that, in spite of changes in the position of the head, the eyes remain fixed on the same point. As the head is turned to the left, the balance of afferent input from the vestibular apparatus on each side is altered. Impulses from the vestibular nuclei activate the ocular muscles, which turn the eyes to the right, and

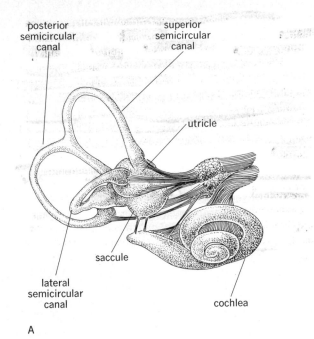

posterior
semicircular
canal

superior
semicircular
canal

utricle

saccule

lateral
semicircular
canal

cochlea

A

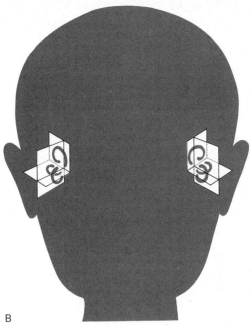

B

FIGURE 16-38

A **Vestibular system.** B **Relationship of the two sets of semicircular canals.**

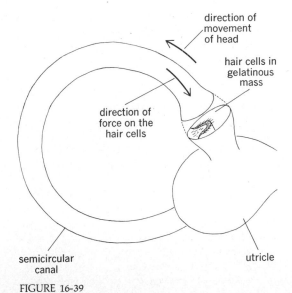

direction of
movement
of head

hair cells in
gelatinous
mass

direction of
force on the
hair cells

semicircular
canal

utricle

FIGURE 16-39

Diagram of a semicircular canal.

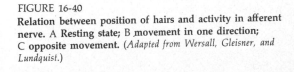

resting activity   stimulation
(depolarization)

inhibition
(hyperpolarization)

discharge rate of vestibular nerve

A   B   C

FIGURE 16-40

Relation between position of hairs and activity in afferent nerve. A **Resting state;** B **movement in one direction;** C **opposite movement.** (*Adapted from Wersall, Gleisner, and Lundquist.*)

direction of movement of head

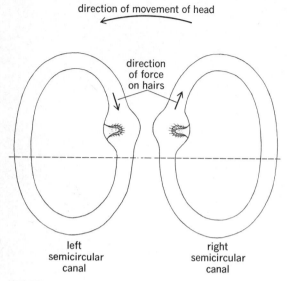

direction
of force
on hairs

left
semicircular
canal

right
semicircular
canal

FIGURE 16-41
Comparison of semicircular canals on opposite sides of the head.

inhibit their antagonists. The eyes turn toward the right as the head turns toward the left, and the net result is that the eyes remain fixed on the point of interest.

Vestibular information is also utilized in reflex mechanisms for maintaining upright posture. In monkeys, cats, and dogs the vestibular apparatus plays a definite role in the postural fixation of the head, orientation of the animal in space, and reflexes accompanying locomotion. However, in man very few postural reflexes are known to depend primarily on vestibular input, despite the fact that the vestibular organs are called the sense organs of balance.

Taste and smell

Receptors sensitive to changes in some chemicals in blood plasma were discussed in Chap. 10. There are also chemoreceptors which respond to certain chemicals in the external environment, i.e., receptors for the sense of taste and the sense of smell. In man, these are much less important than the receptors for vision or hearing although that is certainly not true for most animals. Although taste and smell affect a person's appetite, the initiation of digestion, and the avoidance of harmful substances, they do not exert strong or essential influence.

Taste The specialized receptor organs for the sense of taste are the 10,000 or so *taste buds* which are located on the tongue, roof of the mouth, pharynx, and larynx. The latter receptors are not activated by substances in the mouth but do fire during swallows. Inside the taste buds the receptor cells are arranged like segments of an orange with the multifolded upper surfaces of the receptor cells extending into a small pore at the surface of the taste bud, where they are bathed by the fluids of the mouth (Fig. 16-42).

Taste sensations are traditionally divided into four basic groups: sweet, sour, salt, and bitter, but different types of taste buds or receptor cells which would support this specificity have not been identified. In fact, a single receptor cell can respond in varying degrees to many different chemical substances falling into more than one of the basic categories.

What makes a receptor cell respond? The mechanisms by which the taste receptors are stimulated and action potentials generated are not known. It has been suggested that the first step is a loose binding of the individual molecules of the chemical substance with specific sites on the receptor cell membrane. The fact that a single receptor can be responsive to more than one basic taste quality could be explained if a single receptor

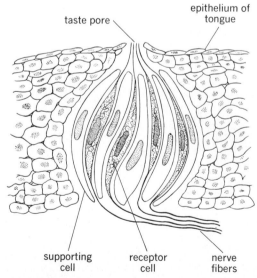

epithelium of
tongue

taste pore

supporting
cell

receptor
cell

nerve
fibers

FIGURE 16-42
Structure and innervation of a taste bud.

cell had several different sites, each capable of binding with a different type of molecule. The theory goes on to suggest that the chemical-substance–receptor-site combination alters the cell membrane, forming pores through which ions move to change the membrane potential of the cell. It is known that the membrane of the receptor cells depolarizes when the cell is stimulated chemically.

Beneath the taste buds lie the nerve fibers which enter the buds to end the receptor cells. One nerve fiber may innervate several receptor cells, and one receptor cell may be innervated by several different neurons. There is clearly no one-to-one relationship by which each receptor cell has a single line into the central nervous system. How can we distinguish so many different taste sensations when the receptor cells lack specificity both in terms of the kind of chemical to which they will respond and the way in which they are connected to the brain? The frequency of action potentials in single nerve fibers increases in response to increasing concentrations of the chemical stimulant; therefore, frequency signals quantity, but what signals the quality? The afferent fibers involved in taste show different firing patterns in response to different substances; e.g., one fiber may fire very rapidly when the stimulatory substance is salt but only sporadically when it is sugar, and another fiber may have just the opposite reaction. This variation in relative sensitivity makes the pattern of firing within a group of neurons meaningful, and awareness of the specific taste of a substance probably depends upon the relative activity in a number of different neurons rather than that in a specific neuron. Identification of the substance is aided by information about its temperature and texture which is transmitted to the central nervous system from receptors on the tongue and surface of the oral cavity. The odor of the substance clearly helps, too, as is attested by the common experience that food lacks taste when one has a stuffy head cold.

Smell The olfactory receptors which give rise to the sense of smell lie in a small patch of mucosa, i.e., membrane which secretes mucus, in the upper part of the nasal cavity (Fig. 16-43A). Because the *olfactory mucosa* is above the path of the main air currents that enter the nose with inspiration, the odorous molecules must either diffuse up to the receptor cells or be drawn up by changes in respiration such as sniffing. The receptor cells (Fig. 16-43B) are really modified neurons having two processes; one passes toward the brain, forming the olfactory nerve, and the other bears many fine cilia and

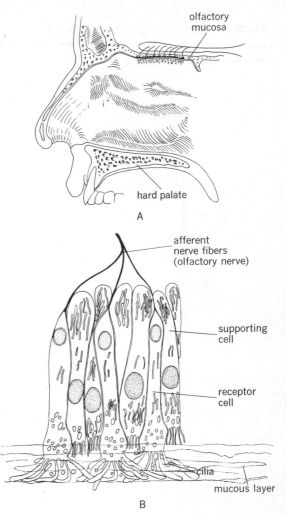

FIGURE 16-43
Location and structure of the olfactory receptors.

extends out to the surface of the olfactory mucosa.

Before an odorous substance can be detected, it must first release molecules which diffuse into the air and pass into the nose to the region of the olfactory mucosa, dissolve in the layer of mucus covering the receptors, establish some sort of relation with the receptor, and depolarize the membrane enough to initiate an action potential in the afferent-nerve fiber. Perhaps the molecule combines with a receptor site on the membrane in such a way that the membrane permeability changes and ions move across the membrane to depolarize the cell, for upon stimulation of the olfactory mucosa with odorous substances, a generator potential can be recorded which changes with stimulus quality and intensity.

The physiological basis for discrimination between the tens of thousands of different odor qualities is speculative. There are no apparent differences in receptor cells, at least on a microscopic level, to account for it, but meaningful differences between receptors are thought to exist at molecular levels. One theory is that odor discrimination depends upon a large number of different types of interactions between the odor-substance molecules and receptor sites. This theory is based on the supposition that the receptor cells probably have 20 or 30 different types of receptor sites, each type capable of interacting with many different odor molecules but responding best to a molecule with a particular size, polarity, and shape. The receptor-site populations vary from one receptor cell to another. For example, one receptor cell may have mainly receptor sites of types A, B, and C, whereas another cell has types B, D, and E. The molecules of the odor substance combine with most types of receptor sites, but the "fit" varies. A good contact occurs when the odor molecule and receptor sites' size, shape, and polarity match; in such a case the resulting depolarization of the receptor cell is large. In cases of poorer fit, the receptor cell still depolarizes but to a smaller degree. All depolarizations occurring together within one cell are summed, forming a generator potential, which determines the firing rate in the afferent-nerve fiber. Receptor cells with different receptor-site populations respond to the same odor substance with different firing rates. Thus, it is the simultaneous yet differential stimulation of many receptor cells that provides the basis for the discrimination. Moreover, olfactory discrimination depends only partially upon the action-potential pattern generated in the different afferent neurons; it also varies with attentiveness, state of the olfactory mucosa (acuity decreases when the mucosa is congested, as in a head cold), hunger (sensitivity is greater in hungry subjects), sex (women in general have keener olfactory sensitivities than men), and smoking (decreased sensitivity has been repeatedly associated with smoking). And just as the awareness of the taste of an object is aided by other senses, so the knowledge of the odor of a substance is aided by the stimulation of other receptors. This is responsible for the description of odors as pungent, acrid, cool, or irritating.

Further perceptual processing

Our actual perception of the events around us often involves areas of brain other than primary sensory cortex. It is apparent that the sensory information of the primary sensory areas achieves further elaboration through the neural activity in *cortical association areas*. These brain areas lie outside the classic primary cortical sensory or motor areas but are connected to them by association fibers (Fig. 16-44). Although it often has not been possible to elucidate the specific roles performed by the association areas, they are acknowledged to be of the greatest importance in the maintenance of higher mental activities in man. This elaboration of sensory information is demonstrated in the following experiments. For example, if areas of primary visual cortex are stimulated (when the brain surface is exposed under local anesthesia during neurosurgical procedures and the patient is awake), the patient "sees" a flash of light. Upon stimulation of the association areas surrounding visual cortex, the patient reports seeing more elaborate visual sensations such as brilliant

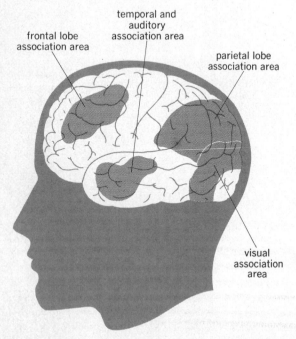

frontal lobe
association area

temporal and
auditory
association area

parietal lobe
association area

visual
association
area

FIGURE 16-44
Areas of association cortex.

colored balloons floating around in an infinite sky. Upon stimulation of association areas still farther from primary visual cortex, the patient might report visual memories which seem to be reenacted before his eyes. Another example of the embellishment of the sensory experience provided by association areas can be found in persons who have undergone removal of parts of cortex because of tumors or accidents. A person who has no primary visual cortex is blind; if a chair is placed in his path, he walks into it because he does not see it. In contrast, a person who has functional primary visual cortex but has no visual association areas sees the chair and therefore does not walk into it, but he is not able to say that the object in his path is a chair or to explain its function. Similarly, a patient with damaged auditory association areas may hear a spoken word but not comprehend its meaning. A patient with damaged parietal association areas can feel a small, cool object in his hand but know neither that it is called a key nor that it is used to open doors. As mentioned earlier, the mechanisms responsible are unknown.

Such further perceptual processing involves arousal, attention, learning, memory, language, and emotions, and it involves comparing the information presented via one sensory modality with that of another. For example, we may hear a growling dog, but our perception of the actual event taking place varies markedly, depending upon whether our visual system detects that the sound source is an angry animal or a loudspeaker.

Another step in the processing of any sensory information is testing the appropriateness of our interpretation. For example, when we see a can on the supermarket shelf whose label contains a picture of pears or tomatoes, we understand that we are looking at cans containing these items. However, if the label pictures a green giant, we do not think that the can contains the flesh of this bizarre creature since we know that that is impossible. We reject that notion and look for further clues to tell us of the can's contents.

Sometimes the objects we view are ambiguous and have more than one logical interpretation. Such a figure is the drawing entitled "My wife and my mother-in-law," illustrated in Fig. 16-45. When first trying to identify such a figure, we pick out one detail, say the curved line at the midportion of the left side of the drawing. If it reminds us of a bent nose, we immediately seek to verify our impression by looking for the expected eyes, mouth, hair, etc. An old woman's face is perceived, but as we look at it, the image suddenly shifts. The line that was a nose

FIGURE 16-45
Ambiguous figure, "My wife and my mother-in-law," in which the young girl's chin is the woman's nose, created by cartoonist W. E. Hill in 1915.

is now a chin; the image of a stylish young woman appears. Since both images are equally plausible, our interpretation of the image shifts back and forth between them. It is an interesting property of our perceptual mechanisms that we see either one or the other of the images; it is impossible to see them as both plausible at the same time.

We put great trust in our sensory-perceptual processes despite the inevitable modifications we know to exist. Some factors known to distort our perceptions of the real world are as follows:

1 Afferent information is distorted by receptor mechanisms and by its processing along afferent pathways, e.g., by accommodation.

2 Such factors as emotions, personality, and social background can influence perceptions so that two people can witness the same events and yet perceive them differently.

3 Not all information entering the central nervous system gives rise to conscious sensations. Actually, this is a very good thing because many unwanted signals, generated

by the extreme sensitivity of our receptors, are canceled out. The afferent systems are very sensitive. Under ideal conditions the rods of the eye can detect the flame of a candle 17 mi away. The hair cells of the ear can detect vibrations of an amplitude much lower than that caused by the flow of blood through the vascular system and can even detect molecules in random motion bumping against the tympanic membrane. Olfactory receptors respond to the presence of only four to eight odorous molecules. It is possible to detect one action potential generated by a pacinian corpuscle. If no mechanisms existed to select, restrain, and organize the barrage of impulses from the periphery, life would be unbearable. Information in some receptors' afferent pathways is not canceled out; it simply lacks the capacity for expression of a conscious correlate. For example, stretch receptors in the muscles detect changes in the length of the muscles, but activation of these receptors does not lead to a conscious sense of anything. Similarly, stretch receptors in the walls of the carotid sinus effectively monitor both absolute blood pressure and its rate of change, but man has no conscious awareness of his blood pressure.

4 We lack suitable receptors for many energy forms. For example, we can have no direct information about radiation and radio or television waves until they are converted to an energy form to which we are sensitive. Many regions of the body are insensitive to touch, pressure, and pain because they lack the appropriate receptors. The brain itself has no pain or pressure receptors, and brain operations can be performed painlessly on patients who are still awake, provided that the cut edges of the sensitive brain coverings are infused with local anesthetic.

However, the most dramatic examples of a clear difference between the real world and our perceptual world can be found in illusions and drug- and disease-induced hallucinations when whole worlds can be created and mistaken for reality but can be proved false by physical measurements.

Any sense organ can give false information; e.g., pressure on the closed eye is perceived as light in darkness, and electric stimulation of any sense organ produces the sensory experience normally arising from the activation of that receptor. Why do such illusions appear? The two bits of false information just mentioned arise because most afferent fibers transmit information about one modality, and action potentials in a given afferent pathway going to a certain area of the brain signal the kind of information normally carried in that pathway. The association of the signal with the particular sensory experience has been in part built into the system during devel-

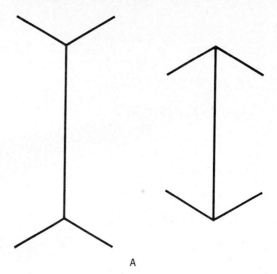

A

FIGURE 16-46
A **Arrow illusion.** B **Basis for the illusion.** (*Photograph of Lincoln Center by Ingrid Froehlich.*)

opment and in part acquired through learning. The explanations in other illusions vary. In the arrow illusion (Fig. 16-46A) the shaft of the arrow with the outgoing fins looks longer than that of the other arrow even though the two shafts are exactly the same length. Like many other illusions this one occurs because our perceptual systems try to group information into meaningful patterns. This particular illusion appears to those of us brought up in a rectangular world of lines and corners, the two-dimensional arrows immediately suggesting the three-dimensional arrangement of a cube in two very familiar situations (Fig. 16-46B). The arrow with outward-going fins suggests the distant corner of a room whereas the arrow with the inward-going fins suggests the nearby corner of a building. We realize that distant objects appear smaller than they really are and mentally adjust for the apparent difference in distance between the two corners. We therefore perceive the arrow with the outward-going fins to be longer than it really is. People brought up in curvilinear societies (round buildings, etc.) do not have this illusion and report that the shafts of both arrows appear to be the same length.

Auditory illusions exist too; for example, it is possible to ''hear'' sounds that do not exist in a sound stimulus of the real world if the missing stimulus could logically be expected from the context. In other words, the probability of a sound stimulus affects its perception. This

B

"filling in" with a bit of illusion is called *auditory induction*. Such sound illusions are generally quite normal and are useful because they compensate for information that would otherwise be lost among the extraneous sounds of our noisy world.

Levels of perception

The two processes of transmitting data through the nervous system and interpreting it cannot be separated. Information is processed at each synaptic level of the afferent pathways. There is no one point along the afferent pathways or one particular level of the central nervous system below which activity cannot be a conscious sensation and above which it is a recognizable, definable sensory experience. Perception has many levels, and it seems that the many separate stages are arranged in a hierarchy, with the more complex stages receiving input only after they are processed by the more elementary systems. Every synapse along the afferent pathways adds an element of organization and contributes to the sensory experience.

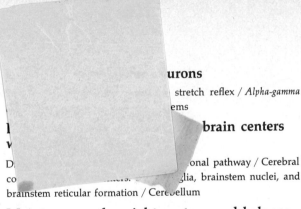
Control of body movement

The execution of a coordinated movement is a complicated process. Consider reaching to pick up an object. The fingers are extended (straightened) and then flexed (bent), the degree of extension depending upon the size of the object to be grasped, and the force of flexion depending upon the weight and consistency of the object. Simultaneously, the wrist, elbow, and shoulder are extended, and the trunk is inclined forward, the exact movements depending upon the distance of the object and the direction in which it lies. The shoulder must be stabilized to support the weight first of the arm and then of the object. Upright posture must be maintained in spite of the body's continually shifting center of gravity.

The building blocks for this seemingly simple action—as for all movements—are active motor units, each comprising one motor neuron together with all the skeletal muscle cells it innervates (Chap. 8). Thus, anything that affects the movement of skeletal muscle does so by means of synaptic input to the motor neurons. Inputs from many sources exert influence upon the activity of the motor neurons, and the precision of coordinated muscle movement in man depends upon the *balance* of their influence. If one system is damaged and its influence is lessened, or if through certain diseases it becomes hyperactive and its influence becomes greater, the *balance* of input to the motor neuron necessary for the best coordination of movement will be destroyed. The clinical terms *spasticity, rigidity,* and *flaccidity* represent abnormal muscle behavior caused by such imbalances. Spastic and rigid muscle movements are stiff and awkward. Flaccid mus-

17

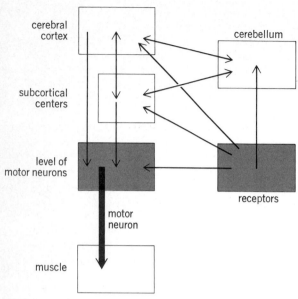

FIGURE 17-1
Diagram of motor-system function.

cles are weak and soft. In both cases the motor units usually continue to function, but they do not operate with the same degree of coordination or under the wide range of conditions typical of a normal motor system. No one source of input to the motor neuron is essential for movement, but to provide the precision and speed of normally coordinated movements the balanced input from all sources is necessary.

The interrelating systems which converge upon the motor neurons to control their activity are the subject of this chapter, and a simplified model of motor-system function follows (Fig. 17-1). Each of the myriad coordinated body movements is characterized by a set of motor-unit activities occurring over space and time. The selection of the specific pattern of neural activity, i.e., the "program" required to achieve a given movement, is the function of higher brain centers. How they do this is unknown. The program in the form of action potentials is transmitted to the motor neurons and on to the muscles. During the movement, various receptors relay information back to the central nervous system, thereby providing moment-to-moment information about the progress of the movement. Any discrepancy between the original program and resulting movements is detected and the program is revised and corrected.

This system requires a "programming device," receptors to measure the performance of the system, and "compensating units" to detect errors between program and performance. The programming is done either by the cerebral cortex together with several subcortical centers or, in many cases, by the subcortical centers themselves. Transmission occurs via pathways which descend in the central nervous system from these centers to the level of the various motor neurons where they play upon locally occurring control mechanisms. The receptors are muscle and joint receptors, skin receptors, vestibular receptors (receptors which detect the position of the head in space or changes in that position), and receptors in the eyes. Compensation is done at many levels of the central nervous system, including the cortex, subcortical centers, cerebellum, and level of the motor neurons.

However, it should be recognized that the system is incomplete as described, for it does not take into account the mechanisms by which the "decision" to make a particular movement is reached. What neural events actually occur in the brain to cause one to "decide" to pick up an object in the first place? Presently we have no insight on this question.

Given such a model, it is difficult to use the word "voluntary" with any real precision. We shall use *voluntary* to refer to those actions which are characterized as follows:

1 These are actions we think about. The movement is accompanied by a conscious awareness of what we are doing and why we are doing it rather than the feeling that it "just happened," a feeling that often accompanies reflex responses.

2 Our attention is directed toward the action or its purpose.

3 The actions are the result of learning. Actions known to have disagreeable consequences are less likely to be performed voluntarily.

In the previous example of reaching to pick up an object, the activation of some of the motor units, such as those actually involved in grasping the object, can be classified clearly as voluntary; but most of the muscle activity associated with the act is initiated without any conscious, deliberate effort. In fact, almost all motor behavior involves both conscious and unconscious components, and the distinction between the two cannot be made easily.

Even a highly conscious act such as threading a needle involves the unconscious postural support of the hand and arm and inhibition of antagonistic muscles (those muscles whose activity would oppose the intended action, in this case the finger extensor muscles which straighten the fingers), and unconscious basic reflexes such as dropping a hot object can be influenced by conscious effort. If the hot object is something that took a great deal of time and effort to prepare, one probably would not drop it but would try to inhibit the reflex, holding on to the object until it could be put down safely. Most motor behavior is neither purely voluntary nor purely involuntary but falls at some point on a spectrum between these two extremes. But even this statement is of little help because patterned muscle movements shift along the spectrum according to the frequency with which they are performed.

For example, when a person first learns to drive a car with standard transmission, stopping is a fairly complicated process involving the accelerator, clutch, and brake. The sequence and force of the various operations depend upon the speed of the car, and their correct implementation requires a great deal of conscious attention. With practice the same actions become automatic. If a child darts in front of the car of an experienced driver, he does not have to think about the situation and decide to remove his foot from the accelerator and depress the brake and clutch. Upon seeing the child, he immediately and automatically stops the car. A complicated pattern of muscle movements is shifted from the highly conscious end of the spectrum over toward the involuntary end by the process of learning.

Whether activated voluntarily or involuntarily, given motor units are frequently called upon to serve many different functions. For example, one demand upon the muscles of the limbs, trunk, and neck is made by postural mechanisms; these muscles must support the weight of the body against gravity, control the position of the head and different parts of the body relative to each other to maintain equilibrium, and regain stable, upright posture after accidental or intentional shifts in position. Superimposed upon these basic postural requirements are the muscle movements associated with locomotion. For these purposes, the muscles must be capable of transporting the body from one place to another under the coordinated commands of neural mechanisms for alternate stepping movements and shifting the center of gravity. And added to the requirements of posture and locomotion can be the highly skilled movements of a ballerina or hockey player. The motor units are activated and the sometimes conflicting demands are settled, usually without any conscious, deliberate effort.

We now turn to an analysis of the individual components of the model for control of the motor system. We begin with local control mechanisms because their activity serves as a base upon which the descending pathways frequently exert their influence.

It should be realized that most information on this model has been obtained by planned experiments on frogs, cats, and monkeys. Similar information about man has been provided chiefly by accidental damage or disease of various parts of the nervous system; only rarely can experiments be performed during neurosurgical procedures. The information on motor control in man is not as explicit as that obtained from animal experiments, but it shows that data from animals can be applied to man only with some reservations. The same neuroanatomical apparatus is present in man, cat, and monkey, but the emphasis sometimes varies. Whenever possible, the physiology of neuromuscular control in man will be discussed.

Local control of motor neurons

Much of the synaptic input to the motor neurons arises from neuronal pathways whose afferents occur at the same level of the central nervous system as the motor neurons. Indeed, some of the neural pathways are activated by receptors in the very muscles controlled by the motor neurons and in other nearby muscles as well as the tendons associated with the muscles. These receptors monitor muscle length and tension, and information concerning these parameters passes via afferent neurons into the central nervous system. This input forms the afferent component of purely local reflexes which provide negative-feedback control over muscle length and tension. In addition, it is transmitted to higher brain centers where it can be integrated with input from other types of receptors.

Length monitoring systems and the stretch reflex

Embedded within skeletal muscle are stretch receptors which are made up of afferent-nerve endings wrapped around modified muscle cells, both partially enclosed in a fibrous capsule. The entire structure is called a *muscle spindle*. The modified muscle fibers within the spindle are known as *spindle fibers;* the typical skeletal muscle cells outside the spindle are the *skeletomotor muscle fibers* (Fig. 17-2).

The muscle spindles are located within the muscle in such a way that passive stretch of the entire muscle pulls on the spindle fibers, stretching them and activating their receptors. Conversely, contraction of the skeletomotor fibers and the resultant shortening of the muscle

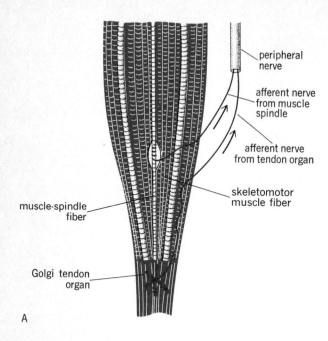

A

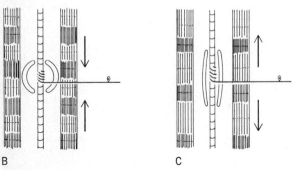

B   C

FIGURE 17-2

A **Diagram of a muscle spindle. B Contraction of the skeletomotor fibers removes tension on the spindle stretch receptors and lowers the rate of firing in the afferent nerve. C Passive stretch of the skeletomotor fibers activates the spindle stretch receptors and causes a higher rate of firing in the afferent nerve.** [*Part A from P. A. Merton, How We Control the Contractions of Our Muscles, Sci. Am.,* **226:** 30, *May* (1972).]

release tension on the muscle spindle and slow down the rate of firing of the stretch receptor. But the muscle spindle is more complicated in that there are different kinds of spindle receptors, one responding to the magnitude of the stretch, another probably to both the absolute magnitude of the stretch and the speed with which it occurs. The importance of the kind of information relayed

by the first receptor is apparent: It tells the central nervous system about the length of the muscle. However, by indicating the rate of change of the muscle length, the second type of receptor allows the central nervous system to anticipate the magnitude of the stretch. If the rate of stretch is increasing very rapidly, the stretch itself cannot stop immediately, and an additional change in length can be predicted. Although these receptors are separate anatomic and physiologic entities, they will be referred to collectively as the *muscle-spindle stretch receptors*.

When the afferent nerves from the muscle spindle enter the central nervous system, they divide into branches which can take several different paths. One group of terminals directly forms (*A* in Fig. 17-3) excitatory synapses upon the motor neurons going back to the muscle that was stretched, thereby completing a reflex arc known as the *stretch reflex*. This reflex is probably most familiar in the form of the knee jerk, which is tested as part of routine medical examinations. The physician taps on the patellar tendon, which stretches over the knee and connects a muscle in the thigh to a bone in the foreleg. As the tendon is depressed, the muscle to which it is attached is stretched, and the receptors within the muscle spindles are activated. Coded information about the change in length of the muscle is fed back to the motor neurons controlling the same muscle. The motor units are excited, and the patient's foreleg is raised to give the familiar knee jerk. The proper performance of the knee jerk tells the physician that the afferent limb of the reflex, the balance of synaptic input to the motor neuron, the motor neuron itself, the neuromuscular junction, and the muscle are functioning normally. The knee jerk nicely illustrates a major physiologic function of the stretch reflex; it permits the muscle to resist any passively induced change in its length and is a good example of a negative-feedback control system.

Because the group of afferent terminals mediating the stretch reflex synapses directly with the motor neurons without the interposition of any interneurons, the stretch reflex is called *monosynaptic*. Stretch reflexes are the only known monosynaptic reflex arcs in man; all other reflex arcs are polysynaptic, having at least one interneuron (and usually many) between the afferent and efferent pathways.

A second group of afferent terminals (*B* in Fig. 17-3) ends on interneurons which, when excited, inhibit the motor neurons controlling antagonistic muscles whose contraction would interfere with the reflex response. For example, the normal response to the knee-jerk reflex is straightening of the knee to extend the foreleg. The antagonists to these extensor muscles are a group of flexor muscles which, when activated, draw the foreleg back and up against the thigh. If both opposing groups of muscles are activated simultaneously, the knee joint is immobilized and the leg becomes a stiff pillar. This is certainly what is required in some situations, but if the foreleg is to be extended from a flexed position, the motor neurons which activate the flexor muscles must be inhibited as the motor neurons controlling the extensor muscles are activated. The excitation of one muscle and the simultaneous inhibition of its antagonistic muscle is called *reciprocal innervation.*

A third group of terminals (*C* in Fig. 17-3) ends on interneurons which, when excited, activate *synergistic muscles,* i.e., muscles whose contraction assists the reflex motion. For example, in the knee jerk, interneurons facilitate motor neurons which control other leg extensor muscles.

A fourth group of afferent terminals (*D* in Fig. 17-3) synapses with interneurons which convey information about the muscle length to areas of the brain dealing with coordination of muscle movement. Although the muscle stretch receptors initiate activity in pathways eventually reaching cerebral cortex, the information relayed by these action potentials does not have a conscious correlate; rather the conscious awareness of the position of a limb or joint comes from the joint, ligament, and skin receptors (Chap. 16).

Alpha-gamma coactivation   The muscle spindles are structurally parallel to the large skeletomotor muscle fibers so that stretch on them is removed when the skeletomotor fibers contract (Figs. 17-2 and 17-4). If the spindle stretch receptors were permitted to shorten at this time they would stop firing action potentials and this important afferent information would be lost. To prevent this, the spindle muscle fibers themselves are frequently made to contract during the shortening of the skeletomotor fibers, thus maintaining tension in the spindle and firing in the receptors. The spindle fibers are not large and strong enough to shorten whole muscle and move joints; their sole job is to produce tension on the spindle stretch receptors. The muscle fibers in the spindles shorten in response to motor neuron activity (Fig. 17-4). The motor neurons which activate the spindle fibers are not, however, the same motor neurons which activate the skeletomotor muscle fibers. The motor neurons controlling the skeletomotor muscle fibers are larger and are classified as *alpha motor neurons;* the smaller neurons whose axons innervate the spindle fibers are known as the *gamma motor*

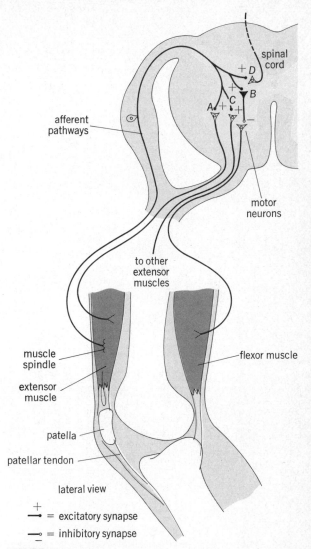

FIGURE 17-3
Terminals of the afferent fiber from the muscle spindle involved in the knee jerk.

neurons. The latter neurons are activated primarily by synaptic input from descending pathways. The overall route—descending pathway, gamma motor neuron, spindle muscle fiber, stretch receptor and afferent neuron, alpha motor neuron—is known as the *gamma loop* (Fig. 17-5).

In many voluntary and involuntary movements alpha and gamma motor neurons are *coactivated,* i.e., fired

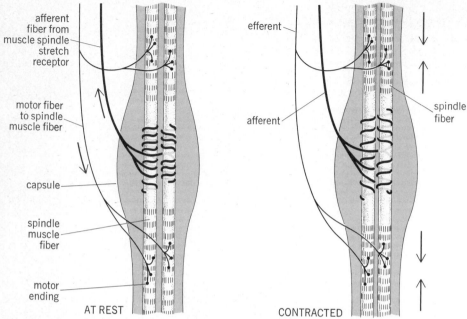

afferent
fiber from
muscle spindle
stretch
receptor

motor fiber
to spindle
muscle fiber

capsule

spindle
muscle
fiber

motor
ending

AT REST

efferent

afferent

spindle
fiber

CONTRACTED

FIGURE 17-4
A **Diagram of muscle-spindle innervation.** B **As the two
striated ends of a spindle fiber contract, they pull on the center
of the fiber and stretch the receptor, which is located there.**
[*From P. A. Merton, How We Control the Contractions of Our
Muscles, Sci. Am.,* **226:** 30, May (1972).]

at almost the same time. To understand the usefulness
of alpha-gamma coactivation, consider picking up a book
whose weight is unknown. Suppose that the initially programmed strength of alpha motor-unit firing is not sufficient to lift the book. The skeletomotor muscle fibers will
be unable to shorten but the spindle fibers, activated
simultaneously by the gamma motor neurons, will shorten
and the spindle receptors will be stretched. By way of
the stretch reflex, the balance of excitatory synaptic input
to the alpha motor units will increase, causing summation
of contraction, recruitment of additional motor units, and
greater muscle tension. Thus the gamma loop provides
a mechanism by which motor commands and muscle
performance can be compared at the local level and
compensation brought about on the spot. Moreover, information that the spindles are longer than expected (i.e.,
that the skeletomotor fibers have not shortened enough)
will be transmitted to those higher brain centers involved
in programming and controlling motor behavior so that
they can alter their output as well.

Alpha-gamma coactivation works the other way
too. If the initial program caused too intense alpha
motor-unit activity, the book would be lifted too rapidly.
The faster-than-expected shortening of the spindle fibers
would remove tension from the spindles, stopping the
receptors' firing. This would reflexly remove a component
of excitatory input from the alpha motor neurons, automatically slowing the muscle movement to a more desirable rate. Thus, coactivation of alpha and gamma
motor neurons can lead to fine degrees of regulation of
muscle activity.

Tension monitoring systems

A second component of the local motor-control apparatus
monitors tension rather than length. The receptors employed in this system are the *Golgi tendon organs,* which
are encapsulated structures located in the tendon near
its junction with the muscle. Endings of afferent-nerve
fibers are wrapped around collagen bundles of the tendon, which are slightly bowed in the resting state. When

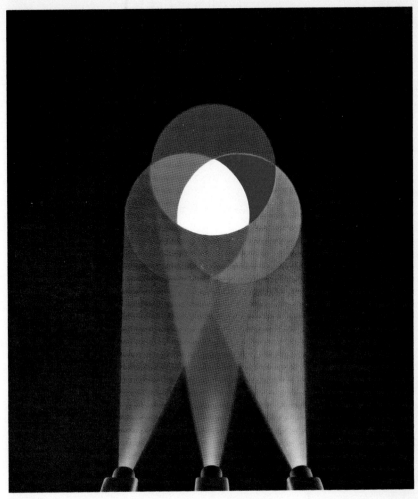

PLATE I

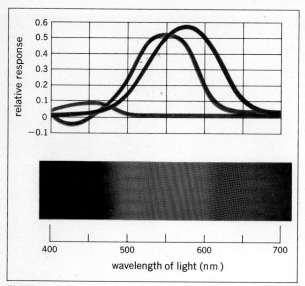

PLATE II

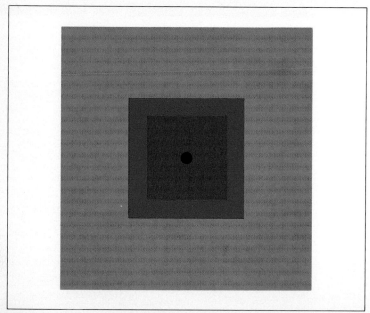

PLATE III

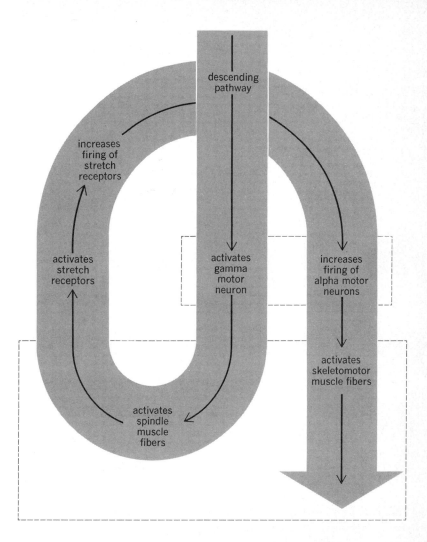

FIGURE 17-5
Gamma loop. The small rectangle at the center indicates those effects occurring in the brainstem or spinal cord; the large rectangle, the effects in the muscle.

the skeletomotor fibers of the attached muscle contract, they pull on the tendon, straightening the collagen bundles, and distorting the receptor endings of the afferent nerves (Fig. 17-2). The receptors fire in relation to the increasing force or tension generated by a contracting muscle. Their activity results in the initiation of IPSPs (inhibitory postsynaptic potentials) in the motor neurons of the contracting muscle (Fig. 17-6). Some of the Golgi tendon organs have high thresholds and respond only when the tension is very high. These high-threshold receptors may function as safety valves, inhibiting the muscle when the force it generates is great enough to damage the limb. The remainder of the Golgi tendon organs

have lower thresholds, comparable to the receptors of the muscle spindle, and they supply the motor control systems with continuous information about the tension generated. This information is necessary for effective movement because a given input to a group of motor neurons does not always provide the same amount of tension. The tension developed by a contracting muscle depends on the velocity of muscle shortening, the muscle length, and the degree of muscle fatigue as well as the number of activated motor neurons and the rate at which they are firing (Chap. 8). Because one set of inputs to the motor neurons can lead to a large number of different tensions, a feedback of information is necessary to inform the motor control systems of the tension actually achieved.

As is true for the reflexes mediated by the spindle stretch receptors, those mediated by the Golgi tendon organs are generally characterized by reciprocal effects on antagonistic muscles. This is accomplished by means of interneurons juxtaposed between the afferent and efferent pathways just as described for the stretch reflex (Fig. 17-6).

Descending pathways and the brain centers which control them

The cerebral cortex, subcortical centers, and cerebellum influence the motor neurons by descending pathways. There are three mechanisms by which these pathways alter the balance of synaptic input converging upon the alpha motor neurons:

1 By synapsing directly upon the alpha motor neurons themselves. This has the advantage of speed and specificity.

2 By synapsing on the gamma motor neurons, which, via the gamma loop, influence the alpha motor neurons. This pathway and number 1 above usually operate together.

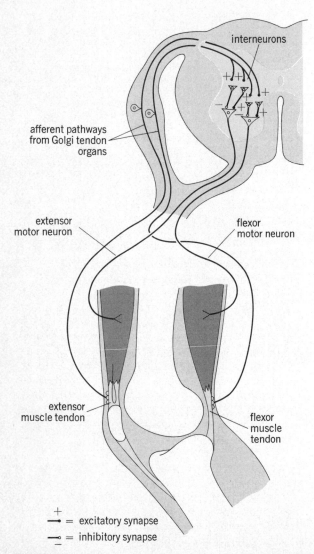

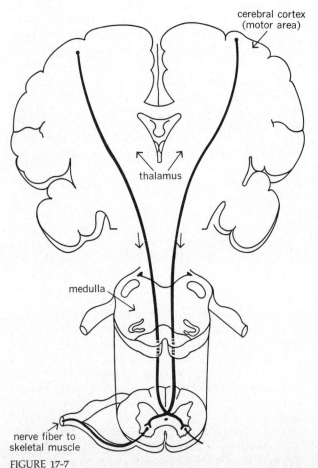

FIGURE 17-6
Golgi tendon organ component of the local control system.

FIGURE 17-7
Diagram of the direct corticospinal pathway.

As described earlier, this has the advantage of maintaining output from the stretch receptors and providing a means for local on-the-spot compensation.

3 By synapsing on interneurons, often the same ones subserving the local reflexes. Although this route is not as fast as directly influencing the motor neurons, it has the advantage of the coordination built into the interneuron network as described earlier (reciprocal innervation).

The degree to which each of these three mechanisms is employed varies, depending upon the nature of the descending pathway, of which there are two major categories: the *direct corticospinal pathway* and the *multineuronal pathway*.

Direct corticospinal pathway

The fibers of the direct corticospinal pathway, as the name implies, have their cell bodies in the cerebral cortex. The axons of these cortical neurons pass directly and without any additional synapsing to end in the immediate vicinity of the motor neurons (Fig. 17-7). The group of fibers innervating muscles of the eye, face, tongue, and throat branch away from these descending pathways in the brainstem to contact motor neurons whose axons travel out with the cranial nerves; the rest descend to their various terminations in the spinal cord, traveling down the spinal cord to innervate the motor neurons controlling the muscles of the extremities which are associated with fine-skilled movements. Near the junction of the spinal cord and brainstem, most of these fibers cross the spinal cord to descend on the opposite side. Thus, the skeletal muscles on the left side of the body are controlled largely by neurons in the right half of the brain. The direct corticospinal pathway is also called the *pyramidal tract* or *pyramidal system*, perhaps because of its shape in some parts of the brain or because it was formerly thought to arise solely from the giant pyramidal neurons of the cortex.

The direct corticospinal pathway is the major mediator of fine, intricate movements. However, it is not the sole mediator of such movements, since surgical section of this pathway in man does not completely eliminate them although it does make them weaker, slower, and less well coordinated. Clearly, the multineuronal pathway must also contribute to the performance of delicate movements.

The fibers of the direct corticospinal pathway end in all three of the ways described, i.e., on alpha motor neurons, gamma motor neurons, and interneurons. In addition they end presynaptically on the central terminals

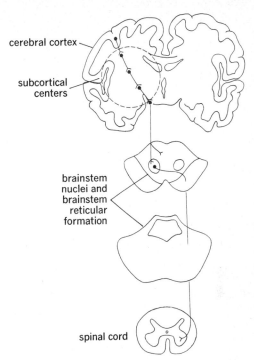

FIGURE 17-8
Diagram of the multineuronal pathway.

of afferent neurons and, through collateral branches, on neurons of the ascending afferent pathways. The overall effect of their input to afferent systems is to limit the area of skin, muscle, or joints allowed to influence the cortical neurons, thereby sharpening the focus of the afferent signal and improving the contrast between important and unimportant information. The collaterals also convey the information that a certain motor command is being delivered and possibly give rise to the sense of effort. Because of this descending (motor) control over ascending (sensory) information, there is clearly no real functional separation of these two systems.

Multineuronal pathway

Some of the neurons of cerebral cortex do not go directly to the region of the motor neurons; rather, they form the first link in the multineuronal pathway which ultimately goes to motor neurons of the brainstem and spinal cord. These cortical neuronal axons pass to the subcortical centers and cerebellum, where they synapse with neurons which, in turn, add their axons to the pathway (Fig. 17-8). The pathway is thus made up of a chain of func-

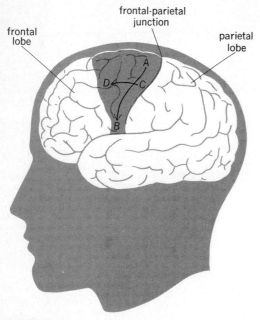

FIGURE 17-9
Regions of the motor cortex.

Cerebral cortex

Many areas of cerebral cortex give rise to the two descending pathways described above, but a large number of the fibers come from the posterior part of the frontal lobe, which is therefore called the *motor cortex* (Fig. 17-9). The function of the neurons in the motor cortex varies with position in the cortex. As one starts at the top of the brain and moves down along the side (*A* to *B* in Fig. 17-9), the cortical neurons lie in such a way that neurons affecting movements of the toes and feet are at the top of the brain, followed (as one moves laterally along the surface of the brain) by neurons controlling leg, trunk, arm, hand, fingers, neck, and face. The size of each of the individual body parts in Fig. 17-10 is proportional to the amount of cortex devoted to its control; clearly, the cortical areas representing hand and face are the largest. The great number of cortical neurons for innervation of the hand and face is one of the factors responsible for the fine degree of motor control that can be exerted over those parts.

The neurons also change character as one explores from the back of the motor cortex forward (*C* to *D* in Fig. 17-9). Those in the back portion, i.e., closest to the junction of the frontal and parietal lobes, mainly contribute to the direct corticospinal pathway. Moving anteriorly (forward) in the motor cortex, this zone of

tionally related neurons, all integrating and transmitting information about control of the skeletal muscles. The final neuron in the chain ends either on interneurons or the gamma motor neurons.

The information conveyed by this pathway (in contrast to the direct corticospinal pathway) continuously changes character as it is transmitted down through the cerebrum and brainstem because cerebellar and other influences are introduced at each synaptic junction between the neurons in the pathway. The multineuronal pathway is responsible for the second type of motor cortex control over muscle movement, i.e., setting into motion patterns of neural interaction which are built into lower control centers of the central nervous system, perhaps in the basal ganglia, brainstem, or spinal cord. For example, the intricate neural mechanisms necessary for walking lie at subcortical levels of the central nervous system; the cortex need only start them off and guide them during the course of the movements.

Despite the distinctions between these pathways, it is wrong to imagine a complete separation of function. All movements, whether automatic or voluntary, require the continual coordinated interaction of both pathways.

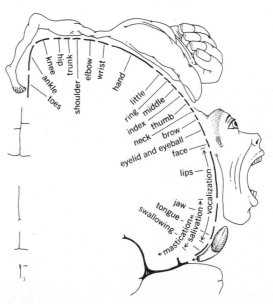

FIGURE 17-10
Arrangement of the motor cortex.

neurons gradually blends into the group which form the first link in the multineuronal pathway. However, none of these cortical neurons functions as an isolated unit. Rather, they are interconnected so that those cortical neurons controlling motor units having related functions fire together. The more delicate the function of any skeletal muscle, the more nearly one-to-one is the relationship between cortical neurons and the muscle's motor units.

Throughout this description we have presented the cerebral cortex as the origin of the descending pathways to motor neurons. But how are the action potentials which travel in these descending pathways initiated? As in most other neurons, the firing pattern of these cells is determined solely by the balance of synaptic activity impinging upon them (there is no evidence for the spontaneous generation of action potentials in these cells). What, then, are the inputs to these cortical cells? First, they receive a large amount of information from the receptors listed earlier in this chapter. Given this input, one might visualize the cortical cells as the integrators of long reflex arcs initiated by the stimulation of receptors, and, indeed, some movement is explainable in these terms. Yet such explanations do not seem to account for the initiation of movement by conscious intention. For example, if the neurons of the motor cortex are stimulated in conscious patients during surgical exposure of the cortex, the patient moves but not in a purposefully organized way. The movement made depends upon the part of the motor cortex stimulated. The patient is aware of the movement but says, "You made me do that," recognizing that it is not a voluntary movement of his own. Stimulation of parts of cortex in the parietal lobe sometimes causes patients to say that they want to make a certain movement, but the movement does not occur.

Someplace in the brain, and it is not known where, immediate afferent information must be associated with data acquired from past experiences and with the neural activity of present feelings (I want to, I feel like doing, etc.). It is the output of this synthesis which ultimately commands the motor cortex.

Finally, the input to the cortex neurons not only must be directed toward some purpose but must select from a variety of ways of achieving that purpose. For a simple example, a rat trained to press a lever whenever a signal light flashes will do so quite consistently but, depending upon its original position in the cage, its movements can be quite varied. If the rat is to the left of the lever, it moves to the right; if it is to the right, it moves to the left. If its paw is on the floor, it raises the paw; if its paw is above the lever, it lowers the paw. The

only consistent act is pressing the lever. The movements performed to achieve the purpose are variable and seem almost inconsequential. Yet, although it is the end result that matters, only the intervening acts or movements can be programmed by the nervous system. How a given program is selected is not known.

Subcortical centers: basal ganglia, brainstem nuclei, and brainstem reticular formation

The multineuronal motor pathway descending from the cortex synapses in the basal ganglia, brainstem nuclei, and brainstem reticular formation. These neuronal clusters, or centers, have important roles in the control of postural mechanisms and coordination of the many simultaneous movements of locomotion. They also serve to correlate fine, detailed voluntary movements with the appropriate postural mechanisms upon which these movements are superimposed. In addition, the basal ganglia serve a special role in the voluntary control of slow, smooth movements. These structures send information back to the motor cortex, which is involved with both fast and slow movements, to modulate its output. These functions become apparent when the basal ganglia are damaged or diseased (for example, Parkinson's disease), as the patient exhibits excessive and disorganized movements and has a marked defect in the voluntary production of smooth motions of different speeds.

Cerebellum

The cerebellum sits on top of the brainstem, as can be seen in Fig. 6-43. It does not initiate movement but acts by influencing other regions of the brain responsible for motor activity. Destruction of the cerebellum does not cause the loss of any specific movement; instead it is associated with a general inadequacy of that movement. The main problems of a person with cerebellar damage are as follows:

1 He cannot perform movements smoothly. If he tries to grasp an object with his hand, the movement is jerky and is accompanied by oscillating, to-and-fro tremors which become more marked as he approaches the object. When his limbs are at rest, they are steady and motionless; but if he tries to move them for any reason, even to help maintain his balance, they go into oscillations which are sometimes quite wild. This oscillating tremor is also known as *intention tremor* and is classically associated with cerebellar damage. It is opposite to the resting tremor that occurs with basal ganglion disease.

2 A person suffering from cerebellar damage walks awk-

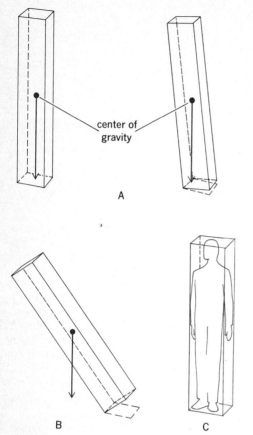

center of
gravity

A

B   C

FIGURE 17-11
The center of gravity is the point in a system at which, if a
string were attached and pulled upward, all the downward
forces due to gravity would be exactly balanced. The center of
gravity must remain within the vertical projections of the base
of the system if stability is to be maintained. A Stable
conditions. B Unstable conditions. C The vertical projections
of the base, in which the center of gravity must remain for
stable posture in man.

wardly with his feet well apart. He has such difficulty maintaining his balance that his gait is reeling and drunken.

3 He cannot start or stop movements quickly or easily and,
if asked to rotate his wrist back and forth as rapidly as
possible, his motions are slow and irregular.

4 He may not be able to combine the movements of several
joints into a single, smooth, coordinated motion. To move
his arm, he might first move his shoulder, then his elbow,
and finally his wrist.

In cases of severe cerebellar damage, the combined difficulties of poor balance and unsteady movements may become so great that the person is incapable
of walking or even standing alone. Since speech depends
on the intricately timed coordination of many muscle
movements, cerebellar damage is accompanied by
speech disturbances. The person with cerebellar damage
shows no evidence whatsoever of sensory or intellectual
deficits.

From this discussion, the reader will deduce that
the cerebellum is involved in the control of muscles utilized both in maintaining steady posture and in effecting
coordinated, detailed movements. It receives input both
from cortex and subcortical centers with information
about what the muscles *should* be doing and from many
afferent systems with information about what the muscles
are doing. If there is a discrepancy between the two, an
error signal is sent from cerebellum to the cortex and
subcortical centers where new commands are initiated
to decrease the discrepancy and smooth the motion.

The cerebellum plays a special role in the control
of rapid movements (and thus serves as the counterpart
to the basal ganglia, which influence slow movements).
Rapid movements are largely "preprogrammed" in their
entirety, rather than being modified during their course
as slower movements are. Such preprogramming requires
calculation of the time needed for the movement (taking
into account the particular amount of muscle force necessary in any special situation) and integration of these
data with information about the moved structure's resting
position and final location. The cerebellum performs
this function and, like the basal ganglia, projects the
information to the motor cortex.

The afferent inputs to the cerebellum come from
the vestibular system, eyes, ears, skin, muscles, joints,
and tendons, i.e., from the major receptors affected by
movement. Inputs from receptors in a single small area
of the body end in the same region of the cerebellum
as do the inputs from the higher brain centers controlling
the motor units in that same area. Thus information from
the muscles, tendons, and skin of the arm arrive at the
same area of cerebellar cortex as the motor commands
for "arm" from the cerebral cortex. This permits the
cerebellum to compare motor commands with muscle
performance.

This completes our analysis of the components of
motor control systems. We now analyze the interactions
of these components in two situations: maintenance of
upright posture and walking.

Maintenance of posture and balance

The skeleton supporting the body is a system of long bones and a many-jointed spine which cannot stand alone against the forces of gravity. Even when held together with ligaments and covered with flesh, it cannot stand erect unless there is coordinated muscular activity. This applies not only to the support of the body as a whole but also to the fixation of segments of the body on adjoining segments, e.g., the support of the head, which is held erect without any conscious effort and without fatigue in a normal awake person. A decrease in postural fixation of the head is seen when someone falls asleep sitting up; his head nods until his chin is on or near his chest.

Added to the problem of supporting his own weight against gravity is that of maintaining equilibrium. Man is a very tall structure balanced on a relatively small base, and his center of gravity is quite high, being situated just below the small of the back. For stability, the center of gravity must be kept within the small area determined by the vertical projection of that base (Fig. 17-11). Yet man is almost always in motion, swaying back and forth and side to side even when he is standing still. Clearly, he often operates under conditions of unstable equilibrium and would be toppled easily by physical forces in the environment if his equilibrium were not protected by reflex postural mechanisms.

The maintenance of posture and balance is accomplished by means of complex counteracting reflexes, all the components of which we have met previously. The efferent arc of the reflexes is, of course, the alpha motor neurons to the skeletal muscle. The major coordinating centers are the basal ganglia, brainstem nuclei, and reticular formation which influence the motor neurons mainly via the descending multineuronal pathway. What is the source of afferent input? One might predict the existence of "center-of-gravity" receptors but, in fact, no such receptors exist. Rather, information about the location of the center of gravity is given by the integration of all the afferent signals from muscles, joints, skin, vestibular system, and eyes. This integration provides the subcortical coordinating centers with a "map" of the position of the whole body in space.

In such a system it is extremely difficult to assign a certain percentage of importance to any one afferent system, even when the exact conditions are specified. However, it does seem that in man, under conditions which permit vision, visual information is probably most important. Yet, so influential are the other inputs and so adaptable is the overall system that a blind person maintains balance quite well, with only slight loss of precision. Moreover, with changing circumstances, the balance of importance between different afferent inputs may change considerably.

Let us take the vestibular system as an example. Despite the fact that the vestibular organs are called the sense organs of balance, persons whose vestibular mechanisms have been completely destroyed may have very little disability in everyday life (one such man was even able to ride a motorcycle). Such persons are not seriously handicapped as long as their visual system, joint position receptors, and cutaneous receptors are functioning. However, they do have difficulty walking in darkness over uneven ground or walking down stairs, where they cannot see a point immediately in front of their feet for visual reference. Thus, in a normal person vestibular input must increase in importance under such circumstances. Finally, vestibular information provides the only clue to orientation with respect to gravity when one is swimming under water, where visual and skin and joint input is confused.

Receptors in the joints and associated ligaments and tendons and their pathways through the nervous system play an important role in the unconscious control of posture and movement and also give rise to the conscious awareness of the position and movement of the joints (Chap. 16). Recall that joint receptors are accurate indicators of movement and position. Skin receptors sensing contact of the body with other surfaces also play a role in regulation of body posture. That both cutaneous and joint-position receptors contribute to postural mechanisms can be shown by the following tests.

Walking in the dark, one's gait is halting and uncertain; stability and confidence are greatly improved by the simple act of running a fingertip along a wall. The fingertip certainly provides no physical support, but it adds significant afferent information to the coordinating centers in the subcortical centers and cerebellum. The joint-position receptors add information too. When a blindfolded person lacking a vestibular system is tilted from an upright position, his trunk does not make appropriate motions to counteract the tilt, but the limbs do so as a result of stimulation of the joint-position receptors in the moving joints. Moreover, the movements are similar to those a normal person would make (but they are not as vigorous; they do not always affect all limbs and may not prevent falling).

Several more examples may further illustrate how integration of more than one input is usually required for appropriate responses to be elicited. Normal subjects

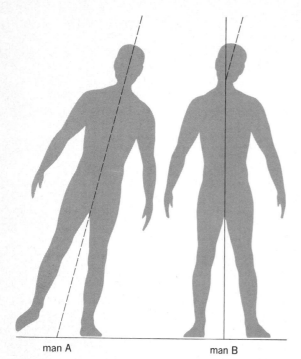

FIGURE 17-12
Interaction of the vestibular and skin, joint, and tendon receptors.

man A   man B

show large postural reflex responses to instability of the base upon which they are seated or standing; but under these conditions, both vestibular receptors and joint-position receptors are excited. If the vestibular system alone is stimulated by moving the head sharply in any direction, there is no reflex movement of the trunk or limbs. It seems, therefore, that the postural reflexes occur only when the *body* is unstable. In the central nervous system the integration of synaptic information determines whether the stimuli from the vestibuli are to excite the postural mechanisms or not, a process which at one time allows the vestibular system to excite these large reflex movements and at another time when they would be inappropriate shuts them out.

The two figures in Fig. 17-12 illustrate this point. Man A is clearly in danger of falling and is in need of muscle movements to keep him upright. Consider the synaptic input to the central nervous system: The vestibular inputs from the two sets of semicircular canals are

different, the inputs from the joints and ligaments of the limbs on the two sides of his body are different, and the joint receptors from the section of vertebral column that runs through his neck indicate that his neck is not bent. The final integration of all these synaptic inputs leads to the initiation of postural reflex responses. Man B is not in danger of falling even though the input from his vestibular systems is exactly the same as that of man A, because the total synaptic input to the central nervous system is quite different. There is no imbalance of input from his limb joint receptors, but there is imbalance from his neck joint receptors. The information in man A indicates instability; that in man B does not. The reactions which would be initiated by vestibular stimulation do not occur unless the input from joint receptors indicates that the body is unstable.

Finally, we must point out the special contribution of the muscle-spindle (stretch) receptors to the maintenance of upright posture against the force of gravity. These length-monitoring receptors provide information, via afferent and ascending pathways, to the higher brain centers but, in addition, they initiate the locally occurring stretch reflex. Imagine a person standing upright; gravity causes his knees to begin to buckle. As this occurs the patellar tendon and extensor muscles are stretched and the stretch reflex is elicited, resulting in increased contraction of the extensor muscles to straighten the knee and prevent him from falling.

From the above description, it might seem (incorrectly) that the stretch reflexes, by themselves, could maintain upright position against the forces of gravity. In fact, their input is effective in enhancing motor activity only when there is a simultaneously occurring facilitation of the alpha motor neurons by descending pathways. Indeed, the providing of such facilitation is a major function of the multineuronal pathway. As might be predicted, this pathway generally facilitates preferentially the motor neurons to the so-called antigravity muscles, for example, the extensor muscles of the legs.

Finally, what happens if all the postural reflexes, once initiated, are unsuccessful and the person loses his balance? If a person is tilted, the arm on the lower side is pulled toward his body, but if the tilting goes so far that he is in danger of losing his balance, the arm on the lower side is quickly extended to break the fall. The first reflex pattern in response to tilt breaks up and a different set of reflexes takes over as soon as there is no hope of restoring balance. Indeed, it is just such a transformation that accounts for stepping in walking.

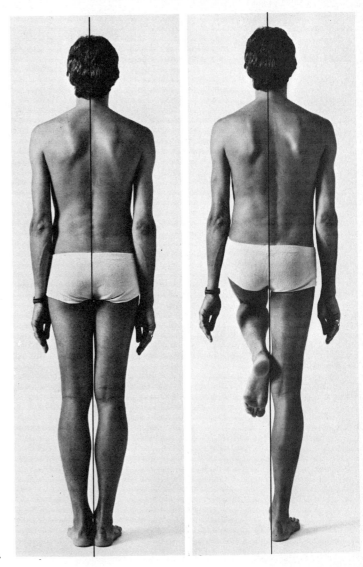

FIGURE 17-13
Postural changes with stepping.
A **Normal standing posture.** B **As the left foot is raised, the whole body leans toward the right so that the center of gravity is vertically above the right foot.**

Walking

Postural fixation of the body is intimately related to the problems of locomotion and maintaining equilibrium in the face of movement. In considering locomotion, the need for some structure or mechanism capable of carrying the body along is added to the basic requirement of anti-gravity support of the body. In walking, the human body is balanced on the very small base provided by one foot.

The weight of the body is supported on each leg alternately, and to accomplish this the body moves from side to side in such a way that the center of gravity is alternately poised over the right and then the left leg. Only when the center of gravity is shifted over the right leg can the left foot be raised from the ground and advanced. As the left foot is lifted, the trunk of the body sways to the right to counterbalance the weight of the left leg (Fig. 17-13). It must be apparent that strict and delicate control of the center of gravity is essential to permit these movements without loss of equilibrium.

Effective stepping can be evoked in newborn infants if the child is held with his feet set on a surface and his legs supporting the weight of his body. The stepping movements can be initiated by simply tilting the child's body forward and rocking him slightly from side to side. Thus, small shifts in the center of gravity start the coordinated, alternate flexion and extension of the leg involved in purposeful stepping.

The stimulus necessary to trigger the stepping is a slight forward tilt of the body. When a child or adult takes a step, the weight of the body is shifted to one foot, and the opposite foot is lifted from the ground. The body is allowed to fall forward and loses its equilibrium until it is caught on the leg which has swung forward. During this process, the center of gravity has moved both sideways and forward from its original position over one leg to a similar position over the other leg. This action is repeated rhythmically, and its continuation depends on both components of the shift in the center of gravity, the forward shift, which causes the fall forward, and the sideways shift, which allows one foot to be lifted and advanced. Thus, there are four necessary components for locomotion: antigravity support of the body, stepping, control of the center of gravity to provide equilibrium,

and a means of acquiring forward motion. All four of these components must be present simultaneously and continuously. It would obviously be futile to apply forward motion without an adequate stepping mechanism. An interesting point is that the forward fall serves to both provide the advance in position and evoke stepping. In fact, if a person leans forward beyond a certain point, he must either take a step or fall, and in this case the stepping is part of a protective reflex.

In persons with diseases of the basal ganglia, disturbances in locomotion often occur, and occasionally the disorders may become so great that the patients become immobilized. Such patients can stand and can make rhythmical, alternate stepping movements, but they cannot walk. One such patient with diseased basal ganglia, who walks only poorly, is able to get along very well if she carries a 14-lb chair in front of her. Her disorder is in the basal ganglion mechanisms which control the forward shift in the center of gravity, and she cannot lean forward to acquire the forward progression necessary for locomotion. The weight of the chair has the effect of bringing the center of gravity forward. Her disorder is only in the ability to tilt her body forward; she does not lack the ability to rock her body from side to side. Other patients have good front-to-back control but lack adequate lateral control.

Consciousness and behavior

Despite the fact that the word behavior is commonly used to refer to those actions which are external, readily visible events, it really denotes the response of individuals to their total environment, both external and internal. Thus behavior may be defined simply as anything a person does, including both somatic and autonomic events. Obviously everything we have described in Part 3 comes under the category of behavior. Chapter 16 introduced the concept of consciousness in terms of perception, and Chap. 17 dealt with it in terms of voluntary and involuntary movements. This chapter discusses consciousness and its relationship to all forms of behavior.

Consciousness

The term *consciousness* includes two distinct concepts, *states of consciousness* and *conscious experience*. The second concept refers to those things of which a person is aware—thoughts, feelings, perceptions, ideas, dreams, reasoning—during any of the states of consciousness. A person's state of consciousness, i.e., whether awake, asleep, drowsy, etc., is defined both by his behavior, covering the spectrum from coma to maximum attentiveness, and by the pattern of brain activity that can be recorded electrically, usually as the electric-potential difference between two points on the scalp. This record is the *electroencephalogram* (EEG).

18

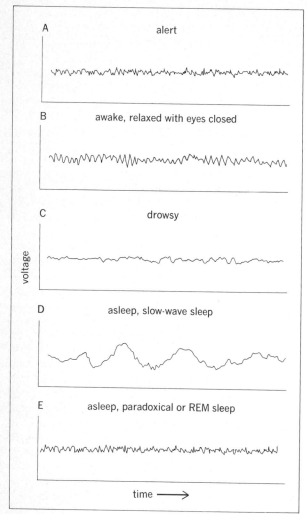

FIGURE 18-1
EEG patterns corresponding to various states of consciousness.

The wavelike pattern of the EEG changes in frequency and amplitude as behavior changes from attentive alertness through quiet resting and drowsiness to sleep (Fig. 18-1). The EEG is produced by the intermittent synchronization of the electric activity of small groups of neurons of the cerebral cortex. The basic units of this electric activity are thought to be individual synaptic potentials (or groups of them) rather than action potentials. These active synapses are driven by some subcortical center, possibly the thalamus. Cortical neurons other than those synchronized at the moment also have fluctuating membrane potentials due to synaptic input, but their activity is not synchronized and, therefore, the potentials tend to cancel each other out. Periodically, new groups of neurons are synchronized.

The EEG is a useful clinical tool because the normal patterns are altered over brain areas that are diseased or damaged. It is also useful in defining states of consciousness, but it is not known what function, if any, this electric activity serves in the brain's task of information processing. Do these electric waves actually influence brain activity or are they merely epiphenomena? (An epiphenomenon is a phenomenon which occurs with an event but is not usually related to it, for example the sound of a baseball bat striking a ball. The sound results from the impact but does not influence how far the ball will travel.) We do not know.

States of consciousness

The waking state and arousal   Behaviorally, the waking state is far from homogeneous, comprising the infinite variety of things one can be doing. The prominent EEG wave pattern of an awake relaxed adult with his eyes closed is a slow oscillation of 8 to 13 Hz (Fig. 18-1B), known as the *alpha rhythm*. Each individual has a characteristic pattern of alpha rhythm, and so does each region of the brain. The alpha rhythms are nearly always larger at the back of the head over the area of visual cortex, and they are also larger when the person has his eyes shut and is not thinking.

When the person is attentive to an external stimulus (or when he thinks hard about something), the alpha rhythm is replaced by lower, faster oscillations (Fig. 18-1A). This transformation is known as *EEG arousal* and is associated with the act of attending to stimuli rather than with the perception itself; for example, if a person opens his eyes in a completely dark room and tries to see, EEG arousal occurs. Also, with decreasing attention to repeated stimuli, the EEG pattern reverts to the alpha rhythm. Thus, the alpha rhythm with its larger and more regular oscillations is associated with decreased levels of attention.

When alpha rhythms are being generated, subjects commonly report that they feel relaxed and happy. A high degree of alpha rhythm is also associated with meditational states. However, people who normally experience high numbers of alpha episodes have not been shown to be psychologically different from others with lower levels, and the relation between brain-wave activity and subjective mood is obscure. People have been trained to increase the amount of alpha brain rhythms by providing them with a feedback signal such as a tone whenever alpha rhythm appears in their EEG. Increasing the number of alpha episodes has been used for the control of some kinds of chronic pain. This is successful in some cases, possibly because (*1*) the alpha training distracts attention from the pain to the feedback signal and inner feelings, (*2*) the patient believes that the method works, (*3*) the relaxation associated with alpha episodes decreases the patient's anxiety, or (*4*) awareness of the possibility of control over the pain changes its meaning and, therefore, the response to it.

Sleep Although average people spend about one-third of their lives sleeping, we know little of the functions served by it. We do know that sleep is an active process and not a mere absence of wakefulness. Moreover, it is not a single simple phenomenon; there are two distinct states of sleep characterized by different EEG and behavior patterns.

The EEG pattern changes profoundly in sleep. As a person becomes drowsy, the alpha rhythm is gradually replaced by irregular, low-voltage potential differences (Fig. 18-1C), and as sleep deepens, the EEG waves become slower, larger, and more irregular (Fig. 18-1D). This *slow-wave sleep* is periodically interrupted by episodes of *paradoxical sleep,* during which the subject still seems asleep but has an EEG pattern similar to that of EEG arousal, i.e., an awake alert person (Fig. 18-1E).

Paradoxical sleep and slow-wave sleep are differentiated by behavioral criteria as well. During slow-wave sleep these criteria are not clear-cut, and it is difficult to tell precisely when a person passes from drowsiness into slow-wave sleep. There is considerable tonus in postural muscles and only a small change in cardiovascular or respiratory activity. The sleeper can be awakened fairly easily during slow-wave sleep, and if he is awakened, he rarely reports dreaming. Slow-wave sleep has a characteristic kind of mentation but is described by subjects as "thoughts" rather than "dreams." The thoughts are more plausible and conceptual, and more concerned with recent events of everyday life and more like waking-state thoughts than are true dreams.

During episodes of paradoxical sleep, on the other hand, the behavioral criteria are precise. At the onset of paradoxical sleep, there is an abrupt and complete inhibition of tone in the postural muscles, although periodic episodes of twitching of the facial muscles and limbs and rapid eye movements behind the closed lids occur. (Paradoxical sleep is therefore also called *rapid-eye-movement* or *REM sleep.*) Respiration and heart rate are irregular, and blood pressure may go up or down. When awakened during paradoxical sleep, 80 to 90 percent of the time a person reports that he has been dreaming.

Continuous recordings show that the two states of sleep follow a regular 30- to 90-min cycle, each episode of paradoxical sleep lasting 10 to 15 min. Thus, slow-wave sleep constitutes about 80 percent of the total sleeping time in adults, and paradoxical sleep about 20 percent. The time spent in paradoxical sleep increases toward the end of an undisturbed night. Normally it is not possible to pass directly from the waking state to an episode of paradoxical sleep; it is entered only after at least 30 min of slow-wave sleep. Thus, a subject wakened at the beginning of every period of paradoxical sleep can be prevented from spending much time in that state although his total sleeping time remains approximately normal. After being deprived of paradoxical sleep for several nights, all subjects spend a greater than usual proportion of time in paradoxical sleep the next time they sleep. Thus, the total number of hours spent in paradoxical sleep tends to remain constant.

What is the functional significance of sleep? What happens to the brain during sleep? We now know that the brain, as a whole, does not rest during sleep and there is no generalized inhibition of activity of cerebral neurons. On the contrary, there is a considerable amount of neuronal activity during slow-wave sleep, and many areas of the brain are more active during paradoxical sleep than they are during waking. The blood flow and oxygen consumption of the brain, signs of its metabolic activity, do not decrease in sleep.

However, during sleep there is a change in distribution or reorganization of neuronal activity, some individual neurons being less active than during waking although the brain as a whole remains relatively active. Although sleep is not a period of generalized rest for the whole brain, it may represent a period of rest for certain specific elements, during which they can replenish substrates necessary for their generation of action potentials. Yet, when isolated neural tissue is exposed to extreme rates of stimulation far exceeding those occurring under physiological circumstances, neurons recover within a period of minutes. Alternatively, it has been suggested that the functional significance of sleep lies not in short-term recovery but in the relatively long-term chemical and structural changes that the brain must undergo to make learning and memory possible.

Thus, to the question: What is the functional significance of sleep? We must answer that we do not know.

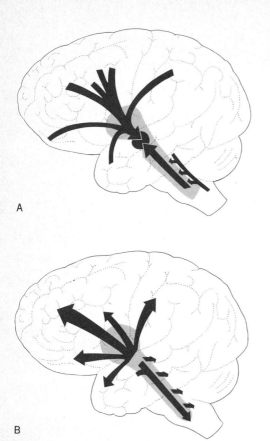

A

B

FIGURE 18-2
A **Convergence of descending, local, and ascending influences upon reticular formation (shaded area). B Projections from reticular formation to spinal cord, brainstem and cerebellum, and cerebrum.** (*Adapted from Livingston.*)

Several answers are possible, but evidence is not sufficient to prove any one.

Role of the reticular formation The prevailing state of consciousness is the resultant of the interplay between three neuronal systems, one causing arousal and the other two sleep, all three of which are parts of the reticular formation. The reticular formation lies in the central core of the brainstem in the midst of the neural pathways ascending and descending between the brain and spinal cord, and the neurons of the reticular formation receive a continuous sample of the neural activity in these pathways, including information from (1) areas of the cerebrum (cerebral cortex, basal ganglia, limbic system, and other regions deep within the cerebrum); (2) spinal cord;

and (3) cerebellum and other brainstem structures (Fig. 18-2A). The output of the reticular formation neurons is determined by these inputs as well as by spontaneous activity generated in the reticular formation itself. The reticular formation projects to (1) the spinal cord, (2) the cerebellum, and (3) the subcortical and cortical areas of the cerebrum. There are also a great many synaptic endings in the brainstem, so that the reticular formation acts upon itself (Fig. 18-2B). Thus, the reticular formation influences and is influenced by virtually all areas of the central nervous system.

However, the reticular formation is not homogeneous, and discrete areas frequently have specific functions. We have already discussed several of the important functions of the reticular formation: It helps to coordinate skeletal muscle activity (Chap. 17); it contains the primary cardiovascular and respiratory control centers (Chaps. 9 and 10); it monitors the huge number of messages ascending and descending through the central nervous system (Chaps. 16 and 17). In this chapter we are concerned with its role in determining states of consciousness.

The reticular activating system In 1934, it was discovered that the EEG of a cerebrum surgically isolated from the spinal cord and lower three-fourths of the brainstem loses the wave patterns typical of an awake animal, indicating that some neural structures within the separated brainstem or spinal cord are essential for the maintenance of a waking EEG. These neural structures lie within the brainstem reticular formation. Electric or chemical stimulation of this area causes EEG arousal, whereas its destruction produces coma and the EEG characteristic of the sleeping state.

This component of the reticular formation is called the *reticular activating system* (RAS). As they pass from the brainstem into the central core of the cerebrum, the neurons of the RAS activate the *diffuse thalamic projection system*. These thalamic neurons synapse in the cortex, but unlike the specific thalamic projections described in Chap. 16, they are not involved in the transmission of information about specific sensory modalities; rather, they carry on the functions of the RAS and maintain the EEG and behavioral characteristics of the awake state.

Single neurons in the reticular formation may be activated by any afferent modality—a flash of light, a ringing bell, a touch on the skin to "arouse" the brain. Human beings are able to perceive a stimulus only when the nervous system is oriented and appropriately receptive toward it, and it is the neurons of the reticular activating system which arouse the brain and facilitate information reception by the appropriate neural structures. However, the sensitivity of this system is selective. A mother may awaken instantly at her baby's faintest whimper whereas she can sleep peacefully through the roar of a jet plane passing overhead.

One of the phenomena leading to selectivity is known as *habituation*. Presentation of a novel stimulus to an awake animal usually leads to an orienting response, during which the animal stops whatever it is doing and looks around or listens intently and the EEG switches from the quiet resting alpha rhythm to that characteristic of arousal. On the other hand, the monotonous repetition of a stimulus of constant strength leads to a progressive decrease in response (habituation). For example, when a loud bell is sounded for the first time, it may evoke a startle response in the animal; but after several ringings, the animal makes progressively less response and eventually may ignore the bell altogether. An extraneous stimulus of another modality or the same stimulus at a different intensity restores the original response (*dishabituation*). Habituation is not due to receptor fatigue or adaptation but is mediated at least in part by the reticular formation.

Although the mechanism of action of centrally acting anesthetics is not known, it has been postulated that some act by interfering with the transmission of neural activity in the reticular formation rather than by blocking the direct transmission of afferent information to the cortex. Also, some stimulating drugs (e.g., the amphetamines) work by enhancing the transmission of nerve impulses through the reticular system.

Parts of the brain other than the reticular system are also important for wakefulness and the alert state. For example, the cortex is necessary for sustained wakefulness, and maintenance of the alert state seems to involve an interplay between cortex and the reticular formation. Electric stimulation of some regions of the cortex activates the reticular formation and awakens or alerts an animal as surely as afferent stimulation does. Finally, certain hypothalamic areas are implicated in the EEG and behavioral aspects of the waking state.

The sleep centers The control of sleep is exerted by two neuronal systems which oppose the tonic activity of the RAS. The two neuronal clusters, one in the central core of the brainstem, the other in the pons, are also part of the reticular formation (Fig. 18-3). The brainstem-core neurons tonically release the transmitter 5-hydroxytryptamine (5-HT, also known as *serotonin*); when serotonin levels become high enough, the neurons of RAS are inhibited. This results in the loss of awake conscious behavior and its EEG manifestations and the replacement

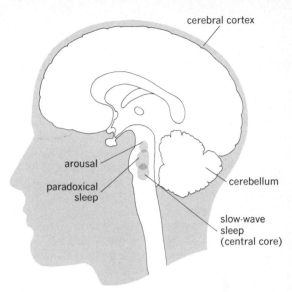

FIGURE 18-3
Brainstem structures involved in arousal, paradoxical sleep, and slow-wave sleep. (*Adapted from Jouvet.*)

of these by the behavior and EEG characteristics of slow-wave sleep.

These brainstem-core neurons also facilitate the sleep center of the pons, whose activity induces paradoxical sleep. Pathways ascending from this paradoxical-sleep center of the pons establish the low-voltage, fast EEG pattern and activate the muscles of the eyes; descending pathways inhibit motor-neuronal activity, which results in loss of muscle tone.

Of great importance is the fact that, while instituting the EEG and behavioral manifestations of paradoxical sleep, the neurons of the sleep center in the pons also influence the central core of the brainstem in a true feedback fashion. We have said that the release of serotonin by these central core neurons causes slow-wave sleep and facilitates the paradoxical-sleep center. It is thought that the paradoxical-sleep-center neurons feed back and stimulate the *uptake* of serotonin by the endings of the brainstem-core neurons. This decrease in free (i.e., extracellular) serotonin concentration lessens the inhibition on the RAS and permits a return of the awake state. Waking continues until sleep is again triggered by the release of sufficient serotonin to inhibit the RAS. Thus, the cycling of the sleeping and waking states of consciousness is due, at least in part, to the slow accumulation and dissipation of chemical transmitters.

As presented above, sleep is basically the result of the cyclical inhibition of the RAS by brainstem-core neurons. However, it should be reemphasized that this inhibition can be overridden by input from afferent pathways or other brain centers so that RAS activity is maintained sufficiently high to keep one awake or interrupt sleep.

When deprived of sleep, we sleep longer at the next sleep cycle to "catch up." However, this recovery period is not directly proportional to the period of sleep deprivation; for example, a good night's sleep of 14 to 16 hr by one subject was sufficient to repay at least the slow-wave component of 300 hr of sleep deprivation (the paradoxical-sleep debt took longer to make up).

Of the two basic components of the sleep-wake cycle, the waking mechanisms seem to be more easily activated than those causing sleep. An example familiar to all parents is that it is much easier to arouse a sleeping child than to get an alert, attentive child to sleep.

Conscious experience

All subjective experiences are popularly attributed to the workings of the mind. This word conjures up the image of a nonneural "me," a phantom interposed between afferent and efferent impulses, with the implication that mind is something more than action potentials and synapses. The truth of the matter is that physiologists and psychologists have absolutely no idea of the mechanisms which give rise to conscious experience. Nor are there even any scientifically meaningful hypotheses concerning the problem.

Conscious experiences are difficult to investigate because they can be known only by verbal report. Such studies lack scientific objectivity and must be limited to man. In an attempt to bypass these difficulties scientists have studied the behavioral correlates of mental phenomena in other animals. For example, a rat deprived of water performs certain actions to obtain it. These actions are the behavioral correlates of thirst. But it must be emphasized that we do not know whether the rat consciously experiences thirst; this can only be inferred from the fact that man is conscious of thirst under the same conditions.

However, one crucial question which cannot be investigated in experimental animals is whether conscious experiences actually influence behavior. Although, intuitively, it might seem absurd to question this, the fact is that the answer is crucial for the development of one's concept of man. It is possible that conscious experience is an epiphenomenon.

Consider the following sequence of events: The ringing of a telephone reminds a student that he had promised to call his mother; he finishes the page he has been reading and makes the call. What causes him to do so? The epiphenomenon view holds that the conscious awareness accompanies but does not influence the passage of information from afferent to motor pathways. Thus, in this view, behavior occurs automatically in response to a stimulus, memory stores supplying the direct link between afferent and efferent activity. In contrast, the processing of the afferent information (the sound of the bell), acting through memory stores, could result in the conscious awareness of his promise, which, in turn, leads to the relevant activity in motor pathways descending from the cortex. There is no way of choosing between these two views at present.

In contrast to this presently unapproachable question, some aspects of conscious experience have yielded, at least in part, to experimentation in man. In general, these can be described by two questions: What is the relationship between the conscious experience and information arriving over the principal afferent pathways? Is there an anatomically distinct area of the central nervous system involved in conscious experience as opposed to those areas of the brain engaged in the unconscious processing of information or the execution of automatic movements?

The first question has been approached by studies performed in conscious human subjects whose brains are exposed for neurosurgery. Tiny electrodes lowered into different areas of the thalamus record the activity of single cells. In certain thalamic cells a given somatic stimulus, such as movement of a joint, regularly produces a given response which contains in coded form the precise location and intensity of the stimulus. The electric response recorded is unchanged regardless of whether the patient is keenly aware of each stimulus or whether his attention is diverted so that he is unaware that he has been stimulated; i.e., the electric activity is the same whether or not information about the stimulus is incorporated into his conscious experience. The thalamic cells which respond in this manner are in nuclei known to form part of the specific afferent pathway. Thus, information relayed in the specific ascending pathways does not necessarily become part of conscious experience.

There are other neurons in nonsensory parts of the thalamus which have quite different properties and patterns of activity which are more closely related to conscious experience. Each of these cells, called *novelty detectors,* responds to input from many parts of the body, but it rapidly *ceases* responding to repeated stimuli of the same kind as the person's attention to them wanes. The firing pattern is more closely related to the degree to which the person is aware of a given stimulus than to precise information about a certain sensory modality.

Other evidence also suggests that conscious experience is determined by central structures as well as by peripheral stimuli. Clinical examples of conscious sensory experiences existing in the absence of neural input are not unusual. For example, after a limb has been amputated, the patient sometimes feels as though it were still present. The nonexistent limb, called a phantom limb, can be the "site" of severe pain.

In this context, it is interesting to study the conscious experiences of persons undergoing periods of sensory deprivation. Student volunteers lived 24 hr a day in as complete isolation as possible—even to the extent that their movements were greatly restricted. External stimuli were almost completely absent, and stimulation of the body surface was relatively constant. At first the students slept excessively, but soon they began to be disturbed by vivid hallucinations which sometimes became so distorted and intense that the students refused to continue the experiment. The neural bases of these hallucinations are poorly understood, but it has been suggested that central structures may generate patterns of activity corresponding to those normally elicited by peripheral stimuli when varied sensory input is absent and that conscious experience is not solely dependent upon the senses.

There seems to be an optimal amount of afferent stimulation necessary for the maintenance of the normal, awake consciousness. Levels of stimulation greater or less than this optimal amount can lead to trances, hypnotic states, hallucinations, "highs," or other altered states of consciousness. In fact, alteration of sensory input is commonly used to induce intentionally such experiences.

The answer to the second question: Is there a specific brain area in which conscious experience resides? can be gleaned from conscious persons undergoing neurosurgical procedures and from persons who have accidental or disease-inflicted damage to parts of the brain; but the answer is still far from clear. We have mentioned that it is possible that conscious experience may be just another aspect of the neural activity in those brain centers which receive and process afferent information. Or, on the other hand, conscious experience may depend upon the transmission of the processed afferent information to special parts of the brain whose function

is to arrive at and release the contents of conscious experience. The only available clues as to how or where this might be done have been gained from inference. For example, with evolution comes (we assume) greater complexity of conscious experience. Since man's brain is distinguished anatomically from that of other mammals by a greatly increased volume of cerebral cortex, this is a logical place to look for the seat of conscious experience. The cortex has been stimulated when patients on the operating table are fully alert and the brain is exposed. If a certain area of association cortex in temporal lobe is stimulated, the subject may report one of two types of changes in his conscious experience. Either he is aware of a sudden change in his interpretation of the present situation, i.e., what he is seeing or hearing suddenly becomes familiar or strange or frightening or coming closer or going away, or he has a sudden flashback or awareness of an earlier experience. Although he is still aware of where he is, an earlier experience comes to him and repeats itself in the same order and detail as the original experience. It may have been a particular occasion when he was listening to music. If asked to do so, he can hum an accompaniment to the music. If in the past he thought the music beautiful, he thinks so again. During such electric stimulation, visual or auditory experiences are recalled only if the patient was attentive to them when they originally occurred. Other experiences which are also part of conscious experience have never been produced by such stimulation. No one has ever reported periods when he was trying to make a decision or solve a problem or add up a row of figures. It is also interesting that no brain area other than temporal association cortex has been found from which complete memories have been activated. However, one cannot assume that association cortex is the site of the stream of consciousness, because stimulation of these cortical neurons causes the propagation of action potentials to many other parts of the brain.

It is best to take a different view of the matter; for, in the words of one famous neurosurgeon, "Consciousness is not something to be localized in space." It is a function of the integrated action of the brain. Sensations and perceptions form part of conscious experience, and yet there is no one point along the ascending pathways or one particular level of the central nervous system below which activity cannot be a conscious sensation and above which it is a recognizable, definable sensory experience. Every synapse along the ascending pathways adds an element of meaning and contributes to the sensory experience.

On the other hand, consciousness and, we presume, the accompanying conscious experiences, are inevitably lost when the function of regions deeper within the cerebrum or the nerve fibers passing to association cortex from the reticular formation are interrupted by injury. Although the matter is far from settled, evidence suggests that neuronal systems in the reticular formation of the brainstem and regions deep within the cerebrum are involved in brain mechanisms necessary for perceptual awareness. It has been suggested that this system of fibers has widespread interactions with various areas of the cortex and somehow determines which of these functional areas is to gain temporary dominance in the on-going stream of the conscious experience.

The concept we want to leave as an answer to the question of where the conscious experience resides is perhaps best presented in the following analogy. In an attempt to say which part of a car is responsible for its controlled movement down a highway, one cannot specify the wheels or axle or engine or gasoline. The final performance of an automobile is achieved only through the coordinated interaction of many components. In a similar way, the conscious experience is the result of the coordinated interaction of *many* areas of the nervous system. One neuronal system would be incapable of creating a conscious experience without the effective interaction of many others.

Motivation and emotion

Motivation

Motivation is presently undefinable in neurophysiological terms, but it can be defined in behavioral terms as the processes responsible for the goal-directed quality of behavior. Much of this behavior is clearly related to homeostasis, i.e., the maintenance of a stable internal environment, an example being putting on a sweater when one is cold. In such homeostatic goal-directed behavior specific bodily needs are being satisfied, the word "needs" having a physicochemical correlate. Thus, in our example the correlate of need is a drop in body temperature, and the correlate of need satisfaction is return of the body temperature to normal. The neurophysiologic integration of much homeostatic goal-directed behavior has been discussed earlier (thirst and drinking, Chap. 11; food intake and temperature regulation, Chap. 13; reproduction, Chap. 14).

However, many kinds of motivated behavior, e.g., the selection of a particular sweater on the basis of style, have little if any apparent relation to homeostasis. Clearly, much of man's behavior fits this latter category. Nonetheless, the generalization that motivated behavior is induced by needs and is sustained until the needs are satisfied is a useful one despite the inability to understand most needs in physicochemical terms.

A concept inseparable from motivation is that of *reward* and *punishment*, rewards being things that organisms work for or things which strengthen behavior leading to them, and punishments being the opposite. They are related to motivation in that rewards may be said to satisfy needs. Many psychologists believe that rewards and punishments constitute the incentives for learning. Because virtually all behavior is shaped by learning, reward and punishment become crucial factors in directing behavior. Although some rewards and punishments have conscious correlates, many do not. Accordingly, much of man's behavior is influenced by factors (rewards and punishments) of which he is unaware.

We have thus far described motivation without regard to its neural correlates. As was true for conscious experience, nothing is known of the mechanisms which underlie the subjective components of this phenomenon, nor is it known how rewards and punishments influence learning and behavior. Present knowledge is limited to recognition of some of the brain areas (and their interconnecting pathways) which are important in motivated behavior.

It should not be surprising that the brain area most important for the integration of motivated behavior related to homeostasis is the hypothalamus, since it contains the integrating centers for thirst, food intake, temperature regulation, and many others. Much information concerning the reinforcing effects of rewards and punishments on hypothalamic function has been obtained through *self-stimulation* experiments, in which an unanesthetized experimental animal regulates the rate at which electric stimuli are delivered through electrodes previously implanted in discrete brain areas. The animal is placed in a box containing a lever it can press (Fig. 18-4). If no stimulus is delivered to the animal's brain when the bar is pressed, he usually presses it occasionally, perhaps out of boredom or curiosity. However, if a stimulus is delivered to the brain as a result of the bar press, a different behavior can result, depending upon the location of the electrodes. If the animal increases his bar-pressing rate above control, the electric stimulus is, by definition, rewarding; if he decreases it, the stimulus is punishing. Thus, the rate of bar pressing is a measure of the effectiveness of the reward (or punishment). Bar pressing that results in self-stimulation of the sensory and motor systems produces response rates not significantly different from the control rate. Brain stimulation through electrodes implanted in certain areas of hypothalamus serves as a positive reward. Animals with electrodes in these areas bar-press to stimulate their brains from 500 to 5,000 times per hour. In fact, electric stimulation of some areas of hypothalamus is more rewarding than external rewards; e.g., hungry rats often ignore available food for the sake of electrically stimulating their brains.

This rewarding effect of self-stimulation is not found in all areas of the hypothalamus but is most closely associated with those areas which normally mediate highly motivated behavior, e.g., feeding, drinking, and sexual behavior. Consistent with this is the fact that the animal's rate of self-stimulation in some areas increases when he is deprived of food; in other areas, it is decreased by castration and restored by administration of sex hormones. Thus, it appears that neurons controlling homeostatic goal-directed behavior are themselves intimately involved in the reinforcing effects of reward and punishment.

Although it has been generally assumed that such feeding and drinking behaviors are associated with the underlying feelings of hunger and thirst, it is puzzling to determine why an animal would self-stimulate in order to experience "thirst" or "hunger." An explanation of the paradox is that the reinforcing effects of self-stimulation are not thirst or hunger sensations per se; rather that stimulation (i.e., activation) of the neural pathways underlying these behaviors is in itself reinforcing and can, therefore, provide the motivation to engage in self-stimulation behavior.

Emotion

Related to motivation are the complex phenomena of *emotion.* Scientists are presently trying to understand the operation of the chain of events leading from the perception of an emotionally toned stimulus, i.e., the subjective feeling of fear, love, anger, joy, anxiety, hope, etc., to the complex display of emotional behavior, and they are beginning to find some answers.

Most experiments point to the involvement of the *limbic system,* which is an interconnected group of brain structures within the cerebrum, including portions of the frontal-lobe cortex, temporal lobe, thalamus, and hypothalamus as well as the circuitous neuron pathways connecting all parts together (Fig. 18-5). Besides being connected with each other, the parts of the limbic system have connections with many other parts of the central

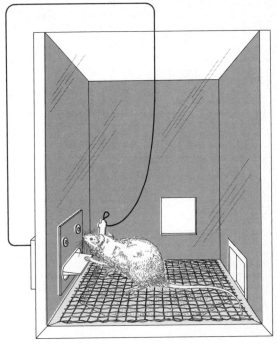

FIGURE 18-4
Apparatus for self-stimulation experiments. (*Adapted from Olds.*)

nervous system. For example, it is likely that information from all the different afferent modalities can influence activity within the limbic system, whereas activity of the limbic system can result in a wide variety of autonomic responses and body movements. This should not be surprising since many emotional feelings are accompanied by autonomically mediated responses such as sweating, blushing, and heart-rate changes, and by somatic responses such as laughing and sobbing.

The limbic system has been studied in experimental animals, using electric stimulation of specific areas within it. The physiological results of these procedures vary markedly but justify the concept that three distinct neural systems mediate the various emotional behaviors. Of course, in these experiments there was no way to assess the subjective emotional feelings of the animals; instead they were observed for behaviors which usually are associated with emotions in man. As different areas of the limbic system were stimulated in awake animals (the stimulus is a small electric current delivered through electrodes previously implanted while the animal was anesthetized), three types of behavior resulted.

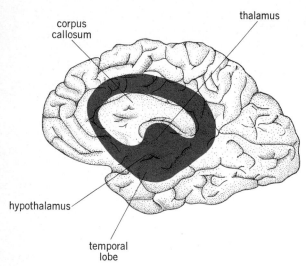

corpus
callosum

thalamus

hypothalamus

temporal
lobe

FIGURE 18-5
Structures of the limbic system are in the shaded area on this section through the midline of the brain. (*View of brain from Curtis, Jacobson, and Marcus, "An Introduction to the Neurosciences," p. 431, W. B. Saunders Company, Philadelphia, 1972.*)

After stimulation of one area the animal actively approaches a situation as though expecting a reward. Stimulation of a second area causes the animal to stop the behavior he is performing, as though he knew it would lead to punishment.

Stimulation of a third area of the limbic system causes the animal to arch its back, puff out its tail, hiss, snarl, bare its claws and teeth, flatten its ears, and strike. Simultaneously, its heart rate, blood pressure, respiration, salivation, and concentrations of plasma epinephrine and fatty acids all increase. Clearly, this behavior typifies that of an enraged or threatened animal. In fact, an animal's behavior can be changed from quiet to savage or from savage to docile simply by electrically stimulating different areas of the limbic system.

Limbic areas have also been stimulated in awake human beings undergoing neurosurgery. These patients, relaxed and comfortable in the experimental situation, report vague feelings of fear or anxiety during periods of stimulation to certain areas even though they are not told when the current is on. Stimulation of other areas induces pleasurable sensations which the subjects find difficult to define precisely.

Surgical damage to parts of the limbic system in experimental animals is another commonly used tool; it leads to a great variety of changes in behavior, particularly that associated with emotion. Destruction of a nucleus in the tip of the temporal lobe produces docility in an otherwise savage animal, whereas surgical damage to an area deep within the brain produces vicious rage in a tame animal; and the rage caused by this lesion can be counteracted by a lesion in the tip of the temporal lobe. A rage response can also be caused by destruction of part of the hypothalamus. Lesioned animals sometimes manifest bizarre sexual behavior in which they attempt to mate with animals of other species; females frequently assume male positions and attempt to mount other animals.

Self-stimulation experiments have shown the presence of reward and punishment responses in various parts of the limbic system. When the electrodes are in certain midline areas, the animal presses the bar once and never goes back, indicating that stimulation of these brain areas has a punishing effect. These are the same brain areas which, when stimulated, give rise to behavioral activity signifying avoidance, rage, or escape. Conversely, self-stimulation of other limbic areas has a strong rewarding effect.

Stimulation of certain hypothalamic areas (like the stimulation of other limbic structures described above) elicits behavior that *seems* to have a strong subjective emotional component; yet if the hypothalamus is isolated from the other portions of the limbic system, the emotional component is lacking. For example, stimulation of a cat's brain can cause enraged, aggressive behavior complete with attack directed at any available object, but as soon as the stimulation ends, the animal immediately reverts to its usual friendly behavior. It seems as though the actions lacked emotionality and purpose, representing only the motor component of the behavior. We use the word "purpose" but could have said that, except for the experimentally induced stimulus, the behavior lacked "motivation."

What is the relationship between the hypothalamus and the rest of the limbic system? The three components of the limbic system described above converge in the medial hypothalamus, which acts as an integrating center. For example, in cats the medial hypothalamus exerts a tonic inhibition on the neural pathways which lead to fight-or-flight behavior. However, upon receipt of appropriate environmental stimuli, the nuclei of the temporal lobe inhibit the medial hypothalamus, thus decreasing its inhibitory influence over the fight-or-flight system and allowing activity in that system to increase. The resulting emotional behavior, then, results from the balance of input to the medial hypothalamic integrating centers. Notice that, although the structures involved in the control of emotional behavior are predominantly located in the limbic system, and (as described in previous chapters) the main controlling centers for consummatory behavior related to homeostasis are located in the hypothalamus, the two meet and interact at the level of the medial hypothalamus.

Finally the subjective aspects, or feelings, that make up part of an emotional experience possibly also involve the cortex, particularly cortex of the frontal lobes, which is implicated because changes in emotional states frequently occur following damage there. These alterations in mood and character are described as fear, aggressiveness, depression, rage, euphoria, irritability, or apathy. There are indications that frontal regions may exert inhibitory influences upon the hypothalamus and other areas of the limbic system. There may be facilitatory frontal regions as well. Anatomical connections between frontal cortex and hypothalamus exist to support the suggested interrelationship of these two areas in motivated and emotional behaviors. Excitatory and inhibitory influences from the limbic system and possibly from non-limbic areas of cortex are of great importance in determining and patterning the level of excitability of hypothalamic and brainstem neurons. How activity in the limbic system is initiated and influences other brain areas is still poorly understood.

Chemical mediators for emotion-motivation

Norepinephrine and other amines play important roles in the mediation of some emotional and behavioral states. For example, norepinephrine is believed to be a neurotransmitter in the pathways subserving rage, for drugs which enhance the effect of norepinephrine also increase rage and those which block it diminish rage behavior. Norepinephrine also seems to operate in the active-approach system. Thus, an animal given a drug that depletes brain norepinephrine stores manifests a decrease in active avoidance of a punishing shock and in rates of self-stimulation, whereas drugs enhancing norepinephrine release have the opposite effect.

The biogenic amines, particularly norepinephrine and dopamine, are also associated with subjective mood. Decreased norepinephrine is associated with depression and increased norepinephrine with elevated mood; furthermore, the antidepressant drugs are thought to work by increasing brain amine concentration.

The biogenic amines are also implicated in the networks subserving learning. This association is not unexpected since we have just stated that they are involved in the neural systems underlying reward and punishment, and many psychologists believe that rewards and punishments constitute the incentives for learning.

Neural development and learning

Evolution of the nervous system

During one of its earliest stages of evolution, the nervous system was probably a simple three-neuron system with a limited number of interneurons interposed between the afferent and efferent nerve cells. The interneuronal component expanded rapidly until it came to be by far the largest part. The interneurons formed networks of increasing complexity, at first involved mainly with the stability of the internal environment and position of the body in space. Those cells with increasing specialization of function came to be localized at one end of the primitive nervous system, and the brain began to evolve.

The brainstem, which is the oldest part of the brain in this evolutionary sense, retains today many of the anatomical and functional characteristics typical of those most primitive brains. With continued evolution, newer, increasingly complex structures were added on top of (or in front of) the older ones. They developed as paired symmetrical tissues, the cerebral hemispheres, and reached their highest degree of sophistication with the formation of the cerebral cortex.

The newer structures served in part to elaborate, refine, modify, and control already existing functions. For example, it is possible to perceive the somatic stimuli of pain, touch, and pressure with only a brainstem, but the stimulus cannot be localized without a functioning cerebral cortex. Perhaps even more important is that these newer, more anterior parts of the brain came to be involved in the perception of goals, the ordering of goal priorities, and the patterning of behaviors to serve in pursuit of these goals.

Development of the individual nervous system

Although all brain cells are present at birth, their rate of growth in size, number of dendrites, degree of myelinization, etc., varies considerably. Notice the changes in visual cortex cells in just the first three months of life (Fig. 18-6) and the gradual change in electric activity (Fig. 18-7). These developmental patterns are reflected in

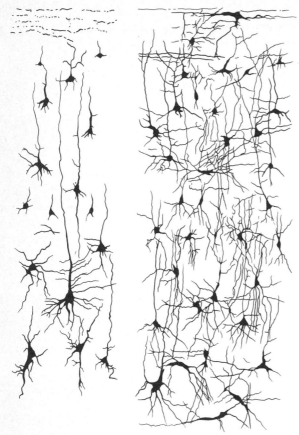

FIGURE 18-6
Visual cortex of a newborn (*left*) **and a three-month-old child**
(*right*). (*Redrawn from J. L. Conel, "The Postnatal Development of the
Human Cerebral Cortex," vol. 1, The Cortex of the Newborn, Harvard
University Press, Cambridge, Mass., 1939.*)

With cortical development, some of the reflexes
whose integrating centers are in subcortical structures
come under at least a degree of cortical control. Examples are the rooting and sucking reflexes present in
infancy. (The stimulus is a touch on the infant's cheek; in
response, the head turns toward the stimulus, the mouth
opens, the stimulating object is taken into the mouth, and
sucking begins.) The reflex is essential for the survival
of the young, but it is superseded by other eating behav-

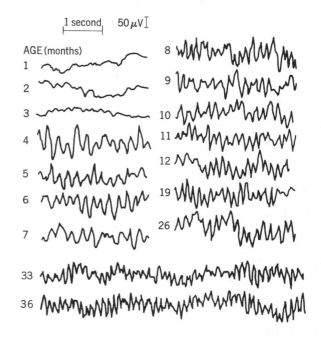

FIGURE 18-7
**EEGs from the same individual, showing the onset of the
alpha rhythm at four months, the attainment of adult wave
frequencies at ten years of age, and little change thereafter.**
(*Redrawn from D. B. Lindsley, Attention, Consciousness, Sleep, and
Wakefulness, in "Handbook of Physiology; Neurophysiology," vol. 3, pp.
1553–1593, American Physiological Society, Washington, D.C., 1960.*)

behavioral changes. It is believed that the cortex is relatively nonfunctional at birth, a notion supported by the
fact that infants born without a cortex have almost the
same behavior and reflexes as shown by a normal infant.
As the cortex develops, it seems gradually to exert an
inhibitory control over the lower (and phylogenetically
older) structures. For example, many of the early motorbehavior patterns are primarily extensor in nature (motions which throughout phylogeny serve supporting and
aggressive functions). Extensor movements are subserved, even in the adult, by subcortical and brainstem
mechanisms. Flexion, largely a cortical phenomenon, can
occur only if extension is inhibited.

iors. As the cerebral hemispheres develop and autonomy increases, allowing the elaboration of hand feeding, the primitive rooting and sucking reflexes are inhibited. The basic reflex arc does not disappear; it is simply inhibited. For example, while examining a patient with damaged frontal lobes of the brain, a physician standing behind the patient and out of his view quietly reached over and touched the patient's cheek. In response, the patient nuzzled the physician's finger until it was in his mouth and even began sucking on it. Upon realizing what he had unconsciously and automatically done, he was quite embarrassed; but his behavior, released from its normal inhibition because of the brain damage, serves as a perfect illustration of the above point.

During development of the brain and spinal cord, nerve growth and synaptic contact occur with remarkable precision and selectivity. Novel experiments by R. W. Sperry demonstrated this fact. The eyes of young frogs were rotated 180° from top to bottom after severing the optic nerve, which connects the eye to the brain. Three questions were asked: (1) Would nerve regrowth occur to produce a confusedly blurred or an orderly visual image, (2) if orderly, would the image be right side up, and (3) if upside down, could proper right-side-up vision be learned? The results showed that when the nerves regrew they established the new synaptic connections in their precisely ordered original position but the new visual image was upside down. Furthermore, the frogs were unable to adapt to the rotated image; for example, they continued to flick their tongues up to catch flies that were below their heads.

How do the right nerve connections get established in the first place? The best answer presently available is the *chemoaffinity theory,* in which the complicated neural circuits are said to grow and organize themselves according to selective attractions between neurons, which are determined by chemical codes under genetic control. Thus, during early stages of their development, neurons acquire individual chemical "identifications" by which they can recognize and distinguish each other. The chemical specificity is precise enough to determine not only the particular postsynaptic cell to which the axon tip will grow but even the precise area on the postsynaptic cell that will be contacted. (Recall that synapses closer to the postsynaptic cell's initial segment have greater influence over its activity.)

Of course, adequate nutritive environments are required to sustain such growth. The level of hormones is also important; for example, decreased thyroid hormone concentrations during development cause a type of mental retardation known as *cretinism.* Although it was formerly believed that the developing fetus "took what it needed" nutritionally from the mother and that, in cases of malnutrition after birth, the brain was "spared" at the expense of the rest of the body, there is increasing evidence that nutritional deprivation both before and after birth has serious and irreversible effects on the chemical and structural maturation of the brain. Intermittent deprivation affects those cells that are still developing at that time more than cells already mature.

Despite the evidence indicating the selective growth and interconnectivity of nerve cells, the nature-nurture controversy, i.e., how much of development is due to genetically determined patterns and how much to experience and subsequent learning, must also be considered, for, in fact, the course of neural development can be altered.

However, the period during which this alteration can occur is genetically built into the neuron's developmental timetable. The modifiability of some neurons is limited to so-called critical periods early in development but in others it persists longer. Thus, some neuronal organizational patterns are highly specified early in development and are unmodifiable thereafter. These periods of maximal structural and functional growth depend upon the availability of proper internal and external environmental conditions for their full development. As an example of such a critical period it was found that the visual systems of sheep dogs deprived of sight in the first 5 weeks of life are anatomically, biochemically, and electrophysiologically retarded; this does not occur if the deprivation occurs from 5 to 10 weeks of age or during a 5-week period in adult dogs.

Other neurons remain uncommitted and are modifiable by function and experience until late in development. But in the case of either limited or lengthy periods of modifiability, it can occur only within the constraints imposed by the neurons' genetic code and its experiential history. The genes specify the capacity whereas the environmental stimuli determine its specific expression and content.

Learning and memory

Learning is the increase in the likelihood of a particular response to a stimulus as the consequence of experience. Rewards or punishments, as mentioned earlier, are crucial ingredients of learning, as is contact with, and manipulation of, the environment. A variety of hypotheses have been advanced recently to explain how individually acquired information may be stored by the brain. The

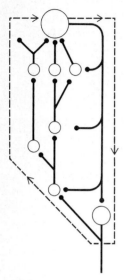

FIGURE 18-8
Reverberating neuronal circuit. Dashed arrow indicates directions of conduction of action potentials.

postulated neural correlate of memory is called the *memory trace.*

The processes involved in laying down the memory trace occur in a matter of minutes or hours, during which the memory is known as *short-term memory.* After this somewhat labile formative period the memory is stored as *long-term memory.* Short-term memory involves cerebral cortex, and long-term memory involves the limbic system; however, there is no exclusive site for memory storage because removal of various parts of the brain does not remove specific memories.

Short-term memory   Short-term memory is a limited-capacity storage process which serves as the initial depository of information. Generally, as new items enter, older ones are displaced, suggesting that the information is organized in a temporal sequence. Information that has entered short-term memory may be forgotten, recycled back through short-term memory by actively rehearsing the information, or transferred to a more durable storage mode. Recycling keeps the item at the front of the temporal sequence so that it will not be forgotten and increases the probability that it will be transferred to long-term storage.

One theory suggests that the memory trace during the early phases of learning is a reverberating neural circuit in which electric activity passes around and

around in closed neuronal loops such as that diagramed in Fig. 18-8. There is certainly evidence for the existence of such neuronal pathways in the brain, and activity once started in such loops could be maintained to keep the "memory" of the input for some time. That such reverberating circuits could be responsible for the temporary storage of acquired information in short-term memory is supported by evidence that conditions such as coma, deep anesthesia, electroconvulsive shock, and insufficient blood supply to the brain, which interfere with the electric activity of the brain, also interfere with the retention of recently acquired information. These same states do not interfere with long-term memory. When a man becomes unconscious from a blow on the head, he often cannot remember anything that happened for about 30 min before he was hit. This phenomenon is called *retrograde amnesia.* The loss of consciousness in no way interferes with memories of experiences that were learned before the period of amnesia.

Long-term memory   The existence of at least two stages of memory, short- and long-term memory, is borne out by the behavior of patients with specific amnesias. Such patients can learn perfectly well and recall the information immediately after it has been presented, but they lose the material much faster than normal people do, particularly if it involves verbal clues. It is as though they had difficulty transferring the information from short- to long-term storage.

Behavioral investigations and common experience indicate that memories of past events and well-learned behavior patterns normally can have very long life-spans. They may be changed or suppressed by other experiences, but, contrary to popular opinion, memories do not usually fade away or decay with time. This stability and durability combined with the fact that removing parts of the brain does not remove specific memories suggest that memory is stored in widespread chemical form or that the memory trace is an alteration in structure of some elements of the brain.

Retrieval of items from short-term memory seems to be much faster than retrieval from long-term memory, perhaps because the stores in long-term memory are so much larger. There are two major theories to explain long-term memory.

Molecular theories   One theory states that large, stable molecules within neurons are changed during learning and that information is stored in the specific configuration of these molecules. The molecules most frequently implicated are RNA and protein. We are already familiar with information storage mechanisms of this kind, e.g., the coding of genetic information by the nucleotide sequences in chromosomal DNA and the transfer of this information to RNA and proteins (Chap. 4). For learning, the question becomes: Are there chemical processes that in some way constitute the store of learned material in a coded form? Studies investigating the transfer of learned experiences by brain extracts in rats suggest that this is so. One group of rats was trained to approach a food cup when they heard a click, and a second group was trained to go to the food cup on a light cue. The rats were then killed and small pieces of brain were removed and extracted chemically. The brain extracts were then injected into other experimental rats which were exposed to the learning situation of approaching the food cup. The rats receiving extract from the click-trained animals tended to respond to the click and not the light, and the light-trained recipients tended to respond to the light and not the click. Moreover, there was no difference between injected animals and controls if the brain tissue was removed immediately after training, supposedly before the memory trace was laid down in a chemically extractable form. The experimental animals had to live approximately 8 hr before the learning achieved a form that could be transferred by injection. Chemical analyses of the extract indicated that the active ingredient was either RNA or protein. The conclusion to be drawn from this type of experiment is intriguing but highly controversial.

Experiments with drugs that block protein synthesis and interfere with the laying down or retention of learned behavior strongly support the hypothesis that RNA or protein synthesis is involved in some aspects of the learning-memory problem. Goldfish were placed in a tank partially divided by a barrier. Learning trials consisted of 20 sec of light followed by 20 sec of light paired with shock. The fish learned to swim over the hurdle from the light to the dark end of the tank to avoid the shock and, when tested 4 days after the learning situation, remembered the experience. If a drug which blocks protein synthesis was injected into the brains of the fish immediately after the learning experience, the fish had no memory of the experience 4 days later. However, if injected 1 hr after the learning situation, they remembered the experience. In other words, the long-term memory of the learning experience was formed during the first hour. Further experiments showed that after fish have been injected with the protein-synthesis blocker, they remember what they learned many days before and they can learn new tasks, but they cannot retain the new lessons.

The blocking agent does not interfere with short-term memory because the fish can perform immediately after they learn the task, but this short-term memory decays gradually over the 2 days following training. The blocking agent does not interfere with long-term memory; it interferes with the *laying down* of long-term memory.

Morphological theories   These theories suggest that changes in the relationship between cells occur during the formation of a memory trace. One theory suggests that modifications of the glial cells surrounding neurons provide the storage of the memory trace. A second theory suggests that new synaptic relationships are established or old synapses become more efficient when new information is transmitted to the brain. These modifications of synaptic relationships could occur with the swelling or shrinking of nerve processes as the result of use or disuse, an increase or decrease in the concentration of synaptic vesicles, changes at the pre- or postsynaptic membranes, etc.

Relationships between cells could also change with the interposition of small, newly formed nerve cell processes between the input and output elements of the nervous system. Indeed, the maturing brain undergoes radical structural transformations after birth: a great outgrowth of dendritic processes of the nerve cells representing an increase in the potential connectivity of neurons; formation of many glial cells, whose processes become interdigitated among the neurons; continuing formation of myelin around neuronal axons and increased density of blood vessels in the brain; and the further multiplication of undifferentiated nerve cells. These morphological changes in the developing brain are to a large extent genetically determined maturational processes. However, studies have suggested that the structural organization of the maturing brain is sensitive to conditions of the physical and social environment of the animal. Rats were raised in either enriched or restricted environments, the enriched animals being housed together in large cages in which they could move about freely and play with various objects and with each other while the restricted animals were raised alone in small cages and deprived of both sensory stimulation and an opportunity for much exercise. At the termination of the experiment, the rats were killed and their brains analyzed. The rats given enriched experiences developed greater weight and thickness of cerebral cortex than rats raised in isolation, and they also had a greater rate of cell proliferation. Even such behavioral variables as handling the animals markedly alters brain development.

Also, since the transmission of neural impulses across synapses requires a chemical step, the experimenters investigated some of the chemical correlates of enriched and restricted environments. They found that the brains (and particularly the cerebral cortices) of the enriched animals had a higher concentration of one of the important chemicals involved in effective synaptic transmission. Brains of adult enriched rats also showed significant increases in cortical weight and synaptic chemical concentration, supporting the argument that these effects are related to experience and, possibly, learning, rather than to genetically determined maturation of the brain.

The two groups of rats were tested to see if the different environmental experiences affected problem-solving ability; in general, depending upon the nature of the problem-solving task, enrichment seemed to improve performance and impoverishment to impair it. The unexpected result of the experiment was that these differences were not long-lasting. Furthermore, when the experiments were tried with monkeys, the results indicated that there were no permanent intellectual effects of prolonged exposure to different qualities of environment. Needless to say, it is risky to extrapolate from data on rats and monkeys to situations involving human beings.

Forgetting   Our understanding of the mechanisms of forgetting, like that of the mechanisms of learning, is still at the level of theories. One such idea is the *interference theory of forgetting,* which states that forgetting results from competition between responses at the time of recall. Competition comes from conflicting information stored both before and after the storage of the particular item that is being recalled. Thus we carry with us (in the form of prior learning) the source of much of our forgetting.

Summary of theories of learning   The molecular and morphological theories are certainly not mutually exclusive because a change in structural interrelationships can occur only through a change in macromolecules such as RNA and protein. However, the steps by which a particular experience results in a specific alteration in RNA and protein are obscure. Hypotheses to suggest how macromolecules can produce neuronal and, ultimately, behavioral changes are also lacking. It is probable that more than one mechanism will be found to be involved in the processes of learning and remembering.

The physical mechanisms ultimately accepted to explain learning must be able to account for the following phenomena:

1 Learning can occur very rapidly. In fact, under some situations, learning can occur in one trial.

2 Learning must be translated from an initial form involving action potentials to some permanent form of storage that can survive deep anesthesia, trauma, or electroconvulsive shock, which disrupt the normal patterns of neural conduction in the brain.

3 Information can be retained over long periods of time during which most components of the body have been renewed many times. Learning and memory must therefore reside either in systems that do not turn over rapidly or in systems which are self-perpetuating.

4 Information can be retrieved from memory stores after long periods of disuse. The common notion that memory, like muscle, always atrophies with lack of use is largely wrong.

5 Learning opposing responses interferes with the memory of initial responses to the same stimulus.

6 When learning a specific task, after an initial period of rapid learning, the process seems to slow down and proceed at a rate that offers diminishing returns.

Language

Language is the only fundamental process in which man is known to differ dramatically from other animals. When human and chimpanzee infants were raised together in human households and subjected to the same linguistic environment, only the human children developed language beyond a few simple words and meaningful gestures. It is precisely because no experimental animal has highly developed language skills that the study of language is so difficult and yet so important. One may hope that an understanding of the mechanisms of language will provide clues to the unique mechanisms of man's brain.

There are no simple anatomical differences between the brains of man and other mammals that could account for language, yet there are subtle differences between the two hemispheres of man's brain related to the fact that in adults language functions occur predominantly in the left hemisphere.

As demonstrated by *aphasias,* i.e., specific language deficits not due to mental defects, language is separable into two components, conceptualization and expression. In forms of aphasia related to conceptualization, the patient cannot understand spoken or written language even though his hearing and vision are unimpaired. In expressive aphasias, the patient is unable to carry out the coordinated respiratory and oral movements necessary for language even though he can move his lips and tongue, understands spoken language, and knows what he wants to say. Expressive aphasias are often associated with an inability to write.

Different areas of cortex are related to specific aspects of language. Areas in the frontal lobe near motor cortex are involved in the articulation of speech, whereas areas in the parietal and temporal lobe are involved in sensory functions and language interpretation. These cortical specializations are not present at birth but are established gradually in childhood during language acquisition. Why language functions localize in the left hemisphere in 96 percent of the population is not known, because during early childhood both hemispheres have language potential. Accidental damage to the left hemisphere of children under two causes no impediment of future language development, and language develops in the intact right hemisphere. Even if the left hemisphere is traumatized after the onset of language, language is reestablished in the right hemisphere after transient periods of loss. The prognosis becomes rapidly worse as the age at which damage occurs increases, so that after the early teens, language is interfered with permanently. The dramatic change in the possibility of establishing language (or a second language) in the teens is possibly related to the fact that the brain attains its final structural, biochemical, and functional maturity at that time. Apparently, with maturation of the brain, language functions are irrevocably assigned and the utilization of language propensities of the right hemisphere is no longer possible.

Recent evidence that language functions reside in the left hemisphere has been obtained from studies on patients whose main commissures (nerve fiber bundles) joining the two cerebral hemispheres have been cut to relieve uncontrollable epilepsy (neuronal activity, which starts at a cluster of abnormal cortical neurons, spreads throughout adjacent cortex, and gives rise to seizures and convulsions). Essentially, this operation leaves two separate cerebral hemispheres with a single brainstem, two separate mental domains within one head. Events experienced, learned, and remembered by one hemisphere remain unknown to the other because the memory processing of one hemisphere is inaccessible to the other. Because language resides in the left hemisphere, only that hemisphere can communicate orally or in writing about its conscious experience. If vision is limited so that only that part of the retina whose fibers pass to the right hemisphere is excited (Fig. 18-9), the left hemisphere, which controls speech, is unaware of the visual experience and the patient is unable to describe it either orally or in writing. On the other hand, if the portion of retina projecting to the left hemisphere is activated, the patient can describe the experience without difficulty. The right

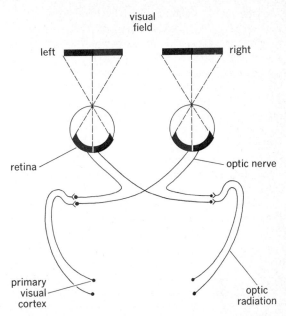

FIGURE 18-9
Visual pathways. Information about objects in the left half of the visual field is projected to the right cerebral hemisphere and vice versa.

silent hemisphere does understand language, however, for if the word for an object is flashed only to that hemisphere and the patient is told to point to the object or demonstrate what the object is used for, the patient complies. Of course, he responds with his left hand since the right hemisphere controls the muscles of the opposite side of the body. Therefore the right hemisphere does have linguistic functions but not those involved in graphic or vocal expression.

The capacity for language is characteristic of man, and language develops provided that there is some stimulation and opportunity. The capacity is at least partially inherited although environmental circumstances are certainly important and can limit language activity. Behavioral characteristics and anatomical correlates of language can thus be identified, but mechanisms by which neural operations result in language are unknown.

Conclusion

Until recently it was commonly thought that control of most complex behavior such as thinking, remembering, learning, etc., was handled almost exclusively by the cerebral cortex. Actually, damage of cortical areas outside of motor and sensory areas produces behavioral results that are subtle rather than obvious (the one exception, language, is highly sensitive to cortical damage), and stimulation of the cortex causes little change in the orientation or level of excitement of the animal.

In general, it is best to consider that particular behavioral functions are not controlled exclusively by any one area of the nervous system but that the control is shared or influenced by structures in other areas. Cortex and the subcortical regions—particularly limbic and reticular systems—form a highly interconnected system in which many parts contribute to the final expression of a particular behavioral performance. Moreover, the nervous system is so abundantly interconnected that it is difficult to know where any particular subsystem begins or ends.

In the early seventeenth century Descartes taught that all things in nature, including man, are machines. The brain's mode of operation was compared to that of a clock. When computers became widely used, the brain was compared to a computer. The most recent analogy compares the brain to a hologram, a new photographic process which records specially processed lightwaves themselves, rather than the image of an object. These widely divergent analogies only emphasize how little we know of how the brain really functions.

References

References for figure adaptations

Adolph, E. F.: "Physiology of Man in the Desert," Interscience, New York, 1947.

Anthony, C. P., and N. J. Kolthoff: "Textbook of Anatomy and Physiology," 8th ed., Mosby, St. Louis, 1971.

Bekesy, G. von, and W. A. Rosenblith: in S. S. Stevens (ed.), "Handbook of Experimental Psychology," Wiley, New York, 1951.

Bloom, W., and D. W. Fawcett: "A Textbook of Histology," 9th ed., Saunders, Philadelphia, 1968.

Boyd, I. A., and T. D. Roberts: *J. Physiol.,* **122:**38 (1953).

Brobeck, J. R., J. Tepperman, and C. N. H. Long: *Yale J. Biol. Med.,* **15:**831 (1943).

Carlson, A. J., V. Johnson, and H. M. Cavert: "The Machinery of the Body," 5th ed., The University of Chicago Press, Chicago, 1961.

Chapman, C. B., and J. H. Mitchell: *Sci. Am.,* May, 1965.

Comroe, J. H.: "Physiology of Respiration," Year Book, Chicago, 1965.

Daughaday, W. H., and D. M. Kipnis: *Recent Prog. Hormone Res.,* **22:**49 (1966).

Deanesley, R., and A. S. Parkes: *J. Physiol.,* **78:**442 (1933).

Dodt, E., and Y. Zotterman: *Acta Physiol. Scand.,* **26:**345 (1952).

Douglas, C. G., and J. S. Haldane: *J. Physiol.,* **38:**401 (1909).

Dubois, E. F.: "Fever and the Regulation of Body Temperature," Charles C Thomas, Springfield, Ill., 1948.

Ganong, W. F.: "Review of Medical Physiology," 4th ed., Lange, Los Altos, Calif., 1969.

Gordon, A. M., A. F. Huxley, and F. J. Julian: *J. Physiol.,* **184:**170 (1966).

Gregory, R. L.: "Eye and Brain: The Psychology of Seeing," McGraw-Hill, New York, 1966.

Guillemin, R., and R. Burgus: The Hormones of the Hypothalamus, *Sci. Am.,* October, 1972.

Guyton, A. C.: "Functions of the Human Body," 3d ed., Saunders, Philadelphia, 1969.

Hensel, H., and K. K. A. Bowman: *J. Neurophysiol.,* **20:**564 (1960).

Hoffman, B. F., and P. E. Cranefield: "Electrophysiology of the Heart," McGraw-Hill, New York, 1960.

Hubel, D. H., and T. N. Wiesel: *J. Physiol.,* **154:**572 (1960); **160:**106 (1962).

Ito, S.: *J. Cell Biol.,* **16:**541 (1963).

Jouvet, M.: *Sci. Am.,* Feb. (1967).

Lambertsen, C. J.: in P. Bard (ed.), "Medical Physiology," 11th ed., Mosby, St. Louis, 1961.

Landaurer, T. K.: "Readings in Physiological Psychology," McGraw-Hill, New York, 1967.

Livingston, R. B.: in G. C. Quarton, T. Melnechuk, and F. W. Schmitt (eds.), "The Neurosciences: A Study Program," Rockefeller University Press, New York, 1967.

Loewenstein, W. R.: *Sci. Am.,* August, 1960.

Mayer, J., N. B. Marshall, J. J. Vitale, J. H. Christensen, M. B. Mashayekhi, and F. J. Stare: *Am. J. Physiol.,* **177:**544 (1954).

Mayer, J., P. Roy, and K. P. Mitra: *Am. J. Nutr.,* **4:**169 (1956).

McNaught, A. B., and R. Callender: "Illustrated Physiology," Williams & Wilkins, Baltimore, 1963.

Nakayama, T., H. T. Hammel, J. D. Hardy, and J. S. Eisenman: *Am. J. Physiol.,* **204:**1122 (1963).

Nathanson, I. T., L. E. Towne, and J. C. Aub: *Endocrinology,* **28:**851 (1941).

Olds, J.: *Sci. Am.,* October, 1956.

Pedersen-Biergaard, K., and M. Tonnesen: *Acta Endocrinol.,* **1:**38 (1948); *Acta Med. Scand.,* **131** (suppl. 213):284 (1948).

Pitts, R. F.: "Physiology of the Kidney and Body Fluids," 2d ed., Year Book, Chicago, 1968.

Purdon-Martin, J.: "The Basal Ganglia and Posture," Lippincott, Phil., 1967.

Rasmussen, A. T.: "Outlines of Neuro-anatomy," 2d ed., Wm. C. Brown Company, Publishers, Dubuque, Iowa, 1943.

Rushmer, R. F.: "Cardiovascular Dynamics," 2d ed., Saunders, Philadelphia, 1961.

Singer, S. J., and Garth L. Nicolson: *Science,* **175:**720 (1972).

Smith, H. W.: "The Kidney," Oxford University Press, New York, 1951.

Steinberger, E., and W. O. Nelson: *Endocrinology,* **56:**429 (1955).

Tepperman, J.: "Metabolic and Endocrine Physiology," 2d ed., Year Book, Chicago, 1968.

Wang, C. C., and M. I. Grossman: *Am. J. Physiol.,* **164:**527 (1951).

Wersall, J., L. Gleisner, and P. G. Lundquist: in A.V.S. de Reuck and J. Knight (eds.), "Myotatic, Kinesthetic, and Vestibular Mechanisms," Ciba Foundation Symposium, Little, Brown, Boston, 1967.

Woodburne, R. T.: "Essentials of Human Anatomy," 3d ed., Oxford University Press, New York, 1965.

Zweifach, B. W.: *Sci. Am.,* January, 1959.

Suggested reading

Like all scientists, physiologists report the results of their experiments in scientific journals. Approximately 900 such journals in the life sciences publish 155,000 papers each year. Every physiologist must be familiar with the ever-increasing number of articles in his own specific field of research, but since these articles are highly technical for the nonexpert, with few exceptions we have not included any below for further reading.

Another type of scientific writing is the review article, a summary and synthesis of the relevant research reports on a specific subject. Such reviews are extremely useful, and many of our suggested readings are of this type. They are published in many different journals, but perhaps the single most valuable source for physiologists is *Physiological Reviews*. Also, in recent years, the American Physiological Society has initiated a massive program of publishing comprehensive reviews in almost all fields of physiology. These reviews are gathered into a series of volumes known collectively as the "Handbook of Physiology." Already published are the sections on Neurophysiology, Circulation, Respiration, Alimentary Canal, Adaptation to the Environment, Adipose Tissue, Endocrinology, and Renal and Electrolyte Physiology.

Another important series is the *Annual Review of Physiology,* which publishes yearly reviews of the most recent research articles in specific fields and provides the most rapid means of keeping abreast of developments in physiology. However, they are generally highly detailed and sometimes difficult to read. There are also *Annual Reviews of Biochemistry, Genetics, Medicine, Pharmacology,* and *Psychology.*

Another type of review which is perhaps of greatest value to the nonexpert is the *Scientific American* article. Written in a manner usually completely intelligible to the layman, these articles, in addition to reviewing a topic, often present actual experimental data so that the reader can obtain some insight into how the experiments were done and the data interpreted. Many articles published in *Scientific American* can be obtained individually as inexpensive offprints from W. H. Freeman and Company, San Francisco, Calif., 94104, which also publishes collections of *Scientific American* articles in a specific area, e.g., psychobiology. Two other magazines which frequently publish excellent reviews of physiology are *New England Journal of Medicine* and *Hospital Practice.*

Another level of scientific writing is the monograph, an extended review of a subject area broader than the review articles described above. Monographs are usually in book form and generally do not attempt to cover all the relevant literature on the topic but analyze and synthesize the most important data and interpretations. At the last level is the textbook, which deals with a much larger field than the monograph or review article; its depth of coverage of any topic is accordingly much less complete and is determined by the audience to which it is aimed. The books and articles listed below comprise only a tiny fraction of the readings available, but they should provide a source of additional information on the topics covered in this book. Their bibliographies will serve as a further entry into the scientific literature, particularly for original research reports.

We have suggested the relative difficulty of entries by asterisks. No asterisk indicates that, after finishing the relevant chapter in this book, the reader should be able to understand the entry with little difficulty. One asterisk indicates that the work is more difficult but should be comprehensible with some effort. Two asterisks denote a work which is highly detailed or technical and may require a strong background in mathematics, chemistry, or biology for a full understanding; this type of reading has been included for students who wish to pursue a subject in depth and are willing to expend the effort required.

We list first a group of books, mainly textbooks, which cover large areas of physiology. Since each can serve as additional reading for many chapters, they are grouped here according to field and are not mentioned again for individual chapters.

Books covering wide areas of physiology

Cell physiology and biochemistry

Baker, Jeffrey J. W., and Garland E. Allen: "Matter, Energy, and Life," 3d ed., Addison-Wesley, Reading, Mass., 1974 (paperback).

Baldwin, E.: "Dynamic Aspects of Biochemistry," 5th ed., Cambridge University Press, New York, 1967.

** Davson, H.: "A Textbook of General Physiology," 4th ed., vols. I and II, Little, Brown, Boston, 1970.

* DuPraw, Ernest J.: "Cell and Molecular Biology," Academic, New York, 1968.

* Giese, Arthur C.: "Cell Physiology," 4th ed., Saunders, Philadelphia, 1973.

* Lehninger, Albert L.: "Biochemistry: The Molecular Basis of Cell Structure and Function," Worth Publishers, Inc., New York, 1970.

Loewy, Ariel G., and Philip Siekevitz: "Cell Structure and Function," 2d ed., Holt, New York, 1970 (paperback).

McElroy, W. D.: "Cell Physiology and Biochemistry," 3d ed., Prentice-Hall, Englewood Cliffs, N.J., 1971 (paperback).

* White, A., P. Handler, and E. Smith: "Principles of Biochemistry," 5th ed., McGraw-Hill, New York, 1973.

Wooldridge, Dean E.: "The Machinery of Life," McGraw-Hill, New York, 1966 (paperback).

Organ system physiology

Astrand, P., and K. Rodahl: "Textbook of Work Physiology," McGraw-Hill, New York, 1970.

* Ganong, W. R.: "Review of Medical Physiology," 6th ed., Lange, Los Altos, Calif., 1963.

* Guyton, A. C.: "Textbook of Medical Physiology," 4th ed., Saunders, Philadelphia, 1971.

** "Handbook of Physiology," Williams & Wilkins, Baltimore, 1959–(continual publication of new volumes).

** Quarton, G. C., T. Melnechuk, and F. O. Schmitt (eds.): "The Neurosciences: A Study Program," Rockefeller University Press, New York, 1967.

* Ruch, T. C., and H. D. Patton: "Medical Physiology and Biophysics," 20th ed., Saunders, Philadelphia, 1973.

** Schmitt, F. O. (ed.): "The Neurosciences: Second Study Program," Rockefeller University Press, New York, 1970.

** Schmitt, F. O., and F. G. Worden (eds.): "The Neurosciences: Third Study Program," M.I.T., Cambridge, Mass., 1974.

Comparative physiology

* Florey, E.: "An Introduction to General and Comparative Physiology," Saunders, Philadelphia, 1966.

** Prosser, C. Ladd (ed.): "Comparative Animal Physiology," 3d ed., Saunders, Philadelphia, 1973.

Anatomy

Anthony, Catherine Parker, and Norma Jane Kolthoff: "Textbook of Anatomy and Physiology," 8th ed., Mosby, St. Louis, 1971.

* Bloom, W., and D. W. Fawcett: "A Textbook of Histology," 9th ed., Saunders, Philadelphia, 1968.

Fawcett, Don W.: "The Cell: An Atlas of Fine Structure," Saunders, Philadelphia, 1966.

* Woodburne, R. T.: "Essentials of Human Anatomy," 5th ed., Oxford University Press, New York, 1973.

History and philosophy of physiology

Bernard, C.: "An Introduction to the Study of Experimental Medicine,'' Dover, New York, 1957 (paperback).

Brooks, C. McC., and P. F. Cranefield (eds.): ''The Historical Development of Physiological Thought,'' Hafner, New York, 1959.

Butterfield, H.: ''The Origins of Modern Science,'' Macmillan, New York, 1961 (paperback).

''The Excitement and Fascination of Science: A Collection of Autobiographical and Philosophical Essays,'' *Annual Reviews,* Palo Alto, Calif., 1966.

Fulton, J. F., and L. C. Wilson (eds.): ''Selected Readings in the History of Physiology,'' 2d ed., Charles C Thomas, Springfield, Ill., 1966.

Harvey, W.: ''On the Motion of the Heart and Blood in Animals,'' Gateway, Henry Regnery, Chicago, 1962 (paperback).

Leake, Chauncey: ''Some Founders of Physiology,'' American Physiological Society, Washington, D.C., 1961.

Taylor, G. R.: ''The Science of Life: A Pictorial History of Biology,'' Panther, London, 1967 (paperback).

Pathophysiology

** Frohlich, Edward D. (ed.): ''Pathophysiology,'' Lippincott, Philadelphia, 1972.

Snively, W. D., Jr., and Donna R. Beshear: ''Textbook of Pathophysiology,'' Lippincott, Philadelphia, 1972.

* Sodeman, William A., and William A. Sodeman, Jr.,: ''Pathologic Physiology: Mechanisms of Disease,'' 5th ed., Saunders, Philadelphia, 1974.

Suggestions for individual chapters

Chapter 1

* Anfinsen, Christian B.: ''The Molecular Basis of Evolution,'' Wiley, New York, 1963 (paperback).

Doty, P.: Proteins, *Sci. Am.,* September, 1967.

Frieden, Earl: The Chemical Elements of Life, *Sci. Am.,* July, 1972.

Kendrew, J. C.: Three-dimensional Study of a Protein, *Sci. Am.,* December, 1961.

Lambert, Joseph B.: The Shapes of Organic Molecules, *Sci. Am.,* January, 1970.

Ritchie-Calder, Lord: Conversion to the Metric System, *Sci. Am.,* July, 1970.

Schmitt, F. O.: Giant Molecules in Cells and Tissues, *Sci. Am.,* September, 1957.

* Speakman, J. C.: ''Molecules,'' McGraw-Hill, New York, 1966 (paperback).

Stein, W. H., and S. Moore: The Chemical Structure of Proteins, *Sci. Am.,* February, 1961.

Chapter 2

Brachet, J.: The Living Cell, *Sci. Am.,* September, 1961.

** Bretscher, Mark S.: Membrane Structure: Some General Principles, *Science,* **181:**662 (1973).

** Dick, D. A. T.: ''Cell Water,'' Butterworth, Washington, D.C., 1966.

Capaldi, Roderick A.: A Dynamic Model of Cell Membranes, *Sci. Am.,* March, 1974.

Fox, C. Fred: The Structure of Cell Membranes, *Sci. Am.,* February, 1972.

Holter, H.: How Things Get into Cells, *Sci. Am.,* September, 1961.

* *Hospital Practice,* **8:**(1973) and **9:**(1974). (One article each month on membrane structure and function.)

** Kotyk, Arnost, and Karel Janacek: ''Cell Membrane Transport,'' Plenum, New York, 1970.

Rustad, R. C.: Pinocytosis, *Sci. Am.,* April, 1961.

** Singer, S. J., and Garth L. Nicolson: The Fluid Mosaic Model of the Structure of Cell Membranes, *Science,* **175:**720 (1972).

** Skou, J. C.: Enzymatic Basis for Active Transport of Na$^+$ and K$^+$ Across Cell Membranes, *Physiol. Rev.,* **45:**596 (1965).

Solomon, A. K.: Pores in the Cell Membrane, *Sci. Am.,* December, 1960.

Solomon, Arthur K.: The State of Water in Red Cells, *Sci. Am.,* February, 1971.

** Stein, W. D.: ''The Movement of Molecules Across Cell Membranes,'' Academic, New York, 1967.

Chapter 3

* Bernhard, Sidney A.: ''The Structure and Function of Enzymes,'' W. A. Benjamin, Inc., New York, 1968 (paperback).

Bolin, Bert: The Carbon Cycle, *Sci. Am.,* September, 1970.

Changeux, J.: The Control of Biochemical Reactions, *Sci. Am.,* April, 1965.

Cloud, Preston, and Aharon Gibor: The Oxygen Cycle, *Sci. Am.,* September, 1970.

* Conn, E. E., and P. K. Stumpf: ''Outlines of Biochemistry,'' 2d ed., Wiley, New York, 1966.

Frieden, E.: The Enzyme-Substrate Complex, *Sci. Am.,* August, 1969.

Green, D. E.: The Mitochondrion, *Sci. Am.,* January, 1964.

Green, D. E.: The Synthesis of Fat, *Sci. Am.,* August, 1960.

Koshland, Daniel E., Jr.: Protein Shape and Biological Control, *Sci. Am.,* October, 1973.

Lehninger, A. L.: How Cells Transform Energy, *Sci. Am.,* September, 1961.

* Lehninger, Albert L.: "Bioenergetics: The Molecular Basis of Biological Energy Transformations," 2d ed., W. A. Benjamin, Inc., New York, 1971 (paperback).

Levine, R. P.: The Mechanism of Photosynthesis, *Sci. Am.,* December, 1969.

Mott-Smith, Morton: "The Concept of Energy Simply Explained," Dover, New York, 1964 (paperback).

** Newsholme, E. A., and C. Start: "Regulation in Metabolism," Wiley, New York, 1973.

Racker, E.: The Membrane of the Mitochondria, *Sci. Am.,* February, 1968.

* Racker, Efraim: The Inner Mitochondrial Membrane: Basic and Applied Aspects, *Hosp. Pract.,* February, 1974.

** Segal, Harold L.: Enzymatic Interconversion of Active and Inactive Forms of Enzymes, *Science,* **180:**25 (1973).

Chapter 4

Allison, Anthony: Lysosomes and Disease, *Sci. Am.,* November, 1967.

Bearn, A. G., and James L. German III: Chromosomes and Disease, *Sci. Am.,* November, 1961.

Clark, B. F. C., and K. A. Marcker: How Proteins Start, *Sci. Am.,* January, 1968.

Crick, F. H. C.: The Genetic Code, *Sci. Am.,* October, 1962; October, 1966.

** Darnell, James E., Warren R. Jelinek, and George R. Molloy: Biogenesis of mRNA: Genetic Regulation in Mammalian Cells, *Science,* **181:**1215 (1973).

Fraenkel-Conrat, H.: "Design and Function at the Threshold of Life: The Viruses," Academic, New York, 1962.

German, James: Studying Human Chromosomes Today, *Am. Sci.,* **58:**182(1970).

Goodenough, Ursula W., and R. P. Levine: The Genetic Activity of Mitochondria and Chloroplasts, *Sci. Am.,* November, 1970.

* Hendler, Richard W.: "Protein Biosynthesis and Membrane Biochemistry," Wiley, New York, 1968.

Hurwitz, J. and J. J. Furth: Messenger RNA, *Sci. Am.,* February, 1962.

* Ingram, V. M.: "The Biosynthesis of Macromolecules," W. A. Benjamin, Inc., New York, 1965 (paperback).

Kornberg, A.: The Synthesis of DNA, *Sci. Am.,* September, 1961.

Mazia, Daniel: The Cell Cycle, *Sci. Am.,* January, 1974.

Miller, O. L., Jr.: The Visualization of Genes in Action, *Sci. Am.,* March, 1973.

Mirsky, A. E.: The Discovery of DNA, *Sci. Am.,* June, 1968.

Neutra, M., and C. P. Leblond: The Golgi Apparatus, *Sci. Am.,* February, 1969.

Nirenberg, M. W.: The Genetic Code, II, *Sci. Am.,* March, 1963.

Ptashne, Mark, and Walter Gilbert: Genetic Repressors, *Sci. Am.,* June, 1970.

Rich, A.: Polyribosomes, *Sci. Am.,* December, 1963.

Temin, Howard M.: RNA-directed DNA Synthesis, *Sci. Am.,* January, 1972.

* Watson, James D.: "Molecular Biology of the Gene," 2d ed., W. A. Benjamin, Inc., New York, 1970 (paperback).

Watson, James D.: "The Double Helix," Atheneum, New York, 1968. (Autobiographical account of Watson's role in discovering the DNA double helix.)

Wessells, N. K., and W. J. Rutter: Phases in Cell Differentiation, *Sci. Am.,* March, 1969.

Yanofsky, C.: Gene Structure and Protein Structure, *Sci. Am.,* May, 1967.

Chapter 5

** Adolph, E. F.: Early Concepts of Physiological Regulations, *Physiol. Rev.,* **41:**737 (1961).

Bernard, C.: "An Introduction to the Study of Experimental Medicine," Dover, New York, 1957 (paperback).

* Bullough, W. S.: "The Evolution of Differentiation," Academic, New York, 1967.

Cannon, W. B.: "The Wisdom of the Body," Norton, New York, 1939 (paperback).

* Elkinton, J. B., and T. S. Danowsky: "The Body Fluids: Basic Physiology and Practical Therapeutics," Williams & Wilkins, Baltimore, 1955.

* Fox, S. W., and K. Dose: "Molecular Evolution and the Origin of Life," Freeman, San Francisco, 1972 (paperback).

Langley, L. L. (ed.): "Homeostasis: Origin of the Concept," Dowden, Hutchinson, & Ross, Inc., Stroudsburg, Pa., 1973.

** Milhorn, H. T., Jr.: "The Application of Control Theory to Physiological Systems," Saunders, Philadelphia, 1966.

** Oparin, A. I.: "The Origin of Life," Dover, New York, 1953 (paperback).

Tustin, A.: Feedback, *Sci. Am.,* September, 1952.

Wald, G.: The Origin of Life, *Sci. Am.,* August, 1954.

* Weiner, N.: Concept of Homeostasis in Medicine, *Trans. Coll. Physicians Phila.,* **20:**87 (1948).

Wolf, A. V.: Body Water, *Sci. Am.,* November, 1958.

Chapter 6

Baker, P. F.: The Nerve Axon, *Sci. Am.,* March, 1966.

* Bodian, D.: The Generalized Vertebrate Neuron, *Science,* **137:**323 (1962).

* Bullock, T. H.: Neuron Doctrine and Electrophysiology, *Science,* **129:** 997 (1959).

Cannon, W. B.: "Bodily Changes in Pain, Hunger, Fear, and Rage," Harper & Row, New York, 1963 (originally published 1915; paperback).

* DeRobertis, E.: Ultrastructure and Cytochemistry of the Synaptic Region, *Science,* **156:**907 (1967).

DiCara, L. V.: Learning in the Autonomic Nervous System, *Sci. Am.,* January, 1970.

Eccles, J. C.: The Synapse, *Sci. Am.,* January, 1965.

** Eccles, J. C.: "The Physiology of Nerve Cells," Johns Hopkins, Baltimore, 1957.

* Eyzaguirre, C.: "Physiology of the Nervous System," Year Book, Chicago, 1969.

Fernstron, J. D., and R. L. Wurtman: Nutrition and the Brain, *Sci. Am.,* February, 1974.

Galambos, R.: "Nerves and Muscles," Science Study Series, Anchor Books, Doubleday, Garden City, N.Y., 1962 (paperback).

* Goodman, L. S., and A. Gilman: Neurohumoral Transmission and the Autonomic Nervous System, in "The Pharmacological Basis of Therapeutics," 4th ed., pp. 402–441, Macmillan, New York, 1970.

* Hodgkin, A. L.: "The Conduction of the Nervous Impulse," Charles C Thomas, Springfield, Ill., 1964.

* Katz, B.: Quantal Mechanism of Neural Transmitter Release, Nobel Prize Lecture, 1970, *Science,* **173:**123 (1971).

* Katz, B.: "Nerve, Muscle, and Synapse," McGraw-Hill, New York, 1966 (paperback).

Katz, B.: How Cells Communicate, *Sci. Am.,* September, 1961.

Keynes, R. D.: The Nerve Impulse in the Squid, *Sci. Am.,* December, 1958.

** Krnjevic, K.: Chemical Nature of Synaptic Transmission in Vertebrates, *Physiol. Rev.,* **54:**418 (1974).

Loewenstein, W. R.: Biological Transducers, *Sci. Am.,* August, 1960.

* Miller, N. E.: Learning of Visceral and Glandular Responses, *Science,* **163:**434 (1969).

* Ochs, S.: "Elements of Neurophysiology," pp. 302–341, Wiley, New York, 1965.

** Ruch, T. C., H. D. Patton, W. Woodbury, and A. L. Towe: "Neurophysiology," chaps. 1 and 2, Saunders, Philadelphia, 1965.

Chapter 7

Davidson, E. H.: Hormones and Genes, *Sci. Am.,* June, 1965.

* Davidson, J. M., and S. Levine: Endocrine Regulation of Behavior, *Ann. Rev. Physiol.,* **35:**375 (1972).

Gillie, R. B.: Endemic Goiter, *Sci. Am.,* June, 1971.

Guillemin, R., and R. Burgus: The Hormones of the Hypothalamus, *Sci. Am.,* October, 1972.

* Lefkowitz, R. J.: Isolated Hormone Receptors, *N. Engl. J. Med.,* **288:**1061 (1973).

* O'Malley, B. W.: Mechanisms of Action of Steroid Hormones, *N. Engl. J. Med.,* **284:**370 (1971).

Pastan, I.: Cyclic AMP, *Sci. Am.,* August, 1972.

Pike, J. E.: Prostaglandins, *Sci. Am.,* November, 1971.

* Schally, A. V., A. Arimura, and A. J. Kastin: Hypothalamic Regulatory Hormones, *Science,* **179:**241 (1973).

** Segal, H. L.: Enzymatic Interconversion of Active and Inactive Forms of Enzymes, *Science,* **180:**25 (1973).

Stent, G. S.: Cellular Communication, *Sci. Am.,* September, 1972.

* Sutherland, E. W.: Studies on the Mechanism of Hormone Action, *Science,* **177:**401 (1972).

* Tepperman, J.: "Metabolic and Endocrine Physiology," 3d ed., Year Book, Chicago, 1973.

* Turner, C. D.: "General Endocrinology," 5th ed., Saunders, Philadelphia, 1971.

Zuckerman, S.: Hormones, *Sci. Am.,* March, 1957.

Chapter 8

* Bendall, J. R.: "Muscles, Molecules and Movement," American Elsevier, New York, 1969.

** Bourne, G. H. (ed.): "The Structure and Function of Muscle," 2d ed., vol. 1 (1972), vols. II and III (1973), vol. IV (1974), Academic, New York.

** Bülbring, Edith, Alison F. Brading, Allan W. Jones, and Tadao Tomita (eds.): "Smooth Muscle," Williams & Wilkins, Baltimore, 1970.

** Close, R. I.: Dynamic Properties of Mammalian Skeletal Muscles, *Physiol. Rev.,* **52:**129 (1972).

** Guth, Lloyd: Trophic Influences of Nerve on Muscle, *Physiol. Rev.,* **44:**645 (1968).

Hoyle, Graham: How Is Muscle Turned On and Off?, *Sci. Am.,* April, 1970.

** Huxley, H. E.: The Mechanism of Muscular Contraction, *Science,* **164:**1356 (1969).

Huxley, H. E.: The Contraction of Muscle, *Sci. Am.,* November, 1958.

Huxley, H. E.: The Mechanism of Muscular Contraction, *Sci. Am.,* December, 1965.

Merton, P. A.: How We Control the Contraction of Our Muscles, *Sci. Am.,* May, 1972.

** Mommaerts, W. F. H. M.: Energetics of Muscle Contraction, *Physiol. Rev.,* **49:**427 (1969).

Murray, John M., and Annemarie Weber: The Cooperative Action of Muscle Proteins, *Sci. Am.,* February, 1974.

* Needham, Dorothy M.: ''Machina Carnis: The Biochemistry of Muscular Contraction in Its Historical Development,'' Cambridge University Press, New York, 1971.

Porter, K. R., and C. Franzini-Armstrong: The Sarcoplasmic Reticulum, *Sci. Am.,* March, 1965.

** Ruegg, Johann Caspar: Smooth Muscle Tone, *Physiol. Rev.,* **51:**201 (1971).

** Sandow, Alexander: Excitation-Contraction-Coupling in Skeletal Muscle, *Pharmacol. Rev.,* **17:**265 (1965).

Satir, P.: Cilia, *Sci. Am.,* February, 1961.

** Weber, Annemarie, and John M. Murray: Molecular Control Mechanisms in Muscle Contraction, *Physiol. Rev.,* **53:**612 (1973).

Chapter 9

Adolph, E. F.: The Heart's Pacemaker, *Sci. Am.,* March, 1967.

* Berne, R. M., and M. N. Levy: ''Cardiovascular Physiology,'' 2d ed., Mosby, St. Louis, 1972.

** Bevegard, B. S., and J. T. Shepherd: Regulation of the Circulation During Exercise in Man, *Physiol. Rev.,* **47:**178 (1967).

* Braunwald, E.: ''The Myocardium: Failure and Infarction,'' HP Publishing Co., New York, 1974.

* Burton, A. C.: ''Physiology and Biophysics of the Circulation,'' 2d ed., Year Book, Chicago, 1972.

Carrier, O., Jr.: The Local Control of Blood Flow: An Illustration of Homeostasis, *Bioscience,* **15:**665 (1965).

Chapman, C. B., and J. H. Mitchell: The Physiology of Exercise, *Sci. Am.,* May, 1965.

** Chien, S.: Role of the Sympathetic Nervous System in Hemorrhage, *Physiol. Rev.,* **47:**214 (1967).

* Folkow, B., and E. Neil: ''Circulation,'' Oxford University Press, Fair Lawn, N.J., 1971.

Mayerson, H.: The Lymphatic System, *Sci. Am.,* June, 1963.

* Rushmer, R. F.: ''Structure and Function of the Cardiovascular System,'' Saunders, Philadelphia, 1972.

Scher, A. M.: The Electrocardiogram, *Sci. Am.,* November, 1961.

Spain, D. M.: Atherosclerosis, *Sci. Am.,* August, 1966.

Wiggers, C. J.: The Heart, *Sci. Am.,* May, 1957.

Wood, J. E.: The Venous System, *Sci. Am.,* January, 1968.

Zweifach, B. J.: The Microcirculation of the Blood, *Sci. Am.,* January, 1959.

Chapter 10

Avery, M. E., N. Wang, and H. W. Taeusch, Jr.: The Lung of the Newborn Infant, *Sci. Am.,* April, 1973.

Baker, P. T.: Human Adaptation to High Altitude, *Science,* **163:**1149 (1969).

* Brewer, G. J., and T. W. Eaton: Erythrocyte Metabolism: Interaction with Oxygen Transport, *Science,* **26:**1205 (1971).

Clements, J. A.: Surface Tension in the Lungs, *Sci. Am.,* December, 1962.

Comroe, J. H., Jr.: The Lung, *Sci. Am.,* February, 1966.

* Comroe, J. H., Jr.: ''Physiology of Respiration,'' Year Book, Chicago, 1965.

* Davenport, H. W.: ''The ABC of Acid-Base Chemistry,'' 6th ed., The University of Chicago Press, Chicago, 1973.

Fenn, W. O.: The Mechanism of Breathing, *Sci. Am.,* January, 1960.

Finch, C. A., and C. Lenfant: Oxygen Transport in Man, *N. Engl. J. Med.,* **286:**407 (1972).

* Lenfant, C., and K. Sullivan: Adaptation to High Altitude, *N. Engl. J. Med.,* **284:**1298 (1971).

* Mitchell, T. H., and G. Blomqvist: Maximal Oxygen Uptake, *N. Engl. J. Med.,* **284:**1018 (1971).

** Morgan, T. E.: Pulmonary Surfactant, *N. Engl. J. Med.,* **284:**1185 (1971).

Perutz, M. F.: The Hemoglobin Molecule, *Sci. Am.,* November, 1964.

Ponder, E.: The Red Blood Cell, *Sci. Am.,* January, 1957.

Smith, C. A.: The First Breath, *Sci. Am.,* October, 1963.

** West, J. B.: Respiration, *Ann. Rev. Physiol.,* **34:**91 (1972).

Chapter 11

* Davenport, H. W.: ''The ABC of Acid-Base Chemistry,'' 6th ed., The University of Chicago Press, Chicago, 1973.

* Elkinton, J. R., and T. S. Danowsky: ''The Body Fluids: Basic Physiology and Practical Therapeutics,'' Williams & Wilkins, Baltimore, 1955.

** Gauer, O. H., and J. P. Henry: Circulatory Basis of Fluid Volume Control, *Physiol. Rev.,* **43:**423 (1963).

** Kuru, M.: Nervous Control of Micturition, *Physiol. Rev.*, **45:**425 (1965).

Merrill, J. P.: The Artificial Kidney, *Sci. Am.*, July, 1961.

Merrill, J. P., and C. L. Hampers: Uremia, *N. Engl. J. Med.*, **282:**953 (1970).

** Peart, W. S.: The Renin-Angiotensin System, *Pharmacol. Rev.*, **17:**143 (1965).

* Pitts, R. F.: "Physiology of the Kidney and Body Fluids," 2d ed., Year Book, Chicago, 1968.

Rasmussen, H.: The Parathyroid Hormone, *Sci. Am.*, April, 1961.

Rasmussen, H., and M. M. Pechet: Calcitonin, *Sci. Am.*, October, 1970.

** Schrier, R. W., and H. E. DeWardener: Tubular Reabsorption of Sodium Ion, *N. Engl. J. Med.*, **285:**1731 (1971).

** Schwartz, I. L., and W. B. Schwartz (eds.): Symposium on Antidiuretic Hormones, *Am. J. Med.*, May, 1967.

Smith, H. W.: "From Fish to Philosopher," Anchor Books, Doubleday, Garden City, N.Y., 1961 (paperback).

Smith, H. W.: The Kidney, *Sci. Am.*, January, 1953.

Solomon, A. K.: Pumps in the Living Cell, *Sci. Am.*, August, 1962.

* Valtin, H.: "Renal Function," Little, Brown, Boston, 1973.

Chapter 12

** Andersson, Sven: Secretion of Gastrointestinal Hormones, *Ann. Rev. Physiol.*, **35:**431 (1973).

** Bortoff, Alexander: Digestion: Motility, *Ann. Rev. Physiol.*, **34:** 261 (1972).

* Brooks, Frank P.: "Control of Gastrointestinal Function: An Introduction to the Physiology of the Gastrointestinal Tract," Macmillan, New York, 1970.

Davenport, Horace W.: Why the Stomach Does Not Digest Itself, *Sci. Am.*, January, 1972.

* Davenport, Horace W.: "Physiology of the Digestive Tract," 3d ed., Year Book, Chicago, 1971.

** Hunt, J. N.: Gastric Emptying the Secretion in Man, *Physiol. Rev.*, **39:**491 (1959).

* Javitt, Norman B., and Charles K. McSherry: Pathogenesis of Cholesterol Gallstones, *Hosp. Pract.*, July, 1973.

Kretchmer, Norman: Lactose and Lactase, *Sci. Am.*, October, 1972.

Neurath, H.: Protein Digesting Enzymes, *Sci. Am.*, December, 1974.

* Phillips, Sidney F.: Fluid and Electrolyte Fluxes in the Gut, *Hosp. Pract.*, March, 1973.

* Stahlgren, Leroy H.: The Dumping Syndrome: A Study of Its Hemodynamics, *Hosp. Pract.*, December, 1970.

** Wilson, T. H.: "Intestinal Absorption," Saunders, Philadelphia, 1962.

Chapter 13

* Adolph, E. F.: "Physiology of Man in the Desert," Interscience, New York, 1947.

Benzinger, T. H.: The Human Thermostat, *Sci. Am.*, January, 1961.

Cannon, W. B.: "The Wisdom of the Body," Norton, New York, 1939 (paperback).

Dole, V. P.: Body Fat, *Sci. Am.*, December, 1959.

Eichenwald, H. F., and P. C. Fry: Nutrition and Learning, *Science,* **163:**644 (1969).

Gray, G. W.: Human Growth, *Sci. Am.*, October, 1953.

Grodsky, G. M.: Insulin and the Pancreas, *Vitamins Hormones,* **28:**37 (1970).

Irving, L.: Adaptations to Cold, *Sci. Am.*, January, 1966.

Mayer, J.: "Overweight: Causes, Cost, and Control," Prentice-Hall, Englewood Cliffs, N.J., 1968.

Mayer, J., and D. W. Thomas: Regulation of Food Intake and Obesity, *Science,* **156:**327 (1967).

* Randle, P. J.: The Interrelationships of Hormones, Fatty Acid and Glucose in the Provision of Energy, *Postgrad. Med. J.,* **40:**457 (1964).

** Randle, P. J.: Insulin, in "The Hormones," vol. 4, Academic, New York, 1964.

* Tepperman, J.: "Metabolic and Endocrine Physiology," 3d ed., Year Book, Chicago, 1973.

* Unger, R. H.: Glucagon: Physiology and Pathophysiology, *N. Engl. J. Med.*, **285:**443 (1971).

Wilkins, L.: The Thyroid Gland, *Sci. Am.*, March, 1960.

Young, V. C., and N. S. Scrimshaw: The Physiology of Starvation, *Sci. Am.*, October, 1971.

Chapter 14

Allen, R. D.: The Moment of Fertilization, *Sci. Am.*, July, 1959.

Csapo, A.: Progesterone, *Sci. Am.*, April, 1958.

Edwards, R. G., and R. E. Fowler: Human Eggs in the Laboratory, *Sci. Am.*, December, 1970.

Klopfer, P. H.: Mother Love: What Turns It On?, *Am. Sci.,* **59:**404 (1971).

* Lloyd, C. W.: "Human Reproduction and Sexual Behavior," Lea & Febiger, Philadelphia, 1964.

* Macleod, J.: The Parameters of Male Fertility, *Hosp. Pract.*, December, 1973.

Masters, W. H., and V. E. Johnson: "Human Sexual Response," Little, Brown, Boston, 1966.

Mittwoch, U.: Sex Differences in Cells, *Sci. Am.,* July, 1963.

Money, J., and A. E. Ehrhardt: "Man and Woman, Boy and Girl. The Differentiation and Dimorphism of Gender Identity from Conception to Maturity," Johns Hopkins, Baltimore, 1973 (paperback).

* Odell, W. D., and D. L. Moyer: "Physiology of Reproduction," Mosby, St. Louis, 1971.

* Patten, B. M.: "Human Embryology," 3d ed., McGraw-Hill, New York, 1968.

* Sherwood, L. M.: Human Prolactin, *N. Engl. J. Med.,* **284:**774 (1971).

* Tepperman, J.: "Metabolic and Endocrine Physiology," 3d ed., Year Book, Chicago, 1973.

* Turner, C. D.: "General Endocrinology," 5th ed., Saunders, Philadelphia, 1971.

* Wilson, J. D.: Recent Studies on the Mechanism of Action of Testosterone, *N. Engl. J. Med.,* **287:**1284 (1972).

** Yates, F. E., S. M. Russell, and J. W. Moran: Brain-adenohypophysial Communication in Mammals, *Ann. Rev. Physiol.,* **33:**393 (1971).

Chapter 15

Abramoff, P., and M. La Via: "Biology of the Immune Response," McGraw-Hill, New York, 1970.

Beer, A. E., and R. E. Billingham: The Embryo as a Transplant, *Sci. Am.,* April, 1974.

Clowes, R. C.: The Molecule of Infectious Drug Resistance, *Sci. Am.,* April, 1973.

Constandinides, P. E., and N. Carey: The Alarm Reaction, *Sci. Am.,* March, 1949.

* Davie, E. W., and O. D. Ratnoff: Waterfall Sequence for Intrinsic Blood Clotting, *Science,* **145:**1310 (1964).

* Deykin, D.: Emerging Concepts of Platelet Function, *N. Engl. J. Med.,* **290:**144 (1974).

Dubos, R.: "Man Adapting," Yale, New Haven, Conn., 1965 (paperback).

Good, R. A., and D. W. Fisher: "Immunobiology," Sinauer, Stamford, 1973.

Holland, J. J.: Slow, Inapparent and Recurrent Viruses, *Sci. Am.,* February, 1974.

Isaacs, A.: Interferon, *Sci. Am.,* May, 1961.

Laki, K.: The Clotting of Fibrinogen, *Sci. Am.,* March, 1962.

Lerner, R. A., and F. J. Dixon, The Human Lymphocyte as Experimental Animal, *Sci. Am.,* June, 1973.

Levey, R. H.: The Thymus Hormone, *Sci. Am.,* July, 1964.

Levine, S.: Stress and Behavior, *Sci. Am.,* January, 1971.

Mason, J. W.: Organization of Psychoendocrine Mechanisms, *Psychosom. Med.,* **30** (II):(1968).

Mayer, M. M.: The Complement System, *Sci. Am.,* November, 1973.

Merigan, T. C.: Host Defenses Against Viral Disease, *N. Engl. J. Med.,* **290:**323 (1974).

Nossal, G. J. V.: How Cells Make Antibodies, *Sci. Am.,* December, 1964.

Notkins, A. L., and H. Koprowski: How the Immune Response to a Virus Can Cause Disease, *Sci. Am.,* January, 1973.

Porter, R. R.: The Structure of Antibodies, *Sci. Am.,* October, 1967.

Ratnoff, O. D.: The Interrelationship of Clotting and Immunologic Mechanisms, *Hosp. Pract.,* April, 1971.

Rensfeld, R. A., and B. D. Kahan: Markers of Biological Individuality, *Sci. Am.,* June, 1972.

Ross, R.: Wound Healing, *Sci. Am.,* June, 1969.

** Sayers, G.: The Adrenal Cortex and Homeostasis, *Physiol. Rev.,* **30:**241 (1905).

Terne, N. K.: The Immune System, *Sci. Am.,* January, 1973.

Weiss, J. M.: Psychological Factors in Stress and Disease, *Sci. Am.,* June, 1972.

Zucker, M. B.: Blood Platelets, *Sci. Am.,* February, 1961.

Chapter 16

* Alpern, M., M. Lawrence, and D. Wolsk: "Sensory Processes," Brooks/Cole, Belmont, Calif., 1967 (paperback).

Attneave, F.: Multistability in Perception, *Sci. Am.,* December, 1971.

Bekesy, G. von: The Ear, *Sci. Am.,* August, 1957.

Bower, T. G. R.: The Object in the World of the Infant, *Sci. Am.,* October, 1971.

Casey, K. L.: Pain: A Current View of Neural Mechanisms, *Am. Sci.,* **61:**194 (1973).

** Daw, N. W.: Neurophysiology of Color Vision, *Physiol. Rev.,* **53:**571 (1973).

Day, R. H.: Visual Spatial Illusions, A General Explanation, *Science,* **175:**1335 (1972).

Deregowski, J. B.: Pictorial Perception and Culture, *Sci. Am.,* November, 1972.

* De Valois, R. L., and G. H. Jacobs: Primate Color Vision, *Science,* **162:**533 (1968).

Gombrich, E. H.: The Visual Image, *Sci. Am.,* September, 1972.

Gordon, B.: The Superior Colliculus of the Brain, *Sci. Am.,* December, 1972.

* Granit, R.: The Development of Retinal Neurophysiol-

ogy, Nobel Prize Lecture, 1967, *Science,* **160:**1192 (1968).

** Granit, R.: "Receptors and Sensory Perception: A Discussion of Aims, Means, and Results of Electrophysiological Research into the Process of Perception," Yale, New Haven, Conn., 1955 (paperback).

Gregory, R. L.: "Eye and Brain: The Psychology of Seeing," 2d ed., World University Library, McGraw-Hill, New York, 1973.

Haber, R. N.: Eidetic Images, *Sci. Am.,* April, 1969.

* Hartline, H. K.: Visual Receptors and Retinal Interactions, Nobel Prize Lecture, 1967, *Science,* **164:**270 (1969).

Hubel, D. H.: The Visual Cortex of the Brain, *Sci. Am.,* November, 1963.

MacNichol, E. F.: Three-pigment Color Vision, *Sci. Am.,* December, 1964.

Michael, C. R.: Retinal Processing of Visual Images, *Sci. Am.,* May, 1969.

Miller, W. H., F. Ratliff, and H. K. Hartline: How Cells Receive Stimuli, *Sci. Am.,* September, 1961.

** Moulton, D. G., and L. M. Beidler: Structure and Function in the Peripheral Olfactory System, *Physiol. Rev.,* **47:**1 (1967).

Noton, D., and L. Stark: Eye Movements and Visual Perception, *Sci. Am.,* June, 1971.

** Oakley, B., and R. M. Benjamin: Neural Mechanisms of Taste, *Physiol. Rev.,* **46:**173 (1966).

Oster, G.: Auditory Beats in the Brain, *Sci. Am.,* October, 1973.

Oster, G.: Phosphenes, *Sci. Am.,* February, 1970.

Pettigrew, J. D.: The Neurophysiology of Binocular Vision, *Sci. Am.,* August, 1972.

Ratliff, F.: Contour and Contrast, *Sci. Am.,* June, 1972.

Robinson, D. A.: Eye Movement Control in Primates, *Science,* **161:**1219 (1968).

Rock, I.: The Perception of Disoriented Figures, *Sci. Am.,* January, 1974.

Rock, I., and C. S. Harris: Vision and Touch, *Sci. Am.,* May, 1967.

Rushton, W. A. H.: Visual Pigments in Man, *Sci. Am.,* November, 1962.

Stent, G. S.: Cellular Communication, *Sci. Am.,* September, 1972.

* Stevens, S. S.: Neural Events and the Psychophysical Law, *Science,* **170:**1043 (1970).

Toates, F. M.: Accommodation Function of the Human Eye, *Physiol. Rev.,* **52:**828 (1972).

* Wald, G.: Molecular Basis of Visual Excitation, Nobel Prize Lecture, 1967, *Science,* **162:**230 (1968).

Warren, R. M., and R. P. Warren: Auditory Illusions and Confusions, *Sci. Am.,* December, 1970.

Warshofsky, F., and S. S. Stevens: "Sound and Hearing," Life Science Library, Time-Life, New York, 1969.

Werblin, F. S.: The Control of Sensitivity in the Retina, *Sci. Am.,* September, 1972.

Young, R. W.: Visual Cells, *Sci. Am.,* October, 1970.

Chapter 17

Evarts, E. V.: Brain Mechanisms in Movement, *Sci. Am.,* July, 1973.

* Eyzaguirre, C.: "Physiology of the Nervous System," Year Book, Chicago, 1969.

** Granit, R.: "The Basis of Motor Control," Academic, New York, 1970.

Lippold, O.: Physiological Tremor, *Sci. Am.,* May, 1971.

** Matthews, P. B. C.: Muscle Spindles and Their Motor Control, *Physiol. Rev.,* **44:**219 (1964).

Merton, P. A.: How We Control the Contraction of Our Muscles, *Sci. Am.,* May, 1972.

* Ochs, S.: "Elements of Neurophysiology," pp. 280–301, 342–363, and 493–526, Wiley, New York, 1965.

** Ruch, T. C., H. D. Patton, J. W. Woodbury, and A. L. Towe: "Neurophysiology," 2d ed., pp. 153–225, 252–300, Saunders, Philadelphia, 1965.

* Sherrington, Sir Charles: "The Integrative Action of the Nervous System," Yale, New Haven, Conn., 1961 (originally published in 1906; paperback).

Wilson, V. J.: Inhibition in the Central Nervous System, *Sci. Am.,* May, 1966.

Chapter 18

Agranoff, B. W.: Memory and Protein Synthesis, *Sci. Am.,* June, 1967.

"Altered States of Awareness," readings from *Scientific American,* Freeman, San Francisco, 1972 (paperback).

Atkinson, R. C., and R. M. Schiffrin: The Control of Short-term Memory, *Sci. Am.,* August, 1971.

Broadbent, D. E.: Attention and the Perception of Speech, *Sci. Am.,* April, 1962.

Butter, C. M.: "Neuropsychology: The Study of Brain and Behavior," Brooks/Cole, Belmont, Calif., 1968 (paperback).

Ceraso, J.: The Interference Theory of Forgetting, *Sci. Am.,* October, 1967.

Corballis, M. C., and I. L. Beale: On Telling Right from Left, *Sci. Am.,* March, 1971.

** Eccles, J. C. (ed.): "Brain and Conscious Experience," Springer-Verlag, New York, 1966.

Fromkin, V. A.: Slips of the Tongue, *Sci. Am.*, December, 1973.

Gazzaniga, M. S.: The Split Brain in Man, *Sci. Am.*, August, 1967.

Geschwind, N.: Language and the Brain, *Sci. Am.*, April, 1972.

Haber, R. N.: How We Remember What We See, *Sci. Am.*, May, 1970.

Hess, W. R.: Causality, Consciousness, and Cerebral Organization, *Science,* **158:**1279 (1967).

Horn, G., S. P. R. Rose, and P. P. G. Bateson: Experience and Plasticity in the Central Nervous System, *Science,* **181:**506 (1973).

Jacobson, M., and R. K. Hunt: The Origins of Nerve-cell Specificity, *Sci. Am.*, February, 1973.

Kimura, D.: The Asymmetry of the Human Brain, *Sci. Am.*, March, 1973.

Lenneberg, E. H.: On Explaining Language, *Science,* **164:**635 (1969).

Luria, A. R.: The Functional Organization of the Brain, *Sci. Am.*, March, 1970.

Oatley, K.: "Brain Mechanisms and Mind," The World of Science Library, Dutton, New York, 1972 (paperback).

* Penfield, W., and L. Roberts: "Speech and Brain-mechanisms," Atheneum, New York, 1966 (paperback).

Premack, A. J., and D. Premack: Teaching Language to an Ape, *Sci. Am.*, October, 1972.

Pribram, K. H.: The Neurophysiology of Remembering, *Sci. Am.*, January, 1969.

"Psychobiology: The Biological Basis of Behavior," readings from *Scientific American,* Freeman, San Francisco, 1972 (paperback).

Rosenzweig, M. R., E. L. Bennett, and M. C. Diamond: Brain Changes in Response to Experience, *Sci. Am.*, February, 1972.

Scott, J. P.: Critical Periods in Behavioral Development, *Science,* **138:**949 (1962).

* Sperry, R. W.: A Modified Concept of Consciousness, *Psychol. Rev.,* **76:**532 (1969).

Tart, C. T. (ed.): "Altered States of Consciousness," Anchor Books, Doubleday, Garden City, N.Y., 1972 (paperback).

Wallace, R. K., and H. Benson: The Physiology of Meditation, *Sci. Am.*, February, 1972.

Wooldridge, D. E.: "Mechanical Man: The Physical Basis of Intelligent Life," McGraw-Hill, New York, 1968 (paperback).

Wooldridge, D. E.: "The Machinery of the Brain," McGraw-Hill, New York, 1963 (paperback).

INDEX

Page references in **boldface** indicate definition or beginning of detailed discussion.